Nanochemistry
Synthesis, Characterization and Applications

T0313236

Editors

Ashutosh Sharma
Tecnologico de Monterrey
School of Engineering and Sciences, Centre of Bioengineering
Mexico

Goldie Oza
National Laboratory for Micro and Nanofluidics (LABMyN)
Centro de Investigación y Desarrollo Tecnológico en Electroquímica (CIDETEQ)
Mexico

CRC Press
Taylor & Francis Group
Boca Raton   London   New York

CRC Press is an imprint of the
Taylor & Francis Group, an **informa** business

A SCIENCE PUBLISHERS BOOK

Cover credit:
1. Victor Merupo, Institute of Electronics, Microelectronics and Nanotechnology, Univ. Lille, ISEN, CNRS, UMR 8520, 59652 Villeneuve d'Ascq, France
2. Brandon Ortiz-Casas, Department of Engineering Science, University of Oxford, Parks Road, Oxford OX1 3PJ United Kingdom
3. Serfio Zamora Zúñiga, Designer and Illustrator, Lima, Perú

First edition published 2023
by CRC Press
6000 Broken Sound Parkway NW, Suite 300, Boca Raton, FL 33487-2742

and by CRC Press
4 Park Square, Milton Park, Abingdon, Oxon, OX14 4RN

© 2023 Taylor & Francis Group, LLC

CRC Press is an imprint of Taylor & Francis Group, LLC

Library of Congress Cataloging-in-Publication Data (applied for)

ISBN: 978-0-367-53444-8 (hbk)
ISBN: 978-0-367-53447-9 (pbk)
ISBN: 978-1-003-08194-4 (ebk)

DOI: 10.1201/9781003081944

Typeset in Times New Roman
by Radiant Productions

Preface

Nanochemistry is a subject that deals with the transformation of the materials when it moves from macro to micro to nanoscale. This resulted in a special phenomenon known as quantum size effects. This effect is responsible for imparting a new degree of freedom to the material, thus consequently leading to altogether new properties in the material. It involves bottom up approach where atoms stick to each other in a very regulated fashion and then stop or restrict the growth of nanoparticles beyond 100 nm. The miniaturization of materials from macro to nanoscale increases the surface to volume ratio. The atoms at the surface of the nanomaterial possess dangling bonds, thus increasing the surface energy. The surface effects are dominantly seen in nanoscale systems. The impurities or dopants decrease the surface energy and this phenomenal repercussion has a greater effect on fuel cells, batteries, catalysis as well as sensing. Further, surface functionalization and passivation and surface functionalization defects also lead to transformation of the properties of the nanomaterials. Hence, self-assembled nanostructures possess plethora of different applications depending upon the tunability and functionalization of the materials. This has been well explained in Chapter 1. In Chapter 2, the surface stabilization and functionalization of metal oxide semiconductors is well explained as they exorbitantly exhibit their role in gas sensing, LEDs, transistors and many more. Such metal oxide nanoparticles possess different physical, chemical, thermal, magnetic and optical properties that is well explained in Chapter 3. Such nanomaterials can be synthesized using physical, chemical or biological methods (Chapters 4, 5 and 6). Further, nanomaterials can be characterized using electron microscopy and SQUID (Chapters 7 and 8).

Chapter 9 explains the different analytical devices formulated from different kinds of nanomaterials such as gold nanoparticles and carbon nanotubes. Such biosensing devices and lab-on-a-chip systems exploit different techniques such as surface enhanced Raman spectroscopy, and optical and magnetic techniques for qualitative and quantitative analysis. Nanoparticles (NPs), especially, organic and inorganic, are used for drug delivery since they improve multiple drug limitations as reduced biodistribution, effectiveness, specificity, bioavailability, and unwanted side effects. These NPs can be designed for passive targeting by taking advantage of enhanced permeation and retention effect in the tumor microenvironment or for an active targeting by modifying their surface with different moieties for responding to endogenous or exogenous stimuli and guide them to the target location and diminish immunogenicity. This is well explained in Chapter 10. Moreover, several applications of nanotechnology, in particular the use of nanomaterials for prevention and treatments of diseases like cancer, infectious diseases and autoimmune diseases, includes the design of efficacious and safe nanomaterial immunotherapies for achieving new immunotherapeutic mechanisms such as molecular and cellular immunotherapies (Chapter 11). On the other hand, cell culture is one of the main milestones in biomedical science. However, tridimensional and dynamical culture models like organs-on-a-chip are promising *in vitro* models since they can mimic the cellular microenvironment and interactions while at the same time monitor the results with different biosensors and keep control of inlets and outlets (Chapter 12).

Chapters 13 and 14 deal with the synthesis of carbon catalysts and graphene and their various applications in different areas such as hydrogen storage (Chapter 15) as well as electrochemical water splitting (Chapter 16). Nanomaterials also have applications in the construction industry during the formulation of cementitious materials for the increment of tensile strength and durability of the building structures (Chapter 17). Last but not the least, Chapter 18 deals with the design of biomimetic devices that can play an important role in the development of brain-computer interface implants and cyborgs. The human-machine hybrid technologies derived from nanomaterials can provide service to human beings with its enhanced speed and strength.

Contents

CHAPTER 1

Self-Assembled Nanostructures

Amreen Khan,[1,2,#] *Manali Jadhav,*[1,2,#] *Nishant Kumar Jain,*[1,#]
Rajendra Prasad[1,3,$] and *Rohit Srivastava*[1,*]

1. Introduction

Self-assembly has been related to nanostructures that are well controlled and tunable, and represent favorable ways for surface functionalization (Kühnle 2009). The process majorly involves somewhat less ordered to more ordered self-assembled nanoarchitectures with minimal energy requirement at final equilibrium configuration (Lombardo et al. 2020). The interplay of molecule-molecule along with molecule-substrate interactions having various molecular building blocks, surface chemistry, and the structure helps to govern molecular self-assembly when considering solid surfaces (Kühnle 2009). Along with this, self-assembly technique has emerged while keeping the nanoparticles' properties intact. Self-assembly can be defined as the process by which molecules or nanoparticles or any other component organize spontaneously through direct specific interactions (Grzelczak et al. 2010). Just like a single nanoparticle displays unique properties, the assembly of such nanoparticles can have collective properties when present in bulk. The assembly of nanoparticles can even be used to enhance the related mechanical properties (Nie et al. 2010).

Various approaches have been explored to design and develop versatile nanostructures with multifunctionality. Among them, self-assembling nanostructures have attracted researchers from diverse fields due to their distinct properties and multitude of applications. Self-assembly is a spontaneous bottom-up process involving specific interactions leading to the formation of nanohybrids (Patil et al. 2008; Zhao et al. 2018). Self-assembly mainly involves either electrostatic forces, host-guest molecular interactions, chemical bonding, non-covalent hydrophobic-hydrophobic or π–π interactions (Zhao et al. 2018, Ji et al. 2019, Li et al. 2012). These self-assembled nanomaterials can be broadly categorized into three types based on their composition and origin which include (1) organic (2) inorganic/hybrid and (3) soft nanohybrids (Nie et al. 2010, Zhao et al. 2018). Organic-derived nanostructures are mainly formed from polymers and small molecules, whereas inorganic nanostructures are usually composites and hybrids. Soft nanohybrids includes DNA and peptides. The self-assembly process is influenced by surrounding media as the crucial interactions between

[1] Department of Biosciences and Bioengineering, Indian Institute of Technology Bombay, Powai, Mumbai – 400076, India.
[2] Centre for Research in Nanotechnology and Science, Indian Institute of Technology Bombay, Powai, Mumbai – 400076, India.
[3] Department of Biotechnology and Food Engineering, Technion – Israel Institute of Technology, Haifa – 3200, Israel.
\# These authors have contributed equally
$ Co-corresponding author: rpmeena@iitb.ac.in
* Corresponding author: rsrivasta@iitb.ac.in

the components occur at the interface because of the freedom of motions of components (Ariga et al. 2019).

The scope of the chapter is to summarize the materials forming self-assembly, their structure, and functionalization. Here, we have tried to capture the basic understanding of nanomaterial chemistry and related morphologies. We have also illustrated a few concepts of nanohybrids with their advantages and disadvantages. The fabrication and modulation of material for multimodal imaging and combined therapy have also been highlighted.

2. Organic self-assembly

The natural tendency of organic material to form diverse structures and functionalize themselves through utilizing self-assembly in an effective biological system has been gaining importance (Stupp et al. 2014). The organic nanomaterials consisting of carbon components are very versatile and offers control over chemical and morphological characteristics (Lombardo et al. 2020). In this section, we have described briefly the major organic molecules stated as polymer and lipid-based self-assembled systems.

2.1 Polymeric self-assembly

Polymeric nanoparticle assembly can be formed by block copolymer with distinguishable hydrophilic and hydrophobic elements. Polymer materials offer a wide range of possible structures influenced by intermolecular interactions sufficient to drive the assembly formation (Rupar et al. 2012). Polyelectrolytes with water-soluble macromolecules have a high charge density to produce a stern double electric layer as per Derjaguin–Landau–Verwey–Overbeek theory. This stern double-layer electric layer serves to increase the electrostatic repulsion against the attractive van der Waal forces. Such sort of electrostatic repulsion and attraction behavior of polyelectrolytes-based micelles is pH-dependent and can also be stated as counterion dependent (Dou at al. 2003). A concept of hierarchically arranged self-assembly of an amphiphilic ABC miktoarm star terpolymer mediated by counterion was revealed by Hanisch et al. The conjugation was extended to explore the polycationic segment. Here, the presence of iodide/triiodide counterions played an important role in forming micro-compartmentalized particles containing fine structures with periodic lamellar morphology (Hanisch et al. 2013). Another aspect is redefining the final self-assembled structure which can't be well predicted and control over final morphology becomes difficult. Ahead of this, research states core-shell micellar nanofiber structure formation with polyvinyl alcohol and polyethylene glycol-block-poly(p-dioxanone) obtained as the final crystallizable core. The approach combined the anisotropic micellization process of the copolymer formation in water and further reassembling into fiber (Zhai et al. 2013). In combination with other molecules like those of biological origin, polymers can act as multivalent counterions for self-assembly and vice versa (Liu et al. 2010). In an example provided in Figure 1, the behavioral characteristic of polymer self-assembly can be easily correlated in different solvents. Likewise, monodispersed micelles of positively charged PEG-*block*-poly(4-vinyl pyridine) (PEG-*b*-P4VP) in acidic solution self-assembled due to the presence of negatively charged circular plasmid DNA which acted as counterion (Zhang et al. 2014). The hierarchical structures of co-assembly of copolymeric micelles can be used for combined therapies with both therapeutic and diagnostic agent carrier systems. Liu et al. produced a multifunctional assembled structure for carrying theranostic agents. The carriers were synthesized by a process of co-assembly of anionic cylinders with poly(acrylamidoethylamine)-*block*-polystyrene cationic spherical micelles stating mix morphologies (Shrestha et al. 2012). Yet another common strategy of effective delivery by controlling particle size is by altering surface chemistry through hydrophilic polymer brush, further reducing the antibodies absorption and targeting specific cells (Champion et al. 2007).

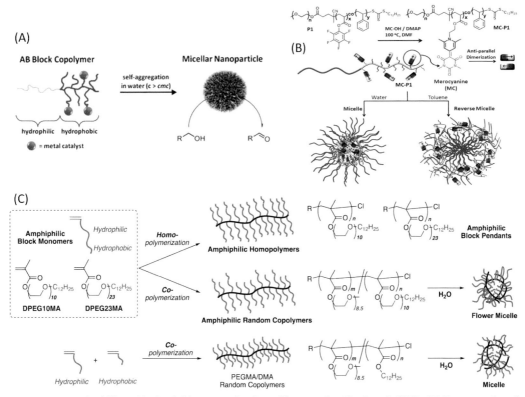

Figure 1. (A) Hydrophilic and hydrophobic groups showing self-aggregation (Sand et al. 2015). (B) Representation of Merocyanine dimerization induced supramolecular cross-linking of micelle in water and reverse micelle formation in toluene (Matsumoto et al. 2019). (C) Strategy design of amphiphilic monomers and copolymers (Rajak et al. 2020).

Another form of material behavior supported by polymers is a hydrogel. They tend to retain water and swell to form 3D structures. Biocompatible, biodegradability and high moisture retaining capability help them to get distinguished as polymeric self-assembled hydrogels. Components like poly (lactic–glycolic acid) (PLGA) and N-isopropylacrylamide (NIPAAm) have been reported for their biopolymeric nature (Vashist et al. 2013). Hydrogels can even be classified as thermoresponsive hydrogels like polymer of PEG–PLGA–PEG triblock and others including poly(ethylene glycol) and poly(L-lactic acid) as drug delivery agents (Lee et al. 2007, Jeong et al. 1997). Another example of self-assembled hydrogel formation is given in Figure 2. A very important requirement of biomedical translation of mouldable and injectable hydrogels is by finely tuning the mechanical properties of hydrogel-based systems. Polymer nanoparticle interaction arising as a result of self-assembled structure doesn't require much complex synthetic approach. A report of a non-covalent interaction between the hydroxypropylmethylcellulose derivatives (HPMC-*x*) and core-shell nanoparticles have been reported. They showed reverse and transient interactions between nanoparticle and chain governing polymer-nanoparticle (NP) self-assembled hydrogel. Poly(ethylene glycol)-*block*-poly(lactic acid) (PEG-*b*-PLA) NP could deliver both molecules of hydrophobic and hydrophilic nature by entrapping them into the gel (Appel et al. 2015).

Molecular self-assembly of polymers, which includes the abrupt formation of supramolecular aggregates and structure in different states, has also been keenly engineered. Due to dynamic chemical structure and the different binding sites, they can be said to be broadly classified depending on the combination as supramolecular engineered polymers which are built by small molecules, a mixture of small molecules and macromolecules, and those which have only macromolecules (Wang et al. 2014). Their unique characteristic has been simplicity in morphology, template ability,

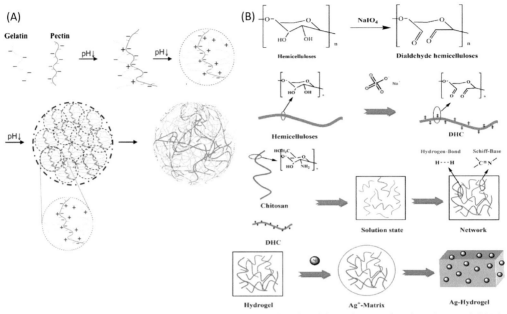

Figure 2. (A) Polymer gelatin and pectin containing self-assembled hydrogel formation at reduced pH (Wu et al. 2015). (B) Self-assembly of chitosan hydrogel and its network (Gun et al. 2015).

functionalization, and responsiveness to be utilized for various biomedical applications (Zhuang et al. 2016). The directionality and spatial geometrical distribution in solvents like water due to hydrogen bonds provide highly ordered structures to get self-assembled ranging from lower to higher units (Okesola et al. 2009). Another technique that has recently advanced in playing a significant role in elucidating the mechanism linked to copolymer self-assembly of the micelle is through simulation (Akcora et al. 2009). Simulation is a theoretical analysis to gain detailed information on experimental studies. In self-assembly, it can be used to study behavioral characteristics and types of copolymer building blocks under different conditions (Zhang et al. 2017). Macromolecular structure prediction, electrostatic interactions, and understanding the charged nanoscale motifs can also be achieved by molecular simulation (Akcora et al. 2019). The dissipative particle dynamics simulation method deals in the study of core-core coupled self-assembly with anisotropic graft copolymer micelle and correlating for experimental data interpretations (Zhuang et al. 2016).

2.1.1 Diagnostics and bioimaging

Visual probes to recognize desired molecular pathways have been of importance and are getting explored in most research areas. Magnetic resonance imaging (MRI), optical and nuclear imaging, all have shown potential in the biomedical field and a similar trend is with polymer-based probes (Wallyn et al. 2019). Polymer-based imaging probes provide advantages over small-molecules due to their integrity, stability, and reduced toxicity (Lu et al. 2007). Apart from this, higher target specificity, mainly while incorporating supramolecular engineered probes, have major advantages of facilitating both diagnosis and treatment. Optical imaging using fluorescence has been widely used when it comes to diagnosis and clinical studies. However, basic hurdles might include absorption, scattering of light, or autofluorescence (Licha et al. 2001). A research group reported supramolecular dot-shaped nanosize-based aggregates on bis(pyrene) derivatives for lysosome-directed imaging. The nano-aggregate represented good pH- and photo-stability with reduced toxicity in physiological conditions. It can be used as a better molecular candidate of fluorescence nanoprobes for live-cell imaging when targeting lysosomes (Wang et al. 2013). Like optical imaging, MRI and CT imaging has also attained prominence recently. Non-ionizing radiation, an easy distinction of target, high spatial and soft-tissue resolution are some advantages of imaging through MRI (Knight et al. 2011). Sensitive supramolecular nuclear probes were designed by Wang et al., where supramolecular nanoparticles of controlled size assisted in microPET/CT imaging while achieving trafficking of the lymph node *in vivo*. Here, the size as the main characteristic modulated the activity of supramolecular nanoparticles as imaging agents (Wang et al. 2009).

2.1.2 Delivery system

Polymeric engineering for therapeutic purposes has escalated tremendously. Few benefits include fast and easy supramolecular precursor synthesis with good biocompatibility and responsiveness (Liechty et al. 2010). A wide variety of different units can also be incorporated to make it more functionalized through multiple building blocks. The manufacturing processes for designing the carrier system is complex enough that hinders the reproducibility and quality-control (Peer et al. 2007). Additionally, building blocks, either polymer or small molecules, have a diverse range of compositions and functionalities required for optics, magnetic and biomedicine, and more. Likewise, amphiphilic polymer with hydrophilic PEGylated compound encapsulated hydrophobic drug photosensitizers has been effective in demonstrating the self-assembly of polymeric nanomicelles (Tu et al. 2011). Various stimuli-responsive drug carriers with enhanced therapeutic efficacy can accomplish active targeting over conventional systems. Polymeric nanocarriers with pH, thermal, light, redox-, enzyme responsiveness have been studied for their drug delivery properties (Fleige et al. 2012). With covalent and non-covalent linkages, controlled assembly and disassembly can be easily achieved for drug delivery. For example, a triple responsive carrier system with a double hydrophilic block of supramolecular having copolymer micelle was synthesized by a research group. Thermal responsive Poly-N-Isopropylacrylamide and pH-responsive poly(dimethylaminoethylmethacrylate) formed the self supramolecular engineered polymers for carrying doxorubicin which is an anticancer drug. The release study stated control of drugs by pH, temperature, and additional competitive molecules in a gradient manner. Also, compared to micelle, it showed covalent linkage, and supramolecules exhibited a faster release profile (Loh et al. 2012).

2.2 Lipid-based self-assembly

Lipids show diversity in their geometry and the type of assembly formed upon coming in contact with water. Structurally, lipids have a hydrophobic head, which is mostly a fatty acyl chain and hydrophilic polar head and hence are categorized under amphiphilic molecules. In an aqueous environment, a compact or internalized part of lipid forms the hydrophobic core and the inner part covers a hydrophilic region. This phenomenon is energetically favorable and can be termed as underlying self-organized molecules when coming in contact with water (Tanford 1974). The geometrical parameter, size, shape, and possible interactions of lipid head or tail are particularly considerate for charged lipids. Likewise, in phosphatidylserine and phosphatidylglycerol, the head groups are distanced enough due to repulsive forces to create a stable structure (Petrache et al. 2004, Brown et al. 2015). Apart from this ion composition of an aqueous solution, temperature and pH may also change the lipids' geometrical parameters to form lipid assembly (Kociurzynski et al. 2015). This causes solid-liquid phase transition where it may be referred to as chain melting transition and varies with different lipids' composition. For example, glycolipids having sugar groups show more solid-phase transition temperature (Kociurzynski et al. 2015). This temperature and degree of packaging in the liquid phase are dependent on the degree and length of fatty acyl chain saturation. Lipids with lower value form low order bilayer phase and with high mobility known as liquid-disordered phase (Van Meer et al. 2008). Cholesterol addition increases acyl chain in liquid disorder membranes (Veatch et al. 2005). Oppositely, bilayers formed by long-chain saturated lipids have higher solid-like phase with low mobility and high order. As the order is high, they are referred to as liquid-ordered phase (Maxfield et al. 2005).

When dealing with drug delivery systems, designing biocompatible and biodegradable nano self-assembly systems is highly important and desirable (Farokhzad et al. 2006). Apart from the polymers described above, liposomes also form composite nanostructures acting as drug carrier in multifunctional plateforms (Torchilin et al. 2005). Liposomes are amphiphilic lipid molecular vesicles of self-assembled structure, in which the hydrophilic agent could be entrapped in the inner portion whereas the hydrophobic pharmaceutical compounds in the liposomal membrane (Xie et al. 2012). As a famous delivery vehicle, liposomes with nanoparticles have higher encapsulation efficiency, demonstrate extended drug release profile and by carrying chemotherapeutic agents, they can be used to strategize nanomedicines (Sengupta et al. 2005).

3. Inorganic/hybrid self-assemblies

Self-assemblies can also be derived from inorganic as well as metallic materials, which may be composed of inorganic material alone or hybrids of inorganic and organic materials (Nei et al. 2010, Sanchez et al. 2005, Rui Hitzky et al. 2003). They are recognized due to their unique optical, magnetic, and electronic properties (Li et al. 2010, Pandey et al. 2016). These properties significantly vary when these materials are brought down from bulk to nanoscale that finds promising clinical applicability. Besides, these properties are tunable and are dependent on the composition, size, shape, and surrounding solvent medium (Pandey et al. 2016, Scholes 2018, Sun et al. 2003, Jain et al. 2007, Wu et al. 2016). Hence, they are widely used for imaging, therapy, and imaging-guided therapy applications. Several major types of inorganic materials used to produce self-assembled structures include gold nanoparticles (NPs), silica NPs, iron oxide NPs, carbon nanotubes (CNTs), graphene, etc. However, these materials are used in conjunction with organic materials to form nanohybrids that provide the necessary interactions for their self-assembly.

Self-assembling gold nanoparticles with functionalized surfaces help in fluorescence sensing applications. They demonstrate tunable optical properties due to which they are used in different applications of sensing, photonics, catalysis, and biotherapy (Miao et al. 2018, Alex et al. 2015). Fu et al. studied self-assembled gold nanoparticles (SAGNPs) for photothermal applications. Polyethylene glycol (PEG) modified by dithiol was used as a crosslinking agent to form SAGNPs presented in Figure 3(A), (B). Results indicated that the nanohybrid uptake improved into cancer cells and glutathione (GSH) responsive bio-disintegration in the tumor microenvironment. Enhanced photothermal performance with enhanced therapeutic efficacy was reported (Fu et al. 2018).

Quesada et al. developed core-shell type nanohybrid material where the redox-responsive amorphous organosilica forms a thin shell (6–10 nm) over the surface of the PLGA nanoparticle core through self-assembly process (Quesada et al. 2013). Silica coating over the PLGA core imparts chemical inertness, porosity, and biocompatibility. The self-assembly was achieved through silsesquioxane and tetraethyl orthosilicate containing a disulfide bridge as shown in Figure 3(D). These organic disulfide bridges act as molecular gates that selectively get cleaved by reducing agents. Such redox-sensitive systems are widely used for slow and sustained delivery of hydrophobic drug molecules. Yang et al. designed core-shell structured self-assemblies of anisotropic rare-earth β-NaREF4 nanocrystals (NC). The self-assembly was driven by selective etching of NC using oleate anion (OA⁻) and highly lattice mismatched driven epitaxial growth (Yang et al. 2019). Further, the nanohybrid surface was functionalized with N-(2-pyridinyl)benzoylacetamide (HPBA), a pH-responsive ligand exhibiting great potential in optical encoding.

Nanostructures based on carbon nanotubes (CNTs) are extensively used in the field of biocatalysis, chemical sensors, hydrogen storage, and power storage (Abdalla et al. 2015, Quesdada et al. 2013, Yang et al. 2019, Ajayan et al. 2001). Li et al. fabricated six various inorganic-organic nanohybrids of metal/metal oxide nano building blocks on CNTs using self-assembly mentioned

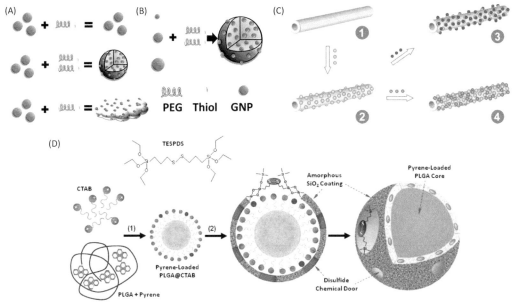

Figure 3. (A) Construction of SAGNPs. (B) Size independent assembly of GNPs (Abdalla et al. 2015). (C) Schematic illustration of the self-assembly schemes for the preparation of nanocomposites: (1) carbon nanotubes (CNTs, primary support), (2) the second phase of nanoparticles are anchored onto CNTs, (3) a third phase of nanoparticles are attached to the surface of the second phase nanoparticles, and (4) the third phase of nanoparticles are also landed onto the CNTs (Li et al. 2007). (D) Synthesis Layout of Pyrene-Loaded PLGA-Organosilica Nanoparticles with Redox-Responsive Disulfide Molecular Gates (Quesada et al. 2013).

in Figure 3(C). In particular, TiO_2/CNTs nanocomposite when used as cathode demonstrated significantly improved electrochemical performance in Li-battery application (Li et al. 2007). Self-assembly resulted in the nanostructures formation at large-scale synthesis with control over particle shape, concentration, and size.

Kabachi et al. reported self-assembled nanostructures formed by electrostatic interaction, utilizing the exfoliated MoS_2 nanoplates of negative charge and iron oxide nanoparticles' (NPs) positive charge having two different sizes (Kabachii et al. 2013). Quaternized poly(2-(dimethylamino)ethyl metacrylate-costearyl metacrylate) (poly(DMAEMQ-co-SMA), an amphiphilic copolymer, was used to functionalie the hydrophobic surface of iron oxide NPs, resulting in improved solubility and imparting the positive charge for self-assembly formation with MoS_2 nanoplates. Results indicated that more MoS_2 was needed to compensate for the charge as a larger fraction of copolymer was retained by larger NPs. Such nanostructures demonstrated high catalytic activity and find their application in the production of hydrogen using solar energy and removing the dangerous pollutants from wastewater. Zheng et al. reported self-assembled three-dimensional graphene nanohybrids demonstrating oxidase-like and intrinsic peroxidase-like activity. The one-pot solvothermal method was used for Fe_3O_4 and Pd nanoparticles' *in situ* synthesis, which was present over the graphene (3DRGO_Fe_3O_4-Pd) (Zheng et al. 2015). Higher catalytic activity was noticed with 3DRGO_Fe_3O_4-Pd Pd containing Fe_3O_4 nanoparticles due to the synergistic effect between them as compared with monometallic loaded nanohybrids. Further, the designed 3DRGO_Fe_3O_4-Pd nanohybrid demonstrated high affinity and catalytic efficiency towards substrate H_2O_2 due to which it was used for the colorimetric detection of urine glucose and GSH with high selectivity and sensitivity. In another experiment, Hussain et al. designed self-assembled core-shell nanohybrids between neutral hematite nanoparticles (HemNPs) and negatively charged carboxylated polystyrene NPs (PSNPs) through heteroaggregation (Hussain et al. 2020). The effect of HemNPs to PSNPize concentration ratio on the size of nanohybrids was studied. Results revealed that with concentration increase of

PSNPs, the size of nanohybrids gets reduced as the homoaggregation of HemNPs is prevented by abundant PSNPs that form the shell, whereas PSNPs' increased size resulted in nanohybrids with the larger size. From the research ongoing for self-assembly to advance inorganic chemistry and related nanohybrids, it can be said to make a huge difference to science in the coming future.

4. Soft nanohybrids

Self-assembly can also be achieved in a solution without using any template or external interference. This type of assembly is favored by bonds having attractive forces such as hydrogen bond, covalent, electrostatic, or dipole-dipole interaction (Nie et al. 2010), while particles can also be assembled with the use of templates such as carbon nanotubes (Correa-Duarte et al. 2005), polymers (Zhang et al. 2006), viruses (Dujardin et al. 2003), or DNA molecules (Aldaye et al. 2008). Likewise, the templates are classified as hard or soft templates. Examples of the hard template can be carbon nanotubes or inorganic nanowires that are chemically functionalized. These hard templates offer a well-crafted structure but lack control over spacing between the deposited nanoparticles, whereas soft templates can be classified as proteins, DNA molecules, or viruses (Nie et al. 2010, Dujardin et al. 2003, Aldaye et al. 2008); these types of templates provide binding sites for nanoparticle attachment.

4.1 DNA nanocages

The emergence of nanotechnology and its contribution towards the discovery of novel materials in biological applications has widely impacted nanomedicine, of which DNA is versatile for the nanoscale structures' self-assembly. The main features which make DNA a suitable nano-assembly are: DNA being double-helical structure has diameter of ~ 2 nm and helical pitch of ~ 3.4 to ~ 3.6 nm, and length of DNA molecule is ~ 50 nm which serves as a self-assembly rigid building block. The DNA base pairing (A:T and G:C) gives direction to the molecular self-assembly pathway. The single-stranded overhangs connect duplexes providing a way for hierarchical assembly (Chandrasekaran et al. 2016).

DNA nanocages serve functions like drug delivery, biosensing, and imaging. DNA-based drug delivery targets the tumor site restricting enzymatic degradation. The two known major DNA nanocages types are pure and hybrid DNA nanostructures with size ranging from 10–100 nm. Pure DNA nanostructures include DNA polyhedrons, nanoribbons, DNA origamis, and nanoflowers, whereas hybrid DNA nanostructures involve hybrid conjugates with DNA like DNA-inorganic nanoparticles, and DNA polymer, and DNA-lipid hybrids (Yang et al. 2020). Many DNA nanocages have been synthesized starting from DNA cubes (Zhang et al. 2009), hinged tetrahedron (Smith et al. 2011) as shown in Figure 4 (A), (B), (C).

DNA nanocarriers function as smart delivery devices for carrying molecular cargos. The conjugate formed by the interaction between inorganic NPs and DNA nanoparticles possesses properties intrinsic to both components. Encapsulation of gold NPs within polyhedral DNA nanocages has been shown in the following Figure 4 (D). This system ensured the controllable release of Au-NPs at the target site proving its potential application for cargo delivery (Zhang et al. 2014). In one of the studies, an assembled antigen-DNA conjugate showed an antibody response that was long lasting against antigen without any reaction with the DNA-assembly.

4.2 Self-assembled peptides

All the 20 amino acids that make peptides are structurally similar but they differ in side groups. The arrangement of these amino acids is based on first-order structures (sequence length), second-order

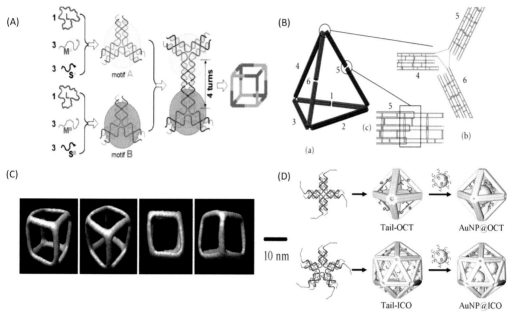

Figure 4. (A) Assembly of DNA cubes from two different component three-point-star tiles (A and B). The separation of any two adjacent tiles is 4 DNA helical turns; thus, the resulting DNA cubes are named 4-turn cubes. Reproduced with permission (Zhang, et al. 2009). (B). Schematic view of DNA origami tetrahedron. (a) In the 3-D representation, each double-helical section is represented by a single cylinder. The six struts are labeled 1 through 6, and each consists of a bundle of six parallelly connected double helices. (b) A two-dimensional diagram of the hinged vertex formed by struts 4, 5, and 6 shows the path of the m13MP18-based scaffold forming the single-stranded connections between bundles and the pattern of staple oligonucleotides around the hinge. (c) Local structure of the scaffold path within strut 5 shows three closely aligned scaffold crossovers stabilized by short staple sections forming the weakened gap. Reproduced with permission (Smith et al. 2011). (C). The 3D reconstruction maps of the DNA cube, reconstructed by imposing a tetrahedron symmetry. Reproduced with permission (Zhang et al. 2009). (D) DNA polyhedra encapsulate gold nanoparticles (AuNPs) to form core-shell structures (AuNP@cages) (top) octahedron (OCT), and (bottom) icosahedron (ICO). When incubating with DNA-AuNPs (solid spheres) that are AuNPs functionalized with complementary DNA single strands, the DNA-AuNPs will be swallowed into the tail-polyhedra driven by maximization of DNA hybridization between the single-stranded tails on DNA polyhedra and the DNA single strands immobilized on DNA-AuNPs. Reproduced with permission (Zhang et al. 2014).

structures (alpha/beta-helix or coiled-coil), and third-order structures (a geometric shape) (Li et al. 2019). Peptide-based nanoparticles are emerging in drug therapy with the fact that peptides can be easily conjugated with other nanoparticles which can help them to rapidly be taken up by the targeting cells. Secondly, peptides are easier to synthesize, have better biocompatibility, bioavailability, and low immunogenicity (Li et al. 2020). Moreover, peptides self-assemble and arrange in the form of nanoparticles, nanofibers, or nanotubes due to their aggregation property (Wang et al. 2016). Some of the functional peptides have been summarized in Table 1.

Peptides can also ensemble through hydrogen bonding, hydrophobic interactions, π–π, or electrostatic interactions. As presented in Figure 5, α-helix, β-sheets, and β-hairpin are the common secondary structures that form self-assembled nanofibers, and nanotubes through non-covalent bonding interactions (Wang et al. 2020, Chen et al. 2019).

One of the reported self -assembled dipeptide Diphelnyalanine, Phe-Phe (FF) has been used for various nanohybrids. It is known as the amyloid-β polypeptide core motif to drive self-assembly in Alzheimer's disease (Görbitz 2006). Among different types of peptide assemblies, "stimuli-responsive" peptide nano-assemblies are more responsive as drug-delivery vehicles, as they show

Table 1. List of self-assembled functional peptides.

Type	Sequence	Function	References
Targeting peptides	ASSLNIA	Treating heart and skeletal muscle	(Yu et al. 2009)
	DDDDDD	Targeting hard tissue (bone and teeth)	(Kasugai et al. 2000)
	RGD	Targeting multiple integrin overexpression tumor cells	(Mas-Moruno et al. 2016)
	KCCYSL	Targeting HER2-positive tumor cells	(Kumar et al. 2007)
Endosomal escape peptides	GLFGAIAGFIENGWEGMIDG	Destructing the membrane of endosomes	(Cheng et al. 2016)
	WEAALAEAALEAALEAALEAA LEAALAALA	Destructing the membrane of endosomes	(Sugahara et al. 2015)
Nuclear localization signal (NLS) peptides	PKKKRKV	Providing intranuclear transport	(Subbarao et al. 1987)
	KRRRR	Nucleocytoplasmic trafficking	(Kim et al. 2012, Leonidova et al. 2014)
Responsive peptides	DEVD	Responsive to caspase-3	(Lu et al. 2014, Song et al. 2018)
	GFLG	Responsive to cathepsin B (CB)	(Yuan et al. 2015)

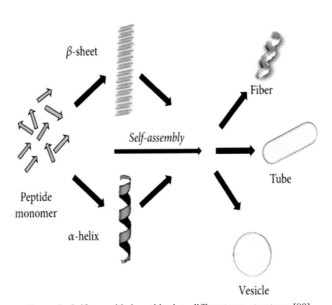

Figure 5. Self-assembled peptides into different nanostructures [89].

drug release by stimuli at the site. Such nano-assemblies can be loaded with nanocarrier and released responding to the environment (Panda et al. 2014).

In some studies, peptide-based hybrid nanostructures are proven to be useful in cancer therapy. In one such study, a peptide-hybrid nano-assembly was made by fabrication of polylactide (PLA) and short peptide sequence VVVVVVKK (V6K2). These were further conjugated with anticancer drugs doxorubicin and paclitaxel. The peptide-based nanoparticle conjugate demonstrated more toxicity to the cancer cells compared to free drug, proving PLA-V6K2 nano-assembly to be efficacious (Jabbari et al. 2013).

5. Conclusion

The actual concept of self-assembled nanoarchitectures could be found in their organization and molecular interplay. As we move towards higher-order, the chances of structural disruption even broaden and can lead to the collapse of assembly. In such cases, the study of energy dissipation is of utmost importance. Another potential feature is a selection of nano self-assembly systems based on the application and carrying capacity. Circumstantial ambiguity lies in the assembly and similarly scaled nanostructure while constructing controlled functionality materials. To be concise, be it organic/inorganic/soft nanohybrids, the conceptual relation of geometry, chemical modification, charge, and synthesis strategy critically defines the spatio-temporal nano self-assembly and its later evolution.

Acknowledgments

We thank the Department of Biotechnology, Government of India. N.K.J. is also grateful for the support of the DST INSPIRE fellowship and IPD fellowship, BSBE, IITB.

References

Abdalla, S., Al-Marzouki, F., Al-Ghamdi, A.A. and Abdel-Daiem, A. 2015. Different technical applications of carbon nanotubes. *Nanoscale Res. Lett.*, 10(1): 1–10.

Ajayan, P.M. and Zhou, O.Z. 2001. Applications of carbon nanotubes. *Carbon Nanotub.*, pp. 391–425.

Akcora, P., Liu, H., Kumar, S.K., Moll, J., Li, Y., Benicewicz, B.C., Schadler, L.S., Acehan, D., Panagiotopoulos, A.Z., Pryamitsyn, V., Ganesan, V., Ilavsky, J., Thiyagarajan, P., Colby, R.H. and Douglas, J.F. 2009. Anisotropic self-assembly of spherical polymer-grafted nanoparticles. *Nat. Mater.*, 8(4): 354–359, doi: 10.1038/nmat2404.

Aldaye, F.A., Palmer, A.L. and Sleiman, H.F. 2008. Assembling materials with DNA as the guide. *Science (80-.)*, 321(5897): 1795–1799, doi: 10.1126/science.1154533.

Alex, S. and Tiwari, A. 2015. Functionalized gold nanoparticles: synthesis, properties and applications-a review. *J. Nanosci. Nanotechnol.*, 15(3): 1869–1894.

Appel, E.A., Tibbitt, M.W., Webber, M.J., Mattix, B.A., Veiseh, O. and Langer, R. 2015. Self-assembled hydrogels utilizing polymer-nanoparticle interactions. *Nat. Commun.*, 6(1): 1–9, doi: 10.1038/ncomms7295.

Ariga, K., Nishikawa, M., Mori, T., Takeya, J., Shrestha, L.K. and Hill, J.P. 2019. Self-assembly as a key player for materials nanoarchitectonics. *Sci. Technol. Adv. Mater.*, 20(1): 51–95, doi: 10.1080/14686996.2018.1553108.

Brown, K.L. and Conboy, J.C. 2015. Phosphatidylglycerol flip-flop suppression due to headgroup charge repulsion. *J. Phys. Chem. B*, 119(32): 10252–10260, doi: 10.1021/acs.jpcb.5b05523.

Champion, J.A., Katare, Y.K. and Mitragotri, S. 2007. Particle shape: A new design parameter for micro- and nanoscale drug delivery carriers. *J. Control. Release*, 121(1-2): 3–9, doi: 10.1016/j.jconrel.2007.03.022.

Chandrasekaran, A.R. and Levchenko, O. 2016. DNA Nanocages. *Chemistry of Materials, American Chemical Society*, 28(16): 5569–5581, doi: 10.1021/acs.chemmater.6b02546.

Chen, J. and Zou, X. 2019. Bioactive Materials Self-assemble peptide biomaterials and their biomedical applications. *Bioact. Mater.*, 4(no. October 2018): 120–131, doi: 10.1016/j.bioactmat.2019.01.002.

cheng Wu, B. and McClements, D.J. 2015. Functional hydrogel microspheres: Parameters affecting electrostatic assembly of biopolymer particles fabricated from gelatin and pectin. *Food Res. Int.*, 72: 231–240, doi: 10.1016/j.foodres.2015.02.028.

Cheng, Y., Huang, F., Min, X., Gao, P., Zhang, T., Li, X., Liu, B., Hong, Y., Lou, X. and Xia, F. 2016. Protease-responsive prodrug with aggregation-induced emission probe for controlled drug delivery and drug release tracking in living cells. *Anal. Chem.*, 88(17): 8913–8919, doi: 10.1021/acs.analchem.6b02833.

Correa-Duarte, M.A., Grzelczak, M., Salgueiriño-Maceira, V., Giersig, M., Liz-Marzan, L.M., Farle, M., Sierazdki, K. and Diaz, R. 2005. Alignment of carbon nanotubes under low magnetic fields through attachment of magnetic nanoparticles. *J. Phys. Chem. B*, 109(41): 19060–19063, doi: 10.1021/jp0544890.

Dou, H., Jiang, M., Peng, H., Chen, D. and Hong, Y. 2003. pH-dependent self-assembly: Micellization and micelle-hollow-sphere transition of cellulose-based copolymers. *Angew. Chemie - Int. Ed.*, 42(13): 1516–1519, doi: 10.1002/anie.200250254.

Dujardin, E., Peet, C., Stubbs, G., Culver, J.N. and Mann, S. 2003. Organization of metallic nanoparticles using tobacco mosaic virus templates. *Nano Lett.*, 3(3): 413–417, doi: 10.1021/nl034004o.

Farokhzad, O.C. and Langer, R. 2006. Nanomedicine: Developing smarter therapeutic and diagnostic modalities. *Adv. Drug Deliv. Rev.*, 58(14): 1456–1459, doi: 10.1016/j.addr.2006.09.011.

Fleige, E., Quadir, M.A. and Haag, R. 2012. Stimuli-responsive polymeric nanocarriers for the controlled transport of active compounds: Concepts and applications. *Adv. Drug Deliv. Rev.*, 64(9): 866–884, doi: 10.1016/j.addr.2012.01.020.

Fu, Y., Feng, Q., Shen, Y., Chen, M., Xu, C., Cheng, Y. and Zhou, X. 2018. A feasible strategy for self-assembly of gold nanoparticles via dithiol-PEG for photothermal therapy of cancers. *RSC Adv.*, 8(11): 6120–6124.

Görbitz, C.H. 2006. The structure of nanotubes formed by diphenylalanine, the core recognition motif of Alzheimer's β-amyloid polypeptide. *Chem. Commun.*, (22): 2332–2334, doi: 10.1039/b603080g.

Grzelczak, M., Vermant, J., Furst, E.M. and Liz-Marzán, L.M. 2010. Directed self-assembly of nanoparticles. *ACS Nano*, 4(7): 3591–3605, Jul. 27, 2010, doi: 10.1021/nn100869j.

Guan, Y., Chen, J., Qi, X., Chen, G., Peng, F. and Sun, R. 2015. Fabrication of biopolymer hydrogel containing Ag nanoparticles for antibacterial property. *Ind. Eng. Chem. Res.*, 54(30): 7393–7400, doi: 10.1021/acs.iecr.5b01532.

Hanisch, A., Gröschel, A.H., Förtsch, M., Löbling, T.I., Schacher, F.H. and Müller, A.H.E. 2013. Hierarchical self-assembly of miktoarm star polymers containing a polycationic segment: A general concept. *Polymer (Guildf)*, 54(17): 4528–4537, doi: 10.1016/j.polymer.2013.05.071.

Hussain, K.A. and Yi, P. 2020. Heteroaggregation of neutral and charged nanoparticles: a potential method of making core-shell nanohybrids through self-assembly. *J. Phys. Chem. C*, 124(35): 19282–19288.

Jabbari, E., Yang, X., Moeinzadeh, S. and He, X. 2013. Drug release kinetics, cell uptake, and tumor toxicity of hybrid VVVVVVKK peptide-assembled polylactide nanoparticles. *Eur. J. Pharm. Biopharm.*, 84(1): 49–62, doi: 10.1016/j.ejpb.2012.12.012.

Jain, P.K., Huang, X., El-Sayed, I.H. and El-Sayed, M.A. 2007. Review of some interesting surface plasmon resonance-enhanced properties of noble metal nanoparticles and their applications to biosystems. *Plasmonics*, 2(3): 107–118.

Jeong, B., Bae, Y.H., Lee, D.S. and Kim, S.W. 1997. Biodegradable block copolymers as injectable drug-delivery systems. *Nature*, 388(6645): 860–862, doi: 10.1038/42218.

Ji, X., Ahmed, M., Long, L., Khashab, N.M., Huang, F. and Sessler, J.L. 2019. Adhesive supramolecular polymeric materials constructed from macrocycle-based host-guest interactions. *Chem. Soc. Rev.*, 48(10): 2682–2697.

Kabachii, Y.A., Golub, A.S., Kochev, S.Y., Lenenko, N.D., Abramchuk, S.S., Antipin, M.Y., Valetsky, P.M., Stein, B.D., Mahmoud, W.E., Al-Ghamdi, A.A. and Bronstein, L.M. 2013. Multifunctional nanohybrids by self-assembly of monodisperse iron oxide nanoparticles and nanolamellar MoS2 plates. *Chem. Mater.*, 25(12): 2434–2440.

Kasugai, S., Fujisawa, R., Waki, Y., Miyamoto, K.I. and Ohya, K. 2000. Selective drug delivery system to bone: Small peptide (Asp)6 conjugation. *J. Bone Miner. Res.*, 15(5): 936–943, doi: 10.1359/jbmr.2000.15.5.936.

Kima, B.K., Kanga, H., Doha, K.O., Lee, S.H., Park, J.W., Lee, S.J. and Lee, T.J. 2012. Homodimeric SV40 NLS peptide formed by disulfide bond as enhancer for gene delivery. *Bioorganic Med. Chem. Lett.*, 22(17): 5415–5418, doi: 10.1016/j.bmcl.2012.07.051.

Knight, J.C., Edwards, P.G. and Paisey, S.J. 2011. Fluorinated contrast agents for magnetic resonance imaging; A review of recent developments. *RSC Adv.*, 1(8): 1415–1425, doi: 10.1039/c1ra00627d.

Kociurzynski, R., Pannuzzo, M. and Böckmann, R.A. 2015. Phase transition of glycolipid membranes studied by coarse-grained simulations. *Langmuir*, 31(34): 9379–9387, doi: 10.1021/acs.langmuir.5b01617.

Kühnle, A. 2009. Self-assembly of organic molecules at metal surfaces. *Curr. Opin. Colloid Interface Sci.*, 14(2): 157–168, doi: 10.1016/j.cocis.2008.01.001.

Kumar, S.R., Quinn, T.P. and Deutscher, S.L. 2007. Evaluation of an 111in-radiolabeled peptide as a targeting and imaging agent for ErbB-2 receptor-expressing breast carcinomas. *Clin. Cancer Res.*, 13(20): 6070–6079, doi: 10.1158/1078-0432.CCR-07-0160.

Lee, P.Y., Cobain, E., Huard, J. and Huang, L. 2007. Thermosensitive hydrogel PEG-PLGA-PEG enhances engraftment of muscle-derived stem cells and promotes healing in diabetic wound. *Mol. Ther.*, 15(6): 1189–1194, doi: 10.1038/sj.mt.6300156.

Leonidova, A., Pierroz, V., Rubbiani, R., Lan, Y., Schmitz, A.G., Kaech, A., Sigel, R.K.O., Ferrarib, S. and Gasser, G. 2014. Photo-induced uncaging of a specific Re(i) organometallic complex in living cells. *Chem. Sci.*, 5(10): 4044–4056, doi: 10.1039/c3sc53550a.

Li, J., Tang, S., Lu, L. and Zeng, H.C. 2007. Preparation of nanocomposites of metals, metal oxides, and carbon nanotubes via self-assembly. *J. Am. Chem. Soc.*, 129(30): 9401–9409.

Li, L., Ma, B. and Wang, W. 2020. Peptide-based nanomaterials for tumor immunotherapy. *Molecules (Basel, Switzerland)*, 26(1). NLM (Medline), Dec. 30, 2020, doi: 10.3390/molecules26010132.

Li, L.L., An, H.W., Peng, B., Zheng, R. and Wang, H. 2019. Self-assembled nanomaterials: Design principles, the nanostructural effect, and their functional mechanisms as antimicrobial or detection agents. *Mater. Horizons*, 6(9): 1794–1811, doi: 10.1039/c8mh01670d.

Li, S., Meng Lin, M., Toprak, M.S., Kim, D.K. and Muhammed, M. 2010. Nanocomposites of polymer and inorganic nanoparticles for optical and magnetic applications. *Nano. Rev.*, 1(1): 5214.

Li, S.-L., Xiao, T., Lin, C. and Wang, L. 2012. Advanced supramolecular polymers constructed by orthogonal self-assembly. *Chem. Soc. Rev.*, 41(18): 5950–5968.

Licha, K., Hessenius, C., Becker, A., Henklein, P., Bauer, M., Wisniewski, S., Wiedenmann, B. and Semmler, W. 2001. Synthesis, characterization, and biological properties of cyanine-labeled somatostatin analogues as receptor-targeted fluorescent probes. *Bioconjug. Chem.*, 12(1): 44–50, doi: 10.1021/bc000040s.

Liechty, W.B., Kryscio, D.R., Slaughter, B.V. and Peppas, N.A. 2010. Polymers for drug delivery systems. *Annu. Rev. Chem. Biomol. Eng.*, 1: 149–173, doi: 10.1146/annurev-chembioeng-073009-100847.

Liu, C., Zhang, K., Chen, D., Jiang, M. and Liu, S. 2010. Transforming spherical block polyelectrolyte micelles into free-suspending films via DNA complexation-induced structural anisotropy. *Chem. Commun.*, 46(33): 6135–6137, doi: 10.1039/c0cc00902d.

Loh, X.J., Tsai, M.H., Del Barrio, J., Appel, E.A., Lee, T.C. and Scherman, O.A. 2012. Triggered insulin release studies of triply responsive supramolecular micelles. *Polym. Chem.*, 3(11): 3180–3188, doi: 10.1039/c2py20380d.

Lombardo, D., Calandra, P., Pasqua, L. and Magazù, S. 2020. Self-assembly of organic nanomaterials and biomaterials: The bottom-up approach for functional nanostructures formation and advanced applications. *Materials (Basel)*, 13(5), doi: 10.3390/ma13051048.

Lu, Y., Sun, W. and Gu, Z. 2014. Stimuli-responsive nanomaterials for therapeutic protein delivery. *Journal of Controlled Release, Elsevier*, 194: 1–19, Nov. 28, 2014, doi: 10.1016/j.jconrel.2014.08.015.

Lu, Z.R., Ye, F. and Vaidya, A. 2007. Polymer platforms for drug delivery and biomedical imaging. *J. Control. Release*, 122(3): 269–277, doi: 10.1016/j.jconrel.2007.06.016.

Mas-Moruno, C., Fraioli, R., Rechenmacher, F., Neubauer, S., Kapp, T.G. and Kessler, H. 2016. αvβ3- or α5β1-Integrin-selective peptidomimetics for surface coating. *Angew. Chemie - Int. Ed.*, 55(25): 7048–7067, doi: 10.1002/anie.201509782.

Matsumoto, M., Takenaka, M., Sawamoto, M. and Terashima, T. 2019. Self-assembly of amphiphilic block pendant polymers as microphase separation materials and folded flower micelles. *Polym. Chem.*, 10(36): 4954–4961, doi: 10.1039/c9py01078e.

Maxfield, F.R. and Tabas, I. 2005. Role of cholesterol and lipid organization in disease. *Nature*, 438(7068): 612–621, doi: 10.1038/nature04399.

Miao, Z., Gao, Z., Chen, R., Yu, X., Su, Z. and Wei, G. 2018. Surface-bioengineered gold nanoparticles for biomedical applications. *Curr. Med. Chem.*, 25(16): 1920–1944.

Nie, Z., Petukhova, A. and Kumacheva, E. 2010. Properties and emerging applications of self-assembled structures made from inorganic nanoparticles. *Nature Nanotechnology*, 5(1): 15–25. Nature Publishing Group, doi: 10.1038/nnano.2009.453.

Okesola, B.O., Wu, Y., Derkus, B., Gani, S., Wu, D., Knani, D., Smith, D.K., Adams, D.J. and Mata, A. 2019. Supramolecular self-assembly to control structural and biological properties of multicomponent hydrogels. *Chem. Mater.*, 31(19): 7883–7897, doi: 10.1021/acs.chemmater.9b01882.

Panda, J.J. and Chauhan, V.S. 2014. Short peptide based self-assembled nanostructures: Implications in drug delivery and tissue engineering. *Polymer Chemistry, Royal Society of Chemistry*, 5(15): 4418–4436, Aug. 07, 2014, doi: 10.1039/c4py00173g.

Pandey, P. and Dahiya, M. 2016. A brief review on inorganic nanoparticles. *J. Crit. Rev.*, 3(3): 18–26.

Patil, A.J. and Mann, S. 2008. Self-assembly of bio-inorganic nanohybrids using organoclay building blocks. *J. Mater. Chem.*, 18(39): 4605–4615.

Peer, D., Karp, J.M., Hong, S., Farokhzad, O.C., Margalit, R. and Langer, R. 2007. Nanocarriers as an emerging platform for cancer therapy. *Nat. Nanotechnol.*, 2(12): 751–760, doi: 10.1038/nnano.2007.387.

Petrache, H.I., Tristram-Nagle, S., Gawrisch, K., Harries, D., Parsegian, V.A. and Nagle, J.F. 2004. Structure and fluctuations of charged phosphatidylserine bilayers in the absence of salt. *Biophys. J.*, 86(3): 1574–1586, doi: 10.1016/S0006-3495(04)74225-3.

Quesada, M., Muniesa, C. and Botella, P. 2013. Hybrid PLGA-organosilica nanoparticles with redox-sensitive molecular gates. *Chem. Mater.*, 25(13): 2597–2602.

Rajak, A., Karan, C.K., Theato, P. and Das, A. 2020. Supramolecularly cross-linked amphiphilic block copolymer assembly by the dipolar interaction of a merocyanine dye. *Polym. Chem.*, 11(3): 695–703, doi: 10.1039/c9py01492f.

Ruiz Hitzky, E. 2003. Functionalizing inorganic solids: towards organic-inorganic nanostructured materials for intelligent and bioinspired systems. *Chem. Rec.*, 3(2): 88–100.

Rupar, P.A., Chabanne, L., Winnik, M.A. and Manners, I. 2012. Non-centrosymmetric cylindrical micelles by unidirectional growth. *Science (80-.)*, 337(6094): 559–562, doi: 10.1126/science.1221206.

Sanchez, C., Soler-Illia, G.J.D.A.A., Ribot, F., Lalot, T., Mayer, C.R. and Cabuil, V. 2001. Designed hybrid organic-inorganic nanocomposites from functional nanobuilding blocks. *Chemistry of Materials*, 13(10): 3061–3083, doi: 10.1021/cm011061e.

Sand, H. and Weberskirch, R. 2015. Bipyridine-functionalized amphiphilic block copolymers as support materials for the aerobic oxidation of primary alcohols in aqueous media. *RSC Adv.*, 5(48): 38235–38242, doi: 10.1039/c5ra05715a.

Scholes, G.D. 2008. Controlling the optical properties of inorganic nanoparticles. *Adv. Funct. Mater.*, 18(8): 1157–1172.

Sengupta, S., Eavarone, D., Capila, I., Zhao, G., Watson, N., Kiziltepe, T. and Sasisekharan, R. 2005. Temporal targeting of tumour cells and neovasculature with a nanoscale delivery system. *Nature*, 436(7050): 568–572, doi: 10.1038/nature03794.

Shrestha, R., Elsabahy, M., Luehmann, H., Samarajeewa, S., Florez-Malaver, S., Lee, N.S., Welch, M.J., Liu, Y. and Wooley, K.L. 2012. Hierarchically assembled theranostic nanostructures for siRNA delivery and imaging applications. *J. Am. Chem. Soc.*, 134(42): 17362–17365, doi: 10.1021/ja306616n.

Smith, D.M., Schüller, V., Forthmann, C., Schreiber, R., Tinnefeld, P. and Liedl, T. 2011. A structurally variable hinged tetrahedron framework from DNA origami. *J. Nucleic Acids*, vol. 2011, doi: 10.4061/2011/360954.

Song, W., Kuang, J., Li, C.X., Zhang, M., Zheng, D., Zeng, X., Liu, C. and Zhang, X.Z. 2018. Enhanced immunotherapy based on photodynamic therapy for both primary and lung metastasis tumor eradication. *ACS Nano*, 12(2): 1978–1989, doi: 10.1021/acsnano.7b09112.

Stupp, S.I. and Palmer, L.C. 2014. Supramolecular chemistry and self-assembly in organic materials design. *Chem. Mater.*, 26(1): 507–518, doi: 10.1021/cm403028b.

Subbarao, N.K., Parente, R.A., Szoka, F.C., Nadasdi, L. and Poneracz, K. 1987. pH-dependent bilayer destabilization by an amphipathic peptide. *Biochemistry*, 26(11): 2964–2972, doi: 10.1021/bi00385a002.

Sugahara, K.N., Scodeller, P., Braun, G.B., de Mendoza, T.H., Yamazaki, C.M., Kluger, M.D., Kitayama, J., Alvarez, E., Howell, S.B., Teesalu, T., Ruoslahti, E. and Lowy, A.M. 2015. A tumor-penetrating peptide enhances circulation-independent targeting of peritoneal carcinomatosis. *J. Control. Release*, 212: 59–69, doi: 10.1016/j.jconrel.2015.06.009.

Sun, Y. and Xia, Y. 2003. Gold and silver nanoparticles: a class of chromophores with colors tunable in the range from 400 to 750 nm. *Analyst*, 128(6): 686–691.

Tanford, C. 1974. Theory of micelle formation in aqueous solutions. *J. Phys. Chem.*, 78(24): 2469–2479, doi: 10.1021/j100617a012.

Torchilin, V.P. 2005. Recent advances with liposomes as pharmaceutical carriers. *Nat. Rev. Drug Discov.*, 4(2): 145–160, doi: 10.1038/nrd1632.

Tu, C., Zhu, L., Li, P., Chen, Y., Su, Y., Yan, D., Zhu, X. and Zhou, G. 2011. Supramolecular polymeric micelles by the host–guest interaction of star-like calix[4]arene and chlorin e6 for photodynamic therapy. *Chem. Commun.*, 47(21): 6063–6065, doi: 10.1039/c0cc05662f.

Van Meer, G., Voelker, D.R. and Feigenson, G.W. 2008. Membrane lipids: Where they are and how they behave. *Nat. Rev. Mol. Cell Biol.*, 9(2): 112–124, doi: 10.1038/nrm2330.

Vashist, A. and Ahmad, H. 2013. Hydrogels: Smart materials for drug delivery. *Orient. J. Chem.*, 29(3): 861–870, doi: 10.13005/ojc/290303.

Veatch, S.L. and Keller, S.L. 2005. Miscibility phase diagrams of giant vesicles containing sphingomyelin. *Phys. Rev. Lett.*, 94(14): 148101, doi: 10.1103/PhysRevLett.94.148101.

Wallyn, J., Anton, N., Akram, S. and Vandamme, T.F. 2019. Biomedical imaging: principles, technologies, clinical aspects, contrast agents, limitations and future trends in nanomedicines. *Pharm. Res.*, 36(6): 1–31, doi: 10.1007/s11095-019-2608-5.

Wang, D., Tong, G., Dong, R., Zhou, Y., Shen, J. and Zhu, X. 2014. Self-assembly of supramolecularly engineered polymers and their biomedical applications. *Chem. Commun.*, 50(81): 11994–12017, doi: 10.1039/c4cc03155e.

Wang, H., Wang, S., Su, H., Chen, K.J., Armijo, A.L., Lin, W.Y., Wang, Y., Sun, J., Kamei, K., Czernin, J., Radu, C.G. and Tseng, H.R. 2009. A supramolecular approach for preparation of size-controlled nanoparticles. *Angew. Chemie - Int. Ed.*, 48(24): 4344–4348, doi: 10.1002/anie.200900063.

Wang, H., Yan, Y.Q., Yi, Y., Wei, Z.Y., Chen, H., Xu, J.F., Wang, H., Zhao, Y. and Zhang, X. 2020. Supramolecular peptide therapeutics: Host – Guest Interaction-Assisted Systemic Delivery of Anticancer Peptides, 2(6): 739–748, doi: 10.31635/ccschem.020.202000283.

Wang, J., Liu, K., Xing, R. and Yan, X. 2016. Peptide self-assembly: Thermodynamics and kinetics. *Chemical Society Reviews. Royal Society of Chemistry*, 45(20): 5589–5604, Oct. 21, 2016, doi: 10.1039/c6cs00176a.

Wang, L., Li, W., Lu, J., Zhao, Y.X., Fan, G., Zhang, J.P. and Wang, H. 2013. Supramolecular nano-aggregates based on bis(pyrene) derivatives for lysosome-targeted cell imaging. *J. Phys. Chem. C*, 117(50): 26811–26820, doi: 10.1021/jp409557g.

Wu, Z., Yang, S. and Wu, W. 2016. Shape control of inorganic nanoparticles from solution. *Nanoscale*, 8(3): 1237–1259.

Xie, W., Xu, G. and Feng, X. 2012. Self-assembly of lipids and nanoparticles in aqueous solution: Self-consistent field simulations. *Theor. Appl. Mech. Lett.*, 2(1): 014004, doi: 10.1063/2.1201404.

Yang, W., Veroniaina, H., Qi, X., Chen, P., Li, F. and Ke, P.C. 2020. Soft and condensed nanoparticles and nanoformulations for cancer drug delivery and repurpose. *Adv. Ther.*, 3(1): 1900102, doi: 10.1002/adtp.201900102.

Yang, Y., Kong, M., Feng, P., Meng, R., Wei, X., Su, P., Cao, J., Feng, W., Liu, W. and Tang, Y. 2019. Self-Assembly of heterogeneous structured rare-earth nanocrystals controlled by selective crystal etching and growth for optical encoding. *ACS Appl. Nano Mater.*, 2(6): 3518–3525.

Yu, C.Y., Yuan, Z., Cao, Z., Wang, B., Qiao, C., Li, J. and Xiao, X. 2009. A muscle-targeting peptide displayed on AAV2 improves muscle tropism on systemic delivery. *Gene Ther.*, 16(8): 953–962, doi: 10.1038/gt.2009.59.

Yuan, Y., Zhang, C.J., Gao, M., Zhang, R., Tang, B.Z. and Liu, B. 2015. Specific light-up bioprobe with aggregation-induced emission and activatable photoactivity for the targeted and image-guided photodynamic ablation of cancer cells. *Angew. Chemie - Int. Ed.*, 54(6): 1780–1786, doi: 10.1002/anie.201408476.

Zhai, F.Y., Huang, W., Wu, G., Jing, X.K., Wang, M.J., Chen, S.C., Wang, Y.Z., Chin, I.J. and Liu, Y. 2013. Nanofibers with very fine core-shell morphology from anisotropic micelle of amphiphilic crystalline-coil block copolymer. *ACS Nano*, 7(6): 4892–4901, doi: 10.1021/nn401851w.

Zhang, C., Ko, S.H., Su, M., Leng, Y., Ribbe, A.E., Jiang, W. and Mao, C. 2009. Symmetry controls the face geometry of DNA polyhedra. *J. Am. Chem. Soc.*, 131(4): 1413–1415, doi: 10.1021/ja809666h.

Zhang, C., Li, X., Tian, C., Yu, G., Li, Y., Jiang, W. and Mao, C. 2014. DNA nanocages swallow gold nanoparticles (AuNPs) to form AuNP@DNA cage core-shell structures. *ACS Nano*, 8(2): 1130–1135, doi: 10.1021/nn406039p.

Zhang, K., Miao, H. and Chen, D. 2014. Water-soluble monodisperse core-shell nanorings: Their tailorable preparation and interactions with oppositely charged spheres of a similar diameter. *J. Am. Chem. Soc.*, 136(45): 15933–15941, doi: 10.1021/ja5099963.

Zhang, Q., Gupta, S., Emrick, T. and Russell, T.P. 2006. Surface-functionalized CdSe nanorods for assembly in diblock copolymer templates. *J. Am. Chem. Soc.*, 128(12): 3898–3899, doi: 10.1021/ja058615p.

Zhang, Q., Lin, J., Wang, L. and Xu, Z. 2017. Theoretical modeling and simulations of self-assembly of copolymers in solution. *Prog. Polym. Sci.*, 75: 1–30, doi: 10.1016/j.progpolymsci.2017.04.003.

Zhao, N., Yan, L., Zhao, X., Chen, X., Li, A., Zheng, D., Zhou, X., Dai, X. and Xu, F.J. 2018. Versatile types of organic/inorganic nanohybrids: From strategic design to biomedical applications. *Chem. Rev.*, 119(3): 1666–1762, doi: 10.1021/acs.chemrev.8b00401.

Zheng, X., Zhu, Q., Song, H., Zhao, X., Yi, T., Chen, H. and Chen, X. 2015. *In situ* synthesis of self-assembled three-dimensional graphene-magnetic palladium nanohybrids with dual-enzyme activity through one-pot strategy and its application in glucose probe. *ACS Appl. Mater. Interfaces*, 7(6): 3480–3491.

Zhuang, Z., Jiang, T., Lin, J., Gao, L., Yang, C., Wang, L. and Cai, C. 2016. Hierarchical nanowires synthesized by supramolecular stepwise polymerization. *Angew. Chemie*, 128(40): 12710–12715, doi: 10.1002/ange.201607059.

CHAPTER 2

Surface Stabilization and Functionalities of Nanostructures

*M. Cruz-Leal, M. Ávila-Gutiérrez and E. Coutino-Gonzalez**

1. Introduction

In the last 30 years, the development of novel technologies based on nanostructures has directed the efforts of the scientific community to unravel the new physicochemical properties of such nanostructures compared to macroscopic materials. Nanotechnology has proven that the development of nanostructures is based on the interconnection of different scientific disciplines such as physics, chemistry, medicine, or engineering, to name a few. Nevertheless, the scientific community currently faces the challenge of understanding the different phenomena at nanoscale (such as the stabilization and functionalization processes), which is directly linked to the rational design of materials for novel applications in different fields.

One of the most relevant materials nowadays are metal oxide semiconductors (MOx's), which display a high versatility making them excellent candidates for their use in different applications such as gas sensors, optoelectronic materials, water splitting, biosensors platforms, smart materials, among others. Nevertheless, to achieve such versatility in MOx's, the development of stabilization and functionalization protocols is of paramount relevance.

This chapter describes several concepts related to the synthesis, characterization, stabilization, and functionalization of pertinent MOx's thin films. Two types of thin film deposition techniques are mainly addressed: chemical vapor deposition (CVD) and ultrasonic spray pyrolysis (USP). Different methods of stabilization and functionalization of surfaces are also brought into focus. Moreover, the development of new strategies to expand the palette of MOx's nanostructures applications is discussed.

1.1 Nanomaterials

Nanomaterials are defined as objects in which at least one dimension is between 1 nm and 100 nm (1 m = 1×10^{-9} nm). They can be in solid form (powders), liquid (colloidal suspensions), and thin films. Up to now, nanomaterials have been involved in numerous applications such as biotechnology, cosmetics, medicine, electronics, among others. This is due to the high

Nanophotonics & Functional Materials Group. Centro de Investigaciones en Óptica A.C., Loma del Bosque 115, Lomas del Campestre León, Guanajuato 37150, México.
* Corresponding author: ecoutino@cio.mx

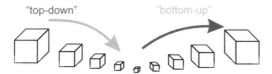

Figure 1. Schematic design of nanomaterial manufacturing methods (top-down/bottom-up).

surface/volume ratio, conferring nanometric-scale materials with peculiar and novel properties such as chemical reactivity, mechanical resistance, enhanced magnetism, or conductivity. Nanomaterials are significantly different from those at the macroscopic scale (bulk size) and can be obtained, in general terms, using two different approaches: top-down (miniaturization) and bottom-up (growth) (see Figure 1).

In the top-down approach, the strategy is to reduce the size of the material by chemical or mechanical methods, for example, through focused ion beam techniques or milling. In contrast, in the bottom-up strategy, the growing of nanostructures is due to the agglomeration of atoms or molecules during the nucleoid formation followed by condensation growth using physical (Iii et al. 2015, Chichkov et al. 1996) or chemical methods (Costanzo et al. 2016) and kinetic synthesis (Vivien et al. 2019).

According to their dimensions, nanomaterials are classified as zero-dimensional (0D), one-dimensional (1D), two-dimensional (2D) and three-dimensional (3D). The 0Ds have three dimensions in the sub-nanometric regime (quantum dots); in the 1Ds, one of their dimensions is outside the nanometric range, this classification includes nanotubes and nanowires; 2D materials possess two dimensions outside the nanometric range, such as thin films or plates. Finally, in 3D materials (also known as superlattices or supercrystals) (Pileni et al. 2014), self-assembly of very well organized nanoparticles gains relevance during the growth of the materials, which can be classified as heterogeneous growth (layer-by-layer deposition of nanoparticles due to evaporation of a good solvent) or homogeneous growth, which form colloidal crystals due to destabilization of a poor solvent (Boles et al. 2016) (see Figure 2).

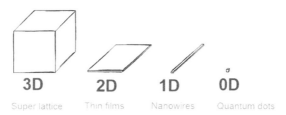

Figure 2. Classification of nanomaterials according to their dimensions.

1.2 Metal oxide semiconductors

Metal oxides (MOx) semiconductors combine an oxide anion with a cation of a single metal and display interesting optical, mechanical, electrical, and chemical properties derived from their characteristics of electronic charge transport, defects in the crystal lattice, and composition. According to classical band theory, the defects introduce discrete energy levels in the band of the material; the majority charge carriers (electrons or holes) are derived in *n*-type (NiO, CuO) or *p*-type (SnO$_2$, In$_2$O$_3$, WO$_3$) semiconductors. Based on their physical properties, MOx´s are classified as (a) non-transitions metal-oxides and (b) transition metal-oxides (Madhuri 2020).

(a) Non-transition metal oxides: A large bandgap separates the filled valence band and empty conduction band. For example, MgO and SiO$_2$.

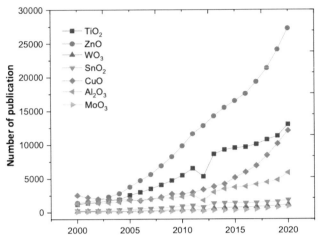

Figure 3. Publications per year related to the most studied MOx from 2000 to 2020. Search string: ALL FILES (TiO$_2$, ZnO, WO$_3$, SnO$_2$, CuO, Al$_2$O$_3$) *Source*: Scopus.

(b) Transition metal oxides are formed by combining a metal transition and oxygen atoms, where *d* orbitals in the metal are partially filled; this characteristic confers them good thermal and electrical conductivity, moreover, transition metal oxides display multiple oxidation states (Ashrit 2017). Examples: ZnO, MnO, WO$_2$, MoO$_3$, CrO$_3$.

Currently, some of the most employed MOx are titanium dioxide (TiO$_2$), zinc oxide (ZnO), tungsten oxide (WO$_3$), tin dioxide (SnO$_2$), copper oxide (CuO), aluminum oxide (Al$_2$O$_3$), and molybdenum oxide (MoO$_3$), to name a few. Figure 3 depicts the evident growth in the number of publications for MOx in the last two decades.

2. Metal oxide semiconductor thin films

Generally, thin films are described as layers of a material deposited on a substrate; the thickness can be in the range of atomic layer to one micrometer. Fabrication of thin films involves bottom-up and top-down methods highly related to physical and chemical deposition techniques (Figure 4). The films exhibit different properties depending on the material, structure, thickness, and deposition conditions. Different characterization techniques are required to unravel the composition, morphology, and structure of thin films; some techniques are briefly described below (Kodigala 2014):

a. Energy dispersive X-ray spectroscopy (EDS): Analysis of the chemical composition in a sample through X-ray emitted by the atoms present in the sample. It is useful for materials composed of elements with atomic numbers higher than carbon.

b. X-ray diffraction (XRD): Provides information on the crystalline structure of the materials and crystallite size.

c. Ellipsometry: Useful technique to determine thin film thickness, growth, surface quality, and optical constant (Hilfiker 2011).

d. Atomic force microscopy (AFM): Employed to quantify roughness in the surface of a material and in the determination of hardness, electrical, and thermal properties (Tararam et al. 2017).

e. Scanning electron microscopy (SEM): Useful technique to visualize structures and morphologies on the surface of a sample.

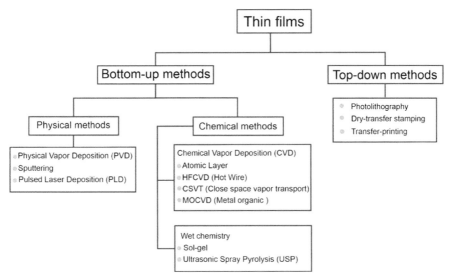

Figure 4. Classification of bottom-up and top-down methodologies to produce thin films, indicating the position of CVD and USP techniques.

f. X-ray photoelectron spectroscopy (XPS): It is employed to investigate the elemental and chemical information and the electronic state of the atoms of a certain material.

g. Transmission electron microscopy (TEM): Provide high-resolution images and quantitative and qualitative chemical information at resolutions below 1 nm.

Thin films are employed in many research fields thanks to their optical, electrical, and physical properties. The growth of thin films can be achieved by different techniques such as sol-gel, ultrasonic spray pyrolysis (USP), sputtering, or chemical vapor deposition (CVD). Every technique produces thin films with specific characteristics, whereby the chosen deposition route depends on the particular needs of the application. In this chapter, two different deposition procedures to fabricate MOx´s thin films, namely chemical vapor deposition (CVD) and ultrasonic spray pyrolysis (USP), will be reviewed.

2.1 Chemical vapor deposition

Chemical vapor deposition (CVD) consists of the dissociation of a material (in solid or liquid phase) activated by heat, light, or plasma. The gas molecules are deposited onto a substrate leading to the growth of solid materials (Pierson 1999). The main advantages of this technique are the versatility for depositing a large variety of elements and compounds, purity of the materials obtained, high reproducibility, uniformity of the films, and a large area of deposit (Frey and Khan 2015). Disadvantages include the use of hazardous gases as precursors, and in some cases, requires highly sophisticated vacuum systems.

CVD systems can display different configurations according to the desired properties of the targeted material. In general, two kinds of CVD reactors are mainly used: hot-wall and cold-wall (Arjmandi-Tash et al. 2017). The first employs a furnace to heat the substrate; this guarantees temperature control, whereas in the second configuration the substrate is locally heated, thus having more control on the deposition rate. The specific characteristics of the synthesized materials and properties are mainly associated with experimental variables, such as precursor materials, atmosphere, temperature, substrate, or reactor type. According to the specific conditions of the process, different CVD variants have been developed and are displayed in Table 1.

Table 1. Main variants of CVD technique and operating principles.

Variant	Operating principle	References
Hot filament chemical vapor deposition (HFCVD)	A filament is heated until chemically decomposed	(Cruz-Leal et al. 2019)
Metal-organic chemical vapor deposition (MOCVD)	Metal-organics are employed as precursors	(Barreca et al. 2003)
Close-space vapor transport (CSVT)	Substrate closely located to the source	(Goiz et al. 2010)
Aerosol-assisted chemical vapor deposition (AACVD)	Precursors are in liquid or gas aerosols	(Stoycheva et al. 2014)
Atomic layer chemical vapor deposition (ALD)	Monolayers of different materials	(Nandi and Sarkar 2014)
Microwave electron cyclotron resonance chemical vapor deposition (MWECR-CVD)	Electron cyclotron resonance by the interaction between microwave and magnetic field	(Zhu et al. 2008)

2.2 Ultrasonic spray pyrolysis

Ultrasonic spray pyrolysis (USP) is one of the most widely used techniques for the deposition of thin films of different metal oxides' nanostructures. Due to its versatility in the homogeneous coating of large deposition substrate areas and the modification of conditions (temperature, atmosphere, frequency, and intensity of ultrasound), the control of film´s morphology and thickness can be nicely adjusted. It is also an excellent alternative for upscaling thin films to an industrial level, reducing production costs compared to other methods in producing thin films (see Figure 5 for a schematic representation of this procedure).

USP technique has also been reported as a method for synthesizing nanoscale materials in powders or particles, having excellent morphological control of nanostructures, and obtaining different uniform particle sizes. Suslick and collaborators reported the formation of nanostructured silica (Bang and Suslick 2010) and carbon (Skrabalak and Suslick 2007), among other materials by USP. Kundu et al. reported the synthesis of $NiCo_2O_4$ nanospheres with high electrochemical responses and employed them as anode materials in supercapacitors or as cathode materials for lithium-ion batteries (Kundu et al. 2018). Meanwhile, Bogovic and collaborators propose a possible model in forming nanoparticles synthesized by USP in their studies on the synthesis of Al_2O_3 nanospheres. This model suggests the formation of nano-objects by evaporation, drying, and calcination of smaller droplets of metal precursors within a solvent (Matula et al. 2013).

USP working principle mainly consists of three steps:

(1) Ultrasound: Its main role is the nebulization of precursors due to the effect of the formation of capillary waves (waves travel along with the interface between two fluids), generating small droplets (sizes average from 1 to 20 μm), and finer sphericity (Chatzikyriakou et al. 2017, Bang and Suslick 2010). Special care should be taken to avoid confusion with the ultrasound used

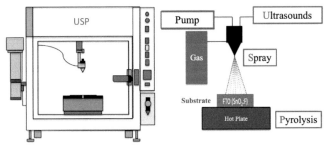

Figure 5. Ultrasonic spray pyrolysis scheme for the deposition of micro-droplets on the formation of thin films.

in sonochemistry, which induces chemical reactions due to the formation of implosive bubbles (acoustic cavitation) (Skrabalak and Suslick 2005, Bang and Suslick 2010).

(2) Spray: Once the micro-droplets have been generated during the nebulization of the precursor, they pass through a nozzle that spreads them uniformly and homogeneously on a selected substrate (polymer, paper, transparent conductive oxides, etc.). The spray must accomplish specific parameters for the control of morphology and thickness in the deposition of thin films.

(3) Pyrolysis: Finally, the substrate is heated (between 50ºC to 1000ºC) during the spreading of the micro-droplets of the precursor. The heating plays two different roles, allowing the evaporation of the solvent where the precursors are contained as organic or aqueous phases and removing the surfactant to have continuous films free of structural fractures. During the calcination, the growth processes of the mesostructured materials take place (Chatzikyriakou et al. 2017).

The fabrication of metal oxides nanostructured thin films using USP has been employed to develop functional materials for different applications due to the diverse physico-chemical characteristics of nanostructures. With the development of optoelectronic devices in past years, several research groups have focused on the innovation of electrochromic materials using MOx, which can absorb ultraviolet (UV), near-infrared (NIR), and visible light. Chatzikyriakou et al. reported the deposition of mesoporous WO_3 films on conductive SnO_2:F (FTO) substrates, obtaining high coloring efficiency due to UV irradiation (Chatzikyriakou et al. 2017). On the other hand, Milliron and co-workers reported the synthesis of tin plasmonic nanocrystals doped with indium oxides deposited on FTO conductive substrates, obtaining active electrochromic materials for near-infrared radiation (1500 to 2000 nm) (Maho et al. 2020). The use of USP facilitates the low-cost production of thin films with high-quality materials. Fauzia and collaborators developed electrodes for UV radiation photodetectors with high sensitivity, responsivity, and detectability, by depositing zinc oxide (ZnO) film on In_2O_3:Sn (ITO) substrates, obtaining a film with irregular crystal size (~ 48 nm), as well as in morphologies, obtaining comparable results to those observed in highly studied electrodes for photodetectors based on noble metals (Au, Ag, Pt) (Putri et al. 2020). Other applications involving the fabrication of nanostructured MOx thin films using ultrasonic spray pyrolysis include the development of materials for hydrogen production through water splitting in solar energy conversion devices. Li and co-workers reported low photoelectrochemical efficiencies in water separation processes using $BiVO_4$ thin films on ITO substrates; however, by doping with WO_3 and conducting thin-film reduction in a hydrogen atmosphere, a high photoelectrochemical efficiency for hydrogen production in water splitting processes was obtained (Li et al. 2010). Table 2 depicts different possible applications of nanostructured metal oxide thin films obtained by USP technique.

In the first part of this section, we addressed USP operation principles, which consists of three steps: the formation of particles by ultrasound, the uniform deposition of the particles via spray spreading, and finally, the calcination of the materials at specific temperatures. However, other parameters should be considered for the deposition of thin films using this technique:

(A) Solution flow rate

An essential factor to consider during the thin film deposition process since the flow rate controls the droplet size in the spray due to the amount of liquid deposited on the substrate (Neagu et al. 2006). Zeggar et al. reported the optical variation of CuO thin films between 1.4 eV – 1.6 eV, where a high flow rate promoted the formation of small CuO crystals (Lamri Zeggar et al. 2015).

(B) Deposition rate on the substrate

This process can significantly influence different aspects during the deposition of metal oxides on substrates, such as the thin film thickness, the size of the deposited nanocrystals, the amount of deposited material, among others. Aoun and co-workers determined the effects caused at different

Table 2. Main applications of MOx thin films produced by USP.

Material	Substrate	Application	Reference
ZnO	Glass substrates	Solar cells	(Aranovich et al. 1979)
SnO_2-Fe_2O_3	Quartz plates	Gas Sensors	(Racheva et al. 1994)
SnO_2:MoO_3	Alumina substrates	Gas Sensors	(Firooz et al. 2010)
SnO_2:Rh	Alumina substrates	Gas Sensors	(Cho et al. 2015)
ZrO_xN_y/ZrO_2	Stainless steel plates	Anti-corrosion	(Cubillos et al. 2015)
TiO_2	Steel substrates	Anti-corrosion	(Yang and Biswas 1999)
RuO_2	Stainless steel plates	Anti-corrosive	(Fugare and Lokhande 2017)
Core-shell Ag/ZnP	Nano-composites	Photocatalysis	(Muñoz-Fernandez et al. 2019)
Composite TiO_2	Graphene oxide	Photocatalysis	(Park et al. 2018)

deposition rates of colloidal solutions on a substrate (between 10 and 35 ml) for the production of ZnO thin films, obtaining diverse structural, optical, and conductivity characteristics and electrical resistance due to the increase of deposited volumetric quantities of colloidal solution on a substrate. (Aoun et al. 2015).

(C) Distance between the nozzle and the substrate

The distance between the colloidal solution nozzle and the substrate (where the particles are deposited) is one of the most important parameters to take into account; distances between 3 to 9 cm have been reported in literature, which can affect the distribution of the particles on the substrate surface. A smaller space favors the obtention of a thicker film by having a total coverage of the deposit on the substrate. On the other hand, depositing at a larger distance will result in a higher evaporation rate of the smaller droplets of the solution during transport to the substrate, resulting in more separated particles. It should be noted that the flow rate of the solution directly influences the distance between the nozzle and the substrate. Neagu et al. showed that the achievement of fracture-free zirconium oxide coatings is related to the distance between the nozzle and the substrate, obtaining a thin film free of fractures with a distance of 57 mm. Meanwhile, depositing at distances less than 40 mm, the films displayed structural fractures (Neagu et al. 2006).

(D) Pyrolysis temperature

It provides control of the solvent evaporation temperature of the solution droplets, the calcination of the surfactant, and the mesoporous or crystalline characteristics of the thin films. Oh et al. reported calcination effects on substrates of Co_3O_4, CuO, and NiO thin films, obtaining different shapes and sizes of crystals, correlating with the material's electrochemical properties (Oh et al. 2007).

MOx semiconductor thin film fabrication requires certain conditions to facilitate their production without being affected by external or internal factors that may impact their structural or chemical properties. Several stabilization methods are carried out to generate high-quality thin films in different MOx materials. In the following section, we will revise certain stabilization parameters for the formation of MOx semiconductor thin films.

2.3 Stabilization

Some materials at the nanoscale tend to change their behavior or properties under different conditions. In this chapter, stabilization can be understood as the conservation of the material's properties when exposed to determined external conditions, maintaining its optimal performance in the application for which it was designed. Many factors can cause oxidation, reduction, crystallinity change, or physicochemical changes in the MOx properties. In this contribution, the emphasis is drawn to environmental conditions, temperature, chemical, and structural stability.

2.3.1 Environmental conditions

Oxygen has an essential role in forming MOx; many studies have been published to describe synthesis methods in controlled conditions where air or water vapor contributed and accelerated the reactions to produce MOx (Malwal and Packirisamy 2018). However, vapor water also can cause cracking, changes in the morphology, and instability in the MOx´s films (Zhang 2021, Saunders et al. 2008). Water vapor also reacts with the surface of MOx in two ways: chemical and physical absorption with MOx, resulting in variation of electrical properties. This behavior is useful in humidity sensors thanks to chemical adsorption that could generate OH groups and electrons related to the conduction band of MOx (Yan et al. 2021).

On the other hand, reactions between oxygen and metals can give origin to different oxides species. For example, in tungsten oxide, partial pressure and temperature induce a change of stoichiometry of the materials evidenced by color change: yellow to WO_3, blue-violet to $WO_{2.9}$, reddish-violet to $WO_{2.7}$, and brown to WO_2 (Lassner and Schubert 1999). Rinaldi et al. reported that more oxygen vacancies within the crystal resulted in darker materials (Figure 6) (Rinaldi et al. 2018).

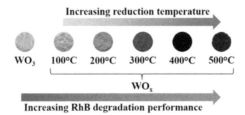

Figure 6. Process of reduction of WO_3 under different temperatures (Reprinted with permission from Rinaldi FG, *Correlations between Reduction Degree and Catalytic Properties of WOx Nanoparticles* (ACS Omega), 2018, 3(8): 8963–8970).

2.3.2 Temperature

Annealing treatments are mainly used to remove organic molecules and evaporate solvents in thin films; this process may also modify the crystallinity, crystallite size, defects, stoichiometry, morphology, and structure of MOx thin films. For example, the annealing of ZnO thin films at 600°C in O_2 induces changes from paramagnetic to ferromagnetic properties (Lim et al. 2021). In WO_3, the annealing treatment has been related to modifications in the morphology, or in the promotion of changes from amorphous to crystalline phase of TiO_2 thin films (RM et al. 2020).

Temperature changes can also promote fractures in thin films, which are caused by dehydration or removal of organic compounds (Roselló-Márquez et al. 2020); for certain applications, it is of paramount importance to avoid this kind of damage in the films (such as in solar cells). One way to protect MOx thin films is by using thermal barrier coatings, which must be thermally stable, chemically inert, resistant to corrosion/oxidation, and adhesive to the substrate; in this context, rare earth metals, zirconates, or mullite coatings are mainly employed (Zazpe et al. 2017).

On the other hand, rapid thermal annealing (RTA) is employed to improve the material's crystallinity by precisely controlling the heating and cooling of the materials. The heating rate is kept very high to minimize any effect of diffusion (RM et al. 2020); moreover, this technique allows the activation of doping elements in MOx, reducing morphology changes, and delaying crystallization.

2.3.3 Surfactants

The use of surfactants is essential for stabilizing microdroplets during deposition processes in the production of thin films. Surfactants must have the ability to prevent agglomeration between particles, with a good dispersion between them due to surfactant-surfactant interactions (steric and electrostatic repulsions) and having excellent solvation with certain solvents. So when the solvent is evaporated, it avoids interaction between particles. Surfactants are involved in the USP calcination process for MOx thin film deposition, where the surfactant is promoted to be eliminated, increasing its critical concentration in the form of micelles on the surface of the microdroplets, controlling the growth of mesostructures (thin films), and varying the thickness, roughness, and porosity. Among the most commonly used surfactants for the production of thin films, we find polyethylene glycol (PEG) (Denayer et al. 2014), cationic surfactants (HTAB), gemini surfactants, anionic surfactants ($NaC_{12}H_{25}SO_4$) (Bertus et al. 2013), copolymers (L62) (Lai et al. 2005) and acetic acid (CH_3COOH) (García-Sánchez et al. 2010) .

In this section, we reviewed some of the essential points for stabilizing colloidal solutions for the deposition of thin films and several topics to be considered to enhance the physicochemical properties of the substrates. In the following sections, we will discuss certain strategies to conduct thin films' functionalization.

2.4 Functionalization

Functionalization of nanostructures can be defined as expanding the material's fundamental properties to add new functionalities, characteristics, or capabilities to obtain certain functions for a possible application. Functionalization can be conducted in several ways, such as chemical modification of the surface, or structural modification, to name a few. This chapter discusses different ways to expand MOx thin film properties using heterostructures and composites.

2.4.1 Heterostructures

Heterostructures are the union of two different materials; one of the main objectives is to develop new structures with improved and tunable physicochemical properties (Vattikuti 2018). The union between semiconductors must have electric equilibrium to generate a flow of charge carriers and align the Fermi levels of both materials, allowing electrons' fluxes from higher energies to lower energy states and inducing the layer depletion, modifying the electrons' flow and electronic properties (Zappa 2018). Electronic properties and structural changes in the heterostructures depend on the properties of the materials combined during this process. The union between different semiconductors can be described as (a) random distribution (doping), (b) a base material and another on the surface, and (c) layer over layer. Some examples of heterostructures formed by two MOx have revealed the enhancement in different applications, such as TiO_2/AgO, and TiO_2/Ag_2O photocatalytic activity for degradation of organic pollutants; the junction *p-n type* semiconductors favored degradation of organic pollutants under ultraviolet and visible irradiation (Xu et al. 2021). In another example, WO_3-Pd demonstrated high efficiency to probe acetone and hydrogen in gas sensors (Chávez et al. 2013). Another field that benefited from the development of heterostructures are Li-ion batteries, for instance, WO_3/ZnO heterostructured electrodes exhibited remarkable reversible discharge capacity, high rate performance, and excellent cycling stability (Tu et al. 2021).

2.4.2 Composites

The development of composite materials can be described as the combination of two or more materials to promote a positive synergy between the properties of each material. A fundamental criterion of composite materials is their architecture, which allows a better efficiency in transporting

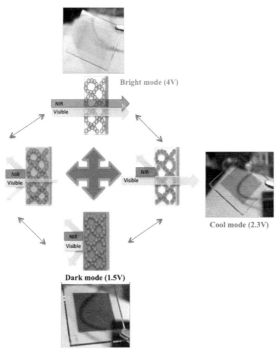

Figure 7. Schematic representation of transparent conducting oxides during the application of current in WO$_3$-NbOx nanocomposite materials (Reprinted with permission from Kim J, *Nanocomposite Architecture for Rapid, Spectrally-Selective Electrochromic Modulation of Solar Transmittance* (Nano Lett., 2015), 15(8):5574–5579).

ions, electrons, or molecular species (Kim et al. 2015). The most studied materials so far are transition metal oxide coatings, including niobium pentoxide (Nb$_2$O$_5$) and tantalum pentoxide (Ta$_2$O$_5$) which, due to their multifunctional properties, have been used in applications such as energy storage (Augustyn et al. 2013), photovoltaic materials (Le Viet et al. 2010), catalysis (Ziolek et al. 2013), and chemical stability (Rani et al. 2014).

These materials (Nb$_2$O$_5$ and Ta$_2$O$_5$) display a dual behavior in their properties when assembled with WO$_3$; they can extend the electrochromic characteristics of tungsten oxides by blocking NIR radiation (Llordés et al. 2013), as well as extending the optical modulation coloration (Wang et al. 2016). Furthermore, both materials have anti-photochromic properties that may be present in amorphous WO$_3$ thin films. Different ways of depositing the aforementioned transition metal oxide coatings have been reported. For instance, Kim et al. reported the formation of WO$_3$ composite thin films, which are responsive to visible light at different voltage loadings (Figure 7). This was achieved by spin coating niobium polyoxometalate oxides (POM) onto an amorphous WO$_3$ thin film, where the WO$_3$ mesopores were filled with POM (Kim et al. 2015).

Wang et al. reported the use of atomic layer deposition (ALD) to deposit a Ta$_2$O$_5$ film on amorphous WO$_3$ thin films to fill the oxygen vacancies located in the crystal structure of WO$_3$ (Wang et al. 2016). By employing this approach, the modulation for the thickness of the thin films was achieved from 0 to 2.5 nm, depending on the required ALD cycles. Transmission electron microscopy (TEM) and energy dispersive X-ray spectroscopy (EDX) characterizations showed that Ta$_2$O$_5$ was completely and uniformly embedded in the tungsten oxide. Saez et al. reported that the pore size of the amorphous WO$_3$ thin films is an important factor in the penetration of transition metal oxide coatings, obtaining poor penetration of Nb$_2$O$_5$ and Ta$_2$O$_5$ by electrodeposition (Saez et al. 2020).

2.4.3 Substrates

This section will discuss the importance of substrates to achieve the fabrication of thin films of MOx with various properties, which are used in multiple technological applications or the innovation of semiconductor materials. For the development of smart materials, substrates are the essential part of the whole system, which must be transparent, electrically conductive, with high morphological uniformity, and high optoelectronic response. In general, these substrates are made of transparent conducting metal oxides (TCO), although other alternatives have been reported (Granqvist 2014). TCO substrates are highly doped thin films generally fabricated by CVD or sputtering methods depending on the type of targeted substrate. For instance, the fabrication of SnO_2:Sb (ATO), and ZnO:Al (AZO) substrates using USP deposition process has been reported (Lee and Park 2006, Marouf et al. 2017). Certain highly doped transparent oxide substrates are listed in Table 3.

Maho et al. conducted a comparative study between ITO and IWO substrates used as substrates in the deposition of WO_3 thin films by USP. The results obtained showed significant differences in terms of morphology between the substrates; meanwhile, ITO substrate showed better resistance, mobility, and better redox reactivity in cyclic voltammetry characterizations; IWO substrates displayed a better coloration efficiency for electrochromic applications (Maho et al. 2017).

Table 3. Types of substrates used in electrochromic materials (Granqvist 2014).

Name TCO	Abbreviation	Composition
Fluorine tin oxide	FTO	SnO_2:F
Indium tin oxide	ITO	In_2O_3:Sn
Indium tungsten oxide	IWO	In_2O_3:WO_3
Indium zinc oxide	IZO	ZnO:In
Indium zinc oxide	IZO	In_2O_3:Zn
Zinc gallium oxide	GZO	ZnO:Ga
Zinc silica oxide	SZO	ZnO:Si
Zinc boron oxide	BZO	ZnO:B
Antimony tin oxide	ATO	SnO_2:Sb
Zinc aluminum oxide	AZO	ZnO:Al
Zinc fluorine oxide	FZO	ZnO:F
Niobium titanate oxide	TNO	TiO_2:Nb
Ternary oxides	AZO/ATO	ZnO:Al/ SnO_2:Sb

3. Applications

Throughout this chapter, we have revised in general terms the great versatility of the physicochemical properties that MOx can display, making them excellent material to be used in cutting-edge applications, as well as in the development of novel strategies to generate smart materials. In this section, we show some of the main application fields of MOx thin films.

3.1 Gas sensor

Gas sensor technology consists of the real-time detection of gaseous compounds. The application extends to monitoring atmospheric pollution, explosive gases, even disease diagnosis. MOx are an important element in the development of the conductometric gas sensor. The main characteristic is the change of conductance of the MOx when exposed to certain gas. These changes are generated by the surface reaction kinetics and electron mobility derived from interactions between gas molecules and the MOx surface.

MOx display different responses to a specific gas; for example, SnO_2 thin films deposited by sputtering were reported as a gas sensor for CO (Chemistry 1996). On the other hand, ZnO and WO_3 have been employed to detect acetone, and TiO_2 displays gas sensing responses to NO_2, O_2, CO_2, and NH_3 (Nunes et al. 2021). In order to improve the conductometric response in gas sensors, MOx are usually doped with certain metals such as Pd, Cr, or Cu.

3.2 *Optoelectronic devices*

Optoelectronic devices consist of electronic materials (usual semiconductors) that absorb different types of radiation between UV and IR regions, with a certain degree of control of the quantum mechanical effects of light (electron-hole photo-generation) (Adams and Barbante 2015). MOx semiconductor thin films have been among the most widely used materials in the development and innovation of optoelectronic devices. These include what is currently called smart materials, i.e., materials that can change their optical properties when irradiated with light.

Currently, WO_3 has been one of the most studied transition MOx's for the development of photochromic/electrochromic devices due to its wide variety of crystal structures, which contain natural cavities; it is an excellent candidate as a receptor for external cations, also called "tungsten bronze". WO_3 is relevant material in developing intelligent windows due to its dual electrochromic/photochromic properties and its electrochemical reversibility, its efficiency in color switching (Kim et al. 2015), and high coloration efficiency (Runnerstrom et al. 2014).

On the other hand, molybdenum oxide (MoO_3) has been studied to a lesser extent than tungsten oxide. However, it also shows excellent photochromic properties and incredible versatility in its perovskite-like crystal structure. Additional materials that have attracted attention due to their photochromic properties are titanium dioxide (TiO_2), vanadium oxide (V_2O_5), and niobium oxide (Nb_2O_5).

3.3 *Photocatalysis*

Photocatalysis is a process heavily employed to remove pollutants from water through photo-induced reactions assisted by a photocatalyst; in this context, semiconductor MOx has displayed high efficiencies in the treatment of polluted industrial wastewater through photocatalytic processes. Several MOx including ZnO, TiO_2, Fe_2O_3, Cu_2O, and WO_3 have been utilized, TiO_2 being one of the most popular MOx due to its high photostability, chemically inert nature, and low cost. For instance, the degradation of methyl orange (Perillo and Rodríguez 2021), polystyrene nanoparticles from aqueous media (Domínguez-Jaimes et al. 2021), and rhodamine B (Dao et al. 2021) has been demonstrated using TiO_2. Moreover, the use of heterostructures such as Ag/ZnO has rendered excellent photocatalytic performances for the degradation of rhodamine B (Ha et al. 2021).

In a general way, this section describes some of the main applications of MOx thin films. Nevertheless, there are many other fields where MOx can be applied. Table 4 displays a summary of MOx thin films' main applications.

4. Conclusions and future perspectives

This chapter focuses on a small part of the fascinating characteristics of transition metal oxide (MOx) thin films, covering certain synthetic procedures, namely chemical vapor deposition (CVD) and ultrasonic spray pyrolysis (USP), and several approaches to enhance their optoelectronic properties through surface stabilization and functionalization methodologies. Different stabilization conditions before, during, and after the deposition process of thin films with different MOx were revised. These

Table 4. Summary of MOx thin films' main applications.

Thin films material	Deposition method	Application	References
WO_3	Sputtering	H_2S gas sensor	(Gullapalli et al. 2010)
TiO_2	Sputtering	NH_3 gas sensor	(Karunagaran et al. 2007)
SnO_2	Spin-coating	H_2 gas sensor	(De et al. 2007)
SnO_2	USP	H_2 gas sensor	(Patil et al. 2009)
SnO_2	Micromachined	CH_4 gas sensor	(Friedberger et al. 2003)
In-doped ZnO	Spray pyrolysis	Solar cells	(Krunks et al. 1999)
ZnO	USP	Solar cells, gas sensors, liquid crystal displays	(Kim et al. 2000)
ZnO	Aerosol assisted chemical vapor deposition (AACVD)	Gas sensing and transparent conductive oxide	(O'Brien et al. 2010)
ZnO	Cathodic electrodeposition from an aqueous solution	Microfluidic devices	(Mei Li et al. 2003)
In_2O_3	Spray pyrolysis	CO, H_2 and ozone gas sensor	(Korotcenkov et al. 2004)
Co_3O_4	Successive Ionic Layer Adsorption and Reaction (SILAR)	Electrochemical supercapacitors for power source applications	(Kandalkar et al. 2008)
Co_3O_4	Usp	NH_3 gas sensing	(Shinde et al. 2006)
MoO_3	Sputtering	Optoelectronic nanodevices	(Navas et al. 2011)
SiO_2	Sol electrophoretic deposition	Flexible panel displays	(Rha et al. 2009)

can ultimately affect the film´s final shape and morphology, avoiding defects in electronic properties, and obtaining fracture-free films due to the use of surfactants. Moreover, possible functionalization protocols involving the use of heterostructures and ligands were addressed, with the ultimate purpose of enhancing or conferring novel functionalities to MOx thin films. Finally, a summary of possible applications where MOx thin films are mainly used is presented. There is still a long way ahead to further explore and develop new strategies to stabilize and functionalize nanostructured materials; however, innovative studies have recently paved the way to achieve such goals. It is expected the scientific community will continue working on this topic, which will ultimately help in the development of emerging technologies to fulfill society's near future necessities.

Acknowledgments

The authors gratefully acknowledge the support provided by CONACYT through grants 2096029 FORDECYT-PRONACES, and CB-A1-S-44458. M. Ávila-Gutiérrez thanks CONACYT for a postdoctoral scholarship (BP-PA-20200709124844902-440361). The authors also thank Dra. Lilian Irais Olvera for the critical revision of the book chapter, and Lisandro Ávila and Vanessa Alfaro for technical support.

References

Adams, F. and Carlo, B. 2015. Nanotechnology and analytical chemistry. pp. 125–147. In Comprehensive Analytical Chemistry. Elsevier Vol. 69 [ed.]

Aoun, Y., Boubaker, B., Said, B. and Brahim, G. 2015. Effect of deposition rate on the structural, optical and electrical properties of zinc oxide (ZnO) thin films prepared by spray pyrolysis technique. *Optik,* 126(20): 2481–84.

Aranovich, J., Armando, O. and Richard, H.B. 1979. Optical and electrical properties of ZnO films prepared by spray pyrolysis for solar cell applications. *Journal of Vacuum Science & Technology*, 16(4): 994–1003.

Arjmandi-Tash, H., Nikita, L., Pauline, M.G. van, D., Jan, A. and Grégory, F.S. 2017. Hybrid cold and hot-wall reaction chamber for the rapid synthesis of uniform graphene. *Carbon*, 118: 438–42.

Ashrit, P. 2017. Introduction to transition metal oxides and thin films. pp. 13–72. *In*: Ghenadii Korotcenkov [ed.]. Transition Metal Oxide Thin Film Based Chromogenics and Devices. Elsevier, Amsterdam, Netherlands.

Augustyn, V., Jérémy, C., Michael, A.L., Jong Woung, K., Pierre Louis, T., Sarah, H.T., Héctor, D.A., Patrice, S. and Bruce, D. 2013. High-rate electrochemical energy storage through Li + Intercalation Pseudocapacitance. *Nature Materials*, 12(6): 518–22.

Bang, J.H. and Kenneth, S.S. 2010. Applications of ultrasound to the synthesis of nanostructured materials. *Advanced Materials*, 22(10): 1039–59.

Barreca, D, Bozza, S., Carta, G., Rossetto, G., Tondello, E. and Zanella, P. 2003. Structural and morphological analyses of tungsten oxide nanophasic thin films obtained by MOCVD. *Surface Science*, 532–535: 439–43.

Bertus, L.M., Faure, C., Danine, A., Labrugere, C., Campet, G., Rougier, A. and Duta, A. 2013. Synthesis and characterization of WO$_3$ thin films by surfactant assisted spray pyrolysis for electrochromic applications. *Materials Chemistry and Physics*, 140(1): 49–59.

Boles, M.A., Michael, E. and Dmitri, V. Talapin. 2016. Self-assembly of colloidal nanocrystals: From intricate structures to functional materials. *Chemical Reviews*, 116(18): 11220–89.

Chatzikyriakou, D., Maho, A., Cloots, R. and Henrist, C. 2017. Ultrasonic spray pyrolysis as a processing route for templated electrochromic tungsten oxide films. *Microporous and Mesoporous Materials*, 240: 31–38.

Chávez, F., Pérez-Sánchez, G.F., Goiz, O., Zaca-Morán, P., Peña-Sierra, R., Morales-Acevedo, A., Felipe, C. and Soledad-Priego, M. 2013. Sensing performance of palladium-functionalized WO$_3$ nanowires by a drop-casting method. *Applied Surface Science*, 275: 28–35.

Chemistry, Theoretical. 1996. The Effect of Pt and Pd Surface Doping on the Response of Nanocrystalline Tin Dioxide Gas Sensors to CO 31: 71–75.

Chichkov, B.N., Momma, C., Nolte, S., von Alvensleben, F. and Tünnermann, A. 1996. Femtosecond, picosecond and nanosecond laser ablation of solids. *Applied Physics A: Materials Science & Processing*, 63(2): 109–15.

Cho, Y.H., Xishuang, L., Yun Chan, K. and Jong, H. Lee. 2015. Ultrasensitive detection of trimethylamine using Rh-doped SnO$_2$ hollow spheres prepared by ultrasonic spray pyrolysis. *Sensors and Actuators, B: Chemical*, 207 (Part A): 330–37.

Chuanbao, T., Zhang, Z., Shao, A., Qi, X., Zhu, C., Li, C. and Yang, Z. 2021. Constructing a directional ion acceleration layer at WO3/ZnO heterointerface to enhance Li-ion transfer and storage. *Composites Part B: Engineering*, 205: 108511.

Costanzo, S., Simon, G., Richardi, J., Colomban, Ph. and Lisiecki, I. 2016. Solvent effects on Cobalt Nanocrystal Synthesis: a facile strategy to control the size of co nanocrystals. *J. Phys. Chem. C*, 120(38): 22054–22061.

Cruz-Leal, M., Goiz, O., Chávez, F., Pérez-Sánchez, G.F., Hernández-Como, N., Santes, V. and Felipe, C. 2019. Study of the thermal annealing on structural and morphological properties of high-porosity A-WO$_3$ films synthesized by HFCVD. *Nanomaterials*, 9(9).

Cubillos, G.I., Bethencourt, M. and Olaya, J.J. 2015. Corrosion resistance of zirconium oxynitride coatings deposited via DC unbalanced magnetron sputtering and spray pyrolysis-nitriding. *Applied Surface Science*, 327: 288–95.

Dao, T.B.T., Ha, T.T.L., Nguyen, T.D., Le, H.N., Ha-Thuc, C.N., Nguyen, T.M.L., Perre, P. and Nguyen, D.M. 2021. Effectiveness of photocatalysis of MMT-supported TiO$_2$ and TiO$_2$ nanotubes for rhodamine B degradation. *Chemosphere*, 280: 130802.

De, G., Kohn, R., Xomeritakis, G. and C.J. Brinker. 2007. Nanocrystalline mesoporous palladium activated tin oxide thin films as room-temperature hydrogen gas sensors. *Chemical Communications*, 18: 1840–42.

Denayer, J., Aubry, P., Bister, G., Spronck, G., Colson, P., Vertruyen, B., Lardot, V., Cambier, F., Henrist, C. and Cloots, R. 2014. Improved coloration contrast and electrochromic efficiency of tungsten oxide films thanks to a surfactant-assisted ultrasonic spray pyrolysis process. *Solar Energy Materials and Solar Cells*, 130: 623–28.

Domínguez-Jaimes, Laura Patricia, Erika Iveth Cedillo-González, Luévano-Hipólito, E., Jawer David Acuña-Bedoya, and Juan Manuel Hernández-López. 2021. Degradation of primary nanoplastics by photocatalysis using different anodized TiO$_2$ structures. *Journal of Hazardous Materials*, 413(July): 125452.

Firooz, Azam, A., Takeo, H., Ali Reza, M., Abbas Ali, K. and Yasuhiro, S. 2010. Synthesis and gas-sensing properties of nano- and meso-porous MoO$_3$-doped SnO$_2$. *Sensors and Actuators, B: Chemical*, 147(2): 554–60.

Foggiato, J. 2015. Chemical vapor deposition of silicon dioxide films. pp. 111–150. *In*: Krishna Seshan [ed.]. Handbook of Thin-Film Technology. Noyes Publication, California, USA.

Frey, H. and Hamid, R. Khan. 2015. Handbook of Thin-Film Technology. Handbook of Thin-Film Technology.

Friedberger, A., Kreisl, P., Rose, E., Müller, G., Kühner, G., Wöllenstein, J. and Böttner, H. 2003. Micromechanical Fabrication of robust low-power metal oxide gas sensors. *Sensors and Actuators, B: Chemical*, 93(1–3): 345–49.

Fugare, B.Y. and Lokhande, B.J. 2017. Study on structural, morphological, electrochemical and corrosion properties of mesoporous RuO$_2$ thin films prepared by ultrasonic spray pyrolysis for supercapacitor electrode application. *Materials Science in Semiconductor Processing*, 71(July): 121–27.

García-Sánchez, Mario, F., Armando Ortiz, Guillermo Santana, Monserrat Bizarro, Juan Peña, Francisco Cruz-Gandarilla, Miguel A. Aguilar-Frutis and Juan C. Alonso. 2010. Synthesis and characterization of nanostructured cerium dioxide thin films deposited by ultrasonic spray pyrolysis. *Journal of the American Ceramic Society*, 93(1): 155–60.

Goiz, O., Chávez, F., Felipe, C., Peña-Sierra, R. and Morales, N. 2010. CSVT as a technique to obtain nanostructured materials: WO_{3-x}. *Journal of Nano Research*, 9: 31–37.

Granqvist, Claes G. 2014. Electrochromics for smart windows: Oxide-based thin films and devices. *Thin Solid Films*, 564: 1–38.

Gullapalli, S.K., Vemuri, R.S., Manciu, F.S., Enriquez, J.L. and Ramana, C.V. 2010. Tungsten oxide (WO_3) thin films for application in advanced energy systems. *Journal of Vacuum Science & Technology A: Vacuum, Surfaces, and Films*, 28(4): 824–28.

Ha, La Phan, P., Tran Hoang The V., Nguyen T. Be Thuy, Cao M. Thi and Pham V. Viet. 2021. Visible-light-driven photocatalysis of anisotropic silver nanoparticles decorated on ZnO nanorods: Synthesis and characterizations. *Journal of Environmental Chemical Engineering*, 9(2): 105103.

Hilfiker, J.N. 2011. *In Situ* spectroscopic ellipsometry (SE) for characterization of thin film growth. pp. 99–151. *In*: Köster Gertjan [ed.]. *In situ* Characterization of thin Film Growth. Woodhead publishing, Cambridge, UK..

Kandalkar, S.G., Gunjakar, J.L. and Lokhande, C.D. 2008. Preparation of cobalt oxide thin films and its use in supercapacitor application. *Applied Surface Science*, 254(17): 5540–44.

Karunagaran, B., Uthirakumar, P., Chung, S.J., Velumani, S. and Suh, E.-K. 2007. TiO2 thin film gas sensor for monitoring ammonia. *Materials Characterization*, 58 (8-9 SPEC. ISS.): 680–684.

Kim, H., Gilmore, C.M., Horwitz, J.S., Piqué, A., Murata, H., Kushto, G.P., Schlaf, R., Kafafi, Z.H. and Chrisey, D.B. 2000. Transparent conducting aluminum-doped zinc oxide thin films for organic light-emitting devices. *Applied Physics Letters*, 76(3): 259–61.

Kim, Jongwook, Gary K. Ong, Yang Wang, Gabriel Leblanc, Teresa E. Williams, Tracy M. Mattox, Brett A. Helms and Delia J. Milliron. 2015. Nanocomposite architecture for rapid, spectrally-selective electrochromic modulation of solar transmittance. *Nano Letters*, 15(8): 5574–79.

Kodigala, S.R. 2014. The role of characterization techniques in the thin film analysis. pp. 67–140. *In*: Thin Film Solar Cells From Earth Abundant Materials. Elsevier, California, USA.

Korotcenkov, G., Brinzari, V., Cerneavschi, A., Ivanov, M., Cornet, A., Morante, J., Cabot, A. and Arbiol, J. 2004. In_2O_3 Films deposited by spray pyrolysis: gas response to reducing (CO, H_2) gases. *Sensors and Actuators, B: Chemical*, 98(2-3): 122–29.

Krunks, K., Bijakina, O., Mikli, V., Varema, T. and Mellikov, E. 1999. Zinc oxide thin films by spray pyrolysis method. *Physica Scripta T*, 79: 209–12.

Kundu, Manab, Gopalu Karunakaran, Evgeny Kolesnikov, Elena Sergeevna Voynova, Shilpa Kumari, Mikhail V. Gorshenkov and Denis Kuznetsov. 2018. Hollow $NiCo_2O_4$ nano-spheres obtained by ultrasonic spray pyrolysis method with superior electrochemical performance for lithium-ion batteries and supercapacitors. *Journal of Industrial and Engineering Chemistry*, 59: 90–98.

Lai, W. H., Shieh, J., Teoh, L.G., Hung, I.M., Liao, C.S. and Hon, M.H. 2005. Effect of copolymer and additive concentrations on the behaviors of mesoporous tungsten oxide. *Journal of Alloys and Compounds*, 396(1-2): 295–301.

Lamri Zeggar, M., Chabane, L., Aida, M.S., Attaf, N. and Zebbar, N. 2015. Solution flow rate influence on properties of copper oxide thin films deposited by ultrasonic spray pyrolysis. *Materials Science in Semiconductor Processing*, 30.

Lassner, E. and Schubert, W.D. 1999. Tungsten compounds and their application. pp. 133–177. *In*: Tungsten. Springer, Boston, USA.

Lee, Seung, Y. and Byung Ok Park. 2006. Structural, electrical and optical characteristics of SnO_2:Sb thin films by ultrasonic spray pyrolysis. *Thin Solid Films*, 510(1-2): 154–58.

Li, Mei, Jin Zhai, Huan Liu, Yanlin Song, Lei Jiang and Daoben Zhu. 2003. Electrochemical deposition of conductive superhydrophobic zinc oxide thin films. *Journal of Physical Chemistry B*, 107(37): 9954–57.

Li, M., Liang, Z. and Liejin, G. 2010. Preparation and photoelectrochemical study of $BiVO_4$ thin films deposited by ultrasonic spray pyrolysis. *International Journal of Hydrogen Energy*, 35(13): 7127–33.

Lim, Weon, C., Jitendra, P., Younghak, K., Jonghan, S.g, Keun Hwa Chae and Tae Yeon Seong. 2021. Effect of thermal annealing on the properties of ZnO thin films. *Vacuum*, 183(January): 109776.

Llordés, A., Guillermo, G., Jaume, G. and Delia J. Milliron. 2013. Tunable near-infrared and visible-light transmittance in nanocrystal-in-glass composites. *Nature*, 500(7462): 323–26.

Madhuri, K.V. 2020. Thermal Protection Coatings of Metal Oxide Powders. Metal Oxide Powder Technologies. INC.

Maho, A., Sylvain, N., Laura, M., Gilles, S., Catherine, H., Rudi, C., Bénédicte, V. and Pierre, C. 2017. Comparison of indium tin oxide and indium tungsten oxide as transparent conductive substrates for WO_3-based electrochromic devices. *Journal of The Electrochemical Society*, 164(2): H25–31.

Maho, A., Camila, A. Saez, C., Meyertons, K.A., Lauren C. Reimnitz, Swagat Sahu, Brett A. Helms and Delia J. Milliron. 2020. Aqueous processing and spray deposition of polymer-wrapped tin-doped indium oxide nanocrystals as electrochromic thin films. *Chemistry of Materials*, 32(19): 8401–11.

Malwal, Deepika and Gopinath Packirisamy. 2018. Recent advances in the synthesis of metal oxide (MO) nanostructures. In Synthesis of Inorganic Nanomaterials, 255–81. *Elsevier*.

Marouf, S., Abdelkrim, B., Kasra, K., Manuel, J.M., Olalla, S., Hugo, Á., Elvira, F. and Rodrigo Martins. 2017. Low-Temperature spray-coating of high-performing ZnO:Al films for transparent electronics. *Journal of Analytical and Applied Pyrolysis*, 127(July): 299–308.

Matula, G., Jelena, B. Srečko, S. and Bernd, F. 2013. Scale up of Ultrasonic Spray Pyrolysis Process for Nano-Powder Production - Part I, 2–5.

Muñoz-Fernandez, L., Alkan, G., Milošević, O., Rabanal, M.E. and Friedrich, B. 2019. Synthesis and characterisation of spherical core-shell Ag/ZnO nanocomposites using single and two – steps ultrasonic spray pyrolysis (USP). *Catalysis Today*, 321-322(July 2017): 26–33.

Nandi, Dip, K. and Shaibal K. Sarkar. 2014. Atomic layer deposition of tungsten oxide for solar cell application. *Energy Procedia*, 54: 782–88.

Navas, I., Vinodkumar, R. and Mahadevan V.P. Pillai. 2011. Self-assembly and photoluminescence of molybdenum oxide nanoparticles. *Applied Physics A: Materials Science and Processing*, 103(2): 373–80.

Neagu, R., Dainius, P., Agnès, P. and Elisabeth, D. 2006. Zirconia coatings deposited by electrostatic spray deposition. Influence of the process parameters. *Surface and Coatings Technology*, 200(24): 6815–20.

Nunes, S., Evelyn, A., Thais C. de Oliveira, Ádamo, E., do Carmo, M., Amanda, A., Coutinho, S., Alan S. dos Santos and Luciana de S. Cividanes. 2021. TiO_2 as a gas sensor: The Novel Carbon Structures and Noble Metals as New Elements for Enhancing Sensitivity – A Review. Ceramics International, March.

O'Brien, Shane, Mark G. Nolan, Mehmet Çopuroglu, Jeff A. Hamilton, Ian Povey, Luis Pereira, Rodrigo Martins, Elvira Fortunato and Martyn Pemble. 2010. Zinc oxide thin films: Characterization and potential applications. *Thin Solid Films*, 518(16): 4515–19.

Oh, Sung Woo, Hyun Joo Bang, Young Chan Bae, and Yang Kook Sun. 2007. Effect of calcination temperature on morphology, crystallinity and electrochemical properties of nano-crystalline metal oxides (Co_3O_4, CuO, and NiO) prepared via ultrasonic spray pyrolysis. *Journal of Power Sources*, 173(1): 502–9.

Park, Joon Kyoo, Seong Ki Lee, and Jae Hoon Kim. 2018. Development of an evaluation method for nuclear fuel debris–filtering performance. *Nuclear Engineering and Technology*, 50(5): 738–44.

Patil, L.A., Shinde, M.D., Bari, A.R. and Deo, V.V. 2009. Highly sensitive and quickly responding ultrasonically sprayed nanostructured SnO_2 thin films for hydrogen gas sensing. *Sensors and Actuators, B: Chemical*, 143(1): 270–77.

Perillo, P.M. and Rodríguez, D.F. 2021. Environmental nanotechnology, monitoring & management photocatalysis of methyl orange using free standing TiO_2 nanotubes under solar light. *Environmental Nanotechnology, Monitoring & Management*, 16(April): 100479.

Pierson, H.O. 1992. Fundamentals of chemical vapor deposition. pp. 17–50. *In*: Handbook of Chemical Vapor Deposition. Noyes Publication, New Mexico, USA.

Pileni, Marie Paule, Davide P. Cozzoli and Nicola Pinna. 2014. Self-assembled supracrystals and hetero-structures made from colloidal nanocrystals. *Crystengcomm*, 16(40): 9365–67.

Putri, N., Cuk, I. and Vivi, F. 2020. ZnO thin films prepared using the ultrasonic spray pyrolysis method for high performance metal oxides-based photoconductors. *Key Engineering Materials*, 860 KEM: 274–81.

Racheva, T.M., Stambolova, I.D. and Donchev, T. 1994. Humidity-sensitive characteristics of SnO_2-Fe_2O_3 thin films prepared by spray pyrolysis. *Journal of Materials Science*, 29(1): 281–84.

Rani, R., Ahmad, S.Z., Anthony P. O'Mullane, Michael W. Austin and Kourosh Kalantar-Zadeh. 2014. Thin films and nanostructures of niobium pentoxide: fundamental properties, synthesis methods, and applications. *Journal of Materials Chemistry*, A2(38): 15683–703.

Rha, S.-K., Chou, T.P., Cao, G., Lee, Y.-S. and Lee W.-J. 2009. Characteristics of silicon oxide thin films prepared by sol electrophoretic deposition method using tetraethyl orthosilicate as the precursor. *Current Applied Physics*, 9(2): 551–555.

Rinaldi, Febrigia, G., Osi, A., Aditya A. Farhan, Tomoyuki, H., Takashi Ogi and Kikuo, O. 2018. Correlations between reduction degree and catalytic properties of WOx nanoparticles. *ACS Omega*, 3(8): 8963–70.

RM, N., Kumaraswamy, G.N., Susheel Kumar Gundanna and Umananda M. Bhatta. 2020. Effect of thermal annealing on structural and electrical properties of TiO_2 thin films. *Thin Solid Films*, 710(September): 138262.

Roselló-Márquez, G., Fernández-Domene, R.M., Sánchez-Tovar, R. and García-Antón, J. 2020. Influence of annealing conditions on the photoelectrocatalytic performance of WO_3 nanostructures. *Separation and Purification Technology*, 238(November 2019): 116417.

Runnerstrom, E.L., Llordés, A., Lounis, S.D. and Milliron, D.J. 2014. Nanostructured electrochromic smart windows: Traditional Materials and NIR-selective plasmonic nanocrystals. *Chem. Commun.*, 50: 10555–10572.

Saez, C., Camila, A., Kristen, M., Sungyeon, H., Andrei, D., Gabriel, L., Gabriel, L. and Delia, J.M. 2020. Direct electrochemical deposition of transparent metal oxide thin films from polyoxometalates. *Chemistry of Materials*, 32(11): 4600–4608.

Saunders, S.R.J., Monteiro, M. and Rizzo, F. 2008. The oxidation behaviour of metals and alloys at high temperatures in atmospheres containing water vapour: A review. *Progress in Materials Science*, 53(5): 775–837.

Shinde, V.R., Mahadik, S.B., Gujar, T.P. and Lokhande, C.D. 2006. Supercapacitive Cobalt Oxide (Co3O4) thin films by spray pyrolysis. *Applied Surface Science*, 252(20): 7487–7492.

Skrabalak, S.E. and Kenneth, S.S. 2005. Porous MoS 2 Synthesized by Ultrasonic Spray Pyrolysis, 9990–91.

Skrabalak, S.E. and Kenneth S. Suslick. 2007. Carbon powders prepared by ultrasonic spray pyrolysis of substituted alkali benzoates. *Journal of Physical Chemistry C*, 111(48): 17807–11.

Stoycheva, T., Annanouch, F.E., Gràcia, I., Llobet, E., Blackman, C., Correig, X. and Vallejos, S. 2014. Micromachined gas sensors based on tungsten oxide nanoneedles directly integrated via aerosol assisted CVD. *Sensors and Actuators B: Chemical*, 198(31): 210–218.

Tararam, R., Garcia, P.S., Deda, D.K., Varela, J.A. and de Lima Leite, F. 2017. Atomic force microscopy: a powerful too for electrical characterization. pp. 37–64. *In*: Nanocharacterization techniques. Elsevier, Amsterdam.

Urban, III, F.K., Hosseini-Tehrani, A., Griffiths, P., Khabari, A., Kim, Y.-W. and Petrov, I. 2002. Nanophase films deposited from a high-rate, nanoparticle beam. *Journal of Vacuum Science & Technology B*, 995: 10–15.

Vattikuti, S.V. Prabhakar. 2018. Heterostructured nanomaterials: Latest trends in formation of inorganic heterostructures. In Synthesis of Inorganic Nanomaterials, 89–120. *Elsevier*.

Viet, A. Le, Jose, R., Reddy, M.V., Chowdari, B.V.R. and Ramakrishna, S. 2010. Nb_2O_5 Photoelectrodes for dye-sensitized solar cells: Choice of the polymorph. *Journal of Physical Chemistry C*, 114(49): 21795–800.

Vivien, Anthony, Maya Guillaumont, Lynda Meziane, Caroline Salzemann, Corinne Aubert, Stéphanie Halbert, Hélène Gérard, Marc Petit and Christophe Petit. 2019. The role of oleylamine revisited: An original disproportionation route to monodispersed cobalt and nickel nanocrystals. *Chemistry of Materials*, acs.chemmater.8b04435.

Wang, Y., Kim, J., Gao, Z., Zandi, O., Heo, S., Banerjee, P. and Milliron, D.J. 2016. Disentangling photochromism and electrochromism by blocking hole transfer at the electrolyte interface. *Chem. Mater.*, 28(20): 7198–7202.

Xu, Wei, Heli Tang, Nan Zhou, Qingyu Zhang, Bo Peng and Yu Shen. 2021. Enhanced photocatalytic activity of TiO_2/Ag_2O heterostructures by optimizing the separation of electric charges. *Vacuum*, May, 110283.

Yan, Meng, Yi Wu, Zhongqiu Hua, Ning Lu, Wentao Sun, Jinbao Zhang and Shurui Fan. 2021. Humidity compensation based on power-law response for MOS sensors to VOCs. *Sensors and Actuators, B: Chemical*, 334(May): 129601.

Yang, G., Biswas, P., Boolchand, P. and Sabata, A. 1999. Deposition of multifunctional titania ceramic films by aerosol routes. *Journal of the American Ceramic Society*, 82(10): 2573–2579.

Zazpe, Raul, Jan Prikryl, Viera Gärtnerova, Katerina Nechvilova, Ludvik Benes, Lukas Strizik, Ales Jäger, Markus Bosund, Hanna Sopha and Jan M. Macak. 2017. Atomic layer deposition Al_2O_3 coatings significantly improve thermal, chemical, and mechanical stability of anodic TiO_2 nanotube layers. *Langmuir*, 33(13).

Zhang, Liang. 2021. The effect of post-metal annealing on the electrical performance and stability of two-step-annealed solution-processed In_2O_3 thin film transistors. *Current Applied Physics*, 23(March): 19–25.

Zhu, Xin Hong, Guang Hua Chen and Mao Sheng Zheng. 2008. Study on an improved MWECR CVD system and preparation of silicon-substrate thin films. *Solar Energy Materials and Solar Cells*, 92(10).

Ziolek, Maria, Izabela Sobczak, Piotr Decyk and Lukasz Wolski. 2013. The ability of Nb_2O_5 and Ta_2O_5 to generate active oxygen in contact with hydrogen peroxide o O. CATCOM 37: 85–91.

CHAPTER 3

Inorganic Nanoparticles
Properties and Applications

Victor Merupo,[1,*] *Jose Carlos Zarate,*[2,*] *Alla Abramova,*[1]
Noé Arjona,[2] *José Herrera-Celis,*[2] *L.G. Arriaga,*[2]
Ashutosh Sharma[3] and *Goldie oza*[2,*]

1. Introduction

Nanoparticles are considered to be intermediate systems—smaller than bulk systems (which obey Newton's laws of motion), and bigger than atoms or a simple molecule (which follow the principles of quantum mechanics). The study of small dimensional materials are of great interest to material scientists. These small objects' typical size ranges from 1 to 100 nm. Nanotechnology is a broad and interdisciplinary area of research that is immensely contributing to the modern science and lead to unprecedented developments in almost every domain. Though the use of NPs has been reported long back, the scientific potentials and their quantum behaviors were introduced to the world by the famous talk from Richard Feynman in 1959 'There's plenty of room at the bottom'.

Though the nano-term was coined by Norio Taniguchi in the year 1974, such systems were already being exploited in all our regular activities. In the 4th century AD, glasses and cups were painted with metal-based paints. Daniel and Astruc have summarized the use of soluble gold from ancient times to the Middle Ages (Daniel and Astruc 2004). Earlier, in and around 4th century BC, soluble gold sols were used in different manufacturing units in Egypt and China (Gao 2014). In the Middle Ages, stained glass windows were seen in many churches, creating impressive artistry. The artisans who were themselves not professional experts were oblivious to nanotechnology but the colorful paintings out of the metal sols were quite fascinating to be seen. Hence, the whole world was already using this technology before the actual coining of the term nano. Academically, these transition metal nanoparticles (MNPs) were explored only in the 20th century. In the seventeenth century, gold sols were exploited to treat different diseases such as epilepsy, tumors, and other venereal diseases (Dykman and Bogatyrev 1997). Later, in 1857, Michael Faraday reduced the

[1] Institute of Electronics, Microelectronics and Nanotechnology, University of Lille, ISEN, CNRS, UMR 8520, 59652 Villeneuve d'Ascq, France.
[2] Laboratorio Nacional de Micro Y Nanofuidica (LABMyN), Centro de Investigación y Desarrollo Tecnológico en Electroquímica (CIDETEQ), Querétaro 76703, México.
[3] Tecnologico de Monterrey, School of Engineering and Sciences, Centre of Bioengineering, Campus Queretaro. Av. Epigmenio González, No. 500, Fracc. San Pablo 76130 Querétaro, México
* Corresponding authors: victorishrayelu@gmail.com; jzarate@cideteq.mx; goza@cideteq.mx

aqueous solution of gold chloride, thus synthesizing gold colloid thin films and study their optical properties that could find reversible color changes when compression was applied on the material (Gao 2014).

In 1908, Gustav Mie explained about the optical properties of the gold sols. This led to the study and description of the plasmonic as well as excitonic properties of different metallic as well as semi-conductor nanoparticles. The research community tried to understand the fundamentals of the nanoparticles based on their different sizes and shapes. They were also involved in synthesis of the particles using different methods, purification and post-synthetic modifications for plethora of applications. The nanomaterials were characterized and studied using different simulation models.

Nanomaterials exhibit many compelling properties mainly due to two physical effects. First, the quantum discrete electronic states, which influences their optics, magnetic and electrical properties. Second, the high surface to volume ratio that enhances their chemical reactivity, thermal and mechanical properties, etc.

Various ways of preparing and engineering nanoparticles were classified into two main approaches, viz., top-down and bottom-up. In the top-down approach, the starting materials are in bulk or macro-sized form, and size-reduction down to nanomaterials follows various physical and/or chemical techniques, whereas in a bottom-up approach, the fundamental particles such as atoms and molecules are bounded together following a so-called self-assembly technique. Top-down approaches consist of physical methods such as lithography, milling, attrition, etc. These methods use either light, electrons or electrostatic forces for the formation of these nanostructures, the only problem being that metallic nanoparticles synthesized through these methods may present undesired imperfections. Bottom-up approaches are considered more precise strategies, different

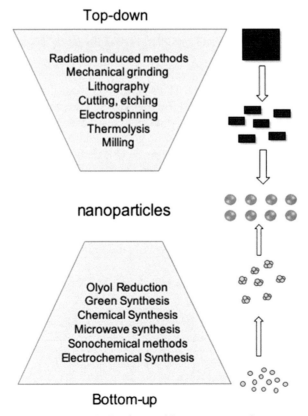

Figure 1. Top-down and Bottom-up approach.

methods include co-precipitation, sol-gel, microemulsions, etc. The bottom-up approach is a far more controlled system in which the precursors form highly specific structures such as rods, wires, spheres, flakes, etc.

2. Properties

2.1 Physical (Structural) properties (dimensions and shapes)

Nanoscale materials behave in different ways in comparison to bulk materials. These special properties are not only confined to different chemical behavior, but to morphological special arrangements with unique and tunable physical and structural properties, mainly related to surface effects. Surface effects are seen (in comparison with bulk) that have less stabilized surface atoms. In small particles, the larger portion of the atoms are on the surface and the average binding energy per atom is also very high (Roduner 2006). The surface-to-volume ratio changes with size reduction, and hence the phase transition temperatures vary (Roduner 2006). Due to large surface area of the material, thermal properties are enhanced due to the occurrence of heat transfers at the surface, and mechanical properties such as strain and elasticity are deemed tunable (Saleh 2020). Thus, physical properties are dependent on conditions such as cristallinity, geometry and surface of the material (Guerra et al. 2019).

Quantum size effects are also introduced in these nano-systems, as delocalized electrons can be described as "particles in a box", for which its electronic behavior (densities of state, energies, etc.) depends on the size of the box (rather than its dimensionality). Therefore, its impact is more related to chemical properties such as catalytic activity, electrical properties and magnetic properties due to the importance of electrons and orbitals influenced by the size effects (Roduner 2006).

The numerous applications given to nanomaterials are supported by their special characteristics; thus, study for enhancing and engineering of these materials is hugely important (Baer et al. 2008). The analysis and understanding of nano-materials requires unique and specific approaches. Molecular dynamics simulations and theory are used to comprehend and predict thermal and mechanical properties in macro and nano-scales (Mohammadi et al. 2020). Thermal analysis techniques are widely used for evaluating a nanomaterial's change of mass, heat flux, change of temperature, mechanical stability with thermogravimetry, differential scanning calorimetry, differential thermal analysis and thermomechanical analysis, respectively (Seifi et al. 2020). Finally, microscope-assisted techniques and theories (i.e., nanoindentation) can help evaluate further mechanical information (elasticity, flexibility) (Reghunadhan et al. 2018) .

According to the classification of dimensionality and shape of nanomaterials (0D, 1D, 2D, 3D) (Sudha et al. 2018), physical properties could be discussed with certain precision.

2.2 Mechanical properties

Mechanical properties of nanomaterials typically depend upon their response, under external stress or load (brittleness, strength, plasticity, toughness, elasticity, etc.). Information on the specific conditions for engineering a material's strength are vital to certain applications. The nature (chemical properties) and preparation (synthesis) are relevant; however, conditions such as grain size, and grain boundary structure have been deemed quintessential to the tunability of mechanical properties (Wu et al. 2020).

Flexibility as a mechanical property is intrinsically related to a bending modulus which indicates bending resistance (Teng et al. 2019), parallel to the Young modulus which characterizes a material's elasticity (deformation resistance) (Callister and Rethwisch 2011). These properties are layer and thickness dependent; therefore, plate-like (2D) nanomaterials are studied, i.e., Young's Modulus will decrease with less than 10 layers, and approach the bulk material constant beyond

20 layers (Sun and Zhang 2003). Layer formation capacity of a nanomaterial is closely related to these properties; thus, stacked monolayered morphology is desired with adequate interlayer distance for functionalization of the material (i.e., ion/electron conductivity) while ensuring uniform distribution without sacrificing the material's stability (Zhu et al. 2018a). For nanotubes, Young's modulus decreases as tube diameter increases (Xiao and Hou 2006), and even these can form multiwalled systems mimicking 2D materials for enhanced mechanical properties (Tao et al. 2015).

2D material arrangements are desired for their flexibility (Teng et al. 2019). Nanosheets have been found to provide mechanical stability as flexible building blocks (Li et al. 2013). Their large aspect ratio (10_6) in comparison with length to thickness, low surface roughness, and even properties such as noncovalent adhesion greatly enhance mechanical stability of these materials. Surprisingly, the mechanical stability of nanosheets decreases as thickness is increased (Fujie et al. 2009).

Some of these layered materials (graphene, h-BN, Zr phosphonate, MoS_2, etc.) show low coefficient friction, which translates into solid-state lubrication. Sliding happens within the layers due to weak interlayer interaction (Van der Waals interactions); therefore, lubricant-substrate interactions may affect the expected behavior (stick-slip and wear) (Spear et al. 2015). Addition to oils in order to obtain 2D nanomaterial based lubricant additives is also explored. Furthermore, stability takes another meaning as these materials have to be tested under extreme conditions (high temperature, high relative motion speed, high vacuum and high pressure), also securing fluid related parameters such as dispersion, in-oil stability, viscosity, etc. (Xiao and Liu 2017).

Hardness and plasticity are also intrinsically related to crystallinity and phase of the material. However, structure and dimensionality can also provide answers towards further improving these characteristics. High-melting point compounds (carbides, nitrides and borides) with 20–35 GPa can be turned to super hard materials with an adequate transition to a nanostructured state (Guo et al. 2013).

In 2D nanomaterials, hardness increases with thickness reduction, while avoiding brittle states (dislocation and crack propagation) (Andrievski 2009). Nanocrystalline metals (Ni- and Co- based) exhibit great hardness-strength relationship and can be used for conventional coarse-grained metals and alloys (Brooks et al. 2008).

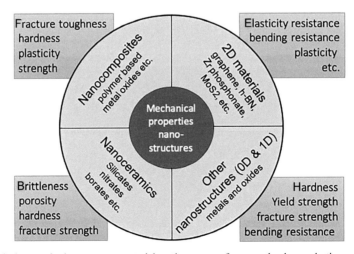

Figure 2. Mechanical properties between nanomaterials acting on a surface, e.g., hardness, elastic modulus, adhesion and friction, and main applications.

2.3 Thermal properties

Thermal applications require the consideration of certain parameters such as thermal conductivity, thermal expansion, thermal stability, etc. Nanomaterials' efficient appliance on insulation, comfort, cooling and energy conversion require previous knowledge and prediction capabilities on heat exchange dynamics and thermal effects and its relation to nanostructuring (Volz 2009).

Thermal conductivity measures a material's capacity to conduct heat (Bird et al. 2007). In a macroscopic system, contribution to this property is deemed by phonon modes with wavelength equal or greater than its macroscopic length scale (using molecular dynamics based on the Boltzmann equation). However, due to anharmonicity in the interaction potential of nanoscopic materials, thermal conductivity is affected. The increase in this property has to do with phonon-phonon scattering reduction as size is decreased (Che et al. 2000). Two-dimensional (2D) structured layered nanomaterials have been studied for this thermal conductivity increase (i.e., 2D-BN nanosheets with 400 W/mK in plane) related to the decrease in thickness (Guerra et al. 2019). Addition of nanoparticles to a base-fluid (nano-fluids) grants enhanced thermal conductivity with correlation to increase in volume fraction, size change, and aspect ratio (i.e., nanorod > nanorhombic > nanosphere, nanocylinder > nanosphere, etc.) (Li 2013, Afrand 2017, Yang et al. 2019).

Thermal expansion coefficient indicates a material's response to temperature in terms of its shape and size (Tipler and Mosca 2003). For nanomaterials, thermal expansion coefficient increases when grain size decreases. Considering surface effects, based on thermodynamical relations, an equation can be used to effectively evaluate this parameter as size varies, and therefore accurately predict expansion behavior (V/V_0) as temperature changes (shown in Eq. n and n+1) (Singh and Singh 2015). Mechanical stability is also closely related to the thermal expansion coefficient, i.e., nanomaterial addition for alloy reinforcement (WS_2 inorganic tubes) improve hardness, tensile strength, etc. (Huang et al. 2018).

$$V = V_0 \left(\frac{1}{1 - \delta_T \alpha_0 (T - T_0)} \right)^{1/\delta_T} \tag{1}$$

where V is volume,

T is temperature and

subscript $_0$ is the reference values of the respective parameters.

α0 is the coefficient of thermal expansion

δT is the Anderson parameter

$$\alpha_0 = \alpha_b \left(1 - \frac{N}{2n} \right)^{-1} \tag{2}$$

where N is the total number of surface atoms and

n is the total number of nanosolids,

α_b is the coefficient of volume thermal expansion of the bulk materials.

Heat capacity is a thermodynamic property, which describes the quantity of heat supplied to a given mass so it produces a unit change in the temperature. Thus, specific heat derives from this property taking into account heat capacity per mass (Tari 2003). Parallel to thermal conductivity, specific heat capacity is only influenced by phonon vibrations, with Debye and Einstein models used for analysis as they take into account vibration modes of inner and surface atoms. This value is apparently higher for cubic and thin particle shapes. Moreover, in terms of size, for nanoparticles larger than 10 nm, the specific heat capacity decreases concerning the nanoparticle diameter due to the quantum size effect. However, at smaller sizes, heat capacity will increase due to quantum size effects (Wang et al. 2006). Effect is also widely harnessed on nanofluids (water-based Al_2O_3 (Zhou and Ni 2008), CuO (Zhou et al. 2010)). Finally, thermal stability, referring to melting point and other phase transitions, vary as well. Melting point decreases as dimensionality and size decreases (surface/volume ratio increase) (Jiang et al. 2003) due to lower stabilization of surface atoms (Roduner 2006). Through various approximations and models, other transitions such as glass transition of polymers can be better understood for designing thermally stable materials (Zhang et al. 2018). Corrosion and wear resistance require testing in order to obtain high-temperature stable materials that can withstand long-term exposure to aggressive environments (Andrievski 2014). It is important to mention that more thermodynamic (rather than thermal) properties also vary in nanomaterials. Physiochemical approximations to nanoscale entropies, enthalpies, free energies, etc., change with size, dimensionality and composition, which can also address endless properties further (Yang and Mai 2014).

2.4 Chemical properties (catalytic)

Inorganic nanocatalysts have received plenty of attention during the last three decades due to their interesting physicochemical properties: high surface area, active sites, better selectivity, high chemical stability, and moderate ease to operate on an industrial scale. They can be applied in energy and environmental applications (oxidation, reduction, pure hydrogen production, fuel cells, etc.). Similarly, these materials can be used in autocatalysis for air cleaning, water purification; and sensors, for detection of harmful or combustible gases or substances in solution, etc.

The exploitation of MNPs has begun since ascertaining the novel catalytic activity of Au nanomaterials in low-temperature CO (carbon monoxide) oxidation (Haruta et al. 1987). Later, in 1941, polymer-protected Pd and Pt nanoparticles for hydrogenation applications were among the earlier reports in the nanocatalysis field (Rampino and Nord 1941). However, it took a long time to focalize the catalysis field on nano and sub-nanostructures. In general, nanoparticles' catalytic properties are enormously dependent on parameters such as size, shape, composition, and morphology. Noble nanoparticles have an extra advantage in photocatalysis applications due to their apparent natural surface plasmonic properties. On the other hand, SCs (semiconductors) and insulator nanostructures (transition metal oxides, borides, sulfides, phosphates, carbides, etc.) present some advantages: selectivity, reactivity, and chemical stability as both photocatalysts and electrocatalysts.

2.4.1 Noble metal catalysts

The plasmonic excitation of noble metal nanoparticles makes them appropriate for various catalytic processes. Noble NPs are highly utilized in electrochemical catalysis and photocatalysis so far due to their LSPR, pairing light flux to the electronic conduction of MNPs, with facile preparation, chemical stability, and high catalytic activities (Wang et al. 2014, Pradeep and Anshup 2009).

The catalytic activity of Au nanostructures has received significant attention for CO oxidation, purification of hydrogen supplies for polymer electrolyte membrane (PEM) fuel cells, and chemical processing (McEwan et al. 2010). Approximately 5 nm-sized Au particles have exhibited excellent catalytic properties in many reactions, such as partial oxidation of hydrocarbons (HCs), water–gas shift reactions, and NO_x reduction (Thompson 2007). Similarly, the studies on CO oxidation on supported Au NPs show that the smaller the size (diameter < 5 nm), the greater activity on various MO (metal oxide) supports. At ambient conditions, tiny gold nanoparticles (2–4 nm in diameter) have largely increased CO oxidation rate by a hundred-fold, in comparison with gold NPs with diameters ranging from 20–40 nm (Ertl 2002).

The transition regime where the intermediate sized (between 2.3 and 1.7 nm) gold nanoparticles exhibit both metallic and molecular behavior demonstrated excellent activity in the oxidation of both CO and alcohols (Zhou et al. 2016). Pt nanoparticle's size effect compared between ~ 30 nm and $1-5$ nm, tested for specific electrocatalytic oxygen reduction reaction (ORR), thus evidencing high efficiencies by sub-nanometric size particles (von Weber and Anderson 2016). Nowadays, sub-nanoparticles of a few dozens of metal atoms (~ 1 nm) have proven highly active in catalytic processes, thanks to modern techniques such as X-ray absorption spectroscopy and aberration corrected electron microscopies.

Some supports can also act synergistically with nanoparticles and enhance catalytic performance. For instance, N-doped carbon supports were outperformed as compared to non-doped carbon supports (Kornienko et al. 2018). Ultra-small (below 2 nm-sized) noble MNPs (Au, Pd, and Pt) highly dispersed on Ni-doped carbon supports have shown extraordinary electrocatalytic activities in the methane oxidation reaction (Liu et al. 2016).

Though particle size is very critical for the reactivity of the nanoparticles, the particle shape counts as well by influencing surface atomic arrangement and exposed facets, active sites, adsorption energies, and reactivity to specific molecules (Gu et al. 2012). Recently, gold particles' activity with similar size distributions (~ 3 nm) was studied using different MO as supports (TiO_2, Fe_2O_3, Al_2O_3, $MgAl_2O_4$, SiO_2). The measured CO oxidation rates for different supports were assigned to the disparities in Au particle shape and their exposed facets on the various oxide supports. The catalytic performance of gold particles supported on Al_2O_3, are more significant than gold particles supported on reducible oxides (TiO_2, Fe_2O_3) (Lopez 2004). For instance, da Silva et al. experimentally proved that Pd NPs exhibited higher electrocatalytic activity on exposure of {100} surface facets in comparison to {111} facets, leading to highest current density, thus causing electrooxidation of formic acid (da Silva et al. 2016).

Likewise, specific multi-metallic nanomaterials can enhance the catalytic performances in comparison to their monometallic counterparts. The compositional variations in multimetallic nanomaterials directly affect their constituents' electronic states. As a consequence, it affects their adsorption energies and catalytic reactivities (Li et al. 2018a). For example, Cu_xPd_{1-x} bimetallic nanocubes (with 20 at% of Cu) showed the best electrocatalytic activity leading to the formic acid oxidation, which can be compared with pure platinum nanocubes of similar sizes (Xu et al. 2010). However, LSPR effect of noble metal NPs can also utilize co-catalysts to improve the performance of SC photocatalysts. Several oxide-based SCs (TiO_2, Bi_2O_3 Fe_2O_3, CeO_2, ZnO, etc.), and chloride and bromide based SCs (AgCl, AgBr) have been explored and studied in detail. A Schottky junction is formed when the metallic nanoparticles are in contact with SCs, thus stimulating fast charge transportation and suppressing charge carrier recombination. This plasmonic effect of NPs (LSPR) contributes significantly to photocatalytic activity (Sarina et al. 2013).

Noble metal NPs deposited on insulator supports (Al_2O_3, ZrO_2, SiO_2, zeolite, etc.) have significantly wider bandgaps where photoexcitation is very limited and not possible under UV light. In this case, noble metal nanoparticles that possess excited electrons due to LSPR react with the reactants and then transform sunlight to chemical energy using photon-activated photocatalytic reaction.

Reasons for the significant photocatalytic activity of noble nanostructures are:

- LSPR of metal NPs acts as antennae to enhance a broad spectral range of UV-Vis light absorption.
- LSPR absorption on the metal cluster surfaces excites electron-hole pairs in the depletion region, which can then activate the surrounded molecules for the desired chemical reactions.

- LSPR polarized the metallic nanoclusters, thus generating electrons and holes on the surface of the metals. These electrons in turn augmented the redox reaction rate as well as mass transfer rates too.
- Plasmonic enhancement of the electric field near metallic NPs surface increases the density of the available conduction electrons much higher than that of any SC's surface.

A majority of photocatalysts are studied in their powder or mobile form to benefit their large surface area, and simplicity of operation. However, it is not easy to retrieve them entirely from the system and may require expensive filtration techniques. Moreover, many mobile nanocatalysts are toxic and difficult to handle in large-scale industrial applications. Therefore, to avoid such issues, recent photocatalysis research focuses on supported/immobile photocatalyst, thus compromising their surface area.

Intrinsic catalytic activity depends on the active sites of catalysts in their electrode form, but their conductivity plays an equally critical role. Graphene and its derivatives are good examples, having high conductivity and good chemical stability, and are suitable to combine with metal or SC nanomaterials to form composite electrocatalysts with further improved activity and stability (Yang et al. 2021).

Kwak et al. reported Au clusters combined with reduced graphene oxide (rGO) in the form of nanocomposites. In the redox process, Au/rGO composites show a significant enhancement of current output and potential difference (13.4 µA and 127 mV at the oxidation peak) compared with that of Au clusters (4.5 µA and 60 mV). Both of these values indicate that the electron transfer process is enhanced by combining Au_{25} with rGO (Kwak et al. 2016).

Noble metals have been extensively studied; however, their high cost limits their use in practical and industrial applications. Consequently, it is necessary to find an alternative where non-noble metals such as MOs, borides, carbides, nitrides and sulfides, etc., have been widely investigated.

2.4.2 Catalytic properties of transition metal oxides

Metal oxides (TiO_2, Bi_2O_3, Fe_2O_3, CeO_2, ZnO, etc.) are earth-abundant and more promising catalysts, and cheaper alternatives than noble metal nanostructures. Moreover, they have been exploited in various electrochemical reactions due to their intrinsic oxidation stability, high-density material defects, and oxygen vacancies, which can benefit catalytic activity.

Similarly, transition MO nanoparticles have been also tested on various supports. Graphene oxide-supported MO nanoparticles exhibited better charge transfer, and synergistic effects in overall catalytic activity, some of them being able to be arranged as one-dimensional (1D) structures such as nanowires, belts, whiskers, and ribbons in order to enhance charge transfer. For instance, SnO2 nanofiber's geometry improves electron transportation by shortening the ions diffusion length and enlarging electrolyte-electrode contact, resulting in high current density.

Transition metal boride (TMB)-based materials (FeB2, Co2B, NixB, etc.) are considered to be efficient electrocatalysts for hydrogen evolution reaction (HER) and oxygen evolution reaction (OER) (Jiang and Lu 2020). Similarly, transition metal carbide and nitride (MCN) nanomaterials are also considered to be efficient electrocatalysts, which is attributed to their fast electron transport, more than enough active sites, and high available surface area. Most MCN catalysts have either a 3D mesoporous structure or ultrathin nanosheet structure, offering highly active site exposure. The strong nano-interface between different domains can promote the formation of excess active sites (Jin et al. 2019). Benefited from this strategy, MCN catalytic activity has been broadly employed in various electrocatalytic reactions such as HER, OER, ORR, CO2RR, etc.

In addition, a more significant density of active sites can also be found in metal sulfide and phosphide nanostructures. For instance, metal-phosphide (Ni2P, Co2P) based nano-catalysts can enhance the intrinsically electrocatalytic activity with their natural high electronegativity behaving as hydrogen transporters in HER. In metal sulfides, MoS2 and WS2 are the most suitable HER electrocatalysts due to their high performance originating from their available surface sites. S-sites behave as active sites, and their S-vacancies can also improve the electrocatalytic HER performances through their H2O association and disassociation process.

In catalysis, a prolific research has been carried out on the supports which hold the transition metal ion catalysts, especially focused on reducing their size down to sub nanometer scale; for example, host ceramic matrices like zeolites, micro structured silica, alumina and other oxides (Cejka et al. 2017). On the other side, carbon based supports with different dimensional materials-1D, 2D and 3D-such as nanotubes, graphene derivatives and metal organic frame works (MOFs) as novel catalyst supports have gotten great attention from the catalysis community (Furukawa et al. 2013). It is noteworthy to mention that metal-organic frameworks (MOFs) are also applied for various electrocatalytic applications. Their reticular synthesis and careful selection allow strong bonding between organic and inorganic domains, resulting in high-grade MOF crystals. These diverse characteristics and MOFs' interior chemical tuning flexibility make them versatile for many catalytic applications such as gas separation, detection, gas storage, synthetic fuel, etc.

2.5 Optical properties

Bulk metals are well known for being good light reflectors due to their high density of charged electrons. When the conductive metal size shrinks down to the nanoscale, like sub-wavelength of incident electromagnetic radiation, the small volume of electrons constrained within the nanostructures indeed absorbs the light and leads to plasmonic effects. That is why when a light shine on metallic nanoparticles, they illuminate in beautiful lustrous colors due to a plasmonic resonance effect. In Middle Age Europe, the ultrafine gold and copper metal particles were utilized by incorporating in the glass to decorate the gothic church windows and artisanal pottery work (Hunt 1976).

The earliest report on the SP (surface plasmon) polaritons was made in 1902 (Wood 1902). Later, Lord Rayleigh interpreted these optical phenomena, and a few years later, Fano consolidated these anomalies with electromagnetic surface waves on selected metallic grating under light illumination ('On the dynamical theory of gratings', 1907) (Fano 1941). In 1950s, the consecutive work of Ritchie, and Powell, and Sawn, proposed and verified the concept of SPs by correlation with an energy loss in thin metallic films (Ritchie 1957, Powell and Swan 1960). A decade later, Otto as well as Kretschmann and Raether almost simultaneously, reported the experimental method to generate SPs of gold and silver nanoparticles for the first time around 1970 (Otto 1968, Kretschmann and Raether 1968).

2.5.1 Surface plasmon polaritons (SPPs)

SPPs are surface-charged electromagnetic waves propagating along the interface of a conductive metal and/or dielectric material coupled to a p-polarized electromagnetic wave. The magnitude of the propagated wave can be determined by the relative permittivity of dielectric material, with nothing but propagating media.

2.5.2 Localized surface plasmon resonance (LSPR)

When light shines on MNPs, the surrounded free-electron cloud oscillates at the incident electric field's frequency, called local surface plasmons. LSPR describes the enhancement of these oscillations when the frequency of electromagnetic radiation matches with the frequency of the electron cloud of metal particles (Baffou and Quidant 2014). Therefore, the dipole-like electron oscillation in LSPR along the associated electromagnetic field's direction makes nanoparticles efficient absorbers with a specific frequency.

The nanoparticles' absorbed energy from the excitation can be further released or transferred afterwards, and radiatively redistributed to the far-field. It can also be released locally as heat dissipation into the metal due to the oscillation damping. This is also known as the photothermal effect of MNPs (Yaqoob et al. 2020).

2.5.3 Applications of surface plasmon polaritons

The capability to manipulate and control the electric field, both within the MNPs and near the confined surface, opens up a realm of possible applications such as SPP waveguides (Maier et al. 2003, Oulton et al. 2008, Weeber et al. 1999) and sources (Lerosey et al. 2009), SERS (Gopinath et al. 2009), optical data processing (O'Connor and Zayats 2010), photovoltaics (Ferry et al. 2008), chemical and bio sensors (Moores et al. 2004, Rosi and Mirkin 2005), etc.

2.5.3.1 Plasmonic waveguides

Optical interconnections in the form of waveguides can provide very high bandwidth in data transmission. SP based photonic circuits are future technologies for achieving high-density photonic integration with small electronics dimensions. SPP waveguides such as Ag spherical nanoparticles arranged in a linear chain fashion, with a distance calculated by Mie theory, have shown light propagation along the chain by pairing the irradiation at parallel polarization (Weeber et al. 1999).

Bozhevolnyi et al. designed plethora of subwavelength waveguide structures, such as V-grooves of Au on Silica substrate, single Y-splitters, and Mach–Zehnder configured interferometers with two consecutive Y-splitters, etc. All these SPPs' structures manifest single-mode with mostly losses of a few dB at telecom wavelengths (Bozhevolnyi et al. 2005).

The surface with low roughness and high-quality patterned noble metal nanostructures are indispensable since SPP propagation loss is directly dependent on it. Recent work shows that the measurement of SPP propagation lengths approaches theoretical values in exhibiting Raman scattering enhancements for sensing applications for perfectly flat and ultra-smooth Au patterns on Si-substrate (Nagpal et al. 2009).

2.5.3.2 Near-field optics

SNOM is an optical imaging technique able to control and observe light at nanometer-scale resolution. The strongly localized SPP modes in nanoscale metal structures essentially slow-down the coupled electromagnetic waves, which means a decrease in the effective wavelength in the midst of the metal nanostructures. Therefore, this confined field of light results in a spatial resolution close to the size of metal nanostructures, and is used to construct nanoscale images.

Ketterson et al. (1995) reported a SNOM based on SPPs looming from the ultrathin metal film surface, and a tungsten probe tip placed in Kretschmann's total internal reflection configuration. As a result, the scattered SPPs are enhanced by the localized surface irregularities and radiate in a conical fashion. In this case, scattered SPPs serve as a probe to shun the complications associated with aperture-based probes, and exploit SPPs to improve the signal and boost the signal's collection efficiency (Kim et al. 1995).

2.5.3.3 Surface-enhanced Raman spectroscopy

Surface-enhanced Raman scattering (SERS) is one of the essential applications of SPPs currently, and was discovered by Jeanmarie and Van Duyne (Jeanmaire and Van Duyne 1977), and Albrecht and Creighton in 1977 (Albrecht and Creighton 1977). Electromagnetic enhancement attributed to the excitation of SPPs is one of the most recognized mechanisms of SERS compared to other mechanisms, such as chemical enhancement or resonance enhancement (Moskovits 1978). Electromagnetic enhancement is nothing but SPP induced enhancement of molecular Raman scattering resulted from the light coupled illuminated metal nanostructures. SERS is known to detect ultra-low concentrated targets with high molecular sensitivity and specificity and provide more information on its chemical structure and conformation (Almehmadi et al. 2019). The SPs of nanostructures can directly affect SERS' sensitivity and selectivity. So far, the most preferred plasmonic nanostructures are Au and Ag as SERS's substrates, definitely due to their higher enhancing factor.

Tian et al. reported SERS effect for the exanimation of organic dye Rhodamine 6G (R6G) as a function of different shapes of gold nanostructures with equivalent dimensions. After running a systematic measurement, the resultant SERS signals' intensity were classified following shape dependence SERS effect: nanospheres < nanosphere aggregates < nanotriangles < nanostars (Tian et al. 2014).

Stamplecoskie et al. described the influence of Ag particle size, with the same Ag concentrate on SERS intensity, in probing R6G dye at a long wavelength excitation (785 nm). The optimal size of Ag NPs was 50–60 nm based on the dynamic depolarization and better absorption on the available surface area at constant Ag concentration (Stamplecoskie et al. 2011).

For example, Xu et al. reported SERS ultrasensitive detection of R6G molecules by using a novel substrate, i.e., porous anode alumina with a periodical aligned silver nanowire arrays. Reported the detection of the Raman signals with a low concentration of 10–13 M under visible light (514.5 nm) excitation (Xu et al. 2009).

2.5.3.4 Data Storage

Future data recording and retrieving mechanisms can use the resonance of SPPs of the metallic nanostructures. Multiplexed optical recording by far/field spectroscopic optical detection technique can allow storing the information beyond 1 Tbit cm/3. In another instance, there are plasmonic based data storage systems where a femtosecond pulsed laser can irradiate a bunch of plasmonic optical disk constituting a collection of plasmonic metal nanostructures. Each nanostructure can change the femtosecond pulse laser spectrum; thus, each structure can store a one/unit bit cell. Thus, the plasmonic based nanostructure based data storage is dependent on the variation of SPP wavelength, polarization as well as spatial dimensions of the individual or multiple nanostructures.

A five-dimensional optical recording studied the peculiarities of longitudinal LSPR of Au nanorods (Figure 3). One-dimensional noble metals (Au nanorods) possess the ability of anisotropic polarization that depends upon their diameter and incident light wavelength. The sample consists of multiple alternative stacks of recording layers (1 micron) and a transparent spacer (10 microns). The recording layer made of spin coated polyvinyl alcohol doped with Au nanorods is divided into domains responsive to three specific wavelengths and polarizations. Therefore, it contains an overall of nine multiplexed states in one single recording layer. The advantage of such a concept is multiplexing in the orthogonal direction and providing multiple recordings in each dimension, and data stability at ambient conditions with numerous readouts (Zijlstra et al. 2009).

2.5.3.5 Solar cells

Solar cells transform light energy directly into electrical energy via the photovoltaic effect. The SPs, including LSPRs and SPPs, have been extensively studied to enhance thin-film solar cells' efficiency. Using plasmonic excited metallic nanostructures, not only can the light be trapped or concentrated at the absorber, but these materials can also serve ingeniously as back contacts or even anti-reflective electrodes in some innovative designs. Plasmonic nanostructures incorporated in the absorber materials can drastically reduce their thickness due to their efficient light absorption compared to the conventional solar cells. Three different configurations of integrated plasmonic structures are utilized to improve overall solar cell efficiency, such as:

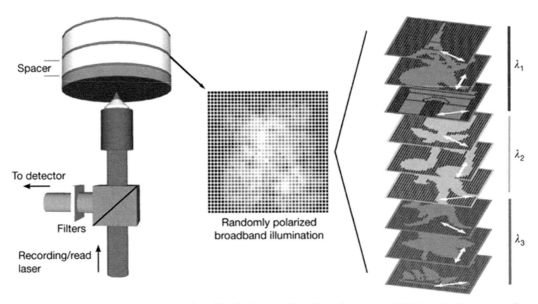

Figure 3. Sample arrangement and patterning of (Left) thin recording films of spin-coated PVA doped with Au nanorods, separated by a transparent pressure-sensitive adhesive. In the recording layers, different images were patterned using various wavelengths ($\lambda 1$–3) and polarizations of the recording laser. When illuminated with unpolarized broadband illumination (Middle), a convolution of all patterns will be observed on the detector. When the adequate polarization and wavelength are chosen (Right), patterns can be read out individually (Reprinted by permission from Macmillan Publishers Limited, Nature, Five-dimensional optical recording mediated by surface plasmons in gold nanorods, Peter Zijlstra, James W.M. Chon & Min Gu, Vol. 459 (2009)).

2.5.3.5.1 Top surface plasmonic structures

They can be configured on the top surface of the solar cell absorber in the form of nanoparticles that could enhance light absorption by scattering the sunlight and reducing the reflection. Plasmonic structures can also be placed in the form of a thin mesh or grid, which eventually serves as a top electrode and aid plasmonic excitation. The MO or dielectric particles can also induce light scattering, but not as good as metal plasmonic elements. For example, high-density gold nanoparticles deposited on the amorphous silicon surface demonstrated an 8.1% increase in energy conversion efficiency (Derkacs et al. 2006). Similarly, photocurrent increment due to SPP was systemically studied with the incorporation of different metallic nanoclusters (Au, Ag and Cu) at the ITO-CuPe interface. Overall photocurrent enhancement followed the sequence of 1.3 (Ag) < 2.2 (Au) < 2.7 (Cu) (Stenzel et al. 1995).

2.5.3.5.2 Embedded plasmonic structures

The intensity of the light absorption in the solar cell can be enhanced by embedding plasmonic structures directly in the absorber layer. Robust light localization around embedded plasmonic elements aid in the generation of photocurrent in the solar cell (Knight et al. 2013).

2.5.3.5.3 Back surface plasmonic structures

Similar to top surface plasmonic structures, they can be configured at the back surface of a solar cell to enhance the light absorption by trapping light and serving as a back electrode. These configurations can benefit both SPP at the metal-absorber interface, and also be coupled to waveguide modes of the absorber. Au convex nanostructures deposited at the back of the perovskite solar cell (PSC) reduced the losses and enhanced light absorption, consequently increasing photocurrent density from

18.63 mA/cm^2 to 23.5 mA/cm^2. Therefore, overall photoconversion efficiency increased from 14.62 to 19.54% (Tooghi et al. 2020).

Generally, photoactive layers must be optically thick enough to absorb most of the incident light and efficiently generate the photoexcited charge carriers. However, thin absorber layers can enhance the conversion efficiency by facilitating adequate photocarrier collection over short distances. Researchers found out that SPPs enabled solar cells can increase light absorption in the active material layer, especially at the SC and metal nanostructures interface, improving the overall efficiency in solar energy devices.

Coupling plasmonic nanostructures within these thin-film photovoltaic devices enable to accomplish higher-efficiency OPVs with PCE $\gg$ 10%. The recent advances on plasmonic-enhanced photovoltaic devices using one-dimensional (1D) and two-dimensional (2D) patterned periodic metal nanostructures are reported elsewhere (Gan et al. 2013).

2.5.3.6 Chemical sensors and biosensors

Continuous demand for reliable, sensitive, and eco-friendly sensors brings on an enormous research interest towards many biological and environmental issues. The first practical sensing application using SPR phenomena was reported in 1983 for biomolecular detection (Liedberg et al. 1983). SPR based bio and chemical sensors have become an essential platform for qualitative and quantitative chemical and biomolecular interactions. Commercial SPR-based biosensors and experimental devices are more often represented in Kretschmann's scheme of plasmon excitation due to their ergonomic planar design and relative simplicity (Kretschmann 1971).

SPR based sensors function in two ways, one being in the change of refractive index on the surface of a metallic film when in contact with the sample (Homola et al. 1999). The characteristics of light waves depend on the refractive index of the dielectric environment, which impacts the propagation length of SPs. Recently, the author reported a second detection mechanism apart from the refractive index change on SPR based sensors. It is based on the surface electron-polaritons' density change, that can be sensitive to the difference in chemical ligands coordination (Daniel and Astruc 2004).

SPP biosensors have many advantages:

1. They are versatile and can be used to detect any chemical or analyte.
2. They require no labels such as radioactive or fluorescent labels to detect the analyte.
3. They have a fast detection response time, and can measure in real-time biomolecular interactions.

In the SPR sensors, any tiny mass, concentration, and refractive index change is tied to the particular immobilized molecules on the exterior of the thin metallic sensor and can be perceived as RU signals with an aid of a transducer. The localized surface plasmon causes electromagnetic enhancement in the tenuous field of the sensor surface (Ekgasit et al. 2004).

Au nanostructures are very commonly used as SPR sensor surfaces due to their versatility and high sensitivity for most analytes. Nano MOs and magnetic nanostructures with significant SPR signal magnification provide monitoring of multiple substances such as drugs, enzymes, hormones as well as bacteria and cells (Herberg et al. 2005).

SPR sensor is constituted of a thin metallic film, on top of which lies another thin dielectric metallic layer. The performance of the sensor purely depends on the thickness of the metal layer and its dielectric constant. Graphene is a one atomic thick carbon layer in the form of a perfectly arranged planar honeycomb structure that has large surface area. Its abundant π conjugation structure renders it a highly suitable candidate for SPR biosensing (Szunerits et al. 2013).

Like graphene, a single atomic thick structure platinum diselenide (PtSe$_2$) has reported remarkable optoelectronic properties. It has drawn significant attention in the fast growing 2D materials field. SPR sensors through 2D layer of PtSe$_2$, graphene, and phosphorene have been studied, proving that only the sensitivities of silver or gold film biochemical sensors for graphene, and phosphorene were observed at 118°/RIU and 130°/RIU, while the sensitivities of the $PtSe_2$-based biochemical sensors touched as high as 162°/RIU (Ag film) and 165°/RIU (Au film) (Jia et al. 2019).

SPR biosensors based on 2D transition metal dichalcogenides (TMDCs), where WS_2 layer coated on Al thin metal film surface, report more than three times angular sensitivity with 7-layers WS_2 coated on 36 nm Al thin film, compared to the typical structure-based single Al thin film, obtaining a maximum phase sensitivity with bilayer WS_2 and 35 nm Al thin-film composite structures (Zhao et al. 2018).

MNPs are excellent materials for sensitive SPR-based biosensors. However, they still presenting some disadvantages, such as the surface inadequacy for biological use, their natural particle aggregation, and the loss of magnetic activity in biological environments. To surmount these problems, MNPs in the form of core-shell (ferrites such as Fe3O4, Fe2O3 as cores with shells containing Au, Ag, or alumina) are synthesized. For instance, Fe3O4@Au as a DNA biosensor (Kouassi and Irudayaraj 2006) have been designed. SPR bio-sensor made of core-shell Fe3O4@Au MNPs were used as agents for the detection of the hepatocellular carcinoma tumor marker, α-fetoprotein (Liang et al. 2012).

2.6 Dielectric properties

The plasmonic structures made of metals suffer from high radiation losses, heating and incompatibility with MOs fabrication processes. Recent developments have been made in nanoscale optical physics, targeting the manipulation of optically induced Mie resonances in dielectric as well as metal or SC nanoparticles with high refractive indices. Such nanostructures have remarkable characteristics known for reduced dissipative losses, and large resonant enhancements in electric and magnetic fields. Resonant dielectric nanostructures offer great promise in the optoelectronics field for various technological applications such as nanoantennae, capacitors, actuators, and sensors.

On the other side, dielectric nanocomposites with a concept of dispersion of nanoparticles in polymer have good compatibility with interfaces. These polymer nanocomposites can cover many applications such as nano-electronic, polymer bionanomaterials, reinforced nano-polymer composites, etc. However, the handling of such diverse technology with desirable dielectric properties is a great challenge, due to the complex and meticulous synthesis techniques. Nanocomposite materials have many advantages in electric and dielectric characterization, such as dielectric confinement effect, surface effect, quantum effect, and volumetric effect. Moreover, nanoparticles are thermodynamically unstable, consequently leading to aggregation. The NP aggregates are thus responsible for the formation of defect centers, enhancement of the local electric field, and decrease of the breakdown strength (Calebrese et al. 2011), but also highly increasing the dielectric losses resulting from DC conduction. Therefore, by blending nanoparticles with many varieties of polymers in the form of nanocomposites, such aggregations could be avoided. In most of the cases, such combinations require the polymer in solution or in melted form in order to disperse the nanostructures homogeneously (Krishnamoorti 2007). The type, size, geometric shape, and dispersion or concentrations of nanoparticles have apparent effects on the overall electric and mechanical properties of nanocomposites.

They can either assist or restrict the free charge mobility and decrease or increase electric insulation to access or limit the generation of mobile charge and the movement of charge carriers in polymer dielectrics. Therefore, the improvement in dispersion of inorganic nanofillers is vital in enhancing the electrical and mechanical properties of nanocomposites (Ng et al. 2001, Zhang et al. 2000).

2.6.1 Dielectric insulators

Blending of nanoparticles into polymeric matrices can control overall dielectric performances with the aim to apply in industrial processes. For instance, nanoparticle composed polymers have better dielectric strength than conventional insulation materials. Furthermore, some of the specific nanomaterials can increase the charging capacity of power dielectric material, thus being used in high-charge storage capacitors. MO nanoparticles like magnesium oxide, fumed silica, barium titanate (BT), alumina, and titanium dioxide can increase the dielectric constant of electrical insulation materials. In contrast, the materials like clay, silica, glass beads nanoparticles decrease dielectric constants with increased volume fraction for variant electrical insulation systems. Similarly, low-loss dielectric materials (Sebastian et al. 2015) have got constant demand in wireless telecommunication applications such as Internet of Things (IoT).

However, the clear relationship between the concentration of filler, and the permittivity of the composite was uncertain (Brosseau et al. 2007), until Maxwell demonstrated an analytical approximation method by considering permittivity and volume fraction of spherical shaped nanoparticles, which are homogeneously distributed in an inclusion polymer matrix (Lombardo 2007). According to the Maxwell-Wagner theory of interfacial polarization, the material's dielectric behavior at low frequency can be represented as two layers. The first layer is a large number of grains, which conducts layers at higher frequencies. The second layer constitutes of grain boundaries, acting as a highly resistive medium at lower frequencies; therefore, as the frequency increases, the constant dielectric increases. At low frequency, generally, (ε') has low values and then rises along with frequency and reaches a constant value at higher frequencies. This rise of (ε') can be correlated to collective contributions of electronic, ionic, and interfacial polarization (Kumar et al. 2006).

The dielectric constant of CdS nanocomposites is larger than the bulk CdS, especially in the low frequency range (Suresh 2014). CdS nanoparticles with particle size around 3.85 nm exhibited better dielectric properties, due to their quantum confinement effect, observed on having a wider bandgap of 2.58 eV with a blue shift of 480 nm in the optical absorption spectrum with respect to the bulk counterpart (2.42 eV).

Mn_3O_4 ferro-nanoparticles demonstrated a high dielectric constant because of their smaller size well-dispersed nanoparticles, and each Mn_3O_4 nanoparticle acts as a nano-dipole under electric fields; therefore, cumulative dipole moment results in a high dielectric constant (Dhaouadi et al. 2012). Nonetheless, nano-grain boundaries play an essential role in transport properties. For instance, ZnO nanofillers in the polymer matrix have shown a high electrical permittivity with high nano-grain boundaries containing many defect structures. The particle size of inorganic nanoparticles in the polymer matrix has quite an influence on the dielectric behavior, where the dielectric properties of nanocrystalline ZnO particles with different particle sizes (d = 22–98 nm) were studied. There are two regions of conduction mechanisms that depend upon the core particle conduction and grain boundary conduction. The 22 nm-sized ZnO nanocomposite has a large volume fraction of interfaces that induce a better barrier hopping conduction and a quantum tunneling mechanism due to the smallest nanoparticles (Parvez Ahmad et al. 2019, Cheng et al. 2018).

BT nanoparticles coated with double layers such as hyperbranched aromatic polyamide (HBP) and PMMA nanocomposites were reported as core@double-shell nanocomposites. These double polymer layer shell has shown higher dielectric constant and low dielectric loss as compared to core@single layer BT@PMMA composites (Xie et al. 2013).

Compared to the PVDF loaded with BT nanofibers (NFs) and BT@TO NFs nanocomposites, the novel design nanocomposites filled with BaTiO3@TiO2@Al2O3 NFs presents low dielectric loss, high breakdown strength, and low leakage current densities. Therefore, a novel core-double shell structure aids the effectivity of polymer nanocomposites for compact and flexible high energy storage devices (Pan et al. 2017).

2.7 *Magnetic properties*

Magnetic properties of nanoparticles:

Magnetic nanoparticles are broadly defined as nanoparticles that can be manipulated using magnetic fields. They are also known as magnetic nanobeads.

2.7.1 *Physical properties of magnetic nanoparticles*

The magnetic moment of a material is obtained by the addition of the individual magnetic moment of particles that have both mass and electric charges. A spinning electric-charged particle results in the generation of a magneton (magnetic dipole). Ferromagnets are materials in which all the magnetons are aligned towards one direction and their resultant moment is different to zero. For example, several crystalline materials exhibit ferromagnetism (Fe, Co, or Ni). If all magnetons presented in a unit volume of ferromagnetic material are aligned in the same direction, it is known as a Weiss domain or magnetic domain. The magnetic behavior of a ferromagnetic material depends on the size of the domain. The size of a ferromagnetic material is reduced to the order of a single magnetic domain and will have a uniform magnetization due to aligned magnetons, whereas in larger particles, the magnetons are pointing to all sorts of random directions due to their multi-domain nature and have non-uniform magnetization. When the size of particle approaches a single domain, due to the discrete nature of electronic band structure, it is possible to observe permanent magnetism in certain metals. For example, Pd nanoparticles are ferromagnetic which is not the case in their bulk form (Sampedro et al. 2003).

Coercivity is the capacity of a ferromagnetic material to resist an external magnetic field without losing its magnetism. Coercivity is denoted by H_C and usually measured in oersted or ampere/meter. The important feature of ferromagnetic materials is their strongly size-dependent coercivities. It has been corroborated that as particle size lessens, the coercivity rises to a maximum and later decreases towards zero. When the magnetic dimension drops down to a critical diameter (Dc) where there is only a single domain, coercivity is maximized. When the size of single-domain particles diminishes below a critical diameter, then coercivity tends to become zero, thus resulting in superparamagnetic in nature (Figure 4). The concept of superparamagnetism of magnetic materials in the nanoscale was first proposed by Frenkel and Doefman in 1930 (FRENKEL and DOEFMAN 1930). They predicted that ferromagnetic NPs with a particle size smaller than single-domain size would display superparamagnetic properties. These superparamagnetic particles are sensitive to thermal fluctuations. Some single domain nanoparticles can behave either as ferromagnetic or superparamagnetic or sometimes both, depending on temperature. This thermal transition from superparamagnetism to ferromagnetism is called blocking temperature (i.e., cobalt ferrites). The advantage of superparamagnetic particles can be employed in biological environments. For example, ferrite oxide-magnetite (Fe_3O_4), and maghemite (Fe_2O_3) are the most studied magnetic materials that can be exploited in the form of superparamagnetic substance (Mikhaylova et al. 2004). The ferromagnetic NPs can be employed to fabricate permanent magnetic arrays for high-density data storage whereas super magnetic NPs possess different applications viz., biological sensing, imaging, and drug delivery.

2.7.2 *Industrial applications*

MNPs are customarily exploited as pigments in porcelain, and paints (Hradil et al. 2003). Along with iron oxide, some ferrites have also been applied as catalysts during the synthesis of ammonia, the desulfurization of natural gas, Fisher-Tropsch synthesis for BT and oxidation of alcohols (Park et al. 2000, Kharisov et al. 2016).

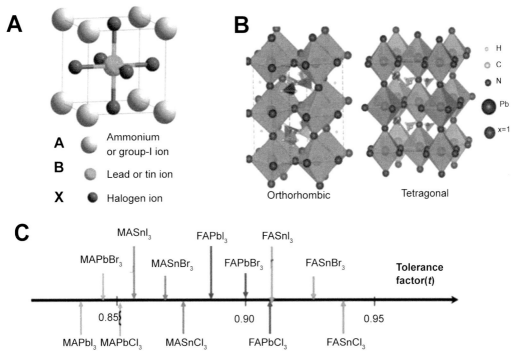

Figure 4. (A) Crystal structure of cubic lead halide perovskite (a phase) (Yin et al. 2015). (B) Crystal structure of the tetragonal crystal system (b) phase and orthorhombic (g) phase of MAPbX3 (Feng and Xiao 2014). (C) Tolerance factors (t) of a series of halide perovskites (Fan et al. 2015).

2.7.2.1 Biomedical applications

Magnetic nanoparticles are widely used in biomedical applications; it can be *in vivo* (inside the body) or *in vitro* (outside the body) (Tartaj et al. 2003). Ferromagnetic iron oxides' nanoparticles have been highly used compared to other materials because of their higher sensitivity and nearly saturated superparamagnetic properties. Consequently, it is easy to track such particles without sacrificing their stability in an aqueous colloidal suspension.

2.7.2.2 *In vivo* applications

For *in vivo* applications, degradation of MNPs is a paramount factor and depends on the core material and the coating material. Further, core-shell nanoparticles are effective in the body since they are highly biocompatible in nature, allows drug delivery and possess very high magnetic properties (KILPATRICK et al. 1997).

To employ the magnetic nanoparticles inside the body, their size and surface functionality play crucial roles. The size of superparamagnetic iron oxide nanoparticles (SPIONs) is around 10 to 40 nm, generally preferred for pro-longed blood circulation and better biodistribution (Lu et al. 2007). On application of an external magnetic field to SPIONs, their magnetic characteristics can be exploited, and magnetic energy can be transferred into SPIONs in the form of heat in order to target tumor tissues locally. This technique is known as hyperthermia and is one of the most successful cancer therapies (Gao et al. 2004) .

2.7.2.2.1 Drug delivery systems

MNPs with the combination of an external magnetic field and/or magnetizable implants on them allows the delivery of nanoparticles *in vivo* to the targeted area. Then, pathogenic cells are fixed by releasing drug at the local site known as magnetic drug targeting technique (Casula et al. 2006). This kind of local site treatments can eliminate side effects and high dosages. The surfaces of these particles are most commonly functionalized with organic polymers and inorganic metals or MOs to generate biocompatibility and suitability for the coupling of multiple bioactive molecules (Gupta et al. 2007). Controlled drug release is one of the recent advances in MNP loaded drug delivery systems (DDS). Combining MRI imaging can permit real-time tumor-tracking, and controlled and stimuli triggered anticancer drug release (Ulbrich et al. 2016).

SPION could serve a critical mission from precise drug delivery to inflammatory sites to maintain adequate concentrations at low cost and with no side effects. Recently, developments in the synthesis and drug delivery efficacity of several hollow structured magnetic MOs were summarized. The drug delivery efficiency has been tested *in vivo* and worked adequately on various antibotics (Cefradine, Vancomycin, enrofloxacin), anticancer drugs (Doxorubicin, cisplatin, camptothecin, paclitaxel, Docetaxel (DOX) and 5-Fluorouracil), nonsteroidal anti-inflammatory drugs (NSAIDs) and Natural drugs ex: RhB or R6G (Gao et al. 2004).

2.7.2.2.2 MRI diagnosis

Magnetic resonance imaging (MRI) is the most dominant non-invasive imaging methods in clinical analysis, based on the relaxation of protons in tissues. MRI is widely used as a diagnostic tool to present high spatial resolution of the tissues. Proton relaxation of accumulated magnetic nanoparticles on specific tissues is compared with that of the surrounding tissues, thus serving as an MRI contrast agent (Sun et al. 2008). This can also serve to monitor physiological and molecular changes to indicate the status of organ functions, blood flow, detection of inflammatory diseases, etc. (Thorek et al. 2006). Iron (Fe) and a few other NPs based on Gd or Mn have been studied as contrast sources in developing new MRI. NPs of Gd2O3, GdF3, and GdPO4 properties have been considered as T1 MRI CAs and tested with other nanostructured materials such as nanoporous silicas and dendrimers perfluorocarbon NPs, and nanotubes. Latest Gd-based NPs have a significantly high improved relaxation due to their ability to stock a high number of paramagnetic ions.

Recently, a novel strategy- grafting/peeling-off surface coating of nanoparticle- has demonstrated adequate activity as magnetic resonance imaging probes can be specifically cleaved by specific enzymes, in an enzyme-rich milieu. Moreover, selective activation of nanoprobes by a specific enzyme renders greater tumor accumulation and enhanced retention time within the tumors. There is a possibility of functionalizing the surface of the nanoprobe with either cleavable or non-cleavable PEG linkers for the construction of active or inactive probes, respectively. To exemplify this aspect, Sun et al. attached cleavable PEG linkers on the fluorescent dyes of the probes; thus, this complex acted as an indicator to track enzymatic activities *in vitro* (Sun et al. 2019).

2.7.2.3 *In vitro* applications

2.7.2.3.1 Bio-separation

MNPs are mainly employed in diagnostic separation, selection, and magnetorelaxometry applications *in vitro*. Moreover, they got considerable attention in the separation of specific biological entities. Performing bio-separation in large volumes is a time-consuming procedure and requires an expensive filtration system, whereas superparamagnetic colloids are ideally optimal for this application due to magnetization being controllable with an external magnetic field (Fatima and Kim 2017).

Optimal particle size is required to have a decent magnetic response and sufficient specific surface area, which directly influences the bio-separation efficiency. Colloidal stability and multifunctionality of magnetic nanocomposites are crucial for better compatibility with various bio agents. Recent advances have been successful in superparamagnetic nanoparticles and nanocomposites' functionalization for separation of proteins, peptides, cells and exosomes as well as blood purification in bio-separation applications (Yang et al. 2020).

2.7.2.3.2 Sensors in detection of bacteria, viruses and proteins

Superparamagnetic nanoparticles (SPIONs) with efficient systems like antibodies, aptamers, can be used as a sensor to detect bacteria, viruses, cells, proteins, etc. (Zhang et al. 2013). The majority of biosensors are based on iron oxides (g-Fe_2O_3 and Fe_3O_4), due to their high active surface, ease of surface modification, high chemical stability, and fast reaction kinetics (Nabaei et al. 2018).

The inherent magnetic properties, like magnetoresistance and giant magnetoresistive effect (GMR), make metallic nanoparticles suitable as magnetic labels indicating molecular interactions. Specific recognition of the biomolecules is an essential requirement of the biosensor. The essential part of bio-recognition is the immobilization of molecules onto magnetic particles. Many bio-recognition mechanisms were applied in sensing. In general, bare MPs are unstable and have low compatibility to biorecognition molecules, consequently limiting their potential in many applications. Therefore, the surface modification strategy of MPs would help colloidal stability and relevant functional groups for the adequate conjugation of biomolecules.

In the previous sections where we have discussed about different properties of nanoparticles, we have dealt with different MNPs. Now we would like to discuss about the two most promising and advanced type of nanoparticles that possess many applications in solar cells, electrical and electronic systems.

3. A special discussion on metal sulfides

Metal sulfides contain either localized and/or itinerant 3d electrons. Covalent bonding is present, resulting in reduced formal charge on transition metals and enabling metal-metal bond formation. Uniqueness to their structural conformation is mainly caused by hybridization between 3s and 3p orbitals with 3d orbitals favoring trigonal-pyramid configuration; S-S bonding may produce molecular anions, and the large polarizability of the anions favor adjacent layer connection via Van der Waals interactions (Rao and Pisharody 1976).

Sulfides are generally softer than their oxidated counterparts (with the exception of PbS). Nonetheless, for this and its layer lattice structure, lubricating properties appear with friction level decreasing at high speeds and loads. Complex sulfides show good pad and disk wear (Melcher and Faullant 2000). Also, these compounds have lower thermodynamical stability than oxides, and lower melting points (rare-earth sulfides being the exception) (Mrowec and Przybylski 1984).

Nanostructured sulfides are sought as the number of active sites and intrinsic activity is related to its morphological characteristics. Hollow metal sulfides have great active site density (Joo et al. 2019), desirable aspect when taking into consideration the great charge capacity metal sulfides show (α-MnS = 900 –1300 mAh/g, FeS = 500 mAh/g, FeS2 = 600 mAh/g, etc.). Thus being adequate enough to be used as electrodes, with their cycling stability being a determinant parameter. Voltage efficiency is also limited to their large band gap hindering any commercial applications (i.e., in lithium-ion batteries) (Xu et al. 2014).

Some sulfides behave as SCs (CdS, PnS, SnS_2, SnS, MoS_2, Bi_2S_3, CuS, etc.), while others behave as metals or even insulators. Therefore, the tuning of its optical properties (i.e., band gap) is achievable as any similar material, through size and morphology modification, surface engineering, doping, etc. (Shen and Wang 2013). Considering previous physical properties, these materials are suitable for flexible optoelectronic applications, as films conserving their crystallinity ensure high conductivities and high charge carrier mobilities (Shinde et al. 2015).

Also, sulfides as energy materials have to deal with high temperature instability, volume expansions and slow reaction kinetics, in some cases due to insufficient interlayer spacing for transport phenomena (Barik and Ingole 2020). Molybdenum and vanadium sulfides are strong candidates for batteries, not only for their capacity, but for their desired 2D layered structure. Graphene-like morphology provides the material with adequate space for electrolyte-ion transportation while shortening transportation of ions and electrons, resulting in high specific capacity and cyclic stability (Geng et al. 2018).

On medical applications, 2D materials are suited for drug delivery, as extensive surface interactions are enabled on a small scale. Mechanical properties permit potential biomedical performance, and even optical-therapy strategies (Zhu et al. 2018b). Usage in photothermal cancer treatment for metal sulfides is viable. Ag_2S, Bi_2S_3, CdS, and CuS NPs coated with proteins have shown high photothermal efficiency, biostability and, furthermore, biocompatibility (Sheng et al. 2018).

3.1 MoS

Molybdenum monosulfide is a rarely studied structure as it can be found at high pressures (120 GPa). At ambient pressure, hexagonal MoS is the most stable with an Imm2 space group (a = 11.660 Å, b = 3.306 Å, c = 5.555 Å) and higher covalent character; however, $Pmn2_1$ (a = 3.013 Å, b = 4.528 Å, c = 4.686 Å), Amm2 (a = 3.148 Å, b = 12.083 Å, c = 5.323 Å), and Pmmn (a = 4.463 Å, b = 3.11 Å, c = 4.381 Å) phases are energetically more stable (metastable) at all pressures (Wei et al. 2017).

DFT has been used to predict monolayered structures, in order to find stable configurations. Buckled configuration is more stable than puckered configuration for MoS, even foretelling its non-magnetic metallic behavior (Pandey and Chakrabarti 2019). Even information such as cluster geometric distribution, electronic structures (numerical electron spin density), and chemical bonding are available. In synthesis, sulfur atoms sequentially occupy the terminal sites, resulting in polisulfidation rather than controlled monosulfidation (Wang et al. 2013). As synthesis methods lack, further properties and applications for this monochalcogenide are missing; however, further possibilities of experimentation and investigation arise.

3.2 MoS$_2$

Molybdenum disulfide is a layered metal sulfide, with generally a laminate structure (~ 0.65 nm (Li and Zhu 2015)), which gives it its lubricating properties (film forming properties) (Rui et al. 2014). It is quite stable as it is not affected by diluted acids or oxygen (Coutinho et al. 2017). It is polytypic: 2H structure follows $P6_3/mmc$ space group (a = b = 3.16 Å, c = 12.29 Å), 3R structure belongs to R3m space group (a = b = 3.16 Å and c = 18.37 Å), and the space group of metastable synthetic 1T is P1 (a = b = 3.36 Å, c = 6.29 Å) (Rui et al. 2014). 2H (hexagonal) is more stable energetically as temperature increases, while 3R (rhombohedral) is better suited for thermal insulation. Their complex dielectric function and absorption are responsive to the plane of polarization of incident light. Through their Brillouin zone, phonon stability is ensured, which indicates dynamical stability (Coutinho et al. 2017).

Metastable MoS$_2$ presents unique physical properties due to its unique structure. Electron hopping between 1T islands show a distinctive metal-like electron transport mechanism. Superconductivity is proven when measuring magnetic and heat capacity (T_c = 4 K); even when ensuring a nanosheet structure, superconducting volume fraction can be completely confirmed (100%). Trimerization of Mo atoms in 1T phases enable ferroelectricity due to spin-orbit coupling effect. Finally, two-spin parallel 4d electrons in $d_{xy,xz,yz}$ orbitals predict ferroelasiticity and may enable ferromagnetism (Zhao et al. 2018a).

As expected, it is a very soft material (1–1.5 Mohs), with a 5.06 g/cm^3 density and a low ionic character (22%). Mechanically, it is highly elastic, has low pre-tension, and for its Young's modulus (0.33 ± 0.07 TPa in 5 to 25 layer nanosheets), it is a suitable alternative to graphene (Castellanos-Gomez et al. 2012). Thermally, it sublimates (theoretically) and oxidates at 450ºC and evaporates/decomposes at 950ºC (Melcher and Faullant 2000) (following MoS_2 to Mo_2S_3 to Mo (Liu et al. 2017)). Electrically, its theoretical capacity of 167–669 mAh/g makes it adequate for energy storage (batteries) (Rui et al. 2014). Mobility could be up to 200 $cm^2/V/s$ (at 25ºC), with current on/off ratio of 1×10^8, following n-type behavior. Its 1.8 eV direct band gap gives a photocurrent response of 7.5 mA/W (at 50V) (Li and Zhu 2015). However, these characteristics are not absolute, as the engineering of different systems and phases of MoS_2 permit the tuning and enhancing further of a variety of properties.

Monocrystalline defect-free MoS_2 possesses high mechanical strength (Young's modulus of 158.2, resulting strength of 27.3 GpA), as previously, the variation from brittle to ductile has been proven to depend on the density of vacancy defects. Therefore, processes such as annealing (annealing temperature and time) greatly change lattice properties, through reconstruction and migration of grain boundaries (GBs) (Wu et al. 2018). Tuning GBs also propitiate ferromagnetic behavior, and presence of Mo-Mo/S-S bonds means unpaired 4d electrons of Mo present in the vicinage of defect rings. Limits are set since a decrease in linear density of homo-elemental bonds means energetic stability increase for GBs (Gao et al. 2017).

Preferred morphology of nanostructured MoS_2 is thermodynamically dependent (pressure-temperature-composition), monolayer 2D structure (nanosheets) being the most thermodynamically favored (in comparison with bi- and multilayer). Also, the increase of layered disulfide is inversely proportional to the number of layers due to surface contributions. As to its preparation, Mo diffusion is the controlling factor since its diffusivity is extremely low in comparison to that of sulfur (Shang et al. 2016).

Nanoribbon structures have also been explored, being non-direct variable bandgap SCs, width and edge atoms dependent. Stiffness is ensured (~ 107 N/m) in the system, and even vacancy defects and adatoms adsorption may induce magnetic moments (Ataca et al. 2011). Zig-zag nanoribbons exhibit magnetic semiconducting behavior, while its armchair counterpart exhibits metallic nonmagnetic behavior. Thus, interest has increased for future research and applications of these highly stable MoS_2 1D structures (Li et al. 2008).

Electrical and energy applications vary from metallic to semiconducting MoS_2. The metallic phase is widely applied in batteries (lithium-ion, sodium-ion, etc.), supercapacitors, and even for hydrogen evolution reaction. Stability has been found to increase on hollow tube assemblies, as the material is easily affected by pH, light, electron beam, heat, etc. (Jiao et al. 2018).

Photocatalytic activity for MoS_2 based catalysts (semiconductive phase) is harnessed on hydrogen production, water remediation (organic pollutants decomposition, inorganic pollutants treatment, and photocatalytic disinfection), and photosynthesis (CO_2 reduction, NH_3 production, and organics production), and also with some photoelectrocatalytic experiments showing improved performance (Li et al. 2018b). As opposed to nanosheets with bandgap tunability being thickness dependent (Lee et al. 2012), NPs have tunable optical properties due to quantum confinement, thus showing excitation dependent emission profiles (with emission lifetime being size dependent), adequate for dye treatment (Bhattacharya et al. 2020).

Ultimately, even MoS_2 is suited for a wide variety of biomedical applications such as functionalization for drug delivery and photothermal cancer treatment. As various pathways have been explored (cellular trafficking, endocytosis/exocytosis, and nanocell activity), NIR laser in tissues requires improvement, as well as MoS_2 *in vivo* stability needs to be ensured (Zhu et al. 2018b).

4. A special discussion on Perovskite oxides

In 1839, calcium titanium oxide ($CaTiO_3$) mineral was discovered by Geologist Gustav Rose in the Ural Mountains of Russia, named perovskite after an eminent Russian mineralogist Lev Perovski (1792–1856) (Johnsson and Lemmens 2007). Perovskites are a class of materials with a similar structural formula ABX_3 (A and B are cations of different valence and ionic radii, and X is an anion). In the cubic unit cell of ABO_3 perovskite, B cation is located at the body center. An octahedron of X anions is located at face-centered positions and surrounded by B ions (Wolfram and Ellialtioglu 2006). In contrast, A cation is located at the cube corner positions (Figure 1). The detailed cell structure is mentioned in the upcoming section. Most of the inorganic perovskites are transition MOs (ABO_3), where X anions are based on oxygen.

Two different types of perovskites exist, i.e., inorganic and hybrid, where hybrid is a mixture of organic and inorganic components.

4.1 Inorganic perovskites

Inorganic perovskites have outstanding magnetic and electrical properties and exhibit different insulating, semiconducting, metallic, or superconducting behaviors. For example, $MgSiO_3$ and $FeSiO_3$ are most abundant in the Earth's crust (Galasso 1969). Some perovskite oxides possess different properties such as insulating, metallic, magnetic and super conducting (Wolfram and Ellialtioglu 2006). Inorganic perovskites got many attentions due to its high performance, low cost, and abundance.

Perovskites are known for various good electric, optical, and magnetic properties. These materials also possess adequate characteristics for numerous applications such as photo and electrochromic catalysis, electronics, energy storage, and acoustics due to their versatile physicochemical properties. They have been employed as active catalysts for CO and HCs oxidation, oxygen reduction reactions, and can also significantly aid in electrochemical applications such as sensing, water-splitting, hydrogen production, and fuel cells.

In the ABO_3 form, an ion A is either an alkali earth metal or lanthanides, and a B ion is generally from transition metal. A is generally larger than B, and O is the oxygen ion with the ratio of 1:1:3 located at face-centered positions (Figure 6) (Johnsson and Lemmens 2007).

Hines et al. reported that the ideal perovskite structure is cubic where corner-linked BO_6 octahedra are linked with interstitial A cations. However, the possibility of various elements and crystalline distortions may endure different structures such as orthorhombic, rhombohedral, hexagonal, and tetragonal from cubic perovskite to orthorhombic. However, all these distorted perovskites hold oxygen coordinations commonly with A and B sites by tilting the BO_6 octahedra and connected A cation displacement (Galasso 1969).

The stabilization of perovskite structure is facile and shows excellent flexibility of composition and various combinations of periodic elements due to their ability to incorporate multiple sizes and charges.

As aforementioned, simple cubic perovskite has an atomic arrangement in a 3D corner-sharing octahedral fashion. Au contraire, layered perovskites include 2D layers of corner-sharing octahedral separated by cations layers. Thus, perovskites and layered perovskites' electronic energy bands are noteworthy, and their structure is singular in specific properties.

Moreover, the substitution of different ions into the A- and/or B-sites would lead to a deviation from an ideal stoichiometry, resulting in tuning optoelectronic properties (Atta et al. 2016).

4.2 Hybrid perovskites

Hybrid halide perovskites (HPs), the mixture of organic-inorganic components, have been at the top in researched optoelectronic materials in recent two decades (Kim et al. 2012). The surprise acceleration in photoelectrical conversion efficiency is considered the future of solar cells with guaranteed 25% solar conversion efficiency lately (Green et al. 2021). Perovskite solar cells (PSCs) have taken over the performance of copper indium gallium diselenide (CIGS) and advances towards commercialization and PSCs play an important role in plethora of applications (Green et al. 2021).

The synthesis of hybrid perovskites is usually achieved by facile chemical routes and in cost-effective methods. Solution-based methods (hydrothermal, solvothermal, etc.) can permit bulk to molecular range covering 3D to 0-dimensional structures (Mitzi 2007).

In the atomic arrangements in hybrid perovskite (ABX_3), the component A is commonly an organic cation such as methylammonium (MA^+) or formamidinium (FA^+), or an atomic cation (often Cs^+) or a mixture, whereas the B component is commonly a divalent metal cation (usually Pb^{2+}, Sn^{2+} or a mixture) and X component is a halide anion (typically Cl^-, I^-, Br^- or a mixture) (Zhao and Zhu 2016, Kitazawa et al. 2002).

Compared to bulk perovskite materials, nanostructured materials such as quantum dots, nanoplatelets, nanowires, and layered 2D structures have exceptional photophysical properties (Zhao and Zhu 2016). The following sections focus on the unique features of hybrid PSCs. We particularly discuss their excellent optoelectronic properties and the impact of their structural changes.

4.3 Tolerance factor of perovskite structure

V.M. Goldschmidt presented the tolerance factor t related to the perovskites crystal structure, which applies to the empirical ionic radii at room temperature

$$t = (R_A + R_B)/(\sqrt{2}(R_X + R_B))$$

where R_A, R_B and R_X are the radius of the A-site cation, B-site cation, and X-site anion, in case of pure inorganic perovskites, where X is ion O^{2-}. Factor t can be used to assess the degree of distortion of perovskite from the ideal cubic structure (Kieslich et al. 2014). In general, the B-site is typically taken by a large atom. Consequently, the A-site must be large enough to reduce the tolerance factor. The tolerance factor for hybrid or inorganic perovskites estimates the ionic radii of the molecular cations to beget perfect perovskite crystal symmetry. From the equation, t will decrease when the ratio of rA/rB diminishes. Based on the tolerance factor value, the variation of perovskite crystalline structure can be estimated. In metal halides, the cubic structure of the perovskite should be close to 1 (between 0.813 and 1.107). On the other side, electroneutrality must not be overlooked, neutral balanced charge is essential to stabilize the perovskite structure. Thus, adding the charges of A as well as B cations would combine together to be equal to one whole charge of X anion, thus balancing the stoichiometry.

Currently, lead and tin HPs have been the most-studied, as mentioned earlier, having MA^+ and/or FA^+ anions for the A site. The tolerance factors for the traditional Pb or Sn HPs are displayed in Figure 6. When large or long-chain (alkylamine) groups occupy the A site, lead HP shifts a 2D layered structure. Similarly, the transition of perovskite structures is temperature-sensitive. For instance, the MAPbI3 perovskite reversible cubic (alpha) to the tetragonal (beta) phase transition happens at 330 K. Similarly, at shallow temperatures tetragonal (beta) phase transits into the orthorhombic (gamma) phase at around 100–150 K; atomic structure is shown in Figure 6 (Zhao and Zhu 2016).

The single-crystal MAPbI3 perovskite with cubic phase structure has a bandgap measured to be 1.51 eV. However, MAPbI3 perovskite PSCs suffer from a comparably large bandgap (1.57 eV) and low charge-carrier transport. Therefore, FAPbI3 perovskite has been envisaged as a substitute, with a cubic phase with a comparatively lower bandgap of 1.48 eV. At the same time, the electronic band structure and, therefore, the absorption coefficient are worthy of comparison. However, they are significantly sensitive to the humidity in the atmosphere that induces a great number of surface defects, reducing the energy barrier for nucleation of the non-perovskite phase. Nanostructured HPs have drawn much attention due to their defect-tolerant structure and facile synthesis in comparison with conventional materials. Their properties and applications in photovoltaics, photocatalysis and photoelectrocatalysis. Nevertheless, HPs suffer from low chemical, thermal, and photostability factors. They are susceptible to photooxidation by exposition to atmospheric oxygen and humidity (Bisquert and Juarez-Perez 2019). HPs are also extremely sensitive to temperature and UV light exposure (Misra et al. 2015). Consequently, enhancing environmental stability is a popular subject in the studies of HPs.

4.4 Tunable light emitting perovskites

Moreover, hybrid perovskites have an exceptional bandgap tunability as compared to other SCs. Simply playing with their physical dimensions and chemical composition can achieve the bandgap engineering and color tunability exhibits from the UV to the NIR spectral regions (390–1050 nm) (Protesescu et al. 2015)

4.5 Bandgap tuning by anion and cation exchange

Ionic radii of commonly used ions in ABX_3 HPs (Shannon 1976, Kieslich et al. 2014) can be studied as follows.

4.5.1 Influence of A-site (cation) modification

From the general band diagram of HPs, A-site cation has no direct contribution to band-edge states. However, their modifications can prevent the geometric distortion of the crystal structure by sharing the corners with PnI6 octahedra in the APnI3 perovskites. Therefore, the bond distance and angle between the Pn^{2+} and X ions depend on the A-site cation size, ultimately affecting the distribution of the band edge states (Amat et al. 2014). For instance, the bandgap of APbI3 perovskite varies depending on the A cation size, such as 1.48 eV, 1.51 eV, and 1.7eV for FA, MA, and Cs, respectively.

4.5.2 Influence of B site (cation) modification

Pb-based perovskite materials were primarily reported due to their high efficiency in the conversion of solar energy. However, researchers are interested in finding an alternative in B-site cation to replace Pd because of the toxicity issues. In that search, Sn has a valence electron configuration as similar to Pd and shows potential to replace in B-site cation in perovskites ($CsSnI_3$, $MASnI_3$, and $FASnI_3$) (Stoumpos et al. 2013). Because of the smaller ionic radius in comparison to Pd^{+2}, bandgaps are narrowed (1.3–1.4 eV). However, photoconversion energy, has been reported to be less than that of compared to Pd-based one (PCE < 12%). It is because they are prone to oxidize easily and convert Sn^{2+} into Sn^{4+} in signification proposition. Plenty of research has been conducted to hinder the oxidation of Sn^{2+}, resulting in a mix of Pd and Sn perovskites which have shown better suppression of Sn^{4+} and reduction of defects and low bandgap around 1.25eV which is ideal for better PCE in solar cells (as high as 24.8%) (Lin et al. 2019).

4.5.3 Influence of X site (anion)-modification

The modification of X-site anions can greatly alter the optoelectronic properties since they directly affect the band edge states. The halide anions with 8 valence electrons change from 3p to 4p to 5p when X changes from Cl to Br to I. According to the ionic radii of X, the octahedral corner sharing with B site cation influences the bandgaps of perovskites.

For example, the HP, the anions including Cl^-, Br^-, and I^- (ionic radii 181, 196, and 220 pm, respectively) bonded with Pb^{2+} in corner sharing octahedral fashion. The decrement in the ionic radius of halide anion causes decrease in their bond length with Pd in octahedral PbX_6, therefore resulting in an increase of the material's bandgap. For example, by substitution of different halides in $MAPbX_3$, their optoelectronic properties were tuned, such as amplified spontaneous emission (ASE) and lasing at wavelengths bewteen 390 nm to 790 nm (Buin et al. 2015, Maculan et al. 2015).

The strategy of using mixed anions such as I and Br was reported $MAPb(I_{1-x}Br_x)_3$ [0 < x < 1] by Noh et al. in order to enhance the perovskite light-harvesting capacity and proved that the ratio of the halides could chemically control the perovskite bandgap engineering (Noh et al. 2013). On the other hand, pseudohalogen polyanions (SCN and BF_4) were used in order to replace the low chemical stable halide ions, especially iodide I^-. For example, BF_4 have been incorporated into $MAPbI_3$, has a negligible optical absorption due to its ionic radii (0.218 nm), and is similar to the I (0.220 nm). Therefore, the mixed pseudohalogen/HPs $MAPbI_{3-x}(BF_4)x$ shows higher order of photoresponse and electrical conductivity due to inclusion of fluorine as compared to pure halide perovskites (Nagane et al. 2014, Hendon et al. 2015).

4.6 Tunable band-gap by nanocrystal shape and size

Compositional changes in perovskites have demonstrated a fine color tuning, where colloidal nanocrystals of size 5–10 nm with mixed-halide (OA: MA)PbX_3 (X = Cl, Br, I, and mixture) report a continuously tunable photoluminescence emission from 385 to 770 nm by changing composition and size (nanostructuring) (Pathak et al. 2015). Kovalenko and co-workers also experimented with $CsPbX_3$ nanoparticles that possess emissions from 410 to 700 nm with a controlled synthesis of monodisperse nano-cubes with crystal sizes varying from 4 to 15 nm (Protesescu et al. 2015).

4.7 Applications of halide perovskite-based composites

Pure HP and HP based composites have been employed in a varied range of applications, such as white LEDs, photoemission, detectors, photocatalysts, photovoltaics and memristors.

4.7.1 LEDs

Current technology of high-quality LEDs is based on III–V direct bandgap SC materials, which is an expensive technique and requires a decent vacuum and high process temperatures. On the other hand, direct-bandgap SC-based perovskites are cheap and facile to synthesize. Their easy color tuning in the visible and infrared regions makes them very capable for various optoelectronic applications such as color displays and lighting and optical communication applications (Pathak et al. 2015, Byun et al. 2016).

The encapsulated HP nanostructures in a polymer matrix manifested a better white light emission performance. Moreover, the polymer encapsulation could avoid oxidation and improve their chemical stability in an open environment (Li et al. 2016).

4.7.2 Photocatalyst and photovoltaic

In general, photocatalysis has attracted significant attention due to its potential applications in CO_2 reduction, water splitting, degradation of organic compounds, etc. HP-nanostructures are best photocatalysts, provided their chemical stability can be enhanced. They possess broader visible light absorption range and carrier diffusion length is also longer. In consequence, lead HP materials can be configured in tandem with MOs or ferrites in order to avoid their degradation in moisture and profit their bandgap and high photovoltage generation.

Recently, photoelectrochemical (PEC) hybrid electrode configuration has been reported, where the MAPbI3 solar cell coupled with the MOs such BiVO4 (Chen et al. 2015) or hematite (Gurudayal et al. 2015) photoanode for solar fuel generation or water splitting applications.

Acknowledgements

Goldie oza and L.G. Arriaga are thankful for the Conacyt Fronteras project 2096029. L.G. Arriaga also expresses his gratitude to the Project No. 316263 (Laboratorio Nacional de Micro y Nanofluidica, 2021) for the financial support granted. Goldie oza is also thankful for the Catedras Conacyt project 746.

References

Afrand, M. 2017. Experimental study on thermal conductivity of ethylene glycol containing hybrid nano-additives and development of a new correlation. *Applied Thermal Engineering*. Elsevier, 110: 1111–1119.

Akbarzadeh, A., Samiei, M. and Davaran, S. 2012. Magnetic nanoparticles: preparation, physical properties, and applications in biomedicine *Nanoscale Research Letters*, 7(1): 144. doi: 10.1186/1556-276X-7-144.

Albrecht, M.G. and Creighton, J.A. 1977. Anomalously intense Raman spectra of pyridine at a silver electrode. *Journal of the American Chemical Society*, 99(15): 5215–5217. doi: 10.1021/ja00457a071.

Almehmadi, L.M., Curley, S.M., Tokranova, N.A., Tenenbaum, S.A. and Lednev, I.K. 2019. Surface enhanced raman spectroscopy for single molecule protein detection. *Scientific Reports*, 9(1): 12356. doi: 10.1038/s41598-019-48650-y.

Amat, A., Mosconi, E., Ronca, E., Quarti, C., Umari, P., Nazeeruddin, Md. K., Grätze, M. and De Angelis, F. 2014. Cation-induced band-gap tuning in organohalide perovskites: Interplay of spin–orbit coupling and octahedra tilting. *Nano Letters*, 14(6): 3608–3616. doi: 10.1021/nl5012992.

Andrievski, R.A. 2009. Brittle nanomaterials: Hardness and superplasticity. *Bulletin of the Russian Academy of Sciences: Physics*. Springer, 73(9): 1222–1226.

Andrievski, R.A. 2014. Review of thermal stability of nanomaterials. *Journal of materials science*. Springer, 49(4): 1449–1460.

Ataca, C., Sahin, H., Akturk, E. and Ciraci, S. 2011. Mechanical and electronic properties of MoS2 nanoribbons and their defects. *The Journal of Physical Chemistry C*. ACS Publications, 115(10): 3934–3941.

Atta, N.F., Galal, A. and El-Ads, E.H. 2016. Perovskite nanomaterials – synthesis, characterization, and applications. In *Perovskite Materials - Synthesis, Characterisation, Properties, and Applications*. InTech. doi: 10.5772/61280.

Baer, D.R., Amonette, J., Engelhard, M., Gaspar, D.J., Karakoti, A., Kuchibhatla, S., Nachimuthu, P., Nurmi, J.T., Qiang, Y., Sarathy, V., Seal, S., Sharma, A., Paul G. Tratnyek and Chong-Min Wang. 2008. Characterization challenges for nanomaterials. *Surface and Interface Analysis: An International Journal Devoted to the Development and Application of Techniques for the Analysis of Surfaces, Interfaces and Thin Films*. Wiley Online Library, 40(3-4): 529–537.

Baffou, G. and Quidant, R. 2014. Nanoplasmonics for chemistry. *Chemical Society Reviews*, 43(11): 3898. doi: 10.1039/c3cs60364d.

Barik, R. and Ingole, P.P. 2020. Challenges and prospects of metal sulfide materials for supercapacitors. *Current Opinion in Electrochemistry*. Elsevier.

Bhattacharya, D., Mukherjee, S., Mitra, R.K. and Ray, S.K. 2020. Size-dependent optical properties of MoS2 nanoparticles and their photo-catalytic applications. *Nanotechnology*. IOP Publishing, 31(14): 145701.

Bird, R.B., Stewart, W.E. and Lightfoot, E.N. 2007. Transport Phenomena, 2nd Edn. New York, NY: John Wilwe & Sons. Inc.

Bisquert, J. and Juarez-Perez, E.J. 2019. The causes of degradation of perovskite solar cells. *The Journal of Physical Chemistry Letters*, 10(19): 5889–5891. doi: 10.1021/acs.jpclett.9b00613.

Bozhevolnyi, S.I., Volkov, V.S., Devaux, E. and Ebbesen, T.W. 2005. Channel plasmon-polariton guiding by subwavelength metal grooves. *Physical Review Letters*, 95(4): 046802. doi: 10.1103/PhysRevLett.95.046802.

Brooks, I., Lin, P., Palumbo, G., Hibbard, G.H. and Erb, U. 2008. Analysis of hardness–tensile strength relationships for electroformed nanocrystalline materials. *Materials Science and Engineering: A*. Elsevier, 491(1-2): 412–419.

Brosseaua, C., NDong, W. and Castel, V. 2007. Electromagnetomechanical coupling characteristics of plastoferrites. *Journal of Applied Physics*, 102(2): 024907. doi: 10.1063/1.2757200.

Buin, A., Comin, R., Xu, J., lp, A.H. and Sargent, E.H. 2015. Halide-dependent electronic structure of organolead perovskite materials. *Chemistry of Materials*, 27(12): 4405–4412. doi: 10.1021/acs.chemmater.5b01909.

Byun, J., Cho, H., Wolf, C., Jang, M., Sadhanala, A., Friend, R.H., Yang, H. and Lee, T.-W. 2016. Efficient visible quasi-2D perovskite light-emitting diodes. *Advanced Materials*, 28(34): 7515–7520. doi: 10.1002/adma.201601369.

Calebrese, C., Le Hui, L., Schadler, L.S. and Nelson, J.K. 2011. A review on the importance of nanocomposite processing to enhance electrical insulation. *IEEE Transactions on Dielectrics and Electrical Insulation*, 18(4): 938–945. doi: 10.1109/TDEI.2011.5976079.

Callister, W.D. and Rethwisch, D.G. 2011. *Materials Science and Engineering*. John wiley & sons NY.

Castellanos-Gomez, A., Poot, M., Steele, G., van der Zant, H., Agraït, N. and Rubio-Bollinger, G. 2012. Elastic properties of freely suspended MoS2 nanosheets. *Advanced Materials*. Wiley Online Library, 24(6): 772–775.

Casula, M.F., Jun, Y.-W., Zaziski, D.J., Chan, E.M., Corrias, A. and Alivisatos, A.P. 2006. The concept of delayed nucleation in nanocrystal growth demonstrated for the case of iron oxide nanodisks. *Journal of the American Chemical Society*, 128(5): 1675–1682. doi: 10.1021/ja056139x.

Cejka, J., Morris, R.E. and Nachtigall, P. (eds.). 2017. *Zeolites in Catalysis*. Cambridge: Royal Society of Chemistry (Catalysis Series). doi: 10.1039/9781788010610.

Che, J., Cagin, T. and Goddard III, W.A. 2000. Thermal conductivity of carbon nanotubes. *Nanotechnology*. IOP Publishing, 11(2): 65.

Chen, Y.-S., Manser, J.S. and Kamat, P.V. 2015. All solution-processed lead halide perovskite-BiVO4 tandem assembly for photolytic solar fuels production. *Journal of the American Chemical Society*. American Chemical Society, 137(2): 974–981. doi: 10.1021/ja511739y.

Cheng, Y., Bai, L., Guang Yu, G. and Zhang, X. 2018. Effect of particles size on dielectric properties of nano-zno/ldpe composites. *Materials*, 12(1): 5. doi: 10.3390/ma12010005.

Ciupagea, L., Andrei, G., Dima, D. and Murarescu, M. 2013. Specific heat and thermal expansion of polyester composites containing singlewall-, multiwall-and functionalized carbon nanotubes. *Digest Journal of Nanomaterials & Biostructures (DJNB)*, 8(4).

Coutinho, S.S., Tavares, M.S., Barboza, C.A., Frazão, N.F., Moreira, E. and Azevedo, D.L. 2017. 3R and 2H polytypes of MoS2: DFT and DFPT calculations of structural, optoelectronic, vibrational and thermodynamic properties. *Journal of Physics and Chemistry of Solids*. Elsevier, 111: 25–33.

da Silva, A.G., Rodrigues, T.S., Taguchi, L.S., Fajardo, H.V., Balzer, R., Probst, L.F. and Camargo, P.H. 2016. Pd-based nanoflowers catalysts: Controlling size, composition, and structures for the 4-nitrophenol reduction and BTX oxidation reactions. *Journal of Materials Science*, 51(1): 603–614. doi: 10.1007/s10853-015-9315-3.

Daniel, M.C. and Astruc, D. 2004. Gold nanoparticles: assembly, supramolecular chemistry, quantum-size-related properties, and applications toward biology, catalysis, and nanotechnology. *Chemical Reviews*. doi: 10.1021/cr030698+.

David, B. 2009. Synthesis, structure, and properties of organic inorganic perovskites and related. *Progress in Inorganic Chemistry*, 96: 1.

Derkacs, D., Lim, S.H., Matheu, P., Mar, W. and Yu, E.T. 2006. Improved performance of amorphous silicon solar cells via scattering from surface plasmon polaritons in nearby metallic nanoparticles. *Applied Physics Letters*, 89(9): 093103. doi: 10.1063/1.2336629.

Dhaouadi, H., Ghodbane, O., Hosni, F. and Touati, F. 2012. Mn3O4 nanoparticles: Synthesis, characterization, and dielectric properties. *ISRN Spectroscopy*, pp. 1–8. doi: 10.5402/2012/706398.

Dykman, L.A. and Bogatyrev, V.A. 1997. Colloidal gold in solid-phase assays. A review. *Biochemistry-New York-English Translation of Biokhimiya*. New York: Consultants Bureau, c1956-, 62(4): 350–356.

Ekgasit, S., Thammacharoen, C., Yu, F. and Knoll, W. 2004. Evanescent field in surface plasmon resonance and surface plasmon field-enhanced fluorescence spectroscopies. *Analytical Chemistry*, 76(8): 2210–2219. doi: 10.1021/ac035326f.

Ertl, G. 2002. Heterogeneous catalysis on atomic scale. *Journal of Molecular Catalysis A: Chemical*, 182: 5–16. doi: 10.1016/S1381-1169(01)00460-5.

Fan, Z., Sun, K. and Wang, J. 2015. Perovskites for photovoltaics: A combined review of organic–inorganic halide perovskites and ferroelectric oxide perovskites. *Journal of Materials Chemistry A*, 3(37): 18809–18828. doi: 10.1039/C5TA04235F.

Fano, U. 1941. The theory of anomalous diffraction gratings and of quasi-stationary waves on metallic surfaces (Sommerfeld's Waves). *Journal of the Optical Society of America*, 31(3): 213. doi: 10.1364/JOSA.31.000213.

Fatima, H. and Kim, K.-S. 2017. Magnetic nanoparticles for bioseparation. *Korean Journal of Chemical Engineering*, 34(3): 589–599. doi: 10.1007/s11814-016-0349-2.

Feng, J. and Xiao, B. 2014. Crystal structures, optical properties, and effective mass tensors of CH 3 NH 3 PbX 3 (X = I and Br) Phases Predicted from HSE06. *The Journal of Physical Chemistry Letters*, 5(7): 1278–1282. doi: 10.1021/jz500480m.

Ferry, V.E., Sweatlock, L.A., Pacifici, D. and Atwater, H.A. 2008. Plasmonic nanostructure design for efficient light coupling into solar cells. *Nano Letters*, 8(12): 4391–4397. doi: 10.1021/nl8022548.

FRENKEL, J. and DOEFMAN, J. 1930. Spontaneous and induced magnetisation in ferromagnetic bodies. *Nature*, 126(3173): 274–275. doi: 10.1038/126274a0.

Fujie, T., Matsutani, N., Kinoshita, M., Okamura, Y., Saito, A. and Takeoka, S. 2009. Adhesive, flexible, and robust polysaccharide nanosheets integrated for tissue-defect repair. *Advanced Functional Materials*. Wiley Online Library, 19(16): 2560–2568.

Furukawa, H., Cordova, K.E., O'Keeffe, M. and Yaghi, O.M. 2013. The chemistry and applications of metal-organic frameworks. *Science*, 341(6149): 1230444. doi: 10.1126/science.1230444.

Galasso, F.S. 1969. Structure, properties and preparation of perovskite-type compounds. pp. 3–49. *In*: Smoluchowski, R. and Kurti, N. (eds.). Pergamon Press, New York.

Gan, Q., Bartoli, F.J. and Kafafi, Z.H. 2013. Plasmonic-enhanced organic photovoltaics: Breaking the 10% efficiency barrier. *Advanced Materials*, 25(17): 2385–2396. doi: 10.1002/adma.201203323.

Gao, N., Guo, Y., Zhou, S., Bai, Y. and Zhao, Z. 2017. Structures and magnetic properties of MoS2 grain boundaries with antisite defects. *The Journal of Physical Chemistry C*. ACS Publications, 121(22): 12261–12269.

Gao, P. 2014. Microwaves in nanoparticle synthesis. Fundamentals and applications. Edited by Satoshi Horikoshi and Nick Serpone. Wiley Online Library.

Gao, X., Cui, Y., Levenson, R.M., Chung, L.W. and Nie, S. 2004. *In vivo* cancer targeting and imaging with semiconductor quantum dots. *Nature Biotechnology*, 22(8): 969–976. doi: 10.1038/nbt994.

Geng, P., Zheng, S., Tang, H., Zhu, R., Zhang, L., Cao, S. and Pang, H. 2018. Transition metal sulfides based on graphene for electrochemical energy storage. *Advanced Energy Materials*. Wiley Online Library, 8(15): 1703259.

Gopinath, A., Boriskina, S.V., Premasiri, W.R., Ziegler, L., Reinhard, B.M. and Dal Negro, L. 2009. Plasmonic nanogalaxies: Multiscale aperiodic arrays for surface-enhanced raman sensing. *Nano Letters*, 9(11): 3922–3929. doi: 10.1021/nl902134r.

Green, M., Dunlop, E., Hohl-Ebinger, J., Yoshita, M., Kopidakis, N. and Hao, X. 2021. Solar cell efficiency tables (version 57). *Progress in Photovoltaics: Research and Applications*, 29(1): 3–15. doi: 10.1002/pip.3371.

Gu, J., Zhang, Y.-W. and Tao, F. (Feng). 2012. Shape control of bimetallic nanocatalysts through well-designed colloidal chemistry approaches. *Chemical Society Reviews*, 41(24): 8050. doi: 10.1039/c2cs35184f.

Guerra, V., Wan, C. and McNally, T. 2019. Thermal conductivity of 2D nano-structured boron nitride (BN) and its composites with polymers. *Progress in Materials Science*, 100: 170–186. doi: https://doi.org/10.1016/j.pmatsci.2018.10.002.

Guo, D., Xie, G. and Luo, J. 2013. Mechanical properties of nanoparticles: Basics and applications. *Journal of physics D: applied physics*. IOP Publishing, 47(1): 13001.

Gupta, A.K., Naregalkar, R.R., Vaidya, V.D. and Gupta, M. 2007. Recent advances on surface engineering of magnetic iron oxide nanoparticles and their biomedical applications. *Nanomedicine*, 2(1): 23–39. doi: 10.2217/17435889.2.1.23.

Gurudayal, Sabba, D., Kumar, M.H., Wong, L.H., Barber, J., Grätzel, M. and Mathews, N. 2015. Perovskite–hematite tandem cells for efficient overall solar driven water splitting. *Nano Letters*, 15(6): 3833–3839. doi: 10.1021/acs.nanolett.5b00616.

Haruta, M., Kobayashi, T., Sano, H. and Yamada, N. 1987. Novel gold catalysts for the oxidation of carbon monoxide at a temperature far below 0°C. *Chemistry Letters*, 16(2): 405–408. doi: 10.1246/cl.1987.405.

Hendon, C.H., Yang, R.X., Burton, L.A. and Walsh, A. 2015. Assessment of polyanion (BF 4⁻ and PF 6⁻) substitutions in hybrid halide perovskites. *Journal of Materials Chemistry A*, 3(17): 9067–9070. doi: 10.1039/C4TA05284F.

Hines, Robert Ian. Atomistic simulation and *ab initio* studies of polar solids. Diss. University of Bristol, 1997.

Homola, J., Yee, S.S. and Gauglitz, G. 1999. Surface plasmon resonance sensors: review. *Sensors and Actuators B: Chemical*, 54(1-2): 3–15. doi: 10.1016/S0925-4005(98)00321-9.

Hradil, D., Grygar, T., Hradilová, J. and Bezdička, P. 2003. Clay and iron oxide pigments in the history of painting. *Applied Clay Science*, 22(5): 223–236. doi: 10.1016/S0169-1317(03)00076-0.

Huang, S.J., Peng, W.Y., Visic, B., Zak, A.M. and Dal Negro, L. 2018. Al alloy metal matrix composites reinforced by WS2 inorganic nanomaterials. *Materials Science and Engineering: A*. Elsevier, 709: 290–300.

Hunt, L.B. 1976. The true story of Purple of Cassius. *Gold Bulletin*, 9(4): 134–139. doi: 10.1007/BF03215423.

Jeanmaire, D.L. and Van Duyne, R.P. 1977. Surface raman spectroelectrochemistry. *Journal of Electroanalytical Chemistry and Interfacial Electrochemistry*, 84(1): 1–20. doi: 10.1016/S0022-0728(77)80224-6.

Jia, Y., Li, Z., Wang, H., Saeed, M. and Cai, H. 2019. Sensitivity enhancement of a surface plasmon resonance sensor with platinum diselenide. *Sensors*, 20(1): 131. doi: 10.3390/s20010131.

Jiang, Q., Zhang, S. and Zhao, M. 2003. Size-dependent melting point of noble metals. *Materials Chemistry and Physics*. Elsevier, 82(1): 225–227.

Jiang, Y. and Lu, Y. 2020. Designing transition-metal-boride-based electrocatalysts for applications in electrochemical water splitting. *Nanoscale*, 12(17): 9327–9351. doi: 10.1039/D0NR01279C.

Jiao, Y., Hafez, A.M., Cao, D., Mukhopadhyay, A., Ma, Y. and Zhu, H. 2018. Metallic MoS2 for high performance energy storage and energy conversion. *Small*. Wiley Online Library, 14(36): 1800640.

Jin, J., Yin, J., Liu, H., Lu, M., Li, J., Tian, M. and Xi, P. 2019. Transition metal (Fe, Co and Ni)−Carbide−Nitride (M−C−N) nanocatalysts: Structure and electrocatalytic applications. *ChemCatChem*, 11(12): 2780–2792. doi: 10.1002/cctc.201900570.

Joo, J., Kim, T., Lee, J., Choi, S.I. and Lee, K. 2019. Morphology-controlled metal sulfides and phosphides for electrochemical water splitting. *Advanced Materials*. Wiley Online Library, 31(14): 1806682.

Kharisov, B.I., Kharissova, O.V. and Méndez, U.O. 2016. *Radiation Synthesis of Materials and Compounds*. CRC press.

Kieslich, G., Sun, S. and Cheetham, A.K. 2014. Solid-state principles applied to organic–inorganic perovskites: New tricks for an old dog. *Chem. Sci.*, 5(12): 4712–4715. doi: 10.1039/C4SC02211D.

Kilpatrick, K.E., WRING, S.A., WALKER, D.H., MACKLIN, M.D., PAYNE, J.A., SU, J.L. and McINTYRE, G.D. 1997. Rapid development of affinity matured monoclonal antibodies using RIMMS. *Hybridoma*, 16(4): 381–389. doi: 10.1089/hyb.1997.16.381.

Kim, H.S., Lee, C.R., Im, J.H., Lee, K.B., Moehl, T., Marchioro, A. and Park, N.G. 2012. Lead iodide perovskite sensitized all-solid-state submicron thin film mesoscopic solar cell with efficiency exceeding 9%. *Scientific Reports* 2(1): 591. doi: 10.1038/srep00591.

Kim, Y.K., Lundquist, P.M., Helfrich, J.A., Mikrut, J.M., Wong, G.K., Auvil, P.R. and Ketterson, J.B. 1995. Scanning plasmon optical microscope. *Applied Physics Letters*, 66(25): 3407–3409. doi: 10.1063/1.113369.

Kitazawa, N., Watanabe, Y. and Nakamura, Y. 2002. Optical properties of CH3NH3PbX3 (X = halogen) and their mixed-halide crystals. *Journal of Materials Science*, 37(17): 3585–3587. doi: 10.1023/A:1016584519829.

Knight, M.W., Wang, Y., Urban, A.S., Sobhani, A., Zheng, B.Y., Nordlander, P. and Halas, N.J. 2013. Embedding plasmonic nanostructure diodes enhances hot electron emission. *Nano letters*. ACS Publications, 13(4): 1687–1692.

Kornienko, N., Zhang, J.Z., Sakimoto, K.K., Yang, P. and Reisner, E. 2018. Interfacing nature's catalytic machinery with synthetic materials for semi-artificial photosynthesis. *Nature Nanotechnology*, 13(10): 890–899. doi: 10.1038/s41565-018-0251-7.

Kouassi, G.K. and Irudayaraj, J. 2006. Magnetic and gold-coated magnetic nanoparticles as a DNA sensor. *Analytical Chemistry*. doi: 10.1021/ac051621j.

Kretschmann, E. and Raether, H. 1968. Radiative decay of non radiative surface plasmons excited by light. *Zeitschrift fur Naturforschung - Section A Journal of Physical Sciences*. doi: 10.1515/zna-1968-1247.

Kretschmann, E. 1971. The determination of the optical constants of metals by excitation of surface plasmons. *Zeitschrift für Physik A Hadrons and Nuclei*. doi: 10.1007/BF01395428.

Krishnamoorti, R. 2007. Strategies for dispersing nanoparticles in polymers. *MRS Bulletin*, 32(4): 341–347. doi: 10.1557/mrs2007.233.

Kumar, A., Singh, B.P., Choudhary, R.N.P. and Thakur, A.K. 2006. Characterization of electrical properties of Pb-modified BaSnO3 using impedance spectroscopy. *Materials Chemistry and Physics*, 99(1): 150–159. doi: 10.1016/j.matchemphys.2005.09.086.

Kwak, K., Azad, U.P., Choi, W., Pyo, K., Jang, M. and Lee, D. 2016. Efficient oxygen reduction electrocatalysts based on gold nanocluster-graphene composites. *ChemElectroChem*, 3(8): 1253–1260. doi: 10.1002/celc.201600154.

Lee, H.S., Min, S.W., Chang, Y.G., Park, M.K., Nam, T., Kim, H. and Im, S. 2012. MoS2 nanosheet phototransistors with thickness-modulated optical energy gap. *Nano Letters*. ACS Publications, 12(7): 3695–3700.

Lerosey, G., Pile, D.F.P., Matheu, P., Bartal, G. and Zhang, X. 2009. Controlling the Phase and Amplitude of Plasmon Sources at a Subwavelength Scale. *Nano Letters*, 9(1): 327–331. doi: 10.1021/nl803079s.

Li, C., Liu, T., He, T., Ni, B., Yuan, Q. and Wang, X. 2018a. Composition-driven shape evolution to Cu-rich PtCu octahedral alloy nanocrystals as superior bifunctional catalysts for methanol oxidation and oxygen reduction reaction. *Nanoscale*, 10(10): 4670–4674. doi: 10.1039/C7NR09669K.

Li, M. 2013. A nano-graphite/paraffin phase change material with high thermal conductivity. *Applied Energy*, 106: 25–30. doi: https://doi.org/10.1016/j.apenergy.2013.01.031.

Li, N., Zhou, G., Li, F., Wen, L. and Cheng, H.M. 2013. A self-standing and flexible electrode of Li4Ti5O12 nanosheets with a N-doped carbon coating for high rate lithium ion batteries. *Advanced Functional Materials*. Wiley Online Library, 23(43): 5429–5435.

Li, X. and Zhu, H. 2015. Two-dimensional MoS2: Properties, preparation, and applications. *Journal of Materiomics*. Elsevier, 1(1): 33–44.

Li, X., Wu, Y., Zhang, S., Cai, B., Gu, Y., Song, J. and Zeng, H. 2016. CsPbX 3 quantum dots for lighting and displays: room-temperature synthesis, photoluminescence superiorities, underlying origins and white light-emitting diodes. *Advanced Functional Materials*, 26(15): 2435–2445. doi: 10.1002/adfm.201600109.

Li, Y., Zhou, Z., Zhang, S. and Chen, Z. 2008. MoS2 nanoribbons: High stability and unusual electronic and magnetic properties. *Journal of the American Chemical Society*. ACS Publications, 130(49): 16739–16744.

Li, Z., Meng, X. and Zhang, Z. 2018b. Recent development on MoS2-based photocatalysis: A review. *Journal of Photochemistry and Photobiology C: Photochemistry Reviews*. Elsevier, 35: 39–55.

Liang, R.P., Yao, G.H., Fan, L.X. and Qiu, J.D. 2012. Magnetic Fe 3O 4@Au composite-enhanced surface plasmon resonance for ultrasensitive detection of magnetic nanoparticle-enriched α-fetoprotein. *Analytica Chimica Acta*. doi: 10.1016/j.aca.2012.05.043.

Liedberg, B., Nylander, C. and Lunström, I. 1983. Surface plasmon resonance for gas detection and biosensing. *Sensors and Actuators*, 4: 299–304. doi: 10.1016/0250-6874(83)85036-7.

Lin, R., Xiao, K., Qin, Z., Han, Q., Zhang, C., Wei, M. and Tan, H. 2019. Monolithic all-perovskite tandem solar cells with 24.8% efficiency exploiting comproportionation to suppress Sn(ii) oxidation in precursor ink. *Nature Energy*, 4(10): 864–873. doi: 10.1038/s41560-019-0466-3.

Liu, B., Yao, H., Song, W., Jin, L., Mosa, I.M., Rusling, J.F. and He, J. 2016. Ligand-free noble metal nanocluster catalysts on carbon supports via "Soft" nitriding. *Journal of the American Chemical Society*, 138(14): 4718–4721. doi: 10.1021/jacs.6b01702.

Liu, F., Zhou, Y., Liu, D., Chen, X., Yang, C. and Zhou, L. 2017. Thermodynamic calculations and dynamics simulation on thermal-decomposition reaction of MoS2 and Mo2S3 under vacuum. *Vacuum*. Elsevier, 139: 143–152.

Lombardo, N. 2007. A two-way particle mapping for calculation of the effective dielectric response of graded spherical composites. *Composites Science and Technology*. doi: 10.1016/j.compscitech.2006.04.009.

Lopez, N. 2004. On the origin of the catalytic activity of gold nanoparticles for low-temperature CO oxidation. *Journal of Catalysis*, 223(1): 232–235. doi: 10.1016/j.jcat.2004.01.001.

Lu, A.-H., Salabas, E.L. and Schüth, F. 2007. Magnetic nanoparticles: synthesis, protection, functionalization, and application. *Angewandte Chemie International Edition*, 46(8): 1222–1244. doi: 10.1002/anie.200602866.

Maculan, G., Sheikh, A.D., Abdelhady, A.L., Saidaminov, M.I., Haque, M.A., Murali, B. and Bakr, O.M. 2015. CH 3 NH 3 PbCl 3 single crystals: Inverse temperature crystallization and visible-blind UV-photodetector. *The Journal of Physical Chemistry Letters*, 6(19): 3781–3786. doi: 10.1021/acs.jpclett.5b01666.

Maier, S.A., Kik, P.G., Atwater, H.A., Meltzer, S., Harel, E., Koel, B.E. and Requicha, A.A. 2003. Local detection of electromagnetic energy transport below the diffraction limit in metal nanoparticle plasmon waveguides. *Nature Materials*, 2(4): 229–232. doi: 10.1038/nmat852.

Mats, J. and Peter, L. 2006. Crystallography and chemistry of Perovskites. *J. Biol. Chem.*, 262: 7486–7491.

McEwan, L., Julius, M., Roberts, S. and Fletcher, J.C. 2010. A review of the use of gold catalysts in selective hydrogenation reactions. *Gold Bulletin*. doi: 10.1007/bf03214999.

Melcher, B. and Faullant, P. 2000. *A comprehensive study of chemical and physical properties of metal sulfides*. SAE Technical Paper.

Mikhaylova, M., Kim, D.K., Bobrysheva, N., Osmolowsky, M., Semenov, V., Tsakalakos, T. and Muhammed, M. 2004. Superparamagnetism of magnetite nanoparticles: dependence on surface modification. *Langmuir*, 20(6): 2472–2477. doi: 10.1021/la035648e.

Misra, R.K., Aharon, S., Li, B., Mogilyansky, D., Visoly-Fisher, I., Etgar, L. and Katz, E.A. 2015. Temperature- and component-dependent degradation of perovskite photovoltaic materials under concentrated sunlight. *The Journal of Physical Chemistry Letters*, 6(3): 326–330. doi: 10.1021/jz502642b.

Mohammadi, K., Madadi, A.A., Bajalan, Z. and Pishkenari, H.N. 2020. Analysis of mechanical and thermal properties of carbon and silicon nanomaterials using a coarse-grained molecular dynamics method. *International Journal of Mechanical Sciences*. Elsevier, 187: 106112.

Moores, A., Goettmann, F., Sanchez, C. and Le Floch, P. 2004. Synthesis and immobilisation on mesoporous materials. *Chemical Communications*, (24): 2842. doi: 10.1039/b412553c.

Moskovits, M. 1978. Surface roughness and the enhanced intensity of Raman scattering by molecules adsorbed on metals. *The Journal of Chemical Physics*, 69(9): 4159–4161. doi: 10.1063/1.437095.

Mrowec, S. and Przybylski, K. 1984. Defect and transport properties of sulfides and sulfidation of metals. *High Temperature Materials and Processes*. De Gruyter, 6(1-2): 1–80.

Nabaei, V., Chandrawati, R. and Heidari, H. 2018. Magnetic biosensors: Modelling and simulation. *Biosensors and Bioelectronics*, 103: 69–86. doi: 10.1016/j.bios.2017.12.023.

Nagane, S., Bansode, U., Game, O., Chhatre, S. and Ogale, S. 2014. CH 3 NH 3 PbI (3−x) (BF 4) x: Molecular ion substituted hybrid perovskite. *Chemical Communications*, 50(68): 9741. doi: 10.1039/C4CC04537H.

Nagpal, P., Lindquist, N.C., Oh, S.H. and Norris, D.J. 2009. Ultrasmooth patterned metals for plasmonics and metamaterials. *Science*, 325(5940): 594–597. doi: 10.1126/science.1174655.

Ng, C.B., Ash, B.J., Schadler, L.S. and Siegel, R.W. 2001. A study of the mechanical and permeability properties of nano- and Micron-Tio 2 filled epoxy composites. *Advanced Composites Letters*, 10(3): 096369350101000. doi: 10.1177/096369350101000301.

Noh, J.H., Im, S.H., Heo, J.H., Mandal, T.N. and Seok, S.I. 2013. Chemical management for colorful, efficient, and stable inorganic–organic hybrid nanostructured solar cells. *Nano Letters*, 13(4): 1764–1769. doi: 10.1021/nl400349b.

O'Connor, D. and Zayats, A.V. 2010. The third plasmonic revolution. *Nature Nanotechnology*, 5(7): 482–483. doi: 10.1038/nnano.2010.137.

On the dynamical theory of gratings. 1907. *Proceedings of the Royal Society of London. Series A, Containing Papers of a Mathematical and Physical Character*. doi: 10.1098/rspa.1907.0051.

Otto, A. 1968. Excitation of nonradiative surface plasma waves in silver by the method of frustrated total reflection. *Zeitschrift für Physik A Hadrons and nuclei*, 216(4): 398–410. doi: 10.1007/BF01391532.

Oulton, R.F., Sorger, V.J., Genov, D.A., Pile, D.F.P. and Zhang, X. 2008. A hybrid plasmonic waveguide for subwavelength confinement and long-range propagation. *Nature Photonics*, 2(8): 496–500. doi: 10.1038/nphoton.2008.131.

Pan, Z., Zhai, J. and Shen, B. 2017. Multilayer hierarchical interfaces with high energy density in polymer nanocomposites composed of BaTiO 3 @TiO 2 @Al 2 O 3 nanofibers. *Journal of Materials Chemistry A*, 5(29): 15217–15226. doi: 10.1039/C7TA03846A.

Pandey, D. and Chakrabarti, A. 2019. Prediction of two-dimensional monochalcogenides: MoS and WS. *Physics Letters A*, 383(24): 2914–2921. doi: https://doi.org/10.1016/j.physleta.2019.06.018.

Park, S-J., Kim, S., Lee, S., Khim, S., Char, K. and Hyeon, T. 2000. Synthesis and magnetic studies of uniform iron nanorods and nanospheres. *Journal of the American Chemical Society*, 122(35): 8581–8582. doi: 10.1021/ja001628c.

Parvez Ahmad, M., Rao, A., Babu, K. and Rao, G. 2019. Particle size effect on the dielectric properties of ZnO nanoparticles. *Materials Chemistry and Physics*, 224: 79–84. doi: 10.1016/j.matchemphys.2018.12.002.

Pathak, S., Sakai, N., Wisnivesky Rocca Rivarola, F., Stranks, S.D., Liu, J., Eperon, G.E. and Snaith, H.J. 2015. Perovskite crystals for tunable white light emission. *Chemistry of Materials*, 27(23): 8066–8075. doi: 10.1021/acs. chemmater.5b03769.

Powell, C.J. and Swan, J.B. 1960. Effect of oxidation on the characteristic loss spectra of aluminum and magnesium. *Physical Review*, 118(3):640–643. doi: 10.1103/PhysRev.118.640.

Pradeep, T. 2009. Noble metal nanoparticles for water purification: A critical review. *Thin Solid Films*, 517(24): 6441–6478. doi: 10.1016/j.tsf.2009.03.195.

Protesescu, L., Yakunin, S., Bodnarchuk, M.I., Krieg, F., Caputo, R., Hendon, C.H. and Kovalenko, M.V. 2015. Nanocrystals of cesium lead halide perovskites (CsPbX3, X = Cl, Br, and I): Novel optoelectronic materials showing bright emission with wide color gamut. *Nano Letters. American Chemical Society*, 15(6): 3692–3696. doi: 10.1021/nl5048779.

Ramanavičius, A., Herberg, F.W., Hutschenreiter, S., Zimmermann, B., Lapënaitë, I., Kaušaitè, A. and Ramanavičienë, A. 2005. Biomedical application of SPR biosensors Biomedical application of surface plasmon resonance biosensors (review). *Acta Medica Lituanica*, 12.

Rampino, L.D. and Nord, F.F. 1941. Preparation of palladium and platinum synthetic high polymer catalysts and the relationship between particle size and rate of hydrogenation. *Journal of the American Chemical Society*, 63(10): 2745–2749. doi: 10.1021/ja01855a070.

Rao, C.N.R. and Pisharody, K.P.R. 1976. Transition metal sulfides. *Progress in Solid State Chemistry*, 10: 207–270. doi: https://doi.org/10.1016/0079-6786(76)90009-1.

Reghunadhan, A., Kalarikkal, N. and Thomas, S. 2018. Mechanical property analysis of nanomaterials. pp. 191–212. *In*: Sneha, M.B., Oluwatobi, S.O., Nandakumar, K. and Sabu, T. (eds.). *Characterization of Nanomaterials*. Elsevier.

Ritchie, R.H. 1957. Plasma losses by fast electrons in thin films. *Physical Review*, 106(5): 874–881. doi: 10.1103/PhysRev.106.874.

Roduner, E. 2006. Size matters: why nanomaterials are different. *Chemical Society Reviews*. Royal Society of Chemistry, 35(7): 583–592.

Rosi, N.L. and Mirkin, C.A. 2005. Nanostructures in biodiagnostics. *Chemical Reviews*, 105(4): 1547–1562. doi: 10.1021/cr030067f.

Rui, X., Tan, H. and Yan, Q. 2014. Nanostructured metal sulfides for energy storage. *Nanoscale*. Royal Society of Chemistry, 6(17): 9889–9924.

Saleh, T.A. 2020. Nanomaterials: Classification, properties, and environmental toxicities. *Environmental Technology & Innovation*, 20: 101067. doi: https://doi.org/10.1016/j.eti.2020.101067.

Sampedro, B., Crespo, P., Hernando, A., Litrán, R., López, J.S., Cartes, C.L. and Vallet, M. 2003. Ferromagnetism in fcc twinned 2.4 nm size pd nanoparticles. *Physical Review Letters*, 91(23). doi: 10.1103/PhysRevLett.91.237203.

Sarina, S., Waclawik, E.R. and Zhu, H. 2013. Photocatalysis on supported gold and silver nanoparticles under ultraviolet and visible light irradiation *Green Chemistry*, 15(7): 1814. doi: 10.1039/c3gc40450a.

Sebastian, M.T., Ubic, R. and Jantunen, H. 2015. Low-loss dielectric ceramic materials and their properties. *International Materials Reviews*, 60(7), pp: 392–412. doi: 10.1179/1743280415Y.0000000007.

Seifi, H., Gholami, T., Seifi, S., Ghoreishi, S.M. and Salavati-Niasari, M. 2020. A review on current trends in thermal analysis and hyphenated techniques in the investigation of physical, mechanical and chemical properties of nanomaterials. *Journal of Analytical and Applied Pyrolysis*. Elsevier, p. 104840.

Shang, S.L., Lindwall, G., Wang, Y., Redwing, J.M., Anderson, T. and Liu, Z.K. 2016. Lateral versus vertical growth of two-dimensional layered transition-metal dichalcogenides: Thermodynamic insight into MoS2. *Nano Letters*. ACS Publications, 16(9): 5742–5750.

Shannon, R.D. 1976. Revised effective ionic radii and systematic studies of interatomic distances in halides and chalcogenides. *Acta Crystallographica Section A*, 32(5): 751–767. doi: 10.1107/S0567739476001551.

Shen, S. and Wang, Q. 2013. Rational tuning the optical properties of metal sulfide nanocrystals and their applications. *Chemistry of Materials*. ACS Publications, 25(8): 1166–1178.

Sheng, J., Wang, L., Han, Y., Chen, W., Liu, H., Zhang, M. and Liu, Y.N. 2018. Dual roles of protein as a template and a sulfur provider: A general approach to metal sulfides for efficient photothermal therapy of cancer. *Small*. Wiley Online Library, 14(1): 1702529.

Shinde, D.V., Patil, S.A., Cho, K., Ahn, D.Y., Shrestha, N.K., Mane, R.S. and Han, S.H. 2015. Revisiting metal sulfide semiconductors: A solution-based general protocol for thin film formation, hall effect measurement, and application prospects. *Advanced Functional Materials*. Wiley Online Library, 25(36): 5739–5747.

Singh, Madan and Singh, Mahipal. 2015. Impact of size and temperature on thermal expansion of nanomaterials. *Pramana*. Springer, 84(4): 609–619.

Spear, J.C., Ewers, B.W. and Batteas, J.D. 2015. '2D-nanomaterials for controlling friction and wear at interfaces. *Nano Today*. Elsevier, 10(3): 301–314.

Stamplecoskie, K.G., Scaiano, J.C., Tiwari, V.S. and Anis, H. 2011. Optimal size of silver nanoparticles for surface-enhanced raman spectroscopy. *The Journal of Physical Chemistry C*, 115(5): 1403–1409. doi: 10.1021/jp106666t.

Stenzel, O., Stendal, A., Voigtsberger, K. and Von Borczyskowski, C. 1995. Enhancement of the photovoltaic conversion efficiency of copper phthalocyanine thin film devices by incorporation of metal clusters. *Solar Energy Materials and Solar Cells*, 37(3-4): 337–348. doi: 10.1016/0927-0248(95)00027-5.

Stoumpos, C.C., Malliakas, C.D. and Kanatzidis, M.G. 2013. Semiconducting tin and lead iodide perovskites with organic cations: Phase transitions, high mobilities, and near-infrared photoluminescent properties. *Inorganic Chemistry*, 52(15): 9019–9038. doi: 10.1021/ic401215x.

Sudha, P.N., Sangeetha, K., Vijayalakshmi, K. and Barhoum, A. 2018. Nanomaterials history, classification, unique properties, production and market. In *Emerging Applications of Nanoparticles and Architecture Nanostructures*. Elsevier, pp. 341–384.

Sun, C., Lee, J.S.H. and Zhang, M. 2008. Magnetic nanoparticles in MR imaging and drug delivery. *Advanced Drug Delivery Reviews*, 60(11): 1252–1265. doi: 10.1016/j.addr.2008.03.018.

Sun, C.T. and Zhang, H. 2003. Size-dependent elastic moduli of platelike nanomaterials. *Journal of applied physics*. American Institute of Physics, 93(2): 1212–1218.

Sun, Z., Cheng, K., Yao, Y., Wu, F., Fung, J., Chen, H. and Cheng, Z. 2019. Controlled nano–bio interface of functional nanoprobes for *in vivo* monitoring enzyme activity in tumors. *ACS Nano*, p. acsnano.8b05825. doi: 10.1021/acsnano.8b05825.

Suresh, S. 2014. Studies on the dielectric properties of CdS nanoparticles. *Applied Nanoscience*, 4(3): 325–329. doi: 10.1007/s13204-013-0209-x.

Szunerits, S., Maalouli, N., Wijaya, E., Vilcot, J.P. and Boukherroub, R. 2013. Recent advances in the development of graphene-based surface plasmon resonance (SPR) interfaces. *Analytical and Bioanalytical Chemistry*, 405(5): 1435–1443. doi: 10.1007/s00216-012-6624-0.

Tao, Y.B., Lin, C.H. and He, Y.L. 2015. Preparation and thermal properties characterization of carbonate salt/carbon nanomaterial composite phase change material. *Energy Conversion and Management*. Elsevier, 97: 103–110.

Tari, A. 2003. *The Specific Heat of Matter at Low Temperatures*. World Scientific.

Tartaj, P., Morales, M., Veintemillas-Verdaguer, S., González-Carreño, T. and Serna, J. 2003. The preparation of magnetic nanoparticles for applications in biomedicine. *Journal of Physics D: Applied Physics*, 36(13): R182–R197. doi: 10.1088/0022-3727/36/13/202.

Teng, C., Su, L., Chen, J. and Wang, J. 2019. Flexible, thermally conductive layered composite films from massively exfoliated boron nitride nanosheets. *Composites Part A: Applied Science and Manufacturing*, 124: 105498. doi: https://doi.org/10.1016/j.compositesa.2019.105498.

Thompson, D.T. 2007. Using gold nanoparticles for catalysis. *Nano Today*, 2(4): 40–43. doi: 10.1016/S1748-0132(07)70116-0.

Thorek, D.L., Chen, A.K., Czupryna, J. and Tsourkas, A. 2006. Superparamagnetic iron oxide nanoparticle probes for molecular imaging. *Annals of Biomedical Engineering*, 34(1): 23–38. doi: 10.1007/s10439-005-9002-7.

Tian, F., Bonnier, F., Casey, A., Shanahan, A.E. and Byrne, H.J. 2014. Surface enhanced Raman scattering with gold nanoparticles: effect of particle shape. *Anal. Methods*, 6(22): 9116–9123. doi: 10.1039/C4AY02112F.

Tipler, P.A. and Mosca, G. 2003. *Physics for Scientists and Engineers, Volume 1: Mechanics, Oscillations and Waves; Thermodynamics*. Macmillan.

Tooghi, A., Fathi, D. and Eskandari, M. 2020. High-performance perovskite solar cell using photonic–plasmonic nanostructure. *Scientific Reports*, 10(1): 11248. doi: 10.1038/s41598-020-67741-9.

Ulbrich, K., Hola, K., Subr, V., Bakandritsos, A., Tucek, J. and Zboril, R. 2016. Targeted drug delivery with polymers and magnetic nanoparticles: Covalent and noncovalent approaches, release control, and clinical studies. *Chemical Reviews*, 116(9): 5338–5431. doi: 10.1021/acs.chemrev.5b00589.

Volz, S. 2009. *Thermal nanosystems and nanomaterials*. Springer Science & Business Media.

von Weber, A. and Anderson, S.L. 2016. Electrocatalysis by mass-selected pt n clusters. *Accounts of Chemical Research*, 49(11): 2632–2639. doi: 10.1021/acs.accounts.6b00387.

Wang, B.-X., Zhou, L.-P. and Peng, X.-F. 2006. Surface and size effects on the specific heat capacity of nanoparticles. *International Journal of Thermophysics*. Springer, 27(1): 139–151.

Wang, B., Wu, N., Zhang, X.B., Huang, X., Zhang, Y.F., Chen, W.K. and Ding, K.N. 2013. Probing the smallest molecular model of MoS2 catalyst: S2 units in the MoSn–/0 (n = 1–5) clusters. *The Journal of Physical Chemistry A*. American Chemical Society, 117(27): 5632–5641. doi: 10.1021/jp309163c.

Wang, J.L., Ando, R.A. and Camargo, P.H.C. 2014. Investigating the plasmon-mediated catalytic activity of AgAu nanoparticles as a function of composition: Are two metals better than one? *ACS Catalysis*, 4(11): 3815–3819. doi: 10.1021/cs501189m.

Weeber, J.C., Dereux, A., Girard, C., Krenn, J.R. and Goudonnet, J.P. 1999. Plasmon polaritons of metallic nanowires for controlling submicron propagation of light. *Physical Review B*, 60(12): 9061–9068. doi: 10.1103/PhysRevB.60.9061.

Wei, Q., Zhang, Q., Yan, H., Zhang, M., Shi, X. and Zhu, X. 2017. Prediction of stable ground-state and pressure-induced phase transition of molybdenum monosulfide. *Materials Science and Engineering: B*. Elsevier, 226: 114–119.

Wolfram, T. and Ellialtioglu, S. 2006. *Electronic and Optical Properties of D -Band Perovskites*. Cambridge: Cambridge University Press. doi: 10.1017/CBO9780511541292.

Wood, R.W. 1902. XLII. On a remarkable case of uneven distribution of light in a diffraction grating spectrum. *The London, Edinburgh, and Dublin Philosophical Magazine and Journal of Science*, 4(21): 396–402. doi: 10.1080/14786440209462857.

Wu, J., Cao, P., Zhang, Z., Ning, F., Zheng, S.S., He, J. and Zhang, Z. 2018. Grain-size-controlled mechanical properties of polycrystalline monolayer MoS2. *Nano letters*. ACS Publications, 18(2): 1543–1552.

Wu, Q., Miao, W.S., Gao, H.J. and Hui, D. 2020. Mechanical properties of nanomaterials: A review. *Nanotechnology Reviews*. De Gruyter, 9(1): 259–273.

Xiao, H. and Liu, S. 2017. 2D nanomaterials as lubricant additive: A review. *Materials & design*. Elsevier, 135: 319–332.

Xiao, S. and Hou, W. 2006. Studies of size effects on carbon nanotubes' mechanical properties by using different potential functions. *Fullerenes, Nanotubes, and Carbon Nonstructures*. Taylor & Francis, 14(1): 9–16.

Xie, L., Huang, X., Huang, Y., Yang, K. and Jiang, P. 2013. Core@Double-shell structured BaTiO 3–Polymer Nanocomposites with High Dielectric Constant and Low Dielectric Loss for Energy Storage Application. *The Journal of Physical Chemistry C*, 117(44): 22525–22537. doi: 10.1021/jp407340n.

Xu, D., Bliznakov, S., Liu, Z., Fang, J. and Dimitrov, N. 2010. Composition-dependent electrocatalytic activity of Pt-Cu nanocube catalysts for formic acid oxidation. *Angewandte Chemie*, 122(7): 1304–1307. doi: 10.1002/ange.200905248.

Xu, W., Zhang, J., Zhang, L., Hu, X. and Cao, X. 2009. Ultrasensitive detection using surface enhanced Raman scattering from silver nanowire arrays in anodic alumina membranes. *Journal of Nanoscience and Nanotechnology*, 9(8): 4812–4816. doi: 10.1166/jnn.2009.1104.

Xu, X., Liu, W., Kim, Y. and Cho, J. 2014. Nanostructured transition metal sulfides for lithium ion batteries: Progress and challenges. *Nano Today*, 9(5): 604–630. doi: https://doi.org/10.1016/j.nantod.2014.09.005.

Xu, Y., Peng, Y., You, T., Yao, L., Geng, J., Dearn, K.D. and Hu, X. 2018. Nano-MoS 2 and graphene additives in oil for tribological applications. pp. 151–191. *In: Nanotechnology in Oil and Gas Industries*. Springer.

Yang, C.C. and Mai, Y.-W. 2014. Thermodynamics at the nanoscale: A new approach to the investigation of unique physicochemical properties of nanomaterials. *Materials Science and Engineering: R: Reports*. Elsevier, 79: 1–40.

Yang, L., Ji, W., Huang, J.N. and Xu, G. 2019. An updated review on the influential parameters on thermal conductivity of nano-fluids. *Journal of Molecular Liquids*, 296: 111780. doi: https://doi.org/10.1016/j.molliq.2019.111780.

Yang, Q., Dong, Y., Qiu, Y., Yang, X., Cao, H. and Yao, W. 2020. Design of functional magnetic nanocomposites for bioseparation. *Colloids and Surfaces B: Biointerfaces*, 191: 111014. doi: 10.1016/j.colsurfb.2020.111014.

Yang, W., Pan, M., Huang, C., Zhao, Z., Wang, J. and Zeng, H. 2021. Graphene oxide-based noble-metal nanoparticles composites for environmental application. *Composites Communications*, 24: 100645. doi: 10.1016/j.coco.2021.100645.

Yaqoob, S.B., Adnan, R., Rameez Khan, R.M. and Rashid, M. 2020. Gold, silver, and palladium nanoparticles: A chemical tool for biomedical applications. *Frontiers in Chemistry*, 8. doi: 10.3389/fchem.2020.00376.

Yin, W.J., Yang, J.H., Kang, J., Yan, Y. and Wei, S.H. 2015. Halide perovskite materials for solar cells: A theoretical review. *Journal of Materials Chemistry A*, 3(17): 8926–8942. doi: 10.1039/C4TA05033A.

Yu, H., Wang, F., Xie, F., Li, W., Chen, J. and Zhao, N. 2014. The role of chlorine in the formation process of "CH 3 NH 3 PbI 3-x Cl x" Perovskite. *Advanced Functional Materials*, p. n/a-n/a. doi: 10.1002/adfm.201401872.

Zhang, L., Dong, W.-F. and Sun, H.-B. 2013. Multifunctional superparamagnetic iron oxide nanoparticles: Design, synthesis and biomedical photonic applications. *Nanoscale*, 5(17): 7664. doi: 10.1039/c3nr01616a.

Zhang, X., Li, W., Wu, D., Deng, Y., Shao, J., Chen, L. and Fang, D. 2018. Size and shape dependent melting temperature of metallic nanomaterials. *Journal of Physics: Condensed Matter*. IOP Publishing, 31(7): 75701.

Zhang, X.W., Pan, Y., Zheng, Q. and Yi, X.S. 2000. Time dependence of piezoresistance for the conductor-filled polymer composites. *Journal of Polymer Science, Part B: Polymer Physics*. doi: 10.1002/1099-0488(20001101)38:21<2739::AID-POLB40>3.0.CO;2-O.

Zhao, W., Pan, J., Fang, Y., Che, X., Wang, D., Bu, K. and Huang, F. 2018a. Metastable MoS2: Crystal structure, electronic band structure, synthetic approach and intriguing physical properties. *Chemistry–A European Journal*. Wiley Online Library, 24(60): 15942–15954.

Zhao, X., Huang, T., Ping, P.S., Wu, X., Huang, P., Pan, J. and Cheng, Z. 2018b. Sensitivity enhancement in surface plasmon resonance biochemical sensor based on transition metal dichalcogenides/graphene heterostructure. *Sensors*, 18(7): 2056. doi: 10.3390/s18072056.

Zhao, Y. and Zhu, K. 2016. Organic–inorganic hybrid lead halide perovskites for optoelectronic and electronic applications. *Chemical Society Reviews*, 45(3): 655–689. doi: 10.1039/C4CS00458B.

Zhou, L.P., Wang, B.X., Peng, X.F., Du, X.Z. and Yang, Y.P. 2010. On the specific heat capacity of CuO nanofluid. *Advances in Mechanical Engineering*. SAGE Publications Sage UK: London, England, 2: 172085.

Zhou, M., Zeng, C., Chen, Y., Zhao, S., Sfeir, M.Y., Zhu, M. and Jin, R. 2016. Evolution from the plasmon to exciton state in ligand-protected atomically precise gold nanoparticles. *Nature Communications*, 7(1): 13240. doi: 10.1038/ncomms13240.

Zhou, S.-Q. and Ni, R. 2008. Measurement of the specific heat capacity of water-based Al 2 O 3 nanofluid. *Applied Physics Letters*. American Institute of Physics, 92(9): 93123.

Zhu, Y., Peng, P., Fang, Z., Yan, C., Zhang, X. and Yu, G. 2018a. Structural engineering of 2D nanomaterials for energy storage and catalysis. *Advanced Materials*. Wiley Online Library, 30(15): 1706347.

Zhu, X., Ji, X., Kong, N., Chen, Y., Mahmoudi, M., Xu, X., Ding, L., Tao, W., Cai, T., Li, Y., Gan, T., Barrett, A., Bharwani, Z., Chen, H. and Farokhzad, O. 2018b. Intracellular mechanistic understanding of 2D MoS2 nanosheets for anti-exocytosis-enhanced synergistic cancer therapy. *ACS Nano. American Chemical Society*, 12(3): 2922–2938. doi: 10.1021/acsnano.8b00516.

Zijlstra, P., Chon, J.W.M. and Gu, M. 2009. Five-dimensional optical recording mediated by surface plasmons in gold nanorods. *Nature*, 459(7245): 410–413. doi: 10.1038/nature08053.

CHAPTER 4

Physical Methods for Synthesis of Nanoparticles

Kailas R. Jagdeo

1. Introduction

Nanoparticle synthesis is the production of particles from 1 to 100 nm in characteristic size which is considered the key factor in the nanotechnology field. On December 29, 1959, Richard P. Feynman, in his well-known talk on "*There is plenty of room at the bottom: An invitation to enter a new field of Physics*" at the California Institute of Technology (Feynman 1960), introduced the idea of miniaturization and that science and technology would be able to manipulate and control individual atoms and molecules leading to a new field in science. Since then, with a new sky touching thought of Feynman, the new ideas and concept about *Nanotechnology* started flourishing. In 1974, the word nanotechnology was first introduced by Norio Taniguchi (Professor, Tokyo University of Science, Japan) to explain semiconductor processes using thin film deposition as well as ion beam milling to exhibit the characteristic control to the range of nanometers. The evolution in the nanotechnology field leads to tremendous applications in various fields like material science, medical science, physical, chemical and biological sciences, bio-technology, pharmaceutical science, food technology, agriculture, energy, environment, etc.

The nano-size particles show important characteristic of increased surface to volume ratio, which leads to various applications in supercapacitors, hydrogen storage, fuel cell, etc. It is reported that the transformation from bulk material to nano-size material shows change in physical and chemical properties, even if having the same composition as bulk material. This encourages the researcher to develop the nano materials with enhanced properties like catalytic activity, electrical conductivity, hardness, optical, mechanical, antibacterial, etc. Due to the interdisciplinary approach of nanotechnology, there are various techniques available to synthesize nanomaterials. Physical, chemical, biological and hybrid techniques of synthesis give different forms of nanomaterials like tubes, rods, wires, dots, clusters, colloids, thin films, etc. The method of synthesis of nanomaterials is selected as required for different applications. Nanoparticles can be prepared by using either top-down or bottom-up approach. Basically, in top-down approach the large size bulk material is simply broken down into nano-size particles whereas in bottom-up approach nanomaterials are synthesized by building of materials from atom by atom, molecule by molecule and cluster by cluster. The approach and technique of synthesis mostly depends on the type of nanomaterials,

DSPM's K. V. Pendharkar College (Autonomous), Dombivli, Maharashtra, India.
Email: kailashjagdeo@gmail.com

materials of interest, quantity and their applications. On the basis of dimensionality, nanomaterials are categorized into zero dimensional (0D), one dimensional (1D), two dimensional (2D) and three dimensional (3D).

This chapter discusses only simple physical methods for synthesis of nanoparticles/nano materials.

2. Ball Milling

Production of nanomaterials using ball mills is the simplest mechanical method. The working principle of ball milling is the impact and size reduction of source material. The impact is caused due to falling of balls from top of the chamber (as shown in Figure 2) which reduces the size of source materials; this continuous process gives nanomaterials. Ball mills consist of a hollow cylindrical shaped chamber with a tight lid which rotates about its axis, as shown in Figure 1. The inner surface of the chamber is made rough by lining with an abrasion-resistant material like Mn (manganese), stainless steel (SS) or rubber. The electromagnet can be placed outside the chamber which exhibits the pulling magnetic force on the source material. By changing the strength of magnetic force, the milling energy can be changed to the required levels. The chamber is partially filled with balls and source material, usually 2:1 mass ratio. Balls are normally made of SS, WC (tungsten carbide), Si_3N_4 (silicon nitride), Al_2O_3 (alumina), ZrO_2 (zirconia), agate or rubber, etc. Heavy balls can be used for increasing impact energy on the source material on collision. It is observed that the larger size balls produce smaller grain size but it may introduce large defects in synthesized nanoparticles. During the process of synthesis of nanoparticles, the temperature of the chamber may rise to the range 100–1100°C. So water cooling or cryo-cooling can be used to dissipate the generated heat. Various types of ball mill are designed as per required applications. Generally, ball mills are categorized as per their mechanical operations into attritor, horizontal, planetary, high energy, or shaker. A specific gaseous atmosphere is required or vacuum is to be maintained during the process of synthesis of nanomaterials in order to obtain pure nanomaterial. High-energy ball mills are normally used for synthesis of nanomaterials, which are available in the form of planetary ball mills, vibrating ball

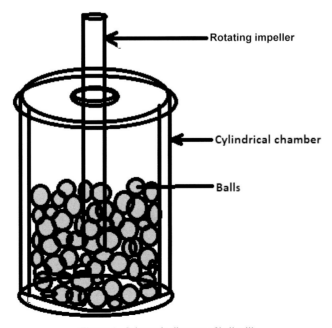

Figure 1. Schematic diagram of ball mill.

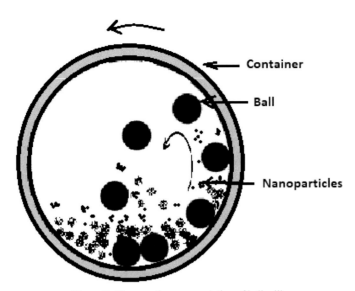

Figure 2. Cross section representation of ball mill.

mills and attrition ball mills. The tumbling mills with low energy are available as per requirements for synthesis of nanomaterials.

Among all these ball mills, planetary mill is widely used for preparation of nanoparticles. In this mill, the chamber with partially filled balls is kept on a specially designed rotating disk and the chamber also rotates around its own axis. As higher distance gives higher kinetic energy to balls and exhibits stronger impact on source material, the size of the chamber of planetary mill is kept larger. This results in increased efficiency of synthesis process of nanomaterials.

In tumbler mill, the partially filled chamber rotates about its longitudinal axis. Here, the diameter of the chamber affects the efficiency of the synthesis process of nanomaterials. The large diameter of chamber allows greater height of balls to fall and consequently higher energy is attained by balls, which results in increased impact on the source material.

In vibratory mills, the partially filled chamber is shaken back and forth at high frequency of vibration. For preparation of nanoparticles, frequency of vibration and amplitude of vibration are key factors but mass of source material also plays a vital role.

Fe nanoparticles (Munoz et al. 2007), CuO nanopowder (Ayoman and Hosseini 2016) and TiO_2 nanoparticles (Carneiro et al. 2014, Damonte et al. 2004) are synthesized by ball milling. Calcium carbonate nanoparticles are synthesized from *Achatina fulica* shell by dry milling followed by wet milling (Gbadeyan et al. 2020). Ball milling of jute fiber wastes is used for preparation of nanocellulose (Baheti et al. 2012) and functionalization of nanocellulose derivatives (Carmen et al. 2019).

3. Electric arc deposition

Electric arc deposition works on the principle of electric arc discharge, which is an electric discharge with the highest current density. The system consists of two electrodes—anode and cathode. Both electrodes are separated by an extremely small distance and kept in a suitable gas atmosphere. An extremely high voltage is maintained across the electrodes, so an extremely high electric field is produced. Due to this electric field, sudden spark is generated which results in extremely high localized temperature that leads to an arc plasma and vaporization of the electrode material. The

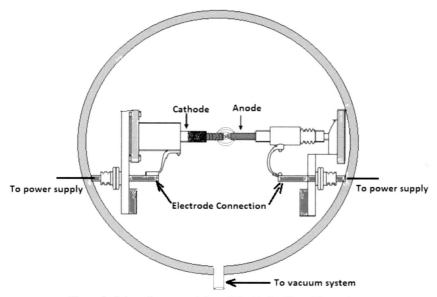

Figure 3. Schematic representation of Electric Arc Deposition system.

vaporization and subsequent cooling of electrode materials produces nanoparticles of it. The vaporization of electrode material works on the principle of thermionic emission.

Arc discharge method became more popular in 1991 when Iijima prepared carbon nanotubes (CNTs) using graphite electrodes (Iijima 1991). Arc discharge direct current (dc) or alternating current (ac) can be used for synthesizing carbon nanotubes and carbon nano fibers but it is reported that dc arc discharge provides higher yields of CNTs, mostly multi-wall CNTs. Single walled CNTs can be prepared using electrodes doped with catalyst particles such as Nickel-Yttrium, Cobalt-Yttrium, Nickel-Cobalt, etc. It is also a powerful technique for preparing graphene using graphite rods.

4. Laser ablation

Laser ablation method is used for obtaining various kinds of nanoparticles and became more popular after using it for synthesis of carbon nanotubes by using graphite target (Smalley et al. 1995). This technique is being used for synthesizing carbon nanotubes, semiconductor quantum dots, nanowires and core shell nanoparticles. The basic principle of nanoparticles' preparation in a background of gas is the nucleation and growth of species generated by laser-vaporization. The rapid quenching of vapor decides the purity of nanoparticles. With extremely rapid quenching of vapors, nanoparticles exhibiting quantum size effect with the range of less than 10 nm can be synthesized.

Laser ablation in general involves the process of laser irradiation, which removes material from the target surface. The basic difference in laser ablation and laser evaporation is that laser ablation uses the non-equilibrium vapor/plasma conditions created at the target surface by intense powerful laser beam, whereas laser evaporation emphasizes the condition of thermodynamic equilibrium; heating and evaporation of material is done at this equilibrium.

Laser ablation system has three important parts:

(a) Powerful laser source like ArF excimer laser, XeCl excimer laser, CO_2 laser, Nd-YAG laser, etc.

(b) Ablation chamber

(c) Vacuum system

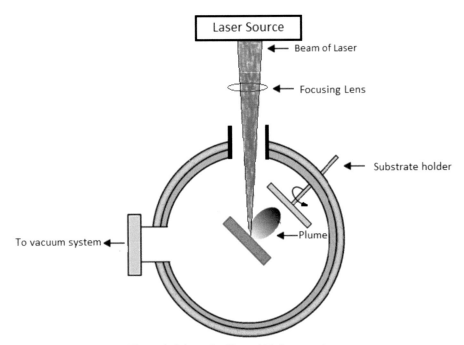

Figure 4. Schematic of Laser Ablation apparatus.

In laser ablation, a focused laser beam from laser source is allowed to incident on the solid target material, fitted in ablation chamber which is filled with ambient media (gas or liquid). It induces the large amount light absorption at surface of target material. So temperature of absorbing material increases, which makes the vaporization of target material into *laser induced plasma plume.* The plume size and emission spectrum of it depends on four factors:

(a) Material of target

(b) Ambient media (gas or liquid)

(c) Ambient pressure

(d) Laser conditions

The vaporized materials in some cases can be condensed into clusters and particles without any chemical reaction, whereas in some cases vaporized material can react with reactants introduced in the ablation chamber and produce new materials. The condensed particle can be either collected through a filter system, which consists of a glass fiber mesh or deposited on a substrate. During the process of vaporization and deposition, essential high vacuum is to be maintained, so High or Ultra-High Vacuum (UHV) vacuum system is attached to ablation chamber. The ablation chamber is evacuated with a vacuum system and can be filled with inert gas or reactive gases as required for synthesis of nanomaterials. Here, gas pressure plays a key role in determining the size of particles and their distribution.

In pulsed laser ablation, a powerful pulsed laser source can be used for vaporizing target material into *plasma plume.* Due to use of a powerful pulsed laser, a tightly confined, both spatially and temporally, *plasma plume* is obtained. But this method produces a small amount of nanoparticles. Iron oxides and strontium ferrite (Shinde et al. 2000), TiO_2 (Harano et al. 2002), ZnO (Mintcheva et al. 2018), silver (Perito et al. 2016) and hydrogenated-silicon (Makimura et al. 2002) nanoparticles are synthesized by using pulsed laser ablation method.

5. Laser pyrolysis

Nowadays, laser pyrolysis method is used for synthesis of nanoparticles. Here, in the pyrolysis chamber, the reactant gases (gaseous-phase precursors) can be introduced with carrier gases like helium or argon where the reactant gases meet the powerful laser beam (CO_2 laser in continuous mode). The high power laser beam (2400 W) generates elevated localized temperatures. As a result, the reactant gas decomposes, which collides and interacts with carrier gas. This triggers the nucleation and growth of nanoparticles. Nanoparticles can be either allowed to grow and deposit on cooled substrate or collected by a catcher, which is equipped with a filter. Here, gas pressure is also the key factor in deciding particle size and particle distribution.

Although the gaseous precursors are used in laser pyrolysis synthesis, solid or liquid can be used in some cases but it involves safety issues, cost of precursor and availability. The liquid precursors with sufficiently high pressure can be delivered as vapors from a bubbler to the reaction zone as small droplets, which rapidly evaporate upon laser heating.

Silicon-Carbide (Ershov et al. 2020), ceramic (D'Amato et al. 2013) and metallic tin-based nanoparticles (Dutu et al. 2015) are synthesized by laser pyrolysis.

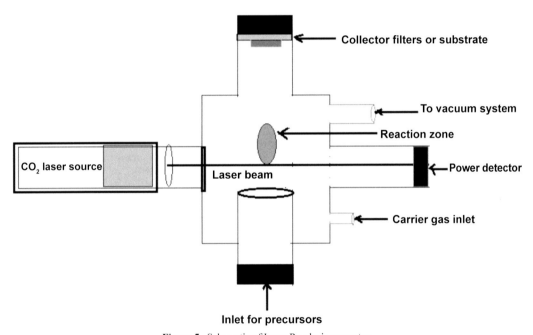

Figure 5. Schematic of Laser Pyrolysis apparatus.

6. Physical Vapor Deposition (PVD)

PVD is the coating technique based on the principle of vaporization of coating material. It involves the transfer of coating material on an atomic level under vacuum conditions. It is being widely used for coating on substrates and obtaining thin films. But the recent technical developments and advances in PVD technique lead to the use of PVD for synthesis of nanostructured coatings, nano-thin films, multilayered nano-thin films and nanocomposites. Normally, the physical method used to generate vapor of coating material is either thermal evaporation or sputtering for deposition on the target substrate. Beside that cathodic arc, ion beam or pulsed laser can also be used for Physical Vapor Deposition. Sputtering is more suitable than thermal evaporation as it gives better bonding strength with the substrate. The PVD follows three steps for synthesis of nanoparticles or nano thin films:

(a) The material which is to be deposited is firstly converted into vapor phase by using physical methods.

(b) The supersaturation of the vapor phase in an inert atmosphere promotes the condensation of metal nanoparticles.

(c) The condensation of vapor on substrate takes place to form nano thin film or consolidation of nanocomposite takes place by thermal treatment under inert atmosphere.

The PVD system consists of filaments or boats of refractory. This is made of metals like tungsten, tantalum and molybdenum. In this refractory, the materials which are to be evaporated are kept in crucibles. A PVD system needs reactive gas or an inert gas with proper pressure for collision with material vapor. It consists of a cold finger, which is connected to a cooling system on which clusters or nanoparticles get condensed. The device consists of scrapers, which are used to scrape nanoparticles and also a special arrangement-piston-anvil in which nanoparticle powder can be compacted.

From high temperature boats of refractory, metals or high pressure metal oxides get evaporated or sublimated. It is observed that near to the boat, density of this evaporated material is notably high and their size normally found in nanometers. These evaporated atoms and clusters can collide with gas molecules and form larger size particles. These particles are forced to move towards the cold finger where they get condensed. Particle size and their distribution depends on the rate of evaporation, gas pressure maintained in the chamber and distance of source of evaporator from cold finger. The particles condensed on the central cooled rod can be scraped and consolidated with a special piston-anvil system to form a pallet. The gases like O_2, H_2 and NH_3 are used which works as reactive gas that can interact with evaporated materials and form oxide, nitride or hydride particles. After obtaining these particles, with appropriate post-treatments one can achieve the desired metal

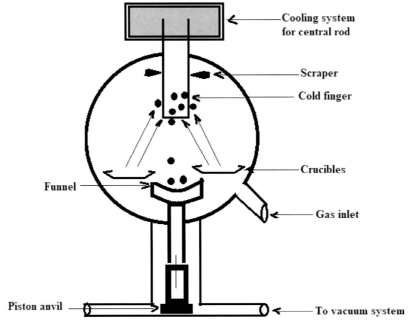

Figure 6. Schematic of (thermal evaporation) PVD.

compound or pure nanoparticles. The gas pressure in the deposition chamber plays a vital role in deciding size, shape and phase of evaporated material.

7. Sputtering

In the sputtering method, vaporization of materials from the surface of the target (solid) is obtained by bombarding high energetic ions of an inert gas (Argon) on the target surface, which causes ejection of atoms and clusters. This ejected material of the target gets deposited on the substrate. Sputtering is mostly used for thin film deposition or for obtaining nano thin film on the substrate.

The sputtering method follows two important steps: the first step is the introduction of controlled inert gas (Ar) into the vacuum chamber and the second step involves electrically energizing the cathode in order to obtain self-sustaining plasma. The surface of the cathode exposed to the substrate acts as a target, which is actually a slab of material to be deposited on the substrate surface. The inert gas atoms may lose their electrons inside plasma and become positively charged ions. These ions are accelerated and directed to strike with sufficient kinetic energy onto the target. Here they lose energy to target atoms and cause the sputtering out of some target atoms/molecules in addition with clusters, ions and secondary electrons. The sputtered material constitutes a vapor stream. This vapor stream traverses the chamber and allows it to hit and stick to the substrate in order to form thin film or surface coating. During the process and before the deposition process, it is necessary to maintain high vacuum or ultra-high vacuum in the chamber.

The basic and commonly used methods for sputter depositions are Direct Current (DC) sputtering, Radio Frequency (RF) sputtering and Magnetron sputtering. The type of sputtering depends upon which type of power is used on the cathodes. For conductive materials, DC power is used whereas for non-conductive materials RF power is suitable. If the reaction gas like O_2 or N_2 is added to Ar gas, the resultant sputtering is known as reactive sputtering. Here with argon, ions of reaction gas are produced. These ions react with the sputtered layer atoms and form the reaction products, which can be deposited on the substrate.

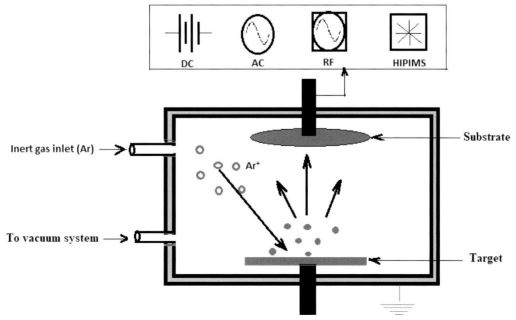

Figure 7. Schematic of Sputtering deposition.

1. DC Sputtering

This is the simplest method of deposition. In this method, the sputter target is connected to the negative terminal of the power source and maintained at high negative voltage. The substrate is connected to the positive terminal of the power source or ground or kept floating. When the required pressure by the vacuum system is obtained in the chamber, inert gas (Ar) can be introduced at suitable pressure, normally less than 10 Pa. As soon as the required high potential is obtained across the anode and cathode, the cathode glows and glow discharge occurs. This results in plasma which contains electrons, ions, neutral atoms and even photons obtained due to collision. Some energetic electrons on impact with inert gas cause gas ionization. A sufficiently large number of ions are produced in the chamber, which get accelerated and cause the sputtering of the target. Further, the deposition of sputtered material on substrate is obtained. The material which is to be deposited on substrate normally decides whether substrate is to be heated or cooled.

2. RF Sputtering

In RF sputtering, AC modulation of power at radio frequencies is used across the terminals of anode and cathode, which alternatively reverses the polarities. This causes an oscillation in electrons, so a sufficient amount of gas ionization is possible. Target itself is biased to negative potential and hence working as cathode. In this sputtering, plasma formed tends to diffuse throughout the whole chamber rather than concentrating around the cathode or target as happens with DC sputtering. Although it is useful for many types of dielectric coating or insulating coatings, it has some disadvantages. The deposition rate in RF sputtering is observed to be less than DC sputtering. Another disadvantage is that the RF power is itself complicated as it requires a high voltage power supply and expensive advanced circuitry, which may cause additional overheating problems.

3. Magnetron Sputtering

DC/RF Magnetron sputtering gives greater sputtering rate than conventional DC/RF sputtering. This sputtering rate can be obtained by providing an additional strong magnet in the DC/RF sputtering system. Here, both electric and magnetic fields acting on electrons make electrons move in helical paths and confine the electrons in the plasma near or at the target surface. These helical moving electrons may ionize more atoms in the gas. These helical moving confined electrons increase the plasma density and deposition rate. It prevents direct damage, which is possibly caused by direct impact of these electrons with substrate or the growing film on substrate.

HIPIMS is High Power Impulse Magnetron Sputtering in which a train of short and intense pulses with low duty cycle is delivered to the cathode. During a pulse, a very high power creates very high dense plasma, which results in a large fraction of ionized sputtered atoms without overheating the target. HIPIMS produces high performance dense coatings with good adhesion, and a uniform film without droplets on complexed shaped substrates.

Copper (Jaiswal et al. 2015), gold (Terauchi et al. 1995) and flower-like tungsten (Acsente et al. 2015) nanoparticles are synthesized by Magnetron Sputtering.

8. Ion beam techniques or ion implantation

Ion implantation is one of the advanced techniques used for synthesis of nanomaterials. Ion implantation assembly typically consists of an ion source with ion gun, ion deflector with accelerator and target chamber. Amount of ions implanted (fluence) per square cm is called dose (ions/cm^2). In ion source with ion gun, ions of desired elements are produced and with specially designed ion gun, ions are ejected. In an ion deflector with an accelerator, ions get electrostatically accelerated and directed towards the target with high energy. In high vacuum target chamber, ions get implanted on the substrate. The low energy (less than 200 eV) as well high energy (few keV to hundreds of keV) ions can be used to obtain nanoparticles. To synthesize nanoparticles, the typical ion beam energies used range from 50 to 150 keV with fluences in range from 10^{16} to 10^{18} ions/cm^2. It is observed that the concentration of implanted atoms which modifies surface composition of substrate increases as fluences increases (William and Poate 1984). The nanostructure modification (Avasthi and Pivin 2010) and shape-modification of patterned nanoparticles (Heo and Gwag 2015) can be possible by ion beam technique. Now, focused ion beam nanofabrication is a promising technique for fabrication of 3D nanostructures and devices (Li et al. 2021).

Reference

Acsente, T., Negrea, R.F., Nistor, L.C., Logofatu, C., Matei, E., Birjega, R. and Dinescu, G. 2015. Synthesis of flower-like tungsten nanoparticles by magnetron sputtering combined with gas aggregation. *The European Physical Journal D*, 69(6): 161.

Avasthi, D.K. and Pivin, J.C. 2010. Ion beam for synthesis and modification of nanostructures, *Current Science*, 98(6): 780–792.

Ayoman, E. and Hosseini, S.G. 2016. CuO nanopowder syntheisized by high-energy ball-milling method and investigation of their catalytic activity on thermal decomposition of ammonium perchlorate particles. *J. Therm Anal Calorim*, 123: 1212–1224.

Baheti, V., Abbasi, R. and Militky, J. 2012. Ball milling of jute fibre wastes to prepare nanocellulose. *World Journal of Engineering*, 9(1): 45–50.

Carmen, C. Piras, Susana Fernandez-Prieto and Wim M. De Borggraeve. 2019. Ball milling: A green technology for preparation and functionalisation of nanocellulose derivatives. *Nanoscale. Adv.*, 1: 937–947.

Carneiro, J.O., Azevedo, S., Fernandes, F., Freitas, E., Pereira, M., Tavares, C.J. and Teixeira, V. 2014. Synthesis of iron-doped TiO$_2$ nanoparticles by ball-milling process: The influence of process parameters on the structural, optical, magnetic, and photocatalytic properties. *Journal of Materials Science*, 49(21): 7476–7488.

D'Amato, R., Falconieri, M., Gagliardi, S., Popovici, E., Serra, E., Terranova, G. and Borsella, E. 2013. Synthesis of ceramic nanoparticles by laser pyrolysis: From research to applications. *Journal of Analytical and Applied Pyrolysis*, 104: 461–469.

Damonte, L.C., Mendoza Zélis, L.A., Marí Soucase, B. and Hernández Fenollosa, M.A. 2004. Nanoparticles of ZnO obtained by mechanical milling. *Powder Technology*, 148(1): 15–19.

Dutu, E., Dumitrache, F., Fleaca, C.T., Morjan, I., Gavrila-Florescu, L., Morjan, I.P. and Vasile, E. 2015. Metallic tin-based nanoparticles synthesis by laser pyrolysis: Parametric studies focused on the decreasing of the crystallite size. *Applied Surface Science*, 336: 290–296.

Ershov, I.A., Iskhakova, L.D., Krasovskii, V.I., Milovich, F.O., Rasmagin, S.I. and Pustovoi, V.I. 2020. Synthesis of silicon-carbide nanoparticles by the laser pyrolysis of a mixture of monosilane and acetylene. *Semiconductors*, 54(11): 1467–1471.

Feynman, R.P. 1960. There's plenty of room at the bottom. *Engineering and Science*, 23(5): 22–36.

Gbadeyan, O.J., Adali, S., Bright, G., Sithole, B. and Onwubu, S. 2020. Optimization of milling procedure for synthesizing nano-CaCO$_3$ from *Achatina fulica* Shell through mechanochemical techniques. *Journal of Nanomaterials* Volume 2020, Article ID 4370172. https://doi.org/10.1155/2020/4370172.

Guo, T., Nikolaev, P., Thess, A., Colbert, D.T. and Smalley, R.E. 1995. Catalytic growth of single-walled nanotubes by laser vaporization. *Chem. Phys. Lett.*, 243: 49–54.

Harana, A., Shimada, K., Okubo, T. and Sadakata, M. 2002. Cystal phases of TiO$_2$ ultrafine particles prepared by laser ablation of solid rods. *J. Nanoparticle Res.*, 4: 215–9.

Heo, K.C. and Gwag, J.S. 2015. Shape-modification of patterned nanoparticles by an ion beam treatment. *Sci. Rep.*, 5: 8523; DOI:10.1038/srep08523 (2015)

Iijima, S. 1991. Helical microtubules of graphitic carbon. *Nature*, 354: 56–58.

Jaiswal, J., Chauhan, S. and Chandra, R. 2015. Influence of sputtering parameters on structural, optical and thermal properties of copper nanoparticles synthesized by dc magnetron sputtering. *Int. J. Sci. Technol. Manage*, 4 (01): 678–688.

Jorge, E. Munoz, Janeth Cervantes, Rodrigo Espaeza and Gerardo Rosas. 2007. Fe nanoparticles synthesized by high-energy ball milling. *J. Nanopart. Res.*, 9: 945–950.

Makimura, T., Mizuta, T. and Murakami, K. 2002. Laser ablation synthesis of hydrogenated silicon nanoparticles with green photoluminescence in the gas phase. *Jpn. J. Appl. Phys.*, 41: L144–L146.

Mintcheva, N., Aljulaih, A., Wunderlich, W., Kulinich, S. and Iwamori, S. 2018. Laser-ablated ZnO nanoparticles and their photocatalytic activity toward organic pollutants. *Materials*, 11(7): 1127.

Perito, B., Giorgetti, E., Marsili, P. and Muniz-Miranda, M. 2016. Antibacterial activity of silver nanoparticles obtained by pulsed laser ablation in pure water and in chloride solution. *Beilstein Journal of Nanotechnology*, 7: 465–473.

Ping, Li, Siyu Chen, Houfu Dai, Zhengmei Yang, Zhiquan Chen, Yasi Wang, Yigin Chen, Wengiang Peng, Wubin Shan and Huigao Duan. 2021. Recent advances in focused ion beam nanofabrication for nanostructures and devices: Fundamental and applications. *Nanoscale*, 13: 1529–1565.

Sharma, Annu, Bahniwal, Suman, Aggarwal, Sanjeev, Chopra, S. and Kanjilal, D. 2011. Synthesis of copper nanoparticles in polycarbonate by ion implantation. *Bull. Mater. Sci.*, 34(4): 645–649.

Shinde, S.R., Kulkarni, S.D., Banpurkar, A.G., Nawathey-Dixit, R., Date, S.K. and Ogale, S.B. 2000. Magnetic properties of nanosized powder of magnetic oxides synthesized by pulsed laser ablation. *J. Appl. Phy.*, 88: 1566–75.

Terauchi, S., Koshizaki, N. and Umehara, H. 1995. Fabrication of Au nanoparticles by radio-frequency magnetron sputtering. *Nanostructured Materials*, 5(1): 71–78.

William, J.S. and Poate, J.M. 1984. Introduction to implantation and beam processing, First Edition, *Academic Press*, eBooK ISBN: 9781483220642, p. 1–11.

CHAPTER 5

Chemical Methods for the Synthesis of Nanomaterials

Jagruti S. Suroshe

1. Introduction

As compared to physical methods, chemical methods provide better control over the characteristic properties of nanocrystallites by tuning the particle size, shape as well as surface modification of nanocrystallites and have been the subject of matter of few reviews and books (Burda et al. 2005, Malik et al. 2010, Lalena et al. 2007). Chemical methods can be categorized as solid phase (mechanical attrition, mechanochemical synthesis), wet chemical or liquid phase (precipitation, micelles or micro-emulsion, solvothermal, electrochemical, sonochemical, microwave synthesis, liquid-liquid interface methods and sol-gel method) and vapor/gas phase processes (chemical vapor deposition, spray pyrolysis, etc.). Among these processes, liquid phase methods are more relevant for the preparation of metal chalcogenides and hence they have been briefly discussed below.

(i) Co-precipitation method

This method involves precipitation of sparingly soluble products induced by chemical reactions such as hydrolysis, complexation, oxidation, reduction or by changing reaction conditions like temperature, concentration, pH, etc. These chemical reactions are governed by nucleation as well as the development process which decides the particle size and morphology of the products (Rane et al. 2018, La Mer and Johnson 1947, Schmidt et al. 2003).

The products obtained from this method are mostly insoluble products obtained under high supersaturation conditions. A huge number of tiny particles will be formed in an important process known as nucleation. The properties, morphology and size of these products will be appreciably influenced by secondary operations like Ostwald ripening and aggregation. The supersaturation state required to persuade precipitation is generally the outcome of a chemical reaction (Rane et al. 2018).

$$X\ Ay^+_{(aq)} + y\ Bx^-_{(aq)} \longrightarrow AxBy_{(s)} \tag{1}$$

Department of Chemistry, DSPM's K. V. Pendharkar College (Autonomous), Dombivli, Thane, MH 421203, India.
Email: skjkbag@gmail.com

The conventional coprecipitation synthetic methods include:

1. Oxides produced from polar and non-polar solutions.
2. Formation of metal chalcogenides by using molecular precursors.
3. Coprecipitation using microwave/ultrasonication
4. Metals obtained using aqueous solutions, by reduction from organic solutions, electrochemical reduction and decay of metal-organic precursors.

Among the earlier methods, this method is used for the synthesis of nanomaterials. However, the method has some advantages and disadvantages. Advantages include simple and fast preparation, low temperature, control over composition, homogeneity and particle size, non-polluting and energy-saving approach. The polymerization of the material may get influenced due to the existence of metallic precursors in the solution. This usually affects the specimens with unacceptable characteristics consisting of less described pore size. Moreover, the above perspective has restricted appropriateness to polymeric supports (White et al. 2009).

Yazid and co-workers synthesized magnetic nanoparticles via the co-precipitation method for effective elimination of bulky metals from synthetic wastewater (Yazid and Joon 2019). Tinwala and co-authors prepared $La_2Ce_2O_7$ nanoparticles by co-precipitation method and obtained nanoparticles that were well characterized by using various characterization techniques like Thermogravimetric analysis (TGA), Electron dispersive spectroscopy (EDS), X-ray diffraction (XRD), Fourier transform infrared spectroscopy (FT-IR) and Transmission electron microscopy (TEM) (Tinwala et al. 2014).

(ii) Solvothermal/Hydrothermal method

The solvothermal method has been a versatile technique to prepare inorganic materials either in non-polar solvent or in the polar solvent at elevated temperatures and pressures in a closed vessel and takes the benefit of increased diffusion of chemicals and solubility of inorganic compounds under these conditions (Rane et al. 2018). The method is addressed as solvothermal method or hydrothermal method depending on the medium (organic solvent or water) in which the reactions are carried out. However, the term solvothermal has also been used in the literature for heating precursors in presence of coordinating solvents. The method offers precise control on the size and appearance of the particles. By altering experimental conditions like duration of reaction, solvent, temperature, surfactant type as well as precursor type, these characteristics can be changed (Lu 2012).

Suroshe et al. synthesized functionalized carbon nanotube/ZnO nanocomposites by solvothermal decomposition method and studied its capacitive behavior by coating it on glassy carbon electrode (Suroshe and Garje 2015). Palve et al. synthesized ZnS nanomaterials from complexes of Zinc(II) thiosemicarbazone and characterized by various characterization techniques (Palve and Garje 2011).

(iii) Sol-gel method

Sol is a suspension of colloidal particles in a solution or liquid. When it is mixed with other liquids, it affects the development of an endless three-dimensional network known as 'gel'. This method usually relates to the hydrolysis as well as condensation of alkoxide precursors or metal alkoxides, directing to the diffusion of oxides in a sol. Then the sol is free from solvent by removing it. More often, H_2O is served as a solvent, although the hydrolysis of precursors can be done using an acid or base. Acid catalysis gives rise to a polymeric gel; however, basic catalysis leads to the development of a colloidal gel (Lam et al. 2008). The properties of the final products get affected by major parameters such as the condensation and hydrolysis rates. Small particles can be observed at gradual as well as the higher governed rate of hydrolysis. Figure 1 demonstrates the route for sol-gel synthesis (Owens et al. 2016). As per the need for a particular application, various steps involved

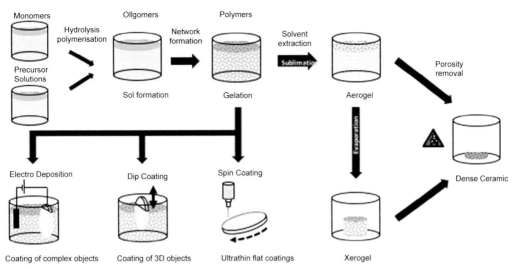

Figure 1. Demonstration of the sol-gel route.

(Reproduced with permission from Owens, Gareth J., Rajendra K. Singh, Farzad Foroutan, Mustafa Alqaysi, Cheol Min Han, Chinmaya Mahapatra, Hae Won Kim and Jonathan C. Knowles. 2016. Sol-gel based materials for biomedical applications. *Progress in Materials Science* 77: 1–79. doi: 10.1016/j.pmatsci.2015.12.001.)

in this process can be changed, elongated or eliminated entirely exclusive of gelation as well as solvation.

This process depends upon the appearance of the sol, i.e., if it is existing in the form of a solution or a suspension of fine particles (Wang 2020). The size of particles is dependent upon the pH, constitution of the solution as well as temperature. In this system, magnetic ordering can consider the obtained phases including the particle volume fraction together with the size distribution along with the dissipation of the particles (Tavakoli et al. 2007). For nanocomposites obtained by the gels, framework conditions, as well as the porosity of the material, are estimated by the rate of hydrolysis along with condensation of the gel precursors and additional redox reactions throughout the stages of gelling and successive heating (Tavakoli et al. 2007, Teja and Koh 2009).

Yahaya and co-workers discussed the preparation of crystalline TiO_2 nanoparticles by the sol-gel process. This process includes two major processes- hydrolysis and condensation. Precursors such as $Ti[OCH(CH_3)_2]_4$, (TTIP), $TiCl_3$, $TiCl_4$, $TiBr_4$ and $Ti(OBu)_4$, are more often used for the preparation of TiO_2 nanomaterials. These precursors are treated with water to carry out hydrolysis reactions and then on condensation, they form a 3-D network (Yahaya et al. 2017).

Reactions for hydrolysis and condensation process are shown below:

Hydrolysis:

$$Ti(OR)_4 \; + \; 4\,H_2O \longrightarrow 2\,Ti(OH)_4 \; + \; 4\,ROH \tag{2}$$

Condensation:

$$Ti(OH)_4 \; + \; Ti(OH)_4 \longrightarrow 2\,TiO_2 \; + \; 4\,H_2O \quad \text{(oxolation)} \tag{3}$$

$$Ti(OH)_4 \; + \; Ti(OR)_4 \longrightarrow 2\,TiO_2 \; + \; 4\,ROH \quad \text{(alcoxolation)} \tag{4}$$

Where, R = ethyl, i-propyl, n-butyl, etc.

(iv) Template method

The template method is very popular for the preparation of functional materials with several nanostructures (Liang et al. 2010, Lou et al. 2008). This is an easy approach for the synthesis of nanomaterials by encouraging the target materials to extend as per the model of the templates. This scheme imparts a simple process for the preparation of nanostructures with expected shape as well as the size and has been broadly used in the fabrication of 1-dimensional nanostructures (Zhao et al. 2008). In most of the cases, templates introduced for the preparation can be categorized into two classes: (1) soft templates, which includes ligands, organogelators, polymers as well as surfactants, (2) hard templates that are either employed in the form of physical scaffolds for the adjacent deposition of expected coating substances or used as structure-specifying layout and as chemical reagents that interact with more reagents to fabricate appropriate nanomaterials (Fan et al. 2008).

In the 1970s, this technique was first introduced by scientist Possin to synthesize nanowires (Possin 1970) and in the 1990s, Martin developed this methodology and proposed the word "template synthesis" (Martin 1994). The often-used templates consist of organized porous membranes developed with anodized Al_2O_3 (Murakami et al. 1999), silica (Kim et al. 2003), nanochannel glass (Berry et al. 1996) and ion-track-etched polymers (Chakarvarti and Vetter 1998).

(v) Electrodeposition method

This method is interesting for forming a metallic layer through the cathodic reduction on the surface in polar otherwise non-polar solvents. Cathode used is the substrate substance that is dipped in a solution consisting of a metal salt to be coated. The solubilized metallic ions are sent towards the cathode followed by reduction to form metals (Anon n.d.). By using the electrodeposition method, Yang and co-authors introduced reduced graphene oxide on carbon fiber electrode to explore its response towards uric acid, dopamine and ascorbic acid. Figure 2 illustrates the synthesis of reduced graphene oxide/carbon fiber composite by electrodeposition method (Yang et al. 2014). With the help of data obtained from various studies, the authors concluded that this composite is an excellent electrode material for biochemical gadgets.

The characteristic merits of the electrodeposition method are that it can allow the conformal coating of materials and one can simply manage the thickness of the coated layers (She et al. 2008).

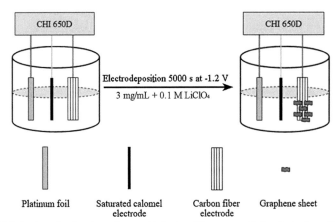

Figure 2. Fabrication of reduced graphene oxide/carbon fiber composite by electrodeposition method.

(Reproduced with permission from Yang, Beibei, Huiwen Wang, Jiao Du, Yunzhi Fu, Ping Yang and Yukou Du. 2014. Direct electrodeposition of reduced graphene oxide on carbon fiber electrode for simultaneous determination of ascorbic acid, dopamine and uric acid. *Colloids and Surfaces A: Physicochemical and Engineering Aspects*, 456(1): 146–52. doi: 10.1016/j. colsurfa.2014.05.029.)

By controlling certain parameters like current, and potential, nanomaterials with desired morphology and required electrocatalytic properties can be synthesized. Experimental results reveal that the crystallite size of nanostructures at greater overvoltage is affected by substrate pore diameters to a small extent. However, this impact is huge at lower overvoltage (Serp et al. 2002). Figure 3 below presents a scanning electron microscope (SEM) image of nano electrodeposited Ni-Co at 80 mA/cm^2 (Paul et al. 2014).

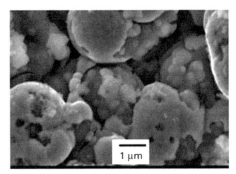

Figure 3. SEM image of nano electrodeposited Ni-Co at 80 mA/cm^2.
(Reproduced with permission from Elsevier, License No. 5114430235058)

(vi) Micro-emulsion method

The method involves the isotropic dispersions of two immiscible liquids like oil with water which can be maintained by surfactants (cationic, anionic or neutral) placed at the interface of oil with water. They can be either water-in-oil (W/O) or oil-in-water (O/W) micro-emulsions. A micelle is a cluster of surfactant molecules dispersed in a liquid where the hydrophilic head point towards the surrounding solvent and the hydrophobic tail is directed inwards to the core of the micelle (oil-in-water). The micelle is referred to as reverse or inverse micelle if the hydrophilic moiety of surfactant points inward to the micelle's center (water-in-oil) (Rane et al. 2018, Pinisetty et al. 2011).

The micro-emulsions allow size-choice development of the particles in droplets of water captured in a hydrocarbon solvent. The characteristic properties are chiefly directed by the water-to-surfactant molar ratio. The size and shape of a micelle is a role of the surfactant molecule's geometry, concentration, along with the pH and ionic strength of the solution (Lu 2012). Finnie and co-workers proposed a system for the synthesis of SiO$_2$ nanoparticles in an acidic and alkaline environment, which are shown below in Figure 4. SiO$_2$ nanoparticles were prepared by interaction of tetramethyl orthosilicate inside the droplets of water of a water-in-oil microemulsion at p^H 1.05 (acidic) along with pH 10.85 (basic). It was observed that with a change in p^H from acidic to basic, the hydrolysis reaction of tetramethyl orthosilicate slows down. Under basic conditions, spheres of ~ 11 nm were formed whereas in an acidic medium ~ 5 nm spheres were obtained which was highly uniform in size (Finnie et al. 2007). The development of particles in microemulsions is controlled by a mechanism similar to that of colloidal suspensions: nucleation with the development of particles takes place by either ripening aggregation or coagulation.

(vii) Chemical Vapor Deposition (CVD)

In this method, vapors of the suitable precursors are put forward into a preheated CVD apparatus where the vapors are adsorbed and thermally decomposed or reacted with other vapors present in the reactor to deposit the required material. Nucleation can either occur in the gaseous phase or on the substrate. In the former case, it is homogeneous while in the latter it is heterogeneous (Martínez and Prieto 2007). Figure 5 shown below illustrates the basic apparatus required for chemical vapor deposition at the laboratory level for coating (Choy 2003).

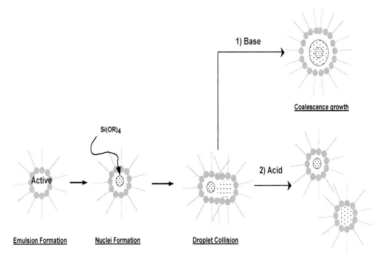

Figure 4. Proposed mechanism for synthesis of SiO_2 nanoparticles in microemulsions under acidic as well as basic conditions.

(Reproduced with permission from American Chemical Society, Langmuir)

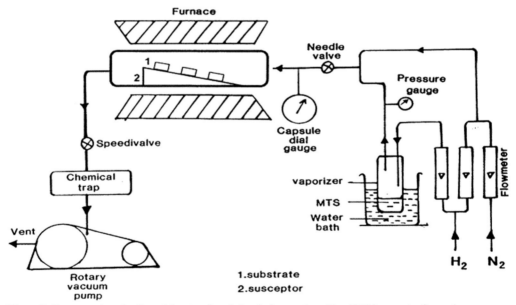

Figure 5. Demonstration of ordinary laboratory-based chemical vapor deposition (CVD) apparatus for coating purposes.

(Reproduced with permission from Choy, K.L. 2003. Chemical vapour deposition of coatings. *Progress in Materials Science*, 48(2): 57–170. doi: 10.1016/S0079-6425(01)00009-3.)

CVD has reactions proceeding the nucleation and development of nanomaterials from the vapor-phase precursors. In the CVD process, at higher temperatures and below nonequilibrium conditions, gas-phase reactions with the development of particles occur. This operation is chiefly related to vapor-phase reactions, nucleation via supersaturated vapors to obtained particles, particle development through vapor condensation and/or heterogeneous chemical interactions, agglomeration via particle-particle collisions promoted through Brownian motion with coalescence or sintering amongst particles. The size of particles between 5–200 nm can be manufactured with the help of this method. Generally, argon or nitrogen can be used as a carrier gas, which helps to carry the volatilized precursors across a heated reaction compartment. A tube-shaped furnace can be employed as a heat source with temperatures up to 1500 K. Carbonyls, chlorides and metal-organic compounds are generally utilized as precursors. Low-temperature synthesis leads to the formation of tiny, porous, and in many examples, glassy nanomaterials. Nucleation of oxides is effectively rapid because of oxidation as well as large concentrations of the precursor vapors.

Salifairus and Rusop synthesized carbon nanotubes (CNTs) using camphor oil above ferrocene and aluminum isopropoxide ($C_9H_{21}O_3Al$) catalyst by the CVD method. They varied deposition temperature from 700ºC to 900ºC and obtained CNTs were well characterized by various characterization techniques (Salifairus and Rusop 2013). Manawi et al. also discussed a review paper that includes the preparation of several carbon nanomaterials like fullerenes, CNTs, carbon nano-onion (CNO), carbide-derived carbon (CDC), carbon nanofibers (CNFs) and graphene by the CVD method (Manawi et al. 2018).

(viii) Polyol method

Out of several methods, to synthesize nanoparticles in solution, the 'polyol method' is simple and can be easily set up; therefore, it is of commercial significance. In this process, the polyol is mentioned as a diol, mostly a 1,2-diol like ethylene glycol along with their derivatives, di-, tri-, tetra- till poly(ethylene glycol), represented as EG, DEG, TEG, TTEG, PEG, etc., respectively, and also to dissimilar isomers of propanediol, butanediol, pentanediol, etc. Substances consisting of higher than two –OH groups like glycerol, carbohydrates and pentaerythritol are also viewed like polyols. The availability of multiple hydroxy groups leads to increase in boiling point and viscosities, which further gives unique properties like reducing and coordinating properties. These properties are very much helpful for the synthesis of nanoparticles for tuning their size and shape (Fievet et al. 2018).

Murray et al. at IBM synthesized FePt nanoparticles by polyol method. Co-reduction of iron pentacarbonyl and Platinum (II) bis (acetylacetonate) at elevated temperatures gives FePt nanoparticles of size 3 nm. Monodispersed magnetic face-centered cubic (fcc) (A1 phase) FePt nanoparticles were synthesized using the same procedure followed by post-synthetic annealing at 500ºC to generate face-centered tetragonal (fct) (L10 phase) FePt crystals (Sun et al. 2000, Yu et al. 2008). Several advantages offered by polyol medium are shown below in Figure 6. Polyol medium has multiple advantages such as high boiling point, reducing medium, high viscosity and its capacity to coordinate metal precursors.

The high boiling point of polyol medium helps to carry out synthesis at comparatively high temperature, confirming crystalline materials can be manufactured. Polyol medium is a reducing medium that prevents the oxidation of synthesized metal particles. It can also coordinate metal precursors and particle surfaces, which further help to minimize coalescence. The high viscosity of the polyol medium facilitates a diffusion-controlled regime for the growth of particles, which ultimately results in structures with controlled morphology (Fievet et al. 2018). FePd nanoparticles can be synthesized with the help of stabilizers such as oleic acid and oleylamine (Chen and Nikles 2002).

(ix) Sonochemical method

Currently, ultrasound has been utilized for the preparation of nanoparticles. It supplies strong mechanical forces, which help to disperse reactants easily. In the course of sonication of liquid, cavitation occurs from the implosive fall down of bubbles, which further leads to the formation of localized hot spots around 5000 K and a lifetime of some nanoseconds or less than that. Therefore, chemical reactions usually occur inside the bubbles. By controlling the ultrasonic irradiation power,

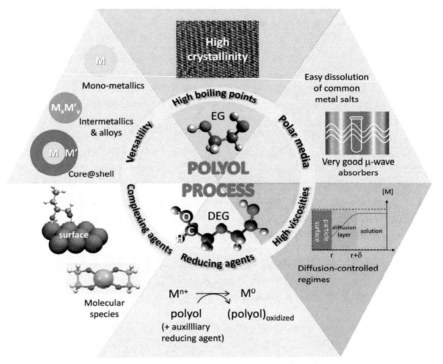

Figure 6. Schematic representation of the merits of the polyol process.

(Reproduced with permission from Royal Society of Chemistry, ISSN-0306-0012)

the shape, as well as size of the nanomaterials, can be governed. A setup for the sonochemical method is illustrated below in Figure 7 (Hujjatul Islam et al. 2019).

The sonochemical method is simple, operating conditions are ambient and someone can easily manage the particle size of desired nanomaterials by introducing precursors of different concentrations in the solution and also by controlling the power of ultrasonic irradiation. Sivakumar and coauthors synthesized nanocrystals of zinc ferrite with the help of the precursors of Zn and Fe acetates along with rapeseed oil (Sivakumar et al. 2006). Cao and coworkers prepared nanocrystalline hydroxyapatite particles with the help of ultrasonic precipitation process using $Ca(NO_3)_2$ and ammonium dihydrogen phosphate as source materials and carbamide (NH_2CONH_2) as precipitating agent (Cao et al. 2005). Bastami et al. prepared Hausmannite (Mn_3O_4) nanoparticles by the sonochemical method in different media under ambient conditions without the addition of any additives. The obtained nanoparticles were well characterized by various characterization techniques. With the change in media, different shapes and sizes of nanomaterials were observed. Using mineral oil, diamond shape nanomaterials with a size ~ 50 nm were observed whereas in vegetable oil, spherical nanoparticles with a size of ~ 7 nm were observed (Bastami and Entezari 2010).

2. Conclusion

In the present chapter, we focused on various chemical methods to synthesize nanomaterials. Every method has certain advantages and some disadvantages. Depending upon the nanomaterial which is to be synthesized, a suitable method can be adopted which helps to synthesize material with good yield; it will be also easy to control the shape and size by changing certain parameters. Here, we successfully demonstrated co-precipitation, solvothermal, sol-gel, template, electrodeposition,

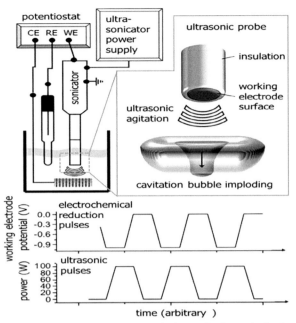

Figure 7. A schematic representation of sonochemical setup for the synthesis of nanomaterials.
(Reproduced with permission from Hujjatul Islam, Md, Michael T.Y. Paul, Odne S. Burheim and Bruno G. Pollet. 2019. Recent developments in the sonoelectrochemical synthesis of nanomaterials. *Ultrasonics Sonochemistry*, 59(April): 104711. doi: 10.1016/j.ultsonch.2019.104711.)

microemulsion, chemical vapor deposition (CVD), polyol and sonochemical method using appropriate examples.

References

Anon. n. d. 2007. Titanium dioxide nanomaterials: synthesis, properties, modifications, and applications. *Chem. Rev.* 107: 2891–2959. doi: 10.1021/cr0500535.

Bastami, T. Rohani and Entezari, M.H. 2010. Sono-synthesis of Mn3O4 nanoparticles in different media without additives. *Chemical Engineering Journal*, 164(1): 261–66. doi: 10.1016/j.cej.2010.08.030.

Berry, A.D., Tonucci, R.J. and Fatemi, M. 1996. Fabrication of GaAs and InAs wires in nanochannel glass. *Applied Physics Letters*, 69(19): 2846–48. doi: 10.1063/1.117338.

Burda, Clemens, Xiaobo Chen, Radha Narayanan and Mostafa A. El-Sayed. 2005. Chemistry and properties of nanocrystals of different shapes. *ChemInform*, 36(27).

Cao, Li Yun, Chuan Bo Zhang and Jian Feng Huang. 2005. Synthesis of hydroxyapatite nanoparticles in ultrasonic precipitation. *Ceramics International*, 31(8): 1041–44. doi: 10.1016/j.ceramint.2004.11.002.

Chakarvarti, S.K. and Vetter, J. 1998. Template Synthesis—a membrane based technology for generation of nano-/micro materials: A review. *Radiation Measurements*, 29(2–6): 149–59. doi: 10.1016/S1350-4487(98)00009-2.

Chen, Min and David E. Nikles. 2002. Synthesis of spherical FePd and CoPt nanoparticles. *Journal of Applied Physics*, 91(10 I): 8477–79. doi: 10.1063/1.1456406.

Choy, K.L. 2003. Chemical vapour deposition of coatings. *Progress in Materials Science*, 48(2): 57–170. doi: 10.1016/S0079-6425(01)00009-3.

Fan, Hai, Yuanguang Zhang, Maofeng Zhang, Xuyang Wang and Yitai Qian. 2008. Glucose-assisted synthesis of CoTe nanotubes *in situ* templated by Te Nanorods. *Crystal Growth and Design*, 8(8): 2838–41. doi: 10.1021/cg7011364.

Fievet, F., Ammar-Merah, S., Brayner, R., Chau, F., Giraud, M., Mammeri, F., Peron, J., Piquemal, J.Y., Sicard, L. and Viau, G. 2018. The polyol process: A unique method for easy access to metal nanoparticles with tailored sizes, shapes and compositions. *Chemical Society Reviews*, 47(14): 5187–5233. doi: 10.1039/c7cs00777a.

Finnie, Kim S., John R. Bartlett, Christophe J.A. Barbé and Linggen Kong. 2007. Formation of silica nanoparticles in microemulsions. *Langmuir*, 23(6): 3017–24. doi: 10.1021/la0624283.

Hujjatul Islam, Md, Michael T.Y. Paul, Odne S. Burheim and Bruno G. Pollet. 2019. Recent developments in the sonoelectrochemical synthesis of nanomaterials. *Ultrasonics Sonochemistry*, 59(April): 104711. doi: 10.1016/j. ultsonch.2019.104711.

Kim, Tae-Wan, In-Soo Park and Ryong Ryoo. 2003. A synthetic route to ordered mesoporous carbon materials with graphitic pore walls. *Angewandte Chemie* 115(36): 4511–15. doi: 10.1002/ange.200352224.

La Mer, Victor, K. and Irving Johnson. 1947. The Determination of the particle size of monodispersed systems by the scattering of light. *Journal of the American Chemical Society*, 69(5): 1184–92. doi: 10.1021/ja01197a058.

Lalena, John N., David A. Cleary, Everett E. Carpenter and Nancy F. Dean. 2007. *Inorganic Materials Synthesis and Fabrication*. Wiley-Interscience, A John Wiley & Sons, Inc., Publication.

Lam, Un Teng, Raffaella Mammucari, Kiyonori Suzuki and Neil R. Foster. 2008. Processing of iron oxide nanoparticles by supercritical fluids. *Industrial and Engineering Chemistry Research*, 47(3): 599–614. doi: 10.1021/ie070494+.

Liang, Hai Wei, Shuo Liu and Shu Hong Yu. 2010. Controlled synthesis of one-dimensional inorganic nanostructures using pre-existing one-dimensional nanostructures as templates. *Advanced Materials*, 22(35): 3925–37. doi: 10.1002/adma.200904391.

Lou, Xiong Wen, Lynden A. Archer and Zichao Yang. 2008. Hollow micro-/nanostructures: Synthesis and applications. *Advanced Materials*, 20(21): 3987–4019. doi: 10.1002/adma.200800854.

Lu, Kathy. 2012. *Nanoparticulate Materials: Synthesis, Characterization, and Processing*. doi: 10.1002/9781118408995.

Malik, Mohammad Azad, Mohammad Afzaal and Paul O'Brien. 2010. Precursor chemistry for main group elements in semiconducting materials. *Chemical Reviews*, 110(7): 4417–46. doi: 10.1021/cr900406f.

Manawi, Yehia M., Ihsanullah, Ayman Samara, Tareq Al-Ansari and Muataz A. Atieh. 2018. A review of carbon nanomaterials' synthesis via the chemical vapor deposition (CVD) Method. *Materials*, 11(5). doi: 10.3390/ma11050822.

Martin, Charles R. 1994. Nanomaterials: A membrane-based synthetic Approach. *Science,* 266(5193): 1961–1966. [doi: 10.1126/science.266. 5193.1961].

Martínez, Agustín and Gonzalo Prieto. 2007. The key role of support surface tuning during the preparation of catalysts from reverse micellar-synthesized metal nanoparticles. *Catalysis Communications*, 8(10): 1479–86. doi: 10.1016/j. catcom.2006.12.025.

Murakami, Hideki, Masao Kobayashi, Hirofumi Takeuchi and Yoshiaki Kawashima. 1999. Preparation of Poly(DL-Lactide-Co-Glycolide) nanoparticles by modified spontaneous emulsification solvent diffusion method. *International Journal of Pharmaceutics*, 187(2): 143–52. doi: 10.1016/S0378-5173(99)00187-8.

Owens, Gareth J., Rajendra K. Singh, Farzad Foroutan, Mustafa Alqaysi, Cheol Min Han, Chinmaya Mahapatra, Hae Won Kim and Jonathan C. Knowles. 2016. Sol-gel based materials for biomedical applications. *Progress in Materials Science*, 77: 1–79. doi: 10.1016/j.pmatsci.2015.12.001.

Palve, Anil M. and Shivram S. Garje. 2011. Preparation of zinc sulfide nanocrystallites from single-molecule precursors. *Journal of Crystal Growth*, 326(1): 157–62. doi: 10.1016/j.jcrysgro.2011.01.087.

Paul, Subir, Sk Naimuddin and Asmita Ghosh. 2014. Electrochemical characterization of Ni-Co and Ni-Co-Fe for oxidation of methyl alcohol fuel with high energetic catalytic surface. *Ranliao Huaxue Xuebao/Journal of Fuel Chemistry and Technology*, 42(1): 87–95. doi: 10.1016/s1872-5813(14)60012-8.

Pinisetty, D., Davis, D., Podlaha-Murphy, E.J., Murphy, M.C., Karki, A.B., Young, D.P. and Devireddy, R.V. 2011. Characterization of electrodeposited bismuth-tellurium nanowires and nanotubes. *Acta Materialia*, 59(6): 2455–61. doi: 10.1016/j.actamat.2010.12.047.

Possin, George E. 1970. A method for forming very small diameter wires. *Review of Scientific Instruments*, 41(5): 772–74. doi: 10.1063/1.1684640.

Rane, Ajay Vasudeo, Krishnan Kanny, Abitha, V.K. and Sabu Thomas. 2018. *Methods for Synthesis of Nanoparticles and Fabrication of Nanocomposites*. Elsevier Ltd.

Salifairus, M.J. and Rusop, M. 2013. Synthesis of carbon nanotubes by chemical vapour deposition of camphor oil over ferrocene and aluminum isopropoxide catalyst. *Advanced Materials Research*, 667: 213–17. doi: 10.4028/www. scientific.net/AMR.667.213.

Schmidt, A., Schneiders, M., Döpfner, M. and Lehmkuhl, G. 2003. Störungskonzepte Für Psychische Probleme Bei Jugendlichen. Pilotstudie Zur Validierung Eines Fragebogens Zu Störungskonzepten Bei Psychischen Problemen von Jugendlichen (SSPJ). *Zeitschrift Fur Kinder- Und Jugendpsychiatrie Und Psychotherapie*, 31(2): 111–21. doi: 10.1024/1422-4917.31.2.111.

Serp, Philippe, Philippe Kalck and Roselyne Feurer. 2002. Chemical vapor deposition methods for the controlled preparation of supported catalytic materials. *Chemical Reviews*, 102(9): 3085–3128.

She, Guangwei, Xiaohong Zhang, Wensheng Shi, Yuan Cai, Ning Wang, Peng Liu and Dongmin Chen. 2008. Template-free electrochemical synthesis of single-crystal CuTe nanoribbons. *Crystal Growth and Design*, 8(6): 1789–91. doi: 10.1021/cg7008623.

Sivakumar, Manickam, Atsuya Towata, Kyuichi Yasui, Toru Tuziuti and Yasuo Iida. 2006. A new ultrasonic cavitation approach for the synthesis of zinc ferrite nanocrystals. *Current Applied Physics*, 6(3): 591–93. doi: 10.1016/j. cap.2005.11.068.

Sun, Shouheng, C.B. Murray, Dieter Weller, Liesl Folks and Andreas Moser. 2000. Monodisperse FePt nanoparticles and ferromagnetic FePt nanocrystal superlattices. *Science*, 287(5460): 1989–1992. doi: 10.1126/Science.287. 5460.1989.

Suroshe, Jagruti S. and Shivram S. Garje. 2015. Capacitive behaviour of functionalized carbon nanotube/ZnO composites coated on a glassy carbon electrode. *Journal of Materials Chemistry A*, 3(30): 15650–60. doi: 10.1039/c5ta01725d.

Tavakoli, A., Sohrabi, M. and Kargari, A. 2007. A review of methods for synthesis of nanostructured metals with emphasis on iron compounds. *Chemical Papers*, 61(3): 151–70. doi: 10.2478/s11696-007-0014-7.

Teja, Amyn S. and Pei Yoong Koh. 2009. Synthesis, properties, and applications of magnetic iron oxide nanoparticles. *Progress in Crystal Growth and Characterization of Materials*, 55(1-2): 22–45. doi: 10.1016/j.pcrysgrow.2008.08.003.

Tinwala, Hozefa, D.V. Shah, Jyoti Menghani and Ranjan Pati. 2014. Synthesis of La2Ce2O7 nanoparticles by co-precipitation method and its characterization. *Journal of Nanoscience and Nanotechnology*, 14(8): 6072–76. doi: 10.1166/jnn.2014.8834.

Wang, Xuanze. 2020. Preparation, Synthesis and Application of Sol-Gel Method University Tutor : Pr. Olivia GIANI Internship Tutor: Mme. WANG Zhen. (October).

White, Robin J., Rafael Luque, Vitaliy L. Budarin, James H. Clark, and Duncan J. Macquarrie. 2009. Supported metal nanoparticles on porous materials. Methods and applications. *Chemical Society Reviews*, 38(2): 481–94. doi: 10.1039/b802654h.

Yahaya, Muhamad Zamri, Mohd Asyadi Azam, Mohd Asri Mat Teridi, Pramod Kumar Singh and Ahmad Azmin Mohamad. 2017. Recent characterisation of sol-gel synthesised TiO2 nanoparticles. *Recent Applications in Sol-Gel Synthesis*. doi: 10.5772/67822.

Yang, Beibei, Huiwen Wang, Jiao Du, Yunzhi Fu, Ping Yang and Yukou Du. 2014. Direct electrodeposition of reduced graphene oxide on carbon fiber electrode for simultaneous determination of ascorbic acid, dopamine and uric acid. *Colloids and Surfaces A: Physicochemical and Engineering Aspects*, 456(1): 146–52. doi: 10.1016/j.colsurfa.2014.05.029.

Yazid, Noraziah Abu and Yap Chin Joon. 2019. Co-precipitation synthesis of magnetic nanoparticles for efficient removal of heavy metal from synthetic wastewater. *AIP Conference Proceedings* 2124(July). doi: 10.1063/1.5117079.

Yu, C.H., Kin Tam and Edman S.C. Tsang. 2008. Chapter 5 chemical methods for preparation of nanoparticles in solution. *Handbook of Metal Physics*, 5: 113–41. doi: 10.1016/S1570-002X(08)00205-X.

Zhao, Yong Sheng, Hongbing Fu, Aidong Peng, Ying Ma, Debao Xiao and Jiannian Yao. 2008. Low-dimensional nanomaterials based on small organic molecules: preparation and optoelectronic properties. *Advanced Materials*, 20(15): 2859–76. doi: 10.1002/adma.200800604.

CHAPTER 6

Bionanofabrication
A Green Approach towards Nanoparticle Synthesis using Plants and Microbes

Annika Durva Gupta[1,a,]* and *Darshana Rajput*[2,b]

1. Introduction

Our nature is full of mysteries which have become a constant source of curiosity and information for us. Nature has figured out how to make the most effective miniaturized practical products in ingenious and stylish ways. Living creatures have devised ingenious ways to produce a wide range of inorganic structures through evolution. Nature knows how to create highly advanced materials that are designed and built to perform complex biological functions as efficiently as possible (Lloyd and Lovley 2001). One such mystery is Quantum Sciences, where the natural operations occur at atomic and sub-atomic levels. It is at this atomic level that the boundaries of Physics, Chemistry and Biology have merged together. At this level, most of the physical laws and chemical principles are transformed to a great scope, giving materials unbelievable properties, which are exploited in material fabrication. Nanotechnology has emerged as a versatile branch of material science which deals with the properties of material at nanometric dimensions (1–100 nm). Nanoparticles of one or more than one dimensions, having the size of 100 nm or less, have enthralled all with their peculiar and interesting properties, as well as a multitude of applications that outperform their bulk equivalents (Daniel and Astruc 2004, Kato 2013). Biology, agriculture, food, telecommunications, cosmetics, electronics, pharmacy, and biomedical and food devices are among these fields. Nanoparticles, especially metallic nanoparticles (MtNPs), have risen to prominence because of their unique physicochemical characters and biotechnological applications (Slavin et al. 2017, Khan et al. 2017).

Physical synthesis, such as thermal decomposition, radiation aided, electrochemical, sonochemical, and microwave aided methods, as well as chemical synthesis, such as reduction of liquids, photochemical and chemical reactions in reverse micelles, are all choices for nanoparticle synthesis (Punjabi et al. 2015). While both chemical and physical methods possess an advantage that the nanoparticle shape and size can be regulated, they have a variety of disadvantages, including

[1] Department of Biotechnology, B.K. Birla College (Autonomous), Kalyan 421304. Dist Thane. Maharashtra, India.
[2] Centro de Investigación y Desarrollo Tecnológico en Electroquímica (CIDETEQ), Querétaro 76703, México.
  Email: drajput@cideteq.mx
* Corresponding author: annikadurve@yahoo.com
[a] ORCID ID-0000-0001-9956-476X
[b] ORCID ID-0000-0002-0373-753X

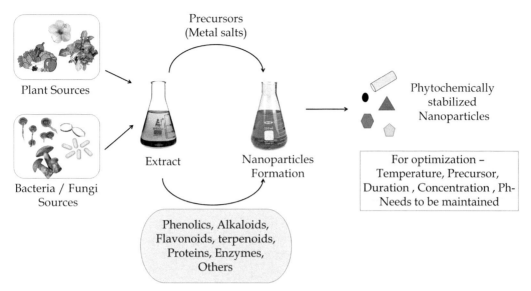

Figure 1. Schematic illustration of biosynthesis of nanoparticles using different plants and microbial sources.

the use of poisonous chemicals, high synthesis costs, and poor yield. MtNPs have hazardous chemicals bound to their surfaces, which may have harmful implications in biomedical applications (Mukherjee et al. 2013, 2015, Patra et al. 2015, Ovais et al. 2018a).

With growing concerns of using physical and chemical methods, research has moved towards the green chemistry route or biological synthesis of MtNPs using various types of microorganisms and plants. The biological method of MtNP production is less expensive, biocompatible, nontoxic, eco-friendly, and comparatively simple (Figure 1) (Mukherjee et al. 2013, 2015, Patra et al. 2015, Ovais et al. 2018a). However, this process results in the generation of polydisperse nanoparticles, due to its diverse photochemistry (Salunke et al. 2014, Ovais et al. 2016) (Figure 1). In this chapter, we will be dealing with the various green eco-friendly mechanisms using microorganisms and plants to produce nanoparticles.

2. Metal Nanoparticles' Synthesis by Microorganisms

Haefeli identified a bacterium isolated from a silver mine as *Pseudomonas stutzeri* AG259, efficient in synthesizing silver (Ag) nanoparticles, as the first bacteria capable of synthesizing silver nanoparticles in 1984 (Haefeli et al. 1984, Venkataraman et al. 2011, Klaus et al. 1999). Nanoparticle biosynthesis utilizing microbial strains such as actinomycetes, microbes, yeast, fungi, marine algae, and viruses has attracted a lot of interest in the field of green and eco-friendly nanotechnology in recent years (Singh et al. 2006). Microorganisms serve as highly effective nanofactories for the environmentally safe and low-cost production of a wide range of metal nanoparticles, including gold (Au), silver, copper (Cu), and palladium (Pd), as well as metallic oxides like titanium oxide (TiO$_2$) and zinc oxide (ZnO). Nanotubes, nanoconjugates, nanorods, and nanowires are examples of nanoscale materials that come in a variety of shapes, sizes, and types (Albanese et al. 2003). Metal nanoparticles have applications in a variety of areas and are showing biocompatibility with minimal toxicity, making them exceptional drug delivery devices and sensor carriers in diagnostic instruments (Sadowski 2010). The reduction of metal in its elemental state, which can either be collected intracellularly or extracellularly, is the key process for microorganism-mediated nanoparticle synthesis (Ahmad et al. 2007, Kalimuthu et al. 2008, Sadowski et al. 2008, Saifuddin et al. 2009, Jain et al. 2011, Janardhanan et al. 2013).

2.1 Bacteria

Bacteria and actinomycetes have evolved to inhabit ecological niches consisting of heavy metal concentrations. Thus, similar mechanisms to tolerate these metals have been selected across different bacterial genera and for different metals (Thakur 2006). Bacteria, which are immune to metals and thrive on them, perform an important role in the biogeochemical cycling of metallic ions. Since the oxidation state of a heavy metal affects its solubility and toxicity, this is a vital consequence of microbial heavy metal resistance (Spain 2003, Malekzadeh et al. 2002). Bacteria and actinomycetes have developed a variety of strategies to combat the toxic effects of metals and metalloids, including aggregation, tolerance, and even biomethylation and transformation to reduce their bioavailability and toxicity.

Metal speciation and transport in the atmosphere are highly influenced by bacterial behaviour.

Different bacteria have different reactions to toxic substances (Thakur 2006). In the rhizosphere, there is increased microbial activity, which could lead to increased trace element aggregation, transformation, degradation, and biomethylation. Biosorption, sulphide deposition, and biotransformation are also established methods for removing harmful heavy metals or metalloids from wastewaters by microbes in the rhizosphere (reduction, volatilization). This mechanism is reliant on the cell's metabolic function, as defined by its intrinsic biochemical and structural properties, physiological and/or genetic adaptation, environmental changes in metal requirements, availability, and toxicity (Cha and Cooksey 1991). Temperature, pH, and biomass concentrations all have an effect on live cells' ability to extract metallic ions from aqueous medium (Chen and Ting 1995).

Metal tolerance in microbes is mediated by phosphate, carbonate, and/or sulphide precipitation, volatilization through methylation or ethylation, physical exclusion of electronegative components in membranes and extracellular polymeric substances (EPS), energy-dependent metal efflux systems, and intracellular sequestration (Gadd 1990, Silver 1996). Protecting a human cell from the toxic effects of the high concentrations of heavy metals is one of the most difficult challenges it faces. By altering the metal ion's redox state, by reduction or the creation of non-toxic complexes such as sulphides and oxides, these effects may be eliminated. The eventual destiny of these metal ions is for them to be converted to a neutral oxidation state and then for each atom to be fabricated into nanoscale particles (Silver 1996).

The intracellular and extracellular methods have been used for metal nanoparticles' synthesis (Figure 2). Extracellular biosynthesis takes place outside the bacterial membrane using techniques like

(a) utilizing bacterial biomass,

(b) utilizing bacterial culture's supernatant and

(c) utilizing extracts free from cell debris.

Synthesis of nanoparticles is favoured, extracellularly over intracellular one, since it does not require complicated downstream processing (Singh et al. 2013). These nanoparticles have been used in a number of uses, the bulk of which are biomedical in nature.

Beveridge's group in 1980 showed that bacteria can synthesize gold nanoparticles and that the cell wall was the site of metal deposition in *Bacillus subtilis* (Beveridge and Murray 1980). His group confirmed that on incubation of bacterial cells with Au^{3+} ions, nano sized gold particles were precipitated within the cells (Southam and Beveridge 1996, Forty and Beveridge 2000). Similar results were obtained by Nair and Pradeep when they incubated lactobacillus strain isolated from butter milk (Nair and Pradeep 2002). They showed that lactic acid bacteria from the whey of buttermilk, when mixed with gold and silver ions together, could form alloys of Au and Ag. Nanoparticles of gold biosynthesized (AuNPs) by *Rhodopseudomonas capsulata* and *Pseudomonas aeruginosa* have been studied and it was seen that pH does perform a crucial role in the shape and

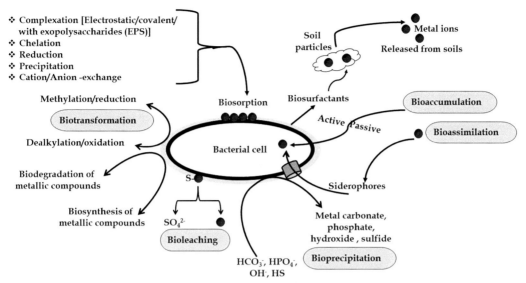

Figure 2. Various bacterial mechanisms for heavy metal resistance.

size of AuNPs. At pH 6, AuNPs with diameters having a range from 20 to 80 nm were developed (Pourali et al. 2017, Singh and Kundu 2014). All metal nanoparticles produced have a size gap, which may be due to the nanoparticles being shaped at different times. Due to restrictions related to particles nucleating within species and released outside the cell, the size of nanoparticles may be restricted (Oza et al. 2012a).

Silver nanoparticles (AgNPs) were biosynthesized in the intracellular periplasmic space by *Bacillus* sp. (Klaus et al. 1999). In one report, two separate isolated strains of *Pseudomonas aeruginosa* were utilized to biosynthesize gold nanoparticles (AuNPs), resulting in AuNPs of various sizes (Pugazhenthiran et al. 2009). *Rhodopseudomonas* capsulata was used to create AuNPs that were spherical (10–50 nm) and triangular plate (50–400 nm) (He et al. 2007). *Serratia ureilytica* was used to produce ZnO nanoflowers, which were then impregnated into cotton fabrics, resulting in antimicrobial activity against *S. aureus* and *E. coli* (Dhandapani et al. 2014). It was noticed that *Lactobacillus plantarum* was utilized to facilitate ZnO nanoparticles' biosynthesis (Selvarajan and Mohanasrinivasan 2013). *Aeromonas hydrophilais* was used to biosynthesize ZnO nanoparticles, which were then tested for antimicrobial properties (Jayaseelan et al. 2012). Halomonas elongate developed triangular shaped nanoparticles of CuO, which were screened for antimicrobial action against *E. coli* and *S. aureus* (Rad et al. 2018). Super paramagnetic iron oxide nanoparticles having a diameter of 29 nm were synthesized by *Bacillus cereus*. These particles had a dose-dependent anti-cancer activity on the 3T3 and MCF-7 cell lines (Fatemi et al. 2018). Bacterial strains were used to make bimetallic Ag-Au nanostructures (Nair and Pradeep 2002).

2.2 Algae

Marine algae, including both micro and macroalgae, have invited substantial attention in the field of nanomanufacturing. Microalgae such as *Chlorella vulgaris* and *Spirogyra insignis*, as well as macroalgae such as *Sargassum wightii* and *Chondrus crispus*, have been used to biosynthesize noble metallic nanoparticles having elevated antimicrobial potential and biocompatibility, which can be used in biomedical tool design, drug distribution, catalysis, and electronics (Govindaraju et al. 2009, Kahzad and Salehzadeh 2020, Castro et al. 2013). Nucleation, regulation of element, and stabilization of nanoparticle shape, facilitated by enzymes in the algal cell, and biological reduction of metallic ions by the functional enzymes are among the mechanisms defined for nanoparticle biosynthesis using these microalgae (Castro et al. 2013).

Scenedesmus sp. was utilized in the biosynthesis of Ag nanoparticles, which resulted in the development of AuNPs with sizes ranging from 3 to 35 nanometers within living cells.

Nucleation points and size stability are thought to be aided by an assortment of stabilizing and reducing agents. The antimicrobial potential of these Ag nanoparticles against *Streptococcus mutans* and *Escherichia coli* was excellent (Jena et al. 2014). *Desmodesmus* sp. was used to make silver nanoparticles of sizes varying from 10 to 30 nanometers. Proteins, polysaccharides, phenolic, and polyphenols compounds have also been implicated in the regulation of particle dimension and stabilization.

Antibacterial activity against *Listeria monocytogenes* and *Salmonella* sp. was observed, as well as antifungal potential against *C. parapsilosis* (Öztürk 2019). The synthesis of AgNPs was performed using *Fusarium oxysporum* cell free culture filtrates, and the results revealed that the highest nanoparticle synthesis rate occurred during the stationary step, when the extracellular enzyme nitrate reductase activity was at its peak. In addition, raising the C:N ratio activated the nitrate reductase enzyme, resulting in the development of small AgNPs with a narrow size distribution (Hamedi et al. 2017). *Laminaria japonica* (Kim et al. 2016), *Gelidium amansii* (Pugazhendhi et al. 2018), *Sargassum plagiophyllum* (Dhas et al. 2014) algal extracts were used to biosynthesize silver nanoparticles. The combination of aqueous silver nitrate with all three algal extracts resulted in the development of 50 nm spherical AgNPs.

Extracts of *Chaetomorpha linum* resulted in the development of AgNP clusters ranging in size from 3–44 nm. The addition of aqueous AgNO3 solutions to entire living cultures of wild type and cell wall deficient *C. reinhardtii* strains resulted in the creation of round-shaped particles having a diameter of less than 10 nm that crystallized in a face centered cubic (FCC) cell lattice structure (Rahman et al. 2019).

Tetraselmis kochinensis was applied for the intracellular synthesis of gold nanoparticles with measurements between the range from 5 to 35 nm and a spherical shape. The metal ions were reduced by enzymes present in the cell wall more than anywhere else, presumably suggesting that they were reduced by cell wall enzymes rather than elsewhere (Senapati et al. 2012). Algal extracts of *Stoechospermum marginatum* (Rajathi et al. 2012), and *Galaxaura elongata* (Abdel-Raouf et al. 2017) were used for the synthesis of gold nanometal. Within the range of 3–95 nm, both extracts gave rise to circular, hexagonal, and triangular forms, with all particles exhibiting strong antibacterial activity. Biosynthesis of AuNPs was also done using *Chlorella pyrenoidusa* extract. The results showed that stable particles with sizes of 25–30 nm formed. The presence of nitrate reductase, which converts auric ions to AuNPs, was indicated by the substantial decrease in nitrate reductase production (Oza et al. 2012b).

Sargassum bovinum has also been used in the biosynthesis of palladium nanoparticles (PdNPs). The bioreduction of Pd2+ ions resulted in octahedral PdNPs with a size range of 5–10 nm. As a hydrogen peroxide tracker, these PdNPs perform admirably (Momeni and Nabipour 2015). The development of spherical PdNPs of 5–20 nm was achieved through a one-step biological synthesis of PdNPs that make use of *Chlorella vulgaris*. At room temperature, the reduction of Pd^{2+} ions into PdNPs took just 10 minutes. The amide and polyol groups found in the algal extract serve as reducing and stabilizing agents, according to FTIR spectra (Arsiya et al. 2017). Extracts of *Padina tetrastromatica* and *Turbinaria conoides* were used as bioreducers to make zinc oxide nanoparticles (ZnO NPs). The role of different functional groups in the development of ZnO NPs was indicated by FTIR analysis of the extract. The antibacterial activity of the crystalline particles obtained was rectangular, pentagonal, and hexagonal in shape, with a size range of 90–120 nm (Shanmugam 2018). Sargassum muticum extract was used to make hexagonal ZnO NPs with a diameter of 30–60 nm. The results show that ZnO NPs have apoptotic, antiangiogenic, and cytotoxic effects at both concentrations and incubation times, indicating that they could be used as a cancer therapy complement (Sanaeimehr et al. 2018).

Diatoms are algae that can be found in the sea or in some other wet area. They are unicellular, with a living component within an amorphous silica shell. Diatoms can range in length from 20 to 200 metres. Polycondensation is prevented by absorbing silicic acid [$Si(OH)_4$] within the cell, which is bound to the cofactor. Silicic acid is stored in the Golgi bodies of cells. Transport vesicles are made of silica. Vesicles are phospholipid bilayers that can absorb a solution or water and aggregate to form silica deposition vesicles (SDV), an organism's mineralized portion (Kulkarni 2015).

The procedures behind biological mineralization or biomineralization, or the deposition of inorganic minerals in living organisms, have been extensively studied in order to create novel nanomaterials. Diatoms are unicellular microalgae made up of frustules, which are biomineralized silica cell walls. Diatom frustules have a 3D-porous micro-nanostructure with a strongly periodic and hierarchical morphology (pennate and centric). Mechanical defense, biological defenses, filtration, UV protection for DNA, and light harvesting optimization are some of the hypotheses regarding their natural functions (Aguirre et al. 2018, Fuhrmann et al. 2004, Losic et al. 2009).

The diatoms have several openings (pores and slits) from which molecules are constantly exchanged with the atmosphere.

The complex patterns and symmetries are unique to each species and are dictated by genetics (Pickett-Heaps et al. 1990). It has been determined which organic molecules are needed for the construction of these diatom walls. Under moderate physiological conditions, diatom silica walls reach a high degree of complexity and hierarchical structure. As a result, the biological processes that produce patterned biosilica are of interest to the rapidly growing field of nanotechnology. From their unique interactions with silaffins and silica, diatoms successfully process silica.

Diatom frustules have many benefits over comparable synthetic mesoporous silica compounds, such as MCM-4, including greater biocompatibility, lower toxicity, and ease of purification (Kröger et al. 2002). The optical and optoelectronic properties of diatom frustules are also fascinating (Fuhrmann et al. 2004, Mazumder et al. 2010). The richness of silanol (Si-OH) groups makes the surface of diatoms readily functionalize (also *in vivo*), enabling maximum use of structural nanopatterning capacity (Li et al. 2014).

Using diatoms to synthesize nanoparticles is a rapidly expanding research area with the goal of fully utilizing the novel properties of diatom frustules, as well as the tremendous ability of the cellular pathway for silica biomineralization, to produce novel functionalized nanomaterials for evolving applications in sensing, photonics, and drug delivery (Panwar and Dutta 2019). The inclusion of the xanthophyll pigment fucoxanthin is another feature of diatom frustule. Many reports have emphasized the successful function of fucoxanthin as a metal ion photo-reducing agent in the stabilization of silver nanoparticles. *In vitro* antimicrobial activity of these silver nanoparticles against *Bacillus stearothermophilus*, *Streptococcus mutans*, and *E. coli* was found (Jena et al. 2015) and their possible utilization in detecting dissolved ammonia in water samples was checked using optical chemosensing (Chetia et al. 2017). Natural silicification was used to produce functionalized biogenic silica (frustules) of 100–200 m size using *Coscinodiscus wailesii* (Li et al. 2014) and frustules were updated with murine monoclonal antibody UN1 using *Coscinodiscus concinnus* Wm (De Stefano et al. 2009). These biogenic silica with photoluminescence activity may be used in the application of immunosensors.

The extracts of *Cosinodiscus argus* and *Nitzschia soratensis* were used to create a multi-layered box series of biogenic silica (frustules) that were bio-mineralized with purified primary rabbit IgG. These functionalized frustules will then be used as a fluorophore-labeled donkey anti-rabbit IgG detection optical immunochip (Kamińska et al. 2017). *Pseudostaurosira trainori* extracts were used to make functionalized biogenic silica (frustules) that were bio-mineralized and integrated with Au nanoparticles. These 4–5 m frustules can be applied in a range of immunosensing functions as well (Zhen et al. 2016).

3. Plants

Plant components for nanoparticle manufacturing have a number of benefits, including ease of preparing, packaging, cost-effectiveness, accelerated growth, reproducible, robust components, environmental friendliness, the avoidance of harsh and poisonous chemicals, and zero environmental pollution (Lade and Shanware 2020). Plant-mediated silver nanoparticle synthesis outperforms chemical and physical approaches and can be readily increased in quantity for maximum production.

In 2003, a paper on plant-facilitated synthesis of silver nanoparticles using *Alfalfa* (Medicago sativa) was published, marking the beginning of plant-mediated nanotechnology. AgNPs have been synthesised from a variety of plant components, including the stem, root, bark, seed, fruit, peel, callus, flower, and leaves (Rajeshkumar and Bharath 2017).

Extracts of black and green tea waste were applied to make gold nanoparticles (AuNPs) and silver nanoparticles (AgNPs). The resulting nanoparticles were found to be stable and to be in the nanoscale range, with AuNPs measuring 10 nm and AgNPs measuring 30 nm (Onitsuka et al. 2019). The synthesized AgNPs had a size distribution ranging from 25 to 85 nm, with an average particle size of 53 nm, and a *C. sinensis* (red tea leaves) extract was also used to create monodispersed AgNPs.

Furthermore, it was discovered that the diameter of a large group of particles was more than 45 nm (Pluta et al. 2017). The starch content in some plants has also been attributed to reduction and formation of stable nanometals. *Rumex dentatus* (Toothed dock) plant extract not only reduced Ag^+ ions but also formed stabilized AgNPs. The particles were in the range of 5–30 nm, enveloped by a thin layer of organic substance, which is characteristic of AgNPs synthesized by plant extracts (El-Shahaby et al. 2013). To synthesise AuNPs, leaf extracts of *Syzygium cumini* (Java Plum) and *Catharanthus roseus* (Madagascar Periwinkle) were used, with the synthesis taking just 15 and 10 minutes, respectively (Lal and Nayak 2012). Plant extracts have also been discovered to have a shape-directing effect on metal nanoparticles, in addition to being easy to synthesise. Single crystalline gold nanotriangles were biosynthesized by changing the volume of *Aloe vera* extract and the carbonyl compounds contained in the extract (Chandran et al. 2006). Water soluble antioxidants, polyphenols (flavonoids), are abundant in plant leaf extract (Pandey and Rizvi 2009).

Green tea extract was recently used to test the cytotoxic and antibacterial effects of Ag^+ ions' reduction and AgNP biosynthesis. The formulation and stabilisation of AgNPs seemed to be mediated by polyphenolic compounds such as catechins found in the extract of *C. sinensis*, which serve as capping and reducing agents (Rolim et al. 2019). The hydroxyl and ketonic groups in phenolic compounds are responsible for metal binding and the chelate effect (Saxena et al. 2012).

Rutile TiO_2 NPs with sizes ranging from 20 to 45 nm are synthesized without the use of a surfactant, mould, or capping agent using the root extract of *Euphorbia heteradena* Jaub. The hydroxyl groups of phenolics present in root extract are liable for the reduction of $TiO(OH)_2$ and also act as capping ligands to the surfaces of TiO_2 NPs, according to FTIR spectroscopic analysis of the extract (Nasrollahzadeh and Sajadi 2015a). Another phenolic agent, rosmarinic acid, was used as a bioreductant for the rapid synthesis of polydispersed, spherical AgNPs in the *Coleus aromaticus* leaf extract (Vanaja et al. 2013). The strong nucleophilic property of the aromatic rings, rather than ideal chelating groups within the molecule, is linked to the chelating ability of phenolic compounds. Active oxygen molecules can scavenge directly from phenolic compounds like flavonoids.

Flavonoids have antioxidant properties, which means they donate electrons or hydrogen atoms to other molecules. Flavonoids and other biochemical agents, which are present in the *F. benghalensis* (Banyan) leaf extract, are not only the source of reducing the AuNPs but also provide the anti-agglomeration ability to nanoparticles (Singh and Jain 2014). Without the use of any additional reducing agents, leaf extracts of *Withania coagulans* (Vegetable rennet) (Atarod et al. 2016) and *Tabebuia berteroi* (Pink trumpet flower) (Vellaichamy and Periakaruppan 2016) reduced Fe^{3+}, Pd^{2+}, and Ag^+ ions to nanometals. The role of flavonoids in the leaf extracts is confirmed by

FTIR analysis, which not only reduce metallic ions but also serve as stabilizing and capping agents. The possible mechanism of palladium nanoparticle (PdNP) formation is given as follows (Atarod et al. 2016) -

$nFlOH + Pd^{+2} \rightarrow nFlO$ (radical) $+ nPd^0$

$nFlO$ (Radical) $+ Pd^{+2} \rightarrow nFlOX + nPd^0$ (Nucleation)

$nPd^0 + Pd^{+2} \rightarrow Pd_n^{+2}$ (Growth)

$Pd_n^{+2} + Pd_n^{+2} \rightarrow Pd_{2n}^{+2n}$

$(Pd_{2n}^{+2n})_n + (FlOH)_n \rightarrow$ Palladium NZV

Flavonol antioxidants such as epicatechin, catechin, and their polymeric derivatives were discovered to reduce Pd^{2+} and Cu^{2+} ions in *Theobroma cocoa* L. seed extract, via the hydrogen donation potential of phenolic antioxidants during a radical mechanism. This culminated in the green synthesis of stable Pd/CuO NPs in aqueous media. The particles were 40 nm in size on average, with a thin layer of phytochemicals on the nanocomposite that avoids particle agglomeration (Nasrollahzadeh et al. 2015). *Ambrosia maritima* (Coastal Ragweed) aqueous leaves' extract was used to biosynthesize AgNPs by reducing silver nitrate (AgNO3).

The transmission electron microscope was used to classify the biosynthesized AgNPs, which revealed they were circular in shape and ranged in size from 25 to 50 nm. The presence of a secondary metabolite, sesquiterpene lactone, and flavonoids in the synthesis and stabilization of these AgNPs is suggested by FTIR study of the extract (El-Kemary et al. 2016). Under surfactant-free conditions, *Ginkgo biloba* L. (Maidenhair tree) leaf extract was used as a stabilizing and reducing agent in the green synthesis of Cu nanoparticles (CuNPs) in the 15 to 20 nm size range. According to an FTIR analysis, the presence of flavonoid and other phenolics in the crude extract of *Ginkgo biloba* L., especially quercetin, may be responsible for metal ion reduction and the formation of metal nanoparticles (Nasrollahzadeh and Sajadi 2015b). The leaf extract of *Psidium guajava* (Common guava) was used to generate spherical and stable silver nanoparticles in less than 10 minutes. The extract's organic molecule content was utilized as a green AgNP capping and reducing agent. Quercetin-3-O—Dxylopyranosid Rutin, isoquercitrin, avicularin, quercetin-3-O—L-arabinoside, quercitrin, quercetin, and kaempferol were included in HPLC study, which may be responsible for the Ag$^+$ ions reduction. Synthesized AgNPs had a typical particle size of 15–20 nm and had excellent antibacterial efficacy (Wang et al. 2018). To make AgNPs, researchers used an aqueous extract of Lagerstroemia speciosa (Pride of India) leaves. We were able to obtain spherical AgNPs with an average size of 12 nm and strong antibacterial action.

FTIR research was used to look at the potential biomolecules involved in reduction and synthesis. The spectrum showed peak at C-X stretching, which is a distinctive trait for alkyl halide functional groups, which may act as capping and reducing agents for the creation of silver nanoparticle (Saraswathi et al. 2017). The electrostatic exchanges between proteins in plant material extract and silver ions have also been suggested as a mechanism for Ag bioreduction.

Proteins reduce Ag$^+$ ions, causing secondary formation changes and the formation of Ag nuclei. Silver nuclei are formed as Ag$^+$ ions are reduced and accumulate at the nuclei, resulting in the creation of AgNPs (Rajeshkumar and Bharath 2017). Within 72 hours, silver ions were reduced in the peels of *Punica granatum* L. fruit, yielding AgNPs with a scale of 5–50 nm. The peel extract's proteins and amino acids formed a coating around the AgNPs, resulting in stable nanoparticles (Al-Othman Monira et al. 2017). The amino and carbonyl groups in amino acids' main chains can bind to metal ions, as can side chains like the nitrogen atoms in histidine's imidazole ring or the carboxyl groups in glutamic and aspartic acid.

Thioethers, thiols, and hydroxyl groups are examples of metal ion binding side chains. The synthesis of AgNPs and AuNPs has been linked to tyrosine residues in peptides or proteins (El-Seedi et al. 2019). AgNPs of 40–50 nm size with strong antibacterial activity against skin resident and transient flora were obtained using *Phoenix sylvestris* L. (Indian date) seed extract to minimize and stabilize Ag^+ ions. The O–H group of phenols, amide groups, and carbonyl groups of proteins present in the seed extract formed a layer of the NPs and serve as potential capping agents to avoid nanometal aggregation and provide stabilization, according to the FTIR range of synthesized NPs (Qidwai et al. 2018). At room temperature, mixing an aqueous mixture of selenium acid (H_2SeO_3) with *C. annuum* extract resulted in the development of highly stable rod-like, quasi-spherical, and spherical selenium/protein composites with dimensions of 100 to 300 nm enclosed by a dense protein coating. Electrophoretic analysis of the crude extract and the protein extracted from the nanocomposites reported a similar 30 kDa protein which is thought to have tyrosine, tryptophan, and phenylalanine residues and might have resulted in the reduction of SeO_3^{2-} ions to Se^0. SDS-PAGE, Cyclic voltammogram (CV), Differential Pulse Voltammetry (DPV) and HPLC results point towards the presence of protein residues such as tyrosine, tryptophan, and phenylalanine and Vitamin C in *C. annuum* extract, which act as stabilizing and reducing agents and convert SeO_3^{2-} ions to Se^0 (Li et al. 2007).

Silver and gold nanoparticles with catalytic activity were generated using aqueous stem extract of *Angelica gigas* (Korean angelica) by green chemistry technique. Field emission gun transmission electron microscope (FEG-TEM), X-ray diffraction (XRD) and particle size analyzer confirmed the crystalline and spherical nature of AgNPs and AgNPs with sizes varying from 40 to 300 nm. FTIR analysis attributed the reduction of metal ions and subsequent stabilization of nanometals to the presence of polypeptides in the extract (Chokkalingam et al. 2019).

The following steps are used in the common procedure for plant-mediated AgNP synthesis: plant materials are collected from different plant sites and washed with detergents before being thoroughly soaked in double distilled water for 2–3 cycles. Farm content that has been washed and dried should be allowed to air dry at room temperature.

Plant materials are weighed and boiled for 10–15 minutes with 100 ml deionized distilled water and subsequently cooled. A nylon mesh cloth was used to filter the solution and held at 4°C. The collected filtrate was then dissolved in an aqueous solution of $AgNO_3$ (1 mM) and stored at room temperature. The production of AgNPs by the interaction of silver metal ions and of plant extract is shown by subsequent colour variations in the reaction mixture. The development of AgNPs can also be verified using a UV-visible spectrophotometer. Metal ion concentration, reaction mixture pH, extract contents reaction, temperature, reaction length, and agitation are both chemical and physical restrictions that influence AgNPs synthesis. The morphology, form, and scale of AgNPs are influenced by factors like metal ion concentration, extract composition, and reaction time (Rajeshkumar and Bharath 2017)

4. General method of biosynthesis of nanoparticles

4.1 Using plants

The metal solution (1 mM) and the plant extract are combined in a particular ratio (Eg 95:5 ml). Microwave oven, autoclave, sonication, heating, boiling, and other techniques can also be used to carry out the reaction. The kinetics of the reaction, which is dependent on which methods were used to carry out the reaction, determines the shape, size, and form or type of nanoparticles produced. A single sound wave, for example, can have a different effect on nanoparticle synthesis than light intensity or colour. The impact of sunlight on the nanoparticle size and shape is unique.

4.2 Using bacterial cells

The bacteria are grown in nutrient broth in an Erlenmeyer flask for 12 hours at 37°C under shaker conditions (150–200 rpm). The culture is combined with the metal salt during the incubation phase. To provide proper stability to the structure, 5–10 g of bacterial mass is needed for the metal ion reduction and capping of synthesised nanomaterials. The biomass can associate with the metal salt over the 120-hour incubation period. The extracellular synthesis of nanomaterials is investigated by looking at the colour shift in the test tube contents.

Following that, the contents of the test tube are centrifuged to isolate the bacterial mass for nanomaterial suspension.

5. Processes of MtNP synthesis by microorganisms

For the biosynthesis of nanomaterials, microbial cells have been extensively used. Microbes such as fungi, bacteria, yeasts, viruses, and actinomycetes can be used as bio factories to reduce metals such as silver, selenium, copper, cadmium, magnetite, gold-silver alloy, palladium, platinum, silica, and various other metals to nanoparticles for use in biological uses (Narayanan and Sakthivel 2010). Despite the fact that a wide range of microbial organisms are capable of forming metal nanostructures, the process of nanoparticle biosynthesis remains unknown.

The catalytic mechanism of the microorganism's isoenzymes can play a role in the biosynthesis of nanoparticles using microorganisms to some degree (He et al. 2008, Ahmed et al. 2003a). Enzymes secreted by microorganisms are responsible for the creation of metal nanostructures, as well as electron transfer to redirect the electron of a reducible agent (such as reductase, reducing sugar, electron donor, etc.) to metal ions. Microbes synthesise these nanoparticles using enzymes formed either extracellularly or intracellularly as a result of a number of bioreduction processes (Figure 3).

5.1 Extracellular enzymes

Extracellular microbial enzymes perform an essential role in the biological synthesis of metal nanoparticles as a reducing agent (Subbaiya et al. 2017). Cofactors such as Nicotinamide Adenine Dinucleotide (NADH) and the reduced form of Nicotinamide Adenine Dinucleotide Phosphate (NADPH) related enzymes also serve as reducing agents by passing electrons from NADH to NADH-dependent enzymes that function as electron carriers, according to research (Bose and Chatterjee 2016). The extracellular biosynthesis of AuNPs by *Rhodopseudomonas capsulata* is aided by the secretion of NADH and NADH-dependent enzymes. *R. capsulate* uses NADH-dependent reductase enzymes to transfer electrons from NADH to perform gold biological reduction of $AuCl_2$. Gold ions accept electrons, reducing Au^{3+} to Au^0 and resulting in the formation of gold nanoparticles (Dhandapani et al. 2017).

A variety of other considerations, such as the precursor concentration, pH, temperature, and reaction time, all play a role in limiting the size of MtNPs. In addition to these enzymes, a host of molecules such as anthraquinones, naphthoquinones, and hydroquinones are involved in the synthesis of MtNPs (Patra et al. 2014). Microorganisms have used variations in biosorption, solubility, extracellular accumulation, metal complexation, toxicity by oxidation-reduction pathways, and the lack of special transporters in the biosynthesis of NPs. Extracellular enzymes like cellobiohydrolase D, acetyl xylan esterase, Beta-glucosidase, and glucosidase are produced by many fungi and perform vital role in the biosynthesis of MtNPs (Ovais et al. 2018b). The fungus-secreted enzyme nitrate reductase aids in the biological reduction and synthesis of MtNPs.

Table 1. Conditions and methods used for synthesis of nanoparticles using microbes and plants.

Sr. No.	Method	Conditions	Other remarks	References
1.	Dark Conditions	Light may oxidize metal (photo-leaching)	Nanoparticles were found to be more stable than particles synthesised with UV radiation.	Lade and Patil 2017, Lade 2017, Li et al. 2011, Rajput et al. 2020
2.	Sunlight	Varying wavelength results in induction or activation of enzymes, functional group of the phytochemicals or secondary metabolites found in the biological extract.	Yields best results, Produces NP quickly	Lade and Patil 2017, Lade 2017
3.	Microwave	Capping agent (reagent) and metallic salts are mixed, microwaved for short time intervals of 20 s for 5 min.	Affects the shape and size NP	Lade 2017, Lade and Shanware 2020
4.	Autoclave	Preheat for 5 min 5–10 psi, flask containing the mixture is kept in the autoclave for 5 mins	The pressure increase is essential for synthesis of nanoparticles where specific pressure and temperature is required	Lade 2017, Lade and Shanware 2020
5.	Sonication	The reaction mixture is exposed to ultrasonic frequencies (> 20 kHz) for a specific period. Can be used for dispensing nanoparticles in liquid solution.	Varying kHz results in formation of varied size and shape of NP	Lade 2017, Lade and Shanware 2020
6.	Heating/ Boiling	The reaction mixture of plant extract and metallic salt is boiled on a burner at a specific temperature (50–100°C) for some time till the solution boils. Cool before use	A colour change in the solution indicates the presence of metal nanoparticles. The colour change may take a few hours or a day depending on the capping and reducing agent involved in the process	Lade 2017, Lade and Shanware 2020, Rajput et al. 2020
7.	Light colour	Blue, green, yellow, red, and orange of 15 W	Light wavelength and colour can be altered for making various shapes/sizes of AgNPs	Lade 2017, Lade and Shanware 2020

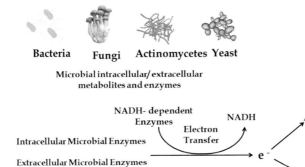

Figure 3. Synthesis of nanoparticles by microbial enzymes and metabolites.

AgNPs are synthesised outside of the cell. Many studies have shown the contribution of nitrate reductase in the extracellular processing of MtNPs (Kumar et al. 2007a,b). Using commercially available nitrate reductase discs, researchers discovered that these NADH-dependent reductase enzymes were involved in the reduction of Ag^+ ions to $Ag(0)$ and the formation of silver nanoparticles (Ingle et al. 2008, Durán et al. 2005). The reduction of gold and silver NPs was carried out using *Fusarium oxysporum.* According to reports, extracellular reductases formed by fungi reduced Au^{3+} and Ag^{1+} to Au–Ag NPs. In addition, shuttle quinone and nitrate-dependent reductases generated by *Fusarium oxysporum* species were utilized in the synthesis of NPs extracellularly (Senapati et al. 2005). Despite the existence of reductase, some plants, such as *Fusarium moniliforme*, were unable to generate AgNPs, meaning that reduction of Ag^{1+} occurs via coupled oxidation-reduction reaction of electron carriers including NADP-dependent nitrate reductase (Durán et al. 2005). In an *in vitro* analysis, *F. oxysporum* nitrate reductase was used to synthesise AgNPs in the absence of oxygen and in the presence of a cofactor (NADPH), a stabiliser protein (phytochelatin), and an electron carrier (4-hydroxyquinoline). Outside of the cell, this fungus developed a significant number of AgNPs (Kumar et al. 2007a, Karbasian et al. 2008). *F. oxysporum*, on the other hand, was shown to be capable of synthesis of semiconductor CdS nanoparticles, extracellularly.

The reductase enzyme obtained from *F. oxysporum* was used to biosynthesize highly luminescent CdSe nanoparticles in this analysis (Kumar et al. 2007b, Ahmad et al. 2002). Other fungal strain's enzymes, such as *Fusarium solani* and *Fusarium semitectum*, were also used in the extracellular processing of AgNPs.

According to the results, specific proteins may be responsible for the reduction of Ag^+, resulting in the production of AgNPs (Basavaraja et al. 2008, Ingle et al. 2009). *Cladosporium cladosporioides* and *Coriolus versicolor* synthesised AgNPs outside of the cell using fungal organic acids, polysaccharides, and proteins. These fungi were found to be capable of synthesizing AgNPs, and the organic acids, polysaccharides, and proteins used in the process influenced the growth and form of the nanocrystals (Balaji et al. 2009). The extracellular biosynthesis of AgNPs was stabilised by fungal proteins after growing *Aspergillus niger* in an AgNO3 solution (Gade et al. 2008). In contrast to other chemical and physical techniques used in the synthesis process, *Aspergillus fumigatus* extracellularly synthesized AgNPs in a fraction of the time (10 mins) (Bhainsa and D'souza 2006). As a result, *A. fumigatus* was discovered to be an excellent applicant for large-scale development of a range of NPs.

In addition, *Penicillium fellutanum* reduced Ag^{1+} ions in a relatively small period of time (10 min). Additional tests showed that the reduction Ag^{1+} ions were due to an enzyme, nitrate reductase (Kathiresan et al. 2009). The release of NADH-dependent enzyme nitrate reductases by *Penicillium brevicompactum* carried out the Ag^{1+} ions reduction (Shaligram et al. 2009). *Sargassum wightii greville*, one of the algae known to biosynthesize NPs, was analysed for its ability to rapidly reduce Au^{3+} ions to form AuNPs having a size of 8–12 nm (Singaravelu et al. 2007). *Chlorella vulgaris*, a filamentous algae, was utilized in the processing of Au nanoparticles, resulting in Au and Au^{+1} nanoparticles (Lengke et al. 2006).

5.2 Intracellular enzymes

Bacteria and fungi, as well as sugar molecules, perform an essential role in the intracellular metal removal mechanism. Following the reduction within the cell, the associations between positively charged groups and intracellular enzymes are used to capture metallic ions from the medium (Dauthal and Mukhopadhyay 2016, Thakkar et al. 2010). The accumulation of MtNPs in the periplasmic region, the cell wall, and the cytoplasmic membrane has been observed.

This may be attributed to metal ion flow through membranes and reduction of the metal using enzymatic processes, which results in the development of MtNPs.

For the intracellular synthesis of AuNPs, alkalo-tolerant (*Rhodococcus* sp.) and alkalo-thermophilic (*Thermomonospora* sp.) actinomycetes were used (Ahmad et al. 2003a,b) for the formation of AuNPs having uniform dimensions. Enzymes found on the mycelia's surface and in the cytoplasmic membrane aided in the reduction of Au^{3+}.

Intracellular reduction and the development of AgNPs is carried out by *Verticillium* biomass.

AgNPs is produced underneath the surface of the cell wall due to bioreduction enzymatically, which is not harmful to the fungi, according to electron microscopic pictures (Mukherjee et al. 2001a). The biosynthesis of AuNPs was carried out in a related manner, with the reducing enzymes coming from the fungus *Verticillium*.

AuNPs were discovered entrapped inside the fungi's cytoplasmic membrane and cell wall, suggesting that Au^{3+} reduction was carried out by reductase enzymes present in that area of the cell (Mukherjee et al. 2001b). AuNPs are precipitated within bacterial cells following incubation with Au3+ ionic solution, according to reports (Southam and Beveridge 1996). When *Pseudomonas stutzeri* (AG259) was subjected to a AgNO3 solution of high concentration, reduction of Ag^{1+} ions took place, and AgNPs were synthesised in the bacterial periplasmic region (Klaus et al. 1999). *Plectonema boryanum*, a filamentous cyanobacterium, was exposed to Au(S2O3)2 3- and AuCl4-solutions, AuNPs were formed at the cell membrane area and gold sulphide was found intracellularly (Lengke et al. 2006). Incubation of *Phanerochaete chrysosporium* in an ionic Au^{3+} solution formed AuNPs having a particle size of 10–100 nm. Laccase and ligninase enzymes were used as extracellular and intracellular reducing agents to reduce Au^{3+} ions, respectively (Sanghi et al. 2011). Other parameters, such as the fungus' incubation age, the incubation temperature, and the concentration of the AuCl4-solution, have shown to have a direct impact on the form of AuNPs. *Shewanella* algae, a mesophilic bacterium, has also shown to be an effective biological reducer of $AuCl^{4-}$ ions to elemental gold. The presence of Au nanoparticles in the periplasmic space of bacteria was discovered thanks to intracellular enzymes (Konishi et al. 2007). *Brevibacterium casei* reduced aqueous mixtures of Au3+ and Ag+ ions to spherical AuNPs and AgNPs, by using intracellular enzymes (Kalishwaralal et al. 2010).

6. Biomolecules involved in nanoparticles synthesis

The procedure for controlling the size of nanoparticles is intriguing in its simplicity.

One of the appropriate pathways is the role of stress proteins such as Glutathione (GSH), PhytoChelatins (PCs), and MetalloThioneins (MTs), as well as superoxide dismutases, catalases, and anti-oxidants such as Vitamin E and Vitamin C. Animals, fungi, algae, some prokaryotes, and in some cases, plants, all contain GSH and metallothioneins (Yadav 2010). In the presence of metals, these molecules are triggered. The introduction of heavy metals into the environment has serious ramifications. In the presence of heavy metals, higher organisms develop cysteine-rich peptides such as PCs, GSH, and MTs, which bind metal ions (such as lead, cadmium, copper, and mercury) and sequester them in biologically inactive ways (Yadav 2010).

Phytochelatins (PCs) are small cysteine-rich polypeptides with the general form (g-Glu-Cys) nGly (Mirza et al. 2014). Phytochelatins are among the most potent metal chelators, capable of neutralising the negative effects of a wide variety of dangerous metals. In the case of heavy metal (HM) toxicity, PCs are synthesised in the cytosol. PC–metal and PC–metalloid complexes are extremely stable in nature, and they shape and sequester themselves in vacuolar compartments where metal toxicity is minimal (Dago et al. 2014, Ray and Williams 2011, Shen et al. 2010). Since phytochelatin (PC) and other metal-binding proteins can bind heavy metals, they have been used to extract them. Phytochelatins are proteins that can form metal complexes with cadmium, copper, silver, lead, and mercury, while metallothioneins are gene-encoded proteins that can bind metals like

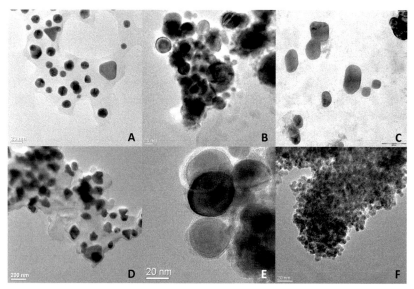

Figure 4. TEM images of biosynthesized nanoparticles using bacteria and plants (Durve 2014).

[A] Gold nanoparticles synthesized by Saffron (*Crocus sativus*) extract.
[B] CdS nanoparticles synthesized by *Pseudomonas aeruginosa* sp.
[C] As nanoparticles synthesized by *Pseudomonas aeruginosa* sp.
[D] Silver nanoparticles synthesized by Green tea (*Camellia sinensis*) leaves extract.
[E] Hg nanoparticles synthesized by *Pseudomonas aeruginosa* sp.
[F] Pb nanoparticles synthesized by *Pseudomonas aeruginosa* sp.

copper, cadmium, and zinc. MTs can aid in (a) maintaining the homeostasis of vital transformation HMs, (b) sequestration of toxic HMs, and (c) defence against intracellular oxidative disruption, despite the fact that their precise physiological function is unknown (Xu et al. 2011). Park et al. (2010) reported on the *in vivo* synthesis of different NPs using recombinant *E. coli* expressing *Arabidopsis thaliana* PCS (AtPCS) and/or *Pseudomonas putida* MT (PpMT). Semiconducting (Cd, Se, Zn, Te), alkali-earth (Cs, Sr), magnetic (Fe, Co, Ni, Mn), noble (Au, Ag), and rare-earth fluorides (Pr, Gd) metals were incubated in different variations of recombinant *E. coli* cells for *in vivo* synthesis. The resulting NPs' optical, magnetic, and physicochemical properties were studied.

The established SeZn and CdTe NPs were of various shapes and sizes, demonstrating that PC (synthesised by AtPCS) and PpMT expressed in cells of *E. coli* could synthesise various semiconducting nanocrystallites *in vivo*. The d lattice structure and size of NPs synthesised *in vivo* by MT and PC vary, meaning that PC and MT have distinct metal binding and assembly properties (Hirata et al. 2005, Cobbett et al. 2002, Robinson et al. 2001, Grill et al. 1985). The binding affinity of MT for Cu was found to be higher than that for Cd or Zn (Coyle et al. 2002). In PC and MT, one Cd ion is seen bound to two and three cysteine residues, respectively (Hirata et al. 2005, Grill et al. 1985). It is possible that coexpression of PpMT and AtPCS in *E. coli* would allow for the synergistic biosynthesis of more diverse metal NPs. Cells that coexpressed or expressed both AtPCS and PpMT simultaneously accumulated NPs. The NPs CdSe, CdZn, SeZn, and CdTe all had well-defined crystalline structures with interplanar lattice lengths, which is a feature of these NPs. Park et al. (2010) showed that coexpression of AtPCS and PpMT results in a wide range of metal NPs. Simultaneous expression of MT and PCS in *E. coli* strengthened the assembly of different metal elements into closely arranged NPs. The size of the metal NPs could be monitored by adapting the concentrations of the supplied metal ions. Semiconducting (Cd, Se, Zn, Te), alkali-earth (Cs, Sr), magnetic (Fe, Co, Ni, Mn), noble (Au, Ag), and rare-earth fluorides (Pr, Gd) metals were incubated in recombinant *E. coli* cells in various variations for *in vivo* synthesis. The optical, magnetic, and physicochemical properties of the resulting NPs were investigated (Toppi and Gabbrielli 1999). PCs are also unable to metabolise or remove Cd. They have been observed to form complexes of Cd-GSH and Cd-PCs in order to effectively sequester Cd within vacuoles (DalCorso et al. 2008, Toppi and Gabbrielli 1999) and to help in Cd transportation via phloem and xylem vessels (Mendoza-Cózatl et al. 2008). The fact that PC growth was related to HM accumulation in both below-ground and above-ground tissues provided further support for PC induction in response to HM stress.

According to study, PCs participate in HM transport (Salt and Rauser 1995), meaning that their detoxifying abilities are secondary or part of a broader process. Despite the fact that PCs have been shown to play a role in HM detoxification and aggregation in higher plants, the development of HM complexes is insufficient to explain the HM or species specificity of hyperaccumulation (Baker et al. 2000). As a result, it is still unclear what role PCs play at the cellular level during the HM-tolerance process, necessitating further investigation.

7. Role of Excreted electron shuttlers in metal ion reduction

The mechanism by which microbes pass electrons to minerals that are poorly soluble has been the focus of extensive research (Turick et al. 2002). Our understanding of cellular transfer derives from a thorough examination of photosynthesis and respiration in both eukaryotic and prokaryotic systems. The structure and role of different membrane-bound proteins involved in electron transport processes is now well understood (Gray and Winkler 1996). During nanoparticle synthesis, electrons may be transferred through low molecular weight redox mediators such as ubiquinol, NADH, or oxygen/superoxide, or by direct interaction between c-type cytochromes redox proteins and the metal ion (Bewley et al. 2013).

In this mode of electron transfer, tiny mobile molecules adept at undergoing redox cycling are used (i.e., an electron shuttle). These molecules serve as terminal electron acceptors, passing electrons from metal-to-metal oxide, which is then reoxidized. In theory, a single shuttle molecule could loop thousands of times and thus have a significant effect on the turnover of the terminal oxidant such as iron. Organic molecules that play an important role as electron shuttlers include humic compounds, quinones, phenazine, and thiol-containing molecules such as cysteine, but anything that is redox-active and has the appropriate redox potential can also be used (Madigan et al. 2000). Several studies have been performed to examine the role of cytochromes and redox mediators in metal nanoparticle extracellular synthesis.

During anaerobic respiration, *Shewanella oneidensis* MR-1 is found to use ferric oxide minerals as terminal electron acceptors, and c-type cytochromes are believed to be active in this electron transfer pathway. Metal-reducing bacteria *S. oneidensis* MR-1 and associated strains *Shewanella* also developed a metal-reducing machinery, or Mtr pathway, for transporting electrons across cell membranes to the surface. The role of c-type cytochromes in moving electrons from quinol at the inner membrane (IM), periplasmic space (PS), and outer membrane (OM) to the metal oxide surface (Fe(III) oxide) has been suggested as a molecular mechanism. CymA, MtrA, MtrB, MtrC, and OmcA are some of the protein components involved in the Mtr pathway.

Outside the bacterial cell membrane, a related mechanism for bacterial metal nanoparticle synthesis has been proposed. CymA is a quinol dehydrogenase from the NapC/NrfH family of inner-membrane tetraheme c-type cytochromes (c-Cyt). The theory is that CymA oxidises the quinol in the inner membrane and passes electrons to MtrA either directly or indirectly through other periplasmic proteins. MtrA is a decaheme c-Cyt found in MtrB, which is a porin-like protein with a trans outer membrane. MtrA transfers electrons from the outer membrane to MtrC and OmcA on the outermost surface, along with MtrB. MtrC and OmcA, two outer-membrane decaheme c-Cyt, are transported from the outer-membrane to the cell surface through the bacterial type II secretion pathway. OmcA and MtrC are terminal reducing agents that can bind to the surface of Fe(III) oxides and directly transfer electrons through the exposed hemes component of these oxides. Flavins secreted by

S. oneidensis MR-1 cells can be used as diffusible factors by OmcA and MtrC to accelerate Fe(III) oxide reduction. OmcA and MtrC, due to their broad redox potentials and extracellular location, can also serve as terminal reducing agents for soluble Fe (III) (Shi et al. 2012).

The molecular structure of *Desulfovibrio vulgaris* NrfH has been determined. According to this model, quinol binds inside the pocket adjacent to heme 1 of *D. vulgaris* NrfH, where it is oxidised (Rodrigues et al. 2006, 2008). Later, the first molecular model of electron transfer across the bacterial outer-membrane was developed (Hartshorne et al. 2009).

As shown in Figure 5, MtrB is a trans outer-membrane spanning-barrel protein that serves as a pocket for MtrA to enter the membrane. MtrAB is an extracellular reductase that acts as a trans outer-membrane electron transport module. MtrA contains a signal peptide that, through the bacterial secretary mechanism, transports the synthesised polypeptide to the periplasmic space. MtrA polypeptides are grouped into two pentaheme domains, each with a sequence similar to that of NrfB from *E. coli*. Since truncated MtrA is expressed in *E. coli* with just one of its pentaheme domains, it folds correctly and has five hemes, indicating that MtrA has two repeated functional domains (Clarke et al. 2008). The electron transfer from the NrfB heme groups that form a molecular wire is facilitated by this sort of heme structure.

In a sample, the molecular structure of MtrF, a MtrC homolog, was determined. The same study discovered molecular structural proof in favour of a terminal reducing agent for *S. oneidensis* MR-1 outer-membrane c-Cyt in Fe(III) oxide reduction (Clarke et al. 2011). A recent study was published to better understand the function of c-type cytochromes in nanoparticle extracellular synthesis. A mutant strain of *Shewanella oneidensis* lacking cytochrome genes (MtrC and OmcA) was used for AgNP biosynthesis.

The mutant strain of *S. oneidensis* produced nanoparticles that were smaller in size and number as compared to the wild-type strain. This indicates that c-type cytochromes aid in electron transfer to extracellular metal ions (Ng et al. 2013). The outer membrane c-type cytochrome protein complexes (ombB, omaB, and omcB) are specifically involved in the extracellular reduction of Fe(III)-citrate and ferrihydrite in a metal-reducing bacterium. The *Geobacter sulfurreducens* PCA strain has been identified (Liu et al. 2014, 2015). Another study on *Geobacter sulfurreducens* AgNP synthesis discovered a similar mechanism of electron transfer through c-cytochromes.

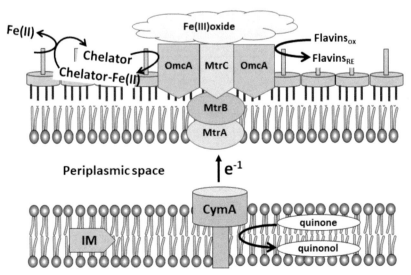

Figure 5. The proposed Metal-reducing (Mtr) extracellular electron transfer pathway of *S. oneidensis* MR-1 which is similar to bacterial metal nanoparticles' synthesis outside the bacterial cell surface (Gahlawat and Choudhury 2019).

Electrons produced during this process were transported through the periplasmic space by Fe2+/hemes of c-cytochromes (Ppc) to Fe3+/hemes of the outer membrane cytochromes (Omc). As a result, electrons were transferred from Omc's protein centre to surface-attached Ag+ ions in order to synthesize AgNPs, while cytochromes were oxidised back to Fe3+/hemes (Vasylevskyi et al. 2017). According to the findings of these studies, multiheme complexes facilitate the transfer of electrons from the inner membrane to the cell wall (or outer membrane) through periplasm, resulting in the extracellular reduction of metal ions into nanoparticles.

Table 2. Applications of nanoparticles in various fields.

Sr. No.	Application	Advantage	References
1.	Antibacterial activity	Biosynthesized AgNPs inhibit nosocomial, bacterial and pathogenic bacteria *viz. Shigella dysentriae, Salmonella infestis* and *Vibrio parahaemolyticus*	(Krithiga et al. 2015, Singh et al. 2016)
2.	Reduction of organic pollutants	AgNPs exhibit effective degradation of organic pollutant 4-NP	(Atarod et al. 2016, Vellaichamy and Periakaruppan 2016)
3.	Cosmetics	AgNPs exhibit good antibacterial activity against acne causing organisms	(Badnore et al. 2019, Qidwai et al. 2018)
4.	Photodynamic therapy	Exposure of AuNPs containing tumour to light at 700-800nm wavelengths swiftly warms up the particles and annihilates the cancerous cells	(Katas et al. 2018)
5.	Drug-delivery agent	AuNPs have huge surface area which helps adsorption of a variety of molecules such as therapeutic agents, polymers, and aiming ligands on their surface	(Katas et al. 2018)
6.	Diagnostics	AuNPs are utilized to identify biomarkers of numerous diseases, along with tumours and contagious agents and even used in home pregnancy detection tests	(Katas et al. 2018)
7.	Sensing	Organic vapours like acetone methanol have been successfully detected using thin gold nanotriangles	(Ankamwar et al. 2005)
8.	Detection	AgNPs have been successfully employed in detecting cysteine in biological samples	(Shen et al. 2016)
9.	Removal of petroleum contaminants	Iron nanoparticles efficiently reduced TPHs from contaminated water and soil samples and in less time as compared to conventional bioremediation process	(Murgueitio et al. 2018)
10.	Photodegradation of dyes	Efficient degradation of methyl red and congo red has been observed with the application of TiO_2 nanoparticles	(Rostami-Vartooni et al. 2016)
11.	Wound healing	AgNPs impregnated in cotton fabric have shown effective antibacterial activity against a range of pathogens	(Nath et al. 2019, Pannerselvam et al. 2017, Shaheen and Abd El Aty 2018, Xu et al. 2017)
12.	PTC	AgNPs provided good protection against contamination as well as improved growth when applied to temporary immersion systems	(Spinoso-Castillo et al. 2017)
13.	Water purification	Small and efficient water purification devices have been developed with the incorporation of AgNPs	(Ghodake et al. 2020)
14.	Agriculture	SiO_2 and AuNPs influenced the growth of lettuce seeds	(Shah and Belozerova 2009)

8. Conclusion

The use of biological agents in nanomaterial synthesis is an environmentally safe, cost-effective, and fast way of producing nanomaterials. Nanomaterials' peculiar properties make them attractive for a wide range of nanotechnology applications. Because of their high stability and biocompatibility, nanomaterials' synthesis using biosynthesis is very unique. It's crucial to understand the various forms of active groups that can be found in nature and bind to the surface of nanoparticles. The existence of these classes, as well as several other physical factors, can influence the scale, form, and morphology of nanoparticles. Using these criteria for biological nanoparticle synthesis will aid in studying the unexplained information of nanoparticle synthesis regulation at the gene stage. In contrast to chemical and physical processes, biological synthesis of nanoparticles is a sluggish process that can take several hours to several days depending on the biological agent. The biosynthesis process would be more appealing if the synthesis time is reduced. The size and monodispersity of the nanoparticle produced are the two most important factors in evaluating nanoparticle synthesis. According to reports, the nanoparticles produced by biological entities decompose after a period of time. As a result, thorough research is needed to fully comprehend these problems. The monodispersity and particle size could be regulated by changing parameters such as microbial cell growth stage (phase), biological entity form, synthesis conditions, pH, growth medium, source compound of target nanoparticle, substrate concentrations, reaction time, temperature, and addition of nontarget ions. Biosynthesis approaches are also advantageous because nanoparticles can be covered with a lipid coating, which provides durability and physiological solubility, which is vital in biomedical functions. There is currently research underway to help pinpoint the compounds that are needed and play a key role in nanoparticle synthesis. This would aid in the development of methods with a short reaction time and a high synthesis efficiency rate.

References

Abdel-Raouf, N., Al-Enazi, N.M. and Ibraheem, I.B.M. 2017. Green biosynthesis of gold nanoparticles using *Galaxaura elongata* and characterization of their antibacterial activity. *Arab. J. Chem.*, 10: S3029–S3039. doi: 10.1016/j.arabjc.2013.11.044.

Aguirre, L.E., Ouyang, L., Elfwing, A., Hedblom, M. and Wulff, A. 2018. Diatom frustules protect DNA from ultraviolet light. doi: 10.1038/s41598-018-21810-2.

Ahari, H. 2017. The use of innovative nano emulsions and nano-silver composites packaging for anti-bacterial properties: An article review. *Iran. J. Aquat. Anim. Heal.*, 3(1): 61–73. doi: 10.18869/acadpub.ijaah.3.1.61.

Ahmad, A., Mukherjee, P., Mandal, D., Senapati, S., Khan, M.I., Kumar, R. and Sastry, M. 2002. Enzyme mediated extracellular synthesis of CdS nanoparticles by the fungus, *Fusarium oxysporum. J. Am. Chem. Soc.*, 124: 12108–12109.

Ahmad, A., Senapati, S., Khan, M.I., Kumar, R., Ramani, R., Srinivas, V. and Sastry, M. 2003a. Intracellular synthesis of gold nanoparticles by a novel alkalotolerant actinomycete, *Rhodococcus* species. *Nanotechnology*, 14: 824.

Ahmad, A., Senapati, S., Khan, M.I., Kumar, R. and Sastry, M. 2003b. Extracellular biosynthesis of monodisperse gold nanoparticles by a novel extremophilic actinomycete, *Thermomonospora* sp. *Langmuir.*, 19: 3550–3553.

Ahmad, R., Minaeian, S., Shahverdi, H.R., Jamalifar, H. and Nohi, A. 2007. Rapid synthesis of silver nanoparticles using culture supernatants of *Enterobacteria*: A novel biological approach. *Process Bioche.*, 42: 919–923.

Albanese, A., Tang, P.S. and Chan, W.C. 2012. The effect of nanoparticle size, shape, and surface chemistry on biological systems. *Annu Rev Biomed Eng.* 14: 1–16. doi: 10.1146/annurev-bioeng-071811-150124.

Al-Othman Monira, R., Abd El-Aziz Abeer, R.M., Mahmoud Mohamed, A. and Hatamleh Ashraf, A. 2017. Green biosynthesis of silver nanoparticles using Pomegranate peel and inhibitory effects of the nanoparticles on aflatoxin production. *Pakistan J. Bot.*, 49(2): 751–756.

Ankamwar, B., Chaudhary, M. and Sastry, M. 2005. Gold nanotriangles biologically synthesized using tamarind leaf extract and potential application in vapor sensing. *Synth. React. Inorganic, Met. Nano-Metal Chem.*, 35(1): 19–26. doi: 10.1081/SIM-200047527.

Arockiya Aarthi Rajathi, F., Parthiban, C., Ganesh Kumar, V. and Anantharaman, P. 2012. Biosynthesis of antibacterial gold nanoparticles using brown alga, *Stoechospermum marginatum* (kützing). Spectrochim. *Acta - Part A Mol. Biomol. Spectrosc.*, 99: 166–173. doi: 10.1016/j.saa.2012.08.081.

Arsiya, F., Sayadi, M.H. and Sobhani, S. 2017. Green synthesis of palladium nanoparticles using *Chlorella vulgaris*. *Materials Letters Elsevier*, 186: 113–115. doi: 10.1016/j.matlet.2016.09.101.

Atarod, M., Nasrollahzadeh, M. and Mohammad Sajadi, S. 2016. Green synthesis of Pd/RGO/Fe3O4 nanocomposite using Withania coagulans leaf extract and its application as magnetically separable and reusable catalyst for the reduction of 4-nitrophenol. *J. Colloid Interface Sci.*, 465: 249–258, doi: 10.1016/j.jcis.2015.11.060.

Badnore, A.U., Sorde, K.I., Datir, K.A., Ananthanarayan, L., Pratap, A.P. and Pandit, A.B. 2019. Preparation of antibacterial peel-off facial mask formulation incorporating biosynthesized silver nanoparticles. *Appl. Nanosci.*, 9(2): 279–287. doi: 10.1007/s13204-018-0934-2.

Baker, A.J.M., McGrath, S.P., Reeves, R.D. and Smith, J.A.C. 2000. Metal hyperaccumulator plants: A review of the ecology and physiology of a biological resource for phytoremedation of metal polluted soils. pp. 85–107. *In*: Terry, N. and Banuelos, G. (eds.). *Phytoremediation of Contaminated Soil and Water*. Michael Lewis, Boca Raton, Fla, USA, 2000.

Bakir, E., Younis, N., Mohamed, M. and El Semary, N. 2018. Cyanobacteria as nanogold factories: Chemical and anti-myocardial infarction properties of gold nanoparticles synthesized by *Lyngbya majuscula*. *Mar. Drugs.*, 16: 217.

Balaji, D., Basavaraja, S., Deshpande, R., Mahesh, D.B., Prabhakar, B. and Venkataraman, A. 2009. Extracellular biosynthesis of functionalized silver nanoparticles by strains of *Cladosporium cladosporioides* fungus. *Colloids Surf. B.*, 68: 88–92.

Basavaraja, S., Balaji, S., Lagashetty, A., Rajasab, A. and Venkataraman, A. 2008. Extracellular biosynthesis of silver nanoparticles using the fungus *Fusarium semitectum. Mater. Res. Bull.*, 43: 1164–1170.

Beveridge, T.J. and Murray, R.G. 1980. Sites of metal deposition in the cell wall of *Bacillus subtilis. J Bacteriol.* 1980 Feb; 141(2): 876–87. doi: 10.1128/jb.141.2.876-887.1980. PMID: 6767692; PMCID: PMC293699.

Bewley, K.D., Ellis, K.E., Firer-Sherwood, M.A. and Elliott, S.J. 2013. Multi-heme proteins: Nature's electronic multi-purpose tool. *Biochim. Biophys. Acta - Bioenerg.* 1827(8-9): 938–948. doi: 10.1016/j.bbabio.2013.03.010.

Bhagat, M., Anand, R., Datt, R., Gupta, V. and Arya, S. 2019. Green synthesis of silver nanoparticles using aqueous extract of *Rosa brunonii* Lindl and their morphological, biological and photocatalytic characterizations. *J. Inorg. Organomet. Polym. Mater.*, 29(3): 1039–1047. doi: 10.1007/s10904-018-0994-5.

Bhainsa, K.C. and D'souza, S. 2006. Extracellular biosynthesis of silver nanoparticles using the fungus *Aspergillus fumigatus*. *Colloids Surf. B.*, 47: 160–164.

Bose, D. and Chatterjee, S. 2016. Biogenic synthesis of silver nanoparticles using guava (*Psidium guajava*) leaf extract and its antibacterial activity against *Pseudomonas aeruginosa*. *Appl. Nanosci.*, 6: 895–901.

Cha, S. and Cooksey, A. 1991. Copper resistance in *Pseudomonas syringae* mediated by periplasmic and outer membrane proteins. *Proceedings of the National Academy of Sciences of the United States of America*, 88: 8915–8919.

Chandran, S.P., Chaudhary, M., Pasricha, R., Ahmad, A. and Sastry, M. 2006. Synthesis of gold nanotriangles and silver nanoparticles using aloe vera plant extract. *Biotechnol. Prog.*, 22(2): 577–583. doi: 10.1021/bp0501423.

Chen, P. and Ting, Y.P. 1995. Effect of heavy metal uptake on the electrokinetic properties of *Saccharomyces cerevisiae*. *Biotechnology Letters*, 17: 107–112.

Chetia, L., Kalita, D. and Ahmed, G.A. 2017. Synthesis of Ag nanoparticles using diatom cells for ammonia sensing. *Sens. Bio-Sensing Res.*, 16: 55–61. doi: 10.1016/j.sbsr.2017.11.004.

Chokkalingam, Mohan, Jahan Rupa, Esrat; Huo, Yue, Mathiyalagan, Ramya, Anandapadmanaban, Gokulanathan, Chan Ahn, Jong, Park, Jin Kyu, Lu, Jing and Yang, Deok Chun. 2019. Photocatalytic degradation of industrial dyes using Ag and Au nanoparticles synthesized from Angelica gigas ribbed stem extracts. Optik, S0030402619305340–. doi: 10.1016/j.ijleo.2019.04.065.

Clarke, Thomas A., Holley, Tracey, Hartshorne, Robert S., Fredrickson, Jim K., Zachara, John M., Shi, Liang, and Richardson, David J. 2008. The role of multihaem cytochromes in the respiration of nitrite in *Escherichia coli* and Fe(III) in *Shewanella oneidensis*. *Biochemical Society Transactions*, 36(5): 1005–1010. doi: 10.1042/BST0361005.

Clarke, T.A., Edwards, M.J., Gates, A.J., Hall, A., White, G.F., Bradley, J., Reardon, C.L., Shi, L., Beliaev, A.S., Marshall, M.J., Wang, Z., Watmough, N.J., Fredrickson, J.K., Zachara, J.M., Butt, J.N. and Richardson, D.J. 2011. Structure of a bacterial cell surface decaheme electron conduit. *Proc. Natl. Acad. Sci. U. S. A.*, 108(23): 9384–9389. doi: 10.1073/pnas.1017200108.

Cobbett, C. and Goldsbrough, P. 2002. Phytochelatins and metallothioneins: roles in heavy metal detoxification and homeostasis. *Annu. Rev. Plant Biol.*, 53: 159–82.

Coyle, P., Philcox, J.C., Carey, L.C. and Rofe, A.M. 2002. Metallothioneins: The Multipurpose Protein. *Cellular and Molecular Life Sciences*, 59: 627–647. https://doi.org/10.1007/s00018-002-8454-2.

DalCorso, G., Farinati, S., Maistri, S. and Furini, A. 2008. How plants cope with cadmium: staking all on metabolism and gene expression. *Journal of Integrative Plant Biology*, 50(10): 1268–1280.

Daniel, M.C. and Astruc, D. 2004. Gold nanoparticles: Assembly, supramolecular chemistry, quantum-size-related properties, and applications toward biology, catalysis, and nanotechnology. *Chemical Reviews*, 104(1): 293–346.

Dauthal, P. and Mukhopadhyay, M. 2016. Noble metal nanoparticles: Plant-mediated synthesis, mechanistic aspects of synthesis, and applications. *Ind. Eng. Chem. Res.*, 55: 9557–9577.

De Stefano, L., Rotiroti, L., De Stefano, M., Lamberti, A., Lettieri, S., Setaro, A. and Maddalena, P. 2009. Marine diatoms as optical biosensors. *Biosens. Bioelectron.*, 24 (6): 1580–1584. doi: 10.1016/j.bios.2008.08.016.

Dhandapani, P., Siddarth, A.S., Kamalasekaran, S., Maruthamuthu, S. and Rajagopal, G. 2014. Bio-approach: Ureolytic bacteria mediated synthesis of ZnO nanocrystals on cotton fabric and evaluation of their antibacterial properties. *Carbohyd. Polym.*, 103: 448–455.

Durán, N., Marcato, P.D., Alves, O.L., De Souza, G.I. and Esposito, E. 2005. Mechanistic aspects of biosynthesis of silver nanoparticles by several *Fusarium oxysporum* strains. *J. Nanobiotechnol.*, 3: 8.

Durve, A. 2014. Bioaccumulation and nanoparticle synthesis of heavy metals by microbial isolates. Ph.D Thesis. University of Mumbai, Mumbai, Maharashtra, India.

El-Kemary, M., Zahran, M., Khalifa, S.A.M. and El-Seedi, H.R. 2016. Spectral characterisation of the Silver nanoparticles biosynthesised using *Ambrosia maritima* plant. *Micro Nano Lett.*, 11(6): 311–314. doi: 10.1049/mnl.2015.0572.

El-Shahaby, O., El-Zayat, M., Salih, E., El-Sherbiny, I.M. and Reicha, F.M. 2013. Evaluation of antimicrobial activity of water infusion plant-mediated silver nanoparticles. *J. Nanomedicine Nanotechnol.*, 4(4). doi: 10.4172/2157-7439.1000178.

El-Seedi, Hesham R., El-Shabasy, Rehan M., Khalifa, Shaden A.M., Saeed, Aamer, Shah, Afzal, Shah, Raza, Iftikhar, Faiza Jan, Abdel-Daim, Mohamed M., Omri, Abdelfatteh, Hajrahand, Nahid H., Sabir, Jamal S.M., Zou, Xiaobo, Halabi, Mohammed F., Sarhan, Wessam and Guo, Weisheng. 2019. Metal nanoparticles fabricated by green chemistry using natural extracts: Biosynthesis, mechanisms, and applications. *RSC Adv.*, 9(42): 24539–24559. doi: 10.1039/c9ra02225b.

Fatemi, M., Mollania, N., Momeni-Moghaddam, M. and Sadeghifar, F. 2018. Extracellular biosynthesis of magnetic iron oxide nanoparticles by *Bacillus cereus* strain HMH1: Characterization and *in vitro* cytotoxicity analysis on MCF-7 and 3T3 cell lines. *J. Biotechnol.*, 270: 1–11.

Fortin, D. and Beveridge, T.J. 2000. Mechanistic routes to biomineral surface development. pp. 7–24. *In*: Bäuerlein, E. (ed.). Biomineralization: From Biology to Biotechnology and Medical Application. Weinheim: Wiley-VCH GmbH.

Fuhrmann, T., Landwehr, S., El Rharbl-Kucki, M. and Sumper, M. 2004. Diatoms as living photonic crystals. *Appl. Phys. B Lasers Opt.*, 78(3-4): 257–260. doi: 10.1007/s00340-004-1419-4.

Gadd, G.M. 1990. Heavy metal accumulation by bacteria and other microorganisms. *Experientia*, 46: 834–840.

Gade, A., Bonde, P., Ingle, A., Marcato, P., Duran, N. and Rai, M. 2008. Exploitation of *Aspergillus niger* for synthesis of silver nanoparticles. *J. Biobased Mater. Biol.*, 2: 243–247.

Gahlawat, G. and Choudhury, A.R. 2019. A review on the biosynthesis of metal and metal salt nanoparticles by microbes. *RSC Adv.*, 9(23): 12944–12967. doi: 10.1039/c8ra10483b.

Ghodake, G., Shinde, S., Saratale, G.D., Kadam, A., Saratale, R.G. and Kim, D.Y. 2020. Water purification filter prepared by layer-by-layer assembly of paper filter and polypropylene-polyethylene woven fabrics decorated with silver nanoparticles. *Fibers Polym.*, 21(4): 751–761. doi: 10.1007/s12221-020-9624-2.

Ghodake, G. and Lee, D.S. 2011. Biological synthesis of gold nanoparticles using the aqueous extract of the brown algae *Laminaria japonica. J. Nanoelectron. Optoelectron.*, 6: 268–271.

Govindaraju, K., Basha, S.K., Kumar, V.G. and Singaravelu, G. 2008. Silver, gold and bimetallic nanoparticles production using single-cell protein (*Spirulina platensis*) Geitler. *J. Mater. Sci.*, 43: 5115–5122.

Govindaraju, K., Kiruthiga, V., Kumar, V.G. and Singaravelu, G. 2009. Extracellular synthesis of silver nanoparticles by a marine alga, Sargassum wightii Grevilli and their antibacterial effects. *J. Nanosci. Nanotechnol.*, 2009 Sep; 9(9): 5497–501.

Gray, H.B. and Winkler, J.R. 1996. Electron transfer in proteins. *Annu. Rev. Biochem.*, 65(1): 537–561. doi: 10.1146/annurev. bi.65.070196.002541.

Grill, E., Winnacker, E.L. and Zenk, M.H. 1985. Phytochelatins: the principal heavy-metal complexing peptides of higher plants. *Science.*, 1985 Nov. 8; 230(4726): 674–6.

Gu, H., Chen, X., Chen, F., Zhou, X. and Parsaee, Z. 2018. Ultrasound-assisted biosynthesis of CuO-NPs using brown alga *Cystoseira trinodis*: Characterization, photocatalytic AOP, DPPH scavenging and antibacterial investigations. *Ultrason. Sonochem.*, 41: 109–119.

Haefeli, C., Franklin, C. and Hardy, K. 1984. Plasmid-determined silver resistance in *Pseudomonas stutzeri* isolated from silver mine. *J. Bacteriol.*, 158: 389–392.

Hamedi, S., Ghaseminezhad, M., Shokrollahzadeh, S. and Shojaosadati, S.A. 2017. Controlled biosynthesis of silver nanoparticles using nitrate reductase enzyme induction of filamentous fungus and their antibacterial evaluation. *Artif. Cells, Nanomedicine Biotechnol.*, 45(8):1588–1596. doi: 10.1080/21691401.2016.1267011.

Hartshorne, R.S., Reardon, C.L., Ross, D., Nuester, J., Clarke, T.A., Gates, A.J., Mills, P.C., Fredrickson, J.K., Zachara, J.M., Shi, L., Beliaev, A.S., Marshall, M.J., Tien, M., Brantley, S., Butt, J.N. and Richardson, D.J. 2009. Characterization of an electron conduit between bacteria and the extracellular environment. *Proceedings of the National Academy of Sciences*, 106(52): 22169–22174.

He, S., Guo, Z., Zhang, Y., Zhang, S., Wang, J. and Gu, N. 2007. Biosynthesis of gold nanoparticles using the bacteria *Rhodopseudomonas capsulata. Mater. Lett.*, 61: 3984–3987.

Hirata, K., Naoki Tsuji and Kazuhisa Miyamoto. 2005. Biosynthetic Regulation of Phytochelatins, Heavy Metal-binding Peptides, 100(6). doi: 10.1263/jbb.100.593.

Husseiny, M., El-Aziz, M.A., Badr, Y. and Mahmoud, M. 2007. Biosynthesis of gold nanoparticles using *Pseudomonas aeruginosa*. Spectrochim. *Acta A Mol. Biomol. Spectrosc.*, 67: 1003–1006.

Ingle, A., Gade, A., Pierrat, S., Sonnichsen, C. and Rai, M. 2008. Mycosynthesis of silver nanoparticles using the fungus *Fusarium acuminatum* and its activity against some human pathogenic bacteria. *Curr. Nanosci.*, 4: 141–144.

Ingle, A., Rai, M., Gade, A. and Bawaskar, M. 2009. *Fusarium solani*: A novel biological agent for the extracellular synthesis of silver nanoparticles. *J. Nanopart. Res.*, 11: 2079.

Jain, N, Bhargava, A., Majumdar, S., Tarafdar, J.C. and Panwar, J. 2011. Extracellular biosynthesis and characterization of silver nanoparticles using *Aspergillus flavus* NJP08: A mechanism perspective. *Nanoscale.*, 3: 635–641.

Janardhanan, A., Roshmi, T., Rintu, T.V., Sonia, E.V., Mathew, J. and Radhakrishnan, E.K. 2013. Biosynthesis of silver nanoparticles by a *Bacillus* sp. of marine origin. *Materials Science-Poland*, 31(2): 173–179.

Jayaseelan, C., Rahuman, A.A., Kirthi, A.V., Marimuthu, S., Santhoshkumar, T., Bagavan, A., Gaurav, K., Karthik, L. and Rao, K.B. 2012. Novel microbial route to synthesize ZnO nanoparticles using *Aeromonas hydrophila* and their activity against pathogenic bacteria and fungi. *Spectrochim. Acta A Mol. Biomol. Spectrosc.*, 90: 78–84.

Jena, J., Pradhan, N., Nayak, R.R., Dash, B.P., Sukla, L.B., Panda, P.K. and Mishra, B.K. 2014. Microalga *Scenedesmus* sp.: A potential low-cost green machine for silver nanoparticle synthesis. *J. Microbiol. Biotechnol.* 2014 Apr; 24(4): 522–33. doi: 10.4014/jmb.1306.06014.

Jena, J., Pradhan, N., Dash, B.P., Panda, P.K. and Mishra, B.K. 2015. Pigment mediated biogenic synthesis of silver nanoparticles using diatom *Amphora* sp. and its antimicrobial activity. *J. Saudi Chem. Soc.*, 19(6): 661–666. doi: 10.1016/j.jscs.2014.06.005.

Kadam, J., Dhawal, P., Barve, S. and Kakodkar, S. 2020. Green synthesis of silver nanoparticles using cauliflower waste and their multifaceted applications in photocatalytic degradation of methylene blue dye and Hg2+ biosensing. *SN Appl. Sci.*, 2(4). doi: 10.1007/s42452-020-2543-4.

Kahzad, N. and Salehzadeh, A. 2020. Green Synthesis of $CuFe_2O_4$@Ag nanocomposite using the *Chlorella vulgaris* and evaluation of its effect on the expression of nora efflux pump gene among *Staphylococcus aureus* Strains. *Biol. Trace Elem. Res.*, 198(1): 359–370. doi: 10.1007/s12011-020-02055-5.

Kalimuthu, K., Babu, R.S., Venkataraman, D., Bilal, M. and Gurunathan, S. 2008. Biosynthesis of silver nanocrystals by *Bacillus licheniformis. Colloids Surf B.*, 65: 150–153.

Kalishwaralal, K., Deepak, V., Pandian, S.R.K., Kottaisamy, M., BarathManiKanth, S., Kartikeyan, B. and Gurunathan, S. 2010. Biosynthesis of silver and gold nanoparticles using *Brevibacterium casei. Colloids Surf. B.*, 77: 257–262.

Kamińska, M. Sprynskyy, Winkler, K. and Szymborski, T. 2017. Ultrasensitive SERS immunoassay based on diatom biosilica for detection of interleukins in blood plasma. *Anal. Bioanal. Chem.*, 409(27): 6337–6347. doi: 10.1007/s00216-017-0566-5.

Karbasian, M., Atyabi, S., Siadat, S., Momen, S. and Norouzian, D. 2008. Optimizing nano-silver formation by *Fusarium oxysporum* PTCC 5115 employing response surface methodology. *Am. J. Agric. Biol. Sci.*, 3: 433–437.

Katas, Haliza, Moden, Noor Zianah, Lim, Chei Sin, Celesistinus, Terence, Chan, Jie Yee, Ganasan, Pavitra and Suleman Ismail Abdalla, Sundos. 2018. Biosynthesis and potential applications of silver and gold nanoparticles and their chitosan-based nanocomposites in nanomedicine. *Journal of Nanotechnology*, 1–13. doi:10.1155/2018/4290705.

Kathiresan, K., Manivannan, S., Nabeel, M. and Dhivya, B. 2009. Studies on silver nanoparticles synthesized by a marine fungus, *Penicillium fellutanum* isolated from coastal mangrove sediment. *Colloids Surf. B.*, 71: 133–137.

Kato, H. 2011. *In vitro* assays: Tracking nanoparticles inside cells. *Nature Nanotechnology*, 6(3): 139–140.

Khan, I., Saeed, K. and Khan, I. 2017. Nanoparticles: Properties, applications and toxicities. *Arab. J. Chem.*, 12(7): 908–931.

Kharissova, O.V., Dias, H.V.R., Kharisov, B.I., Pérez, B.O. and Pérez, V.M.J. 2013. The greener synthesis of nanoparticles. *Trends in Biotechnology*, 2013: 240–248. DOI: 10.1016/j.tibtech.2013.01.003.

Khatoon, N. and Sardar, M. 2017. Efficient removal of toxic textile dyes using silver nanocomposites. *J. Nanosci. Curr. Res.*, 02(03): 2–6. doi: 10.4172/2572-0813.1000113.

Kim, D.Y., Saratale, R.G., Shinde, S., Syed, A., Ameen, F. and Ghodake, G. 2016. Green synthesis of silver nanoparticles using *Laminaria japonica* extract: Characterization and seedling growth assessment. *J. Clean. Prod.*, 172: 2910–2918. doi: 10.1016/j.jclepro.2017.11.123.

Kim, D.H., Gopal, J. and Sivanesan, I. 2017. Nanomaterials in plant tissue culture: The disclosed and undisclosed. *RSC Adv.*, 7(58): 36492–36505. doi: 10.1039/c7ra07025j.

Klaus, T., Joerger, R., Olsson, E. and Granqvist, C.G. 1999. Silver-based crystalline nanoparticles, microbially fabricated. *Proc. Natl. Acad. Sci. USA.*, 96: 13611–13614.

Konishi, Y., Tsukiyama, T., Tachimi, T., Saitoh, N., Nomura, T. and Nagamine, S. 2007. Microbial deposition of gold nanoparticles by the metal-reducing bacterium *Shewanella* algae. *Electrochim. Acta*, 53: 186–192.

Koopi, H. and Buazar, F.A. 2018. Novel one-pot biosynthesis of pure alpha aluminum oxide nanoparticles using the macroalgae *Sargassum ilicifolium*: A green marine approach. *Ceram. Int.*, 44: 8940–8945.

Krithiga, N., Rajalakshmi, A. and Jayachitra, A. 2015. Green Synthesis of Silver Nanoparticles Using Leaf Extracts of *Clitoria ternatea* and *Solanum nigrum* and Study of Its Antibacterial Effect against Common Nosocomial Pathogens.

Kröger, N., Lorenz, S., Brunner, E. and Sumper, M. 2002. Self-assembly of highly phosphorylated silaffins and their function in biosilica morphogenesis. *Science*, 298(5593): 584–586. doi: 10.1126/science.1076221.

Kulkarni, S.K. 2015. Nanotechnology: principles and practices. In Nanotechnology: Principles and Practices. 111–123.

Kumar, S.A., Abyaneh, M.K., Gosavi, S., Kulkarni, S.K., Pasricha, R., Ahmad, A. and Khan, M. 2007a. Nitrate reductase-mediated synthesis of silver nanoparticles from AgNO3. *Biotechnol. Lett.*, 29: 439–445.

Kumar, S.A., Ansary, A.A., Ahmad, A. and Khan, M. 2007b. Extracellular biosynthesis of CdSe quantum dots by the fungus, *Fusarium oxysporum. J. Biomed. Nanotechnol.*, 3: 190–194.

Lade, B.D. 2017. Biochemical and molecular approaches for characterization of wound stress induced antimicrobial secondary metabolites in *Passiflora foetida* linn [Ph. D thesis]. Amravati, MS, India: Biotechnology, Sant Gadge Baba Amravati University.

Lade, B.D. and Patil, A.S. 2017. Silver nano fabrication using leaf disc of *Passiflora foetida* linn. *Applied Nanoscience*, 7(5): 181–119. DOI: 10.1007/s13204-017-0558-y.

Lade, B.D. and Shanware, A.S. 2019. Phytonanofabrication: Methodology and Factors Affecting Biosynthesis of Nanoparticles. DOI: http://dx.doi.org/10.5772/intechopen.90918.

Lade, B.D. and Shanware, A.S. 2020. Phytonanofabrication: Methodology and factors affecting biosynthesis of nanoparticles. *Smart Nanosystems for Biomedicine, Optoelectronics and Catalysis*. 1–17.

Lal, S.S. and Nayak, P.L. 2012. Green synthesis of gold nanoparticles using various extract of plants and spices. *Int. J. Sci. Innov. Discov.*, 2(3): 325–350.

Laura Castro, A.B., María Luisa Blázquez, Jesus Angel Muñoz and Felisa González. 2013. Biological synthesis of metallic nanoparticles using algae. *IET Nanobiotechnol.*, 7(3): 109–16. doi: 10.1049/iet-nbt.2012.0041.

Lengke, M.F., Fleet, M.E. and Southam, G. 2006. Morphology of gold nanoparticles synthesized by filamentous cyanobacteria from gold (I) thiosulfate and gold (III) chloride complexes. *Langmuir*, 22: 2780–2787.

Lengke, M.F., Fleet, M.E. and Southam, G. 2007. Biosynthesis of silver nanoparticles by filamentous cyanobacteria from a silver (I) nitrate complex. *Langmuir*, 23: 2694–2699.

León, E.R., Rodríguez, E.L., Beas, C.R., Plascencia-Villa, G. and Palomares, R.A.I. 2016. Study of methylene blue degradation by gold nanoparticles synthesized within natural zeolites. *J. Nanomater*. doi: 10.1155/2016/9541683.

Li, J. Cai, Pan, J., Wang, Y., Yue, Y. and Zhang, D. 2014. Multi-layer hierarchical array fabricated with diatom frustules for highly sensitive bio-detection applications. *J. Micromechanics Microengineering*, 24(2): 025014, Feb. 2014, doi: 10.1088/0960-1317/24/2/025014.

Li, M., Noriega-Trevino, M.E., Nino-Martinez, N., Marambio-Jones, C., Wang, J., Damoiseaux, R., Ruiz, F. and Hoek, Eric M. V. 2011. Synergistic bactericidal activity of Ag-TiO2 nanoparticles in both light and dark conditions. *Environmental Science & Technology*, 45(20): 8989–8995. DOI: 10.1021/es201675m.

Li, S, Shen, Y., Xie, A., Yu, X., Zhang, X. Yang, L. and Li, C. 2007. Rapid, room-temperature synthesis of amorphous selenium/protein composites using *Capsicum annuum* L. extract. *Nanotechnology*, 18: 40. doi: 10.1088/0957-4484/18/40/405101.

Liu, Yimo, Wang, Zheming, Liu, Juan, Levar, Caleb, Edwards, Marcus J., Babauta, Jerome T., Kennedy, David W., Shi, Zhi, Beyenal, Haluk, Bond, Daniel R., Clarke, Thomas A., Butt, Julea N., Richardson, David J., Rosso, Kevin M., Zachara, John M., Fredrickson, James K. and Shi, Liang. 2014. 2014. A trans-outer membrane porin-cytochrome protein complex for extracellular electron transfer by *Geobacter sulfurreducens* PCA. *Environ. Microbiol. Rep.*, 6(6): 776–785. doi: 10.1111/1758-2229.12204.

Liu, Y., Fredrickson, J.K., Zachara, J.M. and Shi, L. 2015. Direct involvement of ombB, omaB, and omcB genes in extracellular reduction of Fe(III) by *Geobacter sulfurreducens* PCA. *Front. Microbiol.* 6: 1–8. doi: 10.3389/fmicb.2015.01075.

Lloyd, J.R. and Lovely, D.R. 2001. Microbial detoxification of metals and radionuclides. *Current Opinion in Biotechnology*, 12: 248–253.

Losic, D., Mitchell, J.G. and Voelcker, N.H. 2009. Diatomaceous lessons in nanotechnology and advanced materials. *Adv. Mater.*, 21(29): 2947–2958. doi: 10.1002/adma.200803778.

Ghafourian, M., Shahamat, M., Levin, A. and Colwell, R.R. 2002. Uranium accumulation by a bacterium isolated from electroplating effluent. *World Journal of Microbiology and Biotechnology*, 18(4): 295–302.

Maceda, A.F., Ouano, J.J.S., Que, M.C.O., Basilia, B.A., Potestas, M.J. and Alguno, A.C. 2018. Controlling the absorption of gold nanoparticles via green synthesis using *Sargassum crassifolium* Extract. In *Key Engineering Materials; Trans. Tech. Publ.*: Clausthal-Zellerfeld, Germany, 44–48.

Madigan, T., Michael, M.J.M. and Parker, J. 2000. Brock biology of microorganisms, 9th ed. Upper Saddle River, NJ: Prentice Hall.

Malekzadeh, F.A., Farazmand, H., J.F. Moran, Klucas, R.V., Grayer, R.J., Abian, J. and Becan, M. 1997. *Free Radical Bio. Med.*, 22: 861.

Mendoza-Cózatl, D.G., Butko, E., Springer, F., Justin W. Torpey, Elizabeth A. Komives, Julia Kehr and Julian I. Schroeder. 2008. Identification of high levels of phytochelatins, glutathione and cadmium in the phloem sap of *Brassica napus*. A role for thiol-peptides in the long-distance transport of cadmium and the effect of cadmium on iron translocation. *Plant Journal*, 54 (2): 249–259.

Momeni, S. and Nabipour, I. 2015. A simple green synthesis of palladium nanoparticles with sargassum alga and their electrocatalytic activities towards hydrogen peroxide. *Appl. Biochem. Biotechnol.*, 176(7): 1937–1949. doi: 10.1007/s12010-015-1690-3.

Mukherjee, P., Ahmad, A., Mandal, D., Senapati, S., Sainkar, S.R., Khan, M.I., Parishcha, R., Ajaykumar, P., Alam, M. and Kumar, R. 2001a. Fungus-mediated synthesis of silver nanoparticles and their immobilization in the mycelial matrix: A novel biological approach to nanoparticle synthesis. *Nano Lett.*, 1: 515–519.

Mukherjee, P., Ahmad, A., Mandal, D., Senapati, S., Sainkar, S.R., Khan, M.I., Ramani, R., Parischa, R., Ajayakumar, P. and Alam, M. 2001b. Bioreduction of AuCl(4)(-) ions by the fungus, *Verticillium* sp. and surface trapping of the gold nanoparticles formed. *Angew. Chem. Int. Ed.*, 40: 3585–3588.

Mukherjee, S., Vinothkumar, B., Prashanthi, S., Bangal, P.R., Sreedhar, B. and Patra, C.R. 2013. Potential therapeutic and diagnostic applications of one-step in situ biosynthesized gold nanoconjugates (2-in-1 system) in cancer treatment. *RSC Adv.*, 3: 2318–2329.

Mukherjee, S., Dasari, M., Priyamvada, S., Kotcherlakota, R., Bollu, V.S. and Patra, C.R. 2015. A green chemistry approach for the synthesis of gold nanoconjugates that induce the inhibition of cancer cell proliferation through induction of oxidative stress and their *in vivo* toxicity study. *J. Mater. Chem. B.*, 3: 3820–3830.

Murgueitio, E., Cumbal, L., Abril, M., Izquierdo, A., Debut, A. and Tinoco, O. 2018. Green synthesis of iron nanoparticles: Application on the removal of petroleum oil from contaminated water and soils. *Journal of Nanotechnology*. https://doi.org/10.1155/2018/4184769.

Nair, B. and Pradeep, T. 2002. Coalescence of nanoclusters and formation of submicron crystallites assisted by *Lactobacillus* strains. *Cryst. Growth Des.*, 2: 293–298.

Narayanan, K.B. and Sakthivel, N. 2010. Biological synthesis of metal nanoparticles by microbes. *Adv. Colloids Interface Sci.*, 156.

Nasrollahzadeh, M. and Sajadi, S.M. 2015a. Synthesis and characterization of titanium dioxide nanoparticles using *Euphorbia heteradena* Jaub root extract and evaluation of their stability. *Ceram. Int.*, 41(10): 14435–14439. doi: 10.1016/j.ceramint.2015.07.079.

Nasrollahzadeh, M. and Mohammad Sajadi, S. 2015b. Green synthesis of copper nanoparticles using *Ginkgo biloba* L. leaf extract and their catalytic activity for the Huisgen [3+2] cycloaddition of azides and alkynes at room temperature. *J. Colloid Interface Sci.*, 457: 141–147. doi: 10.1016/j.jcis.2015.07.004.

Nasrollahzadeh, M., Sajadi, S.M., Rostami-Vartooni, A. and Bagherzadeh, M. 2015. Green synthesis of Pd/CuO nanoparticles by *Theobroma cacao* L. seeds extract and their catalytic performance for the reduction of 4-nitrophenol and phosphine-free Heck coupling reaction under aerobic conditions. *J. Colloid Interface Sci.*, 448: 106–113. doi: 10.1016/j.jcis.2015.02.009.

Nath, D., Banerjee, P., Ray, A. and Bairagi, B. 2019. Green peptide–nanomaterials; A friendly healing touch for skin wound regeneration. *Adv. Nano Res.*, 2(1): 14–31. doi: 10.21467/anr.2.1.14-31.

Ng, C.K., Sivakumar, Krishnakumar, Liu, Xin, Madhaiyan, Munusamy, Ji, Lianghui, Yang, Liang, Tang, Chuyang, Song, Hao, Kjelleberg, Staffan and Cao, Bin. 2013. Influence of outer membrane c-type cytochromes on particle size and activity of extracellular nanoparticles produced by *Shewanella oneidensis*. *Biotechnol. Bioeng.*, 110(7): 1831–1837. doi: 10.1002/bit.24856.

Onitsuka, S., Hamada, T. and Okamura, H. 2019. Preparation of antimicrobial gold and silver nanoparticles from tea leaf extracts. *Colloids Surfaces B Biointerfaces*, 173: 242–248. doi: 10.1016/j.colsurfb.2018.09.055.

Ovais, M., Khalil, A.T., Raza, A., Khan, M.A., Ahmad, I., Islam, N.U., Saravanan, M., Ubaid, M.F., Ali, M. and Shinwari, Z.K. 2016. Green synthesis of silver nanoparticles via plant extracts: beginning a new era in cancer theranostics. *Nanomedicine.*, 12: 3157–3177.

Ovais, M., Raza, A., Naz, S., Islam, N.U., Khalil, A.T., Ali, S., Khan, M.A. and Shinwari, Z.K. 2017. Current state and prospects of the phytosynthesized colloidal gold nanoparticles and their applications in cancer theranostics. *Appl. Microbiol. Biotechnol.*, 101: 3551–3565.

Ovais, M., A.T. Khalil, N.U. Islam, I. Ahmad, M. Ayaz, M. Saravanan, Z.K. Shinwari, S. Mukherjee. 2018a. Role of plant phytochemicals and microbial enzymes in biosynthesis of metallic nanoparticles. *Appl. Microbiol. Biotechnol.*, 102: 6799–6814.

Ovais, M., Zia, N., Ahmad, I., Khalil, A.T., Raza, A., Ayaz, M., Sadiq, A., Ullah, F. and Shinwari, Z.K. 2018b. Phyto-Therapeutic and nanomedicinal approaches to cure alzheimer's disease: present status and future opportunities. *Front. Aging Neurosci.*, 10.

Oza, G., Pandey, S., Shah, R. and Sharon, M. 2012a. A mechanistic approach for biological fabrication of crystalline gold nanoparticles using marine algae, *Sargassum wightii*, *European Journal of Experimental Biology*, 2(3): 505–512.

Oza, G., Panday, S., Mewada, A., Kalita, G. and Sharon, M. 2012b. Facile biosynthesis of gold nanoparticles exploiting optimum pH and temperature of fresh water algae *Chlorella pyrenoidusa. Library* (Lond)., 3(3): 1405–1412.

Pandey, K.B. and Rizvi, S.I. 2009. Plant polyphenols as dietary antioxidants in human health and disease. *Oxid. Med. Cell. Longev.*, 2(5): 270–278. doi: 10.4161/oxim.2.5.9498.

Pannerselvam, B., Dharmalingam, J., Mukesh, K., Rajenderan, M., Perumal, P., Pudupalayam Thangavelu, K., Kim, H.J., Singh, V. and Rangarajulu, S.K. 2017. An *in vitro* study on the burn wound healing activity of cotton fabrics incorporated with phytosynthesized silver nanoparticles in male *Wistar albino* rats. *Eur. J. Pharm. Sci.*, 100: 187–196. doi: 10.1016/j.ejps.2017.01.015.

Panwar, V. and Dutta, T. 2019. Diatom biogenic silica as a felicitous platform for biochemical engineering: expanding frontiers. *ACS Applied Bio. Materials*, 2(6): 2295–2316. doi: 10.1021/acsabm.9b00050.

Parastoo, P., Seyyed, H.B., Sahebali, M., Tahereh, N., Azadeh, R. and Behrooz, Y. 2017. Biosynthesis of gold nanoparticles by two bacterial and fungal strains, *Bacillus cereus* and *Fusarium oxysporum*, and assessment and comparison of their nanotoxicity *in vitro* by direct and indirect assays. *Electronic Journal of Biotechnology*, 29: 86–93.

Patel, V., Berthold, D., Puranik, P. and Gantar, M. 2015. Screening of cyanobacteria and microalgae for their ability to synthesize silver nanoparticles with antibacterial activity. *Biotechnol. Rep.*, 5: 112–119.

Patra, C.R., Mukherjee, S. and Kotcherlakota, R. 2014. Biosynthesized silver nanoparticles: A step forward for cancer theranostics? *Nanomedicine*, 9: 1445–1448.

Patra, J.K. and Baek, K.H. 2017. Antibacterial activity and synergistic antibacterial potential of biosynthesized silver nanoparticles against foodborne pathogenic bacteria along with its anticandidal and antioxidant effects. *Front. Microbiol.* 8: 1–14. doi: 10.3389/fmicb.2017.00167.

Patra, S., Mukherjee, S., Barui, A.K., Ganguly, A., Sreedhar, B. and Patra, C.R. 2015. Green synthesis, characterization of gold and silver nanoparticles and their potential application for cancer therapeutics. *Mater. Sci. Eng. C*, 53: 298–309.

Pickett-Heaps, J.D., Schmid, A.M. and Edgar, L.A. 1990. The cell biology of diatom valve formation. *Prog. Phycol. Res.*, 7: 1–168.

Pluta, K., Tryba, A.M., Malina, D. and Sobczak-Kupiec, A. 2017. Red tea leaves infusion as a reducing and stabilizing agent in silver nanoparticles synthesis. *Adv. Nat. Sci. Nanosci. Nanotechnol.*, 8(4). doi: 10.1088/2043-6254/aa92b1.

Pugazhendhi, A., Prabakar, D., Jacob, J.M., Karuppusamy, I. and Saratale, R.G. 2017. Synthesis and characterization of silver nanoparticles using *Gelidium amansii* and its antimicrobial property against various pathogenic bacteria. *Microb. Pathog.* 114: 41–45. doi: 10.1016/j.micpath.2017.11.013.

Pugazhenthiran, N., Anandan, S., Kathiravan, G., Prakash, N.K.U., Crawford, S. and Ashokkumar, M. 2009. Microbial synthesis of silver nanoparticles by *Bacillus* sp. *J. Nanopart. Res.* 11: 1811.

Punjabi, K., Choudhary, P., Samant, L., Mukherjee, S., Vaidya, S. and Chowdhary, A. 2015. Biosynthesis of nanoparticles: A review. *Int. J. Pharm. Sci. Rev. Res.*, 30(1): 219–226.

Qidwai, A., Kumar, R. and Dikshit, A. 2018. Green synthesis of silver nanoparticles by seed of *Phoenix sylvestris* L. and their role in the management of cosmetics embarrassment. *Green Chem. Lett. Rev.*, 11(2): 176–188. doi: 10.1080/17518253.2018.1445301.

Rad, M., Taran, M. and Alavi, M. 2018. Effect of incubation time, CuSO4 and glucose concentrations on biosynthesis of copper oxide (CuO) nanoparticles with rectangular shape and antibacterial activity: taguchi method approach. *Nano Biomed. Eng.*, 10: 25–33.

Rahman, A., Kumar, S., Bafana, A., Dahoumane, S.A. and Jeffryes, C. 2019. Biosynthetic conversion of Ag+ to highly Stable Ag0 nanoparticles by wild type and cell wall deficient strains of chlamydomonas reinhardtii. *Molecules*, 24(1). doi: 10.3390/molecules24010098.

Rajeshkumar, S. and Bharath, L.V. 2017. Mechanism of plant-mediated synthesis of silver nanoparticles—A review on biomolecules involved, characterisation and antibacterial activity. *Chem. Biol. Interact.*, 273: 219–227. doi: 10.1016/j.cbi.2017.06.019.

Rajput, D., Paul, S. and Gupta, A.D. 2020. Green synthesis of silver nanoparticles using waste tea leaves. *Adv. Nano Res.*, 3(1): 1–14. doi: 10.21467/anr.3.1.1-14.

Ramakrishna, M., Babu, D.R., Gengan, R.M., Chandra, S. and Rao, G.N. 2016. Green synthesis of gold nanoparticles using marine algae and evaluation of their catalytic activity. *J. Nanostruct. Chem.*, 6: 1–13.

Robinson, N.J., Whitehall, S.K. and Cavet, J.S. 2001. Microbial metallothioneins. *Adv. Microb. Physiol.*, 44: 183–213.

Rodrigues, M.L., Oliveira, T.F., Pereira, I.A.C. and Archer, M. 2006. X-ray structure of the membrane-bound cytochrome c quinol dehydrogenase NrfH reveals novel haem coordination. *EMBO J.*, 25(24): 5951–5960. doi: 10.1038/sj.emboj.7601439.

Rodrigues, M.L., Scott, K.A., Sansom, M.S.P., Pereira, I.A.C. and Archer, M. 2008. Quinol oxidation by c-type cytochromes: structural characterization of the menaquinol binding site of NrfHA. *J. Mol. Biol.*, 381(2): 341–350. doi: 10.1016/j.jmb.2008.05.066.

Rolim, W.R. et al. 2019. Green tea extract mediated biogenic synthesis of silver nanoparticles: Characterization, cytotoxicity evaluation and antibacterial activity. *Appl. Surf. Sci.*, 63: 66–74. doi: 10.1016/j.apsusc.2018.08.203.

Rostami-Vartooni, A., Nasrollahzadeh, M., Salavati-Niasari, M. and Atarod, M. 2016. Photocatalytic degradation of azo dyes by titanium dioxide supported silver nanoparticles prepared by a green method using *Carpobrotus acinaciformis* extract. *J. Alloys Compd.*, 689: 15–20. doi: 10.1016/j.jallcom.2016.07.253.

Sadowski, Z, Maliszewskalh, A., Grochowalska, B., Polowczyk, I. and Koźlecki, T. 2008. Synthesis of silver nanoparticles using microorganisms. *Materials Science-Poland*, 26(2): 419–424.

Sadowski, Z. 2010. Silver nanoparticles, David PP, In Tech. 257–276.

Safavi, K. 2012. Evaluation of using nanomaterial in tissue culture media and biological activity in 2nd international conference on ecological. *Environmental and Biological Sciences*, (EEBS'2012): 5–8.

Sai Saraswathi, V., Kamarudheen, N., BhaskaraRao, K.V. and Santhakumar, K. Phytoremediation of dyes using *Lagerstroemia speciosa* mediated silver nanoparticles and its biofilm activity against clinical strains *Pseudomonas aeruginosa. J. Photochem. Photobiol. B Biol.*, 168: 107–116. doi: 10.1016/j.jphotobiol.2017.02.004. FECHA.

Saifuddin, N, Wonga, C.W. and NurYasumira, A.A. 2009. Rapid biosynthesis of silver nanoparticles using culture supernatant of bacteria with microwave irradiation. *J. Chem.*, 6: 61–70.

Salt, D.E. and Rauser, W.E. 1995. MgATP-dependent transport of phytochelatins across the tonoplast of oat roots. *Plant Physiology*, 107(4): 1293–1301.

Salunke, G.R., Ghosh, S., Kumar, R.S., Khade, S., Vashisth, P., Kale, T., Chopade, S., Pruthi, V. and Kundu, G. 2014. Bellare, Rapid efficient synthesis and characterization of silver, gold, and bimetallic nanoparticles from the medicinal plant *Plumbago zeylanica* and their application in biofilm control. *Int. J. Nanomed.*, 9: 2635.

Sanaeimehr, Z., Javadi, I. and Namvar, F. 2018. Antiangiogenic and antiapoptotic effects of green–synthesized zinc oxide nanoparticles using *Sargassum muticum* algae extraction. *Cancer Nanotechnol.*, 9:3 doi: 10.1186/s12645-018-0037-5.

Sanghi, R., Verma, P. and Puri, S. 2011. Enzymatic formation of gold nanoparticles using *Phanerochaete chrysosporium. Adv. Chem. Eng. Sci.*, 1: 154.

Sanità Di Toppi, L. and Gabbrielli, R. 1999. Response to cadmium in higher plants. *Environmental and Experimental Botany*, 41(2): 105–130.

Satapathy, S. and Shukla, S.P. 2017. Application of a marine cyanobacterium Phormidium fragile for green synthesis of silver nanoparticles. *Indian J. Biotechnol.*, 16: 110–113.

Saxena, A., Tripathi, R.M., Zafar, F. and Singh, P. 2012. Green synthesis of silver nanoparticles using aqueous solution of Ficus benghalensis leaf extract and characterization of their antibacterial activity. *Mater. Lett.*, 67(1): 91–94. doi: 10.1016/j.matlet.2011.09.038.

Selvarajan, E. and Mohanasrinivasan, V. 2013. Biosynthesis and characterization of ZnO nanoparticles using *Lactobacillus plantarum* VITES07. *Mater. Lett.*, 112: 180–182.

Senapati, S., Ahmad, A., M.I. Khan, M.I., Sastry, M. and Kumar, R. 2005. Extracellular biosynthesis of bimetallic Au–Ag alloy nanoparticles. *Small.*, 1: 517–520.

Senapati, S., Syed, A., Moeez, S., Kumar, A. and Ahmad, A. 2012. Intracellular synthesis of gold nanoparticles using alga *Tetraselmis kochinensis. Mater. Lett.*, 79: 116–118, Jul. 2012, doi: 10.1016/j.matlet.2012.04.009.

Shah, V. and Belozerova, I. 2009. Influence of metal nanoparticles on the soil microbial community and germination of lettuce seeds. *Water, Air, and Soil Pollution*, 197(1–4): 143–148. https://doi.org/10.1007/s11270-008-9797-6.

Shaheen, T.I. and Abd El Aty, A.A. 2018. *In-situ* green myco-synthesis of silver nanoparticles onto cotton fabrics for broad spectrum antimicrobial activity. *Int. J. Biol. Macromol.*, 118: 2121–2130. doi: 10.1016/j.ijbiomac.2018.07.062.

Shaligram, N.S., Bule, M., Bhambure, R., Singhal, R.S., Singh, S.K., Szakacs, G. and Pandey, A. 2009. Biosynthesis of silver nanoparticles using aqueous extract from the compactin producing fungal strain. *Process. Biochem.*, 44: 939–943.

Shanmugam, R. 2018. Synthesis of Zinc oxide nanoparticles using algal formulation (Padina tetrastromatica and Turbinaria conoides) and their antibacterial activity against fish pathogens. *Res. J. Biotechnol.*, 13(9): 15–19.

Sharon, M., Mewada, A., Swaminathan, N. and Sharon, C. 2017. Synthesis of biogenic gold nanoparticles and its applications as theranostic agent: A review. *Nanomedicine Nanotechnol. J.*, 1(1): 113.

Shen, Z., Han, G., Liu, C., Wang, X. and Sun, R. 2016. Green synthesis of silver nanoparticles with bagasse for colorimetric detection of cysteine in serum samples. *J. Alloys Compd.*, 686: 82–89. doi: 10.1016/j.jallcom.2016.05.348.

Shi, L., Rosso, K.M., Clarke, T.A., Richardson, D.J., Zachara, J.M. and Fredrickson, J.K. 2012. Molecular underpinnings of Fe(III) oxide reduction by *Shewanella oneidensis* MR-1. *Front. Microbiol.*, 3(FEB): 1–10. doi: 10.3389/fmicb.2012.00050.

Silver, S. 1996. Bacterial resistance to toxic metal ions—A review. *Gene.*, 179: 9–19.

Singaravelu, G., Arockiamary, J., Kumar, V.G. and Govindaraju, K. 2007. A novel extracellular synthesis of monodisperse gold nanoparticles using marine alga, *Sargassum wightii* Greville. *Colloids Surf. B.*, 57: 97–101.

Singh, M., Mallick, A.K., Banerjee, M. and Kumar, R. 2016. Loss of outer membrane integrity in Gram-negative bacteria by silver nanoparticles loaded with *Camellia sinensis* leaf phytochemicals: Plausible mechanism of bacterial cell disintegration. *Bull. Mater. Sci.*, 39(7): 1871–1878. doi: 10.1007/s12034-016-1317-5.

Singh, P. and Jain, S.K. 2014. Biosynthesis of nanomaterials: Growth and properties. *Rev. Adv. Sci. Eng.*, 3(3): 231–238. doi: 10.1166/rase.2014.1066.

Singh, P., Kim, Y.J., Zhang, D. and Yang, D.C. 2016. Biological synthesis of nanoparticles from plants and microorganisms. *Trends Biotechnol.*, 2016 Jul; 34(7): 588–599.

Singh, P.K. and Kundu, S. 2014. Biosynthesis of gold nanoparticles using bacteria. *Proc. Natl. Acad. Sci., India, Sect. B Biol. Sci.*, 84: 331–336. https://doi.org/10.1007/s40011-013-0230-6.

Singh, R., Wagh, P., Wadhwani, S., Gaidhani, S., Kumbhar, A., Bellare, J. and Chopade, B.A. 2013. Synthesis, optimization, and characterization of silver nanoparticles from *Acinetobacter calcoaceticus* and their enhanced antibacterial activity when combined with antibiotics. *Int. J. Nanomed.*, 8: 4277.

Slavin, Y.N., Asnis, J., Häfeli, U.O. and Bach, H. 2017. Metal nanoparticles: Understanding the mechanisms behind antibacterial activity. *J. Nanobiotechnol.*, 15: 65.

Soni, H., Kumar, J.I.N., Patel, K. and Kumar, R.N. 2016. Photocatalytic decoloration of three commercial dyes in aqueous phase and industrial effluents using TiO2 nanoparticles. *Desalin. Water Treat.*, 57(14): 6355–6364. doi: 10.1080/19443994.2015.1005147.

Sonker, A.S., Pathak, J., Kannaujiya, V. and Sinha, R. 2017. Characterization and *in vitro* antitumor, antibacterial and antifungal activities of green synthesized silver nanoparticles using cell extract of *Nostoc* sp. strain HKAR-2. *Can. J. Biotechnol.*, 1: 26–37.

Southam, G. and Beveridge, T.J. 1996. The occurrence of sulfur and phosphorus within bacterially derived crystalline and pseudocrystalline octahedral gold formed *in vitro. Geochim. Cosmochim. Acta*, 60: 4369–4376.

Spain, A. 2003. Implications of microbial heavy metal tolerance in the environment. *Reviews of Undergraduate Research*, 2: 1–6.

Stalin Dhas, T., Ganesh V. Kumar, Karthick, V., Jini K. Angel and Govindaraju, K. 2017. Facile synthesis of silver chloride nanoparticles using marine alga and its antibacterial efficacy. *Spectrochim. Acta - Part A Mol. Biomol. Spectrosc.*, 120: 416–420. doi: 10.1016/j.saa.2013.10.044.

Subbaiya, R., Saravanan, M., Priya, A.R., Shankar, K.R., Selvam, M., Ovais, M., Balajee, R. and Barabadi, H. 2017. Biomimetic synthesis of silver nanoparticles from *Streptomyces atrovirens* and their potential anticancer activity against human breast cancer cells. *IET Nanobiotechnol.*, 11: 965–972.

Swaminathan, S., Murugesan, S., Damodarkumar, S., Dhamotharan, R. and Bhuvaneshwari, S. 2011. Synthesis and characterization of gold nanoparticles from alga *Acanthophora spicifera* (VAHL) Boergesen. *Int. J. Nanosci. Nanotechnol.*, 2: 85–94.

Thakkar, K.N., Mhatre, S.S. and Parikh, R.Y. 2010. Biological synthesis of metallic nanoparticles. *Nanomedicine.*, 6: 257–262.

Thakur, I.S. 2006. Heavy metal pollution, Industrial biotechnology-problems and remedies, I. K International Ltd. publisher, Delhi, 265–287.

Tiwari, D.K., Behari, J. and Prasenjit, S. 2008. Application of activated carbon in waste water treatment. *World Appl. Sci. J.*, 3(3): 417–433.

Tomer, A.K., Rahi, T., Neelam, D.K. and Dadheech, P.K. 2018. Cyanobacterial extract-mediated synthesis of silver nanoparticles and their application in ammonia sensing. *Int. Microbiol.*, 22: 49–58.

Turick, C.E., Tisa, L.S. and Caccavo, F. 2002. Melanin production and use as a soluble electron shuttle for Fe(III) oxide reduction and as a terminal electron acceptor by Shewanella algae BrY. *Appl. Environ. Microbiol.*, 68(5): 2436–2444. doi: 10.1128/AEM.68.5.2436-2444.2002.

Vanaja, M., Rajeshkumar, S., Paulkumar, K., Gnanajobitha, G., Malarkodi, C. and Annadurai, G. 2013. Kinetic study on green synthesis of silver nanoparticles using *Coleus aromaticus* leaf extract. *Pelagia Res. Libr.*, 4(3): 50–55.

Vanaja, M., Paulkumar, K., Baburaja, M., Rajeshkumar, S., Gnanajobitha, G., Malarkodi, C., Sivakavinesan, M. and Annadurai, G. 2014. Degradation of methylene blue using biologically synthesized silver nanoparticles. *Bioinorg. Chem. Appl.*, doi: 10.1155/2014/742346.

Vasylevskyi, M., Kracht, S.I., Corcosa, S., Fromm, P., Giese, K.M. and Füeg, B. 2017. Formation of silver nanoparticles by electron transfer in peptides and c-cytochromes. *Angew. Chem. Int. Ed. Engl.*, 56(21): 5926–5930. doi: https://doi.org/10.1002/anie.201702621e.

Vellaichamy, B. and Periakaruppan, P. 2016. Silver nanoparticle-embedded RGO-nanosponge for superior catalytic activity towards 4-nitrophenol reduction. *RSC Adv.*, 6(91): 88837–88845. doi: 10.1039/c6ra19834a.

Venkataraman, D., Kalimuthu, K., Sureshbabu, R.K.P. and Sangiliyandi, G. 2011. Metal nanoparticles in microbiology. *Springer*, 9: 17–35.

Vijayaraghavan, K., Mahadevan, A., Sathishkumar, M., Pavagadhi, S. and Balasubramanian, R. 2011. Biosynthesis of Au (0) from Au (III) via biosorption and bioreduction using brown marine alga *Turbinaria conoides. Chem. Eng. J.*, 167: 223–227.

Wang, L., Lu, F., Liu, Y., Wu, Y. and Wu, Z. 2018. Photocatalytic degradation of organic dyes and antimicrobial activity of silver nanoparticles fast synthesized by flavonoids fraction of *Psidium guajava* L. leaves. *J. Mol. Liq.*, 263: 187–192. doi: 10.1016/j.molliq.2018.04.151.

Xu, J., Wang, W., Sun, J., Zhang, Y., Ge, Q., Du, L., Yin, H. and Liu, X. 2011. Involvement of auxin and nitric oxide in plant Cd-stress responses. *Plant and Soil*, 346(1): 107–119.

Xu, Q.B., Xie, L.J., Diao, H., Li, F., Zhang, Y.Y., Fu, F.Y. and Liu, X.D. 2017. Antibacterial cotton fabric with enhanced durability prepared using silver nanoparticles and carboxymethyl chitosan. *Carbohydr. Polym.*, 177: 187–193. doi: 10.1016/j.carbpol.2017.08.129.

Yilmaz Öztürk, B. 2019. Intracellular and extracellular green synthesis of silver nanoparticles using *Desmodesmus* sp.: Their antibacterial and antifungal effects. *Caryologia.*, 72(1): 29–43. doi: 10.13128/cayologia-249.

Zhen, L., Ford, N., Gale, D.K., Roesijadi, G. and Rorrer, G.L. 2016. Photoluminescence detection of 2,4,6-trinitrotoluene (TNT) binding on diatom frustule biosilica functionalized with an anti-TNT monoclonal antibody fragment. *Biosens. Bioelectron.*, 79: 742–748.doi: 10.1016/j.bios.2016.01.002.

CHAPTER 7

Electron Microscopy Characterization of Nanoparticles

Diana F. Garcia-Gutierrez[1,2] and *Domingo I. Garcia-Gutierrez*[1,2,*]

1. Introduction

Electron microscopy is the main characterization technique used to study nanostructured materials. Its development was driven by the limitations displayed by visible light microscopy (VLM). In the XIX century, Ernst Karl Abbe published his work related to the limitations of VLM, which are directly related to the diffraction processes involved in the image formation mechanism of such microscopes. Equation 1 shows the mathematical expression, derived by Abbe, related to the minimum distance that can be resolved using VLM:

$$d = \frac{0.61\lambda}{n\,sen\theta} \tag{1}$$

In this equation, d is the resolution of VLM; λ is the visible light wavelength used to illuminate the studied sample; n corresponds to the refraction index where the observation is performed; while θ corresponds to the magnification lens' collection semi-angle. The term in the denominator, $n\,sen\,\theta$, is known as the numerical aperture. The resolution in VLM is defined based on Rayleigh's criterion, as the minimum resolvable distance between two image source points, when the first diffraction minimum of one source point coincides with the maximum of the other. In accordance with Equation 1, in a conventional VLM, under the illumination of blue light ($\lambda = 400$ nm), the numerical aperture reached a value very close to unity several years ago; as a result, the resolution limit reached by conventional VLM is approximately 250 nm. Technical advances in VLM allowed it to reach this theoretical limit around 1930 (Judith et al. 2012), driven mostly by the requirements set by microbiology, and their need to study the inner structure of different cells and microorganisms. Nonetheless, conventional VLM had reached a limit that, according with Abbe himself, human ingenuity would not be capable to surpass.

However, in 1924 French physicist Louis de Broglie published his doctorate dissertation entitled "Research on the theory of the quanta" (Louis de Broglie 1925), where he theorized about the wave nature of electrons, and how these particles show the same dual wave-particle nature

---

[1] Universidad Autónoma de Nuevo León, UANL, Facultad de Ingeniería Mecánica y Eléctrica, FIME, Av. Universidad S/N, Cd. Universitaria, San Nicolás de los Garza, Nuevo León, C.P. 66450, México.

[2] Universidad Autónoma de Nuevo León, UANL, Centro de Innovación, Investigación y Desarrollo en Ingeniería y Tecnología CIIDIT, Apodaca, Nuevo León, C.P. 66628, México.

* Corresponding author: domingo.garciagt@uanl.edu.mx

proposed by Einstein for light, in his seminal work published on 1905 about the photoelectric effect. The first experimental evidence on the wave nature of electrons was observed by C.J. Davisson and L. Germer in 1927. In an experiment investigating the elastic scattering of electrons by the surface of a nickel crystal, the scientists noticed that electrons emerged from the crystal at unexpected angles. These angles corresponded to the angles predicted by the constructive interference condition proposed by Bragg's law. Knowing the crystal interplanar spacing and measuring the diffraction angles, Davisson was able to determine the wavelength of the electrons and found it to coincide with the de Broglie wavelength based on Equation 2.

$$\lambda = \frac{h}{m_e v} \tag{2}$$

In Equation 2, λ is de Broglie's wavelength, h is Plank's constant, m_e is the mass of the electron at rest and v is the velocity of the electron. After this experimental confirmation of his theory, Louise de Broglie was awarded with the Nobel Prize in Physics in 1929 "for his discovery of the wave nature of electrons", as stated on the webpage of the Nobel Prize organization (https://www.nobelprize.org/prizes/lists/all-nobel-prizes-in-physics/).

These advancements in the understanding on the nature of elementary particles, and their behavior, inspired German physicist Ernst Ruska and German engineer Max Knoll to develop a microscope based on electrons. They proposed that a microscope using electrons to illuminate the samples, with wavelengths thousands of times smaller than those of visible light, could provide a much more detailed picture of objects than the images provided by conventional VLM. Based on this idea, they designed and constructed the first transmission electron microscope (TEM) in history in 1933. Ernst Ruska was awarded with the Nobel Prize in Physics in 1986 "for his fundamental work in electron optics, and for the design of the first electron microscope", as stated on the webpage of the Nobel Prize organization (https://www.nobelprize.org/prizes/lists/all-nobel-prizes-in-physics/).

2. Transmission electron microscope

TEM equipment comprises several systems working together to assure a proper performance. The electron gun, or emitter, is the system responsible for producing the electron beam. There are two major commercial types of electron emitters, based on the fundamental phenomenon responsible for the electron emission: thermionic emitters and field effect gun (FEG) emitters. Thermionic emitters are made of tungsten or LaB_6. As suggested by their name, the thermionic emission phenomenon is responsible for the electron emission; hence, they operate at high temperatures ($\sim$ 2800 K for W and $\sim$ 1900 K for LaB_6). Operation vacuums are in the range between $10^{-5} - 10^{-6}$ mbar, while the emission current densities are in the range between $3 - 30$ A/cm^2, with effective source radii between $1{,}500 - 5{,}000$ nm; these two last parameters limit the "smallest" area that can be analyzed to be in the order of several nanometers. On the other hand, FEG emitters are made mostly of tungsten, and there are two types commercially available, Schottky and Cold-FEGs. A promoted tunneling effect, due to the morphology of the emitter's tip and the use of high electric fields, is the mechanism responsible for the electron's emission. They operate at lower temperatures than thermionic emitters ($\sim$ 1800 K for Schottky and $\sim$ 300 K for Cold-FEGs). Operation vacuums for this technology are in the range between $10^{-9} - 10^{-10}$ mbar, while their emission current densities are in the range between $5{,}000 - 17{,}000$ A/cm^2, with effective source radii between 2.5 nm $-$ 15 nm; these two last parameters are responsible for excellent illumination in small regions, allowing the acquisition of high quality high resolution TEM (HRTEM) images. Once the electrons are generated from the emitter, they enter the acceleration ring, where they will be subject to the acceleration voltage, which is a TEM's operational parameter defined by the user. This acceleration voltage will define the final wavelength display by the electron beam, based on the following equation:

$$\lambda = \frac{h}{\sqrt{2m_eVe}\sqrt{1+\dfrac{eV}{2m_ec^2}}} \tag{3}$$

In this expression, λ is de Broglie's wavelength, h is Plank's constant, m_e is the mass of the electron at rest, V is the acceleration voltage, e is the charge of the electron and c is the speed of light in vacuum.

After being accelerated to the desired voltage (velocity), the electrons enter the illumination system. This system is composed of the electromagnetic lenses and apertures that allow the manipulation of the electron beam. The electromagnetic lenses allow to manipulate the electron beam by the generation of a magnetic field produced by an electrical current moving through a copper coil, and the electric filed produced by the voltage applied between two metal plates, known as the polar pieces. The force exerted over the electrons in the traveling beam is described by the Lorenz force (Equation 4).

$$\overline{F} = e(\overline{E} + \overline{v} \times \overline{B}) \tag{4}$$

In Equation 4, F is the total force experienced by the electron beam, e is the charge of the electron, E corresponds to the applied electric field, v is the velocity of electrons and B corresponds to magnetic field applied. As it can be deduced from the Lorenz force equation, applying an electric and magnetic field to a moving electron produces a force that allows its manipulation, whereas the apertures restrict the electron beam's angular spread when entering the electromagnetic lenses.

The vacuum system is very important within a TEM, since electron microscopy relies on the fact that the emitted and accelerated electrons travel in an environment free of anything that could potentially affect their trajectory, while traveling down the emission gun to interact with the sample. The vacuum requirements are mainly defined by the type of the electron source featured by the electron microscope (EM), as previously discussed. All vacuum systems include different types of pumps working together to generate the operational vacuum. In general, modern EM instruments have a mechanical, or roughing, pump to generate the initial vacuum, normally reaching vacuum levels of $\sim 10^{-3}$ Torr. The next vacuum stage is reached with turbomolecular pumps, attaining vacuum levels in the range between $10^{-3} - 10^{-11}$ Torr. In many contemporary thermionic electron source EM, these are the only two pump types forming the vacuum system. However, FEG electron sources equipment always displays a third type of vacuum pump, known as Ion Getter Pumps (IGP). This type of pumps can reach vacuum levels between $10^{-6} - 10^{-12}$ Torr, which are mainly responsible for maintaining the ultra-high-vacuum (UHV) conditions required for proper FEG instruments' operation.

The wave nature of electrons, and its tunable wavelength, is the main reason behind the development of EM. Nevertheless, electrons interact stronger with matter than photons, producing a wide variety of secondary signals not available in VLM. Figure 1 shows a schematic of the main signals generated by the electron scattering of a high energy electron beam interacting with a thin sample (for the rest of the chapter, we'll assume all samples are electron transparent, which is a requirement for a proper analysis in TEM). Each of this signals provide information about different aspects of the analyzed sample, when properly collected and interpreted, these signals provide information about the sample's morphology, chemical composition, crystalline structure, atoms' valence states, atomic chemical bonding interactions, atomic radial distribution functions, optical and electrical properties, among others.

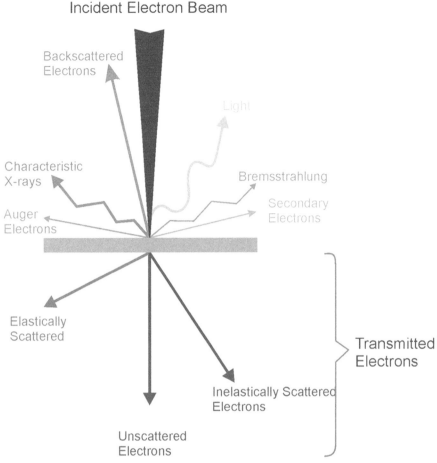

Figure 1. Electron scattering signals produced by the interaction of a high energy electron beam with a thin sample (Garcia 2006).

The different detectors featured by the EM will determine which techniques can be used in that instrument. These set of detectors determine the capabilities displayed by the EM and form the detection system.

In Figure 1 can be observed the main signals generated due to the interaction between a high energy electron beam and a thin specimen. The signals observed "over" the specimen are mainly used in scanning electron microscopes (SEMs), with the exception of the "Characteristic X-Rays", that is equally important in SEM and TEM studies. The signals identified as the "Transmitted Electrons", and showed "below" the specimen, are the main signals used in TEM. Within these signals is contained the information provided by the different TEM techniques available for the characterization of nanoparticles. In the following sections of this chapter, the fundamentals of the main techniques currently used in the characterization of nanoparticles will be briefly described.

3. TEM and HRTEM

Different mechanisms of image formation are related to different TEM techniques. The most basic one relates to the projection mass thickness contrast mechanism (Figure 2a). In this image formation mechanism, the bright areas observed in the TEM images correspond to areas within the specimen

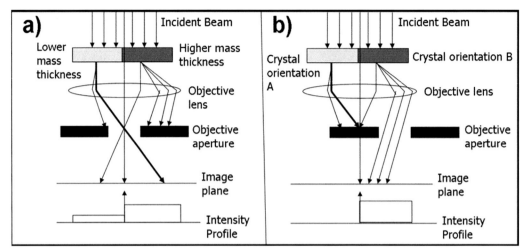

Figure 2. TEM image formation mechanism corresponding to: (a) mass contrast and (b) diffraction contrast (Garcia 2006).

that allow the electron beam transmission through the sample in the forward direction. The images formed through this mechanism correspond to maps of the projected mass thickness within the sample. This image formation mechanism can be easily assimilated if the simple expression used to describe the absorption of radiation by materials (Beer's Law) is remembered.

$$I = I_0 e^{-\left(\frac{\mu}{\rho}\right)\rho t} \tag{5}$$

In Equation 5, I_0 is the original intensity of the electron beam before interacting with the sample, ρ is the sample's density, t is its thickness and μ/ρ is the mass absorption coefficient. This image formation mechanism is predominant in amorphous samples and it is of main significance in biological samples; it exploits, mainly, the unscattered electrons that went through the sample without suffering any interaction (Garcia 2006).

Another basic contrast mechanism for image formation is related to the diffraction contrast mechanism, which is fundamentally important in crystalline samples. A crystalline sample may produce changes in the electron beam trajectory by diffracting the beam off the optic axis of the EM. Then, contrast within the TEM image can be produced between these regions deviating the electron beam and the regions of the sample that do not affect the beam's trajectory (Figure 2b). In this contrast mechanism, the use of an objective aperture is primordial, since this aperture allows to select which diffracted beams will be used for the image formation. A diffraction contrast image can be thought to be something like a map showing the regions within the sample with different crystal orientations. TEM images based on this contrast mechanism will display "bright areas" in the regions where the sample diffracted electrons in the direction selected by the objective aperture, and will display "dark areas" in the regions where the sample diffracted the electron beam away from the objective aperture. One basic approach to describe the constructive interference condition required for diffraction to occur is given by Bragg's Law (Garcia 2006):

$$\lambda = 2d_{hkl}\sin(\theta) \tag{6}$$

In this expression, λ is the wavelength of the radiation, d_{hkl} is the interplanar distance between crystal family planes (hkl) and θ is Bragg's angle, where the electron waves are diffracted. Whenever this relationship between the incident radiation wavelength, interplanar distance within the crystal structure and the orientation of the crystal planes with respect to the electron beam is fulfilled, strong Bragg diffraction spots are expected to appear in the diffraction pattern. Experimentally demonstrated by Davisson, as previously discussed, electron diffraction patterns are produced whenever an electron beam interacts with a crystalline material, as long as the electron beam's characteristic wavelength is in the same order of magnitude as that of the interplanar distances found in the crystalline material. This is always the case for commonly acceleration voltages used in commercial TEMs. Figure 3 shows a typical electron diffraction pattern for a sample of nanoparticles. In this figure, the bright spots correspond to Bragg's spots diffracted to the Bragg angle, θ, measured with respect to the TEM's optical axis, corresponding to the families of planes indicated in the figure; all the Bragg spots at the same distance from the center of the diffraction pattern correspond to the same family of planes.

Diffraction contrast constitutes the basic contrast mechanism for two very important imaging modes in conventional TEM: bright field (BF) and dark field (DF). In BF mode, the TEM image is formed with the sample's regions which are not diffracting the electron beam away from the EM's optic axis while in DF, TEM images show those sample's regions diffracting the electron beam off the optic axis of the electron microscope, such that certain Bragg's spots pass through the objective aperture, located in the EM's objective lens back focal plane, considering that the objective aperture can be manipulated to select the desired Bragg's spots to be used to form the DF image (Williams and Carter 1996). Since diffraction events are one kind of elastic scattering, the discussed

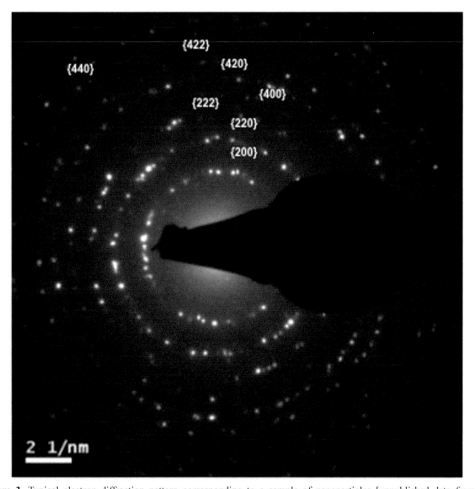

Figure 3. Typical electron diffraction pattern corresponding to a sample of nanoparticles [unpublished data from the chapter's author].

techniques are based on the collection of elastically scattered electrons, however with low scattering angles, corresponding to the Bragg angles. These three contrast imaging mechanisms constitute the basic imaging techniques in conventional TEM.

High resolution imaging (HRTEM) is one of the TEM techniques most used to characterize nanoparticles. In this technique, the imaging contrast mechanism is the phase contrast imaging; for this particular contrast mechanism, it is an important requirement that the sample is considered to be a "phase object", which means that the electron beam won't suffer multiple scattering events within the sample. This translates to a "very thin TEM sample", where single scattering events are assured; once this requirement is fulfilled, the sample can be considered a "weak phase object". This "weak phase object" changes the phase of the electron waves within the electron beam traveling through the sample; however, it does not affect its amplitude. How much the phase of the electron beam would be altered, would be proportional to the crystal potential related to the specimen (phase object) (Garcia 2006).

$$\psi_e = \psi_0 e^{-i\sigma\phi p} \qquad (7)$$

ψ_e is the wave function representing the electron beam at the "lower" surface at the time it "leaves" the specimen, ψ_0 is the wave function of the original incident electron beam, σ is a proportionality constant relating the phase shift and the crystal potential.

$$\phi_P = \int_0^t \phi(x, y, z)dz \qquad (8)$$

φ_p is the projected crystal potential that varies as the position within the crystalline sample changes, $\varphi(x,y,z)$ is the potential energy associated to the electron in the beam traveling through the crystal as a function of its position, while t represents the specimen's thickness (phase object) (Garcia 2006).

The phase contrast image formation mechanism can be realized as follows (Figure 4): the incident electron waves fall upon the sample's surface; this sample is very thin and with an orientation such that the electron waves travel parallel to the atomic sites corresponding to the nuclei. As these electron waves travel through the crystal, those electrons passing through these sites corresponding

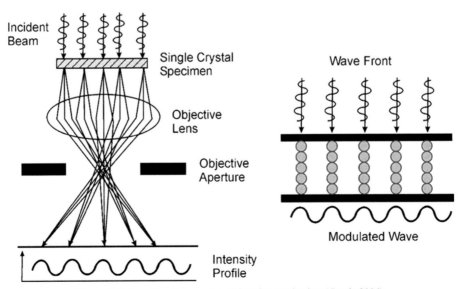

Figure 4. Schematic of phase contrast imaging mechanism (Garcia 2006).

to the position of the atoms' nuclei will be phase delayed relative to the electron waves passing in between the atomic columns. The specimen's periodic crystal potential affects the electron waves' phase passing through it. When the electron waves reach the specimen's exit surface, the electron beam's wavefront displays a phase modulated by the crystal potential of the sample, and these phase modulations transmit information about the sample, such as the atomic positions within the crystal structure. This modulated wavefront interferes with itself as it travels through the electron microscopes' objective lens, carrying this crystallographic information (Garcia 2006).

However, based on the image formation mechanism in electromagnetic lenses, no contrast would be observed in the image plane of an aberration-free electromagentic lens perfectly focused. It is necessary to translate the phase information carried by the electron waves into amplitude information to generate contrast. This procedure could be carried out by slightly defocusing the objective lens. Nonetheless, real electromagnetic lenses display several aberrations, the spherical aberration (C_s) being the most important one. This aberration is responsible for the translation of phase information to amplitude information in focused images; however, it is also responsible for limiting the ultimate achievable resolution in the TEM. Aberrations, defocus and the characteristics of the electromagnetic lenses have a direct effect on the TEM image formed; all possible effects are incorporated to the image in the objective lens' back focal plane, where the amplitude is proportional to the Fourier transformation of the electron wave function at the exit surface of the specimen (Garcia 2006).

$$\chi(g) = (\pi \lambda g^2) \left[\frac{(C_s \lambda^2 g^2)}{2} + \Delta f \right] \tag{9}$$

In Equation 9, χ is known as the contrast transfer function, g is a reciprocal lattice vector associated to a Bragg spot in the diffraction pattern, measured in nm^{-1}, λ is the wavelength of the electron beam, C_s is the spherical aberration displayed by the objective lens and Δf is the defocus value. The electron wave function's Fourier transform at the exit surface of the specimen ($F\{\Psi_e\}$) is multiplied by the factor exp[$-\chi(g)$], which contains the effect the objective lens has on the image formation (aberration, defocus, etc.). However, the important effect is related to the phase contrast; hence, the imaginary term of the Euler expansion of the expression exp[$-\chi(g)$] (sin[$\chi(g)$]) is the only one that matters (Garcia 2006).

$$F\{\psi_e\} \sin \chi(g) \tag{10}$$

The wave function of the electron beam exiting the objective lens' back focal plane is the Fourier transform of the wave function of the electron beam at the exit surface of the specimen, multiplied by the contrast transfer function corresponding to the TEM's objective lens. This wave function will be Fourier transformed, again, at the time it leaves the objective lens, as it propagates to the image plane (Garcia 2006).

The resolution and contrast displayed by TEM phase contrast images are inherently interconnected. Sample thickness variations, or objective lens defocus variations, might produce contrast reversals in phase contrast images; hence, "bright spots" can turn "dark" just by changing the defocus value in the objective lens. Therefore, in order to interpret HRTEM images in a reliable manner, it is necessary to perform image simulations to accurately account for all the factors influencing image formation (Williams and Carter 1996).

4. STEM

Among the TEM techniques that have gained an increasing popularity for the characterization of nanoparticles in the last ten years, the Scanning Transmission Electron Microscopy (STEM) can be highlighted. In this technique, the electron beam is focused on the surface of the sample by the

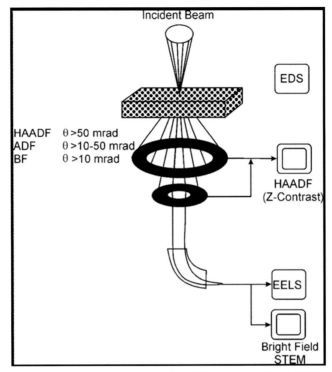

Figure 5. Schematic representation of the STEM technique (Garcia 2006).

objective lens to form an atomic scale probe; this electron probe is "scanned" across the specimen's surface, while all scattered electrons, and secondary signals can be collected by the variety of detectors available in the TEM. In this technique, images are generated by scanning the focused electron beam over the specimen surface, while the scattered or generated signals are collected to form the different types of images or spectra, each signal carrying different information about the sample. On this aspect, there is a great similarity between Scanning Electron Microscopy (SEM) and STEM. Henceforth, a STEM image can be considered as a collection of individual scattering experiments, where different signals can be discriminated by their scattering angle and/or energy loss, generating information about the specimen's structure and/or chemical information that can be collected simultaneously by different detectors (Figure 5) (Garcia 2006).

Electrons scattered elastically are used to form a wide variety of signals when collected by the proper detectors (such as HAADF, ADF, SAED), most of them related to imaging and specimen's crystal structure. On the other hand, electrons scattered inelastically, or the signals produced by such interactions, can be acquired by a different set of detectors (such as EELS and EDXS detectors); these signals provide information about the specimen's chemical, electronic and structural characteristics. Both sets of signals can be acquired simultaneously and in a very efficient manner, allowing the performance of quantitative analyses with spatial resolutions impossible to achieve in other characterization methodologies, making STEM, and their related signals, one of the most powerful analytical techniques available for microanalysis at high-spatial resolution (Browning et al. 1997).

5. HAADF

One of the most important STEM signals used in the characterization of nanoparticles is the High Angle Annular Dark Field (HAADF) technique, also known as Z-contrast. The HAADF images are produced by collecting, with an annular detector, the incoherently elastically scattered electrons that were scattered to high-angle (> 50 mrad) (Figure 5), also known as Howie detector (Howie 1979). The most relevant aspect related to this technique is its chemical sensitivity, since the scattering factors associated to the atoms of different elements are closely related to the nuclear scattering factors, also known as Rutherford-like scattering factors. In this regard, an accurate estimate to determine the scattering cross-section associated to different elements can be obtained from the relativistic Rutherford differential cross-section:

$$\frac{d\sigma(\theta)}{d\Omega} = \frac{\lambda_R Z^2}{64\pi^4 a_0^2 (\sin^2\frac{\theta}{2} + \left(\frac{\theta_0}{2}\right)^2)^2} \tag{11}$$

In Equation 11, σ denotes the atomic interaction cross-section, Ω is the solid angle, λ_R corresponds to the relativistic wavelength of the electrons, Z is the atomic number for the different elements, a_0 corresponds to the Bohr radius, θ denotes the scattering angle, while θ_0 denotes the screening parameter (Williams and Carter 1996).

In STEM mode, particularly in HAADF images, it is important to consider each atom as an independent scatterer; then, any atom's scattering factor can be replaced by the cross section calculated from Equation 11, where a clear $\sim Z^2$ dependence can be observed. Hence, the intensity of the signal observed in this technique is directly proportional to the atomic number of the elements contained in the sample. The spatial resolution in STEM mode techniques is only limited by the probe size, which can be directly understood as the diameter of the electron probe formed by the microscope's illumination system (Browning et al. 1997), which is strongly determined by the coefficients of spherical aberration of the objective lens, C_s, the wavelength of the electron beam, λ, and the lens defocus, Δf.

Compared to phase-contrast methods, such as HRTEM, HAADF is a very attractive technique for composition mapping at atomic resolution and, in general, for high resolution imaging, due to its strong chemical sensitivity and the atomic cross-section scattering, where the atoms within the samples are responsible for the scattering of the electrons, allowing for a direct interpretation of the images, and not requiring the image simulation mentioned in the HRTEM technique.

6. EDXS and EELS

Energy Dispersive X-Ray Spectroscopy (EDXS) and Electron Energy Loss Spectroscopy (EELS) are two techniques highly intertwined since both are related to the same inelastically scattered electron event. One of the most useful inelastic interactions produced between the electron beam and the specimen is the one generated when an incident electron from the beam collides with an inner-shell atomic electron. The total energy during the collision needs to be conserved, as stated by the energy conservation law; hence, the incident electron loses an amount of energy equal to the amount of energy gained by the bombarded inner-shell atomic electron (Garcia et al. 2017). With conventional acceleration voltages used in TEM instruments (80 kV–300 kV), the electron beam energy is high enough to transmit the required energy to the inner-shell atomic electrons to be expelled from the atom, leaving the atom in a highly excited and unstable condition, from which the atom needs to come out as quickly as possible. There are two mechanisms the atoms can use to get rid of this excess of energy and go back to a more stable state. In one mechanism, an outer-shell atomic electron undergoes a downward transition to fill the vacant space left by the expelled inner-shell electron, and the excess energy is released as an electromagnetic wave (X-rays), with an energy equal to the energy difference between the atomic electronic levels involved in the electronic transition, as stated in Bohr's atomic model (Garcia et al. 2017). In the other mechanism, the excited atom releases its excess energy as kinetic energy of another outer-shell electron, from the same energy level originally occupied by the outer-shell electron that made the electronic transition to fill the vacant space left in the inner-shell atomic level. These emitted electrons will have the same energy as the energy difference between the atomic electronic levels involved in the electronic transition. This phenomenon is known as Auger emission, and the emitted electrons through this mechanism are known as Auger electrons. These two energy relaxation mechanisms compete; however, heavier elements tend to use the emission of X-Rays, while lighter elements tend to use the emission of Auger electrons. A graphical description of these mechanisms is illustrated in Figure 6.

As previously discussed, the energy of the X-Rays emitted by the atoms when returning to a more stable configuration from an excited state is equal to the energy difference between the atomic electronic levels involved in the electronic transition; thus, the characteristic X-Rays emitted by the atoms are related to their electronic structure, which is unique in every atom in the periodic table. Hence, every element in the periodic table has its own family of characteristic X-Rays, allowing the identification of the presence of the different elements composing the analyzed sample.

On the other hand, the original electron from the beam colliding to the inner-shell atomic electron, responsible for the generation of the characteristic X-Ray, lost the amount of energy required to free the inner-shell electron from its atom. This electron from the beam no longer has its original energy, associated to the acceleration voltage used in the instrument; instead, this electron loses a specific amount of energy, related to the inelastic scattering event occurring when transmitting through the sample. These inelastic interactions arise primarily from electron-electron interactions, being the inner-shell ionization events previously described the most commonly used in the nanoparticles characterization, particularly to identify their chemical composition. These inelastic electron-electron interactions can provide chemical, structural, electrical, and optical information about the sample. Other important inelastic interactions, in order of importance, are phonon excitations, inter-band and intra-band electron transitions, plasmon excitations, and inner-shell ionizations. Electrons undergoing any of these feasible inelastic interactions produce the Electron Energy Loss (EEL) signal. An EEL spectrum possesses three characteristic regions (Figure 7): the zero-loss peak, which is formed by the transmitted electrons and contains information about the resolution of the spectrum (Ahn et al. 1983, Egerton 1979). Next, the low-loss region (< 50 eV) shows the signals related to the electrons from the beam that suffered an inelastic interaction with the weakly bound outer shell electrons in the sample or were responsible for producing a wide variety of collective or crystal vibrations in the sample, such as phonon vibrations, plasmon oscillations, inter- or intra-band transitions; thus, it can provide information regarding the

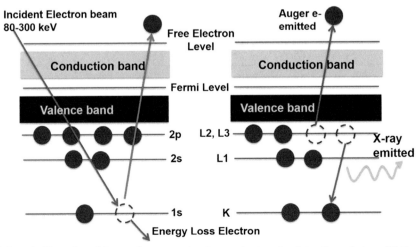

Figure 6. Schematic illustration of the atomic energy relaxation mechanisms involving the emission of X-Rays or Auger electrons [unpublished data from the chapter's author].

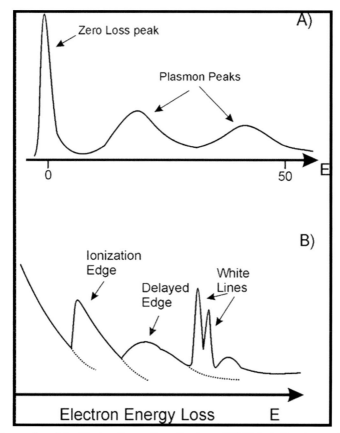

Figure 7. (a) Low-loss region of the Electron Energy Loss Spectrum (EELS), including the zero-loss peak (< 50 eV). (b) High-loss region of the spectrum, considering energies above 50 eV (Garcia 2006).

electronic and optical properties of the sample. Finally, the high loss region of the spectrum (> 50 eV) contains information related to the inelastic interactions of the electron beam with the inner-shell electrons, remembering that this type of interactions convey the information that allows to identify the elements found in the specimen, and how they are bonded. Ionization edges observed in different energies are related to different elements, allowing to identify which elements can be found in the sample, while differences in the fine structure of these ionization edges can be related to the chemical (bonding) state in which the atoms are found within the sample, and structural (atomic arrangement) characteristics of the analyzed sample. Moreover, by selecting electrons with a specific energy loss, with specialized slits, an image can be formed associated with them; in this way, the elemental distribution inside the specimen can be directly visualized (Elemental Mapping).

7. TEM characterization of nanoparticles

Metallic nanoparticles have been widely used in biotechnology applications (Morones et al. 2005, Elechiguerra et al. 2005, Martinez Sanchez et al. 2007). One of the most known cases is the bactericidal effect several metallic nanoparticles have shown, particularly the case of silver (Morones et al. 2005). The antibacterial properties of silver have been known and used by humanity for a very long time; however, nanotechnology allowed a better understanding of this property, and opened new areas that allowed its improvement. The higher reactivity displayed by metallic nanoparticles, related to their higher surface-to-volume ratio along with their crystallographic surface structure, has been proven to affect the bactericidal properties of silver nanoparticles, since nanoparticles in the size range between 1–10 nm have been proven to show an increased interaction with bacteria, mainly due to the fact that small silver nanoparticles are highly faceted, with a high density of (111) facets on their surface. This (111) facets have been observed to strongly interact with the membrane of bacteria, being able to even penetrate within the bacteria (Morones et al. 2005). Another case where the size and surface crystallographic arrangements of silver nanoparticles was determinant in the interaction with another biomolecule or microorganism was in the interaction with the HIV-I virus (Elechiguerra et al. 2005). In this case, silver nanoparticles within this size range attached to the surface of the HIV-I virus; however, the arrangement of the nanoparticles on the surface of the viruses suggest they interact with the virus by preferentially interacting with the glycoprotein gp 120. This conclusion was drawn from two known facts: the center-to-center distance between the nanoparticles coincides with the distance between this type of glycoproteins, and the fact that the gp 120 glycoprotein knobs have exposed sulfur-bearing residues that would be attractive sites for nanoparticle interaction, specially metallic nanoparticles. As a consequence, this interaction between the silver nanoparticles and the gp 120 glycoprotein would inhibit the virus from binding to the cells (Elechiguerra et al. 2005). For these applications, it is crucial to possess the ability to measure, with the highest accuracy, the size of the silver nanoparticles used, and only electron microscopy techniques allow to measure nanoparticles in that size range in a reliable and reproducible manner. Additionally, HRTEM images, in combination with HRTEM image simulations, allow to determine the crystallographic facets found on the surface of the nanoparticles, which has been proven to be very important in determining the interaction between the microorganisms and metallic nanoparticles (Figures 8 and 9).

Specific crystallographic facets forming on the nanoparticles' surface are determined by the nucleation and growth mechanism occurring during their synthesis. One of the most influential factors determining the crystallographic facets found on the surface of nanoparticles is the capping ligand used during the synthesis (Garcia-Gutierrez et al. 2018). In Figure 10, HRTEM images from semiconductor nanoparticles are shown, indicating the presence of two types of crystallographic facets on their surface, (100) and (111) facets. In that particular study, the capping layers based on shorter C-chain carboxylic acids increased the presence of (100) facets on the surface of the nanoparticles, while longer C-chain carboxylic acids increased the presence of (111) facets; these variations had an important effect on the final electronic structure displayed by the semiconductor nanoparticles as well (Garcia-Gutierrez et al. 2018).

Other important applications of metallic and oxide nanoparticles in the biotechnology field are associated to their ability to deliver immunotherapeutic agents to specific target areas, related to their potential for dense and versatile surface functionalization (Evans et al. 2018, Raymond et al. 2012, Burt et al. 2004, Wang et al. 2011, Moreno-Cortez et al. 2015), their capacity to be used for optical or heat-based therapeutic methods, such as hyperthermia induction (Evans et al. 2018, Garza-Navarro et al. 2010, Torres-Martinez et al. 2014, Torres-Martinez et al. 2013, Ravichandran et al. 2015, Ravichandran et al. 2015, Torraya-Brown et al. 2014), and the biolabeling capabilities that have shown several semiconductor nanoparticles (Cheng-An et al. 2007, Gee and Xu 2018, Campargue et al. 2020).

Gold nanoparticles are among the most used when immunotherapeutic applications are intended. This is due to several factors; among them, the following can be highlighted: the wide variety of synthesis methods available, the tunability these methods offer on their final morphological characteristics, along with their optical response, and the flexibility these synthesis methods offer on the gold nanoparticles' surface functionalization (Raymond et al. 2012, Burt et al. 2004, Wang et al. 2011). In all these aspects, the reliable characterization offered by TEM is vital in determining the final characteristics displayed by the gold nanostructures intended for different biotechnological applications. Burt et al. (Burt et al. 2004) developed a synthesis method that used bovine serum albumin protein as the protective agent, to make gold

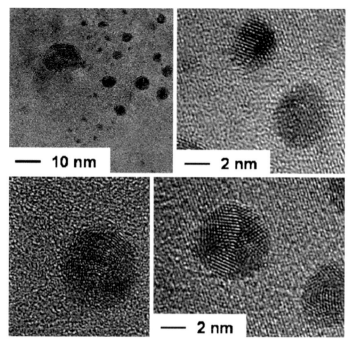

Figure 8. TEM and HRTEM images of Ag nanoparticles. Reprinted from Journal of Alloys and Compounds, 438, R. Martinez-Sanchez et al. Mechanical and microstructural characterization of aluminum reinforced with carbon-coated silver nanoparticles, p. 195–201, Copyright (2021), with permission from Elsevier.

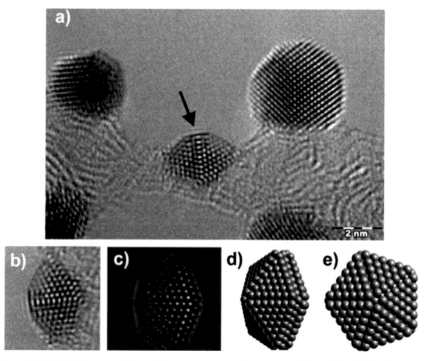

Figure 9. (a) and (b) HRTEM images of bimetallic nanoparticles, (c) image simulation of the HRTEM image; (c) and (d) simulation model showing the proposed structure and morphology of the nanoparticle. Reprinted figure with permission from J.L. Rodriguez-Lopez et al. *Physical Review Letters*, 19: 196102, 2004. Copyright (2021) by the American Physical Society.

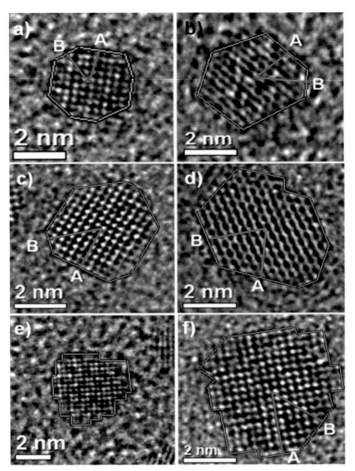

Figure 10. HRTEM images of semiconductor nanoparticles with different capping ligands: (a) and (b) oleic acid; (c) and (d) hexanoic acid; and (e) and (f) myristic acid. D.F. Garcia-Gutierrez, L.P. Hernandez-Casillas, M.V. Cappellari, F. Fungo, E. Martínez-Guerra and D.I. García-Gutiérrez. Influence of the capping ligand on the band gap and electronic levels of Pbs nanoparticles through surface atomistic arrangement determination. *ACS Omega*, 3(1): 393–405 (2018). https://pubs.acs.org/doi/abs/10.1021/acsomega.7b01451. Further permissions related to the material excerpted should be directed to the ACS.

nanoparticles compatible with biotechnology applications. The synthesized gold nanoparticles were characterized by means of TEM techniques. STEM-HAADF studies allowed the easy observation and reliable measurement of the synthesized gold nanoparticles "attached" to the bovine serum albumin protein "film". Figure 11 shows the STEM-HAADF images where gold nanoparticles can be observed as "bright spots", since the intensity in this technique is directly related to the atomic number, Z, of the atoms found in the sample; since gold has a high atomic number (Z = 79), the gold nanoparticles can clearly be discerned from the organic bovine serum albumin protein surrounding them. The synthesized nanoparticles displayed average sizes between 1.4 nm and 1.8 nm; characterization of such small nanoparticles, particularly their size, is more reliable when performed via STEM-HAADF analysis, compared to conventional TEM or HRTEM. The HRTEM studies of the gold nanoparticles showed that the morphology and shape of the nanoparticles with such small average size varied from pentagonal, octahedral, to quasi-spherical (Figure 11 c–e).

Metallic nanostructures, such as gold nanorods covered by a mesoporous silica capping layer, have also been studied as probes for imaging cancer cells (Wang et al. 2011). In this particular system, the fluorescence and Surface Enhanced Raman Scattering (SERS) signals can be generated

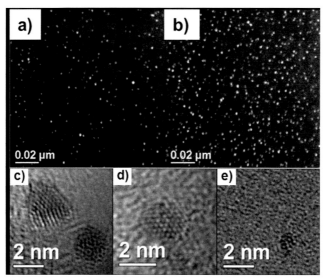

Figure 11. Au nanoparticles synthesized with bovine serum albumin protein as protective agent. (a) and (b) STEM-HAADF images. (c)–(e) HRTEM images showing pentagonal, octahedral, and quasi-spherical Au nanoparticles. Adapted with permission from Justin L. Burt, Claudia Gutierrez-Wing, Mario Miki-Yoshida and Miguel Jose-Yacaman. Noble-metal nanoparticles directly conjugated to globular proteins. *Langmuir*, 20: 11778–11783 (2004). Copyright (2021) American Chemical Society.

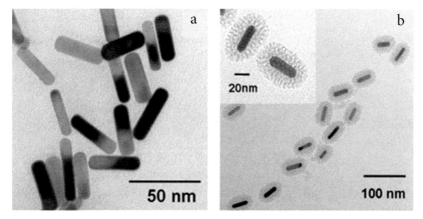

Figure 12. (a) Conventional TEM images of Au nanorods. (b) Conventional TEM images of Au nanorods covered with a mesoporous silica layer. Reprinted from biosensors and bioelectronics, 26, Zhuyuan Wang et al. Dual-mode probe based on mesoporous silica coated gold nanorods for targeting cancer cells. p. 2883–2889, Copyright (2021), with permission from Elsevier.

and processed independently, depending on the excitation wavelength used during the analysis. Moreover, these signals can potentially be used for multiplexed imaging in living cells, compared with traditional imaging probes. In this type of nanostructures, an accurate measurement of the nanorods' dimensions is very important, since the specific dimensions will determine the plasmon excitation energies displayed by the nanorods (Figure 12). Moreover, conventional TEM can also show the mesoporous silica layer covering the gold nanorods, allowing the measurement of the thickness of the capping layer (Figure 12B).

Magnetic nanoparticles have been widely studied for biotechnology applications, such as contrast agent for MRI studies, drug delivery agents, and hyperthermia inductors (Garza-Navarro

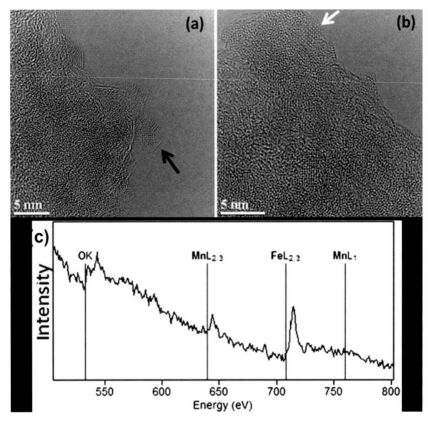

Figure 13. (a) and (b) IC-TEM images of MnFe$_2$O$_3$ nanoparticles. (c) EELS spectrum of nanoparticle indicated in (a) and (b) confirming the desired composition. Adapted with permission from Garza-Navarro, M.A., Torres-Castro, A., Garcia-Gutierrez, D.I., Ortiz-Rivera, L., Wang, Y.C. and Gonzalez-Gonzalez, V.A. Synthesis of spinel-metal-oxide/biopolymer hybrid nanostructured materials. *Journal of Physical Chemistry C*, 114(41): 17574–17579 (2010). Copyright (2021) American Chemical Society.

et al. 2010, Torres-Martinez et al. 2014, Torres-Martinez et al. 2013, Ravichandran et al. 2015, Ravichandran et al. 2015, Torraya-Brown et al. 2014). Hybrid nanostructured materials, combining a magnetic core with an organic shell are the typical structures sought for these kind of applications. For the case of the magnetic core, oxide nanoparticles are the most commonly used, in particular the iron oxide nanoparticles. However, it is important to mention that a reliable characterization of the synthesized iron oxide nanoparticles is extremely important since slight variations in the crystal structure of the nanoparticles can translate into large variations in their magnetic properties. Hence, a reliable methodology to characterize their composition, morphology and crystal structure is vital, with TEM, and their related techniques, being the most used and trusted techniques to perform such characterization.

A hybrid nanostructured material (HNM) based on spinel metal oxide nanoparticles and a biopolymer was synthesized and characterized by Garza-Navarro et al. (Garza-Navarro et al. 2010). In that study, the authors synthesized spinel metal oxide (SMO) nanoparticle of MnF$_2$O$_4$ embedded in a chitosan biopolymer matrix. However, due to the small size of the nanoparticles (size range between 1.8 nm–8 nm) and the chitosan matrix around them, a reliable confirmation of the chemical composition was very challenging. For this case, an image C$_s$ corrector microscope (IC-TEM) was employed to perform the study. Image and probe C$_s$ correctors were developed in the early 2000s, and they allowed to improve the resolution of electron microscopes down to their current limits, close to 50 pm

(Judith et al. 2012). With this improved resolution, and in combination with an EELS analysis, it was possible to confirm the desired composition of the synthesized $MnFe_2O_4$ nanoparticles. In the EELS spectrum shown in Figure 13c, clear oxygen (O), manganese (Mn) and iron (Fe) signals can be detected in the nanoparticles indicated with arrows in the IC-HRTEM images shown in Figures 13a and b.

In a continuation of the previous study, magnetic SMO nanoparticles stabilized in ovoid-like carboxymethyl-cellulose (CMC)/cetyltrimethylammonium-bromide (CTAB) templates were synthesized, and thoroughly characterized via TEM techniques (Torres-Martinez et al. 2013). This HNM was synthesized by controlling the hydrolysis of the inorganic iron salts, used as precursors, into aqueous dissolutions with CMC and CTAB. The competition between the CTAB and SMO nanoparticles to occupy the CMC intermolecular sites close to their carboxylates functional groups, allows to manipulate the size of the SMO nanoparticles and the CMC/CTAB ovoid like templates, just by adjusting the CTAB/SMO ratio, affecting not only the morphology of the HNM, but also its magnetic response (Torres-Martinez et al. 2013). Conventional TEM studies (Bright Field image) allowed the characterization of the shape and size of the organic CMC/CTAB template (Figure 14a), whereas STEM-HAADF studies enabled the observation of small SMO nanoparticles within the CMC/CTAB templates (Figure 14b), even when BF images didn't show the presence of such small nanoparticles within the organic template. This was possible due to the great atomic number difference between the organic CMC/CTAB template, and the composition of the

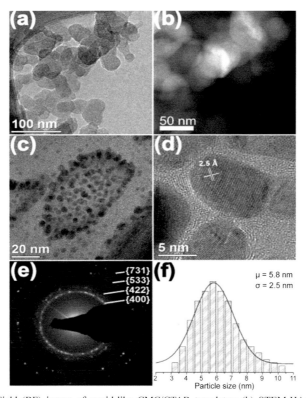

Figure 14. (a) Bright Field (BF) image of ovoid-like CMC/CTAB templates; (b) STEM-HAADF image that reveals the presence of "small" SMO nanoparticles dispersed in the ovoid-like CMC/CTAB templates; (c) BF image of quasi-spherical SMO nanoparticles arranged in an ovoid-like CMC/CTAB template; (d) HRTEM image of a quasi-spherical SMO nanoparticle dispersed in the CMC/CTAB template showed in (c); (e) indexed SAED pattern obtained from the template showed in (c); and (f) particle size distribution of the SMO nanoparticles. Reprinted from Materials Chemistry and Physics, 14, N.E. Torres-Martinez et al. One-pot synthesis of magnetic hybrid materials based on ovoid-like carboxymethyl-cellulose/cetyltrimethylammonium-bromide templates. p. 735–743, Copyright (2021), with permission from Elsevier.

SMO nanoparticles, and the fact that the signal intensity in this technique is directly related to the composition of the sample. In the synthesis conditions that produced "larger" nanoparticles, measuring and characterizing such SMO nanoparticles were not as challenging, their observation being obvious in BF images (Figure 14c). HRTEM images allowed the measurement of the interplanar distances present in the synthesized SMO nanoparticles (Figure 14d) and contributed to finally characterize the crystal structure of the nanoparticles through a selected area electron diffraction (SAED) study, confirming the spinel crystal structure obtained in the synthesized nanoparticles. This is an example of the versatility the TEM offers to characterize nanoparticles. Using one instrument was possible to characterize the organic template of the HNM, at least its size and shape; the size, morphology, composition, and crystal structure of the different magnetic SMO nanoparticles; and how these nanoparticles are distributed within the organic phase to form the HNM. All this information allowed the understanding, interpretation and even tuning of the magnetic properties displayed by the synthesized HNM.

In similar studies, Ravichandran et al. (Ravichandran et al. 2015, Ravichandran et al. 2014) have synthesized cubic spinel cobalt ferrite ($CoFe_2O_4$) nanostructures for possible drug delivery and/ or T_2 MRI contrast agent applications. In this case, the $CoFe_2O_4$ nanostructures displayed a needle-like morphology, with diameters around 15 nm and an aspect ratio close to 3:5. These nano-needles displayed a superparamagnetic behavior, with cell viability results that proved their biocompatibility for possible biotechnology applications. In this case, TEM conventional imaging (BF TEM) was used to study the size and morphology of the $CoFe_2O_4$ nanostructures synthesized; HRTEM results allowed the measurement of interplanar distances displayed by the nano-needles' crystal structure. Also, EDXS results confirmed the chemical composition of the synthesized nanostructures, showing clear signals for the cobalt (Co), iron (Fe) and oxygen (O) in the EDXS spectra acquired, and finally, SAED studies confirmed the cubic spinel cobalt ferrite crystal structure. Examples of all these analyses are shown in Figure 15.

Colloidal quantum dots (CQDs) of semiconductor materials have displayed attractive optical properties, in particular the manipulation of their optical absorbance and emission signals, based on their size and surface chemistry control, as a consequence of quantum confinement effects (Garcia-Gutierrez 2018, Garcia-Gutierrez 2013, Garcia-Gutierrez 2014, Gregg 2003, Wise 2000). When CQDs are photo-excited, electron-hole pairs, known as excitons, are generated. An exciton is a quasiparticle derived as a consequence of the coulombic interaction between a hole in the valence band, or highest occupied molecular orbital (HOMO), and an electron in the conduction band, or lowest unoccupied molecular orbital (LUMO) (Gregg 2003). Upon recombination of these electron-hole pairs, fluorescence light is emitted, where it is possible to manipulate the wavelength of the emitted photon by controlling the size of the CQDs, with smaller CQDs showing smaller wavelengths (Wise 2000). Also, it has been proved that the surface chemistry, and the atomistic surface arrangement, play an important role in determining the final wavelength displayed by photons emitted by CQDs (Garcia-Gutierrez 2018). In this way, it is possible to tune the "color" of the light emitted by CQDs just by changing some of the reaction parameters, such as reaction time, temperature, precursors' ratio, etc. Compared to common organic fluorophores, CQDs offer several interesting advantages, such as a continuous absorption spectra, narrow and symmetric emission spectra, significantly higher fluorescence lifetimes, and a reduced photobleaching (Cheng-An et al. 2007). The most common semiconductor CQDs system studied, regarding their emission properties, are the cadmium chalcogenides (Murray et al. 1993). However, most common CQDs are synthesized in organic solvents; hence, it is imperative to perform ligand exchange procedures that allow replacing the organic soluble capping ligand for an aqueous soluble one (Gee and Xu 2018, Yu et al. 2006). This process also has the goal to increase the biocompatibility of the CQDs, which commonly are composed by elements with a high toxicity, such as Cd and Pb.

One of the potential applications of such CQDs is cellular labeling (Cheng-An et al. 2007). CQDs' surface can be functionalized in order to specifically bound to antibodies targeting structures in cells; in this regard, CQDs can replace organic fluorophores as fluorescence marker of the antibody, taking advantage of all the emission properties previously described for the CQDs. Also, instead of labeling whole cells, specific molecules can be the target, making possible to track the movement of such molecules around the cells. CQDs can also be used to label cells and trace them in cell cultures. Labelling animal tissue is also a possible application of CQDs, aiming to develop the conditions to use them as contrast agents in human beings (Cai et al. 2006). In the particular application in human beings, CQDs have a great advantage, being possible to manipulate their emission wavelength deep into the infrared, just by adjusting their average size. Human skin can absorb light, which is a problem for labeling deep inside tissues; however, for the case of infrared light, human tissue only absorbs moderately this region of the electromagnetic spectrum.

All these possible applications rely on the capability to measure the size of the synthesized CQDs in a reliable and reproducible way. The final emission wavelength sought will determine the

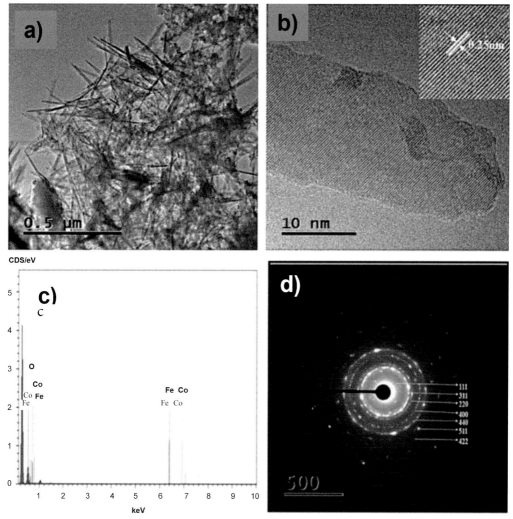

Figure 15. (a) Bright Field TEM image of the $CoFe_2O_4$ nano-needles, (b) HRTEM image of the $CoFe_2O_4$ nano-needles and inset shows interplanar distance appearing in the image. (c) EDXS spectra showing the elemental composition of the $CoFe_2O_4$ nano-needles. (d) Indexed SAED pattern of the $CoFe_2O_4$ nano-needles. Reprinted from Materials Letters, 135, M. Ravichandran et al. One-dimensional ordered growth of magneto-crystalline and biocompatible cobalt ferrite nano-needles, 67–70, Copyright (2021), with permission from Elsevier.

final size synthesized, which will also depend on the chemical composition. For the case of CQDs based on lead chalcogenides, when visible or near infrared emission is sought, their final size is close to 2 nm; measuring such small nanoparticles is more reliable in STEM-HAADF images than in conventional TEM or HRTEM images, as previously stated. The discussed characteristics of such signal reduced the uncertainty at the time of the size measurements in STEM-HAADF images (Figure 16), producing more reliable results. Ultrasmall lead sulfide CQDs were synthesized in the strong quantum confinement regime, displaying absorption and emission signals in the visible and near IR range (Torres-Gomez et al. 2020).

In addition to an accurate measurement of its size, it is also extremely important to determine CQDs' chemical composition, as well as their crystal structure. Again, EDXS and SAED technique prove to be very useful for these purposes. Figure 17 shows the EDXS spectra and SAED pattern acquired for a group of lead selenide CQDs; these results confirmed the expected chemical composition and stoichiometry (PbSe), as well as the expected FCC crystal structure.

Another aspect critical in determining the optical properties of CQDs is their surface capping layer. In general, the characterization of the chemical nature of the layer covering CQDs is performed with the use of FTIR and Raman spectroscopy techniques (Burt et al. 2004, Garza-Navarro et al. 2010, Torres-Martinez et al. 2014, Torres-Martinez et al. 2013, Gee and Xu 2018, Garcia. Gutierrez et al. 2013, Garcia-Gutierrez et al. 2014 Torres-Gomez et al. 2020). Nonetheless, TEM techniques such as STEM-HAADF and EELS can contribute in determining their dimensions and chemical composition (Garcia-Gutierrez et al. 2014). In the case of lead telluride CQDs, it was possible to study the effect of the synthesis conditions on the "final thickness" displayed by the capping layer on these lead chalcogenide CQDs. Aberration corrected STEM (AC-STEM) images allowed the clear observation of the interface between the inorganic crystalline core of the CQDs, and the organic amorphous shell covering them. Additionally, EDXS and EELS results confirmed the presence of lead and tellurium in the core of the CQDs, while also confirming the presence of carbon (C), oxygen (O) and lead (Pb) in the amorphous shell, complementing the FTIR results, leading to the conclusion that the actual capping layer for this kind of CQDs, synthesized by the hot-injection method, display a lead oleate capping layer. For this measurement, the spatial resolution offered by the EELS study, in an AC-STEM instrument, was crucial, along with the high intensity of low atomic number, such as carbon (C) and oxygen (O), in the EELS technique, compared to EDXS. Figure 18 shows the AC-STEM images of the capping layer for different lead telluride CQDs, synthesized under different conditions, and the element line profile study performed, where the composition of the different regions in the CQD can be observed.

Another type of semiconductor nanoparticles with excellent photoemission properties that have gained a lot of attention from the scientific community in recent years are the inorganic perovskites nanoparticles (Manser et al. 2016, Paul 2018, Protesescu et al. 2015, Shamsi et al. 2018). This type of semiconductor nanoparticles exhibits the general formula ABX_3, where A is a cation with valence 1+, normally Cs; B is also a cation with valence 2+, normally Pb; and X is a halogen (Cl, I, Br) with valence 1–. The intense emission displayed by this type of semiconductor nanoparticles can be tuned along the visible region of the electromagnetic spectrum, by changing the nanoparticle size, but mainly by changing the chemical composition, in particular the halogen atom. Studies are being carried out by several research groups all around the world trying to elucidate the nucleation and growth mechanisms of these type of semiconductor nanoparticles. Shamsi et al. (Shamsi et al. 2018) have studied the transformations occurring when using CsX (X = Cl, Br, I) nanocrystals with a cubic crystal structure, and homogeneous size distribution, as the precursor to grow $CsPbX_3$ (X = Cl, Br, I) perovskite nanoparticles. They used HRTEM to characterize the evolution of the perovskite nanoparticles formation. HRTEM studies allowed the determination of the growth mechanism of the perovskite nanoparticles started from the outside of the CsX nanoparticles, promoting the Cs$^+$ cation exchange for Pb^{2+} cations forming, in the early stages of the reaction, a core-shell structure, where

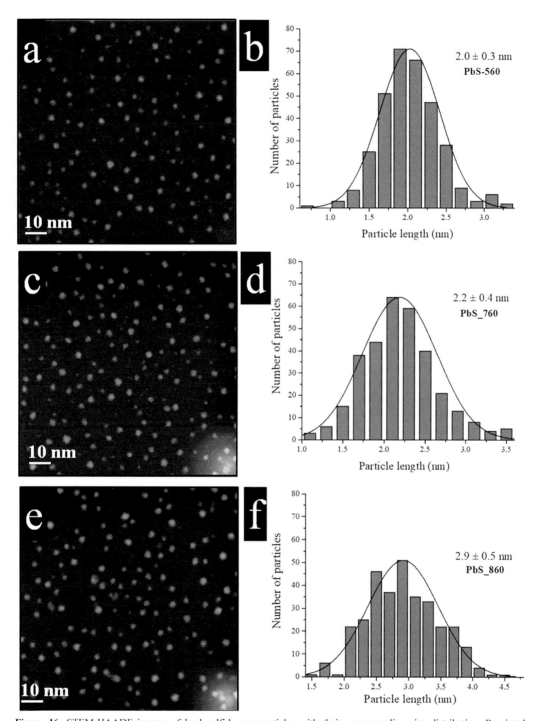

Figure 16. STEM-HAADF images of lead sulfide nanoparticles with their corresponding size distribution. Reprinted from Journal of Alloys and Compounds, 860, Nayely Torres-Gomez et al. Absorption and emission in the visible range by ultra-small PbS quantum dots in the strong quantum confinement regime with S-terminated surfaces capped with diphenylphosphine, 860, Copyright (2021), with permission from Elsevier.

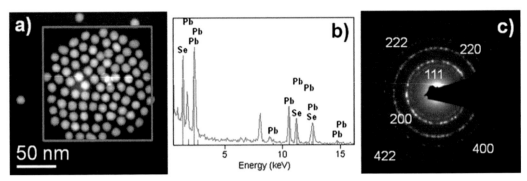

Figure 17. (a) STEM-HAADF image of a group of PbSe CQDs; (b) EDXS spectra acquired from the CQDs; (c) SAED pattern acquired from the CQDs. The gray square marks the area analyzed for the EDXS and SAED studies. "Reprinted by permission from Springer Nature Customer Service Centre GmbH: Springer Nature, *Journal of Nanoparticle Research*; D.I. Garcia-Gutierrez, L.M. De Leon-Covian, D.F. Garcia-Gutierrez, M.A. Garza-Navarro, S. Sepulveda-Guzman. On the role of Pb0 atoms on the nucleation and growth of PbSe and PbTe nanoparticles. *Journal of Nanoparticle Research*, 15(5): 1620 (2013). [COPYRIGHT] (2021).

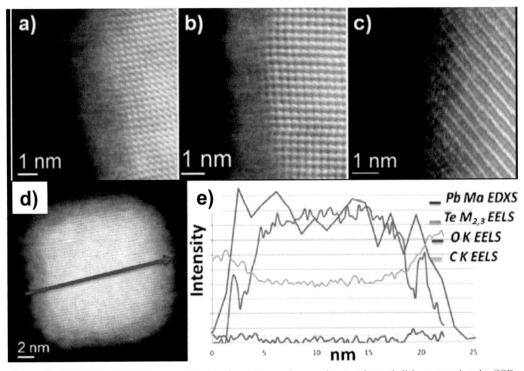

Figure 18. AC-STEM HAADF images of the interface between the organic amorphous shell layer covering the CQDs and the crystalline core: (a) image corresponding to a CQD with higher Pb/Te ratio and longer reaction time; (b) CQD with the same Pb/Te ratio but lower reaction time; and (c) CDQ with lower Pb/Te ratio and lower reaction time. (d) AC-STEM HAADF image showing the CQD analyzed. The line indicates the region where the linescan study was performed, going from left to right. (e) EELS-EDXS line profiles for the Pb Ma EDXS signal and the Te M4,5, O K and C K EELS signals. "Adapted with permission from Domingo I. Garcia-Gutierrez, Diana F. Garcia-Gutierrez, Lina M. De Leon-Covian, Mario T. Treviño-Gonzalez, Marco A. Garza-Navarro, Ivan E. Moreno-Cortez, Rene F. Cienfuegos-Pelaes. Aberration Corrected STEM study of the surface of lead chalcogenide nanoparticles. *Journal of Physical Chemistry C*, 118(38): 22291–22298 (2014). Copyright (2021) American Chemical Society.

the shell was formed by CsPbX$_3$ perovskite, and the core was cubic CsX, as it was observed in the HRTEM images shown in Figure 19.

Finally, advancements in TEM technology electron-optics have reached a resolution limit close to 50 pm in commercially available EM equipment. The availability of C$_s$-correctors for the objective lens, capable of correcting the spherical aberration in the HRTEM images, or in the electron probe in STEM mode, has opened a window to reach resolution limits, even smaller than Bohr's radius, in almost any well-prepared electron microscopy lab in the world. In recent years, the EM community has stopped their attempts to reach smaller resolutions, recognizing the fact that the information that could be acquired from smaller resolutions would be "difficult" to interpret. On the other hand, it has directed their efforts in attempts to study phenomenon, such as the nucleation and growth of nanoparticles, as they occur in real time. The combination of the superb spatial resolution displayed by current EM, along with their increased mechanical, electrical and thermal stability, and the *in situ* capabilities offered by different sample holder technologies, has opened a whole new field within the EM community. The *in situ* or *in-operando* studies represent the frontier in EM nowadays. There are plenty of advantages the biotechnology community could take from this new frontier in EM-from studies to better understand the nucleation and growth mechanisms as the nanoparticles are being synthesized, to the interaction between nanoparticles and microorganisms as it is happening in real time. The possibilities are enormous, just limited by the imagination and creativity of the researchers. As an example, Hutzler et al. (Hutzler et al. 2018) studied the growth mechanism of silver shell on gold nanorods using an *in situ* liquid cell TEM with an advanced architecture. Their advanced design was based on microwells where the liquid was confined between a thin Si$_3$N$_4$ thin film on one side and a few layers of graphene on the other. This well-defined sample thickness, and an ultra-flat cell top, made possible the application of HRTEM analysis combined with analytical TEM techniques, such as EDXS and EELS, in this type of samples and *in situ* studies. This combination of HRTEM data with the chemical information provided by the EDXS results provided completely new insights into the growth of these silver shell gold nanorods. It was shown that silver bromide nanoparticles, formed in the stock solution, play an important role in the exchange of silver ions. HRTEM images, along with EDXS results, as shown in Figure 20, showed that the silver shell growth was correlated to the dissolution of the silver bromide nanoparticles.

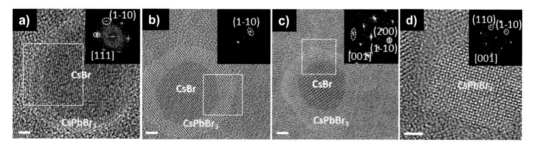

Figure 19. HRTEM images of CsX nanoparticles as CsPbX3 are being synthesized at different stages of the reaction. At early stages of the reaction, (a) a thin CsPbBr3 shell can be observed, with very faint reflections in the Fast Fourier Transform (FFT) inset. (b) and (c) show the increment in the CsPbBr3 shell thickness. (d) At the end of the reaction, the complete nanoparticle has a CsPbX3 composition, along with the crystal structure, as observed in the reflections appearing in the FFT inset. Javad Shamsi, Zhiya Dang, Palvasha Ijaz, Ahmed L. Abdelhady, Giovanni Bertoni, Iwan Moreels, and Liberato Manna. Colloidal CsX (X = Cl, Br, I) Nanocrystals and Their Transformation to CsPbX3 Nanocrystals by Cation Exchange. *Chemistry of Materials.* 30(1): 79–83 (2018). https://pubs.acs.org/doi/abs/10.1021/acs.chemmater.7b04827. Further permissions related to the material excerpted should be directed to the ACS.

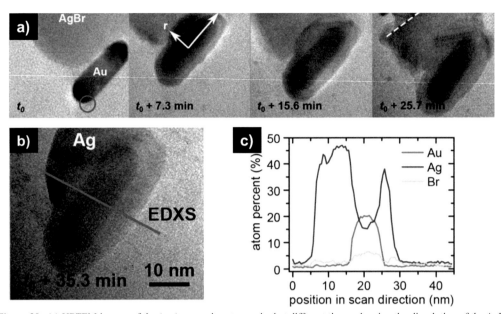

Figure 20. (a) HRTEM images of the *in situ* experiment, acquired at different times, showing the dissolution of the AgBr nanoparticle and (b) the consequent growth of the Ag shell on the Au nanorod. (c) EDXS line profile study showing the composition of the Ag shell on top of the Au nanorod. Andreas Hutzler, Tilo Schmutzler, Michael P. M. Jank, Robert Branscheid, Tobias Unruh, Erdmann Spiecker, and Lothar Frey. Unravelling the Mechanisms of Gold–Silver Core–Shell Nanostructure Formation by in Situ TEM Using an Advanced Liquid Cell Design. *Nano Letters*, 18: 7222–7229 (2018). https://pubs.acs.org/doi/full/10.1021/acs.nanolett.8b03388. Further permissions related to the material excerpted should be directed to the ACS.

8. Conclusions

TEM is the most versatile technique available for the characterization of nanoparticles. Understanding the fundamentals of the different signals offered by this technique is very important to make a better interpretation of the results acquired. A high energy electron beam possesses a small wavelength, determined by the De Broglie wavelength equation, making it possible to analyze objects in the angstrom, even in the picometer scale due to the development of C_s corrector technology. Additionally, the interaction between this high energy electron beam and the analyzed sample produces a wide variety of signals carrying information on the elastic and inelastic interactions the electrons on the beam undergo when traveling through the sample. All the signals related to these electron beam—specimen interactions give rise to the different TEM techniques, which are useful for the study and characterization of nanoparticles for biotechnology applications. It was shown that determining the size, morphology, chemical composition, crystal structure, atomistic surface arrangement and capping layer nature, are among the most important characteristics required to know for a nanoparticle system intended for biotechnology applications. TEM related techniques can provide all the relevant information about these characteristics when the proper signals are correctly analyzed and interpreted. Additionally, the *in situ* and *in-operando* modes of analysis in TEM open a whole new world of opportunities and possibilities in studying the interaction between living systems and nanoparticles.

References

Ahn, C.C., Krivanek, O.L., Burgner, R.P., Disko, M.M. and Swann, P.R. 1983. EELS Atlas Gatan Inc., Pennsylvania.

Alex Gee and Xiaoxue Xu. 2018. Surface functionalisation of upconversion nanoparticles with different moieties for biomedical applications. *Surfaces*, 1: 96–121.

Andreas Hutzler, Tilo Schmutzler, Michael P.M. Jank, Robert Branscheid, Tobias Unruh, Erdmann Spiecker and Lothar Frey. 2018. Unravelling the Mechanisms of Gold–Silver Core–Shell Nanostructure Formation by in Situ TEM Using an Advanced Liquid Cell Design. *Nano Letters*, 18(11): 7222–7229.

Browning, N.D., Wallis, D.J., Nellist, P.D. and Pennycook, S.J. 1997. EELS in the STEM: Determination of materials properties on the atomic scale. *Micron.*, 28(5): 333–348.

Cai, W., Shin, D.-W., Chen, K., Gheysens, O., Cao, Q., Wang, S.X., Gambhir, S.S. and Chen, X. 2006. Peptide-labeled near-infrared quantum dots for imaging tumor vasculature in living subjects. *Nano Lett.*, 6: 669–676.

Cheng-An, J. Lin, Tim Liedl, Ralph A. Sperling, María T. Fernandez-Argüelles, Jose M. Costa-Fernandez, Rosario Pereiro, Alfredo Sanz-Medel, Walter H. Chang and Wolfgang J. Parak. 2007. Bioanalytics and biolabeling with semiconductor nanoparticles (quantum dots). *J. Mater. Chem.*, 17: 1343–1346.

Domingo, I. Garcia-Gutierrez, Diana F. Garcia-Gutierrez, Lina M. De Leon-Covian, Mario T. Treviño-Gonzalez, Marco A. Garza-Navarro, Ivan E. Moreno-Cortez and Rene F. Cienfuegos-Pelaes. 2014. Aberration Corrected STEM study of the surface of lead chalcogenide nanoparticles. *Journal of Physical Chemistry C*, 118(38): 22291–22298.

Egerton, R.F. 1979. K-shell ionization cross-sections for use in microanalysis. *Ultramicroscopy*, 4: 169–179.

Elechiguerra, J.L., Burt, J.L., Morones, J.R., Camacho-Bragado, A., Gao, X., Lara, H.H., Jose Yacaman, M. 2005. Interaction of silver nanoparticles with HIV-1. *J. Nanobiotechnol.*, 3: 6.

Emily Reiser Evans, Pallavi Bugga, Vishwaratn Asthana and Rebekah Drezek. 2018. Metallic nanoparticles for cancer immunotherapy. *Materials Today*, 21(6): 673–685.

Frank, W. Wise. 2000. Lead salt quantum dots: the limit of strong quantum confinement. *Acc. Chem. Res.*, 33(11): 773–780.

Gabriel Campargue, Luca La Volpe, Gabriel Giardina, Geoffrey Gaulier, Fiorella Lucarini, Ivan Gautschi, Ronan Le Dantec, Davide Staedler, Dario Diviani, Yannick Mugnier, Jean-Pierre Wolf and Luigi Bonacina. 2020. Multiorder Nonlinear Mixing in Metal Oxide Nanoparticles. *Nano Lett.*, 20: 8725–8732.

Garcia Gutierrez and Domingo, I. 2006. Transmission electron microscopy characterization of composite nanostructures. Ph.D. Thesis, The University of Texas at Austin, Austin, TX, USA.

Garcia-Gutierrez, D.F., De Leon-Covian, L.M. and Garcia-Gutierrez, D.I. 2017. Electron energy loss spectroscopy. *In*: Wang, Z., Wille, U. and Juaristi, E. (eds.). *Encyclopedia of Physical Organic Chemistry.* https://doi.org/10.1002/9781118468586.epoc4028.

Garcia-Gutierrez, D.F., Hernandez-Casillas, L.P., Cappellari, M.V., Fungo, F., Martínez-Guerra, E. and García-Gutiérrez, D.I. 2018. Influence of the capping ligand on the band gap and electronic levels of PbS nanoparticles through surface atomistic arrangement determination. *ACS Omega*, 3(1): 393–405.

Garcia-Gutierrez, D.I., De Leon-Covian, L.M., Garcia-Gutierrez, D.F., Garza-Navarro, M.A. and Sepulveda-Guzman, S. 2013. On the role of Pb0 atoms on the nucleation and growth of PbSe and PbTe nanoparticles. *Journal of Nanoparticle Research*, 15(5): 1620.

Garza-Navarro, M.A., Torres-Castro, A., Garcia-Gutierrez, D.I., Ortiz-Rivera, L., Wang, Y.C. and Gonzalez-Gonzalez, V.A. 2010. Synthesis of spinel-metal-oxide/biopolymer hybrid nanostructured materials. *Journal of Physical Chemistry C*, 114(41): 17574–17579.

Gregg, B.A. 2003. Excitonic Solar Cells. *The Journal of Physical Chemistry B*, 107(20): 4688–4698.

Howie, A. 1979. Image-Contrast and Localized Signal Selection Techniques. *Journal of Microscopy-Oxford*, 117(SEP): 11–23.

Ivan, E. Moreno-Cortez, Jorge Romero-Garcia, Virgilio Gonzalez-Gonzalez, Domingo I. Garcia-Gutierrez, Marco A. Garza-Navarro and Rodolfo Cruz-Silva. 2015. Encapsulation and immobilization of papain in electrospun nanofibrous membranes of PVA cross-linked with glutaraldehyde vapor. *Materials Science and Engineering* C, 52: 306–314.

Javad Shamsi, Zhiya Dang, Palvasha Ijaz, Ahmed L. Abdelhady, Giovanni Bertoni, Iwan Moreels and Liberato Manna. 2018. Colloidal CsX (X = Cl, Br, I) nanocrystals and their transformation to CsPbX3 nanocrystals by cation exchange. *Chemistry of Materials*, 30(1): 79–83.

Judith, C. Yang, Matthew W. Small, Ross V. Grieshabera and Ralph G. Nuzzo. 2012. Recent developments and applications of electron microscopy to heterogeneous catalysis. *Chem. Soc. Rev.*, 41: 8179–8194.

Justin, L. Burt, Claudia Gutierrez-Wing, Mario Miki-Yoshida and Miguel Jose-Yacaman. 2004. Noble-metal nanoparticles directly conjugated to globular proteins. *Langmuir*, 20: 11778–11783.

Loredana Protesescu, Sergii Yakunin, Maryna I. Bodnarchuk, Franziska Krieg, Riccarda Caputo, Christopher H. Hendon, Ruo Xi Yang, Aron Walsh, and Maksym V. Kovalenko. 2015. Nanocrystals of cesium lead halide perovskites (CsPbX3, X = Cl, Br, and I): Novel optoelectronic materials showing bright emission with wide color gamut. *Nano Letters*, 15(6): 3692–3696.

Louis De Broglie. 1925. Recherches sur la théorie des Quanta. *Ann. Phys.*, 10(3): 22–128.

Manser, J., Christians, J. and Kamat, P. 2016. Intriguing optoelectronic properties of metal halide perovskites. *Chemical Reviews*, 116(21): 12956–13008.

Martínez-Sánchez, R., Reyes-Gasga, J., Caudillo, R., García-Gutierrez, D.I., Márquez-Lucero, A., Estrada-Guel, I., Mendoza-Ruiz, D.C. and José M. Yacaman. 2007. Mechanical and microstructural characterization of aluminum reinforced with carbon-coated silver nanoparticles. *Journal of Alloys and Compounds*, 438(1-2): 195–201.

Morones, J.R., Elechiguerra, J.L., Camacho, A., Holt, K., Kouri, J.B., Tapia J. Ramírez and Yacaman, M.J. 2005. The bactericidal effect of silver nanoparticles. *Nanotechnology*, 16: 2346.

Murray, C.B., Norris, D.J. and Bawendi, M.G. 1993. Synthesis and characterization of nearly monodisperse CdE (E = sulfur, selenium, tellurium) semiconductor nanocrystallites. *J. Am. Chem. Soc.*, 115: 8706– 8715.

Nayely Torres-Gomez, Diana F. Garcia-Gutierrez, Alan R. Lara-Canchea, Lizbeth Triana-Cruz, Jesus A. Arizpe-Zapata, Domingo I. Garcia-Gutierrez. 2020. Absorption and emission in the visible range by ultra-small PbS quantum dots in the strong quantum confinement regime with S-terminated surfaces capped with diphenylphosphine. *Journal of Alloys and Compounds*, 860: 158443.

Nubia, E. Torres-Martínez, Marco A. Garza-Navarro, Raúl Lucio-Porto, Virgilio A. González-González, Alejandro Torres-Castro and Domingo García-Gutiérrez. 2013. One-pot synthesis of magnetic hybrid materials based on ovoid-like carboxymethyl-cellulose/cetyltrimethylammonium-bromide templates. *Materials Chemistry and Physics*, 141(1-2): 735–743.

Paul, T. 2018. Fabrication of all-inorganic CsPbBr3 perovskite nanocubes for enhanced green photoluminescence. *Materials Today: Proceedings*, 5(1): 2234–2240.

Ravichandran, M., Goldie Oza, S. Velumani, Jose Tapia Ramirez, Francisco Garcia-Sierra, Norma Barragán Andrade, Marco A Garza-Navarro, Domingo I. Garcia-Gutierrez, Rafael Lara-Estrada, Emilio Sacristán-Rock and Junsin Yi. 2015. *Cobalt ferrite nanowhiskers as T 2 MRI contrast agent. RSC Advances*, 5(22): 17223–17227.

Ravichandran, M., Goldie Oza, Velumani, S., Jose Tapia Ramirez, Francisco Garcia-Sierra, Norma Barragán Andrade, Marco A. Garza-Navarro, Domingo I. Garcia-Gutierrez and R. Asomoza. 2014. One-Dimensional Ordered growth of magneto-crystalline and biocompatible cobalt ferrite nano-needles. *Materials Letters*, 135: 67–70.

Raymond, P. Brinas, Andreas Sundgren, Padmini Sahoo, Susan Morey, Kate Rittenhouse-Olson, Greg E. Wilding, Wei Deng and Joseph J. Barchi, Jr. 2012. Design and synthesis of multifunctional gold nanoparticles bearing tumor-associated glycopeptide antigens as potential cancer vaccines. *Bioconjugate Chem.*, 23: 1513–1523.

Rodríguez-López, J.L., Montejano-Carrizales, J.M., Pal, U., Sánchez-Ramírez, J.F., Troiani, H.E., García, D., Miki-Yoshida, M. and José-Yacamán, M. 2004. Surface reconstruction and decahedral structure of bimetallic nanoparticles. *Phys. Rev. Lett.*, 92: 196102.

Seiko Toraya-Brown, PhD., Mee Rie Sheen, M.S., Peisheng Zhang, M.D., Lei Chen, B.S., Jason R. Baird, PhD., Eugene Demidenko, PhD., Mary Jo Turk, PhD., Jack P. Hoopes, D.V.M., PhD., Jose R. Conejo-Garcia, M.D., PhD. and Steven Fiering, PhD. 2014. Local hyperthermia treatment of tumors induces CD8 + T cell-mediated resistance against distal and secondary tumors. *Nanomedicine: Nanotechnology, Biology, and Medicine*, 10: 1273–1285.

Torres-Martinez, N., Garza-Navarro, M., García-Gutiérrez, D., González-González, V.A., Torres-Castro, A. and Ortiz-Méndez, U. 2014. Hybrid nanostructured materials with tunable magnetic characteristics. *Journal of Nanoparticle Research*, 16: 2759–2771.

Williams, D.B. and Carter, C.B. 1996. Transmission Electron Microscopy, ed. P. Press. New York: Plenum Publishing Corporation.

Yu, W.W., Chang, E., Drezek, R. and Colvin, V.L. 2006. Water-soluble quantum dots for biomedical applications. *Biochemical and Biophysical Research Communications*, 348(3): 781–786.

Zhuyuan Wang, Shenfei Zong, Jing Yang, Jin Li, and Yiping Cui. 2011. Dual-mode probe based on mesoporous silica coated gold nanorods for targeting cancer cells. *Biosensors and Bioelectronics*, 26: 2883–2889.

CHAPTER 8

Magnetic Characterization of Nanoparticles

Marlene González Montiel

1. Introduction

The magnetic behavior of nanoparticles smaller than 100 mn—dimension similar to the domain wall width (Bean 1955, Kittel 1946)—is usually described, among other factors, by its volume, shape, and embedded matrix because they behave as single domain particles (see Figure 1c). These particles are able to inverse their magnetization merely by means of rotating their macroscopic magnetic moment, a mechanism initially proposed by Stoner-Wohlfarth (Stoner and Wohlafarth 1948), who neglected dipolar interactions and thermal agitation for simplicity's sake. This simple, but useful model, considered almost spherical magnetic nanoparticles of V volume and M_s saturation magnetization with permanent magnetic moment $m = M_s V$ along which the anisotropy energy E_k is minimum as the magnetic moment can be aligned in a privileged magnetization axis (so-called easy axis). This alignment is dependent on the effective anisotropy density constant K (which is proportional to the particle size) and the angle between m and the easy axis θ:

$$E_k = KVsin^2(\theta) \tag{1}$$

At low temperatures (absolute zero) and applied fields lower than the critical field for moment inversion (or reversal), the Stoner-Wohlfarth model suggested the existence of two equilibrium situations in the energy scenery (see Figure 1) in which the m moment can hop between up and down positions with a characteristic relaxation time (τ) because of thermal agitation. This relaxation time is obtained by the Arrhenius-Néel law (Neel 1949) and can be determined by the energy barrier $KV = \Delta E$, the frequency f_0 with the magnitude order of the magnetic moment precession frequency, and the Botzman constant k_B.

$$\tau^{-1} = f_0 \exp\left(\frac{-KV}{k_B T}\right) \tag{2}$$

By using the initial susceptibility, the magnetic response of the nanoparticle system can then be described as the temperature variation of the initial susceptibility, which comprises two different regimes: (1) superparamagnetic particles exhibiting rapid relaxation and (2) thermally stable

CONACyT – National Polytechnic Institute, Center for Applied Science and Advanced Technology, Legaria Unit, Mexico City, Mexico.
Email: mgonzalezmo@conacyt.mx

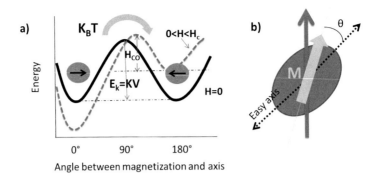

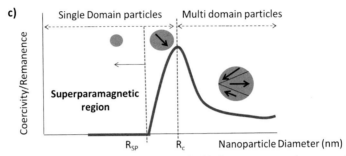

Figure 1. (a) Angular dependence of the energy barrier reversal for ideal superparamagnetic systems. The continuous line is for zero field and the dotted line is for a field lower than the coercive field (see Section 2). Above T_B, the thermal effects allow flips of the magnetic moments between the easy magnetization directions by getting over the energy barriers in zero applied magnetic field and consequently the coercivity $H_C = 0$. (b) The easy rotation axis for a fine particle (see also Section 2). (c) In ideal, non-interacting nanoparticles' systems (superparamagnetic), the particles are very small (typically $R_{SP} < 20$ nm) and belong to a single domain. The critical size value (R_C) depends on the former components, i.e., for Fe_3O_4 nanoparticles, this size is widely estimated to be around 76 nm (Li et al. 2017).

particles exhibiting irreversible magnetic behavior. It must be noticed that, in the Stoner-Wohlfarth model, neither phase transition nor abnormality occurs when the moment of a particle transits from the thermally stable state to the superparamagnetic state.

The characteristic measuring time τ_m (time window) of the applied experimental technique strongly influences the measured magnetic behavior in nanoparticle systems due to its connection to the intrinsic relaxation time τ of the system that is associated with the energy barrier. The time window can vary from slow values—as in magnetometry assessments (usually 100s)—to very fast ones—as in Mössbauer spectroscopy (10^{-8} s). When $\tau_m \gg \tau$, the relaxation is quicker in this time window than the magnetization orientation, which allows the system to reach thermodynamical equilibrium because the nanoparticles are in superparamagnetic regime (Neel 1949). Conversely, when the system relaxation occurs very slowly, $\tau \gg \tau_m$, the quasi-static properties of the system can be seen as ordered magnetic arrangements (the nanoparticles are in the so-called blocked regime). The blocking temperature T_B is then the inflexion transition point from both regimes (slow and quick relaxation) and is influenced by the characteristic measuring time τ_m. T_B is also associated with the energy barrier for reversal, KV, and thereby grows when the particle size upsurges (Knobel et al. 2008). For example, the relation between the critical particle size for a superparamagnetic system and the blocking temperature is directly proportional when $\tau_m = 100$s:

$$T_B \approx {KV}/{25k_B}$$

(3)

Hence, from Equation 3, it can be inferred that bigger particles become superparamagnetic at higher temperatures. Consequently, different particle size in a nanoparticle system will have different blocking temperatures.

In ideal, non-interacting and nanostructured metallic and metal oxide systems, the particles are very small and belong to a single domain (see Figure 1) that is below the critical volume size. For this reason, the thermal energy at which a measurement is done is enough to equilibrate the magnetization of a nanoparticle's assemblage in shorter time than it takes to do the measurement (Neel 1949). As such, particles present a magnetization with long range order, and behave as superparamagnetic.

Nevertheless, in a real sample measurement, the nanoparticle's size is not always monodisperse, as some of the nanoparticles are below the critical size (superparamagnetic) and others remain above the critical size (thermally stable particles). Different particle sizes imply different anisotropy barrier *KV*. For this reason, a detailed information about the particle size distribution, shape and surface contributions in the magnetic system of nanoparticles must be taken into account for real polidisperse or non-spherical (anisotropic) systems.

The role of dipolar interactions (Hansen 1998) between particles must also be mentioned because many studies of magnetic nanoparticles revealed the magnetic behavior of either "weakly" or as "strongly" interacting systems, yet many efforts have been made to synthesize adequate systems in the laboratory that can be used to test and scrutinize the existing theoretical models to fully comprehend the role of dipolar interactions in the description of magnetic nanoparticles (Knobel et al. 2008).

When dealing with real systems, it is very complex to identify the contribution of each effect such as size, shape, and anisotropy origins (magnetocrystalline, shape, stress or surface) in the final measurement; it requires an excellent magnetic nanoparticle synthetic route grade, and more importantly, a basis to elucidate the relationship between their interactions and intrinsic properties. In this regard, samples that are synthesized, nearly monodispersed, and uniformly shaped (Rivas Rojas et al. 2018) may offer a satisfying accord between the known models and the data that is obtained from the magnetic characterization measurements.

This chapter's aim is to expound some of the usual magnetic characterization techniques and the parameters obtained from their analysis to describe magnetic nanoparticles' isotropic systems. As seen below, miscellaneous nanoparticles' systems are discussed to elucidate their best description with the support of recognized author's works, models and approaches known in literature, hoping that this compilation helps the reader to decide on the model/approach which fits the most with his/her research and magnetic data processing.

2. (ZFC-FC) zero field cooling-field cooling measurements

The (ZFC-FC) Zero Field Cooling-Field Cooling is a procedure that consists of magnetization measurements as a function of varying (decreasing or increasing) temperature routine. First, the sample is cooled with non-applied magnetic external field usually from room temperature (300 K), where all nanoparticles regularly show superparamagnetic behavior, to the lowermost accessible temperature (near 1.8 K in Helium liquid Magnetometers). At that moment, a small and constant magnetic external field that is commonly around 8×10^3 *A/m* (≈ 100 *Oe*) is turned on while the sample is heated until it reaches the highest temperature from which it was initially cooled (i.e., 300 K), temperature adequate to detect an early growth and posterior decrease of its magnetization. Finally, the sample is immediately cooled again until it reaches the lowermost temperature while the external applied field is still turned on (see Figures 2, 3 inset and 6).

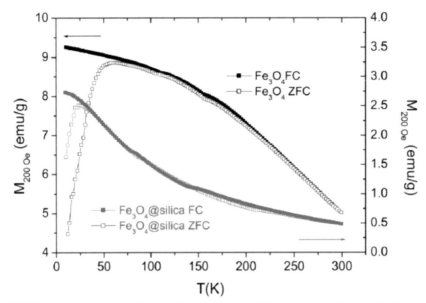

Figure 2. FC/ZFC magnetization curves obtained at H = 200 Oe ≈16 × 10³ A/m in bare Fe_3O_4 and in Fe_3O_4 coated by a silica matrix (Fe_3O_4@silica) nanoparticles. High degree of particle aggregation in the Fe_3O_4 nanopowder (Black curve) is partially reduced by silica coating (Fe_3O_4@silica, gray curve) and can be strongly decreased by dissolution in a host polymer (not shown). Reprinted with permission from Allia et al. 2014.

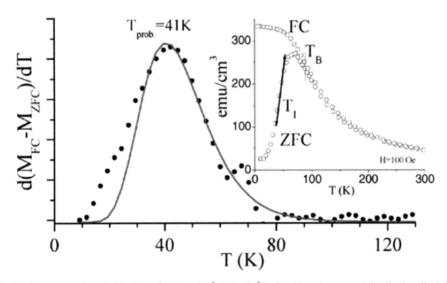

Figure 3. Blocking temperature distribution of [Co λ = 9.7Å/SiO_2 43 Å]$_{20}$ fitted by a log-normal distribution (line) deduced from the ZFC-FC curve (inset), using the $d(M_{FC}-M_{ZFC})/dT$ criterion. Micha et al. pioneer work was one of the first in correlating the particle size radius obtained by Transmission Electronic Microscopy (TEM) and small angle X-ray scattering with the radius obtained by using simple ZFC-FC measurements. The R_{prob} radius was found to be 10 Å greater than that obtained by TEM; the authors ascribed this difference to a magnetic correlation between particles attributed to direct contact or dipolar interactions. Reprinted with permission from Micha et al. 2004.

2.1 Non interacting, monodisperse magnetic particles

In ideal, mono-sized and non-interacting magnetic nanoparticles, the system performs a transition among irreversible and reversible states throughout the ZFC-FC routine. During this transition, a narrow temperature region is expected, while warming with a turned-on applied field, the thermal energy $k_B T$ is firstly smaller than the anisotropy barrier KV and consequently the magnetization remains insignificant. Once $k_B T \cong KV$ (see Equation 2), the magnetization increases rapidly to its value of thermodynamic equilibrium, describing the irreversible/reversible regime transition region owing to the exponential variation of the Néel relaxation time on the temperature (Bean and Livingston 1959). The inflection point (IP) of this increase can be designated as the Blocking temperature T_B in this kind of well isolated systems (Bruvera et al. 2015). This criterion has been reported in several works, where the analyzed systems are considered as monodisperse and each nanoparticle is well separated between them, which results in a ZFC shape with a narrow, symmetrical and well defined peak (Medina et al. 2020).

A good way to differentiate a non-interacting magnetic nanoparticles' arrangement from an interacting one is well exemplified by Allia et al. (Allia et al. 2014), where a characteristic performance (from the ZFC profile) related with an almost non-interacting and monodisperse superparamagnetic nanoparticles system is described for Fe_3O_4 in a silica matrix nanoparticles as depicted in Figure 2. As seen there, the ZFC shape of the peak described by the Fe_3O_4 in a silica matrix nanoparticles corresponds with a narrow, symmetrical and well defined form, whereas for the bare Fe_3O_4 nanoparticles sample, magnetic interactions between particles (or polidisperse size distribution) are anticipated owing to the shape of its ZFC curve (see Figure 2). Here, a wide spread peak is perceived along a wide temperature range, and depending on the author, this maximum value will not necessarily match with the maximum of the ZFC peak, T_{MAX} (see Section 1.2). In this example, the difference between bare Fe_3O_4 and Fe_3O_4 in a silica matrix particles was attributed to the distinct size distributions of the two different magnetic nanoparticles systems (aggregates) and not entirely negligible interactions on the bare Fe_3O_4 nanoparticles. It is also important to notice that the ZFC-FC curve from bare Fe_3O_4 particles presents higher maximum than the Fe_3O_4 in a silica matrix particles. As depicted in Equation 3, bigger particles are expected because they become superparamagnetic at higher temperatures. Thus, $T_{MAX} = T_B$ just when the larger number of nanoparticles within the sample are all blocked. Nevertheless, if the system presents a large size distribution, this method, $T_{MAX} = T_B$, could be the worst to use because the difference between the real T_B and the T_{MAX} can be as big as the double of its number (Bean and Livingston 1959).

2.2 Non interacting, polydisperse magnetic particles

As the particle size is proportional to the anisotropy barrier KV (and size also influences the nanoparticles' relaxation times), a T_B is expected for each different size contribution in a polydisperse nanoparticle system. This section makes this assumption despite the discordance between some articles in the literature, where the use of maximum temperature (T_{MAX}) and the Inflection point (IP) criteria is still reported assuming monodisperse and non-interacting magnetic nanoparticles even in polydisperse systems. The assumption of using different approaches was well defined and defended by expert theoretical and experimental groups around the world to describe more realistic magnetic nanoparticles systems, where polydisperse and interacting samples are more usually obtained in the laboratory. Nevertheless, it is important to remind the reader that monodisperse and non-interacting magnetic nanoparticles are very difficult (but not impossible) to obtain when using good synthetic methods and good experimental mounting of the analyzed samples during measurements.

In this context, Micha et al. (Micha et al. 2004) described the use of magnetic measurements in relation to the magnetic nanoparticle size. In this article, they use an easy method for nanoparticles' size estimation and assume an arrangement of spherical, non-interacting and polydisperse magnetic particles, in which the distribution of the particle size ensemble can be determined and characterized by using ZFC-FC measurements. In this case, the blocking temperature T_B is determined as the ZFC and FC intersection, which is associated with both the effective magnetic anisotropy constant K and to the largest particle R_{MAX} radius by the relation:

$$\frac{4\pi}{3} R_{MAX}^3 = 25 k_B T_B \Big/ K \tag{4}$$

All the ensemble nanoparticles are superparamagnetic above the blocking temperature T_B. Meanwhile, a gradual blocking of minor particles occurs when $T < T_B$ since a critical radius R corresponds to a temperature T agreeing with Equation (4). The projection along the externally applied magnetic field (in this case, Micha et al. used 100 $Oe \approx 8 \times 10^3$ A/m) is provided by the ZFC magnetization measurement of the magnetic moment of randomly frozen particles, for they have greater radius than and the contribution of the superparamagnetic, polarized particles have a radius that is smaller than R. On the other hand, the projection of greater particles (polarized frozen moments) and smaller particles (polarized superparamagnetic ones) is provided by the FC magnetization measurements. Therefore, the blocking distribution of temperatures and the radius distribution can be deduced from the derivative of the difference between ZFC–FC measurements, $d(M_{FC}-M_{ZFC})\big/dT$ (see Figure 3). In Micha et al.'s work, the volume distribution stayed denoted by an asymmetric log-normal distribution (although other distributions can be used as described by El-Hilo (El-Hilo 2012)). Since the FC curve becomes flat at low temperatures, the most probable radius R_{prob} is simply obtained from the $ZFC(T)$ curve's inflexion point at T_1 (Equation 5).

$$\frac{4\pi}{3} R_{prob}^3 = 25 k_B T_1 \Big/ K \tag{5}$$

Above the blocking temperature T_B, the Curie-Weiss law can be applied and the radius R_{CW} can be obtained from the susceptibility variation (Equation 6):

$$M_{CW}(H, T) = \frac{M_S^2 H}{3 k_B (T - \theta_{CW})} R_{CW}^3 \tag{6}$$

where M_S, θ_{CW}, and k_B are the saturation magnetization of the analyzed ion, the Curie-Weiss temperature and the Boltzmann constant, respectively. The difference concerning R_{MAX}, R_{prob}, and R_{CW} radii is influenced by the M_S and K values and the size distribution of the analyzed metal or ion nanoparticles. These assumptions are not attained by supposing strictly monodisperse systems.

A very similar work defending the $d(M_{FC}-M_{ZFC})\big/dT$ criterion was reported by Mamiya et al. (Mamiya et al. 2005). It uses Equation 1 and different theoretical approaches ascribed to the fluctuations of the direction of magnetization that is frozen in a narrow range of temperatures around T_B. Using an excellent theoretical methodology, Mamiya et al. (2005) state that the difference between $\Delta M_{FC}-M_{ZFC}$ in the range between T and $T + \Delta T$ is proportional to the distribution of polydispersive nanoparticles even with an external applied magnetic field H. However, Mamiya et al. (2005) recall that the fluctuations in each particle are really frozen just in the interval among 0.9 T_B and 1.05 T_B. Consequently, the statement of the abrupt particles freezing is an oversimplification for both ideal monodispersive and polydispersive systems when their dynamics are strictly analyzed.

More recently, Bruvera et al. (Bruvera et al. 2015) compared the aforementioned methods for non-interacting particles (Inflection point (IP), the maximum temperature (T_{MAX}) criteria and the derivative $d(M_{FC} - M_{ZFC})/dT$ method) when applied to polysize nanoparticles with log normal distribution. They used the Stoner-Wolfarth model by introducing thermal agitation to numerically estimate their ZFC-FC curves and considering all the particles' easy axes aligned with the same direction of the field. In this work, Bruvera et al. simulated numerous samples with different particle sizes (polysized) by means of diverse parameter sets, fluctuating the scale parameter σ of the log's normal number distribution and the mean magnetic nanoparticles radius. From this, they obtained the T_B distribution from the sum of monosize curves to construct the polysize simulation to later compare the results with the ZFC-FC derivative. As a result, all the simulated cases of $\langle T_B \rangle$ that were compared to the ZFC-FC derivative were identical and remained constant for different scale parameters σ. In contrast, the simulated polysized curve that was compared to the IP value moved towards lower temperatures while the T_{MAX} simulated curve shifted in the opposite direction (see Figure 4).

Additionally, Bruvera et al. tested their simulations by comparing them with experimental measurements from magnetite magnetic nanoparticles in hexane to associate the T_B simulated distribution by means of the ZFC-FC derivative curve with the size information obtained from Transmission Electronic Microscopy (TEM). Figure 5 shows this evaluation with a very good correspondence.

Therefore, after seeing that Bruvera et al. strongly support the validity of Micha-Mamiya's method in polydisperse nanoparticles systems, this section strongly encourages the reader to use the derivative method in polydisperse nanoparticles systems.

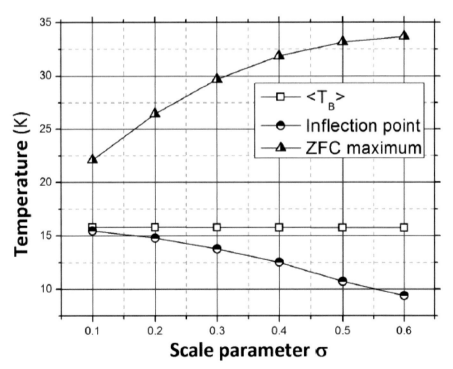

Figure 4. Values of T_B mean (identical to the ZFC-FC derivative), T_{MAX} and IP of simulated curves as a function of the scale parameter σ for 4.5 nm mean radius, 16 KJ/m³ anisotropy constant and 4 K/min heating rate. Reprinted with permission from Bruvera et al. 2015.

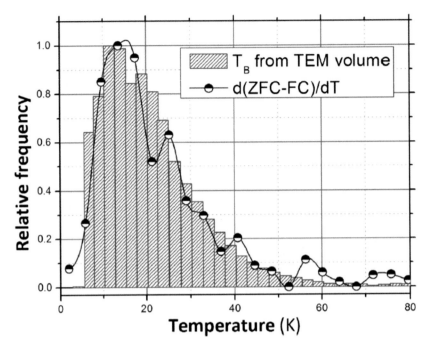

Figure 5. Comparison between the $d(M_{FC}\text{-}M_{ZFC})/dT$ derivative together with T_B distribution from TEM volume obtained by fitting the anisotropy constant K for the maximum coincidence. The translation from TEM volume to T_B was made considering the blocking condition in which the inversion time of the magnetic nanoparticles is approximately equal to the measurement of the magnetization value. Reprinted with permission from Bruvera et al. 2015.

2.3 Interacting magnetic particles

It is well known that, in real nanoparticle systems, a big proportion of the atoms belonging to each nanoparticle are in the surface. In this case, the easier way to diminish that surface energy is by aggregating or agglomerating nanoparticles. The aggregation of a group of nanoparticles is not desirable because the aggregate will behave as a macro particle. Therefore, some external coating agents are usually used to avoid interaction or agglomeration of small nanoparticles that may affect the macroscopic magnetic properties in concentrated systems. The election of a matrix or coating agent will help obtain long distanced nanoparticles with almost negligible interactions between them or controlled dipolar interactions if the particles are embedded in a matrix where the separation between them is small enough.

Well-separated nanoparticles are desirable to avoid interactions between them because the models used for non-interacting superparamagnetic particles are not adequate to describe the results obtained by experimental characterization (Knobel et al. 2008, Moscoso-Londoño et al. 2017, Rivas Rojas et al. 2018), as it was described in the previous section. It is almost impossible to differentiate in real ZFC-FC curves each interaction contribution that is related to the samples with a polydisperse distribution or with different shape (non-spherical), for the behavior of each blocked particle near T_B can be related to mechanisms that are not addressed yet, such as the anisotropy that can be attributed to single or collective origin (Bitoh et al. 1995, Knobel et al. 2008, Mamiya et al. 1998). Therefore, to clarify the nature of the slowdown blocked-particles near T_B, some authors recalled the idea of controlled interacting systems, as will be described in this section.

2.3.1 Dipolar interactions

Mamiya et al. (Mamiya et al. 2004) pointed out that, around the vicinity of T_B, the slowdown mechanism of monodispersive particles with controllable inter-particle interactions must be ascribed to cooperative dipolar interactions, that are able to vary the blocking barrier heights. These interactions influence the transitional region n(TB) between paramagnetic and superparamagnetic states. Therefore, by analyzing the M_{ZFC} and M_{FC} curves, Mamiya et al. constructed a model for the blocking of each spherical particle that results from anisotropy (Bean and Livingston 1959) by introducing an interacting term J_{dd} into each barrier height:

$$E_B = {}^4\!/_3\ K\pi R^3 + \Delta E(J_{dd}) \qquad\qquad (7)$$

where K is the anisotropy constant and R the sphere radius. Since $T_B \propto E_B$, the $\Delta E(J_{dd})$ term is related with the T_B via the $n(T_B)$ distribution variation when the value of J_{dd} increase. In this model, $n(T_B)$ distributes beginning from low T_{BL} to high T_{BH}. Then, the M_{ZFC} and M_{FC} show the features defined in Figure 6a inset: above T_{BH}, both temperatures are equal as the system is sufficiently equilibrated, while below T_{BL}, the M_{ZFC} is different from the M_{FC} as the fluctuations are altogether blocked. Using this model, Mamiya et al. compared the magnetic response through the nT_B distribution of approximately mono-dispersive, magnetic nanoparticle systems that are randomly ordered, which differed merely in their dipolar interactions' strength. Mamiya's group conclusions also demonstrated how dipolar interactions among mono-dispersive, magnetic nanoparticle systems that are randomly ordered can originate a change in the individual blocking mechanism to turn it into a spin-glass-like or any other kind of collective, magnetic spin rotations, depending on the analyzed system, as stated in a very recommendable review from Knobel et al. (Knobel et al. 2008). In this review, Knobel et al. outline how important are controlled dipolar interactions in some nanoparticles assemblies, how their presence has repercussions in the magnetic particles alignment at lower temperatures and in the shape of the nT_B distribution, which advances relevant documentation related with the microscopic mechanisms implicated in the particle magnetic moment rotation.

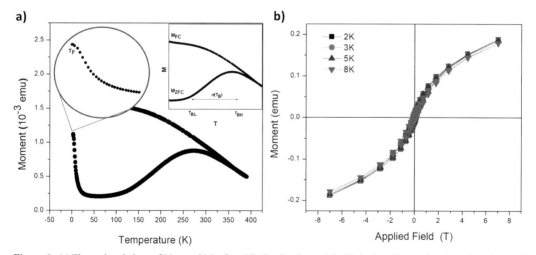

Figure 6. (a) Thermal variations of M_{ZFC} and M_{FC} for a $Ni_{27}Co_{39}Fe_{33}$ layered double hydroxide sample, a broad maximum of the M_{ZFC} is expected for a broad T_B distribution reflecting the distribution of the anisotropy energy barriers. Inset: Example of how the Mamiya model curve supposes individual blocking with $n(T_B)$ distributing from T_{BL} to T_{BH} uniformly. Zoom in: sharp peak at low temperatures, T_F, is ascribed to the freezing of surface spins. (b) The distorted surface-spin state at low temperatures is reflected in a non saturation of the magnetization and changes in the coercive field during the Magnetization versus Applied magnetic field measurements [Unpublished data from the chapter author].

2.3.2 Surface effect interactions

As mentioned in the beginning of the current section, the interactions in nanoparticles' systems can also be attributed to surface effects even when dealing with bigger particles. The surface effects are mainly attributed to the lower coordination number (spin disorder) on the broken exchange bonds in the surface of the particles (Neel 1949, Zysler et al. 2003), which unavoidably causes misalignment of the spins that belong to the surface regarding the orderly spins linked to the core in small size nanoparticles. This spin disorder on the surface might disseminate inside the core of the particle, making the superparamagnetic model no longer valid, where ideal and single domained particles with spins pointing in the same direction with coherent fluctuations are assumed.

For instance, to illustrate the later point, De Biasi et al. (De Biasi et al. 2005) reported a lower magnetization in the surface than in the core when using the Metropolis algorithm and Monte Carlo simulations with amorphous Co-Ni and Fe-Ni-B nanoparticles to provide a clear evidence of how the interactions in the core-surface interface and the superficial anisotropies alter the particle's anisotropy barriers at low temperatures and, therefore, the distribution as well as the collective alignment of the particles at very low temperatures. A similar experimental work is being achieved by M. Gonzalez M. et al. (M. González M. et al. unpublished data) by analyzing different NiCoFe layered double hydroxides, where from analyzing the M_{ZFC} and M_{FC} measurements (see Figure 6a, zoom region), a broad maximum of the M_{ZFC} is expected for a broad T_B distribution associated with the anisotropy energy barriers' distribution. Nevertheless, another but sharper maximum of M_{ZFC} is depicted at much lower temperatures. The observed sharp peak at low temperatures T_F is attributed to the freezing of surface spins in this temperature region, where the surface spin fluctuations delay allows short-range interactions between the spins belonging to the surface forming magnetically correlated spins regions (De Biasi et al. 2002) that lastly lead to distorted and frozen superficial spins states (M. González M. et al. unpublished data, Zysler et al. 2003) reflected in the non saturation of the magnetization and changes in the coercive field during the Magnetization versus Applied magnetic field measurements (see Section 2.1.3 and Figure 6b).

As outlined in this section, very valuable information about the distribution size in nanoparticles systems and some of the mechanisms involved in the particles magnetic moments alignment, attributed to the matrix they are into, can be obtained by a proper and careful analysis of the M_{ZFC} and M_{FC} curves though, as pointed out in the previous paragraph, other measurements are also desirable to complement and achieve additional information concerning the analysed nanoparticles system, as shall be described in Sections 2 and 3.

3. (M Vs H) magnetization Vs applied magnetic field measurements

The (M Vs H) Magnetization Vs Applied Magnetic Field procedure consists of magnetization measurements (M) in a specific temperature (isothermal measurement) while an imposed magnetic field is externally applied (H). The external field variation grows from the null value until it reaches the complete alignment of all nanoparticle magnetic moments in large positive or negative H external field values (usually around 7 T in commercial magnetometers). This is called saturation, which is not always achieved, as described in Section 1.3.2 due to surface effects. The external field variation from positive, then returning to negative values until finally reaching (again) the positive branch saturation is called Hysteresis loop (see Figure 7).

The model of magnetization reversal in non-interacting particles belonging to a single domain that is presented by Stoner-Wohlfart and was described in the previous section offers adequate description about the increased coercivity value below T_B. In this case, the thermal energy (KT) is not able to overcome the magnetic anisotropy (KV), then the direction of each magnetic moment particle starts to fluctuate and rotate from the applied magnetic field backwardly to the nearby easy magnetization axis resulting in a non-null coercivity value. Nevertheless, when the thermal effects

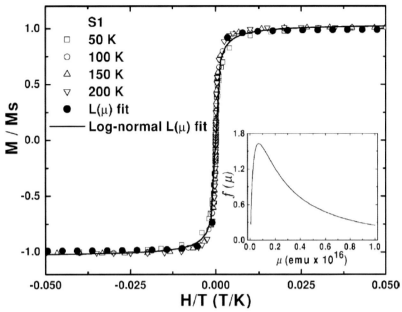

Figure 7. Normalized magnetization as a function of H/T at temperatures of 50,100,150 and 200 K, above T_B, for nanocrystalline Nickel particles. The inset shows the log-normal distribution $f(\mu)$ of the magnetic moments. Reprinted with permission from Fonseca et al. 2002.

can overcome the energy barriers in zero applied magnetic field, a coercivity $H_C = 0$ is expected at temperatures above T_B, since the anisotropy barrier is overwhelmed, which allows the flips of the magnetic moments between the easy magnetization directions (see Figure 1 and 7).

3.1 Langevin function

3.1.1 Non interacting, monodisperse magnetic particles

The magnetization hysteresis (M Vs H) curves for ideal, superpamagnetic, and non-interacting particles that are supposed to rotate coherently usually present a sigmoidal shape with fast saturation, zero coercivity $H_C = 0$, and remanence $M_R = 0$. This happens in thermodynamically equilibrated systems (Tamion et al. 2010) where the anisotropic term is not taken into account since it is assumed to be very low. This statement attends to the classical paramagnetic models, which considers greater magnetic moments (usually thousands of Bohr magnetons) where the highest magnetization M_S value reachable can be described by:

$$M/_{M_S} = \coth\left(\mu H/_{k_B T}\right) - k_B T/_{\mu H} = L\left(\mu H/_{k_B T}\right) \tag{8}$$

where L is the Langevin function. If the Langevin law is valid, the $M/_{M_S}$ versus $H/_T$ curves of a sample at different temperatures must superimpose in a single universal Langevin scaling. A good approximation scaling can be obtained using low fields and temperatures above T_B, where results compatible with the superparamagnetic behavior are expected (Fonseca et al. 2002). For instance, Fonseca et al. reported that the scaling is valid only when the system is comprised of non-interacting particles and spherical in shape (see Figure 7). However, the experimental evidence (Goya et al. 2003b, Vejpravová et al. 2005) shows that in real non-interacting systems with narrow size distribution, the suggested scaling law does not give appropriate results.

3.1.2 Non interacting, polydisperse magnetic particles

As stated in the last section, real nanoparticles' systems are constituted by a collection of variable particle sizes. Consequently, the magnetization of superparamagnetic particles can be best estimated as a weighted sum of Langevin functions (Ferrari et al. 1997):

$$M = \int_0^\infty L\left(\frac{\mu H}{k_B T}\right) M_S \, dx \tag{9}$$

where M_S is nearly temperature independent and is given by:

$$M_S = \int_0^\infty \frac{1}{\sqrt{2\pi}\,\mu\sigma} exp\left[-\frac{ln^2(\mu/H_0)}{2\sigma^2}\right] d\mu \tag{10}$$

where the log-normal distribution width is symbolized by σ and the median of the distribution values is denoted by μ_0 which is associated to the average magnetic moment $<\mu_m>$ by $<\mu_m> = \mu_0\, e^{-\sigma^2/2}$. Figure 7 shows the best fit curves obtained for normal Langevin function (Equation 8) and the sum of Langevin functions (Equation 10) for nanocristalline Ni particles; as can be seen, even if both fits seem to obtain similar results, the greatest dissimilarity comes from the resulting fitted parameters, as the magnetic moment values μ_0 and μ_m. Comparing these results, the (sum of functions) log-normal Langevin function obtained parameters show excellent accordance with the transmission microscopy observations, confirming the best performance of the weighted sum of Langevin functions for explaining real superparamagnetic systems.

3.1.3 Interacting magnetic particles

According to Kechrakos and Trohidou (Kechrakos and Trohidou 2000), dipolar interactions help to reduce the system's magnetic performance, advising that magnetization approaches saturation and remanence slower than the analogous non-interacting superparamagnetic systems (phenomena also observed when surface effects are not neglected). To comprehend the influence of dipolar interactions in diverse nanoparticles systems, Allia et al. (Allia et al. 2001) formulated a phenomenological Interacting Superparamagnetic Model (ISP) that considers interacting (through dipolar long-range fields) nanoparticles' magnetic moments, μ. The total influence of these interacting particles is modeled by using a fictional temperature T*, so-called apparent temperature, which is added to the real temperature in the weighted sum of Langevin function's denominator (Moscoso-Londoño et al. 2017):

$$M(H,T) = n_{CO}\int_0^\infty \mu_{CO}\, L\left(\frac{\mu_{CO}(H-H_C)}{k_B(T+T^*)}\right) f(\mu_{CO}) d\,\mu_{CO} \tag{15}$$

where the subscript 'CO' denotes corrected fit values, n_{CO} is the corrected number of magnetic moments of paramagnetic particles per mass unit, and the T* is linked to the dipolar interaction energy ε_D through $\varepsilon_D = k_B T^*$, which can be calculated by using low field susceptibility measurements.

The later description was formulated by Moscoso-Lodoño et al. (Moscoso-Londoño et al. 2017), who used iron oxide nanoparticles at different concentrations and T* calculated values that fitted their M Vs H curves. As a consequence, they determined the apparent $\langle\mu_{AP}\rangle$ and corrected magnetic moments $\langle\mu_{CO}\rangle$ with values higher than the apparent ones, as predicted by the Kechrakos and Trohidou (Kechrakos and Trohidou 2000) simulations, see Figure 8. All $\langle\mu_{AP}\rangle$ moments were

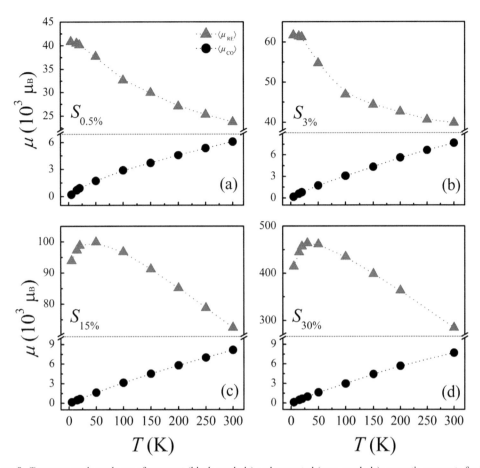

Figure 8. Temperature dependence of apparent (black symbols) and corrected (gray symbols) magnetic moments for iron oxide magnetic nanocomposites using the Interacting Superparamagnetic Model (ISP). Reprinted with permission from Moscoso-lodoño et al. 2017.

of the same order of magnitude for the analyzed samples when using the classical Langevin methodology, while in concentrated systems $\langle \mu_{CO} \rangle$, moments increase from low temperatures until they subsequently reach a decreasing tendency. Near those temperatures, the reached maximum can be described as a constraint on the ISP model, particularly when a substantial proportion of nanoparticles are still blocked.

The Moscoso-Lodoño group unraveled how the ISP model can characterize clusters of different nanoparticles as if they were single and larger magnetic entities per unit volume, a phenomenon that could emerge from interactions amidst and between the clustered nanoparticles. This approach completely agreed with the other characterization techniques used by this group.

A very good work published by Mendoza Zélis et al. (Mendoza Zélis et al. 2017) and Sanchez et al. (Sánchez et al. 2017) also used the aforementioned methods but applied them to different arrays and nanostructured shapes. While these articles are beyond the scope of this chapter, they are strongly recommended for readers working with no-spherical shapes.

3.2 Coercive field non interacting magnetic particles

3.2.1 Non interacting, monodisperse magnetic particles

For ideal superparamagnetic nanoparticles systems, a null coercive field is expected above T_B, as detailed in the beginning of Section 2. However, at low temperatures, the coercivity belonging to a non-interacting and randomly oriented particles system with random distribution of anisotropies is estimated to satisfy the next equation (Bean and Livingston 1959, Linderoth et al. 1993, Torres et al. 2015):

$$H_C(T) = H_{C0}\left[1-\left(\frac{T}{T_B}\right)^{1/2}\right] \tag{11}$$

When supposing T = 0 in the Stoner-Wohlfarth model, the H_{C0} must take the value $H_{C0} = 0.48(2K/M_S)$ and the $H_C(T)$ can be written as (Stoner and Wohlafarth 1948):

$$H_C(T) = 0.48\frac{2K}{M_S}\left[1-\left(\frac{T}{T_B}\right)^{1/2}\right] \tag{12}$$

All the above coercivity expressions contemplate a coherently reversal process of magnetization. However, it does not consider the size distribution because the total H_C of a system does not correspond to a simple superposition of individual particle's coercivities.

As can be appreciated in equations 11 and 12, H_C follows an almost linear dependence with $T^{1/2}$ at low temperatures. Therefore, extrapolating the linear fit interval to T = 0, the values H_{C0} and hence the anisotropy term K can be obtained, respectively. Likewise, the extrapolation of $H_C(T) = 0$ provides the values of T_B, which must be consistent with the values obtained by the M_{ZFC} and M_{FC} measurements (see Section 1), i.e., see Figure 9, where the H_C Vs $T^{1/2}$ linear fit region is plotted for

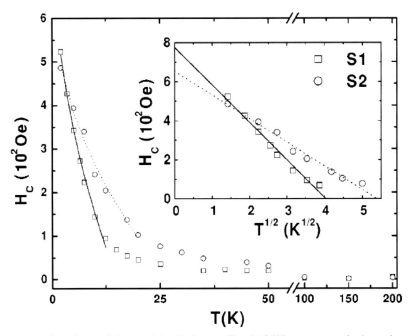

Figure 9. Temperature dependence of the coercivity H_c for two diluted Ni:SiO$_2$ nanocomposites' samples with 1.5 and 5-weight percentage nickel. The inset shows $H_C(T)$ to obey a (T)$^{1/2}$ dependence with H_{C0} of ~ 780 and 650 Oe for samples with 1.5 and 5-weight percentage nickel, respectively. Reprinted with permission from Fonseca et al. 2002.

two diluted Ni:SiO$_2$ nanocomposites' samples with 1.5 and 5-weight percentage nickel from the Fonseca et al. work (Fonseca et al. 2002).

Conversely, fitting the linear region by assuming the anisotropy constant as temperature independent (Torres et al. 2015), the values of K can be obtained by extrapolating T = 0. Nevertheless, it is worth noting that the obtained K values must be used cautiously (Moscoso-Londoño et al. 2017) as Equation 12 is founded on the classic superparamagnetic models and does not take into account important parameters present in real samples' systems, such as the temperature dependence of the magnetic parameters involved (Bean and Livingston 1959), surface effects, particle size dispersion, dipolar interactions effects, or structural disorder effects related to the Verwey transition that may impact the anisotropy determination of the analyzed systems (Yosida and Tachiki 1957).

3.2.2 Non interacting, polydisperse magnetic particles

The $H_c(T)$ behavior that considers particle size distribution was described by Nunes et al. (Nunes et al. 2004) using a phenomenological approach that takes into account the unblocked particles' contributions. Thus, the blocked particles' coercive field is obtained as follows (Knobel et al. 2008, Yosida and Tachiki 1957):

$$H_{CB} = \alpha \frac{2K}{M_S} \left[1 - \left(\frac{T}{\langle T_B \rangle_T} \right)^{1/2} \right] \tag{13}$$

where α takes the 0.48 or 1 values, if the particles are randomly oriented or if their easy axes are almost all oriented, respectively (see Section 2.2.1). As this equation just regards the volume fraction of blocked particles at temperature T, the $\langle T_B \rangle_T$ represents their average blocking temperature.

Therefore, the total description of the particle size distribution can be ascribed to blocked (H_{CB}, Equation 13) and superparamagnetic particles. In this regard, Keller and Luborsky (Kneller and Luborsky 1963) considered the influence of the superparamagnetic and blocked particle contributions when analyzing the magnetization curves as linear for $H < H_{CB}$. Hence, by averaging both contributions, the total coercive field can be calculated:

$$\langle H_C \rangle_T = \frac{M_r(T)}{\chi_S(T) + [(M_r(T)/H_{CB}]} \tag{14}$$

where $M_r(T)$ is the temperature-dependent remanence and $\chi_S(T)$ the superparamagnetic susceptibility in a determined temperature T. The parameters aforementioned in Equation 14 can be acquired by experimental characterization, i.e., the blocking temperature distribution $f(T_B)$ can be obtained from the $M_{FC} - M_{ZFC}$ curves as presented in Section 1. To test this proposition, Nunes et al. used a Cu$_{90}$Co$_{10}$ granular alloy sample that did not follow the H_C linear dependence with $T^{1/2}$. As can be seen in Figure 10, without taking into account the unblocked particles' superparamagnetic susceptibility, they obtained the curves indicated by the dashed and dotted lines by using Equation 13, working well for isolated and narrow distributions (Bean and Livingston 1959). The average blocking temperature reliant on temperature is described by the dashed line, which notably demonstrates the lack of the superparamagnetic correction term. In contrast, the dashed-dotted line was acquired by using Equation 14 and the average blocking temperature for all particles, which resulted in a superb agreement with the experimental data at low temperatures in which the majority of particles remain blocked. Lastly, by using Equation 14 and the average blocking temperature $\langle T_B \rangle_T$ and considering the volume fraction of blocked particles at a determined temperature, they obtained the solid line, which perfectly describes the experimental data in the whole temperature range.

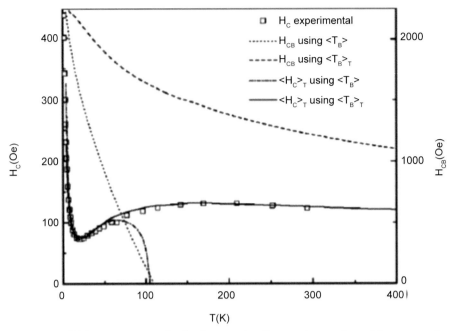

Figure 10. Coercive field Vs Temperature of a $Cu_{90}Co_{10}$ granular alloy sample: experimental (open squares), calculated $<H_c>_T$ (solid and dotted-dashed lines), and H_{CB} (dashed and dotted lines). Reprinted with permission from Nunes et al. 2004.

3.2.3 Interacting magnetic nanoparticles

As isolated nanoparticles tend to aggregate due to steric and surface effects, then dipolar interactions are expected to impact the H_C values. For this reason, these interactions must be taken into account and need to be widely studied when describing more realistic nanoparticle magnetic systems. For example, Neél (Néel 1955) suggested that H_C diminishes (in order to lower the anisotropy energy) with the increase in the "interaction field" or packing fraction ∈ following the relation $H_C = H_{C∞}(1 - ∈)$. In a later paper, Wohfarth (Wohlfarth 1955) demonstrated the existence of a particle orientation dependency on dipolar interactions due to the fact that these interactions are direction-dependent. Consequently, this results in a decrease or increase of H_C depending on the orientation of the particles. Later, more recent works (El-Hilo 2010, El-Hilo and Bsoul 2007, Holmes et al. 2007, Sharif et al. 2007) described an evident dependency on H_C in terms of an increase or decrease in the anisotropy energy due to interaction-induced effects.

As the increase or decrease depends strongly on the studied system, Das et al. (Das et al. 2010) realized that applying interaction induced changes in K could not describe by itself their Nickel nanoparticles' systems but rather can be described by the effect of fluctuating dipolar magnetic fields due to fluctuations in the magnetization of nearest neighbor particles. As a result, they qualitatively described the dynamics of magnetization in the presence of dipolar interactions as a collective effect (Suzuki et al. 2009). These collective dynamics of particle magnetizations that are the result of dipolar interactions, added as an energy term $± E_{dip}$ to the anisotropy energy, were responsible for the increase in H_C in their Nickel nanoparticles' studied systems, see Figure 11.

A recent theoretical work solved by the Monte Carlo Method from Agudelo-Giraldo et al. (Agudelo-Giraldo et al. 2020) also describes how the coercive field and remanence can alter the magnetization processes of a series of nanogranular films. Their simulations also suggest changes

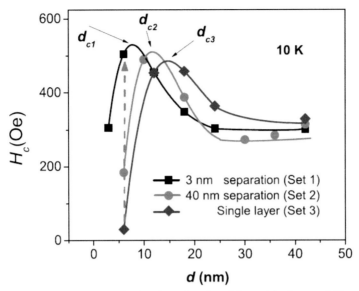

Figure 11. The coercive field H_c at T = 10 K is plotted as a function of the particle diameter d. The peak for each curve separates the single domain (SD) from the multi domain (MD) particles. Particles with diameter higher (smaller) than the peak diameter (d_c) are MD (SD). The critical diameters d_{c1}, d_{c2} and d_{c3} for each sample are identified with arrows. In the SD region (i.e., $d < d_c$), the coercivity increases with increasing dipolar interactions as shown by the vertical dashed arrow. Reprinted with permission from Das et al. 2010.

in H_C and in the anisotropy because of the competition among local and boundary anisotropies, showing how anisotropic local environment and collective phenomena are significant to explain nanomaterials with distorted grain boundaries.

4. Dynamic properties, AC measurements

So far, this chapter has described some of the most suitable techniques to characterize nanomagnetic materials using static measurements. Nonetheless, another useful experimental technique is the AC (Alternating Current) susceptibility measurements, where the excitation field is applied at different frequencies (f) to follow the processes of particle relaxation (see Equation 2). Different applied frequencies imply a variation in the number of particles that can go behind the oscillations, modifying, as a consequence, the magnetic response since the sample magnetization might be delayed from the alternating field. The AC method yields to a susceptibility $\chi = \chi' + \chi''$ built of two main components: (1) χ' the real part and (2) χ'' the imaginary or dynamic part, which is related with dynamic processes that produce spin reversal lags in presence of an alternating external field.

4.1 Relaxation time

Considering small nanoparticles with low anisotropy energy E_a, the nanoparticle magnetic moment can pass the anisotropy energy barrier E_a from one orientation to the other due to thermal agitation, where the characteristic time between the two possible orientations is the so-called Neél relaxation time, τ_N. Writing Equation 2 in a different manner, the Neél relaxation time can be defined as:

$$\tau_N = \tau_0 \, \exp\left(\frac{E_a}{k_B T}\right) \tag{15}$$

Where τ_0 is the characteristic time, which must take values between 10^{-9} and 10^{-11} s according to the superparagmentic theory (Goya et al. 2003a) and $E_a = KV$ is characterized in the simplest case by only one anisotropy axis, as described in the beginning of this chapter.

On the other hand, for unfrozen fluidic solutions in which the nanoparticles are allowed to rotate and, thus, change their orientation and anisotropy axis trying to align with the external magnetic field in order to reduce the magnetic energies, it is necessary to use a different expression to define their relaxation processes. In this regard, the "Brown relaxation" will be the best option because its characteristic time, τ_B, is defined by the liquid viscosity of the matrix in which the sample is contained, η, and the hydrodynamic nanoparticle volume V_H considering the magnetic core and coating (Zhakhovskii and Anisimov 1997):

$$\tau_B = \frac{3\eta V_H}{k_B T} \tag{16}$$

Depending on the system, both relaxation mechanisms can be present above certain temperature (usually in ferrofluids with broad distribution of particle size and mean radii ~ 10 nm) and the resultant characteristic time τ can be obtained by adding both contributions (Rosensweig 2002):

$$\frac{1}{\tau} = \frac{1}{\tau_N} + \frac{1}{\tau_B} \tag{17}$$

Normally, the absorption of the particle system will follow the quickest relaxation processes. Owing to the dissimilar size relations for the two relaxation regimes that are given by Equations 15 and 16, there are different spectral regions in which the two relaxation mechanisms may occur. For example, the Néel relaxation predominates in small particle sizes and measurements with high frequencies and the opposite works for the Brown relaxation, see Figure 12. For this reason, the Brown relaxation is not well applied for hyperthermia experiments (Hergt et al. 2010).

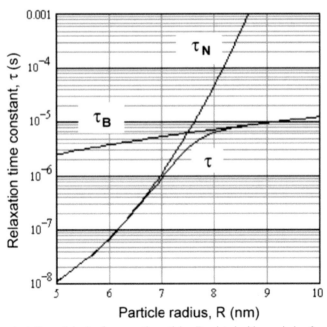

Figure 12. Time constants Vs particle size for magnetic particles. Reprinted with permission from Rosensweig 2002.

As stated in last paragraph, the Neél process is the dominant relaxation detected in small nanoparticles' arrangements (r < 9 nm) when $\tau_N < \tau_B$ (Rosensweig 2002). Considering that the measurement typical time for a vibrating sample magnetometer τ_m is around 0.1 – 100s, then $\tau_m \gg \tau_B$ and the nanoparticles are allowed to freely rotate in the course of the measurement, while for low temperatures' measurements, i.e., frozen water, merely the Neél relaxation is expected as the nanoparticle magnetic moment is in a minimum energy blocked state when $\tau_m < \tau_N$ (Henrard et al. 2019).

The literature usually reports two main methodologies to acquire the relaxation time of a nanoparticle system: the Neél-Arrhenius and the Vogel-Fulcher laws, both of which will be described in the next sections.

4.2 Neél-Arrhenius law

The expected behavior for superparamagnetic systems assumes a modification (shift) of the maximum in temperature (T_{max}) with increasing frequency for both components χ' (real) and χ'' (imaginary) susceptibilities (Dormann et al. 1999), see Figure 13. Nevertheless, this tendency is not always indicative of superparamagnetic behavior. For that reason, a convenient method for classifying the freezing/blocking processes obtained by the dynamic measurements determines the

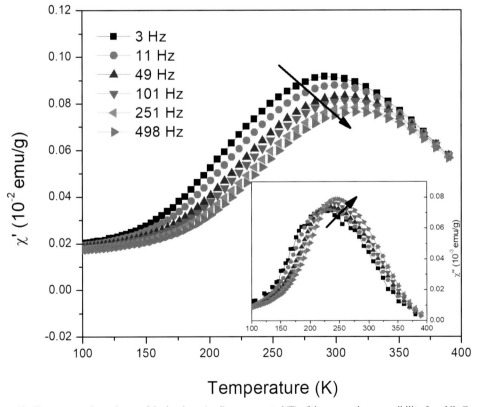

Figure 13. Temperature dependence of the in-phase (real) component χ'(T) of the magnetic susceptibility for a $Ni_{27}Co_{39}Fe_{33}$ layered double hydroxide sample, at different excitation frequencies. Arrows indicate increasing frequencies. Inset: Out of phase (imaginary) component χ''(T). The data were taken with an external magnetic field H = 10Oe ≈ 800 *A/m*. [Unpublished data from the chapter author].

Mydosh parameter Φ (Mydosh 1993), which is an empirical parameter that denotes the relative shift of the temperature per frequency decade:

$$\Phi = \frac{\Delta T_{max}}{T_{max}\,\Delta log_{10}(f)} \tag{18}$$

Where ΔT_{max} is the difference among the T_{max} values measured in the $\Delta log_{10}(f)$ frequency range. For weakly or non-interacting superparamagnetic systems, the expected value for the Mydosh parameter Φ must be ($\Phi \sim 0.1$–0.13) (Masunaga et al. 2009, Vázquez-Vázquez et al. 2011, Zhakhovskii and Anisimov 1997); the values below this interval are characteristic of nanoparticle systems with intermediate interactions (Nadeem et al. 2011) and the values below the interval are related to strong interparticle interacting systems (Moscoso-Londoño et al. 2017) or result from spin-glass like surface behavior ($\Phi \sim 5 \times 10^{-3} - 5 \times 10^{-2}$) (Goya 2002). Consequently, if the analyzed systems show the above mentioned shift in the T_{max} with increasing frequency and a Mydosh parameter value between 0.1–0.13, a thermally activated Néel–Arrhenius fit for superparamagnetic particles must simulate the experimental data.

The Néel–Arrhenius fit model is described by Equations 2 and 15, in which the energy barrier, $E_a = k_{eff}V \sin^2 \theta$ is proportional to the particle volume V at null external magnetic field, k_{eff} is the effective magnetic anisotropy constant and θ is the angle amid the easy magnetization axis and the particle magnetic moment. Then, when plotting $\ln(\tau)$ Vs $1/T_B$ a linear plot is expected. Figure 14 shows the Arrhenius plot obtained for the imaginary component χ'' from the work of Goya et al. (Goya et al. 2003a) in which they used spherical 5 nm magnetite nanoparticles. In this work, they calculated the $k_{eff} = 35.6 \times 10^5 \; erg/cm^3$ and $\tau_0 = 9 \times 10^{-13}$ s values by applying equation 15 and the radii obtained from electronic microscopy data. The resultant effective anisotropy is an order magnitude higher than the first order magnetocrystalline anisotropy from the bulk material. This augmentation

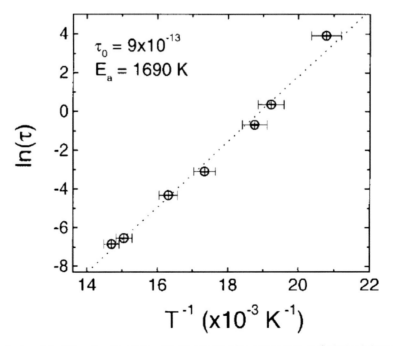

Figure 14. Arrhenius plot of the relaxation time τ Vs inverse blocking temperature T_B^{-1} obtained from the imaginary component χ''(T) for Fe_3O_4 nanoparticles. Dotted line is the best fit with $\tau_0 = 9 \times 10^{-13}$ s and $E_a = 1690$ K. The τ_0 value is slightly smaller than the expected values between 10^{-9} and 10^{-11} s, according to the superparagmentic theory. Reprinted with permission from Goya et al. 2003a.

is usually expected because of smaller particle size, surface effects (Luis et al. 2002), extra intrinsic contributions from the particle anisotropy (shape, magnetostrictive or magnetocrystalline), and interparticle dipolar or exchange interactions, which have direct repercussions on the energy barrier (Goya et al. 2003a). Therefore, the authors point out to be cautious with the values obtained from the Néel–Arrhenius law fitting since they can also lead to obtaining values without physical meaning (Vázquez-Vázquez et al. 2011).

In contrast, when the obtained k_{eff} value is slightly bigger than the bulk one, it is convenient to link this value with the estimated coercive field for spherical particles with magnetocrystalline anisotropy by using the relation $H_C = 2K_{eff}/M_S$ (valid for uniaxial anisotropy particles with no interactions, see Section 2.2.1), whose values must agree with the obtained values from the hysteresis curves at low temperatures.

4.2.1 Neél-Arrhenius modified law-dipolar interactions

For some applications, it is decisive to identify how the superparamagnetic relaxation rate is modified by the magnetic dipolar interactions. Unfortunately, the experiments performed and the theoretical work in the literature provide dissimilar, and inclusively opposing, results because of the many possible origins of the magnetocrystalline anisotropy that predict when the inversion of magnetic moments becomes faster (Hansen 1998, Morup et al. 1995, Mørup and Tronc 1994) or slower (Andersson et al. 1997, El-Hilo et al. 1992) as the dipolar interactions are switched on (Luis et al. 2002). To attest the latter point, Luis et al. followed the theoretical piece of Dormann et al. (Dormann et al. 1988) and made the assumption that, when the magnetic moment of a specified cluster flips, the energy of a neighbor spin changes by an extent equivalent to their common dipolar interaction energy. Hence, before the analyzed spin flips again, the nearest neighbor's magnetization is polarized via a new local dipolar field. Thus, the relaxation time can be expressed as the one used for independent clusters with an extra interaction energy E_{dip} term added in Equation 15:

$$\tau_{\pm} = \tau_0 \, exp\left(\frac{E_a \pm E_{dip}}{k_B T}\right) \tag{19}$$

where the +/– signs are related with the spin reversal direction to the local field (interacting field) produced by the neighbors. An opposed direction is denoted by (+) while in the same direction is denoted by (–), leading to different relaxation times depending on the local field alignment.

This approach has been used in some works (Goya et al. 2003b, Goya 2002) and has given reasonable good fitting parameters. Nevertheless, Vogel-Fulcher is the most commonly reported law for interacting nanoparticle systems, as described below.

4.3 Vogel-Fulcher law

As stated in the previous sections, the Arrhenius law can lead to fitting values without physical meaning (Vázquez-Vázquez et al. 2011), even when using the dipolar term interaction because of the many possible origins of the magnetocrystalline anisotropy. This is specially so at low temperatures when, depending of the analyzed system, some magnetic clusters interact (via dipole, exchange or other forces) so as to have cooperative effects, as evidenced by static magnetic measurements for some interacting systems (see Sections 1 and 2). In this context, the more recommended law to use is the Vogel-Fulcher.

The Vogel-Fulcher law was formulated for interacting nanoparticle systems (likewise Dorman et al. (Dormann et al. 1988, 1999)) assuming that the anisotropy energies were stronger than the interactions between particles (Shtrikman and Wohlfarth 1981). In that sense, the Vogel-Fulcher law introduces a fictitious temperature T_0^{VF} term that is proportional to the dipolar interaction's strength between particles:

$$\tau = \tau_0 \; \exp \left[\frac{K_{eff} V}{k_B (T_{max} - T_0)} \right] \tag{20}$$

where τ_0 is the relaxation time obtained from the Vogel-Fulcher (V-F) law, T_{max} is the temperature indicative of the beginning of the blocking process (i.e., the temperature of the maxima in the peak position in AC susceptibility), as long as $T_{max} \gg \tau_0^{VF}$, and τ is the measuring window. Assuming the pre-exponential factor $\tau_0 = 10^{-9}$s and using the V-F expression, the Vazquez-Vazquez (Vázquez-Vázquez et al. 2011) group obtained fitting T_0 values that indicate strong particle interactions, as T_0 symbolizes the temperature from which the relaxation time diverges.

Also, Moscoso-Lodoño et al. (Moscoso-Londoño et al. 2017) used the V-F law and obtained values around the expected for superparamagnetic nanoparticles' systems (τ_0 around 10^{-9}s and K_{eff} in good agreement with the values extracted from their coercivity analysis) using iron oxide nanoparticles at different concentrations, see Figure 15.

It must be noticed that, in Equation 20, the term $T - T_0$ may resemble the T* fictitious temperature used in the ISP model seen in Section 2.1.3 for magnetic interacting systems; nevertheless, the ISP denominator includes an added *T** term giving different interpretative meaning. In the V-F expression, the temperature T_0 symbolizes that some critical quantity is diverging there (i.e., viscosity or relaxation time). A mixture of both models can be found in the work from Allia and Tiberto (Allia et al. 2011, Allia and Tiberto 2011) in which by modifying the Arrhenius law by adding an extra

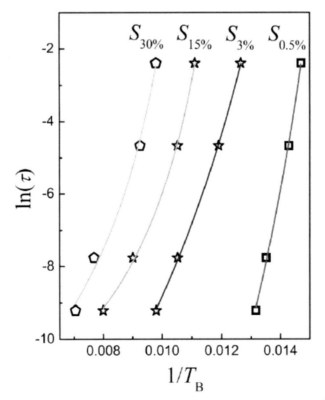

Figure 15. Plots of *In* (τ) Vs 1/T$_{max}^{\chi''}$ (T$_{max}^{\chi''}$ is the peak temperature of the imaginary component at each measured frequency). The full line corresponds to the fit according to Vogel-Fulcher law. Reprinted with permission from Moscoso-lodoño et al. 2017.

term $(T + T^*)$ they explained how anisotropy can drive to colective magnetic interactions that are no negligible. Therefore, the reader must choose very carefully the model that adjusts the best with the hypothesis that best describes its nanoparticle magnetic system.

5. Conclusions

This chapter provides a general overview of the basic techniques used to characterize magnetic nanomaterials by using Vibrating Sample Magnetometers (VSM) and Superconducting Quantum Interference Device (SQUID) magnetometers. These techniques must converge on describing the same phenomena by using the different approaches described in Sections 1, 2 and 3. The non-interacting monodisperse systems are the most simple nanoparticle systems. These systems use simple equations and assumptions to differentiate the superparamagnetic and thermally stable regimes. Nevertheless, the analysis and description complicates when interactions, surface effects, the many possible origins of the magnetocrystalline anisotropy and even viscosity cannot be neglected. As the experimental evidence suggests, cooperative dipolar interactions can vary the blocking barrier heights, the blocking temperature distribution and likewise the heating efficiency during AC magnetic measurements that may have cooperative effects at low temperatures. Therefore, the reader must choose very carefully the model that adjusts the best to the hypothesis that best describes its nanoparticle magnetic system.

Acknowledgment

The author thanks the CONACyT (Mexico) for the Cátedras Reseach-Fellow program and the LNCAE (Laboratorio Nacional de Conversión y Almacenamiento de Energía) for access to its experimental facility.

References

Agudelo-Giraldo, J.D., Moscoso Londoño, O., Velásquez-Salazar, A.A. and Restrepo-Parra, E. 2020. Grain size influence upon magnetic behavior at nanoscale. A computational approach. *Journal of Magnetism and Magnetic Materials*, 515(July). https://doi.org/10.1016/j.jmmm.2020.167296.

Allia, Paolo, Coisson, M., Tiberto, P., Vinai, F., Knobel, M., Novak, M.A. and Nunes, W.C. 2001. Granular Cu-Co alloys as interacting superparamagnets. *Physical Review B - Condensed Matter and Materials Physics*, 64(14): 1444201–14442012. https://doi.org/10.1103/PhysRevB.64.144420.

Allia, Paolo and Tiberto, P. 2011. Dynamic effects of dipolar interactions on the magnetic behavior of magnetite nanoparticles. *Journal of Nanoparticle Research*, 13(12): 7277–7293. https://doi.org/10.1007/s11051-011-0642-2.

Allia, P., Tiberto, P., Coisson, M., Chiolerio, A., Celegato, F., Vinai, F., Sangermano, M., Suber, L. and Marchegiani, G. 2011. Evidence for magnetic interactions among magnetite nanoparticles dispersed in photoreticulated PEGDA-600 matrix. *Journal of Nanoparticle Research*, 13(11): 5615–5626. https://doi.org/10.1007/s11051-011-0249-7.

Allia, P., Barrera, G., Tiberto, P., Nardi, T., Leterrier, Y. and Sangermano, M. 2014. Fe3O4 nanoparticles and nanocomposites with potential application in biomedicine and in communication technologies: Nanoparticle aggregation, interaction, and effective magnetic anisotropy. *Journal of Applied Physics*, 116(11). https://doi.org/10.1063/1.4895837.

Andersson, J., Djurberg, C., Jonsson, T., Svedlindh, P. and Nordblad, P. 1997. Monte Carlo studies of the dynamics of an interacting monodispersive magnetic-particle system. *Physical Review B - Condensed Matter and Materials Physics*, 56(21): 13983–13988. https://doi.org/10.1103/PhysRevB.56.13983.

Bean, C.P. 1955. Hysteresis loops of mixtures of ferromagnetic micropowders. *Journal of Applied Physics*, 26(11): 1381–1383. https://doi.org/10.1063/1.1721912.

Bean, C.P. and Livingston J.D. 1959. Superparamagnetism. *Journal of Applied Physics*, 30(4). https://doi.org/https://doi.org/10.1063/1.2185850.

Bitoh Teruo, Kazuyuki Ohba, Masaki Takamatsu, Takashi Shirane and Susumu Chikazawa. 1995. Field-cooled and zero-field-cooled magnetization of superparamagnetic fine particles in Cu97Co3 alloy: Comparison with spin-glass Au96Fe4 alloy. *J. Phys. Soc. Jpn.*, 1305–1310. https://doi.org/https://doi.org/10.1143/JPSJ.64.1305.

Bruvera, I.J., Mendoza Zélis, P., Pilar Calatayud, M., Goya, G.F. and Sánchez, F.H. 2015. Determination of the blocking temperature of magnetic nanoparticles: The good, the bad, and the ugly. *Journal of Applied Physics*, 118(18). https://doi.org/10.1063/1.4935484.

Das, R.K., Rawal, S., Norton, D. and Hebard, A.F. 2010. A collective dynamics description of dipolar interactions and the coercive field of magnetic nanoparticles. *Journal of Applied Physics*, 108(12). https://doi.org/10.1063/1.3524277.

De Biasi, E., Ramos, C.A., Zysler, R.D. and Romero, H. 2002. Large surface magnetic contribution in amorphous ferromagnetic nanoparticles. *Physical Review B - Condensed Matter and Materials Physics*, 65(14): 1–8. https://doi.org/10.1103/PhysRevB.65.144416.

De Biasi, E., Zysler, R.D., Ramos, C.A., Romero, H. and Fiorani, D. 2005. Surface anisotropy and surface-core interaction in Co-Ni-B and Fe-Ni-B dispersed amorphous nanoparticles. *Physical Review B - Condensed Matter and Materials Physics*, 71(10): 1–6. https://doi.org/10.1103/PhysRevB.71.104408.

Dormann, J.L., Bessais, L. and Fiorani, D. 1988. A dynamic study of small interacting particles: Superparamagnetic model and spin-glass laws. *Journal of Physics C: Solid State Physics*, 21(10): 2015–2034. https://doi.org/10.1088/0022-3719/21/10/019.

Dormann, J.L., Fiorani, D. and Tronc, E. 1999. On the models for interparticle interactions in nanoparticle assemblies: Comparison with experimental results. *Journal of Magnetism and Magnetic Materials*, 202(1): 251–267. https://doi.org/10.1016/S0304-8853(98)00627-1.

El-Hilo, M., O'Grady, K. and Chantrell, R.W. 1992. Susceptibility phenomena in a fine particle system. I. Concentration dependence of the peak. *Journal of Magnetism and Magnetic Materials*, 114(3): 295–306. https://doi.org/10.1016/0304-8853(92)90272-P.

El-Hilo, M. and Bsoul, I. 2007. Interaction effects on the coercivity and fluctuation field in granular powder magnetic systems. *Physica B: Condensed Matter*, 389(2): 311–316. https://doi.org/10.1016/j.physb.2006.07.003.

El-Hilo, M. 2010. Effects of array arrangements in nano-patterned thin film media. *Journal of Magnetism and Magnetic Materials*, 322(9–12): 1279–1282. https://doi.org/10.1016/j.jmmm.2009.06.036.

El-Hilo, M. 2012. Nano-particle magnetism with a dispersion of particle sizes. *Journal of Applied Physics*, 112(10). https://doi.org/10.1063/1.4766817.

Ferrari, E., da Silva, F. and Knobel, M. 1997. Influence of the distribution of magnetic moments on the magnetization and magnetoresistance in granular alloys. *Physical Review B - Condensed Matter and Materials Physics*, 56(10): 6086–6093. https://doi.org/10.1103/PhysRevB.56.6086.

Fonseca, F.C., Goya, G.F., Jardim, R.F., Muccillo, R., Carreño, N.L.V., Longo, E. and Leite, E.R. 2002. Superparamagnetism and magnetic properties of Ni nanoparticles embedded in SiO2. *Physical Review B - Condensed Matter and Materials Physics*, 66(10): 1044061–1044065. https://doi.org/10.1103/PhysRevB.66.104406.

Goya, Gerardo F. 2002. Magnetic dynamics of Zn57Fe2O4 nanoparticles dispersed in a ZnO matrix. *IEEE Transactions on Magnetics*, 38(5 I): 2610–2612. https://doi.org/10.1109/TMAG.2002.803204.

Goya, G.F., Berquó, T.S., Fonseca, F.C. and Morales, M.P. 2003a. Static and dynamic magnetic properties of spherical magnetite nanoparticles. *Journal of Applied Physics*, 94(5): 3520–3528. https://doi.org/10.1063/1.1599959.

Goya, G.F., Fonseca, F.C., Jardim, R.F., Muccillo, R., Carreño, N.L.V., Longo, E. and Leite, E.R. 2003b. Magnetic dynamics of single-domain Ni nanoparticles. *Journal of Applied Physics*, 93(10 2): 6531–6533. https://doi.org/10.1063/1.1540032.

Hansen, M.F. and Mørup, S. 1998. Models for the dynamics of interacting magnetic nanoparticles. *Journal of Magnetism and Magnetic Materials*, 184(3): 262–274. https://doi.org/10.1016/s0304-8853(97)01165-7.

Henrard, D., Vuong, Q.L., Delangre, S., Valentini, X., Nonclercq, D., Gonon, M.F. and Gossuin, Y. 2019. Monitoring of superparamagnetic particle sizes in the langevin law regime. *Journal of Nanomaterials*, pp. 9–11. https://doi.org/10.1155/2019/6409210.

Hergt, R., Dutz, S. and Zeisberger, M. 2010. Validity limits of the Néel relaxation model of magnetic nanoparticles for hyperthermia. *Nanotechnology*, 21(1): 015706. https://doi.org/10.1088/0957-4484/21/1/015706.

Holmes, B.M., Newman, D.M. and Wears, M.L. 2007. Determination of effective anisotropy in a modern particulate magnetic recording media. *Journal of Magnetism and Magnetic Materials*, 315(1): 39–45. https://doi.org/10.1016/j.jmmm.2007.02.200.

Kechrakos, D. and Trohidou, K. 2000. Interplay of dipolar interactions and grain-size distribution in the giant magnetoresistance of granular metals. *Physical Review B - Condensed Matter and Materials Physics*, 62(6): 3941–3951. https://doi.org/10.1103/PhysRevB.62.3941.

Kittel, C. 1946. Theory of the structure of ferromagnetic domains in films and small particles. *Physical Review*, 70(11-12): 965–971. https://doi.org/10.1103/PhysRev.70.965.

Kneller, E.F. and Luborsky, F.E. 1963. Particle size dependence of coercivity and remanence of single-domain particles. *Journal of Applied Physics*, 34(3): 656–658. https://doi.org/10.1063/1.1729324.

Knobel, M., Nunes, W.C., Socolovsky, L.M., De Biasi, E., Vargas, J.M. and Denardin, J.C. 2008. Superparamagnetism and other magnetic features in granular materials: A review on ideal and real systems. *Journal of Nanoscience and Nanotechnology*, 8: 2836–2857.

Li, Q., Kartikowati, C.W., Horie, S., Ogi, T., Iwaki, T. and Okuyama, K. 2017. Correlation between particle size/domain structure and magnetic properties of highly crystalline Fe3O4 nanoparticles. *Scientific Reports*, 7(1): 1–4. https://doi.org/10.1038/s41598-017-09897-5.

Linderoth, S., Balcells, L., Labarta, A., Tejada, J., Hendriksen, P.V. and Sethi, S.A. 1993. Magnetization and Mössbauer studies of ultrafine Fe-C particles. *Journal of Magnetism and Magnetic Materials*, 124(3): 269–276. https://doi.org/10.1016/0304-8853(93)90125-L.

Lois Neel. 1949. Influence des fluctuations thermiques sur l'aimantation de grains ferromagnétiques très fins. *Comptes Rendus Hebdomadaires Des Seances De L Academie Des Sciences*, 228(8): 664–666. HAL Id: hal-02878471. https://hal.archives-ouvertes.fr/hal-02878471.

Luis, F., Petroff, F., Torres, J.M., García, L.M., Bartolomé, J., Carrey, J. and Vaurès, A. 2002. Magnetic relaxation of interacting Co clusters: Crossover from two- to three-dimensional lattices. *Physical Review Letters*, 88(21): 2172051–2172054. https://doi.org/10.1103/PhysRevLett.88.217205.

Mamiya, H., Nakatani, I. and Furubayashi, T. 1998. Blocking and freezing of magnetic moments for iron nitride fine particle systems. *Physical Review Letters*, 80(1):177–180. https://doi.org/10.1103/PhysRevLett.80.177.

Mamiya, H., Ohnuma, M., Nakatani, I. and Furubayashi, T. 2004. Zero-field-cooled and field-cooled magnetization of individual nanomagnets and their assembly. *Physica Status Solidi (A) Applied Research*, 201(15): 3345–3349. https://doi.org/10.1002/pssa.200405512.

Mamiya, H., Ohnuma, M., Nakatani, I. and Furubayashim, T. 2005. Extraction of blocking temperature distribution from zero-field-cooled and field-cooled magnetization curves. *IEEE Transactions on Magnetics*, 41(10): 3394–3396. https://doi.org/10.1109/TMAG.2005.855205.

Masunaga, S.H., Jardim, R.F., Fichtner, P.F.P. and Rivas, J. 2009. Role of dipolar interactions in a system of Ni nanoparticles studied by magnetic susceptibility measurements. *Physical Review B - Condensed Matter and Materials Physics*, 80(18): 1–7. https://doi.org/10.1103/PhysRevB.80.184428.

Medina, M.A., Oza, G., Ángeles-Pascual, A., González, M.M., Antaño-López, R., Vera, A., Leija, L., Reguera, E., Arriaga, L.G., Hernández, J.M.H. and Ramírez, J.T. 2020. Synthesis, characterization and magnetic hyperthermia of monodispersed cobalt ferrite nanoparticles for cancer therapeutics. *Molecules*, 25(19). https://doi.org/10.3390/molecules25194428.

Mendoza Zélis, P., Vega, V., Prida, V.M., Costa-Arzuza, L.C., Béron, F., Pirota, K.R., López-Ruiz, R. and Sánchez, F.H. 2017. Effective demagnetizing tensors in arrays of magnetic nanopillars. *Physical Review B*, 96(17): 1–8. https://doi.org/10.1103/PhysRevB.96.174427.

Micha, J.S., Dieny, B., Régnard, J.R., Jacquot, J.F. and Sort, J. 2004. Estimation of the Co nanoparticles size by magnetic measurements in Co/SiO2 discontinuous multilayers. *Journal of Magnetism and Magnetic Materials*, 272–276(SUPPL. 1), 2003–2004. https://doi.org/10.1016/j.jmmm.2003.12.268.

Mørup, S. and Tronc, E. 1994. Superparamagnetic relaxation of weakly interacting particles. *Physical Review Letters*, 72(20): 3278–3281. https://doi.org/10.1103/PhysRevLett.72.3278.

Morup, S., Bodker, F., Hendriksen, P.V. and Linderoth, S. 1995. Spin-glass-like ordering of the magnetic moments of interacting nanosized maghemite particles. *Physical Review B*, 52(1): 287–294. https://doi.org/10.1103/PhysRevB.52.287.

Moscoso-Londoño, O., Muraca, D., Pirota, K.R., Knobel, M., Tancredi, P., Socolovsky, L.M., Mendoza Zélis, P., Coral, D., Fernández van Raap, M.B., Wolff, U., Neu, V., Damm, C. and de Oliveira, C.L.P. 2017. Different approaches to analyze the dipolar interaction effects on diluted and concentrated granular superparamagnetic systems. *Journal of Magnetism and Magnetic Materials*, 428: 105–118. https://doi.org/10.1016/j.jmmm.2016.12.019.

Mydosh, J.A. 1993. *Spin Glasses : an experimental introduction*. London : Taylor and Francis, 256 p.

Nadeem, K., Krenn, H., Traussnig, T., Würschum, R., Szabó, D.V. and Letofsky-Papst, I. 2011. Effect of dipolar and exchange interactions on magnetic blocking of maghemite nanoparticles. *Journal of Magnetism and Magnetic Materials*, 323(15): 1998–2004. https://doi.org/10.1016/j.jmmm.2011.02.041.

Neel, L. 1949. Théorie du traînage magnétique des ferromagnétiques en grains fins avec applications aux terres cuites. *Ann. Geophys (CNRS)*, 5(99).

Néel, L. 1955. Some theoretical aspects of rock-magnetism. *Advances in Physics*, 4(14): 191–243. https://doi.org/10.1080/00018735500101204.

Nunes, W.C., Folly, W.S.D., Sinnecker, J.P. and Novak, M.A. 2004. Temperature dependence of the coercive field in single-domain particle systems. *Physical Review B - Condensed Matter and Materials Physics*, 70(1): 1–6. https://doi.org/10.1103/PhysRevB.70.014419.

Rivas Rojas, P.C., Tancredi, P., Moscoso Londoño, O., Knobel, M. and Socolovsky, L.M. 2018. Tuning dipolar magnetic interactions by controlling individual silica coating of iron oxide nanoparticles. *Journal of Magnetism and Magnetic Materials*, 451: 688–696. https://doi.org/10.1016/j.jmmm.2017.11.099.

Rosensweig, R.E. 2002. Heating magnetic fluid with alternating magnetic field. *Journal of Magnetism and Magnetic Materials*, 252: 370–374. https://doi.org/https://doi.org/10.1016/S0304-8853(02)00706-0.

Sánchez, F.H., Mendoza Zélis, P., Arciniegas, M.L., Pasquevich, G.A. and Fernández Van Raap, M.B. 2017. Dipolar interaction and demagnetizing effects in magnetic nanoparticle dispersions: Introducing the mean-field interacting superparamagnet model. *Physical Review B*, 95(13): 1–18. https://doi.org/10.1103/PhysRevB.95.134421.

Sharif, R., Zhang, X.Q., Shamaila, S., Riaz, S., Jiang, L.X. and Han, X.F. 2007. Magnetic and magnetization properties of CoFeB nanowires. *Journal of Magnetism and Magnetic Materials*, 310(2 SUPPL. PART 3): 830–832. https://doi.org/10.1016/j.jmmm.2006.10.823.

Shtrikman, S. and Wohlfarth, E.P. 1981. The theory of the Vogel-Fulcher law of spin glasses. *Physics Letters A*, 85(8-9): 467–470. https://doi.org/10.1016/0375-9601(81)90441-2.

Stoner, E.C. and Wohlafarth, E.P. 1948. A mechanism of magnetic hysteresis in heterogeneous alloys. *Philosophical Transactions of the Royal Society of London. Series A, Mathematical and Physical Sciences*, 240(826): 599–642. https://doi.org/10.1098/rsta.1948.0007.

Suzuki, M., Fullem, S.I., Suzuki, I.S., Wang, L. and Zhong, C.J. 2009. Observation of superspin-glass behavior in Fe3O4 nanoparticles. *Physical Review B - Condensed Matter and Materials Physics*, 79(2): 1–7. https://doi.org/10.1103/PhysRevB.79.024418.

Tamion, A., Raufast, C., Hillenkamp, M., Bonet, E., Jouanguy, J., Canut, B., Bernstein, E., Boisron, O., Wernsdorfer, W. and Dupuis, V. 2010. Magnetic anisotropy of embedded Co nanoparticles: Influence of the surrounding matrix. *Physical Review B - Condensed Matter and Materials Physics*, 81(14): 1–6. https://doi.org/10.1103/PhysRevB.81.144403.

Torres, T.E., Lima, E., Mayoral, A., Ibarra, A., Marquina, C., Ibarra, M.R. and Goya, G.F. 2015. Validity of the Néel-Arrhenius model for highly anisotropic CoxFe3-xO4 nanoparticles. *Journal of Applied Physics*, 118(18). https://doi.org/10.1063/1.4935146.

Vázquez-Vázquez, C., López-Quintela, M.A., Buján-Núñez, M.C. and Rivas, J. 2011. Finite size and surface effects on the magnetic properties of cobalt ferrite nanoparticles. *Journal of Nanoparticle Research*, 13(4): 1663–1676. https://doi.org/10.1007/s11051-010-9920-7.

Vejpravová, J., Sechovsky, V., Plocek, J., Nižňansky, D., Hutlová, A. and Rehspringer, J.L. 2005. Magnetism of sol-gel fabricated Co Fe 2 O 4 Si O 2 nanocomposites. *Journal of Applied Physics*, 97(12). https://doi.org/10.1063/1.1929849.

Wohlfarth, E.P. 1955. The effect of particle interaction on the coercive force of ferromagnetic micropowders. *Proceedings of the Royal Society of London. Series A. Mathematical and Physical Sciences*, 232(1189): 208–227. https://doi.org/10.1098/rspa.1955.0212.

Yosida, K. and Tachiki, M. 1957. On the origin of the magnetic anisotropy energy of ferrites. *Progress of Theoretical Physics*, 17(3): 331–359. https://doi.org/10.1143/ptp.17.331.

Zhakhovskii, V.V. and Anisimov, S.I. 1997. Molecular-dynamics simulation of evaporation of a liquid. *Journal of Experimental and Theoretical Physics*, 84(4): 734–745. https://doi.org/10.1134/1.558192.

Zysler, R.D., Romero, H., Ramos, C.A., De Biasi, E. and Fiorani, D. 2003. Evidence of large surface effects in Co–Ni–B amorphous nanoparticles. *J. Magn. Magn. Mater*. https://doi.org/10.1016/S0304-8853(03)00486-4.

CHAPTER 9

Nanostructures in Diagnostics
Bio-sensing and Lab-on-a-chip Systems

Jan-carlo M. Díaz-González[1] and *Jannu R. Casanova-Moreno*[2,*]

1. Introduction

1.1 Dimensionality of nanostructures

Research in nanotechnology has become increasingly popular since it impacts several scientific, technological, and social matters. The nanometric scale, usually considered between 1–100 nm, gives nanomaterials unique properties that are reflected in peculiar and attractive results compared to the bulk materials (Saleh 2020). These nanostructured materials (NMs) can be classified into different categories according to personal interests. For this chapter, NMs will be classified first in terms of their dimensionality and then regarding their properties and function.

In 1995, nanostructured materials were categorized for the first time based on their crystalline shape and chemical composition (Gleiter 1995). This classification, however, was not complete since the relationship between the dimensionality of the nanosized building blocks and the material as a whole was not fully considered. The emergence and exponential creation of a great number of nanomaterials motivated Skorokhod to initially extend Gletier classification and later restructure it, along with Pokropivny. Thus, in 2007, they reported a modified classification in which they coined the terms zero-, one-, two-, and three-dimensional nanostructures, which are widely employed today (Pokropivny and Skorokhod 2007).

Zero-dimensional (0D) nanostructured materials are characterized by having all their dimensions below 100 nm. Nanoparticles are the best example of this class and can have different chemical nature like metallic (Au, Pt, Ag), semiconductor (quantum dots, metal oxides) and magnetic (ferrites, maghemite). Also, the preparation of particles with shapes such as spheres, cubes, and hollow spheres can be found in literature. Meanwhile, the 1D group is formed by wires, tubes, filaments, and fibers, with one of their dimensions not in the nanoscale. In contrast, bidimensional (2D) materials contain only one of their dimensions in the nanometric scale. Some examples of this group are graphene, nanodiscs, nanoprisms, nanoplates, and thin films. Finally, all the dimensions of 3D nanomaterials exceed 100 nm. Nevertheless, they have structural features that are in the nanometric scale (Liu et al. 2011, Singh et al. 2020, Saleh 2020). Nanostructured microspheres, polycrystals,

[1] Centro de Investigación y Desarrollo Tecnológico en Electroquímica, Parque Tecnológico Querétaro, Sanfandila, Pedro Escobedo, Querétaro, 76703, Mexico.

[2] CONACYT – Centro de Investigación y Desarrollo Tecnológico en Electroquímica, Parque Tecnológico Querétaro, Sanfandila, Pedro Escobedo, Querétaro, 76703, Mexico.

* Corresponding author: jcasanova@cideteq.mx

honeycombs, nanocomposites and nanopillar arrays are some examples of this group. Some authors, however, include in this category materials with dimension(s) smaller than 100 nm, but with a notable 3D structure like dendritic particles, and nanobuds (Tiwari et al. 2012). This chapter describes the incorporation of these four NM groups in biosensing and lab-on-a-chip systems.

1.2 Biosensors

Chemical sensors are analytical devices whose response depends on a chemical stimulus (affinity interaction or reaction) by an analyte in a specific manner to detect it qualitatively or quantitatively (Eggins 2007, Ensafi 2019). Biosensors are chemical sensors with a biological element, or a derivative from one, as the sensing element. The concept emerged in 1962, with the glucose biosensor (Clark Jr. and Lyons 1962, Newman and Setford 2006) and was later formalized by the International Union of Pure and Applied Chemistry (Nagel et al. 1992). Biosensors combine a biorecognition element with a transduction method to transform the chemical signal of interest to a measurable, typically electrical, signal (Figure 1).

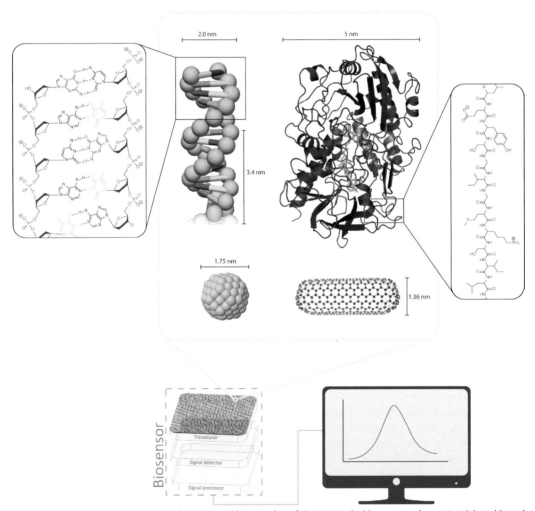

Figure 1. General schematic of nanobiosensors, with examples of the two main bioreceptor classes (nucleic acids and proteins) and the two most reported nanostructures (Au nanoparticles and C nanotubes), all in the same scale. Insets show the molecular structures of double-stranded DNA and part of the primary structure of glucose oxidase. The molecule in green is the enzyme cofactor at the active site.

The coupling gives biosensors the capability to selectively detect analytes of biological origin (proteins, cells, DNA, etc.) and non-biological origin (drugs, toxins, ions, etc.) in simple or complex matrices (Kirsch et al. 2013). Biosensors can be classified by the biorecognition mechanism used, the presence or absence of labels, and the signal transduction type. Biorecognition can be affinity-based, where the binding of biomolecules like nucleic acids and proteins produces the response. Alternatively, catalytic biosensors use the rate of a reaction, typically by an enzyme, as the analytical response. The biorecognition event then must be transduced to be detected, for example, optically, magnetically, electrically or electrochemically as detailed in the following sections. The changes monitored in the transductor can arise directly from physicochemical changes in either the biorecognition element or the molecule to be detected. Sometimes, however, neither of these molecules is suitable for a given transduction mechanism and an appropriate marker or reporter, known as a "label", is added to one of the components. While label-free methods are desirable for simplicity, label-based ones are more generally applicable. At the end, independent of the used strategy, the signal recorded by the device must be related with the analyte concentration in a predictable manner.

1.2.1 Optical transduction

These transductors use changes in the intensity of light at a given wavelength, which arise from the biorecognition event. The most used optical phenomena are absorption, scattering, fluorescence, phosphorescence, polarization, rotation, interference, reflection, and refraction (Farré and Barceló 2020).

Substances that absorb light are known as chromophores and are usually employed either as visual indicators or monitoring their transmittance with a spectrophotometer. Some biomolecules absorb UV or visible light (Figure 2a), making spectrophotometry a standard tool, for example, to determine DNA purity. After absorption, some molecules reemit light at a longer wavelength through fluorescence (Figure 2b). Sensing can be performed by labeling a molecule of interest with a fluorophore, whose intensity will reflect the changes in analyte concentration. Alternatively, a donor fluorophore can transfer energy to an acceptor one via fluorescence (or Förster) resonance energy transfer (FRET) when their separation is smaller than approximately 10 nm. This is a non-radiative dipole-dipole interaction that results in the quenching of the donor and an increase in the acceptor fluorescence (Gauglitz 2005, Bhatt and Bhattacharya 2019). The emission dependence on the donor-acceptor proximity makes it applicable to study intra- or intermolecular interactions, for example, between proteins and cells (Margineanu et al. 2016). Fluorescence has been heavily employed in biosensing, partly due to its high sensitivity (Zhong 2009).

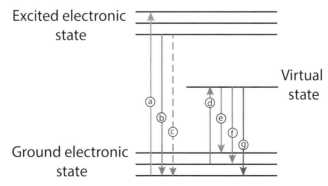

Figure 2. Jabloski diagram showing different processes between energy levels. (a) Absorption, (b) Fluorescence, (c) Non-radiative decay, (d) Raman excitation, (e) Stokes Raman scattering, (f) Elastic (Rayleigh) scattering, (g) Anti-Stokes Raman scattering.

When photons emitted by a laser impinge on a molecule with energy away from any electronic state transition, the molecule will assume a virtual state (Figure 2d). Upon returning to the ground state, it can do so to a vibrational state different from the original one through inelastic scattering. The changes in the vibrational state will cause either a loss (Stokes, Figure 2e) or an increase (anti-Stokes, Figure 2g) in the energy of the inelastically scattered photons. This process is the basis of Raman spectroscopy and is dependent on changes in polarizability of the molecule, therefore providing structural information.

1.2.2 Electrochemical and electrical transduction

An electrochemical biosensor measures the concentration of an electroactive species from variations on the current or potential, for amperometric or potentiometric methods, respectively. The obtained responses are detected on an electronic conductor (frequently referred to as the electrode), which acts as the transducer. In amperometry, the electron transfer processes (redox reactions) result from

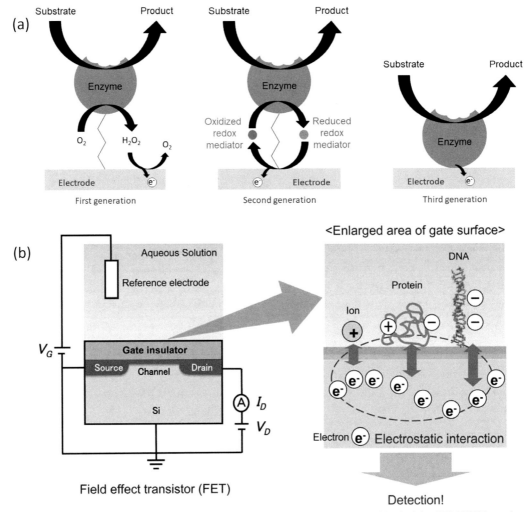

Figure 3. (a) The three different generations of enzymatic electrochemical biosensors. (b) Principle of BioFET biosensing. (Subfigure b republished with permission of The Royal Society of Chemistry, from "Current and emerging challenges of field effect transistor based bio-sensing" Nanoscale, Matsumoto et al. 5, 10702–10718. Copyright (2013); permission conveyed through Copyright Clearance Center, Inc.)

the application of a potential. Electrochemical enzymatic biosensors are the most commercialized example to date. These can be classified in the so-called "generations" according to their electron transfer mechanism (Figure 3a). In the first generation, a reaction subproduct is detected at the electrode, while in the second generation a mediator is added to replace it. In third generation enzymatic biosensors, electrons flow directly from the enzyme cofactor to the electronic conductor or vice versa, depending on whether an oxidation or reduction reaction is employed, respectively. This phenomenon is termed direct electron transfer (DET) and, although it is frequently reported in the literature, the experimental evidence to detect it is still debated (Milton and Minteer 2017, Bartlett and Al-Lolage 2018). Potentiometry measures a potential difference between two electrodes when no current flows through them. Its main application is measuring the activity of ions such as H^+, K^+, Ca^{2+}, Na^+, and Cl^-, through ion selective electrodes (Ding and Qin 2020).

Field effect transistors (FETs) contain a region fabricated in a semiconductor material termed channel, connected to a pair of electrodes termed source (S) and drain (D). A third electrode, known as gate (G), is polarized to create an electric field that affects the conductance between S and D (Figure 3b). Two variations have been used for sensing. In ion sensitive FETs (ISFETs), an ion-sensitive material is used instead of a metallic gate, accumulating charge differently as the ion concentration changes. In biosensor FETs (Bio-FETs), bioreceptors are immobilized on the gate. Interactions with charged molecules create the necessary field changes to obtain a FET response (Pachauri and Ingebrandt 2016, Vu and Chen 2019). FETs are only sensitive to ionic changes within the Debye length, beyond which charges are screened. Because of the high ionic strength of the biological media, its Debye length is in the order of only 1 nm (Schasfoort et al. 1990). Thus, measuring changes in the charge distribution of large molecules (e.g., proteins) immobilized on the surface represents a challenge.

1.3 Lab-on-a-chip

To describe the lab-on-a-chip (LOC) concept, imagine a chemical or biomedical laboratory. All that large equipment with specific functions, located in a space of several square meters, can be replicated in devices, usually called "chips", that measure only a few square centimeters. The miniaturization of the laboratory functions on LOC devices reduces process time, reagent cost, and sample volume while increasing portability. LOCs evolved from the microelectromechanical systems (MEMS) from which they took their micro and nanofabrication techniques. The idea that these devices could be applicable to analytical chemistry was reflected in the coinage of the term "micro total analysis system", or µTAS (Manz et al. 1990). In medical science, the idea is less recent, as it represents a miniaturization of conventional *in vitro* diagnostic devices (Schönberger and Hoffstetter 2016). The LOC structure can be simple or complex according to the needs of a particular application, making their development a multidisciplinary field, with contributions from biology, chemistry, engineering, microfabrication, physics, materials science, microfluidics, etc. Equally, their possible applications are many, their role in two biomedicine fields attracting the most attention. Portable point-of-care (POC) diagnostic devices (Figure 4b) must be considerably simpler to operate than common analytical equipment (e.g., without much sample pretreatment) by non-skilled users. Other desirable qualities are fast analysis time, low cost, and long shelf life (Jung et al. 2015, Primiceri et al. 2018, Mejía-Salazar et al. 2020). Biomimetic organ-on-a-chip (OOC) cell cultures, in contrast, are attractive implementations of LOC platforms for drug research and personalized medicine (Figure 4d). These devices are intended for a lab environment and use by highly trained personnel. As such, they do not require portability and, instead, need high stability during long operation periods (Wu et al. 2020, Azizipour et al. 2020).

Frequently, LOCs have networks of microchannels through which the liquid samples and other reagents flow. These devices usually make use of the behavior fluids present at these scales, including laminar flow (Tarn and Pamme 2014, Zhang and Hoshino 2019). Microfluidic LOC

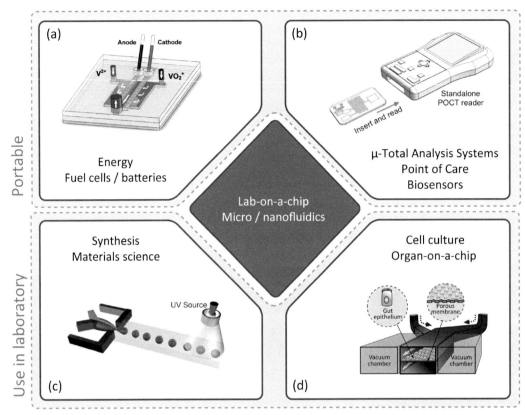

Figure 4. Classification of applications of LOC systems. (Subfigure a was reprinted with permission from Journal of the American Chemical Society, 130, Kjeang et al. "A Microfluidic Fuel Cell with Flow-Through Porous Electrodes", 4000–4006. Copyright (2008) American Chemical Society. Subfigure b was reprinted from Microelectronic Engineering, 132, Jung et al., "Point-of-care testing (POCT) diagnostic systems using microfluidic lab-on-a-chip technologies", 46–57, Copyright (2015), with permission from Elsevier. Subfigure c was republished with permission from The Royal Society of Chemistry, from "Microfluidic synthesis of multifunctional Janus particles for biomedical applications" Lab Chip, Yang et al. 12, 2097–2102, Copyright (2012); permission conveyed through Copyright Clearance Center, Inc. Subfigure d was republished with permission from The Royal Society of Chemistry, from "Human gut-on-a-chip inhabited by microbial flora that experiences intestinal peristalsis-like motions and flow" Lab Chip, Kim et al. 12, 2165–2174, Copyright (2012); permission conveyed through Copyright Clearance Center, Inc.)

platforms are usually fabricated using photolithography, soft-lithography or micromilling, although paper-based devices prepared by inkjet or screen printing are growing in popularity. The interest in nanoscale phenomena and improvements in nanofabrication techniques like e-beam lithography and nanoimprint lithography have prompted the emergence of nanofluidic LOCs. At this scale, van der Waals forces, hydrophobic interactions, electrokinetic effects, and nucleation phenomena dominate the behavior of the flow (Eijkel and Berg 2005, Eijkel et al. 2009, Daiguji 2011).

Optical (Pires et al. 2014), electrochemical (Rackus et al. 2015), and magnetic sensing (Jamshaid et al. 2016) have been incorporated into LOCs. Also, many of the used biosensing strategies employ nanomaterials. Besides the obvious increase in area/volume compared to the bulk materials, nanobiosensors enable unique strategies taking advantage of the fact that the biomolecules themselves are nanostructures (Figure 1). Next, we describe nanomaterials used for biosensing, emphasizing the technologies that have been incorporated in LOCs.

2. 0D Materials

Nanoparticles have been widely used in biosensing, with examples for optical (Kang et al. 2008, Jiang et al. 2013, Ghasemi et al. 2018) and electrochemical (Luo et al. 2006) detection since the 1990s. Besides the *in vitro* applications discussed here, they have been used *in vivo* as agents in imaging, theragnostics and drug delivery (Jiang et al. 2013, Khan et al. 2018).

2.1 Metal nanoparticles

Nanoparticles composed of noble metals, notably gold and silver, have been used in both optical and electrochemical biosensing platforms (Zeng et al. 2011, Vidotti et al. 2011, Doria et al. 2012, Malekzad et al. 2017, Chang et al. 2019).

Most optical sensors using metallic nanoparticles make use of the plasmon resonance that takes place when their conduction electrons oscillate upon being exposed to incident electromagnetic radiation. The radiation wavelength (λ) will therefore cause changes in the transmittance (T), usually called absorption. This quantity ($-\log_{10} T$), however, is more precisely termed extinction, as it includes not only absorption, but also scattering. The extinction cross section (C_E) for a metallic nanosphere of radius r, then, is obtained by adding the cross-sections for each phenomenon (C_A and C_S for absorption and scattering, respectively)

$$C_E = C_A + C_S = k \, \text{Im} \, (\alpha) + \frac{k^4}{6\pi} |\alpha|^2 \tag{1}$$

where k and α are the radiation wave vector and the metal polarizability, respectively. These two quantities are defined as

$$k = \frac{2\pi n_{med}}{\lambda} \tag{2}$$

$$\alpha = 4\pi r^3 \left(\frac{\varepsilon_{part} - \varepsilon_{med}}{\varepsilon_{part} + 2\varepsilon_{med}} \right) \tag{3}$$

where n_{med} is the medium refractive index, ε_{part} the metal dielectric constant, and ε_{med} the medium dielectric constant. Note that dielectric constants for metals are complex numbers (Lakowicz 2005). From Equations 1–3, it can be observed that the absorption and scattering phenomena depend on the nanoparticle size. While absorption is proportional to the cube of the radius, scattering is more readily affected by size, varying in proportion to the sixth power of the radius. Therefore, scattering will dominate on larger particles and absorption on smaller ones. For 22 nm radius gold nanoparticles, for example, most of the extinction is already due to absorption (Messinger et al. 1981).

The quantum model commonly known as "particle in a box" predicts that the electronic transitions redshift for larger "boxes". Indeed, gold nanoparticles (AuNPs) of diameters between 2 and 11 nm, which are most common in biosensing, present a plasmon absorption at wavelengths between 516 and 524 nm, appearing red (Amendola and Meneghetti 2009). Their color has made AuNPs suitable to label biomolecules, evidencing their presence in a particular location. For example, AuNPs have been the most popular label in lateral flow assays like pregnancy tests (Chun 2009, Koczula and Gallotta 2016). AuNPs conjugated to a bioreceptor (commonly an antibody or antigen) interact with antibodies or antigens in the sample and flow along a porous medium through capillary action. A region of the porous medium modified with capture antibodies immobilizes the analyte along with the nanoparticle-bioreceptor conjugate, creating a visibly colored region (typically a line) detectable

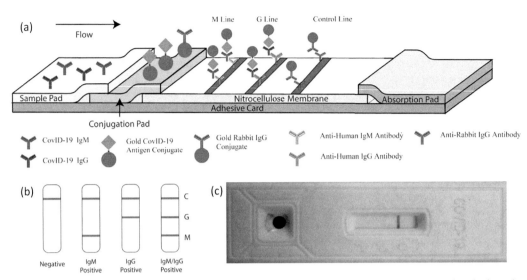

Figure 5. Lateral flow assay for the detection of immunoglobulin M and G antibodies against SARS-CoV-2 virus, the cause of COVID-19 disease. (a) Working principle of the lateral flow immunoassay. (b) Expected gold nanoparticle lines for different test outcomes. (c) COVID-19 IgG/IgM Rapid Test Device commercialized by Hangzhou Realy Tech Co., Ltd. showing a negative result. (Subfigures a and b are reprinted from Journal of Medical Virology, 92, Li et al. "Development and clinical application of a rapid IgM-IgG combined antibody test for SARS-CoV-2 infection diagnosis", 1518–1524, (2020), which is an open access article distributed under the terms of the Creative Commons CC BY license.)

by the naked eye (Figure 5). A control line with a capture strategy that does not require an analyte is used to confirm that flow has gone through the detection areas (O'Farrell 2009). Alternatively, some immunoassays first conjugate the antibodies to small (< 2 nm) AuNPs, later enlarging them to 200 nm by catalytically depositing silver or gold on their surface. This process produces an easily observable colorimetric signal (Lei and Butt 2010).

When AuNPs aggregate, their dipoles interact with the ones of the surrounding particles. Therefore, the side of a particle that is polarized positive will be in proximity with the negatively polarized side of another particle. This plasmon coupling reduces the oscillation frequency and increases the plasmon absorption wavelength (Rechberger et al. 2003). Small gold nanoparticles with usual absorption around 520 nm present a very pronounced redshift to around 620–650 nm, which can be discerned visually. Upon aggregation, initially red suspensions containing such AuNPs become blue/purple. This effect is heavily used to produce a variety of colorimetric assays that change the aggregation state of the nanoparticles when exposed to the analyte. Proteins, nucleic acids, small molecules, whole cells, and even metallic ions have been detected this way (Verma et al. 2015, Poornima et al. 2016, Sabela et al. 2017, Jazayeri et al. 2018, Ghasemi et al. 2018, Gaviña et al. 2018). Silver nanoparticles also change color upon aggregation (yellow to red) but have been less popular, partly because of the poorer stability compared to their gold counterparts (Sabela et al. 2017).

The physicochemical principles behind the aggregation vary widely. A popular approach consists of the formation of cross-linked networks of nanoparticles, very frequently by nucleic acids covalently bound using thiol chemistry (Li et al. 2010). Electrostatic adsorption can also be used to immobilize oligonucleotides, without requiring the target analyte to be thiolated to interact with the nanoparticles. Positively charged capping agents, like polyethyleneimine, can electrostatically interact with anionic DNA and RNA, immobilizing them on the nanoparticle surface (Hakimian et al. 2018).

Commonly, the aggregate size and the inter-particle separation are not controlled (Figure 6a). On the other hand, Guo et al. designed a system in which AuNPs are asymmetrically modified

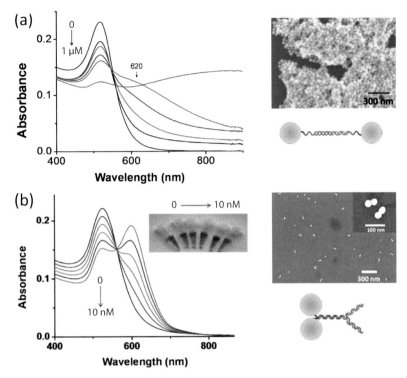

Figure 6. Comparison of two strategies for DNA sensors based on aggregation of AuNPs. In (a), AuNPs modified with two oligonucleotides partially complimentary with the target DNA form large cross-linked aggregates. This results in an increase in absorbance at wavelengths longer than 550 nm without showing a defined peak. In (b), assymetric AuNP modification ensures that only dimers are formed with a specific interparticle gap, resulting in a well defined transition between two conformations. (Adapted with permission from Journal of the American Chemical Society, 135, Guo et al. "Oriented Gold Nanoparticle Aggregation for Colorimetric Sensors with Surprisingly High Analytical Figures of Merit", 12338–12345, Copyright (2013) American Chemical Society.)

with oligonucleotides, which are partly complementary between them and partly complementary to a DNA analyte. As a result, upon interaction with the analyte, the three strands form a Y-shaped structure, holding together a dimer of gold nanoparticles (Figure 6b). Because the aggregate is small, the system presents higher stability than larger aggregates which tend to precipitate. Furthermore, the design ensures a very small distance between the two particles (< 1 nm), enhancing the wavelength shift and therefore improving sensitivity (Guo et al. 2013).

Nanoparticles modified with oligonucleotides can detect more than just nucleic acids. An ingenious example derived from this system by the Mirkin group quantifies copper ions, taking advantage of the "click chemistry" reaction catalyzed by this ion. They employed azide- and alkyne- terminated oligonucleotides immobilized in AuNPs using the shift in the DNA melting temperature to quantify the copper ions (Xu et al. 2010). The use of melting temperature adds a further degree of selectivity of these DNA sensors. Another example of this approach is the single mismatch resolution on a DNA sensor by using a high-fidelity ligase. Both target and mismatch DNA cause AuNP aggregation, but the mismatched strand prevents ligation, resulting in a lower melting temperature (Li et al. 2005).

Metal nanoparticles do not necessarily need to be modified with the biorecognition element. For example, electrostatic immobilization of tyramine on citrate-capped AuNPs has recently been reported for colorimetric sensors. The polymerization of the phenolic hydroxyl moieties in the tyramine is carried out through the reaction with horseradish peroxidase and H_2O_2, causing cross-linking of the nanoparticles and their subsequent aggregation. The analyte is first conjugated with the enzyme catalase, which decomposes the hydrogen peroxide, indirectly preventing the phenol polymerization. This system has been used in both direct (Zheng et al. 2019) and competitive immunoassays (Liang et al. 2018).

Non-crosslinking aggregation employs changes on either the nanoparticle surface charge or the media ionic strength. Changes in the ionic strength can be caused by adding salts (like $MgCl_2$) or by modifying the solution acidity. Since the nanoparticle stability in solution is determined by their inter-particle electrostatic repulsion, high ionic media screens these charges causing aggregation. An interesting work uses oligonucleotide-modified AuNPs complementary to the product of a loop-mediated isothermal amplification (LAMP). Upon increasing the solution ionic strength, particles bound to the amplicon remain stable in solution (red), while aggregation takes place when DNA is either absent or non-complementary (Reuter et al. 2020).

Because of their simple signal detection, colorimetric assays are good candidates for use in POC devices. Visible color changes can be useful qualitatively. Quantitation, on the other hand, requires some sort of (albeit simple) instrumentation. Some approaches have used conventional or smartphone cameras to measure the color change (Lei and Butt 2010, Zheng et al. 2019). Others have gone further and designed portable colorimeters that can measure the absorption in two wavelengths (Zhao et al. 2014, Reuter et al. 2020). In these systems, solutions can be measured using either commercial cuvettes, custom-made microplates or microchannels. Furthermore, nanoparticles can be deposited on paper to perform either spot or lateral flow analysis, prompting systems that not only use the different colors between the dispersed and aggregated nanoparticles, but also their different migration characteristics in the flow (Zhao and van den Berg 2008).

Elastic (Rayleigh) scattering of light by nanoparticles has also been used as an analytical signal. This phenomenon can be measured by using a conventional fluorescence spectrofluorometer with matching excitation and emission wavelengths. As stated in Equations 1–3, scattering is proportional to the sixth power of the particle radius and can be used to monitor processes that involve nanoparticle growth. AuNPs were shown to increase their size when exposed to H_2O_2 in a solution containing $HAuCl_4$ and cetyltrimethylammonium chloride. The H_2O_2 reduces the Au^{3+} only at the gold nanoparticle surface. The enzymatic rate of peroxide-producing oxidases can be followed by monitoring the scattering intensity, allowing the sensing of the enzyme substrates (Shang et al. 2008).

Inelastic scattering of light has also been used in nanobiosensor design as surface enhanced Raman spectroscopy (SERS). Signal enhancement down to the single-molecule detection is caused by a metal structure in close proximity (< 10 nm separation) in two different ways. In the electromagnetic mechanism, excitation and emission rates increase with the square of the local electric field enhancement (E_{Loc}/E). Neglecting the differences between the excitation and emission light frequencies, the enhancement presents a fourth power dependence on E_{Loc}/E (Pilot et al. 2019). In the chemical mechanism, enhancement occurs due to changes in the polarizability of the Raman-active species, including charge transfer with the substrate (Le Ru and Etchegoin 2013, Yaraki and Tan 2020). Label-free SERS use the signals from the analyte and can provide useful structural information but is prone to overwhelming by similar signals from the background. The use of labels enables better quantitation, at the expense of structural information (Li et al. 2020).

Nanomaterial design for SERS focuses on architectures where the electric field is maximized, "hot spots" located on the tips of anisotropic nanostructures and in the gaps between them (Le Ru and Etchegoin 2013). Nanorods, with two plasmon absorptions and strong field enhancement at their ends, have been used as SERS substrates, after various surface modifications (Rekha et al. 2018, Yilmaz et al. 2019). However, reliable selective modification of their ends is an active challenge (Fontana et al. 2016, Fisher et al. 2017) that would allow for consistent end-to-end assembly and reproducible SERS.

Metal colloids are easily adapted to microfluidic SERS. Silver colloids, for example, have been used in a droplet-based platform to detect levofloxacin in urine (Hidi et al. 2016). The analyte

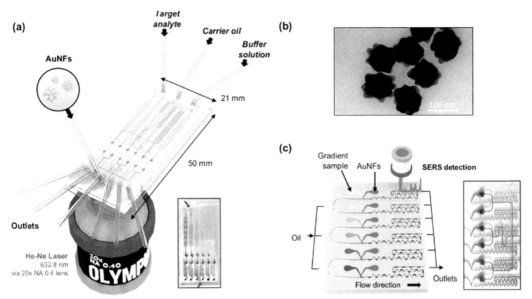

Figure 7. A biphasic droplet-based microfluidic SERS platform. (a) Overview of the device. (b) TEM images of the Au nanoflower (AuNFs) SERS substrate. (c) Detail of the droplet-producing module (Republished with permission of The Royal Society of Chemistry, from "SERS-based droplet microfluidics for high-throughput gradient analysis" Lab Chip, Jeon, et al., 19: 674–681. Copyright (2019); permission conveyed through Copyright Clearance Center, Inc.)

adsorbed on the silver surfaces prompts the aggregation of the nanoparticles creating hot spots (Hidi et al. 2015). However, flowing the sample along the colloid increases random adsorption of SERS structures in the walls. Alternatives include droplet-based assays (Figure 7) and purposely immobilized SERS structures. For example, nanoparticles can be temporarily trapped in a region of the channel. Another option is to fabricate or deposit the nanostructures on the channel surface. Nanosphere lithography has been used to deposit nanotriangle arrays (Camden et al. 2008), whose enhancement factor can be increased coupling nanoparticles to these deposited nanostructures (Li et al. 2013). Raman spectrometry is instrumentally more complex than colorimetry or fluorimetry, causing a delay in SERS point-of-care biosensors. During the last decade, however, several handheld Raman spectrometers have been commercialized, opening the way for more SERS LOCs.

Fluorophores are affected by metallic nanoparticles in a complex manner, quenching their fluorescence in some cases, and enhancing it in others, depending on several factors (Figure 8). Smaller particles favor quenching while larger structures preferentially enhance it, which has been explained in terms of their absorption and scattering processes (Lakowicz 2005). Fluorescence enhancement is maximized at an optimal fluorophore-metal separation; shorter distances cause quenching, while longer separations resemble the fluorescence process without nanoparticles. It is generally accepted that the electric field enhancement around the particle increases the fluorophore excitation rate (Figure 2a). There is some disagreement, however, in its effect on the emission rate. Some people consider that, after the fluorophore transfers energy to the nanoparticles, they can radiate this energy (Figure 2b), effectively increasing the radiative rate (Ribeiro et al. 2017). Others, however, treat this energy transfer as a non-radiative contribution (Figure 2c), and suggest avoiding spectral overlap if enhancement is desired (Kang et al. 2011, Yaraki and Tan 2020). Despite the lack of mechanistic consensus, both fluorescence quenching and enhancement have been used to design biosensors. Quenching-based sensors rely on the initial immobilization of fluorophore-modified biomolecules, typically peptides or oligonucleotides. Upon interaction with the analyte, the fluorophore is released increasing emission. This strategy has been used to detect, for example, enzymes and small molecules (Park et al. 2012, Liu et al. 2014).

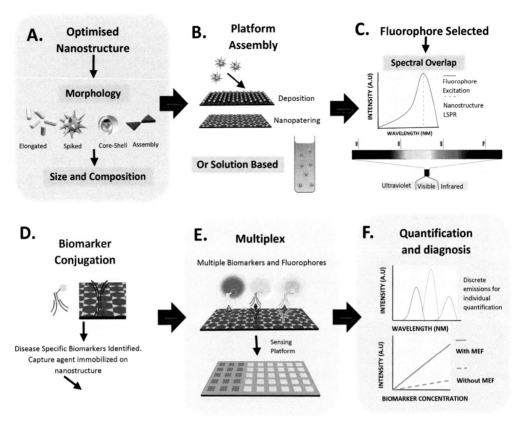

Figure 8. Steps to design a MEF biosensor. (Reproduced from Fothergill et al. "Metal enhanced fluorescence biosensing: from ultra-violet towards second near-infrared window" Nanoscale, 2018, 10: 20914–20929—Published by The Royal Society of Chemistry under a Creative Commons Attribution 3.0 Unported Licence.)

The use of metal enhanced fluorescence (MEF) for actual biomolecule detection was explored only in the last two decades using metal colloids or deposits (Geddes et al. 2005, Jeong et al. 2018, Badshah et al. 2020, Yaraki and Tan 2020). Since a larger enhancement is observed for small quantum yield fluorophores, MEF has been attractive for the sensing of the weak fluorescence of some biomolecules, like nucleotides, without the need for a label. However, most systems rely on fluorescently labelled biomolecules. Due to the highly customizable particle size and shape, MEF systems have been designed for dyes ranging from the UV to the near IR (Fothergill et al. 2018). A common mechanism is the interaction between two biomolecules with affinity, one of which is labeled with the fluorophore and the other with the nanostructure (Figure 8D). MEF systems can be used simply as highly fluorescent labels, requiring constant metal-fluorophore separation for maximum enhancement. A SiO_2 shell has been the preferred separator since it can easily coat metal nanoparticles and can be further functionalized.

The highest change in fluorescence can be obtained by switching from the quenching regime to the enhancement one. An interesting approach first immobilizes the dye directly on the surface of AuNPs, quenching its emission. Separately, silica-coated AgNPs are modified with thiol-terminated groups. The interaction between the two kinds of nanoparticles created "hot spots" in the gap between them, containing the fluorophore. The enhancement factor compared to the quenched state was ~ 100 while the one compared to the free fluorophore was only ~ 5 (Li et al. 2016). Finally, silver, gold and copper nanoclusters smaller than 2 nm are fluorescent under certain conditions. These clusters require capping or templating agents to avoid aggregation. DNA has been used as a templating agent, allowing for the design of bioassays that change fluorescence according to environment modifications by DNA binding events or metal ions (Liu 2014).

Metal nanoparticles (predominantly Au and sometimes Ag) have been used for electrochemical biosensing either bare, coated with their capping agent or functionalized with biomolecules (Rasheed and Sandhyarani 2017, Wongkaew et al. 2019). Current enhancement based simply on the increased electroactive area will not be discussed. AuNPs have been used as labels to electrochemically detect biorecognition events at an electrode surface, using various strategies. They can be oxidatively dissolved, later reducing the ions back to the metallic form on the electrode surface. A related method first deposits Ag on the AuNPs, and Ag electrochemistry is followed instead. Alternatively, captured nanoparticles can electrocatalyze a reaction (like water reduction) that would not produce significant current in an electrode of another material, like carbon (Parolo et al. 2013). Gold nanoparticles can also be modified with electroactive compounds producing a locally high concentration that can easily be detected. Some of these electroactive compounds, like ferrocene, are typically included as a thiolated derivative covalently modifying AuNPs (Takahashi and Anzai 2013). Others are indirectly incorporated, like $[Ru(NH_3)_6]^{3+}$, which intercalates in AuNPs modified with double-stranded DNA.

A different strategy involves using the AuNPs as electron conductors to promote electrochemical reactions in electrode surfaces modified with mixed self-assembled monolayers containing a biomolecule. These layers prevent the electron exchange with electroactive molecules in the solution. Upon a biorecognition event, AuNPs bind to the layer enabling electron transfer to the electrode. Also, AuNPs improve electron transfer between oxidoreductases and electrode surfaces even better than some redox mediators. This has been used to develop sensors to quantify cholesterol, H_2O_2 and glucose (Shumyantseva et al. 2005, Tangkuaram et al. 2007, Sharma et al. 2012). Interestingly, the use of L-serine as a reducing agent during the AuNP synthesis resulted in nanoparticle chains that presented better conductivity and therefore enhanced biocatalytic currents.

2.2 Semiconductor nanoparticles

Semiconductor nanoparticles have been used in biosensing applications, the most common example being the fluorescent quantum dots (QDs). These are typically cadmium chalcogenide (e.g., CdSe) nanoparticles 2–10 nm in diameter. Because of the quantum confinement and the material bandgap, the fluorescence emission wavelength depends on the particle diameter. Quantum dots absorb light in a broadband but emit in narrow peaks that range from the UV to the near IR. Therefore, QDs are suitable for multiplexing assays with a single excitation wavelength. Since QDs display superior photostability than organic fluorophores, they are used to detect ions, proteins, nucleic acids, drugs, etc. (Frasco and Chaniotakis 2009). In their simplest implementation, quantum dots act as labels, indicating the presence of a biological molecule. This function, for example, has been used to detect apoptotic cells by modifying the quantum dots with the protein annexin V (Le Gac et al. 2006). Later, a similar system was applied in a microfluidic device to screen anticancer drug activities in an efficient manner (Zhao et al. 2013). Energy transfer to and from the QDs enables more complex strategies (Figure 9), starting with their use as donors for FRET assays. Also, QDs can transfer their energy to AuNPs, quenching their fluorescence similar to organic dyes. Finally, QD non-radiative excitation can be carried out by chemiluminescence, bioluminescence, electroluminescence all of which have been used in diverse bioassays (Pisanic et al. 2014).

Metal oxides of several metals (Cu, Co, Ni, Mn, Ti, Ag, Va, Zr, Zn, W, etc.) have been used to produce nanoparticles suitable for biosensing (George et al. 2018). A common approach includes immobilization of biomolecules on their surface by either physical or covalent adsorption. Several enzymes, for example, have been immobilized to produce electrochemical and optical biosensing platforms (Shi et al. 2014). These materials are less costly than noble metals and present catalytical or electrocatalytical activity for several reactions of interest, sometimes even offering an (albeit less specific) alternative to enzymes.

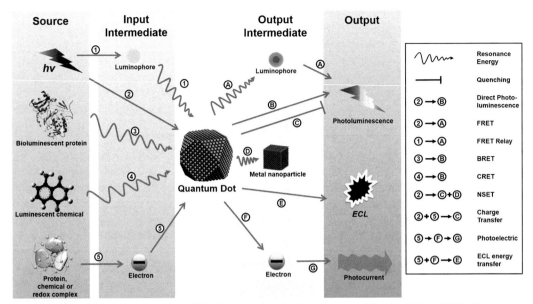

Figure 9. Map of the different processes in which QDs can be involved. Acronyms are defined as follows- FRET: fluorescence resonance energy transfer, BRET: bioluminescence resonance energy transfer, CRET: chemiluminescence resonance energy transfer, NSET: nanosurface energy transfer, ECL: electrochemiluminescence. (Republished with permission of The Royal Society of Chemistry, from "Quantum dots in diagnostics and detection: principles and paradigms" Analyst, Pisanic II et al., 139: 2968–298 Copyright (2014); permission conveyed through Copyright Clearance Center, Inc.)

2.3 Magnetic nanoparticles

Magnetite (Fe_3O_4) and maghemite (γFe_2O_3) nanoparticles are superparamagnetic, remaining magnetized while an external magnetic field is present. This is because their size is smaller than the magnetic domain size, causing the individual magnetic moments of all its atoms to be aligned. Magnetic nano- and microparticles have been used in biosensing and LOCs, both for separation and detection (Jamshaid et al. 2016). In separation, magnetic nanoparticles (MNPs) are functionalized with biorecognition elements with affinity to the analyte. Upon application of a non-uniform magnetic field either through the movement of a magnet or activation of an electromagnet, MNPs in a diamagnetic fluid (like H_2O) are attracted towards the regions of higher magnetic field. This phenomenon is called magnetophoresis and traps magnetic particles while the rest of the molecules can be washed away. Upon removal of the magnetic field, the particles are resuspended and further processed. This batch strategy has its shortcomings, like unwanted entrapment of contaminants in the spaces between the aggregated particles. Also, there is the risk of strong interactions forming between functionalized particles during the aggregation that may hamper their resuspension. Alternatively, MNPs can be sorted in a continuous flow inside microfluidic channels (Alnaimat et al. 2018). A magnetic field gradient normal to the flow results in a sum of velocities in a diagonal direction. Since the deflection is dependent on the particle size, separation can be achieved based on this property. Besides separating magnetically labeled biomolecules and cells, magnetophoresis has also been employed to separate red blood cells, which have intrinsic magnetic properties.

A different strategy uses MNPs as labels, usually detecting them through giant magnetoresistance (GMR) and magnetic relaxation switching assays (MRSw). In GMR, an electrically conductive non-magnetic thin layer is placed between two ferromagnetic ones. If the spins of the ferromagnetic layers are parallel, little spin-dependent electron scattering takes place and the electric resistance is low. When one of the ferromagnetic layers switches its spin due to an external magnetic field, the spins of the two ferromagnetic layers are antiparallel. Increased spin-induced scattering causes

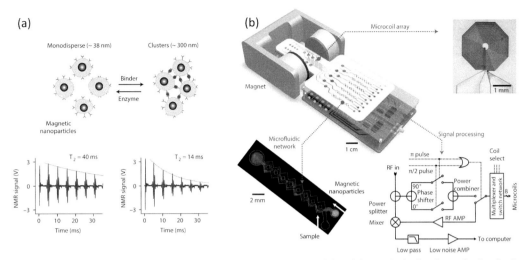

Figure 10. Magnetic relaxation switching (MRSw) bioassays. (a) Principle of the method, showing reduction in the transverse relaxation time upon aggregation of superparamagnetic nanoparticles labeled with antibodies. (b) Integration of the assay in a microfluidic chip consisting of a permanent magnet, metal microcoils for RF excitation and NMR measurements, microchannels with chaotic mixers and the required electronics. AMP stands for amplifier. (Reprinted by permission from Nature, Nature Medicine, 14: 869–874, Chip-NMR biosensor for detection and molecular analysis of cells. Lee at al. Copyright (2008)).

a high resistance that can be monitored externally. These sensors can detect weak magnetic signals and are cheaper and much simpler to include in LOC devices compared to superconducting quantum interference device (SQUID) magnetometers (Shen et al. 2018). Furthermore, cobalt nanoparticles have been embedded in gels, forming nanogranular GMR sensors with the potential to be printed in paper-based biosensors (Eickenberg et al. 2013).

Magnetized nanoparticles create an inhomogeneity on the surrounding magnetic field, which affects the protons in the neighboring water molecules. Their proton magnetic relaxation is characterized by measuring the transverse T_2 relaxation time using nuclear magnetic resonance (NMR) techniques. T_2 decreases when MNPs aggregate, motivating a variety of so-called magnetic relaxation switching (MRSw) bioassays (Figure 10a), favouring aggregation or disaggregation in response to the analyte (Haun et al. 2010). NMR measurements have been integrated into microfluidic portable platforms (Figure 10b), in which patterned coils are used to both emit the required radio frequency (RF) excitation signal and to measure the NMR relaxation response (Lee et al. 2008). A mechanistic study using avidin as a model analyte predicted that this technique could achieve limits of detection of ~ 100 fM before the synthesis of particles with high transverse relaxivity becomes too complex (Min et al. 2012).

3. 1D Materials

3.1 Carbon nanotubes (CNTs)

These nanotubes can be thought of as graphene sheets that have been folded into a cylindrical shape. Their intramolecular interactions are dominated by covalent bonds between sp^2 carbon atoms, while van der Waals forces prevail between neighboring CNTs. The rolling direction (chiral vector) heavily influences their properties (Aqel et al. 2012). Carbon nanotubes consisting of a single layer or graphene are termed single-walled (SWCNT) while those consisting of several concentric layers are called multi-walled (MWCNT). Multi-walled CNTs are highly electrically conductive, while the conductivity of SWCNTs depends on their chiral vector, adopting a conductor, semiconductor, or non-conductor character. Besides their electrical characteristics, CNTs present optical, thermal, and mechanical properties that make them interesting in several fields. A common challenge when designing biosensors that employ CNTs is their hydrophobic nature. Therefore, CNTs are chemically functionalized, also improving their biocompatibility and reducing their toxicity. It is common to find the CNTs by themselves or with a biomolecule, incorporated in different polymers/biopolymers, detergents, solvents, etc. (Pastorin et al. 2005, Tîlmaciu and Morris 2015).

Because of their electrical conductivity and shape, CNTs can serve as electrode "extensions" to achieve direct electron transfer between the electrode and the enzyme active site, which in some cases is deeply buried (Figure 1). Some of the enzymes that, together with CNTs, have been used in biosensors are glucose oxidase (GOx), lactate oxidase, horseradish peroxidase (HRP), xanthine oxidase, glucose dehydrogenase, cholesterol oxidase, alcohol dehydrogenase, and organophosphorus hydrolase (Wu and Hu 2007, Xia and Zeng 2020). It is possible to functionalize the CNTs with carboxyl, amino or sulfhydryl groups, for example, to generate bonds with the biomolecules (Zhou et al. 2019). CNTs modified with more complex groups, like anthracene, aminoethylphenyl, naphthoate, 1-pyrenebutyric acid adamantyl amide, favorably orient the enzyme immobilization for better DET (Meredith et al. 2011, Lalaoui et al. 2016, Gentil et al. 2018). Zhao et al. evaluated the conformational changes of GOx, the most popular enzyme to study DET, when it interacts with MWCNTs. Using circular dichroism, they found a higher amount of β-sheet structure and a lower amount of α-helix structure compared to the free enzyme. These conformational changes could reduce the hindering to the active site, enabling the DET with the CNTs (Zhao et al. 2010). The number of walls in MWCNTs affects the efficiency of the DET process. Using CNTs with predominantly 1, 3, 7 and 12 walls, Liu et al. reported a maximum DET with the triple-walled tubes. They rationalize their observations in terms of the electron tunneling between the walls in the CNT (Liu et al. 2018).

Semiconductor SWCNTS display fluorescence in the near-infrared, with their chiral-dependent bandgap determining their emission wavelength between 900–1600 nm (Li and Shi 2014). SWCNTs offer high photostability and, because of their NIR emission, reduced interference from biological autofluorescence (Diao et al. 2015). Other advantages include the absence of blinking and a large difference between excitation and emission wavelengths. SWCNTs have been applied in optical sensing of proteins, biomarkers, DNA, and small molecules like adenosine 5′-triphosphate (ATP), reactive heteroatom species, H_2O_2 and nitric oxide (Farrera et al. 2017, Hofferber et al. 2020).

SWCNTs have been incorporated in Bio-FET devices, delivering "ultrasensitive" detection, probably because of their high surface-area to volume ratio as well as the fact that all the charge is transported at the surface (Matsumoto and Miyahara 2013, Sadighbayan et al. 2020). CNT FETs have been employed to detect DNA hybridization, as well as to detect and quantify protein and intracellular molecules of biomedical interest (Allen et al. 2007, Liu and Guo 2012). To use CNTs as FET components, they must be arranged between the source and drain electrodes, which demands high fabrication capabilities. Also, only semiconductor CNTs are useful, therefore requiring their separation from highly conductive ones.

The inclusion of CNTs in LOC devices has been extensively reviewed (Choong et al. 2008, Ghasemi et al. 2017). Besides the applications in the sensing stage, they have been incorporated on the walls or forming membranes. CNTs can be grown on substrates that later form part of the chip structure or be deposited after assembly, for example, by drop casting or flowing of CNT-containing solutions or gels through the channels. An interesting use of CNT-modified walls is to obtain superhydrophobic surfaces that minimize the fluid-wall contact, therefore reducing the hydrodynamic resistance (Joseph et al. 2006). Also, the adsorption of organic compounds by CNTs has been exploited to design separation (solid phase extraction and chromatography) stages integrated into the LOC devices (Mogensen and Kutter 2012).

3.2 *Nanowires*

Nanowires (NWs) are cylindrical in shape and usually have diameters of around 10 nm while displaying variable aspect ratios, usually above 1000. They are made of many materials, including metals (Cu, Au, Ag, Ni, Pt), oxides (ZnO, SiO_2, TiO_2), semiconductors (Si, In, Ga, Ge), superconductors (YBCO) and conducting polymers (polypyrrole, polyaniline). Semiconductor nanowires are the most useful as biosensing transducers. In fluorescence biosensors, NWs can improve the sensitivity by an optical fiber effect, in which fluorophore emission is captured by NWs via near field interactions, concentrating and guiding the light. This effect is dependent on the NW diameter and light wavelength and is useful for the collection of low fluorescence signals (Warren-Smith et al. 2010, Verardo et al. 2018). In an interesting example of NWs in fluorescence-based sensors, phosphorylated fluorescein was immobilized on Si NWs. Upon dephosphorylation by alkaline phosphatase fluorescence increases, allowing enzyme quantification (Wang et al. 2014).

NWs have been used as semiconductor channels in FETs, increasing the sensitivity compared to conventional flat FETs (McAlpine et al. 2003). This is partly because of their increased area/volume that allows more charged molecules near the semiconductor. BioFETs have been prepared using NWs, either depositing them on the electrodes or nanofabricating them. While the first option is simpler and cheaper, the second one offers more control and reproducibility. By immobilizing a variety of bioreceptors, molecules such as glucose, cholesterol, uric acid, proteins, and oligonucleotides, have been detected (Ahmad et al. 2018).

There are limited examples of NW in LOCs, most of which are included in a few reviews that also deal with CNT (Lee et al. 2009, Lim et al. 2010, Krivitsky et al. 2012, Adam et al. 2013, Gamal et al. 2015, Liu et al. 2018). A good example incorporates a nanofabricated SiNW FET in PDMS microfluidic chips (Figure 11a-b). The NW is modified with amino terminated molecules

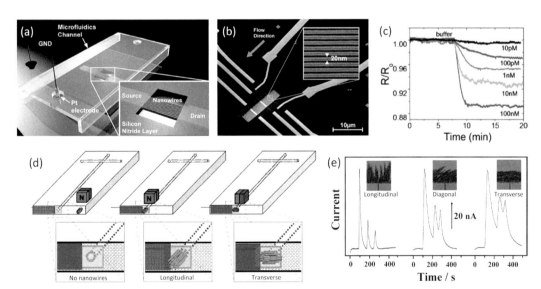

Figure 11. Integration of NWs in microfluidic devices. (a) Scheme of SiNW BioFET inside a PDMS chip. (b) SEM of the SiNW FET array with detail of the 20 nm wide NWs. (c) Changes in the relative resistance as complementary DNA is flown through the microchannel. (d) Magnetically controllable NiNWs placed at the end of a microfluidic electrophoresis separation system. (e) Changes in the electropherograms as the orientation of the nanowires relative to the channel axis is modified. (Subfigures a–c were adapted with permission from Journal of the American Chemical Society, 128, Bunimovich et al. "Quantitative Real-Time Measurements of DNA Hybridization with Alkylated Nonoxidized Silicon Nanowires in Electrolyte Solution", 16323–16331. Copyright (2006) American Chemical Society. Subfigures d-e were adapted with permission from Analytical Chemistry, 79, Piccin et al. "Adaptive Nanowires for Switchable Microchip Devices", 4720–4723. Copyright (2007) American Chemical Society.)

onto which ssDNA was electrostatically adsorbed. Upon flow of the complementary DNA through the chip, the increase in negative charges produced by the DNA hybridization results in a decrease in the FET resistance (Figure 11c) (Bunimovich et al. 2006). Another work demonstrated the use of magnetic NiNWs at the outlet of an electrophoresis microchannel as an "adaptative" material. These NWs were controlled by an external magnet, positioning them in contact with a current collector to use them as an electrode for electrochemical determinations. By rotating the magnet, the NW angle with respect to the channel could be controlled, affecting the separation of the signals in the electropherogram (Figure 11d-e) (Piccin et al. 2007).

4. 2D Materials

4.1 Graphene-based materials

Graphene has very special electronic and optical properties. Graphene oxide (GO) is obtained by chemically oxidizing graphite, exfoliating sheets with a large number of defects and oxygen-containing groups (Compton and Nguyen 2010). In GO, most of the carbon atoms present a sp^3 hybridization and the material is an electrical insulator. GO can be reduced either thermally, chemically, or electrochemically, partially restoring the sp^2 structure while leaving some defects and oxygen-containing groups. Different electrical conductivities can be obtained depending on the C/O atomic ratio (Punckt et al. 2013). These three materials have been integrated into several electrochemical, electrical and optical biosensing schemes (Pumera 2011).

Graphene oxide has been reported to not offer any advantage over graphite in electrochemical sensing (Chua et al. 2011). Reduced graphene oxide (RGO) has displayed better performance, thanks to a balance between the presence of electroactive defect sites and material conductivity. For example, it can resolve the electrochemical oxidation of the four nucleobases, while glassy carbon and graphite fail to properly separate adenine and thymine signals. This can help estimate oligonucleotide composition without the need for a label (Zhou et al. 2009). Graphene has been used in FET biosensors. In a recent example, messenger RNA (mRNA) riboswitches immobilized on graphene were used to detect uncharged purine analogs. Upon binding of the analytes with the riboswitch, its conformation changed, moving the charged phosphate backbone closer to the graphene, shifting the FET transfer curve (Tian et al. 2020). In principle, this strategy could also be applied to conventional DNA aptamers.

Graphene-based materials have been coupled with other nanomaterials to attain further benefits (Parnianchi et al. 2018). For example, decorating graphene with gold nanoparticles helps reduce its agglomeration acting as a nano-spacer, while improving the electrical conductivity and surface area (Khalil et al. 2016). Several graphene-based biosensors employing electrochemical and FET transduction have been successfully incorporated into lab-on-a-chip platforms (Sengupta and Hussain 2019). Furthermore, graphene derivatives have also been used in other components of the chips. For example, a mixture of reduced graphene oxide and PDMS was used to adsorb potentially toxic residues at the chip outlet (Chalupniak and Merkoci 2017).

5. 3D Materials

Asymmetric silicon nanopillar arrays, in which each row is laterally shifted with respect to the previous one, have been used for separation. In this arrangement, smaller objects go straight while larger ones follow the nanopillar shift angle, having been employed to separate exosomes and DNA in microfluidic chips (Wunsch et al. 2016, 2019). Arrays of metal-coated nanopillars have been employed as substrates for surface enhanced Raman spectroscopy, with signal intensity depending on the pillar density and the homogeneity of the hot spots. Some approaches rely on depositing

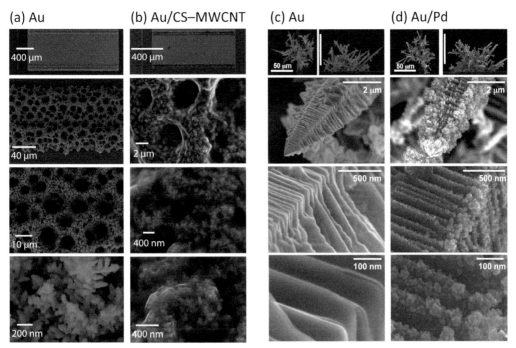

Figure 12. Hierarchical 3D electrodeposited nanomaterials for biosensing. Au nanofoam before (a) and after (b) deposition of chitosan (CS) and MWCNTs. Fractal Au structures before (c) and after (d) Pd deposition. (Subfigures a-b republished with permission of The Royal Society of Chemistry, from "A dual-enzyme, micro-band array biosensor based on the electrodeposition of carbon nanotubes embedded in chitosan and nanostructured Au-foams on microfabricated gold band electrodes" Analyst, Juska et al. 145: 402–414 Copyright (2020); permission conveyed through Copyright Clearance Center, Inc. Subfigures c-d adapted with permission from Soleymani, et al. "Hierarchical Nanotextured Microelectrodes Overcome the Molecular Transport Barrier To Achieve Rapid, Direct Bacterial Detection" ACS Nano, 5: 3360–3366. Copyright (2011) American Chemical Society.")

metal only at the top of the nanopillars, generating "suspended" plasmonic nanoparticles (Magno et al. 2018). Others use the structure in the metal deposits to achieve high EM fields (Chen et al. 2020, Choi et al. 2020). More ambitious strategies use the gaps between the pillars to create the hot spots (Liu et al. 2017). Careful fabrication leads to good control of the gap size to maximize the desired signal enhancement (Li et al. 2014, Kim et al. 2017). Even smaller gaps were obtained by using high aspect ratio nanopillars which, upon solvent evaporation, form structures in which the tips of several nanopillars aggregate in a reproducible fashion. Molecules trapped in the space between the tips generate the SERS signal (Kim et al. 2011, 2012). These strategies are compatible with LOC microfabrication and several examples of integrated devices have been reported (Huang et al. 2015), including a nanopillar array integrated within a centrifugal microfluidic device. Besides acting as a SERS platform, the array separates analytes based on their affinity for the gold structures (Durucan et al. 2018). Although nanosphere lithography has been employed to fabricate arrays of simple features like holes and triangles, complex structures like hole-disc and hole-ring combinations have also been obtained, providing improved SERS enhancement (Ho et al. 2014).

Hierarchical microscale materials with nanoscale features have been particularly useful in electrochemical biosensing. Hollow titania microspheres composed of nanosheets, for example, have been shown to trap enzymes in their interior, using the nanoplate structure as funnels, while allowing the enzyme substrate to move freely (Xie et al. 2011). The Kelley group has developed fractal electrodeposited structures that, when modified with bioreceptors, provide ultrahigh sensitivity in electrochemical biosensing (Figure 12c-d). By extending this technology to the 100 μm scale, they were able to efficiently capture long bacterial RNA, reducing accumulation time by various orders of magnitude (Soleymani et al. 2011). Composite nanomaterials can synergistically employ the properties of each component to improve performance. For example, a bi-enzymatic electrochemical biosensor was recently developed on a nanostructured porous gold electrode (Figure 12a-b). The pores increased the electroactive area and helped in the electrodeposition of a chitosan—MWCNT matrix. GOx and HRP were immobilized in this nanocomposite, achieving excellent analytical performance and remarkable long-term stability (Juska and Pemble 2020).

6. Perspectives

Portable biosensors are here to stay. Although they will never be a replacement for clinical laboratories, POCs will be important not only in healthcare but also in areas such as food, environment, and agriculture (Ragavan and Neethirajan 2019, Farré and Barceló 2020). Microfluidics and related techniques will be instrumental in achieving portable and user-friendly devices. As discussed throughout the previous sections, several nanostructures have proven their potential for biosensing in LOCs. To fully exploit their capabilities, however, several nanostructures can be simultaneously integrated with the microscale device features. One great example of such integration uses MNPs for separation and AuNPs for transduction (Zheng et al. 2019). The authors also employ polystyrene microspheres loaded with catalase flowing the whole system in a microfluidic device (Figure 13). The presence of *E. coli* is quantified employing a smartphone application to extract the hue value of images of the detection chamber. Furthermore, the lower energy consumption requirements of

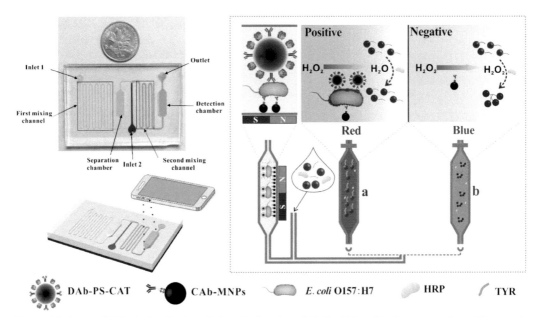

Figure 13. A microfluidic device for the colorimetric detection of *Escherichia coli* using a smartphone. The sample containing the bacteria is first mixed with polystyrene (PS) microbeads modified with a detection antibody (DAb) and the enzyme catalase (CAT), and magnetic nanoparticles with immobilized capture antibodies (CAb). The PS—*E. coli*—MNPs are then separated in a chamber using an external magnetic field. Once released, they move to a second channel in which they mix with H_2O_2, horseradish peroxidase, tyramine (TYR) and gold nanoparticles. In the absence of bacteria, catalase is not present, and the peroxide triggers a phenol polymerization reaction, cross-linking the AuNPs and changing the solution to blue. When *E. coli* is present, the catalase-containing constructs move into the mixing channel, decomposing the peroxide and preventing the aggregation of AuNPs, retaining the red color. (Reprinted from Biosensors and Bioelectronics, 124–125, Zheng et al. "A microfluidic colorimetric biosensor for rapid detection of *Escherichia coli* O157:H7 using gold nanoparticle aggregation and smart phone imaging", 143–149, Copyright (2019), with permission from Elsevier.)

these smaller devices will prompt the integration of energy sources in the same device (Figure 4a), achieving self-powered biosensors (Hickey et al. 2016, Çakıroğlu and Özacar 2018).

Nevertheless, there are several challenges in nanobiosensor design, one of them being its interdisciplinarity. While the synthesis of the nanomaterials is fully dominated by material scientists, the sensing design usually involves the participation of physical, biological, and analytical chemists. A lack of deep understanding in any of these areas can result in "handwavy" explanations that hinder progress in device reliability. While the use of oversimplified cartoons is didactic, it hides the complexity that truly lies in these systems, and that is usually reflected in the uncertainties of fabricated device operation. Acknowledging the effects of heterogeneity, contamination and batch to batch variations will allow more rational design, and ultimately more predictable responses.

The recent COVID-19 pandemic tested the response of research and development institutions to bring applicable ideas to life in an efficient manner. Apart from vaccine development, the spotlight was on the availability of testing methods that could be widely applicable in all social contexts. Soon, publications appeared detailing POC approaches to the detection of either viral RNA or antibodies (Tymm et al. 2020, Hussein et al. 2020). A year after the disease appeared, lateral flow immunoassays (Figure 5) and some integrated microfluidic benchtop equipment like the Cepheid GeneXpert are the only examples massively applied. This showcases the maturity stage of POCs and nanobiosensors and highlights the need for further improvements to achieve the goal of handheld ultrasensitive diagnostics.

Finally, it must be remembered that the relationship between nanostructures and lab-on-a-chip devices is both ways. While this work has focused on the importance of nanostructures as components of biosensing devices, the use of LOC systems to synthesize and characterize nanomaterials has been reported (Figure 4c) with exciting developments that can ultimately influence the future of nanobiosensors (Medina-Sanchez et al. 2012, Krishna et al. 2013).

Acknowledgements

The authors thank the funding support by the Mexican National Council for Science and Technology (CONACYT) through grants 314907 (National Laboratories), A1-S-32500 (Basic Research) and 312687 (COVID). J.M.D. benefited from a CONACYT scholarship for his graduate studies.

References

Adam, T., Hashim, U., Dhahi, T.S. and Leow, P.L. 2013. Nano Lab-on-chip systems for biomedical and environmental monitoring. *African J. Biotechnol.*, 12: 5486–5495.

Ahmad, R., Mahmoudi, T., Ahn, M. and Hahn, Y. 2018. Recent advances in nanowires-based field-effect transistors for biological sensor applications. *Biosens. Bioelectron.*, 100: 312–325.

Allen, B.L., Kichambare, P.D. and Star, A. 2007. Carbon nanotube field-effect-transistor-based biosensors. *Adv. Mater.*, 19(11): 1439–1451.

Alnaimat, F., Dagher, S., Mathew, B., Hilal-Alnqbi, A. and Khashan, S. 2018. Microfluidics based magnetophoresis: a review. *Chem. Rec.*, 18(11): 1596–1612.

Amendola, V. and Meneghetti, M. 2009. Size evaluation of gold nanoparticles by UV-Vis spectroscopy. *J. Phys. Chem. C.*, 113(11): 4277–4285.

Aqel, A., Abou El-Nour, K.M.M., Ammar, R.A.A. and Al-Warthan, A. 2012. Carbon nanotubes, science and technology part (I) structure, synthesis and characterisation. *Arab. J. Chem.*, 5(1): 1–23.

Azizipour, N., Avazpour, R., Rosenzweig, D.H., Sawan, M. and Ajji, A. 2020. Evolution of biochip technology: a review from lab-on-a-chip to organ-on-a-chip. *Micromachines.*, 11(6): 599.

Badshah, M.A., Koh, N.Y., Zia, A.W., Abbas, N., Zahra, Z. and Saleem, M.W. 2020. Recent developments in plasmonic nanostructures for metal enhanced fluorescence-based biosensing. *Nanomaterials.*, 10(9): 22.

Bartlett, P.N. and Al-Lolage, F.A. 2018. There is no evidence to support literature claims of direct electron transfer (DET) for native glucose oxidase (GOx) at carbon nanotubes or graphene. *J. Electroanal. Chem.*, 819: 26–37.

Bhatt, G. and Bhattacharya, S. 2019. Biosensors on chip: a critical review from an aspect of micro/nanoscales. *J. Micromanufacturing.*, 2(2): 198–219.

Bunimovich, Y.L., Shin, Y.S., Yeo, W.S., Amori, M., Kwong, G. and Heath, J.R. 2006. Quantitative real-time measurements of DNA hybridization with alkylated nonoxidized silicon nanowires in electrolyte solution. *J. Am. Chem. Soc.*, 128(50): 16323–16331.

Çakıroğlu, B. and Özacar, M. 2018. A self-powered photoelectrochemical glucose biosensor based on supercapacitor Co(3) O(4)-CNT Hybrid on TiO(2). *Biosens Bioelectron.*, 119: 34–41.

Camden, J.P., Dieringer, J.A., Zhao, J. and Van Duyne, R.P. 2008. Controlled plasmonic nanostructures for surface-enhanced spectroscopy and sensing. *Acc. Chem. Res.*, 41(12): 1653–1661.

Chalupniak, A. and Merkoci, A. 2017. Toward integrated detection and graphene-based removal of contaminants in a lab-on-a-chip platform. *Nano Res.*, 10(7): 2296–2310.

Chang, C.C., Chen, C.P., Wu, T.H., Yang, C.H., Lin, C.W. and Chen, C.Y. 2019. Gold nanoparticle-based colorimetric strategies for chemical and biological sensing applications. *Nanomaterials.*, 9(6): 861.

Chen, K.H., Pan, M.J., Jargalsaikhan, Z., Ishdorj, T.O. and Tseng, F.G. 2020. Development of surface-enhanced raman scattering (SERS)-based surface-corrugated nanopillars for biomolecular detection of colorectal cancer. *Biosensors-Basel.*, 10(11): 13.

Choi, M., Kim, S., Choi, S.H., Park, H.H. and Byun, K.M. 2020. Highly reliable SERS substrate based on plasmonic hybrid coupling between gold Nanoislands and periodic nanopillar arrays. *Opt. Express.*, 28(3): 3598–3606.

Choong, C., Milne, W.I. and Teo, K.B.K. 2008. Review: carbon nanotube for microfluidic lab-on-a-chip application. *Int. J. Mater. Form.*, 1(2): 117–125.

Chua, C.K., Ambrosi, A. and Pumera, M. 2011. Graphene based nanomaterials as electrochemical detectors in lab-on-a-chip devices. *Electrochem. Commun.*, 13(5): 517–519.

Chun, P. 2009. Colloidal gold and other labels for lateral flow immunoassays. pp. 1–19. *In*: Wong, R. and Tse, H. (eds.). Lateral Flow Immunoassay. Humana Press, Totowa, NJ.

Clark, Jr., L.C. and Lyons, C. 1962. Electrode systems for continuous monitoring in cardiovascular surgery. *Ann. N. Y. Acad. Sci.*, 102(1): 29–45.

Compton, O.C. and Nguyen, S.T. 2010. Graphene oxide, highly reduced graphene oxide, and graphene: versatile building blocks for carbon-based materials. *Small.*, 6(6): 711–723.

Daiguji, H. 2011. 4.11 - Nanofluidics. pp. 315–338. *In*: Andrews, D.L., Scholes, G.D. and Wiederrecht, G.P. (eds.). Comprehensive Nanoscience and Technology. Academic Press, Amsterdam, Netherlands.

Diao, S., Hong, G.S., Antaris, A.L., Blackburn, J.L., Cheng, K., Cheng, Z. and Dai, H.J. 2015. Biological imaging without autofluorescence in the second near-infrared region. *Nano Res.*, 8(9): 3027–3034.

Ding, J. and Qin, W. 2020. Recent advances in potentiometric biosensors. *TrAC Trends Anal. Chem.*, 124: 115803.

Doria, G., Conde, J., Veigas, B., Giestas, L., Almeida, C., Assuncao, M., Rosa, J. and Baptista, P.V. 2012. Noble metal nanoparticles for biosensing applications. *Sensors*, 12(2): 1657–1687.

Durucan, O., Wu, K.Y., Viehrig, M., Rindzevicius, T. and Boisen, A. 2018. Nanopillar-assisted SERS chromatography. *Acs Sensors*, 3(12): 2492–2498.

Eggins, B.R. 2007. Chemical sensors and biosensors. Ed. Wiley Blackwell. J. Wiley & Sons, Chichester, West Sussex; Hoboken.

Eickenberg, B., Meyer, J., Helmich, L., Kappe, D., Auge, A., Weddemann, A., Wittbracht, F. and Hütten, A. 2013. Lab-on-a-chip magneto-immunoassays: how to ensure contact between superparamagnetic beads and the sensor surface. *Biosensors*, 3(3): 327–340.

Eijkel, J.C.T. and Berg, A. 2005. Nanofluidics: What is it and what can we expect from it? *Microfluid. Nanofluidics.*, 1(3): 249–267.

Eijkel, J.C.T., Sparreboom, W., Shui, L., Salieb-Beugelaar, G.B. and van den Berg, A. 2009. Nanofluidics: Fundamentals and Applications. *Int. Sol. Stat. Sens. Act. and Micro. Conf. USA* 1561–1565.

Ensafi, A.A. 2019. Chapter 1—An introduction to sensors and biosensors. pp. 1–10. *In*: Ali A. Ensafi (ed.). Electrochemical Biosensors. Elsevier, Amsterdam, Netherlands.

Farré, M., and Barceló, D. 2020. Microfluidic devices: biosensors. pp. 287–351. *In*: Pico, Y. (ed.). Chemical analysis of food (Second Edition). Academic Press, Valencia, Spain.

Farrera, C., Torres Andón, F. and Feliu, N. 2017. Carbon nanotubes as optical sensors in biomedicine. *ACS Nano.*, 11(11): 10637–10643.

Fisher, E.A., Leung, K.K., Casanova-Moreno, J., Masuda, T., Young, J. and Bizzotto, D. 2017. Quantifying the selective modification of Au(111) facets via electrochemical and electroless treatments for manipulating gold nanorod surface composition. *Langmuir.*, 33(45): 12887–12896.

Fontana, J., Charipar, N., Flom, S.R., Naciri, J., Pique, A. and Ratna, B.R. 2016. Rise of the charge transfer plasmon: programmable concatenation of conductively linked gold nanorod dimers. *Acs Photonics.*, 3(5): 904–911.

Fothergill, S.M., Joyce, C. and Xie, F. 2018. Metal enhanced fluorescence biosensing: from ultra-violet towards second near-infrared window. *Nanoscale.*, 10(45): 20914–20929.

Frasco, M.F. and Chaniotakis, N. 2009. Semiconductor quantum dots in chemical sensors and biosensors. *Sensors*, 9(9): 7266–7286.

Gamal, R., Ismail, Y. and Swillam, M.A. 2015. Optical biosensor based on a silicon nanowire ridge waveguide for lab on chip applications. *J. Opt.*, 17(4): 45802.

Gauglitz, G. 2005. Direct optical sensors: principles and selected applications. *Anal. Bioanal. Chem.*, 381(1): 141–155.

Gaviña, P., Parra, M., Gil, S. and Costero, A.M. 2018. Red or blue? Gold nanoparticles in colorimetric sensing. *In*: Rahman, M. and Asiri, A.M. (eds.). Gold Nanoparticles-Reaching New Heights. IntechOpen, London, UK.

Geddes, C.D., Aslan, K., Gryczynski, I. and Lakowicz, J.R. 2005. Metal-enhanced fluorescence sensing. pp. 121–181. *In*: Thompson, R.B. (ed.). Fluorescence Sensors and Biosensors. CRC Press, Boca Raton, FL.

Gentil, S., Carrière, M., Cosnier, S., Gounel, S., Mano, N. and Le Goff, A. 2018. Direct electrochemistry of bilirubin oxidase from magnaporthe orizae on covalently-functionalized MWCNT for the design of high-performance oxygen-reducing biocathodes. *Chemistry* (Easton), 24(33): 8404–8408.

George, J.M., Antony, A. and Mathew, B. 2018. Metal oxide nanoparticles in electrochemical sensing and biosensing: a review. *Microchim. Acta.*, 185(7): 26.

Ghasemi, A., Amiri, H., Zare, H., Masroor, M., Hasanzadeh, A., Beyzavi, A. et al. 2017. Carbon nanotubes in microfluidic lab-on-a-chip technology: current trends and future perspectives. *Microfluid. Nanofluidics*, 21(9): 151.

Ghasemi, A., Rabiee, N., Ahmadi, S., Hashemzadeh, S., Lolasi, F., Bozorgomid, M. et al. 2018. Optical assays based on colloidal inorganic nanoparticles. *Analyst.*, 143(14): 3249–3283.

Gleiter, H. 1995. Nanostructured materials: state of the art and perspectives. *Nanostructured Mater.*, 6(1–4): 3–14.

Guo, L.H., Xu, Y., Ferhan, A.R., Chen, G.N. and Kim, D.H. 2013. Oriented gold nanoparticle aggregation for colorimetric sensors with surprisingly high analytical figures of merit. *J. Am. Chem. Soc.*, 135(33): 12338–12345.

Hakimian, F., Ghourchian, H., Hashemi, A.S., Arastoo, M.R. and Rad, M.B. 2018. Ultrasensitive optical biosensor for detection of MiRNA-155 using positively charged Au nanoparticles. *Sci. Rep.*, 8: 9.

Haun, J.B., Yoon, T.J., Lee, H. and Weissleder, R. 2010. Magnetic nanoparticle biosensors. Wiley Interdiscip. *Rev. Nanobiotechnology*, 2(3): 291–304.

Hickey, D.P., Reid, R.C., Milton, R.D. and Minteer, S.D. 2016. A self-powered amperometric lactate biosensor based on lactate oxidase immobilized in Dimethylferrocene-Modified LPEI. *Biosens. Bioelectron.*, 77: 26–31.

Hidi, I.J., Jahn, M., Weber, K., Cialla-May, D. and Popp, J. 2015. Droplet based microfluidics: spectroscopic characterization of levofloxacin and its SERS detection. *Phys. Chem. Chem. Phys.*, 17(33): 21236–21242.

Hidi, I.J., Jahn, M., Pletz, M.W., Weber, K., Cialla-May, D. and Popp, J. 2016. Toward levofloxacin monitoring in human urine samples by employing the LoC-SERS technique. *J. Phys. Chem. C.*, 120(37): 20613–20623.

Ho, C., Zhao, K. and Lee, T. 2014. Quasi-3D gold nanoring cavity arrays with high-density hot-spots for SERS applications via nanosphere lithography. *Nanoscale*, 6(15): 8606–8611.

Hofferber, E.M., Stapleton, J.A. and Iverson, N.M. 2020. Review—single walled carbon nanotubes as optical sensors for biological applications. *J. Electrochem. Soc.*, 167(3): 37530.

Huang, J.A., Zhang, Y.L., Ding, H. and Sun, H.B. 2015. SERS-enabled lab-on-a-chip systems. *Adv. Opt. Mater.* 3(5): 618–633.

Hussein, H.A., Hassan, R.Y.A., Chino, M. and Febbraio, F. 2020. Point-of-care diagnostics of COVID-19: from current work to future perspectives. *Sensors*, 20(15): 26.

Jamshaid, T., Neto, E.T.T., Eissa, M.M., Zine, N., Kunita, M.H., El-Salhi, A.E. and Elaissari, A. 2016. Magnetic particles: from preparation to lab-on-a-chip, biosensors, microsystems and microfluidics applications. *Trac-Trends Anal. Chem.*, 79: 344–362.

Jazayeri, M.H., Aghaie, T., Avan, A., Vatankhah, A. and Ghaffari, M.R.S. 2018. Colorimetric detection based on gold nano particles (GNPs): an easy, fast, inexpensive, low-cost and short time method in detection of analytes (Protein, DNA, and Ion). Sens. Bio-Sensing Res., 20: 1–8.

Jeong, Y., Kook, Y.M., Lee, K. and Koh, W.G. 2018. Metal enhanced fluorescence (MEF) for biosensors: general approaches and a review of recent developments. *Biosens. Bioelectron.*, 111: 102–116.

Jiang, S., Win, K.Y., Liu, S.H., Teng, C.P., Zheng, Y.G. and Han, M.Y. 2013. Surface-functionalized nanoparticles for biosensing and imaging-guided therapeutics. *Nanoscale*, 5(8): 3127–3148.

Joseph, P., Cottin-Bizonne, C., Benoît, J.M., Ybert, C., Journet, C., Tabeling, P. and Bocquet, L. 2006. Slippage of water past superhydrophobic carbon nanotube forests in microchannels. *Phys. Rev. Lett.*, 97(15): 156104.

Jung, W., Han, J., Choi, J. and Ahn, C.H. 2015. Point-of-Care Testing (POCT) diagnostic systems using microfluidic lab-on-a-chip technologies. *Microelectron. Eng.*, 132: 46–57.

Juska, V.B. and Pemble, M.E. 2020. A dual-enzyme, micro-band array biosensor based on the electrodeposition of carbon nanotubes embedded in chitosan and nanostructured au-foams on microfabricated gold band electrodes. *Analyst.*, 145(2): 402–414.

Kang, H., Wang, L., O'Donoghue, M., Cao, Y.C. and Tan, W. 2008. Nanoparticles for biosensors. pp. 583–621. *In*: Ligler, F.S. and Taitt, C.R. (eds.). Optical Biosensors. Second Ed. Elsevier, Amsterdam, Netherlands.

Kang, K.A., Wang, J.T., Jasinski, J.B. and Achilefu, S. 2011. Fluorescence manipulation by gold nanoparticles: from complete quenching to extensive enhancement. *J. Nanobiotechnology*, 9: 13.

Khalil, I., Julkapli, N.M., Yehye, W.A., Basirun, W.J. and Bhargava, S.K. 2016. Graphene-gold nanoparticles hybrid-synthesis, functionalization, and application in a electrochemical and surface-enhanced raman scattering biosensor. *Materials* (Basel), 9(6): 38.

Khan, H.A., Sakharkar, M.K., Nayak, A., Kishore, U. and Khan, A. 2018. Nanoparticles for biomedical applications: an overview. pp. 357–384. *In*: Narayan, R. (ed.). Nanobiomaterials. Woodhead Publishing, Cambridge, UK.

Kim, A., Ou, F.S., Ohlberg, D.A.A., Hu, M., Williams, R.S. and Li, Z.Y. 2011. Study of molecular trapping inside gold nanofinger arrays on surface-enhanced raman substrates. *J. Am. Chem. Soc.*, 133(21): 8234–8239.

Kim, A., Barcelo, S.J., Williams, R.S. and Li, Z.Y. 2012. Melamine sensing in milk products by using surface enhanced raman scattering. *Anal. Chem.*, 84(21): 9303–9309.

Kim, Y.T., Schilling, J., Schweizer, S.L., Sauer, G. and Wehrspohn, R.B. 2017. Au coated PS nanopillars as a highly ordered and reproducible SERS substrate. *Photonics Nanostructures-Fundamentals Appl.*, 25: 65–71.

Kirsch, J., Siltanen, C., Zhou, Q., Revzin, A. and Simonian, A. 2013. Biosensor technology: recent advances in threat agent detection and medicine. *Chem. Soc. Rev.*, 42(22): 8733–8768.

Koczula, K.M. and Gallotta, A. 2016. Lateral flow assays. pp. 111–120. *In*: Estrela, P. (ed.). Biosensor Technologies for Detection of Biomolecules. Essays in Biochemistry. Portland Press Ltd, Venice, Italy.

Krishna, K.S., Li, Y.H., Li, S.N. and Kumar, C. 2013. Lab-on-a-chip synthesis of inorganic nanomaterials and quantum dots for biomedical applications. *Adv. Drug Deliv. Rev.*, 65(11-12): 1470–1495.

Krivitsky, V., Hsiung, L., Lichtenstein, A., Brudnik, B., Kantaev, R., Elnathan, R., Pevzner, A., Khatchtourints, A. and Patolsky, F. 2012. Si nanowires forest-based on-chip biomolecular filtering, separation and preconcentration devices: nanowires do it all. *Nano Lett.*, 12(9): 4748–4756.

Lakowicz, J.R. 2005. Radiative decay engineering 5: metal-enhanced fluorescence and plasmon emission. *Anal. Biochem.*, 337(2): 171–194.

Lalaoui, N., David, R., Jamet, H., Holzinger, M., Le Goff, A. and Cosnier, S. 2016. Hosting adamantane in the substrate pocket of laccase: direct bioelectrocatalytic reduction of O2 on functionalized carbon nanotubes. *ACS Catal.*, 6(7): 4259–4264.

Le Gac, S., Vermes, I. and van den Berg, A. 2006. Quantum dots based probes conjugated to Annexin V for photostable apoptosis detection and imaging. *Nano Lett.*, 6(9): 1863–1869.

Le Ru, E.C. and Etchegoin, P.G. 2013. Quantifying SERS Enhancements. *Mrs. Bull.*, 38(8): 631–640.

Lee, H., Sun, E., Ham, D. and Weissleder, R. 2008. Chip-NMR biosensor for detection and molecular analysis of cells. *Nat. Med.*, 14(8): 869–874.

Lee, M., Baik, K.Y., Noah, M., Kwon, Y., Lee, J.O. and Hong, S. 2009. Nanowire and nanotube transistors for lab-on-a-chip applications. *Lab Chip.*, 9(16): 2267–2280.

Lei, K.F. and Butt, Y.K.C. 2010. Colorimetric immunoassay chip based on gold nanoparticles and gold enhancement. *Microfluid. Nanofluidics.*, 8(1): 131–137.

Li, C., and Shi, G.Q. 2014. Carbon nanotube-based fluorescence sensors. *J. Photochem. Photobiol. C-Photochemistry Rev.*, 19: 20–34.

Li, J.H., Chu, X., Liu, Y.L., Jiang, J.H., He, Z.M., Zhang, Z.W., Shen, G.L. and Yu, R.Q. 2005. A colorimetric method for point mutation detection using high-fidelity DNA ligase. *Nucleic Acids Res.*, 33(19): 9.

Li, J.Q., Chen, C., Jans, H., Xu, X.M., Verellen, N., Vos, I., Okumura, Y., Moshchalkov, V.V., Lagae, L. and Van Dorpe, P. 2014. 300 mm wafer-level, ultra-dense arrays of au-capped nanopillars with sub-10 Nm Gaps as reliable SERS substrates. *Nanoscale.*, 6(21): 12391–12396.

Li, J.S., Deng, T., Chu, X., Yang, R.H., Jiang, J.H., Shen, G.L. and Yu, R.Q. 2010. Rolling circle amplification combined with gold nanoparticle aggregates for highly sensitive identification of single-nucleotide polymorphisms. *Anal. Chem.*, 82(7): 2811–2816.

Li, M., Cushing, S.K., Liang, H.Y., Suri, S., Ma, D.L. and Wu, N.Q. 2013. Plasmonic nanorice antenna on triangle nanoarray for surface-enhanced raman scattering detection of hepatitis B virus DNA. *Anal. Chem.*, 85(4): 2072–2078.

Li, P., Long, F., Chen, W., Chen, J., Chu, P.K. and Wang, H. 2020. Fundamentals and applications of surface-enhanced raman spectroscopy–based biosensors. *Curr. Opin. Biomed. Eng.*, 13: 51–59.

Li, S., Zhang, T., Zhu, Z., Gao, N. and Xu, Q. 2016. Lighting up the gold nanoparticles quenched fluorescence by silver nanoparticles: a separation distance study. *Rsc Adv.*, 6(63): 58566–58572.

Liang, Y., Huang, X.L., Chen, X.R., Zhang, W.J., Ping, G. and Xiong, Y.H. 2018. Plasmonic ELISA for naked-eye detection of ochratoxin a based on the tyramine-H2O2 amplification system. *Sensors and Actuators B-Chemical.*, 259: 162–169.

Lim, Y.C., Kouzani, A.Z. and Duan, W. 2010. Lab-on-a-chip: a component view. *Microsyst. Technol.*, 16(12): 1995–2015.

Liu, F., Ni, L. and Zhe, J. 2018. Lab-on-a-chip electrical multiplexing techniques for cellular and molecular biomarker detection. *Biomicrofluidics.*, 12(2): 21501.

Liu, J.C., Guan, Z., Lv, Z.Z., Jiang, X.L., Yang, S.M. and Chen, A.L. 2014. Improving sensitivity of gold nanoparticle based fluorescence quenching and colorimetric aptasensor by using water resuspended gold nanoparticle. *Biosens. Bioelectron.*, 52: 265–270.

Liu, J.W. 2014. DNA-stabilized, fluorescent, metal nanoclusters for biosensor development. *Trac-Trends Anal. Chem.*, 58: 99–111.

Liu, L., Zhang, Q., Lu, Y.S., Du, W., Li, B., Cui, Y.S., Yuan, C.S. et al. 2017. A high-performance and low cost SERS substrate of plasmonic nanopillars on plastic film fabricated by nanoimprint lithography with AAO template. *Aip Adv.*, 7(6): 12.

Liu, R., Duay, J. and Lee, S.B. 2011. Heterogeneous nanostructured electrode materials for electrochemical energy storage. *Chem. Commun.*, 47(5): 1384–1404.

Liu, S. and Guo, X. 2012. Carbon nanomaterials field-effect-transistor-based biosensors. *NPG Asia Mater.*, 4(8): e23–e23.

Liu, Y.X., Zhang, J., Cheng, Y. and Jiang, S.P. 2018. Effect of carbon nanotubes on direct electron transfer and electrocatalytic activity of immobilized glucose oxidase. *Acs Omega.*, 3(1): 667–676.

Luo, X.L., Morrin, A., Killard, A.J. and Smyth, M.R. 2006. Application of nanoparticles in electrochemical sensors and biosensors. *Electroanalysis.*, 18(4): 319–326.

Magno, G., Belier, B. and Barbillon, G. 2018. Al/Si Nanopillars as very Sensitive SERS substrates. *Materials* (Basel), 11(9): 9.

Malekzad, H., Zangabad, P.S., Mirshekari, H., Karimi, M. and Hamblin, M.R. 2017. Noble metal nanoparticles in biosensors: recent studies and applications. *Nanotechnol. Rev.*, 6(3): 301–329.

Manz, A., Graber, N. and Widmer, H.M. 1990. Miniaturized total chemical-analysis systems - a novel concept for chemical sensing. *Sensors and Actuators B-Chemical.*, 1(1–6): 244–248.

Margineanu, A., Chan, J.J., Kelly, D.J., Warren, S.C., Flatters, D., Kumar, S., Katan, M., Dunsby, C.W. and French, P.M.W. 2016. Screening for protein-protein interactions using förster resonance energy transfer (FRET) and fluorescence lifetime imaging microscopy (FLIM). *Sci. Rep.*, 6(1): 28186.

Matsumoto, A. and Miyahara, Y. 2013. Current and emerging challenges of field effect transistor based bio-sensing. *Nanoscale.*, 5(22): 10702–10718.

McAlpine, M.C., Friedman, R.S., Jin, S., Lin, K., Wang, W.U. and Lieber, C.M. 2003. High-performance nanowire electronics and photonics on glass and plastic substrates. *Nano Lett.*, 3(11): 1531–1535.

Medina-Sanchez, M., Miserere, S. and Merkoci, A. 2012. Nanomaterials and lab-on-a-chip technologies. *Lab Chip.*, 12(11): 1932–1943.

Mejía-Salazar, J.R., Rodrigues Cruz, K., Materón Vásques, E.M. and Novais de Oliveira, Jr. O. 2020. Microfluidic point-of-care devices: new trends and future prospects for Ehealth diagnostics. *Sensors* (Basel), 20(7): 1951.

Meredith, M.T., Minson, M., Hickey, D., Artyushkova, K., Glatzhofer, D.T. and Minteer, S.D. 2011. Anthracene-modified multi-walled carbon nanotubes as direct electron transfer scaffolds for enzymatic oxygen reduction. *ACS Catal.*, 1(12): 1683–1690.

Messinger, B.J., Vonraben, K.U., Chang, R.K. and Barber, P.W. 1981. Local-Fields at the surface of noble-metal microspheres. *Phys. Rev. B.*, 24(2): 649–657.

Milton, R.D. and Minteer, S.D. 2017. Direct enzymatic bioelectrocatalysis: differentiating between myth and reality. *J. R. Soc. Interface.*, 14(131): 20170253.

Min, C., Shao, H.L., Liong, M., Yoon, T.J., Weissleder, R. and Lee, H. 2012. Mechanism of magnetic relaxation switching sensing. *ACS Nano.*, 6(8): 6821–6828.

Mogensen, K.B. and Kutter, J.P. 2012. Carbon nanotube based stationary phases for microchip chromatography. *Lab Chip*, 12(11): 1951–1958.

Nagel, B., Dellweg, H. and Gierasch, L.M. 1992. Glossary for chemists of terms used in biotechnology - (IUPAC Recommendations 1992). *Pure Appl. Chem.*, 64(1): 143–168.

Newman, J.D. and Setford, S.J. 2006. Enzymatic biosensors. *Mol. Biotechnol.*, 32(3): 249–268.

O'Farrell, B. 2009. Evolution in lateral flow–based immunoassay systems. pp. 1–33. *In*: Wong, R. and Tse, H. (eds.). Lateral Flow Immunoassay. Humana Press, Totowa, NJ.

Pachauri, V., and Ingebrandt, S. 2016. Biologically sensitive field-effect transistors: from ISFETs to NanoFETs. Ed. P Estrela. *Biosens. Technol. Detect. Biomol.*, 60: 81–90.

Park, S.Y., Lee, S.M., Kim, G.B. and Kim, Y.P. 2012. Gold nanoparticle-based fluorescence quenching via metal coordination for assaying protease activity. *Gold Bull.*, 45(4): 213–219.

Parnianchi, F., Nazari, M., Maleki, J. and Mohebi, M. 2018. Combination of graphene and graphene oxide with metal and metal oxide nanoparticles in fabrication of electrochemical enzymatic biosensors. *Int. Nano Lett.*, 8(4): 229–239.

Parolo, C., Medina-Sanchez, M., Monton, H., de la Escosura-Muniz, A. and Merkoci, A. 2013. Paper-based electrodes for nanoparticle detection. *Part. Part. Syst. Charact.*, 30(8): 662–666.

Pastorin, G., Kostarelos, K., Prato, M. and Bianco, A. 2005. Functionalized carbon nanotubes: towards the delivery of therapeutic molecules. *J. Biomed. Nanotechnol.*, 1(2): 133–142.

Piccin, E., Laocharoensuk, R., Burdick, J., Carrilho, E. and Wang, J. 2007. Adaptive nanowires for switchable microchip devices. *Anal. Chem.*, 79(12): 4720–4723.

Pilot, R., Signorini, R., Durante, C., Orian, L., Bhamidipati, M. and Fabris, L. 2019. A review on surface-enhanced raman scattering. *Biosensors*, 9(2).

Pires, N.M.M., Dong, T., Hanke, U. and Hoivik, N. 2014. Recent developments in optical detection technologies in lab-on-a-chip devices for biosensing applications. *Sensors*, 14(8): 15458–15479.

Pisanic, T.R., Zhang, Y. and Wang, T.H. 2014. Quantum dots in diagnostics and detection: principles and paradigms. *Analyst.*, 139(12): 2968–2981.

Pokropivny, V.V. and Skorokhod, V.V. 2007. Classification of nanostructures by dimensionality and concept of surface forms engineering in nanomaterial science. *Mater. Sci. Eng. C.*, 27(5): 990–993.

Poornima, V., Alexandar, V., Iswariya, S., Perumal, P.T. and Uma, T.S. 2016. Gold nanoparticle-based nanosystems for the colorimetric detection of Hg2+ ion contamination in the environment. *Rsc Adv.*, 6(52): 46711–46722.

Primiceri, E., Chiriacò, M.S., Notarangelo, F.M., Crocamo, A., Ardissino, D., Cereda, M., Bramanti, A.P., Bianchessi, M.A., Giannelli, G. and Maruccio, G. 2018. Key enabling technologies for point-of-care diagnostics. *Sensors* (Basel), 18(11).

Pumera, M. 2011. Graphene in biosensing. *Mater. Today*, 14(7-8): 308–315.

Punckt, C., Muckel, F., Wolff, S., Aksay, I.A., Chavarin, C.A., Bacher, G. and Mertin, W. 2013. The effect of degree of reduction on the electrical properties of functionalized graphene sheets. *Appl. Phys. Lett.*, 102(2): 023114.

Rackus, D.G., Shamsi, M.H. and Wheeler, A.R. 2015. Electrochemistry, biosensors and microfluidics: a convergence of fields. *Chem. Soc. Rev.*, 44(15): 5320–5340.

Ragavan, K.V. and Neethirajan, S. 2019. Nanoparticles as biosensors for food quality and safety assessment. pp. 147–202. *In*: López Rubio, A., Fabra Rovira, M.J., martínez Sanz, M. and Gómez-Mascaraque, L.G. (eds.). Nanomaterials for Food Applications. Elsevier, Amsterdam, Netherlands.

Rasheed, P.A. and Sandhyarani, N. 2017. Electrochemical DNA sensors based on the use of gold nanoparticles: a review on recent developments. *Microchim. Acta*, 184(4): 981–1000.

Rechberger, W., Hohenau, A., Leitner, A., Krenn, J.R., Lamprecht, B. and Aussenegg, F.R. 2003. Optical properties of two interacting gold nanoparticles. *Opt. Commun.*, 220(1–3): 137–141.

Rekha, C.R., Nayar, V.U. and Gopchandran, K.G. 2018. Synthesis of highly stable silver nanorods and their application as sers substrates. *J. Sci. Mater. Devices.*, 3(2): 196–205.

Reuter, C., Urban, M., Arnold, M., Stranik, O., Csaki, A. and Fritzsche, W. 2020. 2-LED-MSpectrophotometer for rapid on-site detection of pathogens using noble-metal nanoparticle-based colorimetric assays. *Appl. Sci.*, 10(8): 14.

Ribeiro, T., Baleizao, C. and Farinha, J.P.S. 2017. Artefact-free evaluation of metal enhanced fluorescence in silica coated gold nanoparticles. *Sci. Rep.*, 7: 12.

Sabela, M., Balme, S., Bechelany, M., Janot, J.M. and Bisetty, K. 2017. A review of gold and silver nanoparticle-based colorimetric sensing assays. *Adv. Eng. Mater.*, 19(12): 24.

Sadighbayan, D., Hasanzadeh, M. and Ghafar-Zadeh, E. 2020. Biosensing based on field-effect transistors (FET): Recent progress and challenges. *TrAC Trends Anal. Chem.*, 133: 116067.

Saleh, T.A. 2020. Nanomaterials: classification, properties, and environmental toxicities. *Environ. Technol. Innov.*, 20: 101067.

Schasfoort, R.B.M., Bergveld, P., Kooyman, R.P.H. and Greve, J. 1990. Possibilities and limitations of direct detection of protein charges by means of an immunological field-effect transistor. *Anal. Chim. Acta.*, 238: 323–329.

Schönberger, M. and Hoffstetter, M. 2016. 6 - Emerging trends. pp. 235–268. *In*: Markus Schönberger and Hoffstetter, M. (eds.). Emerging Trends in Medical Plastic Engineering and Manufacturing. William Andrew Publishing, Oxford, UK.

Sengupta, J. and Hussain, C.M. 2019. Graphene and its derivatives for analytical lab on chip platforms. *Trac-Trends Anal. Chem.*, 114: 326–337.

Shang, L., Chen, H.J., Deng, L. and Dong, S.J. 2008. Enhanced resonance light scattering based on biocatalytic growth of gold nanoparticles for biosensors design. *Biosens. Bioelectron.*, 23(7): 1180–1184.

Sharma, S., Gupta, N. and Srivastava, S. 2012. Modulating electron transfer properties of gold nanoparticles for efficient biosensing. *Biosens. Bioelectron.*, 37(1): 30–37.

Shen, H.M., Hu, L. and Fu, X. 2018. Integrated giant magnetoresistance technology for approachable weak biomagnetic signal detections. *Sensors*, 18(1): 20.

Shi, X.H., Gu, W., Li, B.Y., Chen, N.N., Zhao, K. and Xian, Y.Z. 2014. Enzymatic biosensors based on the use of metal oxide nanoparticles. *Microchim. Acta.*, 181(1-2): 1–22.

Shumyantseva, V.V., Carrara, S., Bavastrello, V., Jason Riley, D., Bulko, T.V., Skryabin, K.G., Archakov, A.I. and Nicolini, C. 2005. Direct electron transfer between cytochrome P450scc and gold nanoparticles on screen-printed rhodium-graphite electrodes. *Biosens. Bioelectron.*, 21(1): 217–222.

Singh, V., Yadav, P. and Mishra, V. 2020. Recent advances on classification, properties, synthesis, and characterization of nanomaterials. pp. 83–97. *In*: Srivastava, N., Srivastava, M., Mishra, P. and Gupta, V.K. (eds.). Green Synthesis of Nanomaterials for Bioenergy Applications. Wiley Online Books. Varanasi, India.

Soleymani, L., Fang, Z.C., Lam, B., Bin, X.M., Vasilyeva, E., Ross, A.J., Sargent, E.H. and Kelley, S.O. 2011. Hierarchical nanotextured microelectrodes overcome the molecular transport barrier to achieve rapid, direct bacterial detection. *ACS Nano.*, 5(4): 3360–3366.

Takahashi, S. and Anzai, J. 2013. Recent progress in ferrocene-modified thin films and nanoparticles for biosensors. *Materials* (Basel), 6(12): 5742–5762.

Tangkuaram, T., Ponchio, C., Kangkasomboon, T., Katikawong, P. and Veerasai, W. 2007. Design and development of a highly stable hydrogen peroxide biosensor on screen printed carbon electrode based on horseradish peroxidase bound with gold nanoparticles in the matrix of chitosan. *Biosens Bioelectron.*, 22(9-10): 2071–2078.

Tarn, M.D. and Pamme, N. 2014. Microfluidics. pp. 1–6. *In*: Reference Module in Chemistry, Molecular Sciences and Chemical Engineering. Elsevier.

Tian, M., Li, Z.H., Song, R.H., Li, Y.X., Guo, C.G., Sha, Y.J., Cui, W.L., Xu, S.C., Hu, G.D. and Wang, J.H. 2020. Graphene Biosensor as affinity biosensors for biorecognition between guanine riboswitch and ligand. *Appl. Surf. Sci.*, 503: 8.

Tilmaciu, C. and Morris, M.C. 2015. Carbon nanotube biosensors. *Front. Chem.*, 3(59).

Tiwari, J.N., Tiwari, R.N. and Kim, K.S. 2012. Zero-dimensional, one-dimensional, two-dimensional and three-dimensional nanostructured materials for advanced electrochemical energy devices. *Prog. Mater. Sci.*, 57(4): 724–803.

Tymm, C., Zhou, J.H., Tadimety, A., Burklund, A. and Zhang, J.X.J. 2020. Scalable COVID-19 detection enabled by lab-on-chip biosensors. *Cell. Mol. Bioeng.*, 13(4): 313–329.

Verardo, D., Lindberg, F.W., Anttu, N., Niman, C.S., Lard, M., Dabkowska, A.P., Nylander, T., Månsson, A., Prinz, C.N. and Linke, H. 2018. Nanowires for biosensing: lightguiding of fluorescence as a function of diameter and wavelength. *Nano Lett.*, 18(8): 4796–4802.

Verma, M.S., Rogowski, J.L., Jones, L. and Gu, F.X. 2015. Colorimetric biosensing of pathogens using gold nanoparticles. *Biotechnol. Adv.*, 33(6): 666–680.

Vidotti, M., Carvalhal, R.F., Mendes, R.K., Ferreira, D.C.M. and Kubota, L.T. 2011. Biosensors based on gold nanostructures. *J. Braz. Chem. Soc.*, 22(1): 3–20.

Vu, C.A. and Chen, W.Y. 2019. Field-effect transistor biosensors for biomedical applications: recent advances and future prospects. *Sensors*, 19(19): 22.

Wang, H., Mu, L., She, G., Xu, H. and Shi, W. 2014. Fluorescent biosensor for alkaline phosphatase based on fluorescein derivatives modified silicon nanowires. *Sensors Actuators B Chem.*, 203: 774–781.

Warren-Smith, S.C., Afshar, S. and Monro, T.M. 2010. Fluorescence-based sensing with optical nanowires: a generalized model and experimental validation. *Opt. Express.*, 18(9): 9474–9485.

Wongkaew, N., Simsek, M., Griesche, C. and Baeumner, A.J. 2019. Functional nanomaterials and nanostructures enhancing electrochemical biosensors and lab-on-a-chip performances: recent progress, applications, and future perspective. *Chem. Rev.*, 119(1): 120–194.

Wu, Q., Liu, J., Wang, X., Feng, L., Wu, J., Zhu, X., Wen, W. and Gong, X. 2020. Organ-on-a-chip: recent breakthroughs and future prospects. *Biomed. Eng. Online.*, 19(1): 9.

Wu, Y. and Hu, S. 2007. Biosensors based on direct electron transfer in redox proteins. *Microchim. Acta*, 159(1): 1–17.

Wunsch, B.H., Smith, J.T., Gifford, S.M., Wang, C., Brink, M., Bruce, R.L., Austin, R.H., Stolovitzky, G. and Astier, Y. 2016. Nanoscale lateral displacement arrays for the separation of exosomes and colloids down to 20 Nm. *Nat. Nanotechnol.*, 11(11): 936–940.

Wunsch, B.H., Kim, S.C., Gifford, S.M., Astier, Y., Wang, C., Bruce, R.L., Patel, J. V. et al. 2019. Gel-on-a-Chip: Continuous, velocity-dependent dna separation using nanoscale lateral displacement. *Lab Chip.*, 19(9): 1567–1578.

Xia, H., and Zeng, J. 2020. Rational surface modification of carbon nanomaterials for improved direct electron transfer-type bioelectrocatalysis of redox enzymes. *Catalysts*, 10(12): 1447.

Xie, Q., Zhao, Y.Y., Chen, X., Liu, H.M., Evans, D.G. and Yang, W.S. 2011. Nanosheet-based titania microspheres with hollow core-shell structure encapsulating horseradish peroxidase for a mediator-free biosensor. *Biomaterials*, 32(27): 6588–6594.

Xu, X.Y., Daniel, W.L., Wei, W. and Mirkin, C.A. 2010. Colorimetric Cu2+ detection using dna-modified gold-nanoparticle aggregates as probes and click chemistry. *Small*, 6(5): 623–626.

Yaraki, M.T. and Tan, Y.N. 2020. Metal nanoparticles-enhanced biosensors: synthesis, design and applications in fluorescence enhancement and surface-enhanced raman scattering. *Chem. Asian J.*, 15(20): 3180–3208.

Yilmaz, H., Bae, S.H., Cao, S.S., Wang, Z.Y., Raman, B. and Singamaneni, S. 2019. Gold-nanorod-based plasmonic nose for analysis of chemical mixtures. *Acs Appl. Nano Mater.*, 2(6): 3897–3905.

Zeng, S.W., Yong, K.T., Roy, I., Dinh, X.Q., Yu, X. and Luan, F. 2011. A review on functionalized gold nanoparticles for biosensing applications. *Plasmonics.*, 6(3): 491–506.

Zhang, J.X.J. and Hoshino, K. 2019. Chapter 3 - microfluidics and micro total analytical systems. pp. 113–179. *In*: John, X., Zhang, J. and Hoshino, K. (eds.). Molecular Sensors and Nanodevices (Second Edition). Academic Press, London, UK.

Zhao, C., Zhong, G.W., Kim, D.E., Liu, J.X. and Liu, X.Y. 2014. A portable lab-on-a-chip system for gold-nanoparticle-based colorimetric detection of metal ions in water. *Biomicrofluidics*, 8(5): 9.

Zhao, H., Sun, J., Song, J. and Yang, Q. 2010. Direct electron transfer and conformational change of glucose oxidase on carbon nanotube-based electrodes. *Carbon* N. Y., 48(5): 1508–1514.

Zhao, L., Cao, J., Wu, Z., Li, J. and Zhu, J. 2013. Lab-on-a-chip for anticancer drug screening using quantum dots probe based apoptosis assay. *J. Biomed. Nanotechnol.*, 9(3): 348–356.

Zhao, W.A. and van den Berg, A. 2008. Lab on paper. *Lab Chip.*, 8(12): 1988–1991.

Zheng, L.Y., Cai, G.Z., Wang, S.Y., Liao, M., Li, Y.B. and Lin, J.H. 2019. A microfluidic colorimetric biosensor for rapid detection of *Escherichia coli* 0157:H7 using gold nanoparticle aggregation and smart phone imaging. *Biosens. Bioelectron.*, 124: 143–149.

Zhong, W.W. 2009. Nanomaterials in fluorescence-based biosensing. *Anal. Bioanal. Chem.*, 394(1): 47–59.

Zhou, M., Zhai, Y.M. and Dong, S.J. 2009. Electrochemical sensing and biosensing platform based on chemically reduced graphene oxide. *Anal. Chem.*, 81(14): 5603–5613.

Zhou, Y., Fang, Y. and Ramasamy, R.P. 2019. Non-covalent functionalization of carbon nanotubes for electrochemical biosensor development. *Sensors*, 19(2).

CHAPTER 10

Drug-delivery using Inorganic and Organic Nanoparticles

Juan Luis de la Fuente-Jiménez,[1,] Goldie oza,[2] Brian A. Korgel,[3] Abraham Ulises[2] and Ashutosh Sharma[1,*]*

1. Introduction

The main field of nanomedicine is treating and diagnosing diseases through the improvement and use of nanotechnology and nanomaterials (Li et al. 2019). Nanomaterial dimensions are between 1 and 100 nm, which is an appropriate size range for helping drugs overcome biological barriers and contact the target. Organic and inorganic nanoparticles (NPs) have been developed for these purposes. NPs display a wide range of useful chemical, biological, mechanical, electrical, magnetic, and structural properties, and have been employed for gene and drug intracellular distribution. Some examples of NPs are quantum dots (QD), lipid nanoparticles (LN), dendrimers (den), metallic nanoparticles (MNP), nanogels, carbon nanotubes (CNTs), and polymers, among others (Patra et al. 2018, Su and Kang 2020, Tayo 2017). Moreover, NPs as drug delivery systems (DDS) can improve the efficiency of drugs for anti-microbial, antiviral, and anticancer treatments by using various targeting strategies (Figure 1) (Raza et al. 2019a).

NPs designed for drug delivery systems have unique morphological, physical, and chemical properties, which are affected by their size, shape and surface chemistry, all determining their biocompatibility, immune response evasion, and controlled release of their cargo (Patra et al. 2018, Tayo 2017). The payload can be functional genes, small interfering RNA (siRNA), proteins, imaging agents, and drugs for *in vivo* and *in vitro* therapies (Waehler et al. 2007). For DDS, the configuration of the NPs (e.g., inorganic or organic) and the way the payload is linked to the NP (e.g., matrix system or core-shell system) are both important (Siepmann et al. 2008). NPs are intended to enhance the efficacy and safety profiles of chemotherapeutic drugs or agents and increase the target specificity, through passive or active targeting, while being protected from the surrounding conditions and without disturbing healthy cells or tissue (Iqbal and Keshavarz 2018, Jeong et al. 2007). The nanoscale dimensions of NPs enable efficient infiltration of tissue, guarantee a more efficient supply of drug at the targeted site, and reduce adverse effects (Mirza and Siddiqui 2014).

[1] Tecnologico de Monterrey, School of Engineering and Sciences, Centre of Bioengineering, Campus Queretaro. Av. Epigmenio González, No. 500, Fracc. San Pablo 76130 Querétaro, México.

[2] Centro de Investigación y Desarrollo Tecnológico en Electroquímica, Parque Tecnológico, Querétaro, Sanfandila, Pedro Escobedo, 76703, Querétaro, México.

[3] McKetta Department of Chemical Engineering and Texas Materials Institute, The University of Texas at Austin, Austin, Texas 78712-1062, United States.

* Corresponding authors: juanluisdelafuente420@gmail.com; asharma@tec.mx

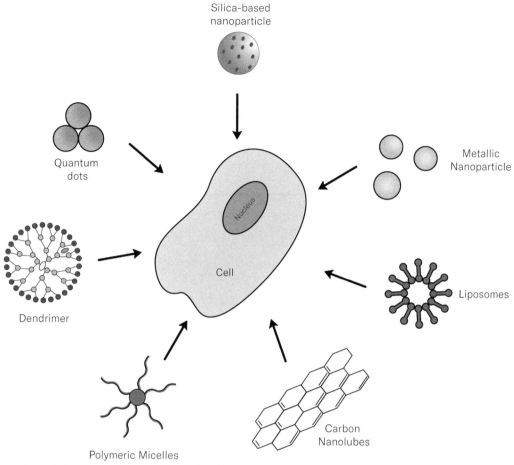

Figure 1. Representation of several organic and inorganic nanoparticles employed for intracellular drug delivery.

NPs can also be designed to respond to endogenous or exogenous stimuli, such as enzymes, pH, glutathione (GSH) or light, temperature, ultrasound (US), among others (Tayo 2017). Stimuli-responsive DDS can extend the capabilities of normal therapeutic treatments. For example, stimuli-responsive polymeric NPs have a key role to play in different tissue or cell sections, intracellular or extracellular, and achieving a precise and efficient payload release (Raza et al. 2019a). Finally, NPs have been applied in the field of theranostics, comprising therapy, diagnosis, and as imaging devices at the same time, especially in cancer (Haba et al. 2007, Koo et al. 2011, Lehner et al. 2013).

2. Drug and nanoparticle protection

The administration of drugs or therapeutic agents to the desired location is one of the main difficulties for achieving an efficient treatment in several diseases. The therapy with these agents has several shortcomings, such as reduced biodistribution, non-selectivity, low bioavailability, unwanted adverse effects, poor specificity, and low effectiveness (Kadam et al. 2012, Majumder and Minko 2020). Likewise, drugs need to overcome several physiological barriers, such as opsonization by the mononuclear phagocyte system and poor absorption and solubility (Blanco et al. 2015, Bonifácio et al. 2014). An encouraging solution for these problems is to combine drugs with NPs that show an enhancement in their biological and physicochemical characteristics and construct a proper targeting pathway for delivering the therapeutic agents into the specific location of the body (Jahan et al. 2017). This targeted application provides for increased drug concentration in the desired location, thus reducing the total dosage needed and reducing overall negative side-effects. The several disadvantages of NPs are poor access after oral administration, intravenous intakes need more drug quantities, lower diffusion capability into the exterior membrane, and undesired side-effects after administration (Patra et al. 2018). Furthermore, NPs help in the evasion of the destruction of drugs in the gastrointestinal tract and protecting non-water-soluble drugs from physical obstacles to reach the desired site (Patra et al. 2018). NPs interact with different drugs via physical interactions such as electrostatic and van der Waals forces or chemical interactions such as covalent and hydrogen bonds (Patra et al. 2018). Covalently attached drugs to NPs have numerous benefits like increased circulation time, less toxicity, more drug stability, and better biocompatibility (Raza et al. 2019a). In addition, contrast agents linked to NPs associated with drugs offer a method of tracking and imaging in *in vivo* systems (Yetisgin et al. 2020). However, the efficiency of the delivery will depend on the shape, size, and chemical/physical properties of the NPs as well as the targeting of the drug via passive or active pathways (Bonifácio et al. 2014).

3. Passive and active targeting

NPs incorporate their payload via two different pathways: passive and self-delivery. In the passive mode, cargo is integrated via hydrophobic interactions in the inner cavity. When NPs are targeted and reach the desired sites, the proposed quantity of the drug is discharged in a hydrophobic location. In self-delivery, the drug molecules are used themselves for building the NPs instead of being only the payload and are directly ligated to the NPs structure for simpler distribution. Nevertheless, in this method the timing is critical since if it is not released at the proper time, the cargo will not reach the target area (Lu et al. 2016).

Besides, an extra important feature is responsible for categorizing the targeting of drugs into passive and active targeting. In passive targeting, the NPs with the payload move inside the bloodstream and are attracted to the target site by their affinity to certain properties like molecular shape, pH, temperature, and accumulate in specific regions taking advantage of the enhanced permeation and retention (EPR) effect as in tumors (Patra et al. 2018, Raza et al. 2019a). The EPR effect is attributed to leaky vasculature, lack of efficient lymphatic drainage, and greater number of blood vessels in tumors as compared to healthy tissue (Figure 2B) (Fang et al. 2011, Tayo 2017). The main disadvantage of passive targeting is the early delivery of the drug even if the NP has an extended biological half-life (Raza et al. 2019a).

For DDS with active targeting, NPs are coupled to moieties that provide NPs with the ability to target and bind specific locations and diminish immunogenicity. NPs can be coated with recognition molecules, including antibodies (Ab), peptides, polysaccharides, polymers, folic acid (FA), among others. These modified NPs can recognize the targets or overexpressed receptor structures in malignant cells or diseased tissue. The key objectives are often unique proteins or antigens expressed on the surface of cancerous cells, as well as receptors and lipid constituents of the cell membrane. After identification and attachment to the cell, NPs can cross the cell membrane, reach the desired site and deliver the payload (Figure 2A) (Patra et al. 2018, Raza et al. 2019a, Tayo 2017). These NPs accumulate in the targeted site, enhance the therapeutic efficiency and reduce side effects from the loaded drugs (Majumder and Minko 2020). For example, hyaluronic acid (HA)—a polysaccharide found in the extracellular matrix—is a well-used ligand that has demonstrated encouraging outcomes to enhance the antitumor activity against several cancers like breast cancer, melanoma stem-like, and lung adenocarcinoma, together with reduction of protein corona development and assisting DDS for retinal gene treatment (Gao et al. 2017, Shen et al. 2015, Wang et al. 2017). NPs can encounter several barriers inside the organism that prevent effective delivery of the therapeutic agent.

A)

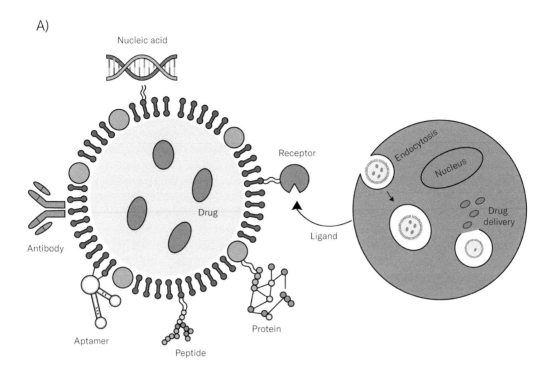

B)

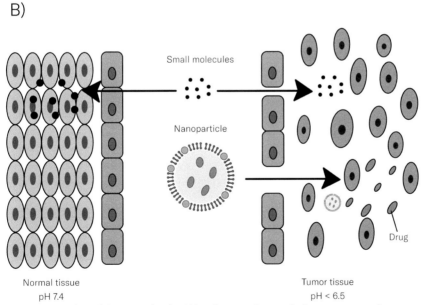

Figure 2. Representation of drug targeting by (A) active targeting employing overexpressed receptor structures of the malignant cells and (B) passive targeting taking advantage of the enhanced permeation and retention effect of tumors.

Non-specific binding to non-targeted cells is one common problem. One solution is conjugation with polyethylene glycol (PEG) to decrease NP interactions with blood cells and reduce undesirable protein adsorption, which is vital for non-fouling NPs in several applications (Tayo 2017).

4. Barriers to the uptake of nanoparticles

NPs have several obstacles once they enter into the bloodstream. They are predisposed to protein opsonization (proteins attach to NP surface as a label for detection from the immune system), reticuloendothelial system (RES), aggregation, and immunogenicity to reach the target site (Sercombe et al. 2015, Yetisgin et al. 2020). For instance, opsonized NPs are released from the bloodstream by filtration from the spleen, kidney, and liver or by phagocytosis. This quick clearance from the body reduces retention time and restricts bioavailability. To circumvent this problem, NPs are coated with PEG, acetyl groups, or other moieties (Shreffler et al. 2019). Additionally, NPs larger than 200 nm are cleared in the RES by phagocytic cells while NPs lesser than 10 nm are cleared by filtration via the kidney. Hence, NPs with < 100 nm have a higher circulation period. Further NPs between 20–200 nm exhibit more accumulation in tumors since they are not identified by the RES or filtrate by physiological systems (Bhatia 2016). Furthermore, the surface charge of the NPs has a key function in their clearance. NPs with a positive surface charge produce a greater immune response compared to negative or neutral NPs. Surface potentials from –10 to +10 mV exhibit less vulnerability to non-specific interactions and phagocytosis (Ernsting et al. 2013). Polymers like PEG, poly-lactic acid (PLA), chitosan, and poly-L-lysine reduce the toxicity of cationic NPs (Patra et al. 2018).

Once the NPs have reached the target area, they can be internalized by cellular uptake as endocytic processes like vesicular internalization, non-phagocytic, and phagocytic pathways. The caveolar- and clathrin-mediated endocytosis and micropinocytosis are commonly studied cellular uptake mechanisms. After cellular uptake, the particles are engulfed by the cell membrane and are channeled to the intracellular endocytic pathway including endosomes and lysosomes. If the target is the nucleus, the NP shape, ligand coating, surface charge, and size need to be well established to guide the NP across the nuclear membrane and achieve the internalization into the nucleus (Doherty and McMahon 2009, Patra et al. 2018, Zhao et al. 2011). Further, nuclear targeting can be realized using TAT peptides conjugated with silica nanoparticles for the delivery of doxorubicin (Pan et al. 2012). TAT peptides can also be conjugated with gold nanoparticles as well as different kinds of polymeric nanoparticles for effective targeting towards the nucleus (Tkachenko et al. 2003). Also, once the NPs have arrived at the targeted location either via passive or active targeting, NPs can experience poor release of drugs into the target site. To overcome this challenge, stimuli-responsive NPs have been developed to provide guided, more efficient and controlled delivery upon several stimuli (Majumder and Minko 2020).

5. Endogenous and exogenous stimuli

Stimuli-responsive NPs stimulate and control the distribution of the payload and decrease adverse effects by shifting the hydrophobic and hydrophilic equilibrium of the NPs structure or by cleaving disulfide bonds offering several biomedical applications (Tayo 2017). These stimuli are either endogenous (internal or biological) or applied externally (Table 1). Stimuli-responsive NPs are

Table 1. Endogenous and Exogenous stimuli.

Endogenous stimuli	Exogenous stimuli
pH-responsive	Temperature-responsive
Redox-responsive	Light-responsive
Enzyme-responsive	Ultrasound-responsive
Temperature-responsive	Magnetic-responsive
Ionic microenvironment-responsive	Electrical-responsive

usually created with hydrophilic or amphiphilic exterior shell and a hydrophobic interior core. The shell is normally an endogenous or exogenous susceptible stimuli-responsive polymer (Majumder and Minko 2020).

6. Endogenous stimuli

Endogenous stimuli are specific for diseased tissue and are useful for boosting drug or agent specificity. This involves pH, temperature, high quantities of enzymes, and increased concentration of GSH due to oxidative stress (Li et al. 2018). The NPs are designed to respond to the endogenous stimuli, thus guiding towards the destruction of the NP structure and ultimately delivering the encapsulated or attached agent (Rasheed et al. 2018). For instance, tumors are considered to have atypical changes in its microenvironment (TME) such as temperature, pH, and oxygen and enzyme levels (Yokoyama 2002). These features can be utilized to stimulate the NPs to disintegrate and released their cargo at the chosen location (Pärnaste et al. 2017).

6.1 pH-responsive NPs

NPs can be coated with polymers that are sensitive to pH changes since they have ionizable pendant groups attached to the hydrophobic chain of the polymer. Thus, they conjugate a category of polyelectrolytes with ionic functional bases (pyridine and amines) that undergoes conformational change with variations in the pH or there are acidic sensitive clusters such as sulfonate and carboxylic acids that break in acidic pH (Deirram et al. 2019, Manchun et al. 2012, Pang et al. 2016). They either donate or accept protons as a consequence of the switching pH (Liu et al. 2017). As a result of these electrostatic alterations in the charges, the polymeric chain is changed and the NP is disrupted. When the net charge of the chain increases, it produces electrostatic repulsion of the chain, creating an alteration in the system from collapsed to expanded state, while when the net charge diminishes it goes from expanded to collapsed state (Tayo 2017). This pH responsive strategy as well as the targeting moieties attached to the NPs helps in the delivery of the payload in the injured tissues as well as in the cell interior. The reason behind this release is that there is minor alteration in the pH of damaged tissues in comparison to the normal ones, which stimulates the pH responsive polymers to release the cargo, especially in the case of chemotherapeutic drug release in tumor microenvironment (Figure 3) (Colson and Grinstaff 2012). The pH of intracellular and extracellular matrix is also different, aiding for more specific targets (Wu et al. 2018).

Two protocols are used for creating NPs with pH-responsiveness like integrating acid-functionalized groups to the polymeric backbone or linking active drugs to the lateral chain of the polymer. These active drugs have the option to begin the pH-responsiveness, making several conformational variations in the polymeric support altering the NPs structure. Numerous polymers have been employed for fabricating pH-responsive NPs such as acrylic acid, PLA, polyketals, acrylonitrile, polycarbonates, polyanhydrides, methacrylic acid, and polycaprolactone (PCL) (Chen et al. 2010, Zhang et al. 2007). These polymers are involved partially or completely in initiating changes in the hydrophilic character as a response to pH-stimuli leading to the destruction of the NP (Raza et al. 2019a). For example, healthy tissues and malignant tissues have different pH values. TME and inflammation sites have a pH of 6.5 to 7.2 as compared to 7.4 of normal tissues. The TME has lower pH since acidic metabolites are generated from the hypoxic surroundings, thus stimulating NPs to release the anti-cancer drugs towards the target site (Figure 3) (Tayo et al. 2015). Additionally, pH-responsive liposomes are applied against Hepatitis-B virus to realize cytosolic cargo delivery (Wu et al. 2018).

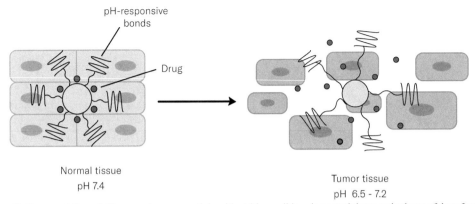

Figure 3. Representation of pH-responsive nanoparticle with stable conditions in normal tissue and release of drug from the nanoparticle at lower pH in tumor tissue.

6.2 Redox-responsive based drug delivery

Redox potential gradients exist around extra- and intracellular surroundings of diseased tissue. The concentration of the reducing agent GSH is two times higher in the nucleus and cytosol than in the extracellular space or endosomes (Cheng et al. 2011, Meng et al. 2009, Schafer and Buettner 2001, Son et al. 2012). The redox potential around intracellular (reductive) and extracellular (oxidative) space is associated with the GSH concentration (Schafer and Buettner 2001). NPs redox-responsiveness is helpful when the payload is targeted inside the cell and is more advantageous for DNA or siRNA distribution (Fleige et al. 2012, Mintzer and Simanek 2009). Moreover, higher amounts of oxidizing agents such as superoxide anions and hydrogen peroxide are the result of cancer cell proliferation (Raza et al. 2019a). One goal is to create glutathione-responsive NPs with acid-labile groups (Cheng et al. 2011). This includes disulfide linkers in the shell of the NP that can be reduced and cleaved to release the drug or gene at the target site within the cell (Cheng et al. 2008, Meng et al. 2009, Son et al. 2012). The disulfide connection in glutathione-responsive block copolymers among hydrophobic and hydrophilic parts generates a micelle configuration called "shell-shedable". At the same time, when these micelles enter the circulation in the extracellular milieu, the disulfide linkers of the NPs are undamaged, but when they enter into the cytosol, following endocytosis inside the malignant cells, the disulfide linkers are reduced to thiol groups and cleaved due to the elevated levels of GSH, leading to the destruction of the NP and eliciting the cytosolic discharge of the payload (Figure 4) (Cheng et al. 2011, Majumder and Minko 2020, Meng et al. 2009). Since tumors have higher GSH values than healthy tissue, NPs with cleavable disulfide bonds can be utilized to target tumors with redox-responsive systems and distribute anti-cancer drugs like trastuzumab, doxorubicin (DOX), and p53 tumor suppressor gene (Kumar et al. 2017, Son et al. 2010).

6.3 Enzyme-responsive NPs

The biocatalytic activity of enzymes present within diseased cells can also be utilized for targeted delivery by creating enzyme-responsive NPs (Rasheed et al. 2018). In every metabolic and biological activity, the dysregulation or regulation of enzymes in the microenvironment at intracellular level have main roles in responsive DDS since they offer advantageous and exclusive characteristics (Hu et al. 2013, Rasheed et al. 2018). The payload is delivered when the polymeric moiety is destroyed by enzymes (Raza et al. 2019a). Glycosidases, phospholipases, and proteases are of high significance because they are regularly overexpressed in malignant tissue as inflammation and cancer (Figure 5) (Basel et al. 2011, Majumder and Minko 2020). Trypsin is one of the main digestive proteinases and has a key role in exciting numerous digestive enzymes. Radhakrishan

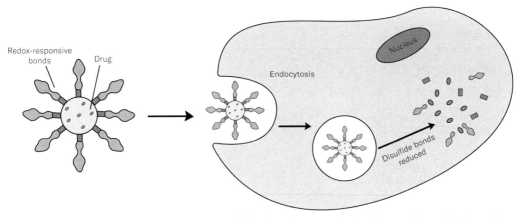

Figure 4. Internalization of redox-responsive nanoparticle. The redox-responsive bonds are stable outside the malignant cells with low levels of glutathione. Once the nanoparticle enters the malignant cell by endocytosis, the levels of glutathione rise, reducing the disulfide bonds to thiol groups and releasing the drug from the nanoparticle.

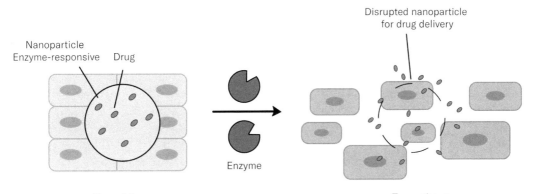

Figure 5. Enzyme-responsive nanoparticle in normal state before enzyme action and disrupted nanoparticle with the enzyme sensitive layer releasing the drug after the enzyme action.

and co-workers (2014) created a nanocapsule for delivering anticancer drugs inside the malignant cells that can be disrupted with the presence of trypsin and deliver the cargo (Radhakrishnan et al. 2014). Moreover, another therapeutic target involved in Alzheimer's disease are oxidoreductases due to their key function in oxidative stress and in cancerous tissues is the extracellular proteolytic enzymes called matrix metalloproteinases which are overexpressed (Eskandari et al. 2019, Kundu and Surh 2010, Shay et al. 2015). In addition, neutrophils are the first immune cells to arrive at the inflammatory sites and secrete a protease called human neutrophil elastase. This protease is employed for controlled DDS and enhances the targeted anti-inflammatory therapies (Aimetti et al. 2009, Korkmaz et al. 2010, Rosales 2018).

6.4 Temperature-responsive NPs

Polymers susceptible to temperature become totally miscible with water/solvent above or below the upper critical solution temperature (UCST) or lower critical solution temperature (LCST), respectively (Bordat et al. 2019, Tayo 2017). These NPs are created to survive at high temperatures by releasing the cargo and elevated temperatures are natural in tumor or diseased tissue areas (Khoee and Karimi 2018). For instance, healthy and damaged tissues have a narrow difference

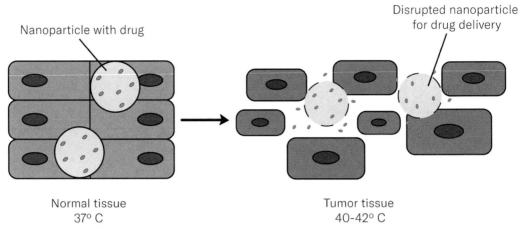

Figure 6. Representation of disrupted temperature-responsive nanoparticle by internal temperature stimulus releasing the drug at tumor tissue areas.

of temperatures. This provides the opportunity to design NPs with a payload that remain stable under normal body temperature and then switch precisely and rapidly their hydrophilicity and hydrophobicity equilibrium to collapse and deliver the agents or genes within the injured tissues in response to the higher temperature (Figure 6) (Cheng et al. 2008, Chilkoti et al. 2002, Liu et al. 2017, Schmaljohann, 2006, Zhang et al. 2005a). For example, temperature-responsive NPs had more delivery of drugs in diseased tissue with temperatures around 40°C as compared to normal tissue at 37°C, thus exhibiting decreased cell viability in cell cytotoxicity assays (Khoee and Karimi 2018).

6.5 *Ionic microenvironment-responsive NPs*

Ionic microenvironment-responsive NPs are created with suspended basic or acidic functional units in the polymer structure, highly manipulating the degree of ionization. Therefore, from the total number of units the NP has, the degree of ionization will depend on releasing the payload (Raza et al. 2019a, Zhang et al. 2005b). An elevated number of acidic entities in the polymeric vehicle enhances the electrostatic repulsion among carboxyl groups with negative charge of diverse chains causing more swelling at a higher pH. Conversely, a high quantity of basic entities like amines can ionize and at low pH exhibit electrostatic repulsion (Zhang et al. 2005b). Thus, these responsive polymeric NPs can either donate or accept protons as a response to ionic strength changes, thus activating the drug delivery. Polymers with LCST changes are more attractive for fabricating ionic microenvironment-responsive NPs (Furyk et al. 2006).

7. Exogenous stimuli

Exogenous stimuli are physical phenomena that depend upon the external milieu of the target area once the loaded NPs reach the diseased site. These stimuli can involve magnetic field, electric field, ultrasound, temperature, or light and are accountable for alterations or destruction of the NP structure leading to the cargo release (Hu et al. 2013, Liu et al. 2017, Rasheed et al. 2018). Hence, in these procedures the outer factors control exactly the payload delivery and can be applied to specific areas in the body to activate the distribution of the drug. Most of these stimuli lead to the production of heat; thus, the design of the thermo-responsive polymeric NPs are of great interest (Raza et al. 2019a, 2019b).

7.1 Temperature-responsive DDS

Temperature-responsive DDS maintain the agent or drug in normal temperature and upon exposure to elevated temperatures, the cargo is released (Liu et al. 2016). DDS are created to respond to external stimuli over the desired site. This rise of temperature stimulates the NP to generate heat and modify the thermo-sensitive material of the DDS in such a way that it ruptures and delivers the cargos at the desired site. This procedure is normally applied for inducing hyperthermia for thermal-based treatments with chemotherapy at the same time. Light, magnetic field, and ultrasound are usually employed for this purpose because they show an accurate control and production viability (Figure 7) (Raza et al. 2019a, Yang et al. 2018). The temperature-responsive polymers, LCST and UCST, respond to variation of temperature. At lower temperatures below LCST, the polymers are accountable for higher swelling due to increase in hydrophilicity. Furthermore, an increase in hydrophilicity is done by a rise of temperature beyond UCST. These changes in DDS hydrophilicity swelling leads to the payload release (Karimi et al. 2015). For example, micelles loaded with DOX and camptothecin grafted with polypyrrole were exposed to the near infrared region (NIR). The absorption produced heat due to photo-thermal influence raising the temperature and stimulating the cargo delivery from the micelles. This heat created an enlargement of temperature-responsive polymer by conversion from hydrophobic to hydrophilic (Yang et al. 2018).

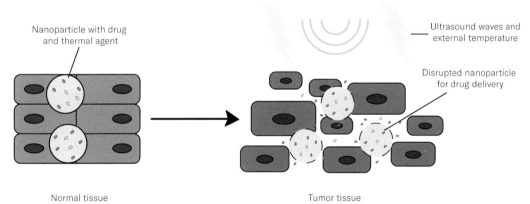

Figure 7. Representation of temperature-responsive nanoparticle with thermal agents in normal tissue and activated by external temperature stimulus disrupting the nanoparticle and releasing the drug at tumor tissue areas.

7.2 Light-responsive NPs

Light-responsive NPs are used in pharmaceutical and biomedical studies to accomplish drug delivery as they have the advantage of superior spatiotemporal control (Lino and Ferreira 2018). Several wavelengths like ultraviolet (UV), visible, and NIR are applied for regulating the cargo release (Figure 8) (Hossion et al. 2013). Nevertheless, due to the low penetration of UV and visible light, they are not considered for therapy purposes *in vivo*, whereas NIR is a valuable light source since it infiltrates deep into tissues being less destructive, with more penetration, and more secure than the UV light to the organism (Saravanakumar and Kim 2014, Xiang et al. 2018). One strategy for constructing a light sensitive NP is by including adequate photochromic functional clusters in block copolymers that switch hydrophobic and hydrophilic equilibrium in the micellar structure once the NP is exposed to the light wavelengths (Babin et al. 2009, Zhao 2012). Photochromic moieties are usually light-responsive polymeric nanomaterials. When light hits the moieties, they experience photochemical changes like photocleavage, photoisomerization, and photodimerization, eventually leading to the disruption of the NP, thus releasing the payload. This light source is easily operated

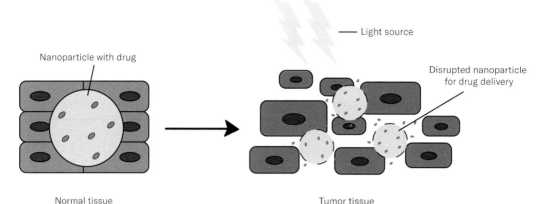

Figure 8. Representation of nanoparticle in stable condition in normal tissue and activated by external lights at tumor areas and disrupting the nanoparticle.

from the exterior of the organism (Fomina et al. 2012, Tomatsu et al. 2011). It is also considerable to use surface plasmon and additional photothermal effects to activate the delivery of the cargo in the desired cells (Zhao 2007). Moreover, the plasmon absorption of NPs can be adjusted for activating them at any required wavelength between UV and NIR region (Tayo 2017). Three approaches exist for drug delivery with NIR:

- Upconverting NPs is a method to initiate elevated energy light-sensitive constituents by the NIR light and transform NIR light to UV or visible light. This method is used for new fluorescence labels for imaging and sensitive bioanalysis. Also, it is beneficial for photopolymerization since after the conversion of NIR to UV or visible light, it activates a cascade of polymerizations (Gwon et al. 2018).

- Two-photon activation: at the same time two identical or different frequency photons excite a particle from its ground state to a greater energy state. This approach needs an excited light source to center small areas to obtain effective instantaneous energy. This technique has been used for imaging and DDS NIR-responsive unit fabrication for a two-photon of NIR absorption with two-photon NIR sensitizing agent (two-photon in NIR equal to one-photon of UV) (Yang et al. 2016b).

- Photo-thermal effect: this technique involves the conversion of light into heat by photo-thermal agents stimulating heat sensitive material, thus disrupting the NPs' structure guiding the drug release. The photo-thermal agent such as IR780 on exposure to a particular wavelength of light transforms into heat, thus weakening the NP platform (Li et al. 2017, Raza et al. 2019).

7.3 Ultrasound-responsive NPs

Ultrasound (US) waves are reviewed as external stimuli since they are generated by radiation forces as well as mechanical and thermal effects, which are accountable for stimulating the DDS for delivering the cargo (Figure 9) (Luo et al. 2017, Paris et al. 2015). Elevated US exposure generates microbubbles that collapse quickly at inertial cavitation boosting the absorbency of the genes or drug delivery into the malignant tissue (Yan et al. 2013). The drawbacks of microbubbles are their big dimensions and small circulation period reducing the clinical uses (Tayo 2017). For instance, US was employed for heating PEGylated mesoporous silica NPs (MSN) with a thermo-responsive linker. The linker was cleaved after 24 hours of US waves treatment guiding the DDS and exhibiting a positive charge on the MSN, thus rising the cellular uptake. This MSN displayed around 50% decrease in cell viability of human osteosarcoma compared with no US exposure (Aryal et al. 2010).

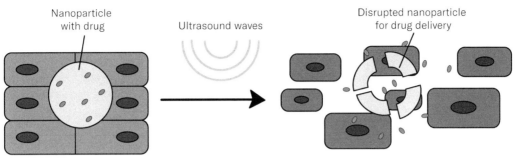

Figure 9. Representation of ultrasound-responsive nanoparticle stable in normal tissue and stimulated with heat and ultrasound vibrations waves at the tumor area for disrupting the nanoparticle and releasing the drug.

7.4 *Magnetic-responsive NPs*

Magnetic field stimuli control the magnetic-responsive DDS since it infiltrates into the body tissue and is employed for magnetic resonance imaging (MRI) (Raza et al. 2019a). This stimulus guides the drug to its target and induces hyperthermia for delivering the payload (Figure 10) (Schleich et al. 2015, Thirunavukkarasu et al. 2018). Hyperthermia-based magnetic NPs are studied for drug delivery purposes. Local hyperthermia gives the possibility of magnetic based imaging due to it being a contrast agent and finally leading to tumor inhibition (Zhou et al. 2018). Thirunavukkarasu and colleagues (2018) showed a difference between the DOX release at different temperatures loaded onto a superparamagnetic iron oxide (Fe_3O_4) nanoparticle (SPION) with temperature sensitive poly-lactic-co-glycolic acid (PGLA) base and the drug alone. The *in vitro* studies showed ~ 39% of drug distribution at 37°C as compared to ~ 57% at 45°C. The temperature of the medium was raised to 5.2°C by employing an alternating magnetic field (AMF) at 4.4 kW displaying an important tumor inhibition besides being an *in vivo* MRI contrast agent (Thirunavukkarasu et al. 2018). Therefore, SPIONs are injected close to the target area and when magnetic fields are applied at the target location, the magnetic force will guide the SPIONs to gather and deliver the cargo (McBain et al. 2008).

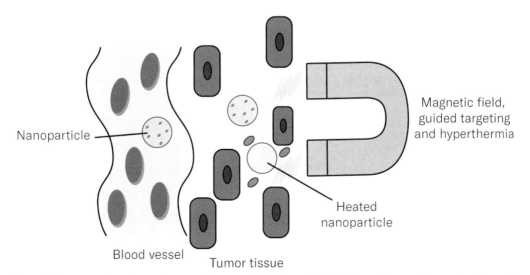

Figure 10. Representation of magnetic-responsive nanoparticle being guided to the tumor tissue area by an external magnetic field and heated to produce magnetic hyperthermia and release the drug.

7.5 Electrical-responsive NPs

After electro-responsive NPs uptake, weak electrical field is employed over the desired area achieving a coordinated on-site drug delivery. Several processes for releasing the payload with electrical stimulation are known such as destroying NPs structure, redox reactions, and incentive of temperature-responsive DDS via generated heat by electrical impulses (Ge et al. 2012, Jeon et al. 2011, Servant et al. 2013). Finally, each stimuli-responsive NP has several challenges and advantages as exhibited in Table 2.

Table 2. Summary of stimuli-responsive nanoparticles' advantages and limitations (Hu et al. 2013, Paris et al. 2015, Rasheed et al. 2018, Su and Kang 2020).

Stimuli-type	Stimuli-responsive	Advantages	Disadvantages
Endogenous	Temperature	• Adjustable drug delivery	• Restricted to only identified target localization • Gathering and clearance in spleen and liver
	Redox	• Localized and quick reply for drug release	• Rapid oxidization and clearance • Restricted specificity to certain microenvironments
	Enzyme	• Localized cargo release • Elevated therapeutic efficiency • Selectivity and sensitivity • Biorecognition • Catalytic efficiency for biomedicine	• Several enzymes with similar cleavage sites • Challenging enzyme-specific substrate creation • Drug delivery less manageable
	pH	• Extended activity • Localized release • Less adverse effects	• Restricted specificity to certain microenvironments • Quick clearance • Uncontrolled payload loading
Exogenous	Temperature	• Adjustable drug delivery	• Restricted to only identified target localization • Gathering and clearance in spleen and liver
	Light	• Elevated efficacy in cargo delivery • Less leakage in the lack of a stimuli	• By-products' adverse effects • Ultraviolet light DNA damages • Possible several administrations of stimuli • Restricted to only identified target localization
	Ultrasound	• On-demand cargo delivery • Non-invasive • Elevated cargo accumulation • Tissue penetration • More safety • Spatiotemporal control	• Restricted to only identified target localization • Creation of free radicals intracellularly as a response to the stimuli • Mechanical induced cell injury
	Magnetic	• Elevated therapeutic efficiency	• High energy spending • Lower therapeutic efficiency • Restricted to only identified target localization • Problematic magnetic nanoparticles manipulation delivery • Target location has to be close to the surface of magnetic force application

8. Organic and inorganic nanoparticles

NPs' size makes them suitable for several agents to overcome biological obstacles and move them to the desired target area, more bioavailability is achieved for poor soluble drugs, and an improved protection from destruction is accomplished. Also, NPs surface can be modified for their effective applications in several medical services, particularly with targeted therapy. These changes functionalize and stabilize the NPs improving their efficiency and letting them to be responsive to internal or external stimuli. Therefore, diverse clinical needs demand different DDS to encounter the particular necessities. These NPs are classified by their structures and compositions into organic and inorganic NPs (Figure 11) (Su and Kang 2020).

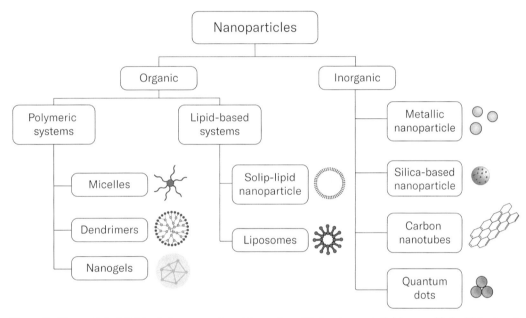

Figure 11. Nanoparticles' division between organic and inorganic and their subsequent divisions used in health treatments and therapeutic applications.

8.1 Organic nanoparticles

Organic NPs are perfect for attaching or encapsulating drugs, genes, or imaging probes and supplying the payload in an organized process into the desired target area. Organic NPs are divided into polymeric systems including micelles, dendrimers, and nanogels and lipid-based systems comprising solid lipid NPs (SLN) and liposomes. The majority of approved NPs by the Food and Drug Administration (FDA) for clinical therapies and treatments are organic NPs (Jahangirian et al. 2017, Xing et al. 2016).

8.1.1 Lipid-based systems

Solid lipid nanoparticles

Solid lipid NPs are aqueous colloidal dispersions of 10 to 100 nm size contained in a solid lipid matrix at body and room temperature. The drug release characteristics are affected by the type of lipid while the biodegradable and biocompatible surfactants enhance their stability (de Jesus and Zuhorn 2015, Majumder and Minko 2020). SLN are multipurpose drug vehicles encapsulating high numbers of hydrophilic, hydrophobic, and lipophilic drugs and nucleic acids (Dolatabadi et al.

2015, Majumder and Minko 2020). SLNs are normally fabricated with an elevated configuration of liquid lipids heated higher than their melting point dissolving the drug in the liquefied lipid. This mixture of drug with melted lipid is distributed in the aqueous phase and heated for achieving a homogenized mixture. Moreover, SLNs can be coated with different moieties to control the target delivery and stimuli-responsive drug discharge exhibiting success in releasing therapeutic agents via pulmonary, intranasal, and oral administration. SLNs increase the oral bioavailability by 3.5- and 11-folds of glibenclamide and efavirenz, correspondingly, compared to drug suspensions (Battaglia et al. 2016, Elbahwy et al. 2017, Gaur et al. 2014, Shen et al. 2015). Likewise, SLNs are employed for topical and ocular drug administration because the benefits of a higher absorption of the NP is achieved on the biological membranes. This way of administration is important since it elevates the therapeutic efficiency by accumulating the cargo locally and decreasing the adverse effects produced by systemic circulation (Mu and Holm 2018). Also, SLNs are employed as DDS to enhance the poor stability of siRNAs and reach the desired location for therapeutic activities (Majumder and Minko 2020). Mu and Holm (2018) gives more information about the SLN design and characterization for drug delivery (Mu and Holm 2018).

Liposomes

Liposomes are spheres constituted by a phospholipid bilayer surrounding completely an aqueous core. Big liposomes have a diameter of 200 nm while the small liposomes between 60 and 80 nm (Joshi et al. 2016). The usual procedure for liposome synthesis is thin layer humidification, motorized stirring, solvent dehydration, solvent insertion, and surfactant solubilization (Kotla et al. 2017). They are arranged in different structures, size, and composition with several lipids and surface changes like attaching ligands or PEG to the surface for more stabilization named PEGylation (Suk et al. 2016, Yetisgin et al. 2020). Moreover, agents are incorporated into the hollow core of the liposome after formation (active) or during formation (passive). These features facilitate the therapeutic treatment since the hydrophilic drugs (e.g., 5-fluro-deoxyuridine and ampicillin) are normally loaded into the aqueous core and the hydrophobic drugs (e.g., amphotericin B and indomethacin) in the hydrocarbon chain, thus increasing their biodistribution (Akbarzadeh et al. 2013, Kotla et al. 2017, Patra et al. 2018). Additionally, several kinds of agents can be incorporated either in numerous aqueous layers or within lipid and aqueous compartments (Figure 12). This enables the delivery of several cargos in sequence with separation of layer among outer shell and inner core (Patil and Jadhav 2014). Nevertheless, payload inside the liposomes are not accessible until they are delivered; hence, it is important to accumulate numerous liposomes in the target site to increase bioavailability (Akbarzadeh et al. 2013).

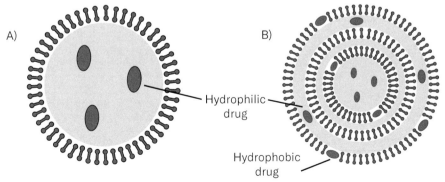

Figure 12. Liposomes used for therapeutic treatments with hydrophilic drugs loaded into the aqueous core and hydrophobic drugs in the hydrocarbon chain. Also, liposomes can be incorporated either in aqueous layers or within lipid and aqueous compartments.

These NPs are noteworthy DDS as the inner aqueous core gives protection to the payload from the surrounding environment and due to the similarity between liposome and cell membrane, it enables the assimilation of the cargo into the cytoplasm (Bozzuto and Molinari 2015, Elegbede et al. 2008). For example, different liposomal shapes like ellipsoids, spheres, and cylinders have been employed with the drug, Doxorubicin (Sumetpipat and Baowan 2014). The first FDA approved NP for therapy were liposomes from lipid systems and micelles. Doxil, a doxorubicin-based liposome, was the initial FDA accepted DDS for treating breast and ovarian cancer and Kaposi's sarcoma (Bulbake et al. 2017, Patra et al. 2018). Furthermore, liposomes have already been combined with numerous stimulators like pH, light, ultrasound, temperature, and enzymatic responses and imaging agents along the target ligand to enhance the efficacy while having theranostic properties (Sercombe et al. 2015, Tayo 2017). Some problems with liposomes moving in the bloodstream are its encounter with low- and high-density lipoproteins reducing their stability and attracting opsonins aiding the RES on identifying and destroying liposomes, but they also have several advantages (Table 3) (Patra et al. 2018).

Polymeric systems

Polymeric nanomaterials have achieved more consideration for drug and gene delivery due to different advantages (Table 3) (Crucho and Barros 2017). Several NPs enter in this category such as nanogels and micelles and their toxicity varies according to the polymer used during the synthesis. Several natural and synthetic polymers are widely employed for NPs' production. Natural polymers like alginate, gelatin, albumin, and chitosan offer an important enhancement in the agent's efficiency and overcome toxic problems. On the other hand, to reduce toxicity and immunogenicity of synthetic polymers as PEG, polyvinyl alcohol, PCL, PGLA, and PLA, the polyester forms are employed (Letchford et al. 2009, Patra et al. 2018). The biodegradability and biocompatibility of polymeric systems depends on which polymer is utilized during synthesis (Patra et al. 2018). Polymeric NPs are divided in two categories according to their composition: (i) nanocapsules have an exclusive polymer membrane that surrounds the therapeutic drug and (ii) nanospheres have a polymeric matrix in which the therapeutic drug is directly distributed throughout the matrix. The size of these NPs varies between 10 and 200 nm for biomedical objectives (Letchford et al. 2009, Patra et al. 2018). Furthermore, these NPs' surface can be changed and functionalized with a recognition ligand increasing the specificity of polymeric NPs for the desired location (Yetisgin et al. 2020). For example, chitosan-based NPs merged with FA and alginate are good carriers of 5-aminolevulinic acid aiding for diagnosis of colorectal cancer and allowing endoscopic fluorescent detection (Yang et al. 2011).

Micelles

Micelles are built of amphiphilic-block copolymers showing a hydrophobic core and a hydrophilic exterior shell, supplying safety from non-specific uptake by the RES guaranteeing more stability (Ahmad et al. 2014). These features help transport several essential water insoluble chemotherapeutic drugs like camptothecin, paclitaxel, docetaxel, and tamoxifen towards their target, while at the same time the exterior shell stabilizes the core, making the entire arrangement water soluble (Maeda 2010, Patra et al. 2018). Micelles are synthesized via precipitation of one block by complementing with a solvent or solvent-based straight ending of polymer trailed by a dialysis process (Miyata et al. 2011, Xu et al. 2013). The size of the micelles varies from 10 to 100 nm, which also increases the capability of taking the advantage of the EPR effect, accumulation in the tumor site, and usually possess a limited dissemination to evade quick renal excretion, thus increasing its tumor tissue concentration (Maeda 2010, Patra et al. 2018). Moreover, surface modification of the micelles aids in protecting the agent or drug from being destroyed by the gastrointestinal system creating an

Table 3. Summary of advantage and disadvantages of organic and inorganic nanoparticles.

Nanoparticle	Nanoparticle type	Advantages	Disadvantages	References
Organic	Lipid-based systems	• Biocompatible • Easy modifiable • Efficient for big payloads carriage • Good safety profile • Tendency to assemble in spleen and liver increasing the concentration and aiding pathogenic disease therapies	• Quick clearance from the RES • Off-target assembly	(Patra et al. 2018, Sercombe et al. 2015, Su and Kang 2020)
	Polymeric systems	• Non-toxic • Biocompatible • Biodegradable • Non-immunogenic • Controlled drug delivery • Enhance drug solubilization • Ease of preparation • Uniformity • Tunable size • Good safety profile	• Positively charged dendrimers or dendrimers with cationic amine groups turns them toxic being an option for a fast clearance and a short circulation period, thus, they are normally modified to decrease this toxicity • Polymer conjugated with proteins decreases the proteins' bioactivity • Nanogels' deformity affects loading capacity, target affinity, and swellability	(Crucho and Barros 2017, Ekladious et al. 2019, Myerson et al. 2019, Su and Kang 2020, Tripathy and Das 2013, Zhu et al. 2019)
Inorganic	Metallic	• Valuable contrast agents • Biocompatibility • Elevated cargo capacity	• Size-dependent toxic effects • Possible bioaccumulation • Simple iron oxide agglomerate and suffer clearance from the RES • Generation of free radicals	(Ali et al. 2016, Han et al. 2007, Patra et al. 2007, Su and Kang 2020)
	Silica-based	• Cost-effective • Complex systems • Thermal stability • Changeable pore size • Co-deliver several agents • Flexible and versatile	• Potential immunogenicity and bioaccumulation in organs	(Chen et al. 2018, Majumder and Minko 2020, Tang et al. 2012)
	Carbon nanotubes	• Easily adopted by cells • Unique thermal and structural properties • Biocompatibility	• Induces oxidative stress in cells • Close link with inflammation, cancer, and fibrosis • Surface charge toxicity	(Dong and Ma 2019, Majumder and Minko 2020, Mohanta et al. 2019, Narei et al. 2018)
	Quantum dots	• Good optical, photoluminescence, and electronic properties • Broad excitation • Elevated brightness • Spectral tunability • Narrow emission spectra • High quantum yield	• Complicated surface chemistry • Moderately large size • Toxicity	(Bilan et al. 2016, Majumder and Minko 2020, Matea et al. 2017)

option for oral distribution of protein drugs like insulin (Veiseh et al. 2014). Drugs are incorporated into polymeric micelles by three processes (Mourya et al. 2011):

- Dialysis: the solution of copolymer and the drug in organic solvent, located in the organic solvent, is merged within a dialysis bag and submerged into water. Then the solvent is exchanged with water, creating the micelle assembly with the drug in solution in a dialysis bag and is dialyzed afterwards.

- Direct dissolution: in the water medium, the drug and amphiphilic copolymer merge together by themselves forming drug-loaded micelles. This method is generally linked with low drug loading.

- Solvent evaporation: a volatile organic solvent dissolves the drug and copolymer, and a thin film of both products is obtained after removing the solvent by evaporation. The polymeric micelles loaded with the drug is the result of the reconstitution of the film with water. Both the cargo and copolymer are dissolved using an organic solvent.

8.1.2 Nanogels

Nanogels are mainly hydrophilic and biocompatible polymeric networks with a diameter smaller than 100 nm produced by electrostatic interactions, covalent or hydrogen bonding, or Van der Waals forces (Dumville et al. 2011, Oh et al. 2008, Skulason et al. 2012). Nanogels possess high hydrophilicity and swelling properties with adaptable size as well as elevated water content permitting them to encapsulate huge quantities of water in aqueous solution (Dumville et al. 2011, Skulason et al. 2012, Tahara and Akiyoshi 2015). They are good DDS options since they entrap drugs, proteins, cytokines, vaccines, and DNA or RNA, reduce drug escape, and are easily administered via mucosal or parenteral ways (Neamtu et al. 2017). In addition, nanogels deliver the payload by passive diffusional discharge or the nanogels' surface structure can be modified for active targeting (Dumville et al. 2011, Skulason et al. 2012).

8.1.3 Dendrimers (den)

Dendrimers are complex, large, three-dimensional, and highly monodispersed NPs. They are synthesized via a repetitive sequence of chemical reactions controlling the size regulated by each generation either by initiating the structure from the core and going to the outside or initiating from the outside of the dendrimer (Figure 13) (Cheng et al. 2011, Paleos et al. 2010). Their usual

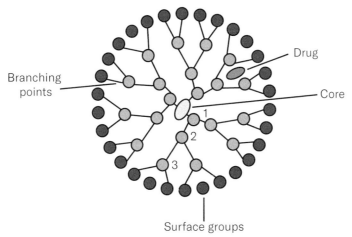

Figure 13. Representation of dendrimer's structure.

size is between 1 and 10 nm and are produced by polymerization in spherical structure guiding the formation of cavities inside the den. These inner core cavities are hydrophobic in nature, thus aiding hydrophobic drug delivery, and possess the capability of entrapping high molecular weight drugs, aiding gene treatments (Chauhan 2018, Pavan et al. 2010, Yetisgin et al. 2020). Three ways exist for loading the payload into the den: covalent conjugation, normal encapsulation, or static interaction (Tripathy and Das 2013). Additionally, den have free end groups, which can be changed or employed to increase the target specificity and distribution of the drug (Yetisgin et al. 2020). den release the cargo by *in vivo* destruction of covalent bonding of the drug due to appropriate enzymes, suitable environment, or variations in it facilitating the administration of the agent (Tripathy and Das 2013). Furthermore, several categories of den exist such as poly-propyleneimine (PPI), core-shell, peptide, chitin, chiral, polyethyleneimine, liquid crystalline, and poly-amidoamine (PAMAM). PAMAM is the most used dDen for biomedical applications such as oral, ocular, and transdermal drug delivery (Kesharwani et al. 2014, Noriega-Luna et al. 2014, Tayo 2017). Moreover, PAMAM is employed for downregulating multidrug resistance in cancer by incorporating siRNAs-related protein (Pan et al. 2019).

8.2 Inorganic Nanoparticles

Inorganic NPs have several intrinsic features like optical, mechanical, electronic, and magnetic properties that are suitable for diagnosis, imaging, and sensing (Dong et al. 2015, Dreaden et al. 2012). Inorganic NPs include gold (Au), silver (Ag), iron oxide, silica NPs, QDs, and CNTs. Nonetheless, only a certain number of inorganic NPs have been recognized for clinical use since the majority remain in clinical trials (Patra et al. 2018).

8.2.1 Metallic nanoparticles

Metallic NPs applied for medical purposes are in the size range of 1–100 nm (Yetisgin et al. 2020). The most reviewed MNPs are Au, Ag, iron (Fe), and copper (Cu), but other types such as palladium (Pd), zinc oxide (ZnO), cerium dioxide (CeO_2), gadolinium (Gd), titanium dioxide (TiO_2), and platinum (Pt) are gaining more attention (Patra et al. 2018). Besides, the surface of MNPs can be modified for decreasing the cytotoxicity of the NPs and enhancing the precision of their targets, biocompatibility, and stability (Tayo 2017, Yetisgin et al. 2020). For example, drugs are physically adsorbed by AuNPs or linked to AuNPs surface through covalent or ionic bonding, delivering them in a stable way via stimuli activation once AuNPs are internalized into cells through receptor-mediated endocytosis (Jia et al. 2017, Kong et al. 2017). AuNPs are also useful for colorimetric biosensing and bioimaging with photoacoustic, two-photon luminescence, and photothermal imaging (Dreaden et al. 2012). Likewise, AuNPs and AgNPs possess a feature, which is known as localized surface plasmon resonance (LSPR) (Patra et al. 2018). When the suitable wavelength of light is applied as an external stimuli, they show photothermal conversion due to the LSPR and generate heat, thus damaging the malignant cells (Yetisgin et al. 2020). In addition, the LSPR properties can be personalized for different application as diagnosis, imaging, photothermal therapy, and electrochemical and optical detection (Singh et al. 2018).

8.2.2 Magnetic nanoparticles

On the other hand, MNPs formed from their oxides like magnetite (Fe_3O_4), maghemite (Fe_2O_3), cobalt ferrite ($CoFe_2O_4$), and chromium dioxide (CrO_2) are interesting since they have the skill of arbitrarily flipping direction of magnetization with the effect of temperature, a characteristic known as superparamagnetism, and they respond to an exterior magnetic field. This magnetic moment provides a robust signal variation in MRI and when they are exposed to an AMF, they generate heat named magnetic hyperthermia, allowing eradication of cancer tumors (Ali et al. 2016, Andocs et al.

2009, Tayo 2017, Yetisgin et al. 2020). Also, these magnetic NPs can be directed to a particular region in the body applying an external magnetic field (Figure 10). A key factor for medical applications is the ratio of stimulated magnetization to the related field called magnetic susceptibility (Cuenca et al. 2006). For instance, SPIONs are extensively used as contrast agents in MRI due to their large magnetic susceptibility (Cuenca et al. 2006). Moreover, the superparamagnetic features enable the stable release of cargo to the target location with a higher accumulation (Fan et al. 2011). Finally, FDA approved ferumoxytol for iron deficiency anemia, blocking tumor development by stimulating pro-inflammatory macrophages polarization (Zanganeh et al. 2016). For more information about magnetic NPs in nanomedicine, check Wu and colleagues' (2019) review (Wu et al. 2019).

8.2.3 Silica-based

Silica-based NPs have several advantages and are an interesting carrier due to their porosity, capacity of functionalization, and surface features, being able to load genes, therapeutic proteins, and drugs (Table 3) (Chen et al. 2018, Tang et al. 2012). Silica NPs have a big surface area wrapped with polar silanol moieties, encouraging water adsorption and enhancing the payload stability (Bharali et al. 2005). This high surface area of silica-based NPs possess a high capacity suitable for loading therapeutic agents and link several ligands such as antibody, peptides, functional groups, among others for active targeting. Moreover, the sealable pores with modifiable size, geometry, and structure enable a homogenous integration of different drugs with diverse properties and sizes (Figure 14) (Majumder and Minko 2020, Tang et al. 2012). Additionally, the pore size is designed to accomplish a continuous release rate and to be covered with stimuli-responsive moieties to increase the drug administration in the targeted location like the supply of the encapsulated β-cyclodextrin in tumor tissue with acidic pH (Mura et al. 2013). Nevertheless, the pore size is highly selective for drug absorption. Silica-based NPs have an open entry for the therapeutic cargo to enter in the body and properly distribute itself in a well-organized fashion. Therefore, these NPs' pore diameter has to be well manufactured and designed for the chosen drug, since molecules superior to the pore diameter will not be efficiently absorbed into the NP (Tang et al. 2012). Tang and associates (2012) possess more information about the synthesis, biocompatibility, and drug delivery of silica-based NPs (Tang et al. 2012).

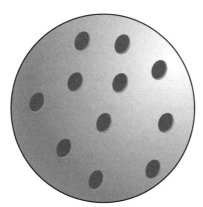

Figure 14. Representation of silica-based nanoparticle.

8.2.4 Carbon nanotubes

Carbon-based nanotubes have tubular structures between 1–100 nm in length and 1 nm in diameter (Figure 11) (Iijima 1991). Carbon nanotubes comprise single-walled nanotubes (SWNTs) (1–2 nm diameter), multi-walled nanotubes (MWNTs) (2 to 6 nm interior and 5 to 20 nm exterior diameter), and C60 fullerenes (Hasnain et al. 2018, Reilly 2007). The firm geometric form and size of carbon nanotubes make them interesting DDS, especially C60 fullerenes and SWNTs, since their inner diameter is equal to half of the average helix diameter of DNA (Reilly 2007). Raw CNTs are hydrophobic; thus, they first need to be modified for biomedical applications and become more biocompatible and water-soluble (Liu et al. 2009). For instance, PEGylation improves the cargo loading size (Liu et al. 2010). A PEGylated CNT with paclitaxel was evaluated in the 4T1 murine breast cancer model that exhibited a higher efficacy in the treatment than paclitaxel alone (Liu et al. 2008). Furthermore, CNTs' functionalization act as vehicles for several therapeutics such as small-molecules, nucleic acids, and peptides, releasing them in several organs depending on the modifications of the CNTs, which can also be responsive to various stimuli (Bianco et al. 2005, Hasnain et al. 2018, Liu et al. 2009). In addition, SWNTs and MWNTs enter the cell by direct insertion or endocytosis via the cell membrane (Yetisgin et al. 2020). CNTs are widely studied for cancer, but they are also studied for avoiding atherosclerosis and crossing the brain-blood barrier (BBB) for neurological diseases (Flores et al. 2020, Gonzalez-Carter et al. 2019).

8.2.5 Quantum dots

Quantum dots have a diameter between 2 and 10 nm with optical properties such as photoluminescence and absorbance, making QDs appropriate for image directed drug delivery purposes. They exhibit an emission in the NIR (< 650 nm) useful for biomedical images (Majumder and Minko 2020). Due to their elevated photo-stability, narrow emission, and bright fluorescence, they are employed for tracking drugs or agents inside the diseased tissue or cell (Bailey et al. 2004, Matea et al. 2017). Besides, achieving multiplex imaging, QDs with the identical light source can give distinct emission colors over a huge spectral range by only changing their core structure, composition, and size (Liu et al. 2010). Moreover, in medicine QDs are reviewed for bioimaging, sensors, and DDS since the exterior aqueous shell is employed for linking the QD with DNA, peptides, or proteins (Iga et al. 2007, Patra et al. 2018). Cai and colleagues (2016) created a responsive ZnO QDs to pH, loaded with DOX, and covered with HA and PEG for targeting HA-receptor CD44 cells. The QDs were secured in physiological pH and DOX was conjugated to PEG and once the QDs entered the TME, the acidic conditions discharged DOX, enhancing the anticancer activity (Cai et al. 2016). Finally, the diverse NPs described previously possess some advantages and limitations as explained in Table 3.

9. Therapeutic applications of nanoparticles

Targeted delivery of therapeutic agents means successfully accumulating the payload in the target location. For this, the cargo needs to elude the immune system, NPs should stay for a small duration of time in the body, and deliver the drug. Approximately 20% of therapeutic DDS under clinical assessment or in clinics were fabricated for cancer, but nanotechnology provides numerous aids for treating chronic human diseases (Patra et al. 2018, Yetisgin et al. 2020). A short list of therapeutic NPs for treatments of several diseases approved by the FDA and European Medicine Agency is shown in Table 4.

9.1 Autoimmune diseases

The two main diseases treated with DDS are acquired immunodeficiency syndrome (AIDS) and rheumatoid arthritis (RA). RA has an unknown cause accountable for the cartilage and bone destruction, affecting nearly 1% of the worldwide population. Therapies for RA need long-term procedures producing systemic side effects (Falconer et al. 2018). DDS release therapeutic agents to inflamed tissue targeted to the synovial membrane. An inhibitor of tumor necrosis factor-α extensively

Table 4. European Medicine Agency (EMA) and Food and Drug Administration (FDA) accepted therapeutic nanoparticles.

Nanoparticle	Purpose	Acceptance	References
Superparamagnetic iron oxide shielded with dextran	Anemia in chronic kidney disease	EMA and FDA	(Bullivant et al. 2013)
Superparamagnetic iron oxide shielded with amino silane	Pancreatic and prostate cancer	EMA	(Weissig et al. 2014)
Polymer-based	Multiple sclerosis	FDA	(Komlosh et al. 2019)
Liposomes	Colorectal and pancreatic cancer	EMA and FDA	(Zhang 2016)
Liposomes	Acute myeloid leukemia	FDA	(Anselmo and Mitragotri 2019)

applied is certolizumab pegol (CZP). CZP combined with PEG increases the half-life and exhibits encouraging outcomes for long-term treatments (Lim et al. 2018). Likewise, liposomal and polymer NPs deliver a target-specific and continuous delivery of anti-HIV drugs for improving the efficacy of the therapies, thus reducing the systemic adverse effects (Herskovitz and Gendelman 2019).

9.2 Cancer

Cancer is one of the biggest causes of death worldwide while chemotherapeutic drugs are the most used treatment for several types of cancer. Nevertheless, these drugs are deficient of aqueous solubility, are non-specific, and toxic to benign cells. Also, cells can become resistant to these drugs which is another important drawback of these drugs (Moorthi et al. 2011, Stavrovskaya 2000). DDS overcome these shortcomings by targeting the cancerous cell directly and delivering the payload in passive or active targeting at a continuous rate increasing the efficiency of the cargo (Mishra et al. 2010). The first DDS for cancer therapy was a PEGylated liposomal formulation of DOX, accomplishing a larger circulation period and avoiding the RES. The size was less than 120 nm, consequently, taking advantage of the EPR effect and gathering in the tumor area and reducing cardiotoxicity (Barenholz 2012). Furthermore, in cancer, theranostic NPs serve to track the location, identifying the stage, and diagnosing the disease plus providing a knowledgeable reaction for therapy (Chen et al. 2014). On the other hand, stimuli-sensitive NPs have also been employed like DOX-based temperature-responsive liposome called thermoDOX and is assessed in phase II breast trials and phase III hepatocellular carcinoma clinical trials (Majumder and Minko 2020).

9.3 Neurodegenerative diseases

Targeted distribution serves for several conditions such as neurodegenerative diseases (NDs) and cancer (Yetisgin et al. 2020). NDs are described through the progressive loss of neuron functions ultimately causing neuronal death. People with ND such as multiple sclerosis (MS), Parkinson's disease (PD), and Alzheimer's disease (AD) have gradual loss of neurons resulting in symptoms related to dementia, memory, and movement (Yetisgin et al. 2020). The current strategies are blocked by the BBB restrictive formation. BBB has an elevated selective semipermeable membrane that allows restricted blood circulation in the brain and inhibits the entrance of almost all the molecules in circulation for keeping the homeostasis in the central nervous system (CNS) (Pardridge 2012). The restrictive formation of the BBB allows only a small number of therapeutic agents to get into the brain; thus, elevated doses are needed that leads to many inimical side effects. DDS are mainly dedicated to cross the BBB and deliver the drugs in the damaged area of the brain (Figure 15) (Wohlfart et al. 2012). Moreover, QDs, lipid-based NPs, and polymeric NPs permit the entrance of cholinesterase inhibitors and N-methyl-D-aspartate to target the key pathological characteristic of AD amyloid-β by crossing the BBB and reducing the side-effects (Yetisgin et al. 2020). Also, a

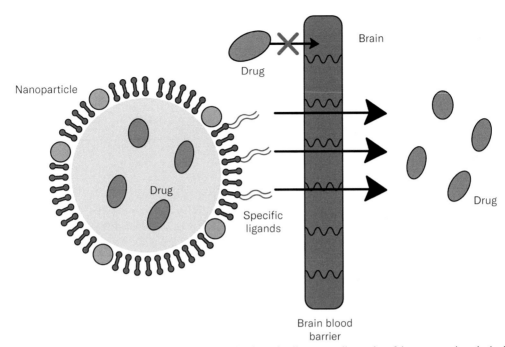

Figure 15. The limiting construction of the brain blood barrier only allows a small quantity of drug to enter into the brain, therefore needing elevated doses having side effects. In addition, nanoparticles are mostly employed to cross the brain blood barrier and deliver the drugs in the injured regions of the brain.

selective degeneration of dopaminergic neurons occurs in PD and targeted delivery with liposomes or polymeric NPs of dopamine is a NP therapeutic approach for PD (Kulkarni et al. 2015). Finally, MS is accountable for motor neuronal damage in charge of voluntary muscle movements such as walking and talking and finally leading to respiratory failure. Loading riluzole, the only agent approved for therapy, into SLN enhances the release of the payload in the CNS (Bondì et al. 2010, Mazibuko et al. 2015). For a more detailed review concerning DDS involvement in BBB, Patel and Patel (2017) can be reviewed (Patel and Patel 2017).

9.4 Cardiovascular diseases

Cardiovascular diseases (CVD) disturb the cardiovascular system, peripheral arteries, and vascular systems of kidney and brain. Despite the new approaches, heart failure affects many people worldwide and is one of the leading reasons of death (Neeland et al. 2018). DDS applied in CVD is more dedicated to targeted delivery and rising bioavailability for vascular restenosis (Yetisgin et al. 2020). SLN and liposome exhibit better oral bioavailability and precise distribution of the cardio-protective polyphenol resveratrol, which has limited water solubility and bioavailability (Neves et al. 2013). Polymeric NPs loaded with vascular endothelial growth factor is an encouraging method to enhance tissue remodeling in acute myocardial ischemic models (Formiga et al. 2010, Simón-Yarza et al. 2013).

9.5 Infectious diseases

Antimicrobial drugs are the main approach to treat infectious diseases, although sometimes the drugs become inadequate once the pathogens have acquired resistance. Therefore, the treatment has to increase the ratio and frequency of the drug, increasing the adverse effects and toxicity.

In addition, numerous pathogens are in latent state or are active, but found intracellularly thus making them difficult to come in contact with the drug (Hillaireau and Couvreur 2009, Sendi and Proctor 2009). DDS overcome these drawbacks and produce formulations to attack several bacteria, parasites, fungi, or viruses (Yetisgin et al. 2020). These DDS can be organic or inorganic NPs. For example, the liposome formulation of ciprofloxacin Lipoquin™ is suggested as a broad-spectrum Ab for infections in lungs and is manufactured for inhalation procedures for continued delivery of up to 24 h and elimination of the systemic effects of the elevated doses of Ab (Yetisgin et al. 2020). Also, several liposomes as vehicles are approved by the FDA for treating fungal or protozoal infections, multiple myeloma, Kaposi's sarcoma, and lung problems (Patra et al. 2018).

9.6 Ocular diseases

Polymeric micelles and liposomes are used for ocular treatments, specially targeting the payload to the proper eye compartment by increasing the corneal permeability and residence period on the tear film (Kulthe et al. 2012, Mandal et al. 2017, Weng et al. 2017). PLGA polymer combined with the drug pranoprofen considerably increases its ophthalmic release, and analgesic and local anti-inflammatory effects of the agent (Abrego et al. 2015). In addition, lipid-based NPs were used to treat glaucoma with brimonidine (Ibrahim et al. 2015).

9.7 Pulmonary diseases

Pulmonary lung diseases such as cystic fibrosis, asthma, pulmonary tuberculosis, chronic obstructive pulmonary disease, and idiopathic pulmonary fibrosis are frequently lethal and without an efficient therapy for repairing lung function (Sugawara and Nikaido 2014, Yetisgin et al. 2020). Actual treatments are applied in lungs locally or systemically through inhalation of the drug. However, the active molecules of the drugs are in aerosol formula leading to elevated lung toxicity due to the burst delivery. Additionally, the size of the aerosol lowers the efficacy of the therapeutic agent since higher than 5 µm and lower than 1 µm cannot go through or remain in the respiratory tract or are suspended in the air and respired. DDS allows to increase the bioavailability of the cargo and its organized release, and lower the frequency of treatment and its dosage (Lim et al. 2016, Yhee et al. 2016). Natural and synthetic polymers are extensively employed in nanomedicine preparations for lung inhalation. For example, PAMAM den were successfully applied for lung inhalation produced with the anti-asthma agent beclomethasone dipropionate (Nasr et al. 2014).

9.8 Regenerative therapy

Regenerative therapy centers on increasing regeneration and repair of tissues with biocompatible materials' utilization by exploiting the natural mechanisms of the cell. One approach for regenerative treatments is with stem cell-based therapy (Yetisgin et al. 2020). The aim is to endorse bone regeneration with therapeutic NPs as natural or synthetic polymers, MNPs, and silica-based NPs. For instance, calcium-phosphate inorganic NPs are applied since they have similarities with the bone (Alves Cardoso et al. 2012, Yetisgin et al. 2020). Furthermore, releasing numerous growth factors by DDS promotes osteoblast for bone formation (Park et al. 2016). Natural and synthetic polymers' NPs with anti-inflammatory drugs are transported into the infected target of large wounds blocking the inflammation (Gonçalves et al. 2015, Warabi et al. 2001).

10. Limitations

Delivery of drugs with large dimensions has several drawbacks like reduced target specificity, solubility, bioavailability, and absorption, possible side effects, and instability (Patra et al. 2018).

DDS could overcome these disadvantages, but their design has some aspects that need to be reviewed and considered before the creation of NPs such as biodegradability, clearance or safer elimination after payload administration, biocompatibility, metabolism, and long-term toxicity (Raza et al. 2019a, Yetisgin et al. 2020). Hence, manufacturing of NPs has to take into consideration the size and its possible relationship with toxicity and initiation of immune responses (Yetisgin et al. 2020). For example, smaller NPs enhance nucleus distribution of the agent, causing systemic problems at cellular level; polymeric NPs have several problems such as immunogenicity, biocompatibility, difficulties for scaling up, cytotoxicity, and organic solvents used for synthesis; and some NPs like den, MNP, micelles, and QDs have the tendency to agglomerate with other circulating NPs producing lower supply of the cargo; consequently, it is necessary to PEGylate these NPs for avoiding this problem (Chang et al. 2013, Yetisgin et al. 2020). Moreover, regarding responsive stimuli NPs, several downsides exist such as premature drug release, and non-targeted areas with similar characteristics as pH and oxidative stress could potentiate the delivery of the payload in off-target regions (Raza et al. 2019a, Su and Kang 2020). Also, a better knowledge of the physiological environment of the diseased and healthy tissue could avoid the enzymatic destruction of the NPs on the route to arrive to the target location (Majumder and Minko 2020).

Likewise, administrative matters for achieving support and employing the DDS is an extra main holdup in the field (Raza et al. 2019a). Very few NPs succeed the trials and become accessible in the market and commercially for the people. This implies more revisions are needed about the toxicity, manufacture, and preparation (Yetisgin et al. 2020). For instance, more studies regarding the chronic toxicity, safety entrée of NPs into the cells, immune compatibility of NPs with macrophages, contact of NPs with blood components, and biocompatibility of a single component do not assure NPs' safety and can provoke pro-inflammatory consequences, activation of complements, and coagulation. Consequently, more toxicity and *in vivo* safety evaluations are necessary in studies before clinical applications (Su and Kang 2020). Finally, large scale production is still limited due to little loading efficacy, purification, and identical production (D'Mello et al. 2017).

11. Conclusions

In summary, drug delivery systems, especially NPs like lipid-based systems, polymeric systems, metallic and iron oxide NPs, silica-based NPs, carbon nanotubes, and quantum dots can help drugs overcome several limitations including poor biodistribution, non-selectivity, low bioavailability, unwanted adverse effects, poor specificity, and low effectiveness. Once the drugs are incorporated into the NPs, these DDS surfaces can be modified for active targeting and be guided to specific cells or damaged tissue and distribute the payload and avoid detection from the immune system. However, the NPs can also use the passive targeting by employing several characteristics the damaged tissue have such as EPR and pH levels. Furthermore, these NPs can be manufactured with stimuli-responsive polymers and take advantage of endogenous stimuli such as pH, temperature, enzyme, and redox or exogenous stimuli such as magnetic, ultrasound, light, and temperature. These stimuli control the direction of the NP to the target location and release the cargo from the NPs by reacting with the different stimuli. Some NPs can answer to a dual stimulus, either exogenous or endogenous, or mixture of both, increasing the target specificity and efficiency for drug delivery (An et al. 2016, Raza et al. 2019b). Zhang and co-workers (2018) display a tumor targeting micelle with double endogenous stimuli, pH and redox (Zhang et al. 2018). Additionally, You and colleagues (2018) created a tumor targeting NP with endogenous (redox) and exogenous (NIR) stimuli (You et al. 2018). Moreover, the future of DDS is leading to multi-therapeutic NPs for treating more than one disease. Currently, several NPs have been approved by the FDA and different NPs are designed for diverse diseases like ocular, infectious, pulmonary, autoimmune, cardiovascular, cancer, and neurodegenerative.

Acknowledgement

We would like to acknowledge Iris Solano López for making the figures using Adobe Illustrator.

References

Abrego, G., Alvarado, H., Souto, E.B., Guevara, B., Bellowa, L.H., Parra, A., Calpena, A. and Garcia, M.L. 2015. Biopharmaceutical profile of pranoprofen-loaded PLGA nanoparticles containing hydrogels for ocular administration. *European Journal of Pharmaceutics and Biopharmaceutics*, 95(February): 261–270. https://doi.org/10.1016/j.ejpb.2015.01.026.

Ahmad, Z., Shah, A., Siddiq, M. and Kraatz, H.B. 2014. Polymeric micelles as drug delivery vehicles. *RSC Advances*, 4(33): 17028–17038. https://doi.org/10.1039/c3ra47370h.

Aimetti, A.A., Tibbitt, M.W. and Anseth, K.S. 2009. Human neutrophil elastase responsive delivery from poly(ethylene glycol) hydrogels. *Biomacromolecules*, 10(6): 1484–1489. https://doi.org/10.1021/bm9000926.

Akbarzadeh, A., Rezaei-Sadabady, R., Davaran, S., Joo, S.W., Zarghami, N., Hanifehpour, Y., Samiei, M., Kouhi, M. and Nejati-Koshki, K. 2013. Liposome: Classification, preparation, and applications. *Nanoscale Research Letters*, 8(1): 1. https://doi.org/10.1186/1556-276X-8-102.

Ali, A., Zafar, H., Zia, M., ul Haq, I., Phull, A.R., Ali, J.S. and Hussain, A. 2016. Synthesis, characterization, applications, and challenges of iron oxide nanoparticles. *Nanotechnology, Science and Applications* 9: 49–67. https://doi.org/10.2147/NSA.S99986.

Alves Cardoso, D., Jansen, J.A. and Leeuwenburgh, S.C. 2012. Synthesis and application of nanostructured calcium phosphate ceramics for bone regeneration. *Journal of Biomedical Materials Research. Part B, Applied Biomaterials*, 100(8): 2316–2326. https://doi.org/10.1002/jbm.b.32794.

An, X., Zhu, A., Luo, H., Ke, H., Chen, H. and Zhao, Y. 2016. Rational Design of Multi-Stimuli-Responsive Nanoparticles for Precise Cancer Therapy. *ACS Nano*, 10(6), 5947–5958. https://doi.org/10.1021/acsnano.6b01296.

Andocs, G., Renner, H., Balogh, L., Fonyad, L., Jakab, C. and Szasz, A. 2009. Strong synergy of heat and modulated electromagnetic field in tumor cell killing. *Strahlentherapie Und Onkologie*, 185(2): 120–126. https://doi.org/10.1007/s00066-009-1903-1.

Anselmo, A.C. and Mitragotri, S. 2019. Nanoparticles in the clinic: An update. *Bioengineering & Translational Medicine*, 4(3): 1–16. https://doi.org/10.1002/btm2.10143.

Aryal, S., Hu, C.M.J. and Zhang, L. 2010. Polymer-cisplatin conjugate nanoparticles for acid-responsive drug delivery. *ACS Nano*, 4(1): 251–258. https://doi.org/10.1021/nn9014032.

Babin, J., Pelletier, M., Lepage, M., Allard, J.F., Morris, D. and Zhao, Y. 2009. A new two-photon-sensitive block copolymer nanocarrier. *Angewandte Chemie - International Edition*, 48(18): 3329–3332. https://doi.org/10.1002/anie.200900255.

Bailey, R.E., Smith, A.M. and Nie, S. 2004. Quantum dots in biology and medicine. *Physica E: Low-Dimensional Systems and Nanostructures*, 25(1): 1–12. https://doi.org/10.1016/j.physe.2004.07.013.

Barenholz, Y. 2012. Doxil® - The first FDA-approved nano-drug: Lessons learned. *Journal of Controlled Release*, 160(2): 117–134. https://doi.org/10.1016/j.jconrel.2012.03.020.

Basel, M.T., Shrestha, T.B., Troyer, D.L. and Bossmann, S.H. 2011. Protease-sensitive, polymer-caged liposomes: A method for making highly targeted liposomes using triggered release. *ACS Nano*, 5(3): 2162–2175. https://doi.org/10.1021/nn103362n.

Battaglia, L., Serpe, L., Foglietta, F., Muntoni, E., Gallarate, M., del Pozo Rodriguez, A. and Solinis, M.A. 2016. Application of lipid nanoparticles to ocular drug delivery. *Expert Opinion on Drug Delivery*, 13(12): 1743–1757. https://doi.org/10.1080/17425247.2016.1201059.

Bharali, D.J., Klejbor, I., Stachowiak, E.K., Dutta, P., Roy, I., Kaur, N., Bergey, E.J., Prasad, P.N. and Stachowiak, M.K. 2005. Organically modified silica nanoparticles: A nonviral vector for *in vivo* gene delivery and expression in the brain. *Proceedings of the National Academy of Sciences of the United States of America*, 102(32): 11539–11544. https://doi.org/10.1073/pnas.0504926102.

Bhatia, S. 2016. Natural polymer drug delivery systems: Nanoparticles, plants, and algae. In *Natural Polymer Drug Delivery Systems: Nanoparticles, Plants, and Algae*. https://doi.org/10.1007/978-3-319-41129-3.

Bianco, A., Kostarelos, K. and Prato, M. 2005. Applications of carbon nanotubes in drug delivery. *Current Opinion in Chemical Biology*, 9(6): 674–679. https://doi.org/10.1016/j.cbpa.2005.10.005.

Bilan, R., Nabiev, I. and Sukhanova, A. 2016. Quantum dot-based nanotools for bioimaging, diagnostics, and drug delivery. *ChemBioChem*, 17(22): 2103–2114. https://doi.org/10.1002/cbic.201600357.

Blanco, E., Shen, H. and Ferrari, M. 2015. Principles of nanoparticle design for overcoming biological barriers to drug delivery. *Nature Biotechnology*, 33(9): 941–951. https://doi.org/10.1038/nbt.3330.

Bondì, M.L., Craparo, E.F., Giammona, G. and Drago, F. 2010. Brain-targeted solid lipid nanoparticles containing riluzole: Preparation. *Characterization and Biodistribution. Nanomedicine*, 5: 25.

Bonifácio, B.V., Silva, P.B. da, Ramos, M.A. dos S., Negri, K.M.S., Bauab, T.M. and Marlus Chorilli. 2014. Nanotechnology-based drug delivery systems and herbal medicines : A review. *International Journal of Nanomedicine*, 9: 1–15.

Bordat, A., Boissenot, T., Nicolas, J. and Tsapis, N. 2019. Thermoresponsive polymer nanocarriers for biomedical applications. *Advanced Drug Delivery Reviews*, 138: 167–192. https://doi.org/10.1016/j.addr.2018.10.005.

Bozzuto, G. and Molinari, A. 2015. Liposomes as nanomedical devices. *International Journal of Nanomedicine*, 10: 975–999. https://doi.org/10.2147/IJN.S68861.

Bulbake, U., Doppalapudi, S., Kommineni, N. and Khan, W. 2017. Liposomal formulations in clinical use: An updated review. *Pharmaceutics*, 9(2): 1–33. https://doi.org/10.3390/pharmaceutics9020012.

Bullivant, J.P., Zhao, S., Willenberg, B.J., Kozissnik, B., Batich, C.D. and Dobson, J. 2013. Materials characterization of feraheme/ferumoxytol and preliminary evaluation of its potential for magnetic fluid hyperthermia. *International Journal of Molecular Sciences*, 14(9): 17501–17510. https://doi.org/10.3390/ijms140917501.

Cai, X., Luo, Y., Zhang, W., Du, D. and Lin, Y. 2016. PH-sensitive ZnO quantum Dots-doxorubicin nanoparticles for lung cancer targeted drug delivery. *ACS Applied Materials and Interfaces*, 8(34): 22442–22450. https://doi.org/10.1021/acsami.6b04933.

Chang, S., Chen, D., Kang, B. and Dai, Y. 2013. UV-enhanced cytotoxicity of CdTe quantum dots in PANC-1 cells depend on their size distribution and surface modification. *Journal of Nanoscience and Nanotechnology*, 13(2): 751–754. https://doi.org/10.1166/jnn.2013.6085.

Chauhan, A.S. 2018. Dendrimers for drug delivery. *Molecules*, 23(4). https://doi.org/10.3390/molecules23040938.

Chen, F., Ehlerding, E.B. and Cai, W. 2014. Theranostic Nanoparticles. *Journal of Nuclear Medicine*, 55(12): 1919–1922. https://doi.org/10.2967/jnumed.114.146019.

Chen, F., Hableel, G., Zhao, E.R. and Jokerst, J.v. 2018. Multifunctional nanomedicine with silica: Role of silica in nanoparticles for theranostic, imaging, and drug monitoring. *Journal of Colloid and Interface Science*, 521: 261–279. https://doi.org/10.1016/j.jcis.2018.02.053.

Chen, W., Meng, F., Cheng, R. and Zhong, Z. 2010. pH-Sensitive degradable polymersomes for triggered release of anticancer drugs: A comparative study with micelles. *Journal of Controlled Release*, 142(1): 40–46. https://doi.org/10.1016/j.jconrel.2009.09.023.

Cheng, C., Wei, H., Shi, B.X., Cheng, H., Li, C., Gu, Z.W., Cheng, S.X., Zhang, X.Z. and Zhuo, R.X. 2008. Biotinylated thermoresponsive micelle self-assembled from double-hydrophilic block copolymer for drug delivery and tumor target. *Biomaterials*, 29(4): 497–505. https://doi.org/10.1016/j.biomaterials.2007.10.004.

Cheng, R., Feng, F., Meng, F., Deng, C., Feijen, J. and Zhong, Z. 2011. Glutathione-responsive nano-vehicles as a promising platform for targeted intracellular drug and gene delivery. *Journal of Controlled Release*, 152(1): 2–12. https://doi.org/10.1016/j.jconrel.2011.01.030.

Chilkoti, A., Dreher, M.R., Meyer, D.E. and Raucher, D. 2002. Targeted drug delivery by thermally responsive polymers. *Advanced Drug Delivery Reviews*, 54(5): 613–630. https://doi.org/10.1016/S0169-409X(02)00041-8.

Colson, Y.L. and Grinstaff, M.W. 2012. Biologically responsive polymeric nanoparticles for drug delivery. *Advanced Materials*, 24(28): 3878–3886. https://doi.org/10.1002/adma.201200420.

Crucho, C.I.C. and Barros, M.T. 2017. Polymeric nanoparticles: A study on the preparation variables and characterization methods. *Materials Science and Engineering C*, 80: 771–784. https://doi.org/10.1016/j.msec.2017.06.004.

Cuenca, A.G., Jiang, H., Hochwald, S.N., Delano, M., Cance, W.G. and Grobmyer, S.R. 2006. Emerging implications of nanotechnology on cancer diagnostics and therapeutics. *Cancer*, 107(3): 459–466. https://doi.org/10.1002/cncr.22035.

de Jesus, M.B. and Zuhorn, I.S. 2015. Solid lipid nanoparticles as nucleic acid delivery system: Properties and molecular mechanisms. *Journal of Controlled Release*, 201: 1–13. https://doi.org/10.1016/j.jconrel.2015.01.010.

Deirram, N., Zhang, C., Kermaniyan, S.S., Johnston, A.P.R. and Such, G.K. 2019. pH-responsive polymer nanoparticles for drug delivery. *Macromolecular Rapid Communications*, 40(10): 1–23. https://doi.org/10.1002/marc.201800917.

D'Mello, S.R., Cruz, C.N., Chen, M.L., Kapoor, M., Lee, S.L. and Tyner, K.M. 2017. The evolving landscape of drug products containing nanomaterials in the United States. *Nature Nanotechnology*, 12(6): 523–529. https://doi.org/10.1038/nnano.2017.67.

Doherty, G.J. and McMahon, H.T. 2009. Mechanisms of endocytosis. *Annual Review of Biochemistry*, 78: 857–902. https://doi.org/10.1146/annurev.biochem.78.081307.110540.

Dolatabadi, J.E.N., Valizadeh, H. and Hamishehkar, H. 2015. Solid lipid nanoparticles as efficient drug and gene delivery systems: Recent breakthroughs. *Advanced Pharmaceutical Bulletin*, 5(2): 151–159. https://doi.org/10.15171/apb.2015.022.

Dong, H., Du, S.R., Zheng, X.Y., Lyu, G.M., Sun, L.D., Li, L.D., Zhang, P.Z., Zhang, C. and Yan, C.H. 2015. Lanthanide Nanoparticles: From design toward bioimaging and therapy. *Chemical Reviews*, 115(19): 10725–10815. https://doi.org/10.1021/acs.chemrev.5b00091.

Dong, J. and Ma, Q. 2019. Integration of inflammation, fibrosis, and cancer induced by carbon nanotubes. *Nanotoxicology*, 13(9): 1244–1274. https://doi.org/10.1080/17435390.2019.1651920.

Dreaden, E.C., Alkilany, A.M., Huang, X., Murphy, C.J. and El-Sayed, M.A. 2012. The golden age: Gold nanoparticles for biomedicine. *Chem. Soc. Rev.*, 41(7): 2740–2779. https://doi.org/10.1039/C1CS15237H.

Dumville, J.C., O'Meara, S., Deshpande, S. and Speak, K. 2011. Hydrogel dressings for healing diabetic foot ulcers. *Cochrane Database of Systematic Reviews*, 9. https://doi.org/10.1002/14651858.cd009101.pub2.

Ekladious, I., Colson, Y.L. and Grinstaff, M.W. 2019. Polymer–drug conjugate therapeutics: advances, insights and prospects. *Nature Reviews Drug Discovery*, 18(4): 273–294. https://doi.org/10.1038/s41573-018-0005-0.

Elbahwy, I.A., Ibrahim, H.M., Ismael, H.R. and Kasem, A.A. 2017. Enhancing bioavailability and controlling the release of glibenclamide from optimized solid lipid nanoparticles. *Journal of Drug Delivery Science and Technology*, 38: 78–89. https://doi.org/10.1016/j.jddst.2017.02.001.

Elegbede, A.I., Banerjee, J., Hanson, A.J., Tobwala, S., Ganguli, B., Wang, R., Lu, X., Srivastava, D.K. and Mallik, S. 2008. Mechanistic studies of the triggered release of liposomal contents by matrix metalloproteinase-9. *Journal of the American Chemical Society*, 130(32): 10633–10642. https://doi.org/10.1021/ja801548g.

Ernsting, M.J., Murakami, M., Roy, A. and Li, S.D. 2013. Factors controlling the pharmacokinetics, biodistribution and intratumoral penetration of nanoparticles. *Journal of Controlled Release*, 172(3): 782–794. https://doi.org/10.1016/j.jconrel.2013.09.013.

Eskandari, P., Bigdeli, B., Porgham Daryasari, M., Baharifar, H., Bazri, B., Shourian, M., Amani, A., Sadighi, A., Goliaei, B., Khoobi, M. and Saboury, A.A. 2019. Gold-capped mesoporous silica nanoparticles as an excellent enzyme-responsive nanocarrier for controlled doxorubicin delivery. *Journal of Drug Targeting*, 27(10): 1084–1093. https://doi.org/10.1080/1061186X.2019.1599379.

Falconer, J., Murphy, A.N., Young, S.P., Clark, A.R., Tiziani, S., Guma, M. and Buckley, C.D. 2018. Review: Synovial cell metabolism and chronic inflammation in rheumatoid arthritis. *Arthritis and Rheumatology*, 70(7): 984–999. https://doi.org/10.1002/art.40504.

Fan, C., Gao, W., Chen, Z., Fan, H., Li, M., Deng, F. and Chen, Z. 2011. Tumor selectivity of stealth multi-functionalized superparamagnetic iron oxide nanoparticles. *International Journal of Pharmaceutics*, 404(1-2): 180–190. https://doi.org/10.1016/j.ijpharm.2010.10.038.

Fang, J., Nakamura, H. and Maeda, H. 2011. The EPR effect: Unique features of tumor blood vessels for drug delivery, factors involved, and limitations and augmentation of the effect. *Advanced Drug Delivery Reviews*, 63(3): 136–151. https://doi.org/10.1016/j.addr.2010.04.009.

Fleige, E., Quadir, M.A. and Haag, R. 2012. Stimuli-responsive polymeric nanocarriers for the controlled transport of active compounds: Concepts and applications. *Advanced Drug Delivery Reviews*, 64(9): 866–884. https://doi.org/10.1016/j.addr.2012.01.020.

Flores, A.M., Hosseini-Nassab, N., Jarr, K.U., Ye, J., Zhu, X., Wirka, R., Koh, A.L., Tsantilas, P., Wang, Y., Nanda, V., Kojima, Y., Zeng, Y., Lotfi, M., Sinclair, R., Weissman, I.L., Ingelsson, E., Smith, B.R. and Leeper, N.J. 2020. Pro-efferocytic nanoparticles are specifically taken up by lesional macrophages and prevent atherosclerosis. *Nature Nanotechnology*, 15(2): 154–161. https://doi.org/10.1038/s41565-019-0619-3.

Fomina, N., Sankaranarayanan, J. and Almutairi, A. 2012. Photochemical mechanisms of light-triggered release from nanocarriers. *Advanced Drug Delivery Reviews*, 64(11): 1005–1020. https://doi.org/10.1016/j.addr.2012.02.006.

Formiga, F.R., Pelacho, B., Garbayo, E., Abizanda, G., Gavira, J.J., Simon-Yarza, T., Mazo, M., Tamayo, E., Jauquicoa, C., Ortiz-de-Solorzano, C., Prósper, F. and Blanco-Prieto, M.J. 2010. Sustained release of VEGF through PLGA microparticles improves vasculogenesis and tissue remodeling in an acute myocardial ischemia-reperfusion model. *Journal of Controlled Release*, 147(1): 30–37. https://doi.org/10.1016/j.jconrel.2010.07.097.

Furyk, S., Zhang, Y., Ortiz-Acosta, D., Cremer, P.S. and Bergbreiter, D.E. 2006. Effects of end group polarity and molecular weight on the lower critical solution temperature of poly(N-isopropylacrylamide). *Journal of Polymer Science, Part A: Polymer Chemistry*, 44(4): 1492–1501. https://doi.org/10.1002/pola.21256.

Gao, X., Zhang, J., Xu, Q., Huang, Z., Wang, Y. and Shen, Q. 2017. Hyaluronic acid-coated cationic nanostructured lipid carriers for oral vincristine sulfate delivery. *Drug Development and Industrial Pharmacy*, 43(4): 661–667. https://doi.org/10.1080/03639045.2016.1275671.

Gaur, P.K., Mishra, S., Bajpai, M. and Mishra, A. 2014. Enhanced oral bioavailability of Efavirenz by solid lipid nanoparticles: *In vitro* drug release and pharmacokinetics studies. *BioMed Research International*, 2014. https://doi.org/10.1155/2014/363404.

Ge, J., Neofytou, E., Cahill, T.J., Beygui, R.E. and Zare, R.N. 2012. Drug release from electric-field-responsive nanoparticles. *ACS Nano*, 6(1): 227–233. https://doi.org/10.1021/nn203430m.

Gonçalves, R.M., Pereira, A.C.L., Pereira, I.O., Oliveira, M.J. and Barbosa, M.A. 2015. Macrophage response to chitosan/poly-(γ-glutamic acid) nanoparticles carrying an anti-inflammatory drug. *Journal of Materials Science: Materials in Medicine*, 26(4): 1–12. https://doi.org/10.1007/s10856-015-5496-1.

Gonzalez-Carter, D., Goode, A.E., Kiryushko, D., Masuda, S., Hu, S., Lopes-Rodrigues, R., Dexter, D.T., Shaffer, M.S.P. and Porter, A.E. 2019. Quantification of blood-brain barrier transport and neuronal toxicity of unlabelled multiwalled carbon nanotubes as a function of surface charge. *Nanoscale*, 11(45): 22054–22069. https://doi.org/10.1039/c9nr02866h.

Gwon, K., Jo, E.J., Sahu, A., Lee, J.Y., Kim, M.G. and Tae, G. 2018. Improved near infrared-mediated hydrogel formation using diacrylated Pluronic F127-coated upconversion nanoparticles. *Materials Science and Engineering C*, 90(February): 77–84. https://doi.org/10.1016/j.msec.2018.04.029.

Haba, Y., Kojima, C., Harada, A., Ura, T., Horinaka, H. and Kono, K. 2007. Preparation of poly(ethylene glycol)-modified poly(amido amine) dendrimers encapsulating gold nanoparticles and their heat-generating ability. *Langmuir*, 23(10): 5243–5246. https://doi.org/10.1021/la0700826.

Han, G., Ghosh, P., De, M. and Rotello, V.M. 2007. Drug and gene delivery using gold nanoparticles. *NanoBiotechnology*, 3(1): 40–45. https://doi.org/10.1007/s12030-007-0005-3.

Hasnain, M.S., Ahmad, S.A., Hoda, M.N., Rishishwar, S., Rishishwar, P. and Nayak, A.K. 2018. Stimuli-responsive carbon nanotubes for targeted drug delivery. In *Stimuli Responsive Polymeric Nanocarriers for Drug Delivery Applications: Volume 2: Advanced Nanocarriers for Therapeutics*. Elsevier Ltd. https://doi.org/10.1016/B978-0-08-101995-5.00015-5.

Herskovitz, J. and Gendelman, H.E. 2019. HIV and the macrophage: From cell reservoirs to drug delivery to viral eradication. *Journal of Neuroimmune Pharmacology*, 14(1): 52–67. https://doi.org/10.1007/s11481-018-9785-6.

Hillaireau, H. and Couvreur, P. 2009. Nanocarriers' entry into the cell: Relevance to drug delivery. *Cellular and Molecular Life Sciences*, 66(17): 2873–2896. https://doi.org/10.1007/s00018-009-0053-z.

Hossion, A.M.L., Bio, M., Nkepang, G., Awuah, S.G. and You, Y. 2013. Visible light controlled release of anticancer drug through double activation of prodrug. *ACS Medicinal Chemistry Letters*, 4(1): 124–127. https://doi.org/10.1021/ml3003617.

Hu, X., Tian, J., Liu, T., Zhang, G. and Liu, S. 2013. Photo-triggered release of caged camptothecin prodrugs from dually responsive shell cross-linked micelles. *Macromolecules*, 46(15): 6243–6256. https://doi.org/10.1021/ma400691j.

Ibrahim, M.M., Abd-Elgawad, A.-E.H., Soliman, O.A.-E. and Jablonski, M.M. 2015. Natural bioadhesive biodegradable nanoparticle-based topical ophthalmic formulations for management of glaucoma. *Translational Vision Science & Technology*, 4(3): 12. https://doi.org/10.1167/tvst.4.3.12.

Iga, A.M., Robertson, J.H.P., Winslet, M.C. and Seifalian, A.M. 2007. Clinical potential of quantum dots. *Journal of Biomedicine and Biotechnology*, 2007. https://doi.org/10.1155/2007/76087.

Iijima, S. 1991. Helical microtubules of graphitic carbon. *Nature*, 354(6348): 56–58. https://doi.org/10.1038/354056a0.

Iqbal, H.M.N. and Keshavarz, T. 2018. Bioinspired polymeric carriers for drug delivery applications. In *Stimuli Responsive Polymeric Nanocarriers for Drug Delivery Applications: Volume 1: Types and Triggers*. Elsevier Ltd. https://doi.org/10.1016/B978-0-08-101997-9.00018-7.

Jahan, S.T., Sadat, S.M.A., Walliser, M. and Haddadi, A. 2017. Targeted therapeutic nanoparticles: An immense promise to fight against cancer. *Journal of Drug Delivery*, 2017(iv): 1–24. https://doi.org/10.1155/2017/9090325.

Jahangirian, H., Lemraski, E.G., Webster, T.J., Rafiee-Moghaddam, R. and Abdollahi, Y. 2017. A review of drug delivery systems based on nanotechnology and green chemistry: Green nanomedicine. *International Journal of Nanomedicine*, 12: 2957–2978. https://doi.org/10.2147/IJN.S127683.

Jeon, G., Yang, S.Y., Byun, J. and Kim, J.K. 2011. Electrically actuatable smart nanoporous membrane for pulsatile drug release. *Nano Letters*, 11(3): 1284–1288. https://doi.org/10.1021/nl104329y.

Jeong, J.H., Kim, S.W. and Park, T.G. 2007. Molecular design of functional polymers for gene therapy. *Progress in Polymer Science (Oxford)*, 32(11): 1239–1274. https://doi.org/10.1016/j.progpolymsci.2007.05.019.

Jia, Y.P., Ma, B.Y., Wei, X.W. and Qian, Z.Y. 2017. The *in vitro* and *in vivo* toxicity of gold nanoparticles. *Chinese Chemical Letters*, 28(4): 691–702. https://doi.org/10.1016/j.cclet.2017.01.021.

Joshi, S., Cooke, J.R.N., Chan, D.K.W., Ellis, J.A., Hossain, S.S., Singh-Moon, R.P., Wang, M., Bigio, I.J., Bruce, J.N. and Straubinger, R.M. 2016. Liposome size and charge optimization for intraarterial delivery to gliomas. *Drug Delivery and Translational Research*, 6(3): 225–233. https://doi.org/10.1007/s13346-016-0294-y.

Kadam, R.S., Bourne, D.W.A. and Kompella, U.B. 2012. Nano-advantage in enhanced drug delivery with biodegradable nanoparticles: Contribution of reduced clearance. *Drug Metabolism and Disposition*, 40(7): 1380–1388. https://doi.org/10.1124/dmd.112.044925.

Karimi, M., Zangabad, P.S., Ghasemi, A. and Hamblin, M.R. 2015. Smart internal stimulus-responsive nanocarriers for drug and gene delivery. In *Smart internal stimulus-responsive nanocarriers for drug and gene delivery* (Issue November). https://doi.org/10.1088/978-1-6817-4257-1.

Kesharwani, P., Jain, K. and Jain, N.K. 2014. Dendrimer as nanocarrier for drug delivery. *Progress in Polymer Science*, 39(2): 268–307. https://doi.org/10.1016/j.progpolymsci.2013.07.005.

Khoee, S. and Karimi, M.R. 2018. Dual-drug loaded Janus graphene oxide-based thermoresponsive nanoparticles for targeted therapy. *Polymer*, 142: 80–98. https://doi.org/10.1016/j.polymer.2018.03.022.

Komlosh, A., Weinstein, V., Loupe, P., Hasson, T., Timan, B., Konya, A., Alexander, J., Melamed-Gal, S. and Nock, S. 2019. Physicochemical and biological examination of two glatiramer acetate products. *Biomedicines*, 7(3). https://doi.org/10.3390/BIOMEDICINES7030049.

Kong, F.Y., Zhang, J.W., Li, R.F., Wang, Z.X., Wang, W.J. and Wang, W. 2017. Unique roles of gold nanoparticles in drug delivery, targeting and imaging applications. *Molecules*, 22(9). https://doi.org/10.3390/molecules22091445.

Koo, H., Huh, M.S., Sun, I.-C., Yuk, S.H., Choi, K., Kim, K. and Kwon, I.C. 2011. *In vivo* targeted delivery of nanoparticles for theranosis. *Accounts of Chemical Research*, 44(10): 1018–1028. https://doi.org/10.1021/ar2000138.

Korkmaz, B., Horwitz, M.S., Jenne, D.E. and Gauthier, F. 2010. Neutrophil elastase, proteinase 3, and cathepsin G as therapeutic targets in human diseases. *Pharmacological Reviews*, 62(4): 726–759. https://doi.org/10.1124/pr.110.002733.

Kotla, N.G., Chandrasekar, B., Rooney, P., Sivaraman, G., Larrañaga, A., Krishna, K.V., Pandit, A. and Rochev, Y. 2017. Biomimetic lipid-based nanosystems for enhanced dermal delivery of drugs and bioactive agents. *ACS Biomaterials Science and Engineering*, 3(7): 1262–1272. https://doi.org/10.1021/acsbiomaterials.6b00681.

Kulkarni, A.D., Vanjari, Y.H., Sancheti, K.H., Belgamwar, V.S., Surana, S.J. and Pardeshi, C.v. 2015. Nanotechnology-mediated nose to brain drug delivery for Parkinson's disease: A mini review. *Journal of Drug Targeting*, 23(9): 775–788. https://doi.org/10.3109/1061186X.2015.1020809.

Kulthe, S.S., Choudhari, Y.M., Inamdar, N.N. and Mourya, V. 2012. Polymeric micelles: Authoritative aspects for drug delivery. *Designed Monomers and Polymers*, 15(5): 465–521. https://doi.org/10.1080/1385772X.2012.688328.

Kumar, B., Jalodia, K., Kumar, P. and Gautam, H.K. 2017. Recent advances in nanoparticle-mediated drug delivery. *Journal of Drug Delivery Science and Technology*, 41: 260–268. https://doi.org/10.1016/j.jddst.2017.07.019.

Kundu, J.K. and Surh, Y.J. 2010. Nrf2-keap1 signaling as a potential target for chemoprevention of inflammation-associated carcinogenesis. *Pharmaceutical Research*, 27(6): 999–1013. https://doi.org/10.1007/s11095-010-0096-8.

Lehner, R., Wang, X., Marsch, S. and Hunziker, P. 2013. Intelligent nanomaterials for medicine: Carrier platforms and targeting strategies in the context of clinical application. *Nanomedicine: Nanotechnology, Biology, and Medicine*, 9(6): 742–757. https://doi.org/10.1016/j.nano.2013.01.012.

Letchford, K., Liggins, R., Wasan, K.M. and Burt, H. 2009. *In vitro* human plasma distribution of nanoparticulate paclitaxel is dependent on the physicochemical properties of poly(ethylene glycol)-block-poly(caprolactone) nanoparticles. *European Journal of Pharmaceutics and Biopharmaceutics*, 71(2): 196–206. https://doi.org/10.1016/j.ejpb.2008.08.003.

Li, H., Yang, X., Zhou, Z., Wang, K., Li, C., Qiao, H., Oupicky, D. and Sun, M. 2017. Near-infrared light-triggered drug release from a multiple lipid carrier complex using an all-in-one strategy. *Journal of Controlled Release*, 261: 126–137. https://doi.org/10.1016/j.jconrel.2017.06.029.

Li, J., Ma, Y.J., Wang, Y., Chen, B.Z., Guo, X.D. and Zhang, C.Y. 2018. Dual redox/pH-responsive hybrid polymer-lipid composites: Synthesis, preparation, characterization and application in drug delivery with enhanced therapeutic efficacy. *Chemical Engineering Journal*, 341: 450–461. https://doi.org/10.1016/j.cej.2018.02.055.

Li, L., Xing, H., Zhang, J. and Lu, Y. 2019. Functional DNA molecules enable selective and stimuli-responsive nanoparticles for biomedical applications. *Accounts of Chemical Research*, 52(9): 2415–2426. https://doi.org/10.1021/acs.accounts.9b00167.

Lim, H., Lee, S.H., Lee, H.T., Lee, J.U., Son, J.Y., Shin, W. and Heo, Y.S. 2018. Structural biology of the TNFα antagonists used in the treatment of rheumatoid arthritis. *International Journal of Molecular Sciences*, 19(3): 1–14. https://doi.org/10.3390/ijms19030768.

Lim, Y.H., Tiemann, K.M., Hunstad, D.A., Elsabahy, M. and Wooley, K.L. 2016. Polymeric nanoparticles in development for treatment of pulmonary infectious diseases. *Wiley Interdisciplinary Reviews: Nanomedicine and Nanobiotechnology*, 8(6): 842–871. https://doi.org/10.1002/wnan.1401.

Lino, M.M. and Ferreira, L. 2018. Light-triggerable formulations for the intracellular controlled release of biomolecules. *Drug Discovery Today*, 23(5): 1062–1070. https://doi.org/10.1016/j.drudis.2018.01.019.

Liu, D., Yang, F., Xiong, F. and Gu, N. 2016. The smart drug delivery system and its clinical potential. *Theranostics*, 6(9): 1306–1323. https://doi.org/10.7150/thno.14858.

Liu, J., Lau, S.K., Varma, V.A., Moffitt, R.A., Caldwell, M., Liu, T., Young, A.N., Petros, J.A., Osunkoya, A.O., Krogstad, T., Leyland-Jones, B., Wang, M.D. and Nie, S. 2010. Molecular mapping of tumor heterogeneity on clinical tissue specimens with multiplexed quantum dots. *ACS Nano*, 4(5): 2755–2765. https://doi.org/10.1021/nn100213v.

Liu, M., Du, H., Zhang, W. and Zhai, G. 2017. Internal stimuli-responsive nanocarriers for drug delivery: Design strategies and applications. *Materials Science and Engineering C*, 71: 1267–1280. https://doi.org/10.1016/j.msec.2016.11.030.

Liu, Z., Chen, K., Davis, C., Sherlock, S., Cao, Q., Chen, X. and Dai, H. 2008. Drug delivery with carbon nanotubes for *in vivo* cancer treatment. *Cancer Research*, 68(16): 6652–6660. https://doi.org/10.1158/0008-5472.CAN-08-1468.

Liu, Z., Tabakman, S.M., Chen, Z. and Dai, H. 2009. Preparation of carbon nanotube bioconjugates for biomedical applications. *Nature Protocols*, 4(9): 1372–1381. https://doi.org/10.1038/nprot.2009.146.

Liu, Z., Sun, X., Nakayama-Ratchford, N. and Dai, H. 2010. Erratum: Supramolecular chemistry on water-soluble carbon nanotubes for drug loading and delivery (ACS Nano (2007) 1 (50-56)). *ACS Nano*, 4(12): 7726. https://doi.org/10.1021/nn103081g.

Lu, H., Wang, J., Wang, T., Zhong, J., Bao, Y. and Hao, H. 2016. Recent progress on nanostructures for drug delivery applications. *Journal of Nanomaterials*, 2016. https://doi.org/10.1155/2016/5762431.

Luo, Z., Jin, K., Pang, Q., Shen, S., Yan, Z., Jiang, T., Zhu, X., Yu, L., Pang, Z. and Jiang, X. 2017. On-demand drug release from dual-targeting small nanoparticles triggered by high-intensity focused ultrasound enhanced glioblastoma-targeting therapy. *ACS Applied Materials and Interfaces*, 9(37): 31612–31625. https://doi.org/10.1021/acsami.7b10866.

Maeda, H. 2010. Tumor-selective delivery of macromolecular drugs via the EPR effect: Background and future prospects. *Bioconjugate Chemistry*, 21(5): 797–802. https://doi.org/10.1021/bc100070g.

Majumder, J. and Minko, T. 2020. Multifunctional and stimuli-responsive nanocarriers for targeted therapeutic delivery. *Expert Opinion on Drug Delivery*, 18(2): 205–227. https://doi.org/10.1080/17425247.2021.1828339.

Manchun, S., Dass, C.R. and Sriamornsak, P. 2012. Targeted therapy for cancer using pH-responsive nanocarrier systems. *Life Sciences*, 90(11-12): 381–387. https://doi.org/10.1016/j.lfs.2012.01.008.

Mandal, A., Bisht, R., Rupenthal, I.D. and Mitra, A.K. 2017. Polymeric micelles for ocular drug delivery: From structural frameworks to recent preclinical studies. *Journal of Controlled Release*, 248: 96–116. https://doi.org/10.1016/j.jconrel.2017.01.012.

Matea, C.T., Mocan, T., Tabaran, F., Pop, T., Mosteanu, O., Puia, C., Iancu, C. and Mocan, L. 2017. Quantum dots in imaging, drug delivery and sensor applications. *International Journal of Nanomedicine*, 12: 5421–5431. https://doi.org/10.2147/IJN.S138624.

Mazibuko, Z., Choonara, Y.E., Kumar, P., du Toit, L.C., Modi, G., Naidoo, D. and Pillay, V. 2015. A review of the potential role of nano-enabled drug delivery technologies in amyotrophic lateral sclerosis: Lessons learned from other neurodegenerative disorders. *Journal of Pharmaceutical Sciences*, 104(4): 1213–1229. https://doi.org/10.1002/jps.24322.

McBain, S.C., Yiu, H.H.P. and Dobson, J. 2008. Magnetic nanoparticles for gene and drug delivery. *International Journal of Nanomedicine*, 3(2): 169–180. https://doi.org/10.2147/ijn.s1608.

Meng, F., Hennink, W.E. and Zhong, Z. 2009. Reduction-sensitive polymers and bioconjugates for biomedical applications. *Biomaterials*, 30(12): 2180–2198. https://doi.org/10.1016/j.biomaterials.2009.01.026.

Mintzer, M.A. and Simanek, E.E. 2009. Nonviral vectors for gene delivery. *Chemical Reviews*, 109(2): 259–302. https://doi.org/10.1021/cr800409e.

Mirza, A.Z. and Siddiqui, F.A. 2014. Nanomedicine and drug delivery: A mini review. *International Nano Letters*, 4(1). https://doi.org/10.1007/s40089-014-0094-7.

Mishra, B., Patel, B.B. and Tiwari, S. 2010. Colloidal nanocarriers: A review on formulation technology, types and applications toward targeted drug delivery. *Nanomedicine: Nanotechnology, Biology, and Medicine*, 6(1): 9–24. https://doi.org/10.1016/j.nano.2009.04.008.

Miyata, K., Christie, R.J. and Kataoka, K. 2011. Polymeric micelles for nano-scale drug delivery. *Reactive and Functional Polymers*, 71(3): 227–234. https://doi.org/10.1016/j.reactfunctpolym.2010.10.009.

Mohanta, D., Patnaik, S., Sood, S. and Das, N. 2019. Carbon nanotubes: Evaluation of toxicity at biointerfaces. *Journal of Pharmaceutical Analysis*, 9(5): 293–300. https://doi.org/10.1016/j.jpha.2019.04.003.

Moorthi, C., Manavalan, R. and Kathiresan, K. 2011. Nanotherapeutics to overcome conventional cancer chemotherapy limitations. *Journal of Pharmacy and Pharmaceutical Sciences*, 14(1): 67–77. https://doi.org/10.18433/j30c7d.

Mourya, V.K., Inamdar, N., Nawale, R.B. and Kulthe, S.S. 2011. Polymeric micelles: General considerations and their applications. *Indian Journal of Pharmaceutical Education and Research*, 45(2): 128–138.

Mu, H. and Holm, R. 2018. Solid lipid nanocarriers in drug delivery: Characterization and design. *Expert Opinion on Drug Delivery*, 15(8): 771–785. https://doi.org/10.1080/17425247.2018.1504018.

Mura, S., Nicolas, J. and Couvreur, P. 2013. Stimuli-responsive nanocarriers for drug delivery. *Nature Materials*, 12(11): 991–1003. https://doi.org/10.1038/nmat3776.

Myerson, J.W., McPherson, O., Defrates, K.G., Towslee, J.H., Marcos-Contreras, O.A., Shuvaev, V.V., Braender, B., Composto, R.J., Muzykantov, V.R. and Eckmann, D.M. 2019. Cross-linker-modulated nanogel flexibility correlates with tunable targeting to a sterically impeded endothelial marker. *ACS Nano*, 13(10): 11409–11421. https://doi.org/10.1021/acsnano.9b04789.

Narei, H., Ghasempour, R. and Akhavan, O. 2018. Toxicity and safety issues of carbon nanotubes. In *Carbon Nanotube-Reinforced Polymers: From Nanoscale to Macroscale*. Elsevier Inc. https://doi.org/10.1016/B978-0-323-48221-9.00007-8.

Nasr, M., Najlah, M., D'Emanuele, A. and Elhissi, A. 2014. PAMAM dendrimers as aerosol drug nanocarriers for pulmonary delivery via nebulization. *International Journal of Pharmaceutics*, 461(1-2): 242–250. https://doi.org/10.1016/j.ijpharm.2013.11.023.

Neamtu, I., Rusu, A.G., Diaconu, A., Nita, L.E. and Chiriac, A.P. 2017. Basic concepts and recent advances in nanogels as carriers for medical applications. *Drug Delivery*, 24(1): 539–557. https://doi.org/10.1080/10717544.2016.1276232.

Neeland, I.J., Poirier, P. and Després, J.P. 2018. Cardiovascular and metabolic heterogeneity of obesity: Clinical challenges and implications for management. *Circulation*, 137(13): 1391–1406. https://doi.org/10.1161/CIRCULATIONAHA.117.029617.

Neves, A.R., Lúcio, M., Martins, S., Lima, J.L.C. and Reis, S. 2013. Novel resveratrol nanodelivery systems based on lipid nanoparticles to enhance its oral bioavailability. *International Journal of Nanomedicine*, 8: 177–187. https://doi.org/10.2147/IJN.S37840.

Noriega-Luna, B., Godínez, L.A., Rodríguez, F.J., Rodríguez, A., Zaldívar-Lelo De Larrea, G., Sosa-Ferreyra, C.F., Mercado-Curiel, R.F., Manríquez, J. and Bustos, E. 2014. Applications of dendrimers in drug delivery agents, diagnosis, therapy, and detection. *Journal of Nanomaterials*, 2014. https://doi.org/10.1155/2014/507273.

Oh, J.K., Drumright, R., Siegwart, D.J. and Matyjaszewski, K. 2008. The development of microgels/nanogels for drug delivery applications. *Progress in Polymer Science (Oxford)*, 33(4): 448–477. https://doi.org/10.1016/j.progpolymsci.2008.01.002.

Paleos, C.M., Tsiourvas, D., Sideratou, Z. and Tziveleka, L.A. 2010. Drug delivery using multifunctional dendrimers and hyperbranched polymers. *Expert Opinion on Drug Delivery*, 7(12): 1387–1398. https://doi.org/10.1517/17425247.2010.534981.

Pan, J., Mendes, L.P., Yao, M., Filipczak, N., Garai, S., Thakur, G.A., Sarisozen, C. and Torchilin, V.P. 2019. Polyamidoamine dendrimers-based nanomedicine for combination therapy with siRNA and chemotherapeutics to overcome multidrug resistance. *European Journal of Pharmaceutics and Biopharmaceutics*, 136(January): 18–28. https://doi.org/10.1016/j.ejpb.2019.01.006.

Pan, L., Qianjun, H., Jianan, L., Yu, C., Linlin, Z. and Jianlin, Shi. 2012. Nuclear targeting via TAT on mesoporous NPs. *Journal of the American Chemical Society*, 134(13): 5722–25. https://doi.org/10.1021/ja211035w.

Pang, X., Jiang, Y., Xiao, Q., Leung, A.W., Hua, H. and Xu, C. 2016. PH-responsive polymer-drug conjugates: Design and progress. *Journal of Controlled Release*, 222: 116–129. https://doi.org/10.1016/j.jconrel.2015.12.024.

Pardridge, W.M. 2012. Drug transport across the blood-brain barrier. *Journal of Cerebral Blood Flow and Metabolism*, 32(11): 1959–1972. https://doi.org/10.1038/jcbfm.2012.126.

Paris, J.L., Cabanas, M.V., Manzano, M. and Vallet-Regí, M. 2015. Polymer-grafted mesoporous silica nanoparticles as ultrasound-responsive drug carriers. *ACS Nano*, 9(11): 11023–11033. https://doi.org/10.1021/acsnano.5b04378.

Park, J.S., Yi, S.W., Kim, H.J., Kim, S.M. and Park, K.H. 2016. Regulation of cell signaling factors using PLGA nanoparticles coated/loaded with genes and proteins for osteogenesis of human mesenchymal stem cells. *ACS Applied Materials and Interfaces*, 8(44): 30387–30397. https://doi.org/10.1021/acsami.6b08343.

Pärnaste, L., Arukuusk, P., Langel, K., Tenson, T. and Langel, Ü. 2017. The formation of nanoparticles between small interfering RNA and amphipathic cell-penetrating peptides. *Molecular Therapy - Nucleic Acids*, 7(June): 1–10. https://doi.org/10.1016/j.omtn.2017.02.003.

Patel, M.M. and Patel, B.M. 2017. Crossing the blood–brain barrier: Recent advances in drug delivery to the brain. *CNS Drugs*, 31(2): 109–133. https://doi.org/10.1007/s40263-016-0405-9.

Patil, Y.P. and Jadhav, S. 2014. Novel methods for liposome preparation. *Chemistry and Physics of Lipids*, 177: 8–18. https://doi.org/10.1016/j.chemphyslip.2013.10.011.

Patra, H.K., Banerjee, S., Chaudhuri, U., Lahiri, P. and Dasgupta, A.K. 2007. Cell selective response to gold nanoparticles. *Nanomedicine: Nanotechnology, Biology, and Medicine*, 3(2): 111–119. https://doi.org/10.1016/j.nano.2007.03.005.

Patra, J.K., Das, G., Fraceto, L.F., Campos, E.V.R., Rodriguez-Torres, M.D.P., Acosta-Torres, L.S., Diaz-Torres, L.A., Grillo, R., Swamy, M.K., Sharma, S., Habtemariam, S. and Shin, H.-S. 2018. Nano based drug delivery systems: Recent developments and future prospects. *Journal of Nanobiotechnology*, 16(1): 71. https://doi.org/10.1186/s12951-018-0392-8.

Pavan, G.M., Posocco, P., Tagliabue, A., Maly, M., Malek, A., Danani, A., Ragg, E., Catapano, C.v. and Pricl, S. 2010. PAMAM dendrimers for siRNA delivery: Computational and experimental insights. *Chemistry - A European Journal*, 16(26): 7781–7795. https://doi.org/10.1002/chem.200903258.

Radhakrishnan, K., Tripathy, J., Gnanadhas, D.P., Chakravortty, D. and Raichur, A.M. 2014. Dual enzyme responsive and targeted nanocapsules for intracellular delivery of anticancer agents. *RSC Advances*, 4(86): 45961–45968. https://doi.org/10.1039/c4ra07815b.

Rasheed, T., Bilal, M., Abu-Thabit, N.Y. and Iqbal, H.M.N. 2018. The smart chemistry of stimuli-responsive polymeric carriers for target drug delivery applications. In *Stimuli Responsive Polymeric Nanocarriers for Drug Delivery Applications: Volume 1: Types and Triggers*. Elsevier Ltd. https://doi.org/10.1016/B978-0-08-101997-9.00003-5.

Raza, A., Rasheed, T., Nabeel, F., Hayat, U., Bilal, M. and Iqbal, H.M.N. 2019a. Endogenous and exogenous stimuli-responsive drug delivery systems for programmed site-specific release. *Molecules*, 24(6): 1–21. https://doi.org/10.3390/molecules24061117.

Raza, A., Hayat, U., Rasheed, T., Bilal, M. and Iqbal, H.M.N. 2019b. "Smart" materials-based near-infrared light-responsive drug delivery systems for cancer treatment: A review. *Journal of Materials Research and Technology*, 8(1): 1497–1509. https://doi.org/10.1016/j.jmrt.2018.03.007.

Reilly, R.M. 2007. Carbon nanotubes: Potential benefits and risks of nanotechnology in nuclear medicine. *Journal of Nuclear Medicine*, 48(7): 1039–1042. https://doi.org/10.2967/jnumed.107.041723.

Rosales, C. 2018. Neutrophil: A cell with many roles in inflammation or several cell types? *Frontiers in Physiology*, 9(FEB): 1–17. https://doi.org/10.3389/fphys.2018.00113.

Saravanakumar, G. and Kim, W.J. 2014. Stimuli-Responsive Polymeric Nanocarriers as Promising Drug and Gene Delivery Systems. https://doi.org/10.1007/978-94-017-8896-0_4.

Schafer, F.Q. and Buettner, G.R. 2001. Redox environment of the cell as viewed through the redox state of the glutathione disulfide/glutathione couple. *Free Radical Biology and Medicine*, 30(11): 1191–1212. https://doi.org/10.1016/S0891-5849(01)00480-4.

Schleich, N., Danhier, F. and Préat, V. 2015. Iron oxide-loaded nanotheranostics: Major obstacles to *in vivo* studies and clinical translation. *Journal of Controlled Release*, 198: 35–54. https://doi.org/10.1016/j.jconrel.2014.11.024.

Schmaljohann, D. 2006. Thermo- and pH-responsive polymers in drug delivery. *Advanced Drug Delivery Reviews*, 58(15): 1655–1670. https://doi.org/10.1016/j.addr.2006.09.020.

Sendi, P. and Proctor, R.A. 2009. Staphylococcus aureus as an intracellular pathogen: The role of small colony variants. *Trends in Microbiology*, 17(2): 54–58. https://doi.org/10.1016/j.tim.2008.11.004.

Sercombe, L., Veerati, T., Moheimani, F., Wu, S.Y., Sood, A.K. and Hua, S. 2015. Advances and challenges of liposome assisted drug delivery. *Frontiers in Pharmacology*, 6(DEC): 1–13. https://doi.org/10.3389/fphar.2015.00286.

Servant, A., Bussy, C., Al-Jamal, K. and Kostarelos, K. 2013. Design, engineering and structural integrity of electro-responsive carbon nanotube-based hydrogels for pulsatile drug release. *Journal of Materials Chemistry B*, 1(36): 4593–4600. https://doi.org/10.1039/c3tb20614a.

Shay, G., Lynch, C.C. and Fingleton, B. 2015. Moving targets: Emerging roles for MMPs in cancer progression and metastasis. *Matrix Biology*, 44–46: 200–206. https://doi.org/10.1016/j.matbio.2015.01.019.

Shen, H., Shi, S., Zhang, Z., Gong, T. and Sun, X. 2015. Coating solid lipid nanoparticles with hyaluronic acid enhances antitumor activity against melanoma stem-like cells. *Theranostics*, 5(7): 755–771. https://doi.org/10.7150/thno.10804.

Shreffler, J.W., Pullan, J.E., Dailey, K.M., Mallik, S. and Brooks, A.E. 2019. Overcoming hurdles in nanoparticle clinical translation: The influence of experimental design and surface modification. *International Journal of Molecular Sciences*, 20(23): 1–25. https://doi.org/10.3390/ijms20236056.

Siepmann, F., Herrmann, S., Winter, G. and Siepmann, J. 2008. A novel mathematical model quantifying drug release from lipid implants. *Journal of Controlled Release*, 128(3): 233–240. https://doi.org/10.1016/j.jconrel.2008.03.009.

Simón-Yarza, T., Formiga, F.R., Tamayo, E., Pelacho, B., Prosper, F. and Blanco-Prieto, M.J. 2013. PEGylated-PLGA microparticles containing VEGF for long term drug delivery. *International Journal of Pharmaceutics*, 440(1): 13–18. https://doi.org/10.1016/j.ijpharm.2012.07.006.

Singh, P., Pandit, S., Mokkapati, V.R.S.S., Garg, A., Ravikumar, V. and Mijakovic, I. 2018. Gold nanoparticles in diagnostics and therapeutics for human cancer. *International Journal of Molecular Sciences*, 19(7). https://doi.org/10.3390/ijms19071979.

Skulason, S., Holbrook, W.P., Thormar, H., Gunnarsson, G.B. and Kristmundsdottir, T. 2012. A study of the clinical activity of a gel combining monocaprin and doxycycline: A novel treatment for herpes labialis. *Journal of Oral Pathology and Medicine*, 41(1): 61–67. https://doi.org/10.1111/j.1600-0714.2011.01037.x.

Son, S., Singha, K. and Kim, W.J. 2010. Bioreducible BPEI-SS-PEG-cNGR polymer as a tumor targeted nonviral gene carrier. *Biomaterials*, 31(24): 6344–6354. https://doi.org/10.1016/j.biomaterials.2010.04.047.

Son, S., Namgung, R., Kim, J., Singha, K. and Kim, W.J. 2012. Bioreducible polymers for gene silencing and delivery. *Accounts of Chemical Research*, 45(7): 1100–1112. https://doi.org/10.1021/ar200248u.

Stavrovskaya, A.A. 2000. Cellular mechanisms of multidrug resistance of tumor cells. *Biochemistry (Moscow)*, 65(1): 95–106.

Su, S. and Kang, P.M. 2020. Recent advances in nanocarrier-assisted therapeutics delivery systems. *Pharmaceutics*, 12(9): 1–27. https://doi.org/10.3390/pharmaceutics12090837.

Sugawara, E. and Nikaido, H. 2014. Properties of AdeABC and AdeIJK efflux systems of *Acinetobacter baumannii* compared with those of the AcrAB-TolC system of *Escherichia coli*. *Antimicrobial Agents and Chemotherapy*, 58(12): 7250–7257. https://doi.org/10.1128/AAC.03728-14.

Suk, J.S., Xu, Q., Kim, N., Hanes, J. and Ensign, L.M. 2016. PEGylation as a strategy for improving nanoparticle-based drug and gene delivery. *Advanced Drug Delivery Reviews*, 99(3): 28–51. https://doi.org/10.1016/j.addr.2015.09.012.

Sumetpipat, K. and Baowan, D. 2014. Three model shapes of Doxorubicin for liposome encapsulation. *Journal of Molecular Modeling*, 20(11): 1–11. https://doi.org/10.1007/s00894-014-2504-1.

Tahara, Y. and Akiyoshi, K. 2015. Current advances in self-assembled nanogel delivery systems for immunotherapy. *Advanced Drug Delivery Reviews*, 95: 65–76. https://doi.org/10.1016/j.addr.2015.10.004.

Tang, F., Li, L. and Chen, D. 2012. Mesoporous silica nanoparticles: Synthesis, biocompatibility and drug delivery. *Advanced Materials*, 24(12): 1504–1534. https://doi.org/10.1002/adma.201104763.

Tayo, L.L., Venault, A., Constantino, V.G.R., Caparanga, A.R., Chinnathambi, A., Ali Alharbi, S., Zheng, J. and Chang, Y. 2015. Design of hemocompatible poly(DMAEMA-co-PEGMA) hydrogels for controlled release of insulin. *Journal of Applied Polymer Science*, 132(32): 1–12. https://doi.org/10.1002/app.42365.

Tayo, L.L. 2017. Stimuli-responsive nanocarriers for intracellular delivery. *Biophysical Reviews*, 9(6): 931–940. https://doi.org/10.1007/s12551-017-0341-z.

Thirunavukkarasu, G.K., Cherukula, K., Lee, H., Jeong, Y.Y., Park, I.-K. and Lee, J.Y. 2018. Magnetic field-inducible drug-eluting nanoparticles for image-guided thermo-chemotherapy. *Biomaterials*, 180: 240–252. https://doi.org/10.1016/j.biomaterials.2018.07.028.

Tkahenko, A.G., Huan, X., Donna, C., Wilhelm, G., Joseph, R., Miles, F.A., Stefan, F. and Daniel, L.F. 2003. Multifunctional gold nanoparticle-peptide complexes for nuclear targeting. *Journal of the American Chemical Society*, 125(16): 4700–4701. https://doi.org/10.1021/ja0296935.

Tomatsu, I., Peng, K. and Kros, A. 2011. Photoresponsive hydrogels for biomedical applications. *Advanced Drug Delivery Reviews*, 63(14-15): 1257–1266. https://doi.org/10.1016/j.addr.2011.06.009.

Tripathy, S. and Das, M.K. 2013. Dendrimers and their applications as novel drug delivery carriers. *Journal of Applied Pharmaceutical Science*, 3(9): 142–149. https://doi.org/10.7324/JAPS.2013.3924.

Veiseh, O., Tang, B.C., Whitehead, K.A., Anderson, D.G. and Langer, R. 2014. Managing diabetes with nanomedicine: Challenges and opportunities. *Nature Reviews Drug Discovery*, 14(1): 45–57. https://doi.org/10.1038/nrd4477.

Waehler, R., Russell, S.J. and Curiel, D.T. 2007. Engineering targeted viral vectors for gene therapy. *Nature Reviews Genetics*, 8(8): 573–587. https://doi.org/10.1038/nrg2141.

Wang, T., Hou, J., Su, C., Zhao, L. and Shi, Y. 2017. Hyaluronic acid-coated chitosan nanoparticles induce ROS-mediated tumor cell apoptosis and enhance antitumor efficiency by targeted drug delivery via CD44. *Journal of Nanobiotechnology*, 15(1): 1–12. https://doi.org/10.1186/s12951-016-0245-2.

Warabi, S., Tachibana, Y., Kumegawa, M. and Hakeda, Y. 2001. Dexamethasone inhibits bone resorption by indirectly inducing apoptosis of the bone-resorbing osteoclasts via the action of osteoblastic cells. *Cytotechnology*, 35(1): 25–34. https://doi.org/10.1023/A:1008159332152.

Weissig, V., Pettinger, T.K. and Murdock, N. 2014. Nanopharmaceuticals (part 1): Products on the market. *International Journal of Nanomedicine*, 9: 4357–4373. https://doi.org/10.2147/IJN.S46900.

Weng, Y., Liu, J., Jin, S., Guo, W., Liang, X. and Hu, Z. 2017. Nanotechnology-based strategies for treatment of ocular disease. *Acta Pharmaceutica Sinica B*, 7(3): 281–291. https://doi.org/10.1016/j.apsb.2016.09.001.

Wohlfart, S., Gelperina, S. and Kreuter, J. 2012. Transport of drugs across the blood-brain barrier by nanoparticles. *Journal of Controlled Release*, 161(2): 264–273. https://doi.org/10.1016/j.jconrel.2011.08.017.

Wu, K., Su, D., Liu, J., Saha, R. and Wang, J.P. 2019. Magnetic nanoparticles in nanomedicine: A review of recent advances. *Nanotechnology*, 30(50). https://doi.org/10.1088/1361-6528/ab4241.

Wu, W., Luo, L., Wang, Y., Wu, Q., Dai, H. bin, Li, J.S., Durkan, C., Wang, N. and Wang, G.X. 2018. Endogenous pH-responsive nanoparticles with programmable size changes for targeted tumor therapy and imaging applications. *Theranostics*, 8(11): 3038–3058. https://doi.org/10.7150/thno.23459.

Xiang, J., Tong, X., Shi, F., Yan, Q., Yu, B. and Zhao, Y. 2018. Near-infrared light-triggered drug release from UV-responsive diblock copolymer-coated upconversion nanoparticles with high monodispersity. *Journal of Materials Chemistry B*, 6(21): 3531–3540. https://doi.org/10.1039/c8tb00651b.

Xing, H., Hwang, K. and Lu, Y. 2016. Recent developments of liposomes as nanocarriers for theranostic applications. *Theranostics*, 6(9): 1336–1352. https://doi.org/10.7150/thno.15464.

Xu, W., Ling, P. and Zhang, T. 2013. Polymeric micelles, a promising drug delivery system to enhance bioavailability of poorly water-soluble drugs. *Journal of Drug Delivery*, 2013(1): 1–15. https://doi.org/10.1155/2013/340315.

Yan, F., Li, L., Deng, Z., Jin, Q., Chen, J., Yang, W., Yeh, C. K., Wu, J., Shandas, R., Liu, X. and Zheng, H. 2013. Paclitaxel-liposome-microbubble complexes as ultrasound-triggered therapeutic drug delivery carriers. *Journal of Controlled Release*, 166(3): 246–255. https://doi.org/10.1016/j.jconrel.2012.12.025.

Yang, G., Liu, J., Wu, Y., Feng, L. and Liu, Z. 2016. Near-infrared-light responsive nanoscale drug delivery systems for cancer treatment. *Coordination Chemistry Reviews*, 320-321: 100–117. https://doi.org/10.1016/j.ccr.2016.04.004.

Yang, J., Zhai, S., Qin, H., Yan, H., Xing, D. and Hu, X. 2018. NIR-controlled morphology transformation and pulsatile drug delivery based on multifunctional phototheranostic nanoparticles for photoacoustic imaging-guided photothermal-chemotherapy. *Biomaterials*, 176: 1–12. https://doi.org/10.1016/j.biomaterials.2018.05.033.

Yang, S.J., Lin, F.H., Tsai, H.M., Lin, C.F., Chin, H.C., Wong, J.M. and Shieh, M.J. 2011. Alginate-folic acid-modified chitosan nanoparticles for photodynamic detection of intestinal neoplasms. *Biomaterials*, 32(8): 2174–2182. https://doi.org/10.1016/j.biomaterials.2010.11.039.

Yetisgin, A.A., Cetinel, S., Zuvin, M., Kosar, A. and Kutlu, O. 2020. Therapeutic nanoparticles and their targeted delivery applications. In *Molecules* (Vol. 25, Issue 9). https://doi.org/10.3390/molecules25092193.

Yhee, J., Im, J. and Nho, R. 2016. Advanced therapeutic strategies for chronic lung disease using nanoparticle-based drug delivery. *Journal of Clinical Medicine*, 5(9): 82. https://doi.org/10.3390/jcm5090082.

Yokoyama, M. 2002. Gene delivery using temperature-responsive polymeric carriers. *Drug Discovery Today*, 7(7): 426–432. https://doi.org/10.1016/S1359-6446(02)02216-X.

You, C., Wu, H., Wang, M., Gao, Z., Sun, B. and Zhang, X. 2018. Synthesis and biological evaluation of redox/NIR dual stimulus-responsive polymeric nanoparticles for targeted delivery of cisplatin. *Materials Science and Engineering C*, 92: 453–462. https://doi.org/10.1016/j.msec.2018.06.044.

Zanganeh, S., Hutter, G., Spitler, R., Lenkov, O., Mahmoudi, M., Shaw, A., Pajarinen, J.S., Nejadnik, H., Goodman, S., Moseley, M., Coussens, L.M. and Daldrup-Link, H.E. 2016. Iron oxide nanoparticles inhibit tumour growth by inducing pro-inflammatory macrophage polarization in tumour tissues. *Nature Nanotechnology*, 11(11): 986–994. https://doi.org/10.1038/nnano.2016.168.

Zhang, H. 2016. Onivyde for the therapy of multiple solid tumors. *OncoTargets and Therapy*, 9: 3001–3007. https://doi.org/10.2147/OTT.S105587.

Zhang, K., Luo, Y. and Li, Z. 2007. Synthesis and characterization of a pH- and ionic strength-responsive hydrogel. *Soft Materials*, 5(4): 183–195. https://doi.org/10.1080/15394450701539008.

Zhang, L., Qin, Y., Zhang, Z., Fan, F., Huang, C., Lu, L., Wang, H., Jin, X., Zhao, H., Kong, D., Wang, C., Sun, H., Leng, X. and Zhu, D. 2018. Dual pH/reduction-responsive hybrid polymeric micelles for targeted chemo-photothermal combination therapy. *Acta Biomaterialia*, 75: 371–385. https://doi.org/10.1016/j.actbio.2018.05.026.

Zhang, W., Shi, L., Wu, K. and An, Y. 2005a. Thermoresponsive micellization of poly(ethylene glycol)-b-poly(N-isopropylacrylamide) in water. *Macromolecules*, 38(13): 5743–5747. https://doi.org/10.1021/ma0509199.

Zhang, R., Tang, M., Bowyer, A., Eisenthal, R. and Hubble, J. 2005b. A novel pH- and ionic-strength-sensitive carboxy methyl dextran hydrogel. *Biomaterials*, 26(22): 4677–4683. https://doi.org/10.1016/j.biomaterials.2004.11.048.

Zhao, F., Zhao, Y., Liu, Y., Chang, X., Chen, C. and Zhao, Y. 2011. Cellular uptake, intracellular trafficking, and cytotoxicity of nanomaterials. *Small*, 7(10): 1322–1337. https://doi.org/10.1002/smll.201100001.

Zhao, Y. 2007. Rational design of light-controllable polymer micelles. *Chemical Record*, 7(5): 286–294. https://doi.org/10.1002/tcr.20127.

Zhao, Y. 2012. Light-responsive block copolymer micelles. *Macromolecules*, 45(9): 3647–3657. https://doi.org/10.1021/ma300094t.

Zhou, X., Wang, L., Xu, Y., Du, W., Cai, X., Wang, F., Ling, Y., Chen, H., Wang, Z., Hu, B. and Zheng, Y. 2018. A pH and magnetic dual-response hydrogel for synergistic chemo-magnetic hyperthermia tumor therapy. *RSC Advances*, 8(18): 9812–9821. https://doi.org/10.1039/c8ra00215k.

Zhu, Y., Liu, C. and Pang, Z. 2019. Dendrimer-based drug delivery systems for brain targeting. *Biomolecules*, 9(12): 1–29. https://doi.org/10.3390/biom9120790.

CHAPTER 11

Engineering Immunity to Disease Using Nanotechnology

Michelle Z. Dion[1,2,3] and *Dr. Natalie Artzi*[2,3,4,5,*]

1. Introduction

Disease and its associated pathologies result from the inability of the immune system to appropriately respond to pathogens, commensal microorganisms, and the body's own cells (healthy or compromised). In the case of cancer and infectious disease, initial immune responses that can control disease burden may become dysregulated and ineffective as the disease evolves, resulting in immune escape and disease progression. For autoimmune and inflammatory diseases, the immune system fails to sustain tolerance to self-antigens or commensal antigens, resulting in immune effector-mediated damage to healthy cells and tissue. Even appropriate immune responses can be detrimental to wound healing and to the outcome of therapeutic interventions, such as antibody therapy and transplantation. A common theme across diseases is that traditional clinical strategies use blunt approaches that do not address the underlying immunopathology of the disease.

To address this gap, immunotherapies have become a pillar in the treatment of cancer and other diseases. Immunotherapy seeks to generate or restore a state where therapeutic immune responses can be achieved by strategically modulating one or more stages in the development of protective effector immunity or of immune tolerance to treat disease. This includes therapeutic strategies that engineer the phenotype and repertoire of T cells and B cells and that deliver cytokines, small molecules, and antibodies to alter immune cell signaling, activation, and trafficking. The immunotherapy arsenal consists of vaccines; immunomodulatory molecules such as cytokines and small molecule adjuvants; adoptive cell therapies; antibodies; and oncolytic viruses (Table 1). The strategies and components of this arsenal are continuously expanding and have been used to treat a variety of diseases. In cancer and infectious disease, effective immunotherapy requires that the proposed intervention results in the successful induction, maintenance, and resolution of a specific, high quality immune response to the diseased cells or pathogen at the desired site within the body that can overcome disease evolution and immune escape (Eppler and Jewell 2019). In hyperactive

[1] Harvard-MIT Division of Health Sciences & Technology, Massachusetts Institute of Technology, Cambridge, MA 02139, USA.

[2] Institute for Medical Engineering and Science, Massachusetts Institute of Technology, Cambridge, MA 02139, USA.

[3] Department of Medicine, Engineering in Medicine Division, Brigham & Women's Hospital, Harvard Medical School, Boston, MA 02115, USA.

[4] Wyss Institute for Biologically Inspired Engineering, Harvard University, Boston, MA 02115, USA.

[5] Broad Institute of Harvard and MIT, Cambridge, MA, 02139, USA.

* Corresponding author: nartzi@bwh.harvard.edu

Table 1. Immunotherapy Types and Challenges. A summary of the major immunotherapy types, their current disease indications, and their associated challenges. (Sources: Waldman et al. (2020). A guide to cancer immunotherapy: from T cell basic science to clinical practice. *Nat. Rev. Immunol.*, 20(11), 651–668; Cifuentes-Rius et al. (2020). Inducing immune tolerance with dendritic cell-targeting nanomedicines. *Nat. Nanotechnol.*, 16(1): 37–46; Gong et al. (2021). Nanomaterials for T-cell cancer immunotherapy. *Nat. Nanotechnol.*, 16(1): 25–36; Goradel et al. (2021). Oncolytic virotherapy: Challenges and solutions. *Curr. Probl. Cancer*, 45(1): 100639.)

Strategy	Description	Uses	Challenges
Antibodies	Therapeutic immune protein macromolecules used for their exquisite potency and specificity for their target antigens and because of their Fc effector functions that interface with Fc receptors on immune cells; often used to target checkpoint inhibitors in cancer immunotherapy	Across all disease types	- Toxicity, including anti-drug antibodies and autoimmune responses to checkpoint blockade therapy - Response durability - Response rates and biomarkers of response
Vaccines	The therapeutic or prophylactic introduction of foreign or self-antigens to the body to produce protective cellular and humoral immune responses or to induce tolerance	Infectious Diseases, Cancer, Autoimmune Disease	- Antigen prediction or generation, particularly for MHC II peptides - Cold chain supply - Disease evolution - Patient-specific design of antigen cocktail may be required
Adoptive and engineered immune cell therapy	Immune cells that have been engineered for enhanced therapeutic efficacy by the introduction of a synthetic receptor or signaling system, like a chimeric antigen receptor; also, the use of autologous or allogenic cell transplants, such as those used in stem cell, bone marrow, and islet cell transplants	Cancer, transplantation, Type I Diabetes	- Penetration, activity, and durability in solid tumors - Expansion *in vivo* - Immune escape/antigen loss - Manufacturing complexity, variability, and cost - Off-target toxicity and the specificity of targeted antigens for diseased tissue - Control of pharmacokinetics *in vivo*
Small molecule immunomodulators	The use of chemokines, cytokines, interleukins, small molecule immune signaling agents, and synthetic small molecules to modify the localization, composition, or activation of a microenvironment, tissue, or organ's immune infiltrate	Cancer, autoimmune disease, transplantation, IBD	- Toxicity - Pharmacokinetics/Biodistribution - Usually require combination therapy to achieve efficacy
Oncolytic Virus Therapy	The use of modified viruses that specifically infect and destroy tumor cells while sparing healthy cells; used to stimulate a proinflammatory tumor microenvironment through immunogenic cell death	Cancer	- Toxicity - Potential to induce immunosuppression - Poor uptake and replication within tumor cells, reducing efficacy - Poor tumor microenvironment penetration and persistence - Off-target infection of healthy cells

and autoimmune diseases, the goal of immunotherapy is to dampen or control hyperactive immune responses to promote tolerance and tissue regeneration. In comparison to traditional therapies, immunotherapeutic approaches can offer numerous advantages including specificity (ability to generate immune responses specific to disease- or self-antigens while limiting bystander damage), durability (ability to generate robust responses with the capacity for immune memory that prevent disease occurrence or recurrence regardless of the time or location), and adaptability (the ability of the immune system to alter its repertoire of cells and their specificity through epitope spreading or the like to protect against disease evolution) (Galluzzi et al. 2018). Immunotherapy can also generate immune responses with abscopal effects that allow them to control or eliminate disease at sites in the body disseminated from the treatment site, making it advantageous for the treatment of metastatic cancer. Because of these advantages, immunotherapeutic strategies are being explored across disease types and have garnered many regulatory body approvals in recent years (Dobosz and Dzieciatkowski 2019).

Despite these successes, the clinical scope of immunotherapy remains limited. Current challenges to effective therapy include off-target toxicities, interpatient differences in efficacy and durability, expensive and complex manufacturing procedures, and insufficient control of disease immune escape (Galluzzi et al. 2018). Thus, strategies that provide more precise control over the timing, the location, and the context in which immunomodulatory signals are presented to specific immune cells and tissues have the potential to significantly improve immunotherapy outcomes by overcoming these challenges. Nanomaterials offer a highly tunable mechanism by which to address these challenges given their proven ability to alter the delivery profile of therapy across time and space to improve the safety and potency of therapies. Additionally, nanomaterials have novel mechanisms of action that can synergize with molecular and cellular immunotherapies. In this chapter, we briefly introduce foundational concepts in immunology to provide insight into how the immune system functions and the ways in which it can be modified to achieve improved outcomes in immunotherapy. We then discuss design considerations for developing nanomaterials for immunotherapy based on their range of tunable physiochemical and surface properties. Finally, we highlight how these features are being integrated into the design of nanomaterials to enable novel mechanisms of action to enhance the delivery, potency, and durability of molecular and cellular immunotherapies.

2. Foundational concepts in immunotherapy

The immune system is composed of an expansive repertoire of organs, cells and molecular agents that interact through dynamic signaling pathways to initiate, maintain, and resolve immune function. The body relies on this intricate and dynamic system to protect against disease by recognizing and eliminating foreign pathogens and unhealthy cells while sparing healthy tissue. The ability of effector components of the immune system to differentiate between healthy and diseased host cells relies on the integration of surface and secreted molecule signaling between host and effector immune cells. Cells presenting self-antigen and expressing co-inhibitory molecules instruct the immune system not to attack, while those presenting foreign or disease antigens and expressing co-stimulatory molecules prime the immune system for attack. Traditionally, immunity is divided into the innate and adaptive immune systems, which are differentiated by the timing of their response following pathogenic assault and their capacity to generate memory to encountered antigens (Abbas et al. 2018). The functions of the innate and adaptive immune system are bridged by the activity of specialized innate immune cells known as antigen-presenting cells. Innate and adaptive immunity work cooperatively to respond to and eliminate disease (Figure 1). In this section, you will be introduced to the foundational concepts underpinning innate and adaptive immunity and how these understandings are used to drive the design of efficacious immunotherapies in different disease states. We will provide background on the key molecules, cell types, organs, and responses involved in generating efficacious immune responses. This section will provide context for our later discussion of nanomaterial-based applications for immunotherapy in the proceeding sections.

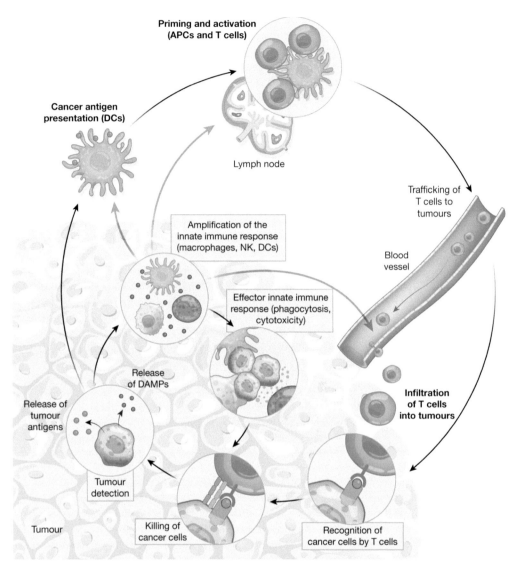

Figure 1. The role of innate and adaptive immunity in the cancer immune cycle. The combined, coordinated action of innate and adaptive immunity allows for the discrimination between healthy and diseased cells or pathogenic agents. The cells of the innate immune system detect the pathogen directly through cell surface interactions or indirectly through damage associated molecular patterns released by dying or infected cells. Innate immune cells then can stimulate an adaptive immune response by presenting antigens-derived from the disease microenvironment to T cells in the lymph nodes. They can also carry out effector mechanisms such as phagocytosis to attempt to eliminate the pathogenic agent directly. Adaptive immune cells that are activated in the lymph nodes then enter the blood and traffic to the disease site, where they need to extravasate and infiltrate the disease microenvironment. Once there, T cells recognize their cognate antigen displayed on diseased cells and eliminate these cells. This creates additional antigens and DAMPs, restimulating the immune cycle until all disease has been eliminated or exhaustion and immune escape occurs. (Reprinted with permission from Demaria et al. (2019). Harnessing innate immunity in cancer therapy. *Nature*, 574(7776): 45–56.)

2.1 Innate immunity

The innate immune system is responsible for generating rapid responses to infectious agents and disease through the recognition of conserved molecular patterns and pathogen components, and generally does not develop immune memory within its cellular repertoire (Abbas et al. 2018). The short-half life and limited proliferation of these cells enable their responses to be quickly resolved to prevent tissue damage from prolonged exposure to their effector activities (Eppler and Jewell 2019). The key cells of the innate immune system are macrophages, monocytes, neutrophils, mast cells, and natural killer cells. Innate immune cells circulate in the blood and are present in tissues as tissue-resident cells to protect against infection and disease. Specialized innate immune cells known as antigen-presenting cells (APCs), which include macrophages and dendritic cells (DCs), are highly phagocytic and continually sample their environment to surveil for pathogens. These cells detect pathogens using pattern recognition receptors (PRRs) that recognize pathogen-associated molecular patterns (PAMPs), which are conserved structural motifs or molecules common to bacteria and viruses that are not usually found in healthy human tissue. PRRs include toll-like receptors (TLRs) and stimulator of interferon genes protein (STING) (among others) and sense a wide variety of microbial and viral products intracellularly and extracellularly including bacterial lipopeptides and lipopolysaccharides (LPS), flagellin, and single-stranded and double-stranded DNA and RNA. In response to PAMPs, APCs become activated and secrete inflammatory cytokines to recruit additional immune cells to the site of disease. These cells can also express cytotoxic molecules to destroy pathogens or damaged tissue and can phagocytose or internalize diseased cells, pathogens, or debris for degradation (Figure 1). Importantly, antigen uptake also drives APC maturation and migration to secondary lymphoid organs, such as the spleen and the lymph nodes, where adaptive immune cells (T cells and B cells) await APC activation to generate cytotoxic cellular and humoral immune responses. APCs must display the cognate antigen of the B cell or T cell (or the specific antigen that the T cell receptor (TCR) or B cell receptor (BCR) can recognize) to activate it. As a result, APCs, particularly DCs, bridge innate and adaptive immune responses. This process will be further discussed in the next section (Adaptive immunity). The innate immune system also contains cytotoxic cells called natural killer cells which are responsible for eliminating host cells that lack MHC-I expression (MHC-I is often downregulated by tumor cells and virally infected cells so that they can avoid killing by adaptive immune cells) (Demaria et al. 2019). Together, these capabilities allow innate immune cells to identify and eliminate pathogens and trigger inflammatory and adaptive immune responses.

2.2 Adaptive immunity

In comparison to the innate immune system, the adaptive immune system responds more slowly to primary exposures (7–14 days) but can generate expansive immune cell repertoires upon antigen challenge that can develop into memory cells that can persist for extended periods of time, including through the life of a person (Abbas et al. 2018). Adaptive immune cells are generated and developed in the bone marrow and thymus before circulating to secondary lymphoid organs, such as the lymph nodes and spleen, for further activation and instruction by APCs. Because they require activation by APCs, adaptive immune cells are slower to respond to pathogens and disease than innate immune cells. The key adaptive immune system cells are T cells and B cells and their associated subtypes, each of which have different effector functions in response to antigen recognition.

B cells recognize soluble or cell-surface-derived antigens and secrete antibodies, which are immune macromolecules capable of directly neutralizing pathogens and that mediate antibody-dependent cellular cytotoxicity (ADCC). Antibodies are involved in immune complement activation and play a critical role in the humoral response to vaccination as the titer, affinity, epitope specificity, valency, and effector mechanisms of the generated antibodies determine the immune system's ability to neutralize pathogens (Clem 2011). To produce specific, long-lasting antibodies that protect against disease, B cells must undergo several processes. Class switching is one of these processes, which produces antibodies with different structural features, or classes, that enable a range of immunological functions required to clear pathogens. Another important process is affinity maturation, which is mediated by helper T cells (discussed in the next paragraph) and occurs in specialized domains of the lymph nodes called germinal centers. In this process, B cells iteratively generate antibodies with higher affinity and selectivity for their target antigen. B cells in germinal centers can further differentiate into plasma cells or memory B cells. Together, these processes ensure that a more potent and specific antibody repertoire with diverse effector functions is generated to neutralize infections.

T cells complement the function of B cells by coordinating adaptive immune responses and executing cytotoxic response. T cells are mainly divided into CD4+ (helper T cell) and CD8+ (cytolytic T cell) subtypes. CD4+ T cells are largely responsible for orchestrating downstream responses and secrete cytokines that mediate their effector functions. They are critical in the phagocytic elimination of microbes by macrophages, in activating other leukocytes such as neutrophils and eosinophils, and in the proliferation and differentiation of other T cells and B cells (Tay et al. 2020). They respond primarily to extracellular pathogens and are activated by antigens presented on major histocompatibility complex II (MHC-II) molecules on APCs. In contrast, CD8+ T cells are activated by antigens presented on MHC-I molecules. They are primarily responsible for killing infected or diseased (tumor) cells that display foreign antigens and for eliminating intracellular pathogens through the release of granules following target cell engagement (Waldman et al. 2020). Certain DC subsets can load exogenous antigen peptides on MHC-I and MHC-II complexes through a process known as cross-presentation, priming antigen-specific immune responses in both CD4 and CD8 T cells. Cross-presentation is crucial in both mounting an immune response against cancer and infectious diseases, as well as in controlling immune reactivity and inducing tolerogenic responses (Abbas et al. 2018). This allows for both intracellular and extracellular disease antigens to be presented to immune cells, giving access to a broader range of epitopes.

The interactions between adaptive immune cells and APCs, particularly DCs, are critically important to the proper generation and regulation of adaptive immune responses and rely on the presence of three signals. DCs present antigens to T cells using MHC-I and MHC-II complexes (signal 1), with the outcome of antigen recognition being determined by the nature of the expressed co-receptors and cytokines (signals 2 and 3) (Figure 2). The engagement of co-stimulatory or co-inhibitory molecules at the immune synapse during antigen presentation governs the type, specificity, and duration of the immune response (Friedl et al. 2005). Secreted cytokines are considered the third signal involved in immune synapse signaling and affect the ability of adaptive immune cells to sustain effector functions. This is a mechanism of immune tolerance where the context of antigen presentation (whether with the expression of costimulatory molecules and cytokines or inhibitory co-receptors and cytokines) licenses the functional activity of T cells, making this interaction a critical target in immunotherapy. This complements central tolerance, which occurs during T cell development, to limit self-reactivity to prevent the development of autoimmune diseases. APC negative feedback signaling mechanisms can also use this process to restrain immune responses to prevent tissue damage from chronic immune activation, as is seen in cancer (Peng et al. 2020, Oh et al. 2020). To explain further, APCs continually uptake self-antigens during immunosurveillance but are unable to activate adaptive immune cells due to low levels of antigen presentation and the lack of immunostimulatory cue expression. When an APC senses danger (PAMPs) in addition to antigen, it will express co-stimulatory surface molecules and cytokines that trigger T cell activation via integration of the surface exposed and soluble inflammatory cues. The presentation of antigen to T cells concurrently with tolerogenic signals or in the chronic absence of costimulatory signals can result in the production of T regulatory cells (Tregs), in T cell anergy (a functionally inactive, non-proliferative, nonresponsive state), or in T cell deletion (Friedl et al. 2005). Tregs play an important role in immune tolerance as they suppress the activation and effector functions of other T cells (Abbas et al. 2018). As a result, Tregs have significant influence over the maintenance and duration of immune responses to tumors and to self-antigens in autoimmune disease (Friedl et al. 2005, Cifuentes-Rius et al. 2020). Tregs can amplify tolerogenic responses in bystander immune cells through the production of immunosuppressive cytokines. Due to the impact of these processes on

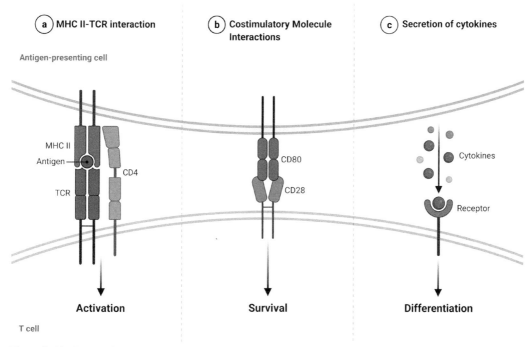

Figure 2. The Immune Synapse. The interactions between soluble proteins and cell surface receptors on T cells and antigen-presenting cells govern the outcome of T cell activation. The three signals that antigen-presenting cells can give to T cells are: (1) the antigen-MHC complex; (2) co-stimulatory receptors; (3) immuno-instructive cytokines. (Adapted from "Three Signals Required for T Cell Activation", by BioRender.com (2021). Retrieved from https://app.biorender.com/biorender-templates.)

the generation of protective immunity or of immune tolerance, the molecules, cells, tissues, and processes involved in immune synapse interactions are powerful targets for immunotherapeutic modulation (Figure 3). Following immune synapse signaling, activated T cells undergo rapid proliferation, exponentially expanding the availability of antigen-specific effector cells. These cells then enter the circulation and traffic to sites of disease following chemokine gradients generated by innate immune cells. If the activated T cell recognizes its antigen, it will carry out its effector mechanisms to induce elimination of the pathogenic agent or cell.

Together, B cells and T cells represent a diverse repertoire of cells that execute highly specific and potent immune responses using a range of effector mechanisms. The cooperative, multi-component processes required to generate and regulate effective adaptive immune responses illustrate the complexity of this system.

2.3 The influence of non-immune components on immune responses

In addition to modulating the functions and components of the immune system discussed above, immunotherapy strategies can also modify the activity and composition of non-immune components, such as diseased cells, stromal tissue, the vasculature, and the microbiome. Diseased cells, such as cancer cells, are eliminated by the immune system based on their susceptibility to immune cell killing and recognition (antigen expression) through a process known as immunoediting (Dunn et al. 2004). This places a negative selection pressure on diseased cells that promotes the growth and evolution of clones or variants that can escape the immune system. Mechanisms of immune evasion include subversion of the immune response by expression of surface or secreted immune regulatory molecules such as PD-L1, the loss of immunogenic antigen expression, and the downregulation of

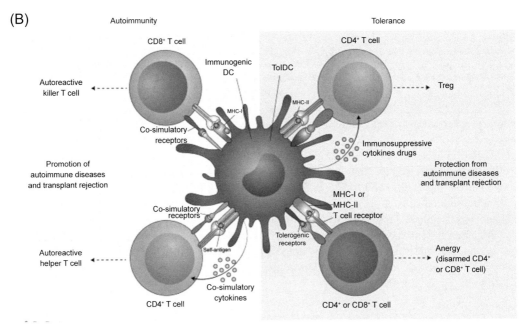

Figure 3. The Impact of Antigen-Presenting Cell and T Cell Interactions on the Development of Antigen-Specific Immunity in Cancer and Autoimmune Disease. (A) The presentation of tumor antigen with co-stimulatory surface receptors and cytokines results in T cell activation, proliferation, and migration to the tumor microenvironment, where the T cell can carry out its effector function against tumor cells displaying the cognate antigen. (B) The presentation of self-antigen with co-stimulatory receptors and cytokines results in the activation of autoreactive CD8+ cytotoxic T cells and CD4+ helper T cells that promote autoimmunity. The presentation of self-antigen with co-inhibitory receptors and immunosuppressive cytokines or in the chronic absence of co-stimulatory signals can result in the production of T regulatory cells or anergy. (Figure 3A: Reprinted with permission from BioRender. Figure 3B: Reprinted with permission from Cifuentes-Rius et al. (2020). Inducing immune tolerance with dendritic cell-targeting nanomedicines. *Nat. Nanotechnol.*, 16(1): 37–46.)

expression of antigen-presenting proteins, such as MHC molecules (Galluzzi et al. 2018). This can result in a rapid increase in disease burden and disease dissemination, overwhelming the immune system and causing extensive tissue damage. The creation of an immunosuppressive disease microenvironment by these cells can further inhibit the induction of immunity by immunotherapy or the activity and durability of the generated response.

Bystander cells and tissues of the body are also gaining recognition as significant influencers of immunity including stromal cells and the gut microbiome (Krausgruber et al. 2020, Sepich-Poore et al. 2021). Non-immune, non-disease microenvironment components, such as the blood and lymphatic vasculatures, stromal cells, nerve cells, and the extracellular matrix (ECM), can influence the efficacy of immunotherapy (Galluzzi et al. 2018). The integrity of the blood vasculature and the molecular expression profile of the blood vasculature influences the trafficking of immune cells and therapy to the disease site. The blood vasculature also regulates the access of oxygen and key nutrients to immune cells and others in the disease microenvironment. The lymphatic vasculature, or lymphatics, provide a channel for activated antigen-presenting cells to leave peripheral tissues and traffic to the local draining lymph nodes to prime T cells. Macromolecules, such as antigens, pathogens, and cytokines, can also be transported to the lymph nodes from peripheral tissues via the lymph, which can further induce adaptive immune cell responses in the lymph nodes. Within the microenvironment, the composition and density of the extracellular matrix can influence the penetration of immune cells and nanoparticle therapy in addition to the migration and spread of diseased cells. Finally, recent work has shown that the gut microbiome, or the collection of microorganisms that live in the human digestive tract, and its metabolites can modulate the immune system and alter the efficacy of immunotherapy in cancer and other diseases (Round and Mazmanian 2009, Gopalakrishnan et al. 2017, Matson et al. 2018, Halfvarson et al. 2017).

While incomplete, this is illustrative of the numerous points of potential intervention beyond components of the immune system for nanotechnology-mediated immunotherapy. Each disease will have unique features of its cells, microenvironment, baseline immune response, and systemic sequalae that will influence the quality of the incited immune response to targeted interventions. This provides extensive opportunities to employ new and innovative nanotechnology-based immunotherapy approaches as discussed in the following section.

3. Designing nanomaterials for immunotherapy

For immunotherapy to be successful, it needs to be able to induce a timely response at specific tissues with sufficient breadth, specificity, and durability to eliminate disease in cancer and infectious disease or to control aberrantly activated effector immune mechanisms in autoimmune and inflammatory diseases (Irvine and Dane 2020, Cifuentes-Rius et al. 2020). Furthermore, the immunotherapeutic strategy needs to be dynamic, capable of countering disease evolution and preventing disease immune escape and recurrence. As a result, multi-faceted approaches requiring dynamic control over the spatiotemporal kinetics of combination therapies are likely required to balance durable immunotherapeutic efficacy with safety. Given their tunability, nanomedicines offer a unique opportunity to enable these features for immunotherapy without inducing toxicities.

Nanomaterials have been studied extensively for the treatment of disease because of their ability to alter the biodistribution, tissue accumulation, and cell uptake profiles of therapy by passive and active mechanisms; to prevent drug degradation; to enable unique combination therapies by reducing toxicities; and to control the temporal release of therapy by enabling controlled, pulsatile, or stimuli-responsive release (Goldberg 2019, Song et al. 2019, Irvine and Dane 2020, Gong et al. 2021). These mechanisms of action are enabled by the highly tunable features of nanomedicines, such as their modifiable physiochemical properties and nano-biological surface interfaces (Eppler and Jewell 2019). These properties are often referred to as the nanotechnology toolbox (Song et al.

2019, Gong et al. 2021) and encompass approaches such as surface, size, shape, and stiffness modification; controlled, activatable, and/or sequential release; and selection of nanomaterial type and structure (Figure 4). This toolbox is continually expanding to include new functions and enable new therapeutic strategies. In addition to the mechanisms previously discussed, new strategies are emerging that allow for nanomedicines to overcome physiological barriers and to recapitulate complex cell-cell interactions more effectively (Eppler and Jewell 2019, Gong et al. 2021). Nanomaterials have also given us more sophisticated control over the *in vivo* delivery and activity of genetic, molecular, and cell engineering therapies that are increasingly expanding our ability to engineer biology for therapeutic purposes (Ditto et al. 2009, Shim and Kwon 2012, Tong et al. 2019). Finally, nanomaterials are enhancing our ability to monitor disease and host responses to therapy *in vivo* across time and spatial scales (Zhong et al. 2019).

The following section will provide an overview of considerations for designing nanomaterials for immunotherapy before we describe specific examples of studies that used nanomaterials to enable novel immunotherapeutic applications in the final sections of this chapter. From this section, you will understand the influence of material properties on the delivery and therapeutic performance of immunotherapeutic nanomedicines. Further, this section is meant to provide an overview of the immuno-supportive mechanisms of action associated with each property or feature. More extensive discussions of these fundamental mechanisms can be found in the literature (Look et al. 2010,

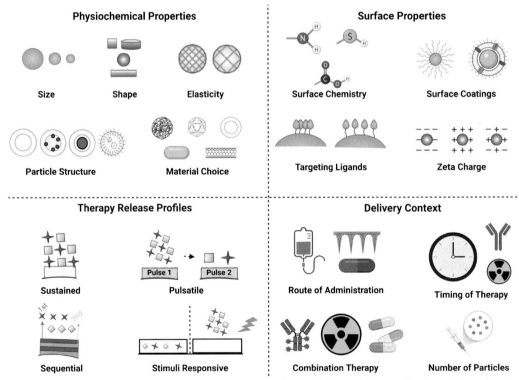

Figure 4. The Tunable Properties of Nanomaterials. Nanomaterials have highly tunable properties that can be used to alter the drug delivery (biodistribution, pharmacokinetics, cell uptake profile, etc.) and therapeutic (efficacy, immune profiling) profiles of immunotherapy. These properties include physiochemical properties (material choice, particle structure, size, shape, elasticity); surface properties (charge; chemistry; coatings; targeting ligands-multivalency, density, types); release properties (sustained; stimuli-responsive; programmable); and their delivery context (number of particles; route of administration; timing of administration; combinations). (Adapted from Song, W., Anselmo, A. C., & Huang, L. (2019). Nanotechnology intervention of the microbiome for cancer therapy. Nat Nanotechnol, 14(12), 1093–1103. and Gong, N., Sheppard, N. C., Billingsley, M. M., June, C. H., & Mitchell, M. J. (2021). Nanomaterials for T-cell cancer immunotherapy. Nat Nanotechnol, 16(1), 25–36.)

Eppler and Jewell 2019, Goldberg 2019, Nam et al. 2019, Shields et al. 2019, Cifuentes-Rius et al. 2020, Irvine and Dane 2020, Gong et al. 2021).

3.1 Physiochemical properties

The physiochemical properties of nanoparticles include material type, size, shape, deformability, and particle structure. Each of these features can be designed to influence the resultant bioavailability, biodistribution, cell uptake, and immune response profiles of the nanomedicine. While the general implications of each feature are discussed, it should be noted that the incited immune responses and delivery profiles associated with a given immunotherapeutic nanomedicine result from the combined features of the particle and can also be influenced by the given application.

3.1.1 Material type and particle structure

Given its central role in the formation of nanoparticles and their structure, material type has a significant influence over the physiochemical properties and biological profiles (distribution, pharmacokinetics, mechanisms of action, invoked immune responses) of the nanoparticle. Further, while materials were previously thought to be immunologically inert, recent studies have illustrated that materials can generate a range of immunomodulatory effects with the invoked responses varying greatly in their type, duration, and magnitude depending on the context of the studied response (Table 2; Shields et al. 2019). The immunomodulatory properties of nanoparticle materials have inspired interest in using these materials in many novel applications including as drug-free immunotherapies (Casey et al. 2019, Saito et al. 2019) and as adjuvants in minimalist nanovaccines (Luo et al. 2017). Further, recent advances in the computational design of programmable biological materials have inspired the use of immunotherapies, such as antibodies, proteins, and DNA or RNA adjuvants, as the building blocks of carrier-free or polymer-free nanoparticles (King et al. 2014, Zhang et al. 2014, Divine et al. 2021). These applications are discussed in greater detail in the subsection of this chapter, Direct Immunomodulation through Novel Material Design.

Beyond its immunomodulatory effects, material type determines different nanomaterial mechanisms of action that can synergize with immunotherapy. For example, certain materials, such as inorganic materials and metal-organic frameworks, can act as radiosensitizers or photothermal sensitizers and/or are magnetic, and, as a result, can enhance immunogenic cell death by localizing and potentiating treatment with external energy sources (Irvine and Dane 2020). These properties can also allow these materials to release cargo in response to stimuli or home to specific environments and be retained there when different physical fields are applied (Sanz-Ortega et al. 2019). This endows nanomaterial-based immunotherapy with higher delivery specificity than free immunotherapies. Magnetic properties of materials have also been used to induce surface receptor clustering on nanoparticles to mimic the dynamics of surface proteins and receptors found at immunological synapses to enhance immune signaling with T cells and the resultant activation and proliferation of these cells (Kosmides et al. 2018). These properties of material type enable these materials to act imaging contrast agents. For example, Zhong et al. recently used erbium-based rare-earth nanoparticles as near-infrared-IIb molecular imaging probes to study the cellular responses of the tumor microenvironment and secondary lymphoid organs to immunotherapy in live animals (Zhong et al. 2019). As the development of efficacious immune responses is reliant on the coordinated spatiotemporal modification of immune cells, this nanoparticle material-based property can provide critical information about the biological action of therapies driving patient responses in real-time by allowing scientists and clinicians to monitor immune cell migration and activation across time and correlate it with therapeutic outcomes, delivery profiles, and treatment schedules. Materials based on quantum dots (Bruns et al. 2017) and carbon nanotubes (Fadel and Fahmy 2014), among others, can also serve as theranostic particles. Additionally, material type can influence the therapy's release kinetics from the particle and the nanoparticle's biodegradation properties. A common example is the use of polyester-based nanoparticles to control drug release by altering polymer properties that affect its degradation by esterase-mediated hydrolysis, such as polymer chain length/molecular weight, polymer hydrophobicity, and end-capping (Makadia and Siegel 2011). Polymer architecture and chemistry can also influence the material's ability to deliver cargo, such as genetic materials and small molecular innate immune receptor agonists, intracellularly (Irvine and Dane 2020). Given the emergence of therapeutic genetic engineering technologies that can engineer biology to treat disease with molecular and cellular precision, an understanding of the impact of nanomaterial properties on organ-specific, cell type-specific, and subcellular delivery and trafficking is of critical interest for the safe and effective implementation of these therapies in patients.

Particle material also dictates particle chemistry and therefore particle structure. Particle structure impacts numerous drug delivery parameters including the types of therapies that can be delivered, the number and amount of therapies that can be loaded, and the way these therapies are presented to the immune system or body. Core-shell particles, for example, have multiple compartments that can enable the encapsulation of therapy into their core, incorporation of therapy into their shells, or conjugation or absorption of therapy to their surface. This enables the delivery of therapies with divergent properties, such as peptide antigens and some small molecule adjuvants, to the same cell or tissue (Irvine and Dane 2020). Particle structure also impacts the nature (chemistry, surface area, etc.) of the nanomaterial-biological interface and the biological stability or durability of the particle. In one example, Moon et al. engineered interbilayer-crosslinked multilamellar vesicles (ICMVs) to create lipid drug carriers that exhibit enhanced extracellular stability compared to multilamellar vesicles and liposomes but that degraded rapidly in response to high concentrations of lipases intracellularly (Moon et al. 2011). When used to co-entrap antigens and an immunostimulatory adjuvant, the structural properties of ICMVs generated enhanced antibody titers in comparison to MLVs and liposomes and resulted in increased antigen-specific humoral and T cell responses following repeat administration, providing a distinct advantage over viral-based vaccine vectors.

While non-exhaustive, these studies highlight the myriad of ways material type and particle structure can impact the design of nanoparticle immunotherapies and their applications and can result in distinct synergistic mechanisms of action.

3.1.2 Size

Because many biological barriers have size-dependent permeability, a nanoparticle's size has a significant impact on its biological transport, and therefore its ability to reach different subcellular, cellular, and tissue immune compartments. One of the most common barriers to nanoparticle delivery is the vascular endothelium. The ability of nanoparticles to cross endothelial barriers is dependent on both active and passive transport mechanisms and the physiology and pathophysiology of the vasculature. The physiology, such as the presence of fenestrations and the size of cell-cell junctions, of the tissue-associated vasculature and its alteration in disease dictates its permeability to nanomedicines and the design of nanoparticle properties that can enable efficient extravasation from the blood (Poon et al. 2020). For example, nanoparticles have been thought to take advantage of a passive transport phenomena specific to tumor vasculature pathophysiology (the Enhanced Permeability and Retention effect or EPR) by which long-circulating nanoparticles leverage increased fluid leakage from poorly formed vasculature to enter the tumor tissue (Prabhakar et al. 2013). While the clinical applicability of the EPR is the focus of intense debate, it is illustrative of how physiological and pathophysiological features of biological barriers can influence the design of nanoparticle size. Similar design considerations can be drawn for surface modification, such as targeting moieties, that can take advantage of differential surface receptor expression by diseased vasculature to enable active transport effects. Nanoparticle size can also impact the particle's ability to transport through the tissue stroma, which is composed of the extracellular matrix (ECM) and connective tissue cells (Chen et al. 2019). Components of the ECM include proteoglycans and fibrous proteins. The composition and density of the ECM can vary between tissues and disease states, which can alter the pore size and steric hindrances within this matrix, affecting the diffusion of nanoparticles of different sizes and charges (Netti et al. 2000). This can impact the ability of nanoparticles to reach their targeted cell types. At a cellular level, size impacts the nanoparticle cell internalization method, with particles < 100 nm in diameter being preferentially internalized by clathrin-mediated endocytosis, while larger particles require phagocytosis or micropinocytosis (Jiang et al. 2008). Nanoparticle shape can also influence these processes.

In immunotherapy applications, size has a critical influence on nanoparticle delivery to the lymph nodes and lymphatic trafficking (Schudel et al. 2019). Nanoparticles smaller than 50 nm in diameter selectively accumulate in draining lymph nodes whereas larger particles rely on migratory DC-dependent transport of particles to the draining lymph nodes to incite an immune response (Reddy et al. 2007, Liu et al. 2014). As a result, the direct lymph node processing of smaller particles could lead to more rapid and enhanced activation of cellular and humoral immune responses than larger particles that are dependent on cell migration to the lymph nodes. Furthermore, nanoparticle size can dictate particle access to different immune cells within the lymph nodes as the reticular network restricts the access of materials in the lymph from reaching the paracortex in a size-dependent fashion (Schudel et al. 2019). This can influence the immune response generated due to nanoparticle association with different lymph node immune cell populations. For example, nanoparticles that were similar in size to viruses were found to predominantly interact with B cells within the lymph nodes (Manolova et al. 2008). Therefore, immunotherapy nanomedicines that target interactions or processes occurring in specific subregions of the lymph node should include size optimization in their design strategy. Specific strategies that have been used to target the lymph nodes and its subregions are discussed further in this chapter's subsection, Targeting Tissues that Regulate Immunity.

In summary, nanoparticle size has a major influence over the particle biodistribution and transport kinetics, including in critical immune infrastructure like the lymphatics and lymph nodes. Nanoparticle size also contributes to particle geometry, including surface area and volume, which impact cell uptake, molecular and cellular interactions, tolerability, and the amount of therapy loaded.

3.1.3 Shape

Like size, nanoparticle shape has a significant influence over the nanoparticle's biological transport, biodistribution, and cell interactions. Particle shape can influence the surface area of the particle for a given volume, which can impact the adhesion dynamics of the particle due to the formation and maintenance of multivalent interactions with cell surface proteins (targeting, uptake) and circulating proteins (protein corona formation) (Kapate et al. 2021). Nanoparticle shape influences the fluid flow dynamics of particles within biological fluids, like the blood. In the blood vasculature, nanoparticles with non-spherical shapes can drift because of their asymmetry, which influences particle margination and, as a result, particle plasma half-life and biodistribution (Tan et al. 2012). Nanoparticle shape also influences the cell-particle interaction dynamics. Different types of innate immune cells demonstrate preferential nanoparticle uptake that is shape-dependent, with spherical nanoparticles being more preferentially taken up by macrophages and rod- and disk-shaped particles being more preferentially internalized by neutrophils (Safari et al. 2020). This is influenced by the internalization mechanisms of the cell. Non-phagocytic cells, such as endothelial and epithelial cells, use different mechanisms to uptake extracellular materials, such as endocytosis and pinocytosis (Kapate et al. 2021). Particles with lager aspect ratios exhibited higher uptake rates and improved cytosolic drug delivery in these cells (Parakhonskiy et al. 2015). Importantly, the shape-dependent difference in macrophage phagocytosis also impacts the biodistribution of these particles as it reduces the liver clearance rate of non-spherical particles from circulation in comparison to spherical particles, enhancing their bioavailability to other organs (Decuzzi et al. 2010, Arnida et al. 2011). Non-spherical particles also have less lung alveolar macrophage uptake, which was shown to reduce adverse nanoparticle-mediated cardiopulmonary reactions because of delayed particle recognition following intravenous administration in pigs (Wibroe et al. 2017). Following cell internalization, nanoparticle shape can also influence intracellular trafficking, as non-spherical particles were shown to increase nanoparticle import through the nuclear envelope, an important consideration for genetic cargo (Hinde et al. 2016). Finally, shape is known to impact the immune response to nanoparticles. Tazaki et al. found that gold nanoparticles with different shapes (sphere, rod, cubic) generated different APC cytokine secretion profiles and antibody titers following vaccination in mice (Tazaki et al. 2018). In a separate study, Kumar et al. demonstrated that antigen-conjugated spherical polystyrene particles generated an antigen-specific immune response that was biased towards Th1 with a lower IgG1/IgG2a antibody ratio while the rod-shaped particles' response was biased towards Th2 and had a higher IgG1/IgG2a antibody ratio (Kumar et al. 2015). In summary, nanoparticle shape has a significant influence over the fluid flow transport and cell-interaction dynamics of nanoparticles that can mediate shape-dependent biodistribution and immune activation profiles.

3.1.4 Elasticity

The impact of nanoparticle elasticity is not as well studied as the other nanoparticle physiochemical properties as it is more difficult to modulate (in isolation) across particle types. Elastic modulation is most often achieved by changing the particle crosslinking densities in gel-like particles. A study by Guo et al. examined the role of elasticity in nanolipogel particle uptake by different cell types and found that softer nanoparticles accumulated more in tumors, compared to their less deformable counterparts which accumulated in the liver (Guo et al. 2018). This resulted from an elasticity-dependent shift in particle internalization pathways where softer particles were able to use more energy-efficient fusion mechanisms and harder particles used endocytosis. Cell membranes can less efficiently wrap softer nanoparticles as these particles can deform and prevent the formation of adequate attachments, reducing cell internalization (Anselmo and Mitragotri 2017). This is further exacerbated if the particle has a large particle edge with which the cell is interacting. Similarly, forces driving ligand-receptor interactions between nanoparticles and cells can cause more significant deformation of softer particles, reducing their cell binding (Hui et al. 2020). Anselmo et al. demonstrated similar results for polyethylene glycol-based hydrogel nanoparticles and showed that harder nanoparticles are more rapidly and more frequently phagocytosed by immune cells (Anselmo et al. 2015). Together, this drives the reduced uptake of softer particles by circulating immune cells and liver macrophages *in vivo*, increasing their blood circulation time and tissue targeting potential. A nanoparticle's deformability can also influence its ability to squeeze between cell-cell gaps and through biological matrices, such as during extravasation (Song et al. 2021). This feature could be used to mediate enhanced delivery if the particle can squeeze through tight junctions and reform without losing its cargo. These studies demonstrate that nanoparticle biodistribution, pharmacokinetics, cell uptake, and cell surface interactions can be optimized by tuning the particle's elasticity.

3.2 Surface chemistry and surface modifications

Nanoparticles interact with many different fluids, molecules, cells, and tissue structures during their time in the body. These interactions are referred to as nanomaterial-biological interactions and they dictate many of the therapeutic outcomes of nanomedicines (Poon et al. 2020). The nanoparticle surface is the main interface of interaction between nanomaterials and the body's components. As a result, numerous studies have focused on engineering the particle surface by altering its chemistry, coating it with biological membranes, and modifying it with targeting ligands to achieve desired molecular and cellular interactions and biodistribution profiles.

Surface properties can impact the adsorption of proteins from biological fluids onto the nanoparticle, influencing the type, diversity, and conformation of proteins within the protein corona and its associated biological and immune responses (Lundqvist et al. 2008). The characteristics of the protein corona can influence the biodistribution of the nanoparticle, based on the particle's interaction with macrophages in the liver, APCs in the spleen, and other components of the reticuloendothelial system (Poon et al. 2020). For example, hydrophilic and inert coatings, such as poly(ethylene) glycol (PEG) coatings, are frequently used to reduce the formation of protein coronas on nanoparticle surfaces, which can minimize nanoparticle recognition by macrophages and other phagocytes (Stolnik et al. 1994). Despite these coatings, however, protein coronas generally still form and become a new interface through which the nanoparticles interact with cells and tissues, ultimately impacting their *in vivo* fate (cell uptake, biodistribution, immune response) (Vincent et al. 2021). The protein corona may also influence the efficacy of other surface modification approaches, such as targeting moieties (Mirshafiee et al. 2013, Salvati et al. 2013). Recent studies have shown that surface chemistry can mediate the adsorption of complement proteins and of albumin, as well as the folding state of these adsorbed proteins, which influenced particle opsonization, uptake by innate immune cells, and clearance (Moghimi and Szebeni 2003, Chen et al. 2016, Nguyen and Lee 2017, Ren et al. 2019, Vincent et al. 2021). Further, a recent study showed that variances in plasma composition between patients can cause differences in the formed protein coronas and the resultant nanoparticle-immune cell interaction profiles (Ju et al. 2020). As a result, strategies that can engineer the composition of the protein corona *in situ* or *ex vivo* to achieve specific nanoparticle properties are of intense interest.

The propensity of nanoparticles to form protein coronas has been leveraged for antigen capture and has been widely used in *in situ* vaccination strategies as described later in this chapter. In a seminal study, Min et al. leveraged this propensity of nanoparticles to form protein coronas in biological fluids to engineer nanoparticles that, when injected intratumorally, efficiently captured tumor-associated antigens released by radiation therapy and enhanced the presentation of these antigens by APCs (Min et al. 2017). The particles' surface chemistry impacted the composition and diversity of the formed protein corona, whereas unmodified particles and DOTAP nanoparticles captured the widest variety of proteins, including cancer neoantigens and DAMPs. When combined with checkpoint blockade immunotherapy, this strategy improved survival and tumor control in both primary treated and metastatic untreated lesions via the abscopal effect. Overall, this study demonstrated a new mechanism by which nanoparticle surface chemistry can influence protein corona characteristics to enhance the combination of standard therapies (like chemotherapy and radiotherapy) with immunotherapy and augment *in situ* vaccination strategies.

While not directly probing the mechanism of the nano-biological interactions, other studies have reported that particle surface charge can influence cell uptake propensity and immune cell activation, with cationic particles demonstrating higher uptake and being more inflammatory than anionic particles (He et al. 2010, Fromen et al. 2016). Kranz et al. found that negatively charged lipid nanoparticles could enhance the delivery efficiency of RNA to lymphoid organs and APCs, particularly splenic dendritic cells, following intravenous administration (Kranz et al. 2016). In a separate study, Cheng et al. showed that lipid nanoparticles could be targeted to select organs (besides the liver) and cells by altering the amount and charge of different lipid "SORT" molecules, which influenced the generated particles' global pKa (Cheng et al. 2021). While not exhaustive, these examples illustrate how nanoparticle surface properties can influence particle biodistribution, pharmacokinetics, and cell interactions.

To achieve biological interactions with high specificity and high affinity, active targeting agents, such as antibodies, antibody fragments, ligands, DNA aptamers, and peptides, can be chemically conjugated or indirectly bound to the surface of the nanoparticle (Bertrand et al. 2014). In immunotherapy, this is an attractive approach because many immune cell types and activation states are classified based on surface marker expression, which can be targeted by these molecules to enhance cell and activation state delivery selectivity. Numerous studies have used this functionality to target nanoparticles to immune cells *in vivo* (Ramishetti et al. 2015, Schmid et al. 2017, Smith et al. 2017, Kedmi et al. 2018, Veiga et al. 2018, Zhang et al. 2019a). Further, antibodies, which are commonly used targeting agents because of their high specificity and affinity, can also be engineered to interact with innate immune surface receptors through their Fc regions, resulting in specific effector functions. Molecular engineering approaches can be further appropriated to generate targeting agents with specific structures and activities and that can leverage different conjugation chemistries. The number of moieties, their distribution, and their density on the particle surface impacts the observed binding affinity and selectivity of the nanoparticle construct for its targeted moiety (Bertrand et al. 2014, Tjandra and Thordarson 2019). Depending on the properties of the targeting agent, these features can also affect the duration and magnitude of the invoked signaling cascade, which can be important if the agent is also being used for immune modulation or other therapeutic purposes. However, these interactions are ligand-target pair specific as the conformational binding dynamics of different moieties can influence the receptor antagonism, agonism, or neutral responses.

Multivalency is used to enhance the discrimination of active targeting agents for different cell types through super-selectivity (Tjandra and Thordarson 2019). Multivalent binding is the formation of multiple bonds between nanoparticle surface molecules and their receptors on the target cell or biological entity. Super-selectivity is the ability of multivalent constructs to obtain near on-off binding to their targets expressing a certain number of receptors compared to monovalent ones (Liu et al. 2020). This, in theory, allows cancer cells to be discriminated from healthy cells when nanoparticles are used to target receptors overexpressed by cancer cells (Carlson et al. 2007). However, studies have suggested that the application of this approach in the *in vivo* setting is complex, with scavenger endothelial cells and macrophages acting as cellular reservoirs for nanoparticles (Miller et al. 2015, Hayashi et al. 2020) and macrophages taking up more cancer cell targeted nanoparticles than tumor cells (Turk et al. 2004). Further, a study by Korangath et al. suggests that antibody-targeted nanoparticle retention in immunocompetent mice is a function of nanoparticle-immune cell interactions rather than antibody-cell receptor interactions (Korangath et al. 2020). A more thorough study of the influence of targeting molecule characteristics (Fc type, ligand type, multivalency), receptor type, conjugation chemistry and specificity, and nanoparticle properties (size, material type) is needed before this can be drawn as a general conclusion regarding nanoparticle active targeting. While macrophage phagocytosis can be exploited based on the physiochemical properties of the particle, as previously discussed, these findings suggest that the nanotechnology-biological interactions regarding receptor-targeting (and others) need to be more carefully investigated. Recently, using statistical mechanical modeling and experimental trials, Liu et al. described the combinatorial entropy-related phenomena of range selectivity where, under certain conditions, multivalent entities bind only to other entities that have a receptor density within a given span (Liu et al. 2020) (Of note, this phenomenon can also occur for monovalent constructs, but it is described here given its importance to understanding multivalent biological interactions). The authors suggested that, biologically, this phenomenon could be used by cells to unbind nanoparticle constructs from their surface receptors by increasing the number of receptors available or to avoid recognition by other cell types. Together, these studies illustrate the need to carefully study the intricacies of nanotechnology-biological interactions, including those governing active targeting. Beyond cell type targeting, surface modification with biologically active agents can be used to control the presentation, diversity, and valency of immunomodulatory signals presented to immune cells by nanomedicines (Irvine and Dane 2020). This concept and its applications are discussed further later in this chapter (Improved immune signaling and immune interactions using multivalency).

Coating nanoparticles with cell membranes or creating nanoparticles from cell membrane-derived vesicles is another method to appropriate bioderived entities to recapitulate complex biological interactions. In comparison to active targeting strategies, these strategies can easily incorporate multiple surface molecules and leverage inherent cell-cell interactions and biological functions of the cell from which the membrane was derived. However, this approach gives less control over the number and the identity of the surface moieties and may require *ex vivo* manipulation of donor cells. Genetic engineering can enhance control over certain characteristics of the derived cell membrane or vesicle, such as surface protein expression, to achieve tailored functions. As demonstrated later in this chapter, the creation of nanoparticle-cell hybrids using these strategies has resulted in nanoparticle immunotherapeutic strategies with novel functions.

3.3 Stimuli-responsive delivery and controlled release

Stimuli-responsive materials have allowed for the controlled alteration of particle physiochemical properties and the release of surface coatings and targeting agents to allow for improved nanoparticle maneuvering of different biological microenvironments. Additionally, this feature can program nanoparticles to release therapies in response to endogenous alterations in the target microenvironments or in response to exogenously provided cues and be used to cleave prodrugs (Goldberg 2019). As a result, novel immunomodulatory drug release profiles (such as pulsatile release) can be achieved in addition to improved delivery to specific intracellular compartments and anatomical locations. This can be important in controlling the delivery of cytotoxic agents, such as chemotherapy and antimicrobial agents, which can synergize with immunotherapy by causing immunogenic cell death, by eliminating immune-evasive clones or variants, and/or by reducing disease burden. Careful attention should be paid, however, to the kinetics of these therapies' release profiles, as prolonged exposure to subtherapeutic levels can enhance the development of therapeutic resistance and burst release can obliterate bystander immune cells in the disease microenvironment, potentially limiting its therapeutic synergy with immunotherapy and safety (Kirtane et al. 2021). External stimuli, such as those derived from photothermal and ultrasonic modalities, can also provide therapeutic effects. Examples of different stimuli include pH, temperature, enzymes (endogenous host/microbiome-derived and CRISPR-Cas9), redox potential, light, ultrasonic waves, and electrical and magnetic fields (Mura et al. 2013). In a related feature, the chemistry of the nanoparticle and its stimuli-responsiveness can allow for programmed release of therapy with different kinetics, including sustained, sequential, and pulsatile profiles. This can widen the therapeutic window, impact the nature of the generated immune responses, and reduce the dosing frequency of therapy (Goldberg 2019). For example, vaccination with poly(lactide-co-glycolide) nanoparticles was shown to generate superior humoral and cellular responses than vaccination with liposomal particles due to differences in antigen release kinetics (Demento et al. 2012). As this example illustrates, one of the largest contributions these features provide to nanomaterials-based immunotherapy is the ability to better control the spatiotemporal kinetics of the therapy or therapies, allowing for dynamic alterations in immune signaling, and in the magnitude and duration of the incited immune responses.

3.4 Delivery context

While not a direct parameter of the nanotechnology itself, the delivery context of the immunomodulatory nanoparticle therapy, including the number of particles given, its route of administration, and the timing of its administration relative to other therapies, can influence the design of the nanoparticle, its fate in the body, and the outcome of immunotherapy. Ouyang et al. made the important discovery that a dose threshold exists for enhancing nanoparticle accumulation in tumors (Ouyang et al. 2020). The authors showed that doses over 1 trillion nanoparticles overwhelmed Kupffer cell uptake rates, causing a nonlinear decrease in liver clearance and prolonging nanoparticle circulation. This enhanced tumor delivery efficiency to up to 12% in comparison to the median delivery of 0.7% of injected dose typically seen using nanoparticle-mediated delivery (Wilhelm et al. 2017). While applied to chemotherapy here, this concept is critical for the design of systemic nanomedicine therapies that target tissues outside the organs of the reticuloendothelial system. This study is an important example of how nanomedicine design extends beyond particle attributes alone to achieve effective delivery and therapeutic outcomes.

The route of administration determines the molecules, cells, tissues, and organs to which the nanomedicine will be exposed and the biological barriers that will impede its delivery to the target site. It determines which components the nanoparticle will interact with and in what order the nanoparticle will encounter them as it travels to its target site (or otherwise). This has a significant influence on many nanomaterial-biological interactions, like protein corona formation, as different protein types and compositions are encountered in the blood than in the interstitial fluid, cerebrospinal fluid, saliva, and mucosa (Nguyen and Lee 2017). These interactions dictate the nanomaterial's fate in the body (biodistribution, pharmacokinetics) and its ability to perform its therapeutic function effectively (Poon et al. 2020). This, in turn, influences the propensity of the nanomedicine to accumulate in off-target sites, in part dictating the development of adverse side effects. In the context of immunotherapy, local administration has been investigated as it can confine potent therapies to the target site, enhancing their interactions with these cells and tissues and minimizing their exposure to off-target tissues (Goldberg 2019, Irvine and Dane 2020). Francis et al. demonstrated that checkpoint blockade therapy was more efficacious when delivered locoregionally to tumors using materials than when given systemically because of increased therapy accumulation in target tissues (tumor, tumor-draining lymph nodes) (Francis et al. 2020). Nanomaterials can be used to increase the retention of therapies at target sites by engineering their size and surface chemistry to interact with components of the microenvironment, such as the extracellular matrix (ECM) (Engin et al. 2017). Systemic therapy via intravenous injection (IV) is more clinically translatable and can reach sites that are not readily accessible for local injection, while transdermal and intramuscular delivery are frequently used for the delivery of vaccines. In a recent study using synthetic nanoparticle vaccines linking neoantigen peptides and a TLR7/8 agonist, Baharom et al. showed that route of administration altered the quality of the immune response generated in response to nanoparticle-mediated vaccination because of the duration of antigen presentation by DCs (Baharom et al. 2020). IV vaccination resulted in more stem-like T cells than subcutaneous injection, which generated more effector cells. When combined with checkpoint blockade, therapeutic IV vaccination led to improved antitumor responses in comparison to subcutaneous vaccination. Another study by Hunter et al. reported that route of administration also impacted the efficacy of a tolerogenic vaccine against experimental autoimmune encephalomyelitis because of its effect on the internalization of nanoparticles by different DC subtypes (Hunter et al. 2014). IV delivery again showed improved outcomes in comparison to subcutaneous, intraperitoneal, and oral administration routes.

One important consideration in selecting the route of administration for a nano-immunotherapy is understanding the differences between the immune response the therapy generates in different tissues. It is increasingly recognized that immune responses are tissue-specific (Hu and Pasare 2013, Horton et al. 2019) and, as a result, the impact and response generated by one route of administration may significantly differ from another if it accumulates in lymphoid tissue in one area of the body as opposed to another. For example, Hiltensperger et al. showed that helper T cells primed in the inguinal lymph nodes versus the mesenteric gut draining lymph nodes have distinct functional phenotypes and infiltration patterns in central nervous system autoimmunity while a study by Esterházy et al. demonstrated that the lymph nodes of the same organ can have distinct immunological function, with proximal lymph nodes in the gut mediating tolerogenic responses and distal lymph nodes mediating pro-inflammatory responses (Hiltensperger et al. 2021, Esterházy et al. 2019). As a result, the immunobiology of different organs and disease pathologies are important considerations when designing nanomaterial-based immunotherapy systems.

Finally, when designing a nanomaterial immunotherapy strategy, one must consider its context in relation to therapies used in current clinical practice and within proposed combinations. Often, it is desirable to situate novel immunotherapies within the context of currently used clinical therapies, as you can build upon the known efficacy of these therapies, more readily enabling clinical translation. However, for example, conventional cancer therapies, like chemotherapy and surgery, can have local and systemic immunological effects that can blunt the efficacy of some immunotherapies (Galluzzi et al. 2018). Modifying the timing or delivery of these therapies in relation to nanoparticle immunotherapy (or in their release from combination chemoimmunotherapy nanoparticle systems) can determine whether its impact on the resultant immunological response is detrimental, neutral, additive, or synergistic (Silvestrini et al. 2017). Furthermore, certain therapies, such as anti-vascular endothelial growth factor antibodies, ultrasound and photothermal therapy, can be used to precondition target microenvironments to enhance nanoparticle therapy delivery (Mead et al. 2019, Chen et al. 2019). Conversely, nanoparticle immunotherapy can also be used to alter the target tissue's immunobiology prior to conventional therapy. In one example, gold nanoparticles were conjugated to TNF-a (a vascular targeting agent) and administered prior to photothermal therapy in tumor-bearing rodents (Shenoi et al. 2011). These particles induced vascular damage and significant decreases in blood flow, recruiting inflammatory immune cells to the tumor and significantly enhancing photothermal therapy. These studies demonstrate how the therapeutic context of nanoparticle therapy can influence immunotherapeutic outcomes.

3.5 Summary

Taken together, these sections provide a foundation for understanding the influence of nanoparticle properties on nanotechnology-mediated immunotherapeutic delivery and therapeutic outcomes. The next sections will describe applications in which these nanomaterial engineering principles were leveraged to develop innovative therapy for cancer, infectious diseases, and autoimmune diseases.

4. Leveraging nanomaterial properties for enhanced immunotherapy

Nanomaterials can be designed with enhanced properties that achieve more specific delivery across multiple length scales, more potent immunomodulatory properties, and that direct the spatiotemporal action of combination therapy to enable the design of safe and efficacious immunotherapeutic treatments. In this section, we discuss the ways in which the different intrinsic properties of nanomaterials, such as material type, stimuli responsiveness, and surface modification, have been leveraged to modulate immune cells and their interactions with other cells and microorganisms.

4.1 Direct immunomodulation through novel material design

Recent studies have illustrated that not all materials are immunologically inert, as previously thought, but instead can generate potent immunomodulatory responses. A summary of the broad classification of the anti-inflammatory or pro-inflammatory properties of different material types is given in Table 2. However, many of these immune responses depend on a range of material-based and application-based factors, and the magnitude of the immune response varies greatly between some materials. Regardless, the ability for a material to modulate the immune system is attractive as it can minimize the number of components needed within a given nanomaterial immunotherapy

Table 2. The Immunological Effects of Materials. A broad categorization of the immunomodulatory effects of common materials used in nanotechnology. It should be noted that many of these observed immune responses depend on a range of material-based and application-based factors, and the magnitude of the immune response varies greatly between some materials and studies. As a result, this should be used as a guide for material selection, but individual application and context screening should be completed for a given indication. Materials that are immunologically inert or have not been studied for their immunomodulatory properties are excluded. (Source: Adapted from Shields et al. (2019). Materials for Immunotherapy. *Adv. Mater.*, 32(13): 1901633.)

Material Category	Anti-Inflammatory	Inflammatory
Natural	Chitosan, Hyaluronic acid (> 1000 kDa), Nanocellulose	Alginate, Hyaluronic acid (< 10 kDa), Pullulan, Collagen, Nucleic acids, Cationic lipids
Synthetic	N/A	Polyesters (PLGA, PCL), PAMAM dendrimers
Inorganic	Carbon nanotubes, Gold	Iron oxide, Mesoporous silica, Alum, Graphene, Titanium Dioxide, Silver

formulation. This has resulted in the design of nanomaterials and nanoparticle compositions with immunomodulatory effects.

Traditional polymer-based materials have been explored as drug-free strategies for modulating immune responses. Casey et al. developed cargo-free nanoparticles to control inflammation resulting from sepsis (Casey et al. 2019). Here, the authors showed that the immunomodulation achieved by the particles, composed of poly(lactic-co-glycolic acid) and poly(lactic acid), was dependent on numerous particle properties including molecular weight, composition, and charge. Particles prepared with a negatively charged surfactant (poly(ethylene-alt-maleic acid), PEMA) significantly blunted immune cell responses to intracellular and extracellular TLR stimulation, whereas particles prepared with a neutral-charged surfactant had limited activity. Mice treated with the PEMA particles survived longer in an endotoxemia model. In a separate study, Saito et al. investigated how polymer molecular weight and composition influenced the activity of drug-free poly(DL-lactide-co-glycolide) and poly(DL-lactide) nanoparticles in EAE (Saito et al. 2019). Nanoparticles composed of high molecular weight PLG resulted in improved disease control in comparison to low molecular weight PLG particles and PDLA particles because of increased particle interaction with immune cells in circulation that restrained these cells entrance into the central nervous system, lowering inflammation. Together, these studies show how particle material, composition, and preparation can be used to generate immunomodulatory responses using only polymeric materials.

Several novel synthetic materials have also been developed that have intrinsic immune adjuvant properties in addition to their delivery properties to enable minimalist nanovaccines (Figure 5A; Luo et al. 2017). Luo et al. were the first to demonstrate that a synthetic material could simultaneously agonize innate immune STING pathway signaling (which promotes immune activation, inflammation, and antigen cross-presentation) and efficiently deliver tumor antigens to the cytosol of DCs (Luo et al. 2017). To identify this polymer, the authors screened a library of pH-sensitive polymers for their ability to incite cytotoxic lymphocyte and helper T cell responses (Figure 5B-C). The transition pH (pH 6.9, early endosomal pH) and polymer architecture (seven-membered ring side chain) dictated the efficacy of the induced T cell response. Further, the polymer's transition pH also determined its ability to buffer the luminal pH of the endosome to drive cytosolic delivery of its encapsulated antigens via the proton sponge effect. (According to the Proton Sponge Effect, the polymer's absorption and accumulation of protons in the endosome increases the membrane potential and drives the diffusion of chloride into the endosome, increasing osmotic pressure until the endosome's lipid bilayer ruptures and releases its contents into the cell (Akinc et al. 2005).) These properties allowed for potent STING-dependent inhibition of tumor growth in multiple murine models and, when combined with checkpoint blockade therapy, led to 100% survival in a

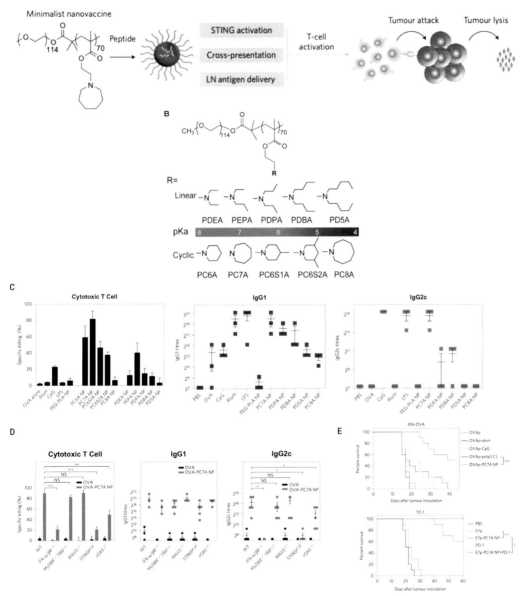

Figure 5. Minimalist Nanovaccines Using Inherently Immunostimulatory Materials. (A) Schematic of the design of a minimalist nanovaccine for cancer immunotherapy. (B) Depiction of the characteristics of the ultra pH-sensitive polymer library screened for their capability to incite cellular and humoral immune responses *in vivo*. (C) Quantitative comparison of the cytotoxic cellular and antibody humoral immune responses incited by the different polymeric nanoparticles loaded with OVA-antigen in comparison to responses derived from vaccination with mixtures of OVA and common immune adjuvants (L to R: CTL response, IgG1 titer, IgG2c titer). (D) The cytotoxic cellular and antibody humoral immune responses incited by the minimalist PC7A OVA nanoparticle vaccine are dependent on STING pathway signaling (L to R: CTL response, IgG1 titer, IgG2c titer). (E) The minimalist PC7A OVA nanoparticle or PC7A E7p nanoparticle vaccines significantly enhances survival in two murine cancer models, with combination with aPD-1 checkpoint blockade therapy resulting in complete curative survival in the TC-1 model. (Reprinted with permission from Luo et al. (2017). A STING-activating nanovaccine for cancer immunotherapy. *Nat. Nanotechnol.*, 12(7): 648–654.)

TC-1 tumor model (Figure 5D-E). This group also showed that this nanovaccine bound STING in a non-competitive site to traditional agonists, such as cGAMP, allowing for STING activation in tumor tissue derived from cGAMP-resistant patient populations (Li et al. 2021). A separate study by another group used a three-dimensional multicomponent reaction system to screen over 1,000 ionizable lipid nanoparticle formulations to identify mRNA delivery vehicles that activated potent immune responses (Miao et al. 2019). The lipids with the best performances each contained common features, including a cyclic amine head group. Like Luo et al.'s polymer, these formulations enhanced vaccine-mediated anti-tumor efficacy via activation of the STING pathway. Nanomaterials have also been developed that agonize other innate immune receptors, such as TLRs (Mosquera et al. 2019). Additional strategies have investigated the incorporation of known immunomodulatory materials directly into the polymer backbone. In a recent study, Davenport Huyer et al. created a library of polyesters that incorporated itaconate, a power anti-inflammatory molecule that inhibits succinate dehydrogenase in macrophages impacting the generation of reactive oxygen species, into the polymer backbone and assessed their ability to recapitulate the anti-inflammatory and antibacterial properties of the parent molecule (Davenport Huyer et al. 2020). These itaconate polymers were able to rapidly resolve inflammation in comparison to silicone control polymers in an *in vivo* model of biomaterial-driven inflammation.

Another strategy is to use the immunomodulatory therapy itself to form the nanoparticle material. This is commonly achieved using therapies that are themselves biological materials, such as DNA and antibodies. Antibodies are y-shaped immune proteins that interact with their target antigens through complementarity-determining regions (CDRs) in their fragment antigen binding domains (Fabs) and with receptors on immune cells through their constant regions (Fc). Molecular engineering techniques can be used to develop antibodies with exquisite selectivity and affinity for their target antigens. The framework residues of the Fc regions can also be engineered using molecular techniques, beyond the differences seen between different classes of antibodies, which allow them to execute a wide range of effector functions. Antibodies are widely used immunotherapies due to these antigen-neutralizing and immunomodulatory features and can be easily manufactured using genetic engineering techniques. Divine et al. used computational techniques to design proteins that can drive the self-assembly of antibodies or Fc-fusion proteins into nanocages with controlled valency and geometry (Divine et al. 2021). Cages were designed from 2, 6, 12, or 30 antibodies without the need for chemical modification of the antibody or the incorporation of polymers. These structures enhanced antibody-dependent signaling in different applications. In cancer therapy, nanocages formed with a death-receptor targeting antibody could induce apoptosis of tumor cells, which mediated immunogenic cell death. In infectious disease, this strategy was able to neutralize pseudo-virus when assembled from antibodies against the SARS-Cov2 spike protein. This strategy has many advantages over traditional nanoparticle antibody conjugates, such as ease of manufacturing due to self-assembly, controllable antibody orientations, and programmable symmetric higher order structures. Given the modular nature of antibodies, this strategy can be applied in multiple ways to enhance antibody signaling and therapeutic efficacy. In a similar manner, DNA can also be organized into programmable structures using DNA origami (Zhang et al. 2014). The application of programmable protein and DNA nanostructures is discussed further in Section 4.3, where these structures are used to present antigens to APCs with defined density and valencies. In a separate strategy, an enzymatic rolling circle amplification method was used to generate a DNA nanococoon from CpG oligodeoxynucleotides, which are innate immune TLR9 agonists, that encapsulated a checkpoint blockade antibody and triglycerol monostearate (TGMS) nanoparticles containing a restriction enzyme (Wang et al. 2016). This nano-immunotherapy was used to prevent post-surgical tumor recurrence. Post-surgical inflammation triggers a nanomaterial degradation cascade where inflammation-related enzymes cleave the ester linkages of the TGMS nanoparticles to release their encapsulate DNA-cutting restriction enzyme, which in turn degrades the nano-cocoon, enabling the concurrent sustained release of checkpoint blockade therapy and CpG. The sustained, local release of combination immunotherapy significantly improved outcomes in metastatic cancer models following primary tumor resection.

4.2 Targeting tissues that modify immunity

Lymphoid organs are delivery sites of interest for immunotherapy because they contain high concentrations of immune cells and are often the locations where immunological important processes are carried out such as lymphocyte and leukocyte development and activation. Nanomaterials have been used to enhance the delivery of immunotherapies to the lymph nodes for vaccine development and cancer therapy. Targeting therapeutics to the lymph nodes is an attractive strategy as it can be used to address lymphocyte and sentinel lymph node metastatic malignancies and latent viral reservoirs as well as promote vaccine responses in prophylactic and therapeutic settings for both immunogenic and tolerogenic applications (Schudel et al. 2020). After subcutaneous, intradermal, intratumoral, or intramuscular injection, lymphatic trafficking is largely mediated by two pathways: (1) cell-mediated transport, where APCs uptake the particles at the injection site and home to the lymph nodes to induce adaptive immune responses; (2) direct lymphatic drainage, where particles drain to the lymph nodes directly via the lymphatics (Irvine et al. 2020). Particles can also diffuse into the blood, limiting their exposure to the lymph nodes and potentially causing systemic adverse effects.

Studies have examined the impact of size, charge, deformability, and targeting ligands on the delivery of nanoparticles to the lymph nodes (Schudel et al. 2019). In a seminal study, Reddy et al. developed antigen-conjugated poly(propylene sulfide) nanoparticle vaccines to target lymph-node resident dendritic cells and activate these cells via *in situ* complement activation (Reddy et al. 2007). Importantly, the authors showed that the lymph node delivery efficiency of these nanoparticle vaccines was critically dependent on particle size, where 100 nm particles were less efficient than 25 nm particles. This difference led to demonstrable differences in the generation of humoral and cellular immunity. As a result, many studies have targeted this smaller particle size directly or through stimuli-responsive alterations in size to enhance lymph node drainage. In a unique study, Song et al. developed a biomimetic strategy where they engineered the deformability of albumin-stabilized vaccine nanoemulsions (~ 330 nm) to mimic the deformability of immunocytes to enable both direct lymph node targeting and antigen-presenting cell-mediated transport of these particles (Song et al. 2021). The deformability of the droplets allowed for direct lymph node migration while the size allowed for some of the particles to form an antigen depot at the injection site. Nanoemulsions with and without adjuvant co-delivery (mono-phosphoryl lipid A) significantly improved survival and delayed tumor growth in mice, with 100% survival seen out to Day 50 for adjuvanted nanoemulsions.

In a different strategy, Schudel et al. developed a dynamic nanoparticle approach using programmable degradable linkers to enable delivery of therapy to specific lymph node structures and their associated lymphocyte subpopulations (Figure 6A) (Schudel et al. 2020). Delivering therapy to specific cells within lymph nodes is difficult to achieve due to anatomical barriers that regulate the passage of materials based on physiochemical properties, such as size. To overcome these barriers and achieve intra-lymph-node-specific delivery, the authors conjugated cargo to thiolated poly(propylene sulfide) nanoparticles using thiol-reactive oxanorbornadiene linkers, where the linker half-lives are dependent on the oxanorbornadiene substituents (Figure 6B). Particle size was selected for efficient transport to the lymph nodes while linker half-life was selected depending on the targeted cells. The use of the OND linker nanoparticle system led to increased cell uptake of small molecule cargos in cells of the lymph node paracortex (Figure 6C), with peak uptake of cargo in these cells corresponding to the half-life of the OND linker (Figure 6D). This approach could allow for enhanced targeting of immunomodulatory small molecules to different areas of the lymph nodes to enhance the generation of adaptive immune responses. As you will see

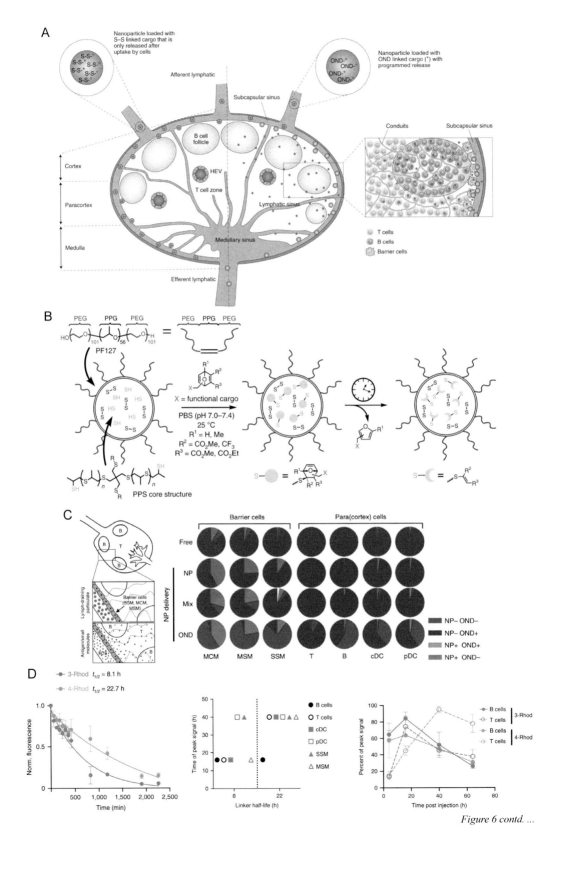

Figure 6 contd. ...

in additional sections, scientists and engineers have tuned the physiochemical and surface properties of nanomedicines to reach a wide range of lymphoid organs.

In a separate approach, immunity can be modulated by targeting non-lymphoid tissues that are known to exert immunomodulatory effects. These include the gut microbiome and disease microenvironment, among others. As the disease microenvironment location varies greatly with different types of disease and within the same families of disease, we will focus our discussion on targeting the body's microbiome. The microbiome is the collection of microorganisms found on the surface of barrier tissues in the skin, lungs, mouth, throat, gut, and vagina. The disease microenvironment can also have its own associated microbiome that influences treatment efficacy and patient outcomes (Riquelme et al. 2019, Nejman et al. 2020). Dysbiosis of the gut microbiome impacts the immunopathology of many diseases including cancer (Gopalakrishnan et al. 2017, Matson et al. 2018) and IBD (Round and Mazmanian 2009, Halfvarson et al. 2017). Most notably, in cancer immunotherapy, recent studies have highlighted its contribution to the poor response rates seen in patients treated with checkpoint blockade immunotherapy (Gopalakrishnan et al. 2017, Matson et al. 2018). Nanomaterials can be used to enhance targeting of therapy to different regions within the colon and to directly modify metabolic and immunologic functions of the gut microbiome (Song et al. 2019), such as their production of short chain fatty acids (SCFAs) that influence the activity of the immune system (Scheppach et al. 1995, Parada Venegas et al. 2019).

Studies have used the innate ability of materials, such as dietary fibers, to directly modify the gut microbiome to enhance immunomodulatory therapy outcomes in cancer and IBD. Han et al. showed that the oral administration of a colon-retentive inulin gel could modify the gut microbiome *in situ* by increasing the number of anti-tumor commensal microorganism and their production of immunomodulatory SCFAs (Han et al. 2021). This therapy synergized with checkpoint inhibitor therapy and induced notable changes in the local and systemic immune response, including the establishment of stem-like T cells within the TME and of systemic memory T cells. This study provides a workflow for how materials can be screened for their microbiome-modulating activity and formulated into gels using materials chemistry to retain in specific locations in the gut. In a related work by the same group, Lee et al. described a nanomedicine created from hyaluronic acid and bilirubin to treat colitis that accumulated in inflamed colon tissue, regulated the response of pro-inflammatory macrophages, and modulated the gut microbiome to increase its diversity and richness (Lee et al. 2019). The authors showed that the efficacy of this therapy was in part mediated by its ability to modify the gut microbiome as depletion of commensal bacteria with broad-spectrum antibiotics partly abrogated its therapeutic efficacy. In another study, Mosquera et al. demonstrated that the gut microbiome can modulate the immune activity and efficacy of nanovaccines in mice with metabolic syndrome, which is prevalent in approximately 25% of the world's population (Mosquera et al. 2019, Saklayen 2018). In this study, they showed that disease-altered TLR5 innate

---

Figure 6 contd. ...

Figure 6. Leveraging Programmable Nanoscale Chemistry for Enhanced Delivery to Lymph Node Subregions. (A) Schematic illustrating the difference when an intracellular redox sensitive linker nanoparticle is used for delivery to lymph node subregions in comparison to the programmable release from oxanorbornadiene (OND) linker particles developed by Schudel et al. (B) Schematic depicting the preparation of poly(propylene sulfide) nanoparticles with OND linkers that release their cargo using a retro-Diels-Alder mechanism, with linker half-life being determined by the OND substituents. (C) Cell uptake percentage by cell type and lymph node region of AF647-nanoparticle or OND-cargo or both following administration of free cargo (Free), NPs labeled with AF-647 only (NP), AF-647 labeled NPs mixed with free cargo (Mix), and NPs labeled with both AF-647 and OND-conjugated cargo (OND). MCM = medullary cord macrophages; MSM = medullary sinus macrophages; SSM = subcapsular sinus macrophages; cDC = conventional dendritic cell; pDC = plasmacytoid dendritic cell. (D) Influence of OND linker half-life on the peak fluorescence signal time and duration in different draining lymph node cell populations following OND nanoparticle treatment. (Figure 6A: Reprinted with permission from Porter and Trevaskis. (2020). Targeting immune cells within lymph nodes. *Nat. Nanotechnol.*, 15(6): 423–425. Figure 6B-D: Reprinted with permission from Schudel et al. (2020). Programmable multistage drug delivery to lymph nodes. *Nat. Nanotechnol.*, 15(6): 491–499.)

immune sensing of the gut microbiota diminished germinal center responses to poly(lactic-co-glycolic acid) nanovaccines. Interestingly, the PLGA vaccines also impacted the composition of the gut microbiome. To rescue the low immune response observed for nanovaccines, the authors developed an immunostimulatory pyridine-poly(hydroxyethyl methacrylate) nanogel vaccine, which functions through an alternative TLR, that resulted in stronger germinal center responses than adjuvant-supplemented PLGA vaccine. This demonstrates how material's choice can influence gut microbiome composition and overcome dampened immune responses to vaccines resulting from disease pathology.

Phage-guided nanomedicines can also be used to both target and modulate specific bacterial species in the microbiome. A unique study by Zheng et al. used phage-guided delivery of abiotic-biotic chemotherapy-loaded nanoparticles to treat colorectal cancer in a multi-faceted approach (Zheng et al. 2019). Bacteriophages are viruses that propagate within bacteria with high specificity, allowing them to target individual species of bacteria without infecting other bacteria or mammalian cells, making them desirable therapeutic agents for tailored microbiome modification. The authors isolated and selected a bacteriophage from human saliva based on its ability to target *F. nucleatum*, which is known to drive the formation of a highly chemo-resistant and immunosuppressive tumor microenvironment (Yu et al. 2017). The isolated phage was functionalized with azide moieties to enable biorthogonal click-chemistry conjugation of DBCO-modified chemotherapy-loaded nanoparticles. Furthermore, the authors tailored the design of the covalently linked nanoparticles to have multifaceted anti-tumor activity. They screened eight different types of materials for their ability to support the proliferation of commensal anti-tumor microbes as a prebiotic therapy before selecting dextran. Irinotecan was also encapsulated within the dextran nanoparticles, enhancing its delivery specificity. When given orally or intravenously to mice bearing orthotopic colorectal tumors, the phage-targeted nanomedicine showed enhanced tumor growth control and survival as well as reduced chemotherapy-associated side effects. Together, this work demonstrates how phage-guided nanomaterials can enable combination delivery of therapeutic agents to disease sites by targeting tumor-associated bacteria, with a unique focus on how the choice of nanoparticle material and carrier agents can be used to enhance anti-tumor immunity by therapeutically modifying the gut microbiome. Similar strategies could be used to modify the gut microbiome to enhance immunity in other diseases and to directly treat inflammatory disease like IBD.

4.3 Improved immune signaling and immune interactions using multivalency

The ability of nanoparticles to display multiple copies of the same ligand and multiple copies of different ligands have allowed nanoparticles to effectively target specific cells, mimic cell-cell immune interactions, and bridge multiple cell types (such as immune and cancer cells) to enhance immune signaling and immunotherapeutic outcomes. Nanoparticle multivalency can also control the way that antigen is presented to the immune system for processing in nanovaccines.

While many studies have focused on how nanoparticle multivalency can influence extracellular or cell surface interactions as described earlier in this chapter and below, few studies have focused on the ways polymer multivalency can engineer intracellular signaling. A recent proof-of-concept study by Li et al. demonstrated how polymer multivalency can be used to induce the assembly of macromolecular complexes of an intracellular signaling protein, intracellular stimulator of interferon genes protein (STING), through polyvalent phase condensation (Li et al. 2021). In signaling pathways, the assembly of membrane-less macromolecular complexes through phase condensation can lead to enhanced sensitivity to small changes in the microenvironment, impacting the regulation of signal transduction (Banani et al. 2017). The PC7A polymer used in this study was a pH-sensitive polymer containing seven-membered cyclic tertiary amine structures, which together allow for its cellular internalization and activation of STING. Luo et al. had previously synthesized this polymer to act concurrently as a nano-vaccine delivery vehicle and immune adjuvant. Here, the multivalency of the PC7A polymer led to distinct spatiotemporal STING activation profiles in comparison to cGAMP, an endogenous cyclic dinucleotide STING agonist that can induce STING oligomerization upon binding. The duration of STING activation and the related cytokine production and the robustness of the induced innate immune response was dependent on the length of the polymer, whereas longer polymers with higher valency (to a certain extent) induced higher levels of STING activation compared to polymers of lower valency and to cGAMP. While these alterations in spatiotemporal signaling dynamics did not lead to significant changes in survival in comparison to cGAMP therapy, the ability to engineer biological processes that are regulated in part by polyvalent phase condensation, such as gene expression and signal transduction, using polymers as therapeutics holds great promise in immunotherapy.

Nanoparticle multivalency has been used to enhance cell-cell interactions through the development of nanomaterial-based bispecific T cell and NK cell engagers (nanoengagers), allowing for enhanced cytotoxic activity of adaptive and innate immune cells against tumors (Yuan et al. 2017, Au et al. 2020). Nanoengagers act as a bridge between immune cells and tumor cells to induce tumor cell cytotoxicity that is MHC-I-independent, addressing issues with loss of antigen or MHC expression that can result in immune escape (Galluzzi et al. 2018). Further, multivalent contact between the nanoengager and both cell types can enhance apparent binding affinity because of avidity, resulting in improved establishment of each pair of cells, which can improve cell killing. Different combinations of antibody or targeting moieties can be further used to generate different cell type pairings to enhance their interactions. Immunomodulatory agents, such as cytokines, can also be encapsulated within the core of the nanoengagers and released over time to enhance immune cell recruitment or to further stimulate the desired immune action or cell function. While promising, nanoparticle architectures or chemistries that provide enhanced control over the orientation and localization of the conjugated antibodies or targeting agents could significantly improve this strategy. Future studies will also have to demonstrate the superiority of this approach to molecular engineering and bioconjugation strategies that achieve the same function, as these approaches are more readily manufacturable and translatable.

In a separate application, multivalency has been used in the design of nanovaccines. Nanoparticle architecture and conjugation chemistry influence the ways in which the antigens are transported, processed and displayed following immunization, resulting in immune responses of different potency to nanovaccines (Eppler and Jewell 2019). Importantly, nanoparticles can deliver multiple copies of a given antigen simultaneously and allow for display of antigens in their native conformation, which can enhance the specificity and magnitude of the induced immune response. Numerous nanoparticle techniques have been used to study the connection between antigen display and the incited immune response. Tokatlian et al. found that innate immune recognition pathway receptors mediated the rapid complement-dependent transport of separate nanoparticles made from two different heavily glycosylated HIV immunogens to the follicular dendritic cell network (Tokatlian et al. 2018). This resulted in immunogen concentration in lymph node GCs and enhanced antibody responses. The same trafficking was not seen with the soluble antigens. Thus, this study demonstrates the important of glycosylation and multivalency on nanovaccine design. In another study, Veneziano et al. used the ability of DNA origami to generate precise programmable structures to study the effect of antigen nanovaccine design principles (Veneziano et al. 2020). Applying this platform to a clinical vaccine immunogen for HIV, the authors were able to elucidate design principles for both antigens (such as antigen copy number, affinity, and nearest-neighbor antigen spacing) and nanoparticles (stiffness, shape) that led to maximum B cell activity. This study illustrates a workflow for investigating the impact of different antigens, their combination, and nanoparticle properties on immune responses to determine optimal design parameters for nanovaccines.

Beyond design principles, multivalent nanoparticles have been developed to study whether particles containing the receptor binding domain from four to eight animal betacoronaviruses (mosaic particles) can broadly protect against future coronaviruses that evolve from cross-species transmissions in mice in comparison to a homotypic particle, soluble antigen vaccination, and natural infection (Cohen et al. 2021). Immunization using a boosting schedule with either nanoparticle formulation resulted in antibodies that could bind and neutralize heterologous coronaviruses whereas soluble immunization and natural infection produced weak heterologous responses, if any. These data demonstrate the influence of antigen multivalency on the induction of cross-reactive antibody responses. Further, mosaic nanoparticles showed enhanced generation of neutralizing antibodies against the coronavirus subgenus, sarbecovirus, which are known to cross over to humans more readily (Boni et al. 2020). This work demonstrates the use of nanoparticle multivalency to co-display multiple types of one antigen and of multiple antigens to induce protective immune responses that combat infectious disease evolution.

Together, these studies demonstrate how nanoparticle and polymer multivalency can be used to enhance and control immune signaling and immune interactions at the subcellular, intracellular, and intercellular levels.

4.4 Enabling novel vaccination strategies

Vaccines were the first therapies to use immunoengineering mechanisms to enable their therapeutic effects. However, traditional methods of delivering vaccines using free, soluble mixtures of antigens and adjuvants suffer from poorly controlled spatiotemporal kinetics and cumbersome manufacturing (Irvine et al. 2020). Additionally, bioinformatic methods to predict the most effective cocktail of antigens for established disease (therapeutic vaccination) or future infections (prophylactic) can be costly and may not provide sufficient coverage of the disease epitopes to enable disease eradication (Fang et al. 2014). Given their ability to incorporate cargo with very different properties into one medicine, to enhance delivery to lymphoid cells and tissue, and to control the presentation of these immunogenic molecules to APCs, nanotechnology-based vaccines have been extensively explored as alternatives to traditional soluble vaccines. The foundational aspects of nanoparticle-mediated vaccination have been reviewed thoroughly in the literature and the reader is directed to those articles for additional reference (Zhao et al. 2014, Pati et al. 2018, Irvine et al. 2020). Here, we will cover the ways in which nanotechnology-mediated vaccines are revolutionizing this therapeutic approach using biomimetic and genetic engineering strategies to achieve improved outcomes and manufacturability.

As highlighted by the Moderna and Pfizer-BioNTech vaccines for the SARS-COV2 virus, lipid nanoparticle-based (LNP) delivery of mRNA vaccines is an effective strategy for the induction of therapeutic prophylactic adaptive immune responses (Polack et al. 2020, Baden et al. 2021). mRNA vaccines take advantage of the transient expression profile of cytosolically delivered mRNA to mediate controlled, prolonged exposure of antigens to immune cells within an immunostimulatory context (Pardi et al. 2018). Certain mRNAs can also play an immune adjuvant role in these vaccines in addition to serving as the antigen source. As a result, this strategy has traditionally been applied in the setting of costimulatory vaccines for the prevention of infectious disease or for the treatment of cancer responses (Pardi et al. 2018). However, it is beginning to be explored in autoimmune and autoinflammatory diseases. Krienke et al. developed a novel non-inflammatory, tolerizing mRNA LNP vaccine delivering the self-antigen, myelin oligodendrocyte glycoprotein (MOG), for the treatment of EAE (Krienke et al. 2021). In this setting, vaccination is designed to suppress antigen-specific immune responses, in part, by activating regulatory T cells rather than priming cytolytic responses by CD8+ cytotoxic T cells. To counter the immunostimulatory adjuvant activity of mRNA, the authors used a less immunogenic form of mRNA that could not activate TLRs in their vaccine. Mice vaccinated with mRNA LNP vaccines encoding MOG but not those vaccinated with an irrelevant antigen were protected against EAE when vaccinated before disease onset (prophylactic) and showed delayed disease progression when vaccinated early in disease onset (therapeutic). The induction of antigen-specific regulatory T cells and their immunosuppression of bystander effector T cells mediated therapeutic efficacy rather than the deletion of pre-existing autoreactive T cells. Further, the authors demonstrated that the vaccine did not induce antigen-specific autoantibodies and that a cocktail of antigens could enhance therapeutic outcomes. While numerous questions remain, including the duration of the induced immunoregulatory effects, the safety and efficacy of repeat exposure to autoantigens, and the choice of antigen cocktail for a given disease, this study demonstrates the proof-of-concept use of mRNA vaccines in autoimmune diseases to expose antigen to the immune system in the absence of co-stimulatory signals for the generation of tolerogenic, therapeutic immune responses. Additional studies into the impact of LNP formulation adjuvant activity and the type of antigen encoded on the induced immune response could provide additional design principles for engineering both proinflammatory and tolerogenic mRNA vaccines.

While mRNA vaccination has been shown to be safe and effective in some disease settings, it is still burdened by the need to identify the antigen or antigens required to induce effective immune responses. It is also limited by the amount of therapeutic cargo that can be loaded in a particle, reducing the number of antigens that can be included. One method to address difficulties in antigen prediction and the incorporation of multiple disease antigens is to use nanoparticles cloaked in cell-membrane coatings derived from diseased or infected cells. Cell membrane coatings from diseased cells, such as tumor cells, can retain a wide array of surface-displayed antigens from their source cells, providing optimal target antigens for nanovaccines without requiring costly bioinformatics strategies to identify antigens or multivalent nanoparticle functionalization to deliver antigen cocktails (Fang et al. 2014). A seminal study by Fang et al. leveraged the rich antigenic information derived from diseased cell membranes to generate tumor-specific immunity using a nanovaccine compromised of adjuvant-loaded nanoparticles coated with cancer cell membranes (Fang et al. 2014). In this study, the cancer cell membrane coatings also improved drug delivery through homotypic tumor targeting. The same group later engineered cancer cell membrane nanoparticle coatings that expressed a co-stimulatory marker to enable direct simulation and priming of antigen-specific T cells without the need for antigen-presenting cells (Jiang et al. 2020). This bypasses the need for antigen uptake and cross-presentation by APCs if the nanomedicine can be effectively targeted to lymphoid tissues, increasing the potential for therapeutic efficacy and efficiency. A similar approach where host cell membranes are synthetically modified to express co-inhibitory molecules could be used to engineer immune tolerance in autoimmune diseases such as type 1 diabetes and multiple sclerosis. Cytokine signals could also be incorporated within the nanoparticle core to further enhance or inhibit immune cell priming using this strategy. One inherent drawback of this approach is that it involves external manipulation of a patient's diseased tissue or cells to generate the nanoparticle coating.

In contrast, *in situ* vaccination, or the ability to promote antigen-specific adaptive immune responses with memory through the generation of antigens from diseased tissues in the body directly and without external manipulation, has many advantages over traditional vaccine approaches. Because antigens are generated from diseased tissue within the body, this strategy reduces the dependency of the therapy on neoantigen prediction or on the isolation and identification of patient-specific antigens. This makes therapy more cost effective, convenient, and efficient than alternative strategies. Furthermore, *in situ* vaccination may improve therapeutic efficacy by more effectively addressing tumor heterogeneity, both within a given tumor (intrapatient) and between patients (interpatient). However, there is potential for immune escape if disease subclones aren't susceptible to or are not reached by the method that is used to generate antigens and if their epitope profile is not covered by other susceptible subclones. In this approach, an immunogenic cell death (ICD) inducing strategy is used to generate tumor antigens and release DAMPs, stimulating the uptake and cross-presentation of the released antigens by APCs. Often, a small molecule immune adjuvant is provided to further stimulate APCs and to help revert the immunosuppressive nature of the tumor microenvironment while checkpoint blockade therapy may be included to block immune escape and prevent the inhibition of T cell activity. Nanoparticles are used to mediate the spatiotemporal kinetics of this combination therapy and to enhance antigen uptake and cross-presentation by DCs. Together, this nanomaterial-mediated strategy seeks to enhance anti-tumor immunity by enabling the coordinated and efficient execution of a multi-step immune cascade *in situ*.

However, one inherent challenge in this approach is that individual steps in the immune cascade can require different nanomaterial designs (size, charge, etc.) to navigate their different processes (i.e., APC internalization vs. lymph node draining). To address this issue, Qin et al. engineered a novel thermo-responsive nanoparticle-based gel system by mixing alpha-cyclodextrin with dendrimer nanoparticles conjugated to CpG and a reductively cleavable mannopyranoside immune cell targeting agent (Qin et al. 2020). To mediate ICD and antigen release, they encapsulated immunogenic doxorubicin chemotherapy and the photosensitizing agent, indocyanine green (ICG). Each of these therapies served an additional purpose-ICG potentiated photothermal therapy upon 808-nm laser irradiation, inducing tumor cell apoptosis and enhancing thermal degradation of the gel to release doxorubicin and CpG nanoparticles while chemotherapy was included to enhance cytotoxicity at the tumor core where laser irradiation efficiency is poor (Figure 7A). Furthermore, CpG dendrimer nanoparticles released from the gel were able to capture released antigens via electrostatic adsorption, drain to the lymph nodes more efficiently due to their small size and neutral charge, and co-deliver the adsorbed antigen and conjugated adjuvant under intracellular reductive conditions (Figure 7B-C). The thermal-responsive and reduction-responsive nature of this nanoparticle-based gel system allowed for the generation of materials with different properties that collectively enhanced therapeutic outcomes by maximizing the efficiency of each step in the immune cascade. Furthermore, combination therapy with immune checkpoint blockade resulted in enhanced survival in postresection models by inducing tumor-specific immune responses against recurrence and metastasis (Figure 7D-F).

Additional studies have used nanoparticles to mediate genetic engineering approaches *in vivo* for the execution of novel *in situ* vaccination strategies. In one study, Ngamcherdtrakul et al. delivered CpG adjuvant in combination with silencing RNA against signal transducer and activator of transcription 3 (STAT3) using mesoporous silica nanoparticles to induce *in situ* tumor vaccination (Ngamcherdtrrakul et al. 2021). In multiple tumor-associated immune cells, STAT3 mediates multiple mechanisms of immunosuppression; it can cause immunogenic cell death when silenced in cancer cells (Kortylewski et al. 2009). As a result, siRNA therapy against STAT3 is a multifaceted immunomodulatory approach. This nanovaccine resulted in enhanced curative survival in murine models of melanoma, breast, and colon cancers when combined with checkpoint blockade therapy. It was also shown to be well-tolerated in nonhuman primates, accelerating translation to the clinic. Another approach by Li et al. used an ICD-inducing lipid nanoparticle formulation to deliver self-replicating RNA encoding the cytokine interleukin-12 (IL-12) fused to lumican, an extracellular-matrix binding protein which enhanced cytokine tumor retention (Li et al. 2020). IL-12 is a cytokine that regulates Th1 responses and promotes the expansion, survival, and activity of cytotoxic T cells and natural killer cells. The authors showed that an ionizable lipid formulation could induce ICD of cancer cells while mediating RNA delivery, which resulted in innate immune stimulation of TLR3 to the RNA and enhanced anti-tumor T cell responses because of IL-12 expression and retention. The design and use of a replicon mediated enhanced IL-12 expression in tumor tissue that necessitated the use of a collagen-binding protein to prevent dissemination of IL-12 into the blood to counter the potential for the development of systemic toxicities. Importantly, large established tumors and their distal untreated metastases in several immunocompetent mouse models were completely eradicated in mice treated with a single injection of this therapy. Other payloads could be incorporated into the replicons to adapt this approach to target other pathways. This approach demonstrates the significant potential for immunotherapeutic approaches combining synthetic biology, genetic engineering, and materials.

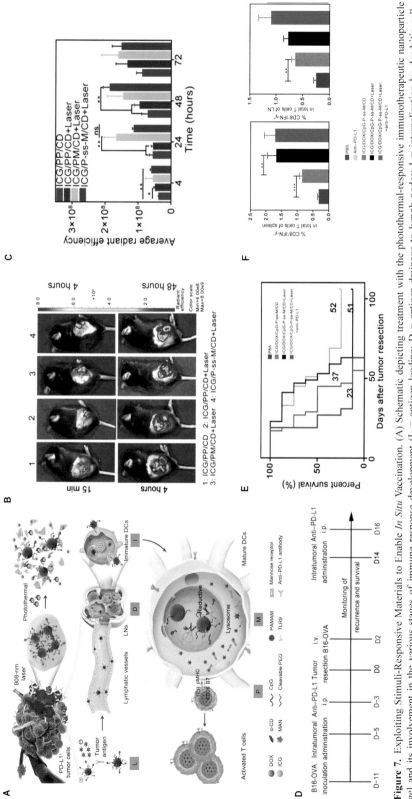

Figure 7. Exploiting Stimuli-Responsive Materials to Enable *In Situ* Vaccination. (A) Schematic depicting treatment with the photothermal-responsive immunotherapeutic nanoparticle gel and its involvement in the various stages of immune response development (L = antigen loading; D = antigen drainage to lymph nodes; I = internalization by dendritic cells; M = dendritic cell maturation for costimulatory molecule expression; P = antigen presentation to T cells by dendritic cells). (B) *In vivo* imaging of tumor-bearing mice treated with the gel over time following tumor irradiation illustrating the transfer of material signal from tumor microenvironment to the lymph nodes (Red circles = sub-iliac lymph nodes). (C) *Ex vivo* semi-quantitative fluorescent signal evaluation of the tumor draining lymph nodes over time. (D) Schematic of the *in vivo* study timeline for treatment with the photothermal-responsive immunotherapeutic nanoparticle gel in a melanoma B16F10 resection model. (E) Survival of mice following the outlined treatments in the resection model study. (F) Differences in the percentages of IFN-g expressing T cells in the spleen and the tumor draining lymph nodes following treatment with therapy in the resection model. (Reprinted with permission from Qin et al. (2020). A tumor-to-lymph procedure navigated versatile gel system for combinatorial therapy against tumor recurrence and metastasis. *Sci. Adv.*, 6(36).)

5. Nanomaterial-cell immunotherapies

The large, diverse repertoire of immune cells and their associated functions are a powerful resource for immunotherapy in the treatment of disease. Directing the precise location, activation state, functions, and interactions of these cells is integral to achieving safe and efficacious immunotherapy. In this section, we will discuss the ways in which nanomaterials and immune cells are interfacing to achieve these outcomes.

5.1 Altering immune cell trafficking and disease microenvironment infiltration

An important area of research is how to control immune cell migration as the location of the immune cells influence the processes under which they'll go, their activity, and immunotherapeutic outcomes. The movement of immune cells (and other cells) is regulated in part by their integration of chemotactic and physical microenvironment signaling cues via their dynamic expression and regulation of their surface receptors. Chemotactic cues, such as chemokines, are used to create protein and chemical gradients between immune tissues, disease sites, and the blood to mediate immune cell migration. The maturation state of the immune cell determines whether it will express the required surface molecules to drive chemotaxis. An important chemokine-receptor pair is the proinflammatory chemokine CXCL10 and its receptor, CXC receptor 3 (CXCR3). CXCL10 expression is induced upon activation of multiple innate immune inflammatory pathways. CXCL10 serves many immunostimulatory roles including acting as a chemoattractant for many immune cell types, such as T cells and dendritic cells. The ability to recruit these cells to the disease microenvironment is important as recent studies have shown that the efficacy of immune checkpoint blockade in cancer therapy may rely on T cell responses generated outside the tumor microenvironment (Yost et al. 2021). Zhao et al. used erythrocyte-anchored nanoparticles entrapping the chemokine CXCL10 to target metastatic tumor cells in the lungs for immunotherapy delivery. The PLGA nanoparticle chemistry (lactic acid to glycolic acid ratio, end group type) was engineered to allow the particles to non-covalently attach to the cell surface and to deposit and accumulate in the lungs in response to physio-mechanical stresses exerted by the lung capillary endothelium (Zhao et al. 2020). The particles were further modified with ICAM-1 binding antibodies to promote particle adhesion to the lung endothelium, extending the retention time of the nanoparticles in the lungs from 20 minutes to over six hours *in vivo*. Lung deposition of the CXCL10-loaded nanoparticles generated a chemokine gradient between the metastases and the blood, driving enhanced immune cell tumor infiltration that led to *in situ* immunization and induced systemic immunity (from a metastatic site) and the formation of immune memory. Another study delivered DNA encoding chemokine receptor 7 (CCR7) and protein antigens to DCs using a mannose-targeted micelle to drive DC chemotaxis to the lymph nodes following antigen uptake for efficient T cell priming (Yang et al. 2018). This strategy overcame tumor-derived immunosuppression in the lymph nodes, which prevents efficient trafficking of APCs to these tissues to incite an immune response. Together, these studies illustrate how nanoparticle delivery of chemotactic molecules can influence immune cell migration to immune and pathological tissues.

In addition to the receptors found on immune cells, the surface molecules found on the endothelial cells of lymphoid tissue and of the blood and lymphatic vasculatures play a key role in mediating immune cell migration and activation. The surface receptors of endothelial cells in lymphoid tissues are responsible for sensing signals from circulation and expressing surface molecules, like integrin receptors, that modify the ingress and egress of immune cells from these tissues. Recently, Krohn-Grimberghe et al. demonstrated that nanoparticles delivering small interfering RNA (siRNA) to bone-marrow endothelial cells to silence genes involved in regulating the release of hematopoietic cells and leukocytes from the bone marrow could improve healing and outcomes after myocardial infarction (Krohn-Grimberghe et al. 2020). The authors screened *in vivo* formulations with different PEG surface coatings (density and molecular weight) and PEG lipid architectures to identify lipid-polymer hybrid nanoparticles that enhanced delivery of siRNA to the bone marrow niche endothelial cells. These alterations to the PEG surface chemistry potentially reduced nanoparticle uptake in other organs with higher systemic circulation perfusion such as the lungs. Using this strategy, the authors showed that they could both inhibit and stimulate immune cell release from the bone marrow following therapy, leading to improved outcomes in a model of myocardial infarction when inhibitory signaling was used. This strategy could be used to silence other receptors, signaling molecules, and cell-cell interaction molecules expressed by or on the endothelium to modulate the ability of these cells to interact with and respond to molecular and cellular components in the circulation. This study demonstrates how nanoparticle gene therapy can be targeted to cells in tissues and organs involved in the regulation of immune cell trafficking within the body, altering the ability of these cells to reach sites of inflammation and disease, with potential to treat a range of disease types.

The interaction between surface molecules on endothelial vasculature cells and circulating lymphocytes also can determine the ability of lymphocytes to home to specific tissues. To induce gut-specific tropisms, vitamin A-derived retinoic acid, a natural product of gut-resident DCs, can be delivered to lymphocytes to drive the surface expression of the gut-homing C-C chemokine receptor 9 (CCR9) and the a4b7 integrin, which binds to an addressin expressed specifically by gut endothelial cells (Iwata et al. 2004). Xia et al. used retinoic acid as an adjuvant for oil-in-polymer vaccine capsules to drive peripheral dendritic cell upregulation of CCRR9 and homing to gut mucosal tissues following intramuscular injection for enhanced systemic and gastrointestinal responses (Xia et al. 2018). The same group later enhanced the efficacy of this vaccine strategy by developing a polymer/lipid nanoparticle that improved dendritic cell migration to the peripheral lymph nodes to prime gut-specific T cells rather than their direct homing to gut tissue (Du et al. 2019). These particles co-delivered retinoic acid, CpG oligodeoxynucleotides, and antigens, where the inclusion of CpG in the vaccine formulation induced draining lymph node homing of DCs and the amplification of lymphocyte activation and homing receptor switching. This effect, which they termed the draining lymph node-amplifying effect, boosted systemic responses, such as IgG secretion and T cell activation, in peripheral tissues and invoked enhanced T cell homing and antigen-specific IgA levels in gut mucosal tissue in response to EV71 vaccination. This strategy could be used to mediate enteric therapeutic vaccination for other diseases, such as IBD and gastrointestinal cancers.

Physical barriers found in healthy and pathologic tissues can also prevent the migration of immune cells into disease microenvironments. In cancer, stromal cells can upregulate their expression of extracellular matrix (ECM) components forming a fibrotic barrier to immune cell (and nanoparticle) infiltration (Netti et al. 2000). To address this issue, Huang et al. developed pancreatic tumor-targeting calcium phosphate liposome core shell particles delivering a-mangostin (an antifibrotic therapy that decreases collagen deposition) and a DNA plasmid encoding LIGHT (a cytokine that increases T cell recruitment) (Huang et al. 2020). By concurrently reducing the fibrotic physical barrier preventing tumor infiltration using a-mangostin and enhancing immune cell recruitment by generating chemoattractants using LIGHT, this nanomedicine enabled immune cell infiltration and enhanced checkpoint blockade therapy. Interestingly, this therapy also induced the formation of intratumoral tertiary lymphoid structures (TLSs). These structures, which consist of T cells, B cells, DCs, and other APCs, can locally prime and activate adaptive immune cells within the disease microenvironment, enhancing endogenous and therapy-induced immune responses. TLSs have recently emerged as a biomarker of checkpoint blockade therapy responses in many different solid tumors (Sautès-Fridman et al. 2019, Helmink et al. 2020); however, the mechanisms by which TLSs are induced and maintained are not currently known nor is it known how their associated immune processes and responses differ from those generated in the lymph nodes and spleen (Sharma et al. 2021). As a result, strategies that enhance lymphocyte recruitment and induce the formation of TLSs are of intense interest to the development of more efficacious cancer immunotherapy strategies.

5.2 Cell-nanomaterial hybrids for improved immunotherapy

Natural materials have evolved to achieve biologically distinct functions that are difficult to fully recapitulate with synthetic materials. Bioderived and biomimetic materials seek to leverage this specificity to create new capabilities for nanomedicines. One recent biomimetic strategy has been to use cell membranes to cloak synthetic nanoparticles with surface coatings that recapitulate the complex interfaces found naturally and that reduce exposure of the synthetic particle to the immune system (Hu et al. 2015). These cell coatings have been derived from erythrocytes, immune cells, tumor cells, and platelets, among others. The generated core-shell particles can be superior to cell-membrane derived vesicles alone due to their improved *in vivo* stability (Zhang et al. 2019b). Cell membrane coatings initially emerged as a strategy to enhance nanoparticle drug delivery by reducing immune cell recognition and clearance of circulating particles. In 2011, Hu et al. first demonstrated the use of a top-down approach to generate erythrocyte-coated polymeric nanoparticles in which the membrane components retained their structure and functional activity. The authors used the erythrocyte membrane coating to disguise nanoparticles from circulating phagocytic immune cells to extend their plasma half-life. More recently, a pH-responsive erythrocyte membrane-coated nanogel was synthesized to deliver combination chemoimmunotherapy of paclitaxel and IL-2 with enhanced tumor accumulation and potent antineoplastic activity (Song et al. 2017). Paclitaxel was encapsulated within the nanogel and promoted the ICD of tumor cells. IL-2 is an FDA-approved cytokine therapy that supports the survival and proliferation of cytotoxic immune effector cells. Here, the membrane coating also acted as the delivery material for IL-2 as the membrane allowed for increased adsorption of the protein on the surface of the particles (nanosponge activity) and provided glycoproteins that could directly bind to IL-2. Membrane-coatings have also been derived from tumor-associated macrophages (TAMs) and natural killer cells to enhance nanoparticle tumor penetration and tumor-cell homing by improving nanoparticle immune compatibility in the tumor microenvironment (Deng et al. 2018, Chen et al. 2021, Zhang et al. 2021). A recent study showed that this property enabled improved photodynamic immunotherapeutic activity of TAM membrane-coated particles loaded with a photosensitizer (Chen et al. 2021). Furthermore, the TAM membrane coating depleted colony stimulating factor 1, an immunosuppressive molecule secreted by tumor cells to reduce macrophage activation, enhancing the polarization of macrophages to a more inflammatory state. The ability of cell membrane-coatings to sequester tumor-secreted or pathogen-secreted immunomodulatory factors can be used to modulate pathologic cell-cell communications impacting host immunity across many disease states.

Cell membrane coatings can also be used to enhance interactions between nanoparticles and specific cell types as these coatings retain a variety of cell-cell interaction molecules, such as adhesion ligands, and other functional proteins that allow them to strongly interact with and target specific cell types. Parodi et al. coated nanoporous silicon particles with leukocyte membranes to avoid immune cell clearance in circulation and to enhance preferential binding to and transport of nanoparticles across inflamed endothelium into tumors (Parodi et al. 2012). Platelet membrane-coated nanoparticles were developed to mimic the ability of platelets to selectively adhere to damaged vasculature and certain pathogens (Hu et al. 2015). In addition to reducing particle uptake by phagocytic cells, the platelet membrane coating prevented nanoparticle-mediated complement activation and achieved better therapeutic efficacy compared to noncoated nanoparticles and erythrocyte-coated nanoparticles in mouse models of coronary restenosis and systemic bacterial infection. The ability to tailor nanoparticle-immune complement interactions can be exploited to enhance or reduce innate immune activation. Recently, platelet membrane-cloaked nanoparticles containing the innate immune TLR agonist, resiquimod (R848), were delivered intratumorally to enhance solid tumor immunotherapy (Bahmani et al. 2021). Like platelets, host cells can also have adhesive interactions with pathogens and microbes. Angsantikul et al. cloaked PLGA nanoparticles encapsulating clarithromycin in gastric epithelial cell membranes to specifically target *Helicobacter pylori* (a bacteria that can cause chronic inflammation and increase the risk of developing gastric ulcers and cancers) to leverage the adhesive interaction between the bacterial pathogen and the host cells (Angsantikul et al. 2018). This resulted in improved *H. pylori* killing in an *in vivo* model of *H. pylori* infection as compared to untargeted nanoparticles, free antibiotics, and untreated control. The biomimetic strategy of leveraging the adhesive interactions between host cells and pathogenic agents to improve nanomedicine targeting through host cell membrane coatings could be used to treat a variety of infectious diseases and to modulate the gut microbiome to improve the outcomes of immunotherapy. Finally, hybrid membrane coatings derived from multiple cells have also been developed to take advantage of the distinct functionalities derived from the unique surface molecules present on different cell types. A dual membrane coating derived from platelets and cancer stem cells was used to enhance immunogenic cell death-inducing photothermal therapy in squamous cell carcinoma by using homotypic adhesion molecules derived from tumor cells and antiphagocytic surface molecules derived from platelets to achieve better tumor accumulation of iron oxide nanoparticles (Bu et al. 2019).

Nanoparticles can also be used as cell-mimetic decoys for viral, fungal, bacterial, and parasitic immunotherapy. These decoys can be used to decrease viral load in acute infections by preventing viruses from interacting with host cells and facilitating their uptake by immune cells or removal from the body. Viral infection, for example, is highly dependent on the ability of viral capsid proteins, such as the spike protein, to bind to surface molecules on host cells. By cloaking nanoparticles in the membranes of host cells targeted by a given virus, the nanoparticles can detain viruses from infecting host cells. This approach was used by Wei et al. to generate CD4+ T cell mimicking nanoparticles that prevented viral envelope protein-mediated T cell killing and enabled host cell viral neutralization mechanisms for effective HIV immunotherapy (Wei et al. 2018). Difficulties in overcoming the genetic diversity of viruses, like HIV, could potentially be overcome using this approach. Another decoy approach used erythrocyte membrane-coated nanoparticles to sequester pathogen-derived exotoxins to prevent red blood cell lysis to treat bacterial and fungal infections (Hu et al. 2013a). Hu et al. showed that the nanoparticle-trapped pore-forming toxins could result in nanoparticle-mediated *in situ* vaccination (Hu et al. 2013b). This allows many advantages over traditional vaccine approaches, including that the toxins are displayed to the immune system in their natural setting and conformation. Using a non-biomimetic approach, Sigl et al. developed programmable self-assembling icosahedral shells from DNA that were able to trap hepatitis B virus particles and adeno-associated viruses (AAVs) *in vitro* and that could neutralize infectious AAVs exposed to human cells (Sigl et al. 2021). Given the modularity of this approach, other binders could be assessed easily and targeting moieties such as antibodies and aptamers could be incorporated to enhance affinity and specificity through multivalency. Together, these results show how nanomaterial strategies can be used to interact with and neutralize pathogenic threats.

5.3 Enabling biomimetic strategies through cell-based delivery

The use of cells as nanomaterial carriers enables a novel range of biomimetic functions that can enhance the efficacy of nanoparticle-based immunotherapy. Using red blood cells (RBCs) as carriers, Ukvide et al. mimicked the innate function of RBCs to capture circulating blood pathogens and present them to immune cells in the spleen by designing erythrocyte-based nanoparticles for the delivery of adjuvant-free nanovaccines (Ukidve et al. 2020). Spleen delivery was engineered, in part, by increasing the antigen-conjugated nanoparticle loading on the surface of the RBCs to induce erythrocyte stiffening and deformability. These changes in the cell membrane increased the nanoparticle resistance to being dislodged by shearing from the lung capillaries and prompted their increased capture in the spleen, potentially due to the accelerated clearance of RBCs. This biomimetic approach showed utility as a prophylactic vaccine for tumor prevention that could be extended to other disease types. Platelets also have several targeting and biological mechanisms that make them useful as drug delivery carriers such as their specific adherence to damaged vascular endothelium and their ability to secrete vesicles after activation (Wang et al. 2017). Lv et al. leveraged these functions when they used platelets as a drug delivery vector for photothermal block copolymer nanoparticles and the innate immune adjuvant R837 (Lv et al. 2021). Damaged tumor-associated endothelium attracted therapy-bearing platelets to initially aggregate at the tumor site. Nanoparticle-enhanced photothermal treatment of the lesion further induced vessel damage driving a positive feedback cascade to further enhance nanoparticle-bearing platelet accumulation at the target site (Figures 8A-B). The hyperthermia treatment also induced immunogenic cell death, which synergized with the R837 therapy. Furthermore, activated platelets released nanosized vesicles capable of penetrating deep into the tumor to mediate photothermal nanoparticle and immunostimulant delivery (Figure 8C). This multi-faceted strategy was safe and efficacious in multiple models of murine cancer (Figure 8D). Wang et al. also used the postoperative wound homing and activation-stimulated particle secretion abilities of platelets to enhance delivery of surface-conjugated aPD-1 (Wang et al. 2017). This strategy prolonged the survival time of tumor-bearing mice and prevented postoperative tumor recurrence and metastasis.

5.4 Nanomaterial backpacks for enhanced cellular immunotherapy

In addition to enhancing drug delivery, the linking of nanomaterials to adoptive cell or engineered cell therapies can result in effective combination immunotherapy. The nanocarriers and their conjugation chemistry are engineered such that therapy is released at a controlled rate to stimulate the cell in a pseudo-autocrine fashion or in response to environmental or external stimuli at the target site such that cell activity is induced or enhanced at a specific location or to modify the local microenvironment (Irvine and Dane 2020). Directly linking the nanomedicines to the cell therapies allows the biodistribution of the molecular payload to mirror that of the cellular therapy, allowing for improved synergy and lower doses in comparison to free drug combination therapy. This significantly reduces the potential for off-target toxicity of potent immunomodulatory drugs, like IL-15, that are lethal when administered freely (Berger et al. 2009). This is an alternative strategy to genetically engineering the cell to express the therapeutic payload of interest. In comparison to genetic engineering, the use of nanoparticle strategies to facilitate combination therapy between cellular therapy and molecular therapy (cytokines, bispecific T cell engagers, etc.) allows for control over the dose and release rate of therapy. One limitation of this strategy is that it involves *ex vivo* manipulation of immune cells, whereas *in vivo* genetic engineering strategies using nanomaterials (discussed later in this chapter) are being developed that could allow for more facile manufacturing processes, while achieving similar efficacy. However, such *in vivo* engineering strategies remain in earlier clinical development and are limited in the amount of genetic cargo that can be loaded.

Two challenges with cellular immunotherapy are the poor expansion of adoptive or engineered cells *in vivo* and their inability to maintain activity in the immunosuppressive tumor microenvironment (Gong et al. 2021). Cellular backpacks have been used to co-deliver cytokines to stimulate adoptive cell proliferation and small molecule immune and metabolic adjuvants to support T cell activity in immunosuppressive disease microenvironments. In a seminal paper, Tang et al. created a carrier-free, redox-responsive nanogel by crosslinking IL-15 super agonist complexes using disulfide bonds and anchored this nanogel to the CD45 surface receptor of T cells (Tang et al. 2018). The redox-responsive disulfide bonds allowed for release of the cytokine payload following an increase in the reduction potential of the cell membrane because of TCR engagement by antigen-MHC complexes in the tumor, coupling drug release to tissue-specific signaling. Conjugation of the nanocarriers to the CD45 surface receptor allowed these materials to traffic to the immune synapse

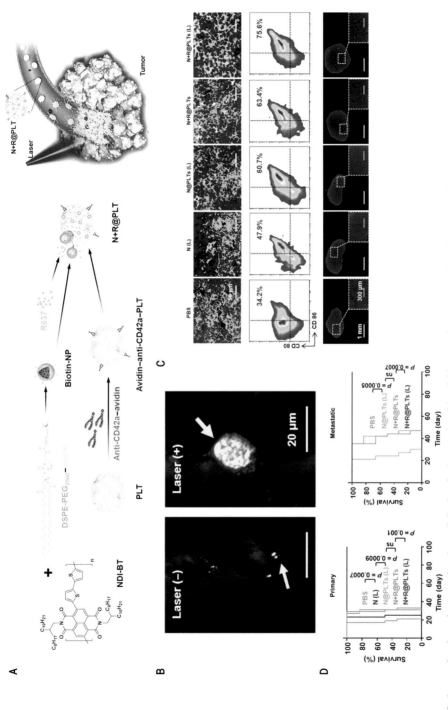

Figure 8. Leveraging the endogenous functions of platelets for improved multimodal immunotherapy: (A) Schematic demonstrating the generation of photoimmunotherapy nanoparticle-loaded platelets (NP-platelet) and their enhanced homing following laser irradiation due to platelet aggregation at sites of damaged vasculature for improved immunotherapeutic outcomes. (B) *In vivo* multiphoton imaging of NP-platelet aggregation in the tumor following laser irradiation. (C) Alterations in the incited immune response in the tumor draining lymph nodes following therapy (top: representative immunohistochemistry fluorescence images; middle: flow cytometry analysis of dendritic cell maturation; bottom: representative IHC analysis for Ki67 expression; blue, cell nuclei; red, Ki67). (D) Survival curves demonstrating enhanced survival for the combination therapy NP-platelets in a primary and a metastatic model of breast cancer (4T1). (Reprinted with permission from Lv et al. (2021). Near-infrared light–triggered platelet arsenal for combined photothermal-immunotherapy against cancer. *Sci. Adv., 7*(13).)

during T cell activation, demonstrating the importance of anchor molecule selection. When injected intravenously into mice bearing melanoma tumors, the nanomaterial backpack therapy led to a significant increase in adoptive T cell expansion in comparison to free IL-15SA and significantly reduced the toxicity associated with systemic delivery of supporting cytokine therapy. A similar strategy used cellular backpacks for the pseudo-autocrine delivery of interferon gamma to adoptively transferred macrophages, which allowed these cells to maintain their proinflammatory phenotype deep within the immunosuppressive milieu of the tumor when the backpacked cells were injected intratumorally (Shields et al. 2020). The therapy also repolarized endogenous tumor-associated macrophages to a proinflammatory state via a bystander effect, removing a significant source of microenvironment immunosuppression. This led to a heightened anti-tumor response characterized by smaller tumor burden with reduced metastases. To combat the metabolic dysfunction of T cells in the hostile, immunosuppressive TME, Zhang and coworkers used avasimibe-loaded liposomal backpacks to maintain engineered T cell activation and function in the metabolically restricted tumor microenvironments of melanoma and glioblastoma (Hao et al. 2020). Avasimibe inhibits cholesterol esterification increasing the concentration of cholesterol in the T cell membrane, which impacts immune synapse formation and signaling and enhances T cell activation, potentiating the anti-tumor response (Yang et al. 2016). Treatment with backpacked T cells increased the antitumor efficacy of therapy in glioblastoma and melanoma tumor-bearing mice, including complete eradication of orthotopic glioblastoma in 60% of mice treated with this therapy compared to 0% for all other therapies. In a separate study, liposomes encapsulating an A2a adenosine receptor antagonist were conjugated to the surface of engineered T cells to prevent hypofunction due to T cell inactivation by adenosine, an immunosuppressive molecule found abundantly in the TME due to its derivatization from extracellular adenosine triphosphate (ATP) by ecto-nucleases (Siriwon et al. 2018). Future strategies using nanoparticle cell backpacks could enable adoptive cell monitoring and tracking and allow for *in situ* modulation of engineered cells to combat disease evolution.

5.5 Engineering immune cells with enhanced functions in vivo

Numerous studies have shown that non-viral nanomaterials can enhance the intracellular delivery of genetic therapies to the cytosol and nucleus *in vivo* (Ditto et al. 2009, Oberli et al. 2016, Shae et al. 2019). Nanomaterial properties such as charge and polymer buffering capacity can be used to enhance the cytosolic delivery of cargo by rupturing the endosome (Eppler and Jewell 2019). A unique application that this enables is the *in vivo* engineering of cellular immunotherapies. In a proof-of-concept study, Smith et al. demonstrated the efficacy of this approach by engineering and expanding circulating host T cells *in vivo* using polymeric anti-CD3e fragment antibody-targeted nanoparticles that co-delivered DNA plasmids encoding a chimeric antigen receptor (CAR) and a piggyBac transposase (Smith et al. 2017). The polymer was further functionalized with peptides that enhanced the nuclear import and gene transfer of the genetic cargo. Targeting of the nanoparticles to circulating T cells with the aCD3 Fab'2 increased accumulation of the nanoparticles in lymphoid organs, such as the spleen, lymph nodes, and bone marrow, and decreased accumulation in the liver. Further, the combination of the nanoparticle and synthetic biology design elements engineered CAR expression in endogenous T cells that persisted for weeks and achieved efficacy comparable to that of CAR T cells produced via standard *ex vivo* modification methods in a mouse model of leukemia. Further work is needed to determine whether this method can achieve sufficient expansion and activity of engineered T cells for application to solid tumors. Despite targeting of T cells with anti-CD3e moieties, one challenge that remains is reducing the uptake of nanomedicines in off-target circulating cells (and their associated toxicities). In a follow-up work, the same group described the delivery of mRNA encoding CAR or T cell receptors using polymeric nanoparticles to transiently engineer disease-specific T cells *in vivo* (Parayath et al. 2020). mRNA has many advantages over DNA-based systems including that it is transiently expressed without the need for genomic integration and has higher target protein expression efficiency, which together results in reduced risks for genotoxicity and provides greater control over therapy pharmacokinetics. In several cancer mouse models, including solid tumors, these nanoparticles achieved similar therapeutic outcomes to *ex vivo* engineered T cells. This study demonstrates how the choice of genetic material cargo can influence the duration and potential safety of *in vivo* engineered cells. Together, these studies demonstrate how nanoparticle-mediated delivery of genetic material can engineer effective T cell therapies *in vivo*. This strategy provides more timely treatment options for patients using traditional intravenous administration to deliver cellular therapy and is more cost effective than existing *ex vivo* engineered cell manufacturing strategies, making it well positioned to transform the implementation of engineered cell immunotherapies clinically.

Besides introducing antigen-specific effector moieties, nanoparticle-mediated genetic engineering approaches can be applied to alter the functional phenotype and activity of circulating cells. Cell therapies both benefit from and can be derided by their ability to integrate complex signaling from healthy and pathological biologic environments, as these signals can enhance or prevent their therapeutic activities. In cancer, tumor-associated macrophages often adopt a pro-tumorigenic, anti-inflammatory phenotype due to immunosuppressive signaling in the tumor microenvironment, causing further immunosuppression and disease progression (Matovani et al. 2002). One therapeutic strategy is to repolarize these cells *in vivo* to a phenotype that promotes anti-tumor immunity. Zhang et al. used mannose-targeted polymeric nanoparticles to deliver mRNA encoding transcription factors (IRF5 and its activating kinase IKKB) that reprogrammed tumor-associated macrophages to an anti-tumor phenotype *in vivo* without causing systemic toxicities or disrupting immune homeostasis (Zhang et al. 2019a). Reduced metastatic burden and increased survival were seen when this strategy was used to treat metastatic disseminated ovarian cancer using intraperitoneal delivery and disseminated pulmonary melanoma using intravenous administration. These outcomes resulted from increased tumor-associated macrophage polarization to an inflammatory phenotype and from an increase in the focal clustering of lymphocytes infiltrating or surrounding the tumors, suggesting that the induced macrophage polarization changes can alter lymphocyte migration and tumor infiltration to enhance therapeutic efficacy. In an alternative strategy, Ledo et al. used siRNA-encapsulating nanoparticles releasing a macrophage chemoattractant, C-C chemokine ligand 2, to preferentially lure myeloid cells to be transfected by the particles, allowing for efficient knockdown of an immunosuppressive gene, *C/EBPb*, in these cells (Ledo et al. 2019). Beyond myeloid cells, Dan Peer and colleagues have demonstrated in multiple studies the use of lipid nanoparticles for the *in vivo* genetic modulation of T cells and tumor cells for therapeutic purposes (Ramishetti et al. 2015, Kedmi et al. 2018, Veiga et al. 2018). In one study, using systemically administered anti-CD4 antibody targeted lipid nanoparticles, this group was able to efficiently modulate T cell function across several lymphoid organs including the spleen, lymph nodes, and bone marrow (Ramishetti et al. 2015). They showed that the efficacy of this therapy was dependent on efficient nanoparticle internalization soon after therapy administration. Together, these approaches highlight the ability of nanoparticle-based gene therapies to transfect and modify immune cells *in vivo* to achieve therapeutic outcomes. Given the complex genetic, environmental, and pathologic interactions impacting immune cell phenotype and function in disease, this approach offers a powerful method to engineer therapeutic responses directly using current and emerging genetic engineering technologies.

6. Outlook and conclusions

The generation of efficacious immune responses requires the proper coordination of multiple steps across different locations in the body and at defined timescales to generate therapeutic effects. Building off the use of nanomaterials to deliver conventional therapies, such as chemotherapy and antimicrobial agents, the first generation of nanomaterial immunotherapies has leveraged material-based systems to alter immunotherapy biodistribution and pharmacokinetics of monotherapies and combination therapies to improve safety and efficacy. This includes tailored nanoscale chemistry formulations and carrier-based delivery approaches that enhance delivery of therapy to specific organs and cells, allowing for targeted genetic and molecular modulation of these cells and tissues for generation of effective immune responses. It also includes the use of nanoparticle structure to enable combination delivery of immunomodulatory therapies with intrinsically different properties and the use of polymer chemistry to control the spatiotemporal kinetics of therapy delivery. Other approaches seek to simplify the formulation components of nanomaterial immunotherapies by using immunomodulatory polymers or by programming self-assembling materials from immunomodulatory therapies like antibodies to enhance the translatability and reduce the manufacturing complexity of nanoscale immunotherapies.

However, the inherent outcomes of immunotherapy are more complex than conventional therapies as immunotherapy seeks to engineer a long-lasting, *living* response by reprogramming immune cells and other microenvironment components and generating memory, while conventional approaches have focused on the diseased cell or pathogen alone, aiming to induce targeted diseased cell or microbial pathogen death, itself a difficult task. As a result, effective immunotherapy requires controlled modulation of multiple steps in the initial signaling process as well as the ability to respond to different dynamic regulatory mechanisms enacted by immune and other host cells and to counter different evolutionary immune escape mechanisms enacted by diseased cells in response to therapy. This will require the design of more dynamic and precise nanomaterial delivery systems. Further, to leverage the specificity of the adaptive immune response, the provided or generated antigens and the resultant immune cell repertoire need to provide sufficient coverage of disease clonal or variant epitopes (genetic heterogeneity) and their evolution such that disease is eradicated or controlled by the induced response and that this response is durable to prevent disease recurrence. For example, in *in situ* vaccination approaches, are the elicited antigens sufficiently immunogenic and shared and displayed by the population of cancer cells that were not susceptible to immunogenic cell death? Will the efficacy and breadth of bystander activation be sufficient to control or eliminate diseased cells not sharing these antigens? While these are questions of disease immunobiology, their answers govern the required drug delivery profiles of therapy, such as the targeted cell uptake profile and spatiotemporal kinetics of therapy. Finally, heightened control over the site of nanotherapeutic action at the subcellular to organ scale is required of the next generation of nanomaterial immunotherapies to reduce toxicity and improve efficacy. Further study into the impact of route of administration, administration timing, and repeat dosing of antigens and immunomodulatory molecules on the nature, extent, breadth, and durability of the invoked immune response will also complement efforts centered on altering nanomaterial biodistribution. Together, these illustrate the on-going challenges associated with engineering controlled, durable, and safe immunomodulation using nanomaterial therapies to eradicate or control disease.

To meet these challenges, the next generation of nanomaterial immunotherapies has begun to explore the design of dynamically responsive materials to overcome biological barriers and the design of combination approaches that integrate biological entities (cell membrane coatings, biological carrier-based delivery, etc.) and/or synthetic biology engineering with nanomaterials as illustrated in the preceding sections. To further address issues in overcoming biological barriers, a deeper understanding of nanomaterial-biological interactions is required, as these will determine how material-based strategies can be designed to leverage these interactions to enable novel applications or to avoid these interactions to increase the efficiency and specificity of delivery. Efforts to understand protein corona formation and active targeting agent interactions discussed earlier in this chapter underscore some of the ongoing research efforts in this area. Further insight into the immunobiology of disease, including disease-specific molecular and cellular compositions, the spatial arrangements of these components in the disease microenvironment, the dynamics of these populations across time and therapy, and the intricate signaling networks between these components and with other systemic immunomodulators (like the gut microbiome), enabled by advances in single cell omics and multiplexed bioimaging technologies will provide more precise resolution of potential targets and strategies for nanomaterial immunotherapies to engineer in a given disease state and patient population. These technologies can also be used to enhance our understanding of the complex immunobiological responses invoked by nanomaterial-biological interactions. Nanomaterials, in turn, can be further developed as imaging agents and diagnostic tools to complement the function of these omics and imaging technologies and to allow for *in vivo* monitoring of immune responses to therapy across time and anatomical locations. Advances in these areas will enhance our ability to engineer biology (including immunology) with molecular and cellular precision, which can be used to impart supraphysiological functions or to better control endogenous agents, improving our ability to combat disease evolution and immunotherapeutic outcomes. As a result, the ability of nanomaterials to precisely deliver genetic immunoengineering therapies or to control and support the function of engineered biological therapies is of critical importance. The continued interplay between materials and synthetic biology therapies will directly enable the therapeutic application of these technologies in patients across a broader spectrum of diseases.

In summary, the tunable properties of nanomaterials have been leveraged to increase the efficacy, safety, and durability of immunotherapies by altering their delivery profile across time and space and by enabling novel mechanisms of action that can synergize with molecular and cellular immunotherapies. The future development of nanomaterial immunotherapies relies on our ability to understand and engineer nanomaterial-biological interactions throughout the nanomaterial's time in the body and within the context of disease pathology and immunobiology. This will require continued synergy between the fields of materials engineering, synthetic biology, basic science, and clinical medicine to elucidate fundamental principles and innovative applications that drive the efficacy and safety of these nanomaterial immunotherapies. Ultimately, this research will provide new diagnostic and therapeutic options to patients across disease types, delivering on nanotechnology's promise to engineer precision delivery for precision medicine in immunotherapy applications and beyond.

Acknowledgments

This work was supported by the Massachusetts Institute of Technology School of Engineering Evergreen Fund Graduate Innovation Fellowship and by the National Science Foundation Graduate Research Fellowship. Any opinions, findings, and conclusions or recommendations expressed in this work are those of the authors and do not necessarily reflect the views of the National Science Foundation.

References

Abbas, A.K., Lichtman, A.H., Pillai, S. and Baker, D.L. 2018. *Cellular and Molecular Immunology*. Elsevier.

Akinc, A., Thomas, M., Klibanov, A.M. and Langer, R. 2005. Exploring polyethyleneimine-mediated DNA transfection and the proton sponge hypothesis. *J. Gene Med.*, 7(5): 657–663.

Angsantikul, P., Thamphiwatana, S., Zhang, Q., Spiekermann, K., Zhuang, J., Fang, R.H., Gao, W., Obonyo M. and Zhang, L. 2018. Coating nanoparticles with gastric epithelial cell membrane for targeted antibiotic delivery against helicobacter pylori infection. *Adv. Ther. (Weinh)*, 1(2): 1800016.

Anselmo, A.C., Zhang, M., Kumar, S., Vogus, D.R., Menegatti, S., Helgeson, M.E. and Mitragotri, S. 2015. Elasticity of nanoparticles influences their blood circulation, phagocytosis, endocytosis, and targeting. *ACS Nano.*, 9(3): 3169–3177.

Anselmo, A.C. and Mitragotri, S. 2017. Impact of particle elasticity on particle-based drug delivery systems. *Adv. Drug Deliv. Rev.*, 108: 51–67.

Arnida, Janát-Amsbury, M.M., Ray, A., Peterson, C.M. and Ghandehari, H. 2011. Geometry and surface characteristics of gold nanoparticles influence their biodistribution and uptake by macrophages. *Eur. J. Pharm. Biopharm.*, 77(3): 417–423.

Au, K.M., Park, S.I. and Wang, A.Z. 2020. Trispecific natural killer cell nanoengagers for targeted chemoimmunotherapy. *Sci. Adv.*, 6(27).

Baden, L.R., El Sahly, H.M., Essink, B., Kotloff, K., Frey, S., Novak, R., Diemert, D., Spector, S., Rouphael, N., Creech, C.B., McGettigan, J., Khetan, S., Segall, N., Solis, J., Brosz, A., Fierro, C., Schwartz, H., Neuzil, K., Corey, L., Gilbert, P., Janes, H., Follmann, D., Marovich, M., Mascola, J., Polakowski, L., Ledgerwood, J., Graham, B.S., Bennett, H., Pajon, R., Knightly, C., Leav, B., Deng, W., Zhou, H., Han, S., Ivarsson, M., Miller, J. and Zaks, T. for the COVE Study Group. 2021. Efficacy and safety of the mRNA-1273 SARS-CoV-2 vaccine. *N. Engl. J. Med.*, 384(5): 403–416.

Baharom, F., Ramirez-Valdez, R.A., Tobin, K.K., Yamane, H., Dutertre, C.-A., Khalilnezhad, A., Reynoso, G.V., Coble, V.L., Lynn, G.M., Mule, M.P., Martins, A.J., Finnigan, J.P., Zhang, X.M., Hamerman, J.A., Bhardwaj, N., Tsang, J.S., Hickman, H.D., Ginhoux, F., Ishizuka, A.S. and Seder, R.A. 2020. Intravenous nanoparticle vaccination generates stem-like TCF1+ neoantigen-specific CD8+ T cells. *Nat. Immunol.*, 22(1): 41–52.

Bahmani, B., Gong, H., Luk, B.T., Haushalter, K.J., DeTeresa, E., Previti, M., Zhou, J., Gao, W., Bui, J.D., Zhang, L., Fang, R.H. and Zhang, J. 2021. Intratumoral immunotherapy using platelet-cloaked nanoparticles enhances antitumor immunity in solid tumors. *Nat. Commun.*, 12(1).

Banani, S.F., Lee, H.O., Hyman, A.A. and Rosen, M.K. 2017. Biomolecular condensates: Organizers of cellular biochemistry. *Nat. Rev. Mol. Cell Biol.*, 18: 285–298.

Berger, C., Berger, M., Hackman, R.C., Gough, M., Elliott, C., Jensen, M.C. and Riddell, S.R. 2009. Safety and immunologic effects of IL-15 administration in nonhuman primates. *Blood*, 114(12): 2417–2426.

Bertrand, N., Wu, J., Xu, X., Kamaly, N. and Farokhzad, O.C. 2014. Cancer nanotechnology: The impact of passive and active targeting in the era of modern cancer biology. *Adv. Drug Deliv. Rev.*, 66: 2–25.

Boni, M.F., Lemey, P., Jiang, X., Lam, T.T.-Y., Perry, B., Castoe, T.A., Rambaut, A. and Robertson, D.L. 2020. Evolutionary origins of the SARS-CoV-2 sarbecovirus lineage responsible for the COVID-19 pandemic. *Nat. Microbiol.*, 5(11): 1408–1417.

Bruns, O.T., Bischof, T.S., Harris, D.K., Franke, D., Shi, Y., Riedemann, L., Bartelt, A., Jaworski, F.b., Carr, J.A., Rowlands, C.J., Wilson, M.W.B., Chen, O., Wei, H., Hwang, G.W., Montana, D.M., Coropceanu, I., Achorn, O.B., Kloepper, J., Heeren, J., So, P.T.C., Fukumura, D., Jensen, K.F., Jain, R.K. and Bawendi, M.G. 2017. Next-generation *in vivo* optical imaging with short-wave infrared quantum dots. *Nat. Biomed. Eng.*, 1(4). https://doi.org/10.1038/s41551-017-0056.

Bu, L.L., Rao, L., Yu, G.T., Chen, L., Deng, W.W., Liu, J.F., Wu, H., Meng, Q.-F., Guo, S.-S., Zhao, X.-Z., Zhang, W.-F., Chen, G., Gu, Z., Liu, W. and Sun, Z.-J. 2019. Cancer stem cell-platelet hybrid membrane-coated magnetic nanoparticles for enhanced photothermal therapy of head and neck squamous cell carcinoma. *Adv. Funct. Mater.*, 29(10): 1807733.

Carlson, C.B., Mowery, P., Owen, R.M., Dykhuizen, E.C. and Kiessling, L.L. 2007. Selective tumor cell targeting using low-affinity, multivalent interactions. *ACS Chem. Biol.*, 2(2): 119–127.

Casey, L.M., Kakade, S., Decker, J.T., Rose, J.A., Deans, K., Shea, L.D. and Pearson, R.M. 2019. Cargo-less nanoparticles program innate immune cell responses to toll-like receptor activation. *Biomaterials*, 218: 119333.

Chen, C., Song, M., Du, Y., Yu, Y., Li, C., Han, Y. et al. 2021. Tumor-associated-macrophage-membrane-coated nanoparticles for improved photodynamic immunotherapy. *Nano Lett.*, 21(13): 5522–5531.

Chen, F., Wang, G., Griffin, J.I., Brenneman, B., Banda, N.K., Holers, V.M., Backos, D.S., Wu, L., Moghimi, S.M. and Simberg, D. 2016. Complement proteins bind to nanoparticle protein corona and undergo dynamic exchange *in vivo*. *Nat. Nanotechnol.*, 12(4): 387–393.

Chen, Y., Liu, X., Yuan, H., Yang, Z., von Roemeling, C.A., Qie, Y., Zhao, H., Wang, Y., Jiang, W. and Kim, B.Y.S. 2019. Therapeutic remodeling of the tumor microenvironment enhances nanoparticle delivery. *Adv. Sci. (Weinh)*, 6(5): 1802070.

Cheng, Q., Wei, T., Farbiak, L., Johnson, L.T., Dilliard, S.A. and Siegwart, D.J. 2020. Selective organ targeting (SORT) nanoparticles for tissue-specific mRNA delivery and CRISPR–Cas gene editing. *Nat. Nanotechnol.*, 15(4): 313–320.

Clem, A.S. 2011. Fundamentals of vaccine immunology. *J. Glob. Infect. Dis.*, 3(1): 73.

Cifuentes-Rius, A., Desai, A., Yuen, D., Johnston, A.P. and Voelcker, N.H. 2020. Inducing immune tolerance with dendritic cell-targeting nanomedicines. *Nat. Nanotechnol.*, 16(1): 37–46.

Cohen, A.A., Gnanapragasam, P.N., Lee, Y.E., Hoffman, P.R., Ou, S., Kakutani, L.M., Keeffe, J.R., Wu, H.-J., Howarth, M., West, A.P., Barnes, C.O., Nussenzweig, M.C. and Bjorkman, P.J. 2021. Mosaic nanoparticles elicit cross-reactive immune responses to zoonotic coronaviruses in mice. *Science*, 371(6530): 735–741.

Davenport Huyer, L., Mandla, S., Wang, Y., Campbell, S.B., Yee, B., Euler, C., Lai, B.F., Bannerman, D., Lin, D.S.Y., Montgomery, M., Nemr, K., Bender, T., Epelman, S., Mahadevan, R. and Radisic, M. 2020. Macrophage immunomodulation through new polymers that recapitulate functional effects of itaconate as a powerhouse of innate immunity. *Adv. Funct. Mater.*, 31(6): 2003341.

Decuzzi, P., Godin, B., Tanaka, T., Lee, S.-Y., Chiappini, C., Liu, X. and Ferrari, M. 2010. Size and shape effects in the biodistribution of intravascularly injected particles. *J. Control Release*, 141(3): 320–327.

Demaria, O., Cornen, S., Daëron, M., Morel, Y., Medzhitov, R. and Vivier, E. 2019. Harnessing innate immunity in cancer therapy. *Nature*, 574(7776): 45–56.

Demento, S.L., Cui, W., Criscione, J.M., Stern, E., Tulipan, J., Kaech, S.M. and Fahmy, T.M. 2012. Role of sustained antigen release from nanoparticle vaccines in shaping the T cell memory phenotype. *Biomaterials*, 33(19): 4957–4964.

Deng, G., Sun, Z., Li, S., Peng, X., Li, W., Zhou, L., Ma, Y., Gong, P. and Cai, L. 2018. Cell-membrane immunotherapy based on natural killer cell membrane coated nanoparticles for the effective inhibition of primary and abscopal tumor growth. *ACS Nano*, 12(12): 12096–12108.

Ditto, A.J., Shah, P.N. and Yun, Y.H. 2009. Non-viral gene delivery using nanoparticles. *Expert Opin. Drug Deliv.*, 6(11): 1149–1160.

Divine, R., Dang, H.V., Ueda, G., Fallas, J.A., Vulovic, I., Sheffler, W., Saini, S., Zhao, Y.T., Morawski, P.A., Jennewein, M.F., Homad, L.J., Wan, Y.-H., Tooley, M.R., Seeger, F., Etemadi, A., Fahning, M.L., Lazarovits, J., Roederer, A., Walls, A.C., Stewart, L., Mazloomi, M., King, N.P., Campbell, D.J., McGuire, A.T., Stamatatos, L., Ruohola-Baker, H., Mathieu, J., Veesler, D. and Baker, D. 2021. Designed proteins assemble antibodies into modular nanocages. *Science*, 372(6537).

Dobosz, P. and Dzieciątkowski, T. 2019. The intriguing history of cancer immunotherapy. *Front Immunol.*, 10.

Du, Y., Xia, Y., Zou, Y., Hu, Y., Fu, J., Wu, J., Gao, X.-D. and Ma, G. 2019. Exploiting the lymph-node-amplifying effect for potent systemic and gastrointestinal immune responses via polymer/lipid nanoparticles. *ACS Nano.*, 13(12): 13809–13817.

Dunn, G.P., Old, L.J. and Schreiber, R.D. 2004. The three Es of cancer immunoediting. *Annu. Rev. Immunol.*, 22(1): 329–360.

Engin, A.B., Nikitovic, D., Neagu, M., Henrich-Noack, P., Docea, A.O., Shtilman, M.I., Golokhvast, K. and Tsatsakis, A.M. 2017. Mechanistic understanding of nanoparticles' interactions with extracellular matrix: The cell and immune system. *Part Fibre Toxicol.*, 14(1).

Eppler, H.B. and Jewell, C.M. 2019. Biomaterials as tools to decode immunity. *Adv. Mater*, 32(13): 1903367.

Esterházy, D., Canesso, M.C., Mesin, L., Muller, P.A., de Castro, T.B., Lockhart, A., ElJalby, M., Faria, A.M. and Mucida, D. 2019. Compartmentalized gut lymph node drainage dictates adaptive immune responses. *Nature*, 569(7754): 126–130.

Fadel, T.R. and Fahmy, T.M. 2014. Immunotherapy applications of carbon nanotubes: From design to safe applications. *Trends Biotechnol.*, 32(4): 198–209.

Fang, R.H., Hu, C.-M.J., Luk, B.T., Gao, W., Copp, J.A., Tai, Y., O'Connor, D.E. and Zhang, L. 2014. Cancer cell membrane-coated nanoparticles for anticancer vaccination and drug delivery. *Nano Lett.*, 14(4): 2181–2188.

Francis, D.M., Manspeaker, M.P., Schudel, A., Sestito, L.F., O'Melia, M.J., Kissick, H.T., Pollack, B.P., Waller, E.K. and Thomas, S.N. 2020. Blockade of immune checkpoints in lymph nodes through locoregional delivery augments cancer immunotherapy. *Sci. Transl. Med.*, 12(563).

Friedl, P., den Boer, A.T. and Gunzer, M. 2005. Tuning immune responses: diversity and adaptation of the immunological synapse. *Nat. Rev. Immunol.*, 5(7): 532–545.

Fromen, C.A., Rahhal, T.B., Robbins, G.R., Kai, M.P., Shen, T.W., Luft, J.C. and DeSimone, J.M. 2016. Nanoparticle surface charge impacts distribution, uptake, and lymph node trafficking by pulmonary antigen-presenting cells. *Nanomedicine*, 12(3): 677–687.

Galluzzi, L., Chan, T.A., Kroemer, G., Wolchok, J.D. and López-Soto, A. 2018. The hallmarks of successful anticancer immunotherapy. *Sci. Transl. Med.*, 10(459).

Goldberg, M.S. 2019. Improving cancer immunotherapy through nanotechnology. *Nat. Rev. Cancer.*, 19(10): 587–602.

Gong, N., Sheppard, N.C., Billingsley, M.M., June, C.H. and Mitchell, M.J. 2021. Nanomaterials for T-cell cancer immunotherapy. *Nat. Nanotechnol.*, 16(1): 25–36.

Gopalakrishnan, V., Spencer, C.N., Nezi, L., Reuben, A., Andrews, M.C., Karpinets, T.V., Prieto, P.A., Vicente, D., Hoffman, K., Wei, S.C., Cogdill, A.P., Zhao, L., Hudgens, C.W., Hutchinson, D.S., Mazo, T., Petaccia de Macedo, M., Cotechini, T., Kumar, T., Chen, W.S., Reddy, S.M., Szczepaniak Sloane, R., Galloway-Pena, J., Jiang, H., Chen, P.L., Shpall, E.J., Rezvani, K., Alousi, A.M., Chemaly, R.F., Shelburne, S., Vence, L.M., Okhuysen, P.C., Jensen, V.B., Swennes, A.G., McAllister, F., Marcelo Riquelme Sanchez, E., Zhang, Y., Le Chatelier, E., Zitvogel, L., Pons, N., Austin-Breneman, J.L., Haydu, L.E., Burton, E.M., Gardner, J.M., Sirmans, E., Hu, J., Lazar, A.J., Tsujikawa, T., Diab, A., Tawbi, H., Glitza, I.C., Hwu, W.J., Patel, S.P., Woodman, S.E., Amaria, R.N., Davies, M.A., Gershenwald, J.E., Hwu, P., Lee, J.E., Zhang, J., Coussens, L.M., Cooper, Z.A., Futreal, P.A., Daniel, C.R., Ajami, N.J., Petrosino, J.F., Tetzlaff, M.T., Sharma, P., Allison, J.P., Jeng, R.R. and Wargo, J.A. 2017. Gut microbiome modulates response to anti–PD-1 immunotherapy in melanoma patients. *Science*, 359(6371): 97–103.

Goradel, N.H., Baker, A.T., Arashkia, A., Ebrahimi, N., Ghorghanlu, S. and Negahdari, B. 2021. Oncolytic virotherapy: Challenges and solutions. *Curr. Probl. Cancer*, 45(1): 100639.

Guo, P., Liu, D., Subramanyam, K., Wang, B., Yang, J., Huang, J., Auguste, D.T. and Moses, M.A. 2018. Nanoparticle elasticity directs tumor uptake. *Nat. Commun.*, 9(1).

Halfvarson, J., Brislawn, C.J., Lamendella, R., Vázquez-Baeza, Y., Walters, W.A., Bramer, L.M., D'Amato, M., Bonfiglio, F., McDonald, D., Gonzalez, A., McClure, E.E., Dunklebarger, M.F., Knight, R. and Jansson, J.K. 2017. Dynamics of the human gut microbiome in inflammatory bowel disease. *Nat. Microbiol.*, 2(5).

Han, K., Nam, J., Xu, J., Sun, X., Huang, X., Animasahun, O., Achreja, A., Jeon, J.H., Pursley, B., Kamada, N., Chen, G.Y., Nagrath, D. and Moon, J.J. 2021. Generation of systemic antitumour immunity via the *in situ* modulation of the gut microbiome by an orally administered inulin gel. *Nat. Biomed. Eng.*, 5: 1377–1388.

Hayashi, Y., Takamiya, M., Jensen, P.B., Ojea-Jiménez, I., Claude, H., Antony, C., Kjaer-Sorensen, K., Grabher, C., Boesen, T., Gilliland, D., Oxvig, C., Strahle, U. and Weiss, C. 2020. Differential nanoparticle sequestration by macrophages and scavenger endothelial cells visualized *in vivo* in real-time and at ultrastructural resolution. *ACS Nano*, 14(2): 1665–1681.

Hao, M., Hou, S., Li, W., Li, K., Xue, L., Hu, Q., Zhu, L., Chen, Y., Sun, H., Ju, C. and Zhang, C. 2020. Combination of metabolic intervention and T cell therapy enhances solid tumor immunotherapy. *Sci. Transl. Med.*, 12(571).

He, C., Hu, Y., Yin, L., Tang, C. and Yin, C. 2010. Effects of particle size and surface charge on cellular uptake and biodistribution of polymeric nanoparticles. *Biomaterials*, 31(13): 3657–3666.

Helmink, B.A., Reddy, S.M., Gao, J., Zhang, S., Basar, R., Thakur, R., Yizhak, K., Sade-Feldman, M., Blando, J., Han, G., Gopalakrishnan, V., Xi, Y., Zhao, H., Amaria, R.N., Tawbi, H.A., Cogdill, A.P., Liu, W., LeBleu, V.S., Kugeratski, F.G., Patel, S., Davies, M.A., Hwu, P., Lee, J.E., Gershenwald, J.E., Lucci, A., Arora, R., Woodman, S., Keung, E.Z., Gaudreau, P.-O., Reuben, A., Spencer, C.N., Burton, E.M., Haydu, L.E., Lazar, A.J., Zapassodi, R., Hudgens, C.W., Ledesma, D.A., Ong, S., Bailey, M., Warren, S., Rao, D., Krijgsman, O., Rozeman, E.A., Peeper, D., Blank, C.U., Schumacher, T.N., Butterfield, L.H., Zelazowska, M.A., McBride, K.M., Kalluri, R., Allison, J., Petitprez, F., Herma Fridman, W., Sautes-Fridman, C., Hacohe, N., Rezvani, K., Sharma, P., Tetzlaff, M.T., Wang, L. and Wargo, J.A. 2020. B cells and tertiary lymphoid structures promote immunotherapy response. *Nature*, 577(7791): 549–555. https://doi.org/10.1038/s41586-019-1922-8.

Hiltensperger, M., Beltrán, E., Kant, R., Tyystjärvi, S., Lepennetier, G., Domínguez Moreno, H., Bauer, I.J., Grassman, S., Jarosch, S., Schober, K., Buchholz, V.R., Kenet, S., Gasperi, C., Ollinger, R., Rad, R., Muschaweckh, A., Sie, C., Aly, L., Knier, B., Garg, G., Afzali, A.M., Gerdes, L.A., Kumpfel, T., Franzenburg, S., Kawakami, N., Hemmer, B., Busch, D.H., Misgeld, T., Dornmair, K. and Korn, T. 2021. Skin and gut imprinted helper T cell subsets exhibit distinct functional phenotypes in central nervous system autoimmunity. *Nat. Immunol.*, 22(7): 880–892.

Hinde, E., Thammasiraphop, K., Duong, H.T., Yeow, J., Karagoz, B., Boyer, C., Gooding, J.J. and Gaus, K. 2016. Pair correlation microscopy reveals the role of nanoparticle shape in intracellular transport and site of drug release. *Nat. Nanotechnol.*, 12(1): 81–89.

Hu, C.-M.J., Zhang, L., Aryal, S., Cheung, C., Fang, R.H. and Zhang, L. 2011. Erythrocyte membrane-camouflaged polymeric nanoparticles as a biomimetic delivery platform. *Proc Natl Acad Sci U S A*, 108(27): 10980–10985.

Hu, C.-M.J., Fang, R.H., Copp, J., Luk, B.T. and Zhang, L. 2013a. A biomimetic nanosponge that absorbs pore-forming toxins. *Nat. Nanotechnol.*, 8(5): 336–340.

Hu, C.-M.J., Fang, R.H., Luk, B.T. and Zhang, L. 2013b. Nanoparticle-detained toxins for safe and effective vaccination. *Nat. Nanotechnol.*, 8(12): 933–938.

Hu, W. and Pasare, C. 2013. Location, location, location: Tissue-specific regulation of immune responses. *J. Leukoc. Biol.*, 94(3): 409–421.

Hu, C.-M.J., Fang, R.H., Wang, K.-C., Luk, B.T., Thamphiwatana, S., Dehaini, D., Nguyen, P., Angsantikul, P., Wen, C.H., Kroll, A.V., Carpenter, C., Ramesh, M., Qu, V., Patel, S.H., Zhu, J., Shi, W., Hofman, F.M., Chen, T.C., Gao, W., Zhang, K., Chien, S. and Zhang, L. 2015. Nanoparticle biointerfacing by platelet membrane cloaking. *Nature*, 526(7571): 118–121.

Huang, Y., Chen, Y., Zhou, S., Chen, L., Wang, J., Pei, Y., Xu, M., Feng, J., Jiang, T., Liang, K., Liu, S., Song, Q., Jiang, G., Gu, X., Zhang, Q., Gao, X. and Chen, J. 2020. Dual-mechanism based CTLs infiltration enhancement initiated by Nano-sapper potentiates immunotherapy against immune-excluded tumors. *Nat. Commun.*, 11(1).

Hui, Y., Yi, X., Wibowo, D., Yang, G., Middelberg, A.P., Gao, H. and Zhao, C.-X. 2020. Nanoparticle elasticity regulates phagocytosis and cancer cell uptake. *Sci. Adv.*, 6(16).

Horton, B.L., Fessenden, T.B. and Spranger, S. 2019. Tissue site and the cancer immunity cycle. *Trends Cancer*, 5(10): 593–603.

Hunter, Z., McCarthy, D.P., Yap, W.T., Harp, C.T., Getts, D.R., Shea, L.D. and Miller, S.D. 2014. A biodegradable nanoparticle platform for the induction of antigen-specific immune tolerance for treatment of autoimmune disease. *ACS Nano.*, 8(3): 2148–2160.

Irvine, D.J. and Dane, E.L. 2020. Enhancing cancer immunotherapy with nanomedicine. *Nature. Rev. Immunol.*, 20(5): 321–334.

Irvine, D.J., Aung, A. and Silva, M. 2020. Controlling timing and location in vaccines. *Adv. Drug Deliv. Rev.*, 158: 91–115.

Iwata, M., Hirakiyama, A., Eshima, Y., Kagechika, H., Kato, C. and Song, S.-Y. 2004. Retinoic acid imprints gut-homing specificity on T cells. *Immunity*, 21(4): 527–538.

Jiang, W., Kim, B.Y., Rutka, J.T. and Chan, W.C. 2008. Nanoparticle-mediated cellular response is size-dependent. *Nat. Nanotechnol.*, 3(3): 145–150.

Jiang, Y., Krishnan, N., Zhou, J., Chekuri, S., Wei, X., Kroll, A.V., Yu, C.L., Duan, Y., Gao, W., Fang, R.H. and Zhang, L. 2020. Engineered cell-membrane-coated nanoparticles directly present tumor antigens to promote anticancer immunity. *Adv. Mater.*, 32(30): 2001808.

Ju, Y., Kelly, H.G., Dagley, L.F., Reynaldi, A., Schlub, T.E., Spall, S.K., Bell, C.A., Cui, J., Mitchell, A.J., Lin, Z., Wheatley, A.K., Thurecht, K.J., Davenport, M.P., Webb, A.J., Caruso, F. and Kent, S.J. 2020. Person-specific biomolecular coronas modulate nanoparticle interactions with immune cells in human blood. *ACS Nano.*, 14(11): 15723–15737.

Kambayashi, T. and Laufer, T.M. 2014. Atypical MHC class II-expressing antigen-presenting cells: can anything replace a dendritic cell? *Nat. Rev. Immunol.*, 14(11): 719–730.

Kapate, N., Clegg, J.R. and Mitragotri, S. 2021. Non-spherical micro- and nanoparticles for drug delivery: Progress over 15 years. *Adv. Drug Deliv. Rev.*, 177: 113807.

Kedmi, R., Veiga, N., Ramishetti, S., Goldsmith, M., Rosenblum, D., Dammes, N., Hazan-Halevy, I., Nahary, L., Leviatan-Ben-Arye, S., Harlev, M., Behlke, M., Benhar, I., Lieberman, J. and Peer, D. 2018. A modular platform for targeted RNAi therapeutics. *Nat. Nanotechnol.*, 13(3): 214–219.

King, N.P., Bale, J.B., Sheffler, W., McNamara, D.E., Gonen, S., Gonen, T., Yeates, T.O. and Baker, D. 2014. Accurate design of co-assembling multi-component protein nanomaterials. *Nature*, 510(7503): 103–108.

Kirtane, A.R., Verma, M., Karandikar, P., Furin, J., Langer, R. and Traverso, G. 2021. Nanotechnology approaches for global infectious diseases. *Nat. Nanotechnol.*, 16(4): 369–384.

Korangath, P., Barnett, J.D., Sharma, A., Henderson, E.T., Stewart, J., Yu, S.-H., Kadala, S.K., Yang, C.-T., Caserto, J.S., Hedayati, M., Armstrong, T.D., Jaffee, E., Gruettner, C., Zhou, X.C., Fu, W., Hu, C., Sukumar, S., Simons, B.W. and Ivkov, R. 2020. Nanoparticle interactions with immune cells dominate tumor retention and induce T cell–mediated tumor suppression in models of breast cancer. *Sci. Adv.*, 6(13).

Kortylewski, M., Swiderski, P., Herrmann, A., Wang, L., Kowolik, C., Kujawski, M., Lee, H., Scuto, A., Liu, Y., Yang, C., Deng, J., Soifer, H.S., Raubitschek, A., Forman, S., Rossi, J.J., Pardoll, D.M., Jove, R. and Yu, H. 2009. *In vivo* delivery of siRNA to immune cells by conjugation to a TLR9 agonist enhances antitumor immune responses. *Nat. Biotechnol.*, 27(10): 925–932.

Kosmides, A.K., Necochea, K., Hickey, J.W. and Schneck, J.P. 2018. Separating T cell targeting components onto magnetically clustered nanoparticles boosts activation. *Nano Lett.*, 18(3): 1916–1924.

Kranz, L.M., Diken, M., Haas, H., Kreiter, S., Loquai, C., Reuter, K.C. et al. 2016. Systemic RNA delivery to dendritic cells exploits antiviral defence for cancer immunotherapy. *Nature*, 534(7607): 396–401.

Krausgruber, T., Fortelny, N., Fife-Gernedl, V., Senekowitsch, M., Schuster, L.C., Lercher, A., Nemc, A., Schmidl, C., Rendeiro, A.F., Bergthaler, A. and Bock, C. 2020. Structural cells are key regulators of organ-specific immune responses. *Nature*, 583(7815): 296–302.

Krienke, C., Kolb, L., Diken, E., Streuber, M., Kirchhoff, S., Bukur, T., Akilli-Ozturk, O., Kranz, L.M., Berger, H., Petschenka, J., Diken, M., Kreiter, S., Yogev, N., Waisman, A., Kariko, K., Tureci, O. and Sahin, U. 2021. A noninflammatory mRNA vaccine for treatment of experimental autoimmune encephalomyelitis. *Science*, 371(6525): 145–153.

Krohn-Grimberghe, M., Mitchell, M.J., Schloss, M.J., Khan, O.F., Courties, G., Guimaraes, P.P., Rohde, D., Cremer, S., Kowalski, P.S., Sun, Y., Tan, M., Webster, J., Wang, K., Iwamoto, Y., Schmidt, S.P., Wojtkiewicz, G.R., Nayar, R., Frodermann, V., Hulsmans, M., Chung, A., Hoyer, F.F., Swirski, F.K., Langer, R., Anderson, D.G. and Nahrendorf, M. 2020. Nanoparticle-encapsulated siRNAs for gene silencing in the haematopoietic stem-cell niche. *Nat. Biomed. Eng.*, 4(11): 1076–1089.

Kumar, S., Anselmo, A.C., Banerjee, A., Zakrewsky, M. and Mitragotri, S. 2015. Shape and size-dependent immune response to antigen-carrying nanoparticles. *J. Control Release*, 220: 141–148.

Ledo, A.M., Sasso, M.S., Bronte, V., Marigo, I., Boyd, B.J., Garcia-Fuentes, M. and Alonso, M.J. 2019. Co-delivery of RNAi and chemokine by polyarginine nanocapsules enables the modulation of myeloid-derived suppressor cells. *J. Control. Release*, 295: 60–73.

Lee, Y., Sugihara, K., Gillilland, M.G., Jon, S., Kamada, N. and Moon, J.J. 2019. Hyaluronic acid–bilirubin nanomedicine for targeted modulation of dysregulated intestinal barrier, microbiome and immune responses in colitis. *Nat. Mater*, 19(1): 118–126.

Li, S., Luo, M., Wang, Z., Feng, Q., Wilhelm, J., Wang, X., Li, W., Wang, J., Cholka, A., Fu, Y.-X., Sumer, B.D., Yu, H. and Gao, J. 2021. Prolonged activation of innate immune pathways by a polyvalent STING agonist. *Nat. Biomed. Eng.*, 5(5): 455–466.

Li, Y., Su, Z., Zhao, W., Zhang, X., Momin, N., Zhang, C., Wittrup, K.D., Dong, Y., Irvine, D.J. and Weiss, R.. 2020. Multifunctional oncolytic nanoparticles deliver self-replicating IL-12 RNA to eliminate established tumors and prime systemic immunity. *Nat. Cancer*, 1(9): 882–893.

Liu, H., Moynihan, K.D., Zheng, Y., Szeto, G.L., Li, A.V., Huang, B., Van Egeren, D.S., Park, C. and Irvine, D.J. 2014. Structure-based programming of lymph-node targeting in molecular vaccines. *Nature*, 507(7493): 519–522.

Liu, M., Apriceno, A., Sipin, M., Scarpa, E., Rodriguez-Arco, L., Poma, A., Marchello, G., Battaglia, G. and Angioletti-Uberti, S. 2020. Combinatorial entropy behaviour leads to range selective binding in ligand-receptor interactions. *Nat. Commun.*, 11(1).

Look, M., Bandyopadhyay, A., Blum, J.S. and Fahmy, T.M. 2010. Application of nanotechnologies for improved immune response against infectious diseases in the developing world. *Adv. Drug Deliv. Rev.*, 62(4-5): 378–393.

Lundqvist, M., Stigler, J., Elia, G., Lynch, I., Cedervall, T. and Dawson, K.A. 2008. Nanoparticle size and surface properties determine the protein corona with possible implications for biological impacts. *Proc. Natl. Acad. Sci. USA*, 105(38): 14265–14270.

Luo, M., Wang, H., Wang, Z., Cai, H., Lu, Z., Li, Y., Du, M., Huang, G., Wang, C., Chen, X., Porembka, M.R., Lea, J., Frankel, A.E., Fu, Y.-X., Chen, Z.J. and Gao, J. 2017. A STING-activating nanovaccine for cancer immunotherapy. *Nat. Nanotechnol.*, 12(7): 648–654.

Lv, Y., Li, F., Wang, S., Lu, G., Bao, W., Wang, Y., Tian, Z., Wei, W. and Ma, G. 2021. Near-infrared light–triggered platelet arsenal for combined photothermal-immunotherapy against cancer. *Sci. Adv.*, 7(13).

Makadia, H.K. and Siegel, S.J. 2011. Poly Lactic-co-Glycolic Acid (PLGA) as biodegradable controlled drug delivery carrier. *Polymers (Basel)*, 3(3): 1377–1397.

Manolova, V., Flace, A., Bauer, M., Schwarz, K., Saudan, P. and Bachmann, M.F. 2008. Nanoparticles target distinct dendritic cell populations according to their size. *Eur. J. Immunol.*, 38(5): 1404–1413.

Mantovani, A., Sozzani, S., Locati, M., Allavena, P. and Sica, A. 2002. Macrophage polarization: Tumor-associated macrophages as a paradigm for polarized M2 mononuclear phagocytes. *Trends Immunol.*, 23(11): 549–555.

Matson, V., Fessler, J., Bao, R., Chongsuwat, T., Zha, Y., Alegre, M.-L., Luke, J.J. and Gajewski, T.F. 2018. The commensal microbiome is associated with anti–PD-1 efficacy in metastatic melanoma patients. *Science*, 359(6371): 104–108.

Mead, B.P., Curley, C.T., Kim, N., Negron, K., Garrison, W.J., Song, J., Rao, D., Wilson Miller, G., Mandell, J.W., Purow, B.W., Suk, J.S., Hanes, J. and Price, R.J. 2019. Focused ultrasound preconditioning for augmented nanoparticle penetration and efficacy in the central nervous system. *Small*, 15(49): 1903460.

Miao, L., Li, L., Huang, Y., Delcassian, D., Chahal, J., Han, J., Shi, Y., Sadtler, K., Gao, W., Lin, J., Doloff, J.C., Langer, R. and Anderson, D.A. 2019. Delivery of mRNA vaccines with heterocyclic lipids increases anti-tumor efficacy by STING-mediated immune cell activation. *Nat. Biotechnol.*, 37(10): 1174–1185.

Miller, M.A., Zheng, Y.-R., Gadde, S., Pfirschke, C., Zope, H., Engblom, C., Kohler, R.H., Iwamoto, Y., Yang, K.S., Askevold, B., Kolishetti, N., Pittet, M., Lippard, S.J., Farokhzad, O.C. and Weissleder, R. 2015. Tumour-associated macrophages act as a slow-release reservoir of nano-therapeutic Pt(IV) pro-drug. *Nat. Commun.*, 6(1).

Min, Y., Roche, K.C., Tian, S., Eblan, M.J., McKinnon, K.P., Caster, J.M., Chai, S., Herring, L.E., Zhang, L., Zhang, T., DeSimone, J.M., Tepper, J.E., Vincent, B.G., Serody, J.S. and Wang, A.Z. 2017. Antigen-capturing nanoparticles improve the abscopal effect and cancer immunotherapy. *Nat. Nanotechnol.*, 12(9): 877–882.

Mirshafiee, V., Mahmoudi, M., Lou, K., Cheng, J. and Kraft, M.L. 2013. Protein corona significantly reduces active targeting yield. *Chem. Commun. (Camb)*, 49(25): 2557.

Moghimi, S.M. and Szebeni, J. 2003. Stealth liposomes and long circulating nanoparticles: Critical issues in pharmacokinetics, opsonization and protein-binding properties. *J. Lipid Res.*, 42: 463–478.

Moon, J.J., Suh, H., Bershteyn, A., Stephan, M.T., Liu, H., Huang, B., Sohail, M., Luo, S., Um, S.H., Khant, H., Goodwin, J.T., Ramos, J., Chiu, W. and Irvine, D.J. 2011. Interbilayer-crosslinked multilamellar vesicles as synthetic vaccines for potent humoral and cellular immune responses. *Nat. Mater*, 10(3): 243–251.

Mosquera, M.J., Kim, S., Zhou, H., Jing, T.T., Luna, M., Guss, J.D., Reddy, P., Lai, K., Leifer, C.A., Brito, I.L., Hernandez, C.J. and Singh, A.. 2019. Immunomodulatory nanogels overcome restricted immunity in a murine model of gut microbiome–mediated metabolic syndrome. *Sci. Adv.*, 5(3).

Mura, S., Nicolas, J. and Couvreur, P. 2013. Stimuli-responsive nanocarriers for drug delivery. *Nat. Mater*, 12(11): 991–1003.

Nam, J., Son, S., Park, K.S., Zou, W., Shea, L.D. and Moon, J.J. 2019. Cancer nanomedicine for combination cancer immunotherapy. *Nat. Rev. Mater*, 4(6): 398–414.

Nejman, D., Livyatan, I., Fuks, G., Gavert, N., Zwang, Y., Geller, L.T., Rotter-Maskowitz, A., Weiser, R., Mallel, G., Gigi, E., Meltser, A., Douglas, G.M., Kamer, I., Gopalakrishnan, V., Dadosh, T., Levin-Zaidman, S., Avnet, S., Atlan, T., Cooper, Z.A., Arora, R., Cogdill, A.P., Khan, M.A.W., Ologun, G., Bussi, Y., Weinberger, A., Lotan-Pompan, M., Golani, O., Perry, G., Rokah, M., Bahar-Shany, K., Rozeman, E.A., Blank, C.U., Ronai, A., Shaoul, R., Amit, A., Dorfman, T., Kremer, R., Cohen, Z.R., Harnof, S., Siegal, T., Yehuda-Shnaidman, E., Gala-Yam, E.N., Shapira, H., Baldini, N., Langille, M.G.I., Ben-Nun, A., Kaufman, B., Nissan, A., Golan, T., Dadiani, M., Levanon, K., Bar, J., Yust-Katz, S., Barshack, I., Peeper, D.S., Raz, D.J., Segal, E., Wargo, J.A., Sandbank, J., Shental, N. and Straussman, R. 2020. The human tumor microbiome is composed of tumor type–specific intracellular bacteria. *Science*, 368(6494): 973–980.

Netti, P.A., Berk, D.A., Swartz, M.A., Grodzinsky, A.J. and Jain, R.K. 2000. Role of extracellular matrix assembly in interstitial transport in solid tumors. *Cancer Res.*, 60: 2497–2503.

Ngamcherdtrakul, W., Reda, M., Nelson, M.A., Wang, R., Zaidan, H.Y., Bejan, D.S., Hoang, N.H., Lane, R.S., Luoh, S.-W., Leachman, S.A., Mills, G.B., Gray, J.W., Lund, A.W. and Yantasee, W. 2021. *In Situ* tumor vaccination with nanoparticle Co-delivering CpG and STAT3 siRNA to effectively induce whole-body antitumor immune response. *Adv. Mater*, 2100628.

Nguyen, V.H. and Lee, B.-J. 2017. Protein corona: a new approach for nanomedicine design. *Int. J. Nanomed.*, 12: 3137–3151.

Oberli, M.A., Reichmuth, A.M., Dorkin, J.R., Mitchell, M.J., Fenton, O.S., Jaklenec, A., Anderson, D.G., Langer, R. and Blankschtein, D. 2016. Lipid nanoparticle assisted mRNA delivery for potent cancer immunotherapy. *Nano. Lett.*, 17(3): 1326–1335.

Oh, S.A., Wu, D.-C., Cheung, J., Navarro, A., Xiong, H., Cubas, R., Totpal, K., Chiu, H., Wu, Y., Comps-Agar, L., Leader, A.M., Merad, M., Roose-Germa, M., Warming, S., Yan, M., Kim, J.M., Rutz, S. and Mellman, I. 2020. PD-L1 expression by dendritic cells is a key regulator of T-cell immunity in cancer. *Nat. Cancer*, 1(7): 681–691.

Ouyang, B., Poon, W., Zhang, Y.-N., Lin, Z.P., Kingston, B.R., Tavares, A.J., Zhang, Y., Chen, J., Valic, M.S., Syed, A.M., MacMillan, P., Couture-Senecal, J., Zheng, G. and Chan, W.C.W. 2020. The dose threshold for nanoparticle tumour delivery. *Nat. Mater*, 19(12): 1362–1371.

Parada Venegas, D., De la Fuente, M.K., Landskron, G., Gonzalez, M.J., Quera, R., Dijkstra, G., Harmsen, H.J.M., Nico Faber, K. and Hermoso, M.A. 2019. Short Chain Fatty Acids (SCFAs)-mediated gut epithelial and immune regulation and its relevance for inflammatory bowel diseases. *Front. Immunol.*, 10.

Parakhonskiy, B., Zyuzin, M.V., Yashchenok, A., Carregal-Romero, S., Rejman, J., Möhwald, H., Parak, W.J. and Skirtach, A.G. 2015. The influence of the size and aspect ratio of anisotropic, porous CaCO3 particles on their uptake by cells. *J. Nanobiotechnol.*, 13(1).

Parayath, N.N., Stephan, S.B., Koehne, A.L., Nelson, P.S. and Stephan, M.T. 2020. *In vitro*-transcribed antigen receptor mRNA nanocarriers for transient expression in circulating T cells *in vivo*. *Nat. Commun.*, 11(1).

Pardi, N., Hogan, M.J., Porter, F.W. and Weissman, D. 2018. mRNA vaccines—a new era in vaccinology. *Nat. Rev. Drug Discov.*, 17(4): 261–279.

Parodi, A., Quattrocchi, N., van de Ven, A.L., Chiappini, C., Evangelopoulos, M., Martinez, J.O., Brown, B.S., Khaled, S.Z., Yazdi, I.K., Enzo, M.V., Isenhart, L., Ferrari, M. and Tasciotti, E. 2012. Synthetic nanoparticles functionalized with biomimetic leukocyte membranes possess cell-like functions. *Nat. Nanotechnol.*, 8(1): 61–68.

Pati, R., Shevtsov, M. and Sonawane, A. 2018. Nanoparticle vaccines against infectious diseases. *Front Immunol.*, 9.

Peng, Q., Qiu, X., Zhang, Z., Zhang, S., Zhang, Y., Liang, Y., Guo, J., Peng, H., Chen, M., Fu, Y.-X. and Tang, H. 2020. PD-L1 on dendritic cells attenuates T cell activation and regulates response to immune checkpoint blockade. *Nat. Commun.*, 11(1).

Polack, F.P., Thomas, S.J., Kitchin, N., Absalon, J., Gurtman, A., Lockhart, S., Perez, J.L., Marc, G.P., Moreira, E.D., Zerbini, C., Bailey, R., Swanson, K.A., Roychoudhury, S., Koury, K., Li, P., Kalina, W.V., Cooper, D., Frenck, R.W., Hammitt, L.L., Tureci, O., Nell, H., Schaefer, A., Unal, S., Tresnan, D.B., Mather, S., Dormitzer, P.R., Sahin, U., Jansen, K.U. and Gruber, W.C. 2020. Safety and efficacy of the BNT162b2 mRNA Covid-19 vaccine. *N. Engl. J. Med.*, 383(27): 2603–2615.

Poon, W., Kingston, B.R., Ouyang, B., Ngo, W. and Chan, W.C. 2020. A framework for designing delivery systems. *Nat. Nanotechnol.*, 15(10): 819–829.

Porter, C.J. and Trevaskis, N.L. 2020. Targeting immune cells within lymph nodes. *Nat. Nanotechnol.*, 15(6): 423–425.

Prabhakar, U., Maeda, H., Jain, R.K., Sevick-Muraca, E.M., Zamboni, W., Farokhzad, O.C., Barry, S.T., Gabizon, A., Grodzinski, P. and Blakey, D.C. 2013. Challenges and key considerations of the enhanced permeability and retention effect for nanomedicine drug delivery in oncology. *Cancer Res.*, 73(8): 2412–2417.

Qin, L., Cao, J., Shao, K., Tong, F., Yang, Z., Lei, T., Wang, Y., Hu, C., Umeshappa, C.S., Gao, H. and Peppas, N.A. 2020. A tumor-to-lymph procedure navigated versatile gel system for combinatorial therapy against tumor recurrence and metastasis. *Sci. Adv.*, 6(36).

Ramishetti, S., Kedmi, R., Goldsmith, M., Leonard, F., Sprague, A.G., Godin, B., Gozin, M., Cullis, P.R., Dykxhoorn, D.M. and Peer, D. 2015. Systemic gene silencing in primary T lymphocytes using targeted lipid nanoparticles. *ACS Nano.*, 9(7): 6706–6716.

Reddy, S.T., van der Vlies, A.J., Simeoni, E., Angeli, V., Randolph, G.J., O'Neil, C.P., Lee, L.K., Swartz, M.A. and Hubbell, J.A. 2007. Exploiting lymphatic transport and complement activation in nanoparticle vaccines. *Nat. Biotechnol.*, 25(10): 1159–1164.

Ren, J., Cai, R., Wang, J., Daniyal, M., Baimanov, D., Liu, Y., Yin, D., Liu, Y., Miao, Q., Zhao, Y. and Chen, C. 2019. Precision nanomedicine development based on specific opsonization of human cancer patient-personalized protein coronas. *Nano Lett.*, 19(7): 4692–4701.

Riquelme, E., Zhang, Y., Zhang, L., Montiel, M., Zoltan, M., Dong, W., Quesada, P., Sahin, I., Chandra, V., San Lucas, A., Scheet, P., Xu, H., Hanasah, S.M., Feng, L., Burks, J.K., Do, K.-A., Peterson, C.B., Nejman, D., Tzeng, C.-W. D., Kim, M.P., Sears, C.L., Ajami, N., Petrosino, J., Wood, L.D., Maitra, A., Straussman, R., Katz, M., White, J.R., Jenq, R., Wargo, J.A. and McAllister, F. 2019. Tumor microbiome diversity and composition influence pancreatic cancer outcomes. *Cell*, 178(4).

Round, J.L. and Mazmanian, S.K. 2009. The gut microbiota shapes intestinal immune responses during health and disease. *Nat. Rev. Immunol.*, 9(5): 313–323.

Safari, H., Kelley, W.J., Saito, E., Kaczorowski, N., Carethers, L., Shea, L.D. and Eniola-Adefeso, O. 2020. Neutrophils preferentially phagocytose elongated particles—An opportunity for selective targeting in acute inflammatory diseases. *Sci. Adv.*, 6(24).

Saito, E., Kuo, R., Pearson, R.M., Gohel, N., Cheung, B., King, N.J.C., Miller, S.D. and Shea, L.D. 2019. Designing drug-free biodegradable nanoparticles to modulate inflammatory monocytes and neutrophils for ameliorating inflammation. *J. Control Release*, 300: 185–196.

Saklayen, M.G. 2018. The global epidemic of the metabolic syndrome. *Curr. Hypertens Rep.*, 20(2).

Salvati, A., Pitek, A.S., Monopoli, M.P., Prapainop, K., Bombelli, F.B., Hristov, D.R., Kelly, P.M., Aberg, C., Mahon, E. and Dawson, K.A. 2013. Transferrin-functionalized nanoparticles lose their targeting capabilities when a biomolecule corona adsorbs on the surface. *Nat. Nanotechnol.*, 8(2): 137–143.

Sanz-Ortega, L., Rojas, J.M., Marcos, A., Portilla, Y., Stein, J.V. and Barber, D.F. 2019. T cells loaded with magnetic nanoparticles are retained in peripheral lymph nodes by the application of a magnetic field. *J. Nanobiotechnol.*, 17(1).

Sautès-Fridman, C., Petitprez, F., Calderaro, J. and Fridman, W.H. 2019. Tertiary lymphoid structures in the era of cancer immunotherapy. *Nat. Rev. Cancer*, 19(6): 307–325.

Scheppach, W., Bartram, H.P. and Richter, F. 1995. Role of short-chain fatty acids in the prevention of colorectal cancer. *Eur. J. Cancer*, 31(7-8): 1077–1080.

Schmid, D., Park, C.G., Hartl, C.A., Subedi, N., Cartwright, A.N., Puerto, R.B., Zheng, Y., Maiarana, J., Freeman, G.J., Wucherpfennig, K.W., Irvine, D.J. and Goldberg, M.S. 2017. T cell-targeting nanoparticles focus delivery of immunotherapy to improve antitumor immunity. *Nat. Commun.*, 8(1).

Schudel, A., Francis, D.M. and Thomas, S.N. 2019. Material design for lymph node drug delivery. *Nat. Rev. Mater.*, 4(6): 415–428.

Schudel, A., Chapman, A.P., Yau, M.-K., Higginson, C.J., Francis, D.M., Manspeaker, M.P., Avecilla, A.R.C., Rohner, N.A., Finn, M.G. and Thomas, S.N. 2020. Programmable multistage drug delivery to lymph nodes. *Nat. Nanotechnol.*, 15(6): 491–499.

Sepich-Poore, G.D., Zitvogel, L., Straussman, R., Hasty, J., Wargo, J.A. and Knight, R. 2021. The microbiome and human cancer. *Science*, 371(6536).

Shae, D., Becker, K.W., Christov, P., Yun, D.S., Lytton-Jean, A.K., Sevimli, S., Ascao, M., Kelley, M., Johnson, D.B., Balko, J.M. and Wilson, J.T. 2019. Endosomolytic polymersomes increase the activity of cyclic dinucleotide STING agonists to enhance cancer immunotherapy. *Nat. Nanotechnol.*, 14(3): 269–278.

Sharma, P., Siddiqui, B.A., Anandhan, S., Yadav, S.S., Subudhi, S.K., Gao, J., Goswami, S. and Allison, J.P. 2021. The next decade of immune checkpoint therapy. *Cancer Discov.*, 11(4): 838–857.

Shenoi, M.M., Shah, N.B., Griffin, R.J., Vercellotti, G.M. and Bischof, J.C. 2011. Nanoparticle preconditioning for enhanced thermal therapies in cancer. *Nanomedicine*, 6(3): 545–563.

Shields, C.W., Wang, L.L.W., Evans, M.A. and Mitragotri, S. 2019. Materials for immunotherapy. *Adv. Mater.*, 32(13): 1901633.

Shields, C.W., Evans, M.A., Wang, L.L.-W., Baugh, N., Iyer, S., Wu, D., Zhao, Z., Pusuluri, A., Ukidve, A., Pan, D.C. and Mitragotri, S. 2020. Cellular backpacks for macrophage immunotherapy. *Sci. Adv.*, 6(18).

Shim, M.S. and Kwon, Y.J. 2012. Stimuli-responsive polymers and nanomaterials for gene delivery and imaging applications. *Adv. Drug Deliv. Rev.*, 64(11): 1046–1059.

Sigl, C., Willner, E.M., Engelen, W., Kretzmann, J.A., Sachenbacher, K., Liedl, A.Kolbe, F., Wilsch, F., Ali Aghvami, S., Protzer, U., Hagan, M.F., Fraden, S. and Dietz, H. 2021. Programmable icosahedral shell system for virus trapping. *Nat. Mater.*, 20: 1281–1289.

Silvestrini, M.T., Ingham, E.S., Mahakian, L.M., Kheirolomoom, A., Liu, Y., Fite, B.Z., Tam, S.M., Tucci, S.T., Watson, K.D., Wong, A.W., Monjazeb, A.M., Hubbard, N.E., Murphy, W.J., Borowsky, A.D. and Ferrara, K.W. 2017. Priming is key to effective incorporation of image-guided thermal ablation into immunotherapy protocols. *JCI Insight*, 2(6).

Siriwon, N., Kim, Y.J., Siegler, E., Chen, X., Rohrs, J.A., Liu, Y. and Wang, P. 2018. CAR-T cells surface-engineered with drug-encapsulated nanoparticles can ameliorate intratumoral T-cell hypofunction. *Cancer Immunol. Res.*, 6(7): 812–824.

Smith, T.T., Stephan, S.B., Moffett, H.F., McKnight, L.E., Ji, W., Reiman, D., Bonagofski, E., Wohlfahrt, M.E., Pillai, S.P.S. and Stephan, M.T. 2017. *In situ* programming of leukaemia-specific T cells using synthetic DNA nanocarriers. *Nat. Nanotechnol.*, 12(8): 813–820.

Song, Q., Yin, Y., Shang, L., Wu, T., Zhang, D., Kong, M., Zhao, Y., He, Y., Tan, S., Guo, Y. and Zhang, Z. 2017. Tumor microenvironment responsive nanogel for the combinatorial antitumor effect of chemotherapy and immunotherapy. *Nano. Lett.*, 17(10): 6366–6375.

Song, T., Xia, Y., Du, Y., Chen, M.W., Qing, H. and Ma, G. 2021. Engineering the deformability of albumin-stabilized emulsions for lymph-node vaccine delivery. *Adv. Mater*, 2100106.

Song, W., Anselmo, A.C. and Huang, L. 2019. Nanotechnology intervention of the microbiome for cancer therapy. *Nat. Nanotechnol.*, 14(12): 1093–1103.

Stolnik, S., Dunn, S.E., Garnett, M.C., Davies, M.C., Coombes, A.G., Taylor, D.C., Irving, M.P., Purkiss, S.C., Tadros, T.F. and Davis, S.S. 1994. Surface modification of poly(lactide-co-glycolide) nanospheres by biodegradable poly(lactide)-poly(ethylene glycol) copolymers. *Pharm. Res.*, 11: 1800–1808.

Tan, J., Shah, S., Thomas, A., Ou-Yang, H.D. and Liu, Y. 2012. The influence of size, shape and vessel geometry on nanoparticle distribution. *Microfluidics Nanofluidics*, 14(1-2): 77–87.

Tang, L., Zheng, Y., Melo, M.B., Mabardi, L., Castaño, A.P., Xie, Y.-Q., Li, N., Kudchodkar, S.B., Wong, H.C., Jeng, E.K., Maus, M.V. and Irvine, D.J. 2018. Enhancing T cell therapy through TCR-signaling-responsive nanoparticle drug delivery. *Nat. Biotechnol.*, 36(8): 707–716.

Tay, R.E., Richardson, E.K. and Toh, H.C. 2020. Revisiting the role of CD4+ T cells in cancer immunotherapy—new insights into old paradigms. *Cancer Gene The.r*, 28(1-2): 5–17.

Tazaki, T., Tabata, K., Ainai, A., Ohara, Y., Kobayashi, S., Ninomiya, T., Orba, Y., Mitomo, H., Nakano, T., Hasegawa, H., Ijiro, K., Sawa, H., Suzuki, T. and Niikura, K. 2018. Shape-dependent adjuvanticity of nanoparticle-conjugated RNA adjuvants for intranasal inactivated influenza vaccines. *RSC Adv.*, 8(30): 16527–16536.

Tjandra, K.C. and Thordarson, P. 2019. Multivalency in drug delivery—when is it too much of a good thing? *Bioconj. Chem.*, 30(3): 503–514.

Tokatlian, T., Read, B.J., Jones, C.A., Kulp, D.W., Menis, S., Chang, J.Y., Steichen, J.M., Kumari, S., Allen, J.D., Dane, E.L., Liguori, A., Sangesland, M., Lingwood, D., Crispin, M., Schief, W.R. and Irvine, D.J. 2018. Innate immune recognition of glycans targets HIV nanoparticle immunogens to germinal centers. *Science*, 363(6427): 649–654.

Tong, S., Moyo, B., Lee, C.M., Leong, K. and Bao, G. 2019. Engineered materials for *in vivo* delivery of genome-editing machinery. *Nat. Rev. Mater.*, 4(11): 726–737.

Turk, M.J., Waters, D.J. and Low, P.S. 2004. Folate-conjugated liposomes preferentially target macrophages associated with ovarian carcinoma. *Cancer Lett.*, 213(2): 165–172.

Ukidve, A., Zhao, Z., Fehnel, A., Krishnan, V., Pan, D.C., Gao, Y., Mandal, A., Muzykantov, V. and Mitragotri, S. 2020. Erythrocyte-driven immunization via biomimicry of their natural antigen-presenting function. *Proc. Natl. Acad. Sci. U S A*, 117(30): 17727–17736.

Veiga, N., Goldsmith, M., Granot, Y., Rosenblum, D., Dammes, N., Kedmi, R., Ramishetti, S. and Peer, D. 2018. Cell specific delivery of modified mRNA expressing therapeutic proteins to leukocytes. *Nat. Commun.*, 9(1).

Veneziano, R., Moyer, T.J., Stone, M.B., Wamhoff, E.-C., Read, B.J., Mukherjee, S., Shepher, T.R., Das, J., Schief, W.R., Irvine, D.J. and Bathe, M. 2020. Role of nanoscale antigen organization on B-cell activation probed using DNA origami. *Nat. Nanotechnol.*, 15(8): 716–723.

Vincent, M.P., Bobbala, S., Karabin, N.B., Frey, M., Liu, Y., Navidzadeh, J.O., Stack, T. and Scott, E.A. 2021. Surface chemistry-mediated modulation of adsorbed albumin folding state specifies nanocarrier clearance by distinct macrophage subsets. *Nat. Commun.*, 12(1).

Waldman, A.D., Fritz, J.M. and Lenardo, M.J. 2020. A guide to cancer immunotherapy: From T cell basic science to clinical practice. *Nat. Rev. Immunol.*, 20(11): 651–668.

Wang, C., Sun, W., Wright, G., Wang, A.Z. and Gu, Z. 2016. Inflammation-triggered cancer immunotherapy by programmed delivery of CpG and anti-PD1 antibody. *Adv. Mater*, 28(40): 8912–8920.

Wang, C., Sun, W., Ye, Y., Hu, Q., Bomba, H.N., and Gu, Z. 2017. *In situ* activation of platelets with checkpoint inhibitors for post-surgical cancer immunotherapy. *Nat. Biomed. Eng.*, 1(2).

Wei, X., Zhang, G., Ran, D., Krishnan, N., Fang, R.H., Gao, W., Spector, S.A. and Zhang, L. 2018. T-cell-mimicking nanoparticles can neutralize HIV infectivity. *Adv. Mater.*, 30(45): 1802233.

Wibroe, P.P., Anselmo, A.C., Nilsson, P.H., Sarode, A., Gupta, V., Urbanics, R., Szebeni, J., Hunter, A.C., Mitragotri, S., Mollnes, T.E. and Moghimi, S.M. 2017. Bypassing adverse injection reactions to nanoparticles through shape modification and attachment to erythrocytes. *Nat. Nanotechnol.*, 12(6): 589–594.

Wilhelm, S., Tavares, A.J., Dai, Q., Ohta, S., Audet, J., Dvorak, H.F. and Chan, W.C. 2016. Analysis of nanoparticle delivery to tumours. *Nat. Rev. Mater.*, 1(5).

Xia, Y., Wu, J., Du, Y., Miao, C., Su, Z. and Ma, G. 2018. Bridging systemic immunity with gastrointestinal immune responses via oil-in-polymer capsules. *Adv. Mater.*, 30(31): 1801067.

Yang, W., Bai, Y., Xiong, Y., Zhang, J., Chen, S., Zheng, X., Meng, X., Li, L., Wang, J., Xu, C., Yan, C., Wang, L., Chang, C.Y.C., Chang. T.-Y., Zhang, T., Zhou, P., Song, B.-L., Liu, W., Sun, S.-C., Liu, X., Li, B.-L. and Xu, C. 2016. Potentiating the antitumour response of CD8+ T cells by modulating cholesterol metabolism. *Nature*, 531(7596): 651–655.

Yang, X., Lian, K., Meng, T., Liu, X., Miao, J., Tan, Y., Yuan, H. and Hu, F. 2018. Immune adjuvant targeting micelles allow efficient dendritic cell migration to lymph nodes for enhanced cellular immunity. *ACS Appl. Mater. Interfaces*, 10(39): 33532–33544.

Yost, K.E., Chang, H.Y. and Satpathy, A.T. 2021. Recruiting T cells in cancer immunotherapy. *Science*, 372(6538): 130–131.

Yu, T.C., Guo, F., Yu, Y., Sun, T., Ma, D., Han, J. et al. 2017. Fusobacterium nucleatum promotes chemoresistance to colorectal cancer by modulating autophagy. *Cell*, 170(3).

Yuan, H., Jiang, W., von Roemeling, C.A., Qie, Y., Liu, X., Chen, Y., Wang, Y., Wharen, R.E., Yun, K., Bu, G., Knutson, K.L. and Kim, B.Y.S. 2017. Multivalent bi-specific nanobioconjugate engager for targeted cancer immunotherapy. *Nat. Nanotechnol.*, 12(8): 763–769.

Zhang, F., Parayath, N.N., Ene, C.I., Stephan, S.B., Koehne, A.L., Coon, M.E., Holland, E.C. and Stephan, M.T. 2019a. Genetic programming of macrophages to perform anti-tumor functions using targeted mRNA nanocarriers. *Nat. Commun.*, 10(1).

Zhang, J., Chen, C., Li, A., Jing, W., Sun, P., Huang, X., Liu, Y., Zhang, S., Du, W., Zhang, R., Liu, Y., Gong, A., Wu, J. and Jiang, X. 2021. Immunostimulant hydrogel for the inhibition of malignant glioma relapse post-resection. *Nat. Nanotechnol.*, 16(5): 538–548.

Zhang, Q., Jiang, Q., Li, N., Dai, L., Liu, Q., Song, L., Wang, J., Li, Y., Tian, J., Ding, B. and Du, Y. 2014. DNA origami as an *in vivo* drug delivery vehicle for cancer therapy. *ACS Nano*, 8(7): 6633–6643.

Zhang, Y., Chen, Y., Lo, C., Zhuang, J., Angsantikul, P., Zhang, Q., Wei, X., Zhou, Z., Obonyo, M., Fang, R., Gao, W. and Zhang, L. 2019b. Inhibition of pathogen adhesion by bacterial outer membrane-coated nanoparticles. *Angew. Chem. Int. Ed.*, 58(33): 11404–11408.

Zhao, L., Seth, A., Wibowo, N., Zhao, C.-X., Mitter, N., Yu, C. and Middelberg, A.P.J. 2014. Nanoparticle vaccines. *Vaccine*, 32(3): 327–337.

Zhao, Z., Ukidve, A., Krishnan, V., Fehnel, A., Pan, D.C., Gao, Y., Kim, J., Evans, M.A., Mandal, A., Guo, J., Muzykantov, V.R. and Mitragotri, S. 2020. Systemic tumour suppression via the preferential accumulation of erythrocyte-anchored chemokine-encapsulating nanoparticles in lung metastases. *Nat. Biomed. Eng.*, 5(5): 441–454.

Zheng, D.-W., Dong, X., Pan, P., Chen, K.-W., Fan, J.-X., Cheng, S.-X. and Zhang, X.-Z. 2019. Phage-guided modulation of the gut microbiota of mouse models of colorectal cancer augments their responses to chemotherapy. *Nat. Biomed. Eng.*, 3(9): 717–728.

Zhong, Y., Ma, Z., Wang, F., Wang, X., Yang, Y., Liu, Y., Zhao, X., Li, J., Du, H., Zhang, M., Cui, Q., Zhu, S., Sun, Q., Wan, H., Tian, Y., Liu, Q., Wang, W., Garcia, K.C. and Dai, H. 2019. *In vivo* molecular imaging for immunotherapy using ultra-bright near-infrared-IIb rare-earth nanoparticles. *Nat. Biotechnol.*, 37(11): 1322–1331.

CHAPTER 12

Basics of Organ-On-A-Chip Technology

Brandon Ortiz-Casas[1],* and *Gloria Cristina Enríquez Cortina*[2]

1. Introduction to tissue engineering and organs-on-a-chip

With more than a century of being researched and applied, cell culture has been one of the main milestones within biomedical science, after providing a plethora of insights about fundamental cell and tissue biology. Besides, cell culture has offered a surfeit of *in vitro* models for various pharmacological and toxicological studies, becoming, along with animal models, a standard within pharmaceutical preclinical research. However, as the development of larger numbers of conventional drugs, biopharmaceuticals, and (more recently) nanopharmaceuticals are facing stricter regulations and clinical studies, the transition from preclinical models to humans started presenting enormous difficulties (Van Norman 2019, Hingorani et al. 2019). Due to the limitations of two-dimensional (2D) cultures mimicking precise phenotypes of human diseases, a paradigm shift in the use of test models is required (Langhans 2018). Accordingly, tissue engineering has offered several tridimensional and dynamical culture models capable of providing more valuable insights than their bidimensional counterparts. Among these models, organs-on-a-chip (OOCs) stand as highly propitious *in vitro* models, due to their ability to mimic the cellular microenvironment and interorgan interactions while keeping control of inlets, outlets, and monitoring through various biosensors (Zhang et al. 2018). In this chapter, a detailed explanation of the main components of an OOC will be outlined, in addition to a few potential applications in basic and clinical research. Furthermore, some biological key considerations will be offered for a successful chip design, in addition to the abundance of potentially useful materials and fabrication technologies. Finally, there will be a more detailed analysis of particular biomimetic models and a discussion about the main challenges these will face in their adoption in industry and the clinic.

2. Overview of organ-on-a-chip technology

An organ-on-a-chip is a tissue engineering technology that emulates the structural and/or functional microenvironmental characteristics of a tissue, organ, or inter-organ system that employs on-chip

[1] Department of Engineering Science, University of Oxford, Parks Road, Oxford OX1 3PJ United Kingdom.
[2] School of Medicine and Healthcare, Tecnológico de Monterrey, Mexico City Campus, Calle Puente 222, Coapa, Arboledas, del Sur, Tlalpan, 14380, Mexico City, Mexico.
* Corresponding author: brandon.ortizcasas@kellogg.ox.ac.uk

microfluidic and microfabrication technologies. In contrast with conventional bidimensional or tridimensional cell cultures, OOCs are *ad hoc* engineered systems that consider a diversity of physical (spatial localization and confinement, shear stress, scaffold stiffness, etc.), chemical (nutrients, hormones, and growth factors), and biological (use of different cells in co-culture) features all at once, achieving a closer degree of fidelity of an *in vivo* model (Zhang et al. 2018, Sosa-Hernández et al. 2018). Furthermore, OOCs open the possibility of generating "more human" alternatives than traditional *in vivo* models like mice, offering greater flexibility regarding the cell phenotype and the anatomical and physiological ratios of organ sizes and flows (Aziz et al. 2017, Van Den Berg et al. 2019). Representatively, OOC technology can be illustrated by four constituents:

Structural and connective tissue components

The structural and connective tissue component (or "stroma") mimics the interface composed of epithelium, connective tissue, blood vessels, and other elements that do not play a direct functional role in the system (Sontheimer-Phelps et al. 2019). This constituent is relevant to mimic because it represents both an interface and a barrier for nutrients, hormones, drugs, and toxins (Figure 1). Compared to 2D culture, where all cells have contact with external molecules homogeneously and indiscriminately, stromal interfaces and barriers play a vital role in pharmacokinetics and cytotoxic profiles due to the heterogeneous distribution and diffusion (Miao et al. 2015, Hua et al. 2017). The correct modeling of these components might play an important duty in developing strategies to pass through nearly impenetrable barriers, such as the blood-brain barrier and the fibrotic stroma of pancreatic cancer (Thomas and Radhakrishnan 2019, Zakharova et al. 2020).

Functional tissue component

Contrary to the structural and connective components of the culture, the functional component (or "parenchyma") is made up of cells that emulate an intrinsic physiological activity of the mimicked organ. Invariably of its nature, these cells are usually very dependent on three-dimensionality, cell density, heterotypic and homotypic interactions, and mechanical, electrical, or biochemical stimuli

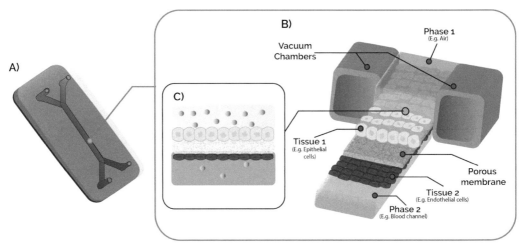

Figure 1. Traditional organ-on-a-chip design for interface modeling. (A) The system is composed of two different substrates with two main microfluidic channels, with their respective inlets and outlets. Every substrate will have a channel representing a biological phase, denoted with either a blue or red color in the figure. (B) Each channel will perfuse the necessary components for the culture of a specific tissue of that phase, which will be separated by a thin, porous membrane. The phases will also be surrounded by vacuum chambers that might simulate mechanical stress or stretching. (C) The obtained system mimics the biological interface in a dynamical and three-dimensional microenvironment, allowing the study of the interaction between phases and the transfer of various components, such as hormones, nutrients, and drugs.

that will allow the best imitation of specific phenotypes (Huh et al. 2011, Bhatia and Ingber 2014, Kaarj and Yoon 2019). The more biologically faithful the OOC, the more predictive power the model will have in experimentation. For this component of the OOC, synergy with other technologies such as organoids is particularly useful (Takebe et al. 2017).

Microfluidic and micromachined components

Arguably, the insertion of microfluidics into tissue engineering has been one of the main achievements in the field overall. Without microfluidics, OOC technology would probably never have proliferated. In this field, volumes are handled within the micro/nanoscale (generally bounded in the 10^{-9} to 10^{-6} L range) due to their strict laminar flow regime, practically devoid of turbulence (Sosa-Hernández et al. 2018). This phenomenon is easily understood when analyzing the intrinsic low Reynolds Number—a dimensionless numerical ratio between inertial forces (responsible for vortices and eddy currents) to viscous forces (which force the flow to stay smooth), with the latter being the predominant force (Rehm et al. 2008). Laminar flows are scarce on the macroscale and present peculiar characteristics such as lack of mixing between parallel flows and the presence of surface tension and interfacial charges. These phenomena can be wisely used for the generation of concentration gradients, the creation of discrete emulsions and droplets, and other several events that would be extremely difficult to replicate on a macro scale (Wang et al. 2017, Ji et al. 2018, Rosser et al. 2019). For the flow control within the chip, micropumps and microvalves are classically used. Besides, the presence of surface charges on the microchannel walls also opens the possibility of employing electroosmotic flows, extending the flexibility of flow control (Kaarj and Yoon 2019). Notwithstanding the advantages of microfluidics, chip design is not a trivial task, and it should always take a careful approach when designing the inlets, outlets and internal microchannels and compartments that will satisfy the requirements of the system, such as the input of nutrients and growth factors, the output of waste, and internal channels for inter-organic interaction. Moreover, other components for mechanical and/or electrical stimulation may be inserted into the chip design, to assist cell differentiation and correct functioning of the parenchymal culture (Kaarj and Yoon 2019).

Sensing and monitoring components

Ultimately, without the capacity of monitoring and quantifying relevant data and experimental parameters, organs-on-a-chip would only remain as elegant (and quite expensive) proofs of concept. The insertion of biosensors allows the quantification and real-time monitoring of important culture parameters, such as the amount of oxygen (both dissolved and gaseous), pH, temperature, glucose and lactate concentrations, cytokines levels, mechanical and electrical activity, and other countless biomarkers (Shaegh et al. 2016, Ferrari et al. 2020). Besides cell metabolism and standard cell culture parameters, other output signals related to tissue integrity can be studied. An important example of the latter is the measurement of the trans-epithelial electrical resistance (TEER), which is related to the integrity and permeability of the tissue endothelium (Odijk et al. 2015)

3. Applications for organs-on-a-chip

The high-throughput nature of organs-on-a-chip makes them attractive systems for various biomedical areas, ranging from basic research to potential medical applications. For an organ-on-a-chip to be relevant, the tissue interface, the parenchyma, the microchannel system, and the sensors must be optimal for the intended purpose. For instance, a chip that may perfectly emulate the physiology of an organ might not be the best model for drug testing if it lacks important stromal interfaces for pharmacokinetic profiles. Before starting any design, the application must be well outlined.

Biological modeling and biomimetics

Due to the difficulty of studying deeply and in real-time the plethora of physiological mechanisms in humans, a considerable part of our understanding of some biochemical and molecular processes might be markedly limited due to our dependence on animal and traditional cell culture models. On the other hand, some physiological mechanisms are massively complex to study *in vivo*. Therefore, the use of OOCs has had recent contributions in elucidating important knowledge about these processes. A particular example to highlight is the work of Shik Mun et al., who modeled a pancreas-on-a-chip model to study the relationship between the cystic fibrosis transmembrane conductance regulator protein and cystic fibrosis-related diabetes. For this model, the researchers isolated and cultured pancreatic ductal epithelial cells and pancreatic islets from patients with pancreatitis, and studied the interaction between both cells, finding a considerable decrease in insulin production (Shik Mun et al. 2019). In parallel, Trapecar et al. developed a multiple interconnected human microphysiological system (MPS) capable of mimicking the complex gut-liver-immune axis for the study of the interaction between organs in complex pathogenesis. In this biomimetic system, an ulcerative colitis (UC) gut model was connected to a healthy liver model, with circulating T cells between the systems. Thusly, this *ex vivo* model was utilized to elucidate the inflammatory regulation processes between T cells and short-chain fatty acids derived from the microbiome (Trapecar et al. 2020).

Drug discovery and screening

Although the new generations of drugs have become more advanced, the number of new pharmaceuticals available on the market has not grown proportionally. For instance, nanomedicine has presented a variety of nanosystems capable of improving the solubility, adsorption, cellular uptake, pharmacokinetics, efficacy, and tissue selectivity of certain drugs, while also decreasing their toxicity. Nonetheless, only a discreet number of them have reached clinical use. In 2017, the FDA had only approved 50 nanodrugs for medical use in the US (Ventola 2017). Adverse drug events are another clinical threat, since several negative medical outcomes cannot be predicted in traditional preclinical models. Therefore, the need for advanced *in vitro* models that will better represent the pharmacological effects opens a crucial opportunity for organs-on-a-chip. Due to the dense three-dimensional nature of some of these templates, OOCs offer a meaningful opportunity to assess the efficacy of pharmacological strategies to cross diverse biological barriers, such as the stromal barriers of tumors (Tanaka and Kano 2018, Sontheimer-Phelps et al. 2019). Added to this, the ability of OOCs to control gradients of nutrients, growth factors, and hormones allows the biomodelling of intricate, spatiotemporal-varying processes such as hormonal cycles. An illustrative example is the work of Xiao and coworkers, who presented a "menstrual-cycle-on-a-chip" capable of mimicking the whole hormonal profile. Being a system capable of evaluating an exclusive female phenotype at specific timepoints, it offers a suitability that would be practically impossible to replicate in traditional cultures (Xiao et al. 2017).

Toxicological screening

Organs-on-a-chip also offers a unique opportunity in the development of various toxicological tests. In contrast to pharmacological tests, toxicological tests are not feasible to escalate to clinical trials, making the findings of advanced *in vitro* models much more valuable (Pridgeon et al. 2018). A clear example of these applications is in the evaluation of air pollutants. In this regard, the team of Xu et al. proposed a 3D lung model to evaluate the effect of particulate matter smaller than 2.5 micrometers. In their system proposal, they cultured endothelial and epithelial cells in a chip design analogous to the alveolar-blood barrier, of which they were able to assess the inflammatory response, generation of reactive oxygen species (ROS), cytotoxicity, and changes in tissue permeability (Xu et al. 2020). Another example is presented by Bovard et al., who designed a model for assessing acute and chronic aerosol exposure. Their bi-organic lung/liver was composed of hepatic spheroids and human bronchial epithelial cells cultured with the air-liquid interface method. To test this system, aflatoxin B1 was supplied to observe the role of liver cells in reducing the cytotoxicity of the *in vitro* model (Bovard et al. 2018).

Regenerative medicine

The evolution of regenerative medicine and tissue engineering has gone hand in hand from the beginning. Attain this objective, OOC technology presents an invaluable opportunity to emulate and study processes such as neuron recovery, tissue regeneration, and cellular differentiation, besides being a potentially ideal element for quality assurance in regenerative medicine trials (Harink et al. 2013, Haring et al. 2017, Pasqualini et al. 2017). As an example, Agrawal, Aung, and Varghese published their progress in skeletal muscle *in vitro* modeling, proposing a system capable of emulating a muscle injury model. Apart from being a useful proof-of-concept, relevant mechanical data such as the elastic modulus, strain profile, and muscle passive tension could be computed. After characterizing the model, they tested it with a cardiotoxin to evaluate structural and mechanical changes. Among these changes, some important events were observed such as cytoskeleton disruption and a decrease in passive tension (Agrawal et al. 2017). Finally, it should not be ignored that OOCs may have a primary role in tissue transplantation. Due to their ability to differentiate cells to desired phenotypes, it might be possible in the future a synergistic combination of OOCs with bioreactor technology to generate customizable tissue, avoiding the use of risky allografts, and offering a more advanced version of current *in vitro* autografts.

4. Biological and physical features of an organ-on-a-chip device

Organs-on-a-chip emerges from the interdisciplinary approach between areas such as medicine, biotechnology, materials science, microfluidics, and micromachining. Therefore, the success of any model will require a comprehensive and strenuous approach. The OOC can simulate the physiological environment of a certain organ function such as regulation of concentration gradients, shear force, cell patterning, tissue-boundaries, and tissue-organ interactions. To do so, it is necessary to deeply understand the organ in its two major states: homeostasis and pathology (Beckwitt et al. 2018). The responses of the body to external and internal stimulus is highly complex, since they depend on the interplay of components from lower levels of biological organization including cells, proteins, and genes. Thus, the study of physiological events by using only *in vivo* models is difficult and considerably inefficient. For decades, traditional two-dimensional cell culture systems represented an accurate method for cell biology research, bringing light to cell metabolism knowledge. Nevertheless, 2D systems are not suitable to study physiological responses of tissues/organs, intra-organ interactions, and microenvironmental factors, requiring replication and verification on *in vivo* models. On the other hand, species differences between animal models and humans represent a pivotal key feature that determines the lack of reproducibility in human experiments. The latter could be attributed to the inadequate understanding of the human tissue environment and, ergo, the inaccurate predictions of the overall effects. OOC technology avoids these obstacles by giving an accurate physiological platform that resembles more accurately the human systems (Md Ali et al. 2018). As a rule of thumb, the more reliable a model will be, the more predictive power it will possess. However, the practicality must also be considered, based on clear objectives of the biomimetic system. Making a more detailed system may yield greater fidelity (although not in all cases), but at the same time, it also may present substantial complexity in the design, assembly, maintenance, and scaling to an industrial level.

5. Biological features

Biological scaling

Among some of the basic biological parameters to consider is the correct biological scaling of the organ or inter-organic system. Despite the lack of scientific consensus on how scaling should be realized for microengineered tissues, there are a series of proposals to respect anatomical and physiological relationships such as the classical allometric scaling, and the organ mass-residence time approach (Moraes et al. 2013, Wikswo et al. 2013). It must also be taken into consideration that, while respecting the mass and volume/area ratios, considerable changes in cell metabolism between *in vitro* and *in vivo* models may distort these relationships, forcing the researcher to perform a series of empirical tests (Moraes et al. 2013). Also, maintaining the tissue-to-liquid ratio and the flow rates as analogous to physiological conditions as possible is highly desirable (Maschmeyer et al. 2015).

Fidelity of cell type

One more decision to consider when mimicking is the selection of the cells to be cultured. The use of primary cells, human pluripotent stem cells (hPSCs), induced pluripotent stem cells (iPSCs), or immortalized cell lines can have several advantages and disadvantages. Some of these are summarized in Table 1.

Table 1. Advantages and disadvantages of different cell types in OCC design.

Type of cultured cell	Advantages	Disadvantages
Mature Primary culture	- Best representation of the native cell phenotype, physiology, and/or pathophysiology - Ability to make customizable *in vitro* models	- Difficulty in extracting them from the patient in several cases - Low availability of cells - Low or no replication in some cell types - Limited life span - Loss of the desired phenotype or dedifferentiation after a short time of being cultured
Adult Stem Cells	- Conservation of genotype and mutations - Ability to make customizable *in vitro* models	- Low frequency
Induced Pluripotent Stem Cells (iPSCs)	- Conservation of genotype and mutations - Ability to make customizable *in vitro* models	- Low reprogramming efficiency - Risk of tumorigenesis and genetic abnormalities
Immortalized cell lines	- Ease of growing - Unlimited life span	- Severe variations in genotype and phenotype

Cell patterning

Besides the cell type alone, the interactions between different cells, their spatial distribution, and organization also dictate the correct function of the whole system. For this purpose, microfluidics enables the devising of *in vitro* chip models with the correct cell distribution. In this regard, different strategies can be used such as modification of surfaces, arrangement, and 3D printing. An appreciable method based on three-dimensional printing was developed by Li et al., who yielded a method to achieve directed topological modifications by using heterotypic cell patterning seeded on glass chips. By using this method, it is possible to study epithelial-mesenchymal interactions

(Li et al. 2015). This method is useful to trace cutaneous epithelial-mesenchymal interaction and other types of cell patterning.

6. Physical features

Fluid shear force

The dynamics of cell culture are determined by the microfluidic flows, facilitating nutrient distribution and waste products disposal. In the OOC, this dynamic environment represents more accurately the *in vivo* features than that one established on a static culture. Besides, organ polarity also can be achieved by fluid shear stress. Moreover, OOC can elicit the activation of different surface molecules that are of primary importance for signaling cascades through physical pressure on the physiological function of endothelial cells (Wojciak-Stothard and Ridley 2003).

Concentration gradient

A stable biomolecule gradient results if the laminar flow is controlled spatially and temporally. Many important biological processes that rely on the concentration gradient of certain molecules, such as angiogenesis, invasion, and migration. Microfluidics can be used to modify the flow speed and channel shape and size by using microvalves and micropumps to resemble three-dimensional (3D) molecular concentration gradient (Figeys and Pinto 2000, Haeberle and Zengerle 2007).

Dynamic mechanical stress

In the human body, the lungs, the cardiovascular system, and the bones are subjected to pressure fluctuations. Thus, the skeletal muscle, blood vessels, bones and cartilage are in constant mechanic stress. Microfluidics allows the use of fenestrated membranes to create cycles of differential mechanical stress rates. The mechanical stress is a critical condition during distinct physiological processes (Chin et al. 2007).

7. Materials in tissue engineering and organ-on-a-chip

The selection of the materials for the OOC modeling represents another critical point. There is no single kit of materials for all cases. Consequently, their selection must be selected based on the requirements of the scaffolds, the microfluidic systems, the nature of the chemicals to be infused, the nature of stress to be applied to the cells, and the type of needed sensors, among other circumstantial needs. Throughout this chapter, the groups of materials with the highest degree of adoption will be reviewed and described according to their respective advantages, drawbacks, and circumstantial applications where they can offer the best utility.

Materials for cellular arrangement and scaffolding

The construction of our cellular scaffold is one of the main success factors in organ-on-a-chip design. As another useful rule of the thumb in tissue engineering, the more similar the cellular scaffold is to the native extracellular matrix (ECM), the closer it will be to obtain the desired cellular phenotype. Different cell populations will require distinct microenvironmental conditions, so it is highly desirable to consider parameters such as the ideal stiffness, porosity, or the presence of ECM-analogous motifs when selecting our materials.

Ceramics

Ceramic materials have a considerable history within tissue engineering due to their stiffness, biological inertness, resorb ability, and biocompatibility (Baino et al. 2015). However, precisely due to their high stiffness and low elasticity, the use of these materials rarely reaches other uses than those related to bones and teeth. Materials such as hydroxyapatite, zirconium, alumina, calcium phosphates, and glass-based ceramics are among the most common examples in tissue engineering. Therefore, the presence of these within bone, dental and musculoskeletal chip models is appreciable in the literature, gaining in recent years considerable interest in its use for stem cell culture. One of the most recent and interesting cases to highlight the latter is the paper of Sieber et al., who developed a bone marrow system ideal for the culture of hematopoietic stem and progenitor cells (HSPCs). As a cellular scaffold, the researchers used a ceramic structure composed of zirconium oxide coated with hydroxyapatite. This cultured scaffold was installed inside a microfluidic system and cultured for several weeks, yielding a fibronectin network and the expression of stem cell factor and other bone marrow-related genes. A Colony Forming Unit -Granulocyte, Erythrocyte, Monocyte, Megakaryocyte (CFU-GEMM) was later built to evaluate the differentiation ability of the cultured HSPCs, finding diverse colony-forming-unit populations related to the progenies of these cells (Sieber et al. 2018). Another approach with potential uses for organs-on-a-chip is presented by Díaz Lantada and collaborators, who manufactured an interesting monolithic chip made of alumina and produced by lithographic methods. This multiwell and multichannel chip showed a certain level of suitability for the culture of HSPCs, even though their pluripotency and other relevant characteristics were analyzed with subsequent tests. Overall, it seems that the use of ceramic materials will keep a special place for bone and dental models, offering not only the right stiffness, porosity, and compatibility to support musculoskeletal tissue, but also in maintaining the pluripotential of hematopoietic stem cells (Díaz Lantada et al. 2017).

Natural polymers

Nature has an enormous number of polymers that sustain life as it is known. Even the extracellular matrix itself offers an important clue about the composition that can be employed for scaffolding. Describing all the natural polymers would require a lengthy description that falls outside the purposes of this chapter. Still, a few polymers can be highlighted in this category such as collagen, fibronectin, alginate, gelatin, agar, hyaluronic acid, cellulose, agarose, and chitosan, whose uses have been appreciable in the development of various engineered systems. One of the advantages among some of these polymers, such as the case of collagen, gelatin, and fibronectin, is the presence of arginine-glycine-aspartic acid (RGD) motifs. RGD motifs are the minimal essential cell adhesion peptide sequence and are strongly associated with increased cell adhesion and integrin affinity (D'Souza et al. 1991, Takahashi et al. 2007, Taubenberger et al. 2010). Also, these polymers generally present good biocompatibility and do not usually release harmful residues after degradation. However, it is also necessary to emphasize the drawbacks that natural polymers have overall, as are their poor mechanical properties and difficulty in remaining structurally stable under physiological conditions (Kondiah et al. 2020). Also, a few of them have poor processability in some manufacturing technologies such as electrospinning. Even though many of these polymers tend to have limited properties without some type of chemical modification, mixing them in different ratios can lead to the generation of various "bio inks" that can be utilized in 3D bioprinting technologies (Liu et al. 2018). In summary, natural polymers generally offer a compatible material for cell scaffolding, with a high degree of cell adhesion. Even so, some drawbacks can be compensated through the mixture with other materials, or through chemical modifications to create a series of semi-synthetic derivatives.

Artificial polymers

The use of artificial polymers is traceable to the beginnings of animal cell culture, where polystyrene use has stood for decades. Although their degree of adoption in OOC technology is lower than their natural counterparts, there are areas of opportunity when these materials meet noticeable properties for the culture model. In addition to the latter, the polymer must meet some basic requirements for its biomedical use, such as good biocompatibility and biodegradability. For applications strictly related to the design of cell scaffolding and tissue engineering, some synthetic polymers such as poly (lactic acid) (PLA), poly (vinyl alcohol) (PVA), polycaprolactone (PCL), polylactic-co-glycolic acid (PLGA), polyurethane (PU) and polyethylene glycol (PEG) have offered a variety of approaches that could translate into more robust applications in organ-on-a-chip technology. Among some of the advantages of artificial polymers are good mechanical strength, chemical inertness, ease of manipulating their geometries, controllable degradation rates, and tunable chemical structures (Guo and Ma 2014). However, most artificial polymers present serious problems such as poor cell adhesion, while some release acidic products when degraded (Stratton et al. 2016). Still, the use of synthetic polymers may hold promise possibilities in a handful of cases. One of them is their use within composites in conjunction with natural polymers and other substances to make more advanced scaffolds. A classic example is the synthesis of hydrogels composed of PEG and fibrinogen for 3D cellular scaffolding (Almany and Seliktar 2005). Another of the most attractive uses of synthetic polymers is the synthesis of ultrafine, microporous membranes that are generally much more complicated to construct and maintain using natural polymers. This is exemplified by the team of Pensabene et al., who built an ultrathin semipermeable membrane composed of poly (L-lactic acid) for biomimicry of the cell basement membrane. Using laser machining, a micropore pattern with constant diameters of 2 μm was obtained in the membrane and subsequently characterized with scanning electron microscopy (Pensabene et al. 2016). In brief, the adoption of these materials in biomimetic systems offers considerable opportunities, such as composite synthesis that will maximize their mechanical properties, and the construction of ultrathin structures that require higher levels of stiffness or more attractive Young's moduli than their natural counterparts can offer.

Semi-synthetic polymers

Through some chemical modifications, some natural polymers can be manipulated to improve some of their properties to the needs of the model. Among these materials, gelatin methacryolyl (GelMA) is arguably the most relevant semi-synthetic material for tissue engineering, due to its tunable mechanical properties, its considerable biocompatibility, and ease of extrusion. To prepare GelMA, natural gelatin is reacted with methacrylic anhydride, which will start substituting the amine groups of the amino acid residues by methacryoyl groups. The total percentage of converted amine groups to methacryoyl groups (known as the degree of substitution) can be chemically controlled, allowing the tunability of a few physical and chemical properties (Yue et al. 2015). Due to the gelatin within its composition, GelMA presents RGD motifs that also confer strong cell adhesion. Another advantage of this hydrogel is its flexibility to be adapted into various fabrication methods, such as photo-pattering, micromolding, and 3D bioprinting (Occhetta et al. 2013, Pepelanova et al. 2018, Ding et al. 2019). Even so, GelMA, like any other hydrogel, also has considerable disadvantages, such as poor resolution in contrast to stiffer materials. Besides, to crosslink the material to form the hydrogel, the composition requires a photoinitiator and UV radiation exposure. The latter, if repeated continuously, can affect cell viability and the preservation of the desired phenotype. Consequently, the search for similar hydrogels that can avoid UV curing is a valuable goal (Hölzl et al. 2016). Other semi-synthetic products such as chitosan derivatives have gained some relevance within tissue engineering and drug delivery (Kim et al. 2008). However, their participation in the generation of

OOC systems seems to be limited. Ultimately, GelMA and other semi-synthetic polymers offer enormous flexibility when developing cell scaffolds, positioning themselves as promising materials for biomimetic models. GelMA can be used with different degrees of substitution and in conjunction with various other components to create an invaluable amount of bioinks. However, its drawbacks must be considered when designing *in vitro* models, to avoid UV overexposure and non-printability ranges.

Nanomaterials and nanocomposites

The recent adoption of nanomaterials to biomedical applications has opened one of the most exciting opportunities in the search for new materials for tissue engineering. Some of these nanostructures possess exceptional mechanical properties, while others resemble the extracellular matrix in terms of its structural conformation. On the other hand, some nanomaterials can tune their properties through surface functionalization, making several of them quite adaptable for different uses. Among all nano-scaled options, nanometric versions of polymers and ceramics have begun to be explored in the development of various scaffolds, in addition to those of metallic and carbon-based composition. On one side, one of the most practical approaches to date is the creation of new nanocomposites, where materials such as nano-hydroxyapatite (nano-HA) crystals, nano-TiO2, and nano calcium phosphate have been implemented for advanced versions of cellular scaffolds (Fan et al. 2016, Singh et al. 2019, Liu et al. 2020). Further, some other unconventional nanostructures have been used to create cell culture scaffolds with some added properties. For example, multi-walled carbon nanotubes (MWCNTs) have been used in nanocomposites in conjunction with gelatin, chitosan, and GelMA, improving not only some mechanical properties but also some electrical properties like electrical conductivity. The latter has gained considerable relevance in cardiac tissue engineering since some culture systems enriched with MWCNTs have shown improvement in cardiac cell growth, differentiation, and regeneration (Gorain et al. 2018). Alternatively, enormous interest has been generated around the use of nanofibers, due to their similarity to the diameters and matrix porosities of the ECM. Additionally, the high surface area/volume ratio of nanofibers is advantageous for cell attachment. For the synthesis of these nanofibers, electrospinning is an easy and favorable technique capable of using some natural and synthetic polymers, with an appreciable number of reported applications in the literature (Mehta and Pawar 2018). Overall, although no nanomaterial will work as a "jack-of-all-trades", nanotechnology offers a huge range of structures with different properties, geometries, and dimensionalities that can complement the ideal architecture for optimal cell culture. The rational and systematic selection of all scaffolding materials, as well as the necessary manufacturing methods, are important keys in obtaining a cellular phenotype sufficiently satisfactory to the needs of the *in vitro* model.

8. Materials for chip design

Chip design is the heart of operations, so selecting a suitable material is just as important as scaffolding design. However, chip design is different from scaffolding since the required materials should be adapted for microfluidic applications. Among some desirable characteristics are the ease of generating microfluidic channels, good channel resolution, optical transparency for optical biosensors and microscopes, thermostability, gas permeability, and lack of absorption of the infused chemicals. Just as in cellular scaffolding, it is highly recommended that the manufacturing methods will be considered as well, as they may have an impact on industrial scalability. If good planning is done regarding the selection of materials and manufacturing methods, one-step organ-on-a-chip fabrication is possible (Lee and Cho 2016).

Glass and silicon chips

The use of glass and silicon dates back to the creation of the first microfluidic systems, and although their use has become more discreet as newer materials were inserted, there are still circumstances in which these materials may still be relevant. Glass maintains some attractive properties for microfluidic applications, such as chemical inertness, good optical transparency, good channel resolution, hydrophilicity, thermal conductivity and high resistant to small molecules' absorption. For this reason, some recent OOC proposals have been made with glass-based substrates. For example, the team of Hirama and collaborators prepared an OOC device composed entirely of glass to overcome some of the intrinsic limitations that elastomer-based chips have, such as the absorption of hydrophobic molecules and background fluorescence (Hirama et al. 2019). However, one of the main disadvantages of glass in microfluidics is its high cost, which not only hinders its transition to the industry but also its use in prototyping. On the other hand, silicone has also been a considerably forgotten material for microfluidics, given its high costs due to the expensive maintenance of the clean rooms needed to work it correctly. Another drawback is its optical opacity, making the use of fluorescence infeasible. Even so, silicon has allowed a whole revolution in the field of microfabrication, being the main substrate for the construction of important technologies such as microelectromechanical systems (MEMS), which have contributed to the development of a technology related to organs-on-a-chip: the lab-on-a-chip (LOC) technology (Azizipour et al. 2020). The insertion of silicon-based semiconductor systems, MEMS, and LOCs into organ-on-a-chip platforms presents a very attractive opportunity for the development of more advanced platforms, which is why some scientists in recent years have experimented with silicon for the creation of OOCs with some embedded microsystems. Exemplifying it, da Ponte et al. proposed a silicon OOC, in which a complementary metal oxide semiconductor (CMS) was integrated into the chip for the construction of an *in situ*, real-time temperature sensor (da Ponte et al. 2021). Even though both glass and silicone are expensive materials for microfluidics, they still have relevance in developing OOC technology for creating advanced proof-of-concepts. Furthermore, given the success of the semiconductor industry, it might be possible to extrapolate silicon chips to commercialization if the market niche requires them in high quantities.

Elastomers

With the adoption of elastomers, and especially, with the adoption of polydimethylsiloxane (PDMS), microfluidics reached a new milestone towards low-cost device manufacturing. PDMS has positioned itself as the most important standard in microfluidics, standing out from other materials due to its optical transparency, flexibility, simplicity and low cost of operation, gas permeability, and cross-linkability with a myriad of materials and other PDMS structures (Torino et al. 2018). Besides, it is possible to make membranes with micrometric porosities that can be easily inserted into the chips through crosslinking (Quirós-Solano et al. 2018). Arguably, the history of organs-on-a-chip would not have progressed to current standards without PDMS, due to the enormous dependence that microfluidics has had on the latter. However, PDMS is not a perfect material, and it also has a few drawbacks that limit its almost universal use. For example, PDMS tends to absorb hydrophobic molecules, which greatly limits the number of substances that can flow through their channels, including some biomolecules and some drugs of hydrophobic nature. Being also a polymer that allows gas permeability, PDMS microchips can also suffer from water evaporation, compromising some culture systems where the flow input is not constant (Toepke and Beebe 2006). Finally, the geometry of the microchannels can be compromised if the system is under high pneumatic pressure (Raj et al. 2018). Due to these substantial drawbacks, there have been considerable efforts trying to find materials "beyond PDMS" (Carlborg et al. 2011), although none have been presented to date with the same or greater quantity of desirable properties. Besides PDMS, some other elastomers have been involved in the development of microfluidic chips. For instance, tetrafluoroethylene-propylene (FEPM) has been proposed recently for organ-on-a-chip design, as an alternative capable of flowing hydrophobic drugs that PDMS would typically absorb (Sano et al. 2019).

Thermoplastic polymers

Thermoplastics are highly malleable materials at the right temperatures and their escalation to the industry is immensely easy. These polymers usually have very low costs, both in production and as raw materials, and several of them have good optical transparency, mechanical stability, and ease for microchannel machining (Gencturk et al. 2017). Because of all this, some thermoplastics have been tested in microfluidic systems and organs-on-a-chip as alternatives to glass, silicon, and PDMS. Polymers like polyetheretherketone (PEEK) have been explored for OOC design, as is the case with the mentioned work of Bovard et al., where this polymer was tested for the development of a lung/liver chip for toxicological screening (Bovard et al. 2018). PEEK is an inert polymer, with considerable biocompatibility, and it is a non-absorbent of small molecules, in contrast to PDMS. However, PEEK is a particularly expensive thermoplastic in high volumes, as seen in the prosthetics industry (Haleem and Javaid 2019). Another detail to consider is that this and other thermoplastics require a different type of equipment and methods to operate them efficiently since the nature of its malleability does not make them compatible with most techniques and equipment of the previously mentioned materials. Therefore, its use could be further explored for prototyping and creation of proofs of concepts, and perhaps its addition to the industry could be carefully considered if its properties justify their selection and do not hinder the production chain.

Hydrogels

Although their use has not been as popular as PDMS, hydrogels in microfluidic applications have been gaining attention in recent years. As previously reviewed, hydrogels are usually biocompatible materials, with a practical resemblance to the extracellular matrix, and with some tunable mechanical properties in several cases. Moreover, hydrogels are materials that can absorb water and swell without being solubilized and can change from swell to shrunk state if the correct stimuli are performed. With these properties, different microfluidic and biomimetic systems based on hydrogels have been presented in the literature as very particular alternatives to conventional microfluidic materials. As a recent case to highlight, the model of Nie et al. proposed a completely hydrogel-based alternative for the engineering of vascularized systems, utilizing a simple manufacturing method. In this work, two hydrogel bulks with microgrooves and composed of alginate, gelatin, and/or GelMA were constructed by a layer-by-layer process and subsequently crosslinked to form a bulk with microchannels inside it. This vessel-on-a-chip was tested with human umbilical vein endothelial cells (HUVEC), which were cultivated inside the channels. As a proof-of-concept, this system showed beneficial properties for cells, which presented adequate spreading morphology along the substrate and a high survival rate after the LIVE/DEAD assay (Nie et al. 2018). However, given the poor mechanical properties of hydrogels overall, the assembly of much more complex structures remains a challenge in OOC design.

9. Sensor selection

Many of the organs-on-a-chip reported in the literature rely heavily on cell viability, proliferation, and cytotoxicity assays, making them impractical to carry out several comprehensive and quantitative analyses that do not compromise the operating system. Therefore, the insertion of sensors that monitor real-time the culture parameters and metabolic activities is highly desirable. It must be considered that, during manufacturing, some sensors will have to be built together with the rest of the OOC chip, so the selection of materials and manufacturing processes should preferably be abreast with the rest of the segments. Similarly, the correct selection of the materials, the physicochemical principles of the sensors, and optimal designing will have an important role in their analytical parameters. Among all the types of sensors used in the literature, electrochemical and optical sensors are the most widely used for organ-on-a-chip applications (Ferrari et al. 2020), representing a practical classification for this chapter.

Electrochemical sensors

Electrochemical sensors are attractive due to their ease of being miniaturized and their high sensitivity. Consequently, they have been applied in OOC technology for monitoring parameters such as dissolved oxygen, pH, substrate, and metabolite levels, and TEER measurement for tissue integrity and permeability. For oxygen measurement, different electrochemical sensors have been proposed, although the work of Moya et al. can be highlighted due to the ease and practicality with which they embedded them *in situ*. In their work, the team microfabricated a set of three amperometric oxygen sensors on an ultrathin, porous membrane of a liver-on-a-chip model for local monitoring of oxygen gradients. For this, they printed a photoresist SU-8 primer layer on the porous membrane, and later a set of gold and silver electrodes through inkjet printing technology, without having any negative effect on diffusion (Moya et al. 2018). Moreover, electrochemical sensors mainly stand out in the detection of metabolites and substrates, due to their ease of being extrapolated to enzyme-based sensors. A relevant case of the latter is shown by Patrick Misun et al., who proposed the use of enzyme-based biosensors based on platinum electrodes functionalized with glucose or lactate oxidase enzymes. These biosensors were successful in detecting and recording in real-time the levels of these metabolites on a 3D spheroid hanging-drop culture system, achieving limits of detection below 10 µM (Misun et al. 2016). One more example to highlight is the work of Khalid et al., who installed in a lung-on-a-chip system a set of impedimetric sensors for pH and TEER monitoring in real-time. Using these sensors, the team was able to perform cytotoxicity tests using doxorubicin and docetaxel while monitoring and recording the outputs. The results showed a gradual pH decrease due to the acidification of the media, while the frequency values of TEER impedance were used to compute a cell index, which was used to assess cell viability in the chip. A decrease in the cell index was found in all tested models, finding a correlation with the results obtained by a LIVE/DEAD cell viability assay (Khalid et al. 2020).

Optical sensors

Optical sensors are another family of sensors with multiple applications within tissue engineering. Like electrochemical sensors, there are notable approaches for glucose and lactate detection and monitoring, along with other detection parameters such as pH. A compelling work that uses several external optical sensors was introduced by Shaegh and coworkers, who proposed a platform with optical sensors for monitoring multiple parameters of an infused culture media. With this technology, it was possible to monitor pH and oxygen through low-cost devices, which offer an efficient and user-friendly system to assess microfluidic-scale bioreactors (Shaegh et al. 2016). Differently, Bavli et al. designed a platform to study the mitochondrial dysfunction of a liver-on-a-chip, using both electrochemical sensors for glucose and lactate monitoring, as optical sensors for real-time monitoring of oxygen, showing the potential of combining both types of sensors given some of their intrinsic advantages (Bavli et al. 2016). Although there is no rule of thumb about what sensor should be installed, since different electrochemical and optical sensors will have their respective advantages and disadvantages, the technology can be chosen based on how familiar the researcher is with it. Even so, the design and fabrication of the sensor should always be considered to maximize their use within the OOC. Therefore, priority should be given to those devices that can be easily inserted or embedded during chip manufacturing, measure the desired signals locally, and present the desired analytical parameters.

10. Fabrication methods for tissue engineering and organs-on-a-chip

Fabrication methods must be evaluated in parallel with the selected materials since different substances can only be handled by certain methods and equipment. Consequently, after selecting the correct range of materials for the chip, design planning should be done to simplify manufacturing as much as possible. Thus, complicated, and illogical workflows that will potentially affect our models should be avoided, such as the abuse of UV and heat curing steps. Regardless of the used materials, it is strongly recommended that the fabrication and manufacturing methods would also consider industrial scaling since the commercial viability of the chip can remarkably depend on the latter. The simpler and more direct the manufacturing chain is established, the more likely the chip will meet commercialization.

Lithographic methods

Lithography, especially within the context of microfabrication, can be generalized as a series of methods that transfer a geometric pattern to our desired structure. Among these methods, we can pigeonhole two particularly important for OOCs: photolithography and soft lithography.

Photolithography

Photolithography is a classic technique within the microfabrication field and is still a standard in some niches such as semiconductor and DNA microarrays industries. This top-down approach uses a photomask, which will transfer a specific pattern to a substrate through the selective passage of light. This light pattern will be imprinted in a material called photoresist, which will be dispersed as a layer on the substrate. Depending on the photoresist, the imprinted pattern on it will be vulnerable or resistant to certain etching processes, allowing the addition of a successive layer uniquely on the removed sections. This process, which has been adapted primarily for silicon wafers, is useful for pattering microchannels and for constructing other microcomponents such as MEMs and embedded sensors. There are discrete but interesting examples reporting its use for organ-on-a-chip technology (Huh et al. 2013, Kimura et al. 2018). Nevertheless, despite being a standard technology in some industries, photolithography has serious drawbacks. The need for expensive clean rooms, equipment, and a potentially large quantity of photomasks makes its use unfeasible if mass production is not met. In other cases, any potential application must be substituted, or as a minimum, should be as limited or outsourced as possible.

Soft lithography

In contrast to photolithography, soft lithography usually prescinds from etching and deposition processes, which generally involve highly aggressive chemicals. Instead, soft lithography uses elastomer seals and molds that will transmit a pattern to the desired structure. In contrast to conventional photolithography, soft lithography allows greater flexibility in material's selection (including some biocompatible), can allow the creation of curved arrangements, and does not require clean rooms as soon as the patterned master is obtained. Due to the popularity of PDMS chips in microfluidics, soft lithography is arguably the most widely used technique to produce organs-on-a-chip (Wang et al. 2015). However, it also has a few drawbacks. With soft lithography, it is not possible to yield the same level of alignment and uniformity that can be obtained with photolithography, so it cannot be used as a total substitute for creating the whole myriad of micro and

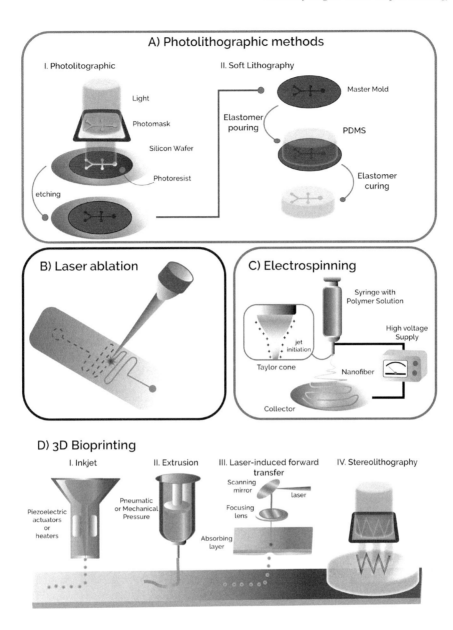

Figure 2. Most used production technologies for the construction of scaffolds and microfluidic systems in tissue engineering and organs-on-a-chip. (A) Lithographic methods: I. The photolithographic process uses a photomask and UV light to insert a pattern on a substrate (usually a Silicon Wafer) on a substrate covered with a photoresist, which after an etching process, will leave the desired pattern. II. Soft lithography uses a master mold with the desired pattern in conjunction with an elastomer (e.g., PDMS). After going through a curing process, the elastomer will acquire the desired pattern. (B) Laser-based technologies irradiate a laser pulse that removes a solid fraction of the substrate surface for the generation of the desired pattern. (C) In electrospinning, the system will be subjected to high voltages, charging the material electrostatically and promoting a stretched geometry known as the Taylor cone. Under suitable experimental conditions, the drop will be executed in a constant flow, forming a nanofiber that will be deposited in a collector. (D) I. Bioprinting inkjet technology makes use of heaters or piezoelectric actuators for the formation of bioink droplets. II. Extrusion bioprinting uses pneumatic pressure through compressed air, mechanical pressure through a piston, or a rotary screw to release a continuous filament of the bioink. III. Laser-induced forward transfer bioprinting makes use of a laser beam that interacts with an absorbing layer on a substrate, producing heat and promoting the deposition of the bioink on the substrate. IV. Stereolithography uses a light-sensitive bioink and exposure to either a laser pattering or a light projection with a distinctive pattern.

nanocomponents. Also, the generation of any master generally requires other lithographic methods, such as photolithography or electron-beam lithography. Although this is not a direct disadvantage, it must always be considered whenever new masters with new patterns are needed (Sahin et al. 2018). Outside these few disadvantages, soft lithography is a firm and reliable technology for the chip design in OOC technology, and one that should be strongly considered when selecting the manufacturing methods.

Laser-based methods

The combination of laser beam technology with computer numerical control (CNC) machines has offered a steadfast technology able to generate micrometric patterns and structures for a surfeit of applications. In principle, laser-based techniques are contactless processes that allow the creation of patterns through laser ablation and cutting, which can be used on a huge variety of materials if they do not produce harmful gases or dust. Fortunately, various materials can be patterned using laser ablation and cutting, positioning this type of method as attractive alternatives in chip design. Among the literature, some papers have already used laser methods for organ-on-a-chip design, highlighting its potential in research and prototyping (Kojic et al. 2019, Sun et al. 2020, Bakhchova et al. 2020). However, there are some major disadvantages to consider from an engineering perspective. These methods are also highly energy-dependent, making their use expensive due to their large power consumption on an industrial scale. Furthermore, the pattering speed and etching rate can vary between materials, achieving exceedingly slow and impractical production times in some cases. However, laser-based methods can be useful prototyping tools for OOC design, and its extrapolation to the industry might be achievable in specific cases.

Electrospinning

The electrospinning technique has become one of the most preferred to produce nanofiber-based structures, making its use appealing for tissue engineering. The electrospinning principle relies on the charge and ejection of a polymer solution in the form of solid and thin fibers, with which structural matrices with high surface-to-volume ratios can be structured. Due to their similarity with the extracellular matrix, this technique has been extensively explored to create different scaffolds, with some of these coupled to a few organs-on-a-chip (Yang et al. 2018, Kim et al. 2019). However, electrospinning also has a few disadvantages, such as the complex control of structural pores if there is a lack of experience on the technique. Due to the latter, some of the obtained matrices usually present poor cellular infiltration and migration (Dahlin et al. 2011). Besides, the process requires the use of solvents whose residues can affect the tissue culture. Still, electrospinning is a simple and generally inexpensive process with good potential for creating cellular scaffolds, if there is a strong knowledge of the technique to yield adequate fiber diameters and porosities.

3D Bioprinting methods

The adoption of 3 D printing to the biomedical field is an appreciable milestone, and the introduction of bioprinting, which allowed the insertion of cells and components within the same "bioinks", is a milestone on its own. 3D printing is a bottom-up technique based on the superposition of successive layers ideal for prototyping, but with high compatibility with industrial scalability. Further, it is also ideal for creating personalized and customizable devices, and it has become an increasingly recurrent technique in the design of OOCs. From 3D bioprinting, an appreciable number of techniques have emerged, with their respective advantages and disadvantages.

Inkjet-based 3D Bioprinting

As one of the pioneering technologies in bioprinting, inkjet printing has evolved rapidly in a series of subtypes with the same purpose: the ejection of droplets to form the desired pattern. For example, in Drop-on-Demand bioprinting, droplets are injected by the action of a heater or piezoelectric actuator, while in Electrohydrodynamic Printing, droplets emerge from the application of high voltages (Gudapati et al. 2016). Inkjet 3D printing is compatible with other non-biological materials, which is ideal for developing other components such as sensors, and it also maintains considerable advantages such as high-speed printing, low cost, and digital-level control (Murphy and Atala 2014). Although inkjet technology has some downsides to consider, such as inhomogeneity in the size and directionality of the droplets and inconsistency with cell encapsulation, this technology has considerable success in OOC design, especially in the design of thin structures such as vascularized models (Bishop et al. 2017).

Extrusion-based 3D bioprinting

This technology is based on the extrusion of a bioink through the pneumatic or mechanical pressure using pressurized air, pistons, or screws. Extrusion bioprinting is widely used due to its ideality to yield high cell densities, fine control over the bioink deposition, and appreciable structural integrity (Bishop et al. 2017). Nonetheless, this technology generally maintains lower resolutions than other printing technologies, and without optimal imprinting conditions for cells, part of the cell population can enter apoptosis (Leberfinger et al. 2017, Cidonio et al. 2019). Further, the deposition speed can decrease as more bioinks in separate syringes are added, slowing down the process and compromising the final structure in a few cases. To overcome the latter, the team of Liu et al. developed a fast, multi-material extrusion system, where bioinks are continuously extruded through a single printhead (Liu et al. 2017).

Laser-induced forward transfer bioprinting

Laser-assisted technologies are another valuable alternative within the bioprinting field. Concretely, laser-induced forward transfer (LIFT) technology uses a pulsed laser beam that will be directed to a transparent substrate known as the donor slide (also known as ribbon), which will be coated in the bottom with an energy-absorbing layer (to favor the expansion of cavitation bubbles), and a bioink layer (with the necessary component of polymers, hydrogels, and cells). The expansive action of these bubbles will eject droplets from the bioink layer, allowing micropatterning with a high level of precision. In addition to the last-mentioned, laser-assisted bioprinting has other appealing advantages such as the possibility of working with high viscosity bioinks, which would generally be difficult to work with other technologies, along with high printing resolutions that reach the micrometer scale (Serra et al. 2010). Nonetheless, this technology is usually expensive, the manufacturing process is slow, and the generated heat may affect cell viability. With all these considerations, laser-based bioprinting is shown as an attractive technology for the design of highly complex microstructures, positioning itself as a first-choice tool in cases where other bioprinting technologies have limitations. On the other hand, it would be favorable to avoid its use in the construction of bulky structures, for which it would be better to complement with more appropriate manufacturing technologies.

Stereolithography bioprinting

Like other lithographic methods, stereolithography uses specific light wavelengths in certain patterns to generate highly detailed structures. In this technology, the light will harden a resin or hydrogel through a photopolymerization reaction. This radiation will first solidify the material at the height of the printing plate and will continue to solidify the desired structure, layer by layer, as the plate moves through the Z-axis. This technology has many advantages, making it a promising technology for tissue engineering. In the first instance, stereolithography is not limited by the viscosity of the bioink as other nozzle-based approaches, it does not produce mechanical stress to the cells of the composition, and it also presents a high printing resolution (Kačarević et al. 2018). Nonetheless, just as LIFT bioprinting, the necessary optical equipment is expensive. Moreover, stereolithography bioprinting is limited to light-curing resins, which not only partially conditions the flexibility of the ink composition but can also cause genetic damage to cells due to the wavelengths of the beams. Notwithstanding, some recent approaches have been presented to solve this last problem, through the adoption of light-curing processes that require wavelengths closer to the visible spectrum (Hoffmann et al. 2017, Sakai et al. 2018, Wang et al. 2018). It should be noted that the presented bioprinting techniques, as well as several of the other manufacturing methodologies, are not mutually exclusive for tissue engineering and organ-on-a-chip design, opening the opportunity to implement several methodologies in the same manufacturing chain. Finally, several of these methodologies can be coupled in the same system, maximizing the benefits, and offsetting the intrinsic drawbacks of each one.

11. Bioengineered systems for organs-on-a-chip technology

An extremely extensive variety of biomimetic systems can be found reported in the literature, ranging from simple vascularized systems to multi-organic and body-on-a-chip proofs-of-concept. However, the detailed understanding of the anatomy and physiology of each tissue and organ is pivotal for the generation of more reliable systems with greater predictive capacity. In this section, some of the most common systems will be presented.

Cardiac

The heart is a potent biological pump that oversees blood delivering and recycling from and towards all extra-cardiac organs. Thus, the importance of managing heart injuries and diseases is imperative for the development of regenerative medicine. There are some difficulties in cardiac system engineering, such as the limited proliferation rate of the cardiomyocytes, and other factors that allow their native alignment. Therefore, cardiomyocytes need to be co-cultured with several cellular and molecular factors to successfully resemble the anatomical cardiac design (Van Berlo and Molkentin 2014). Also, the synchronic beating capacity of cardiomyocytes is lost rapidly inside *in vitro* cultures due to the isolation processes, lack of matrix compatibility, or mismatching matrix properties. On the other hand, drug-induced cardiotoxicity is another pillar of cardiac research. More than 15% of drug recalls are induced by cardiotoxicity, and another important percentage has been retracted from the market primarily due to inconsistent *in vitro* assays' results that imprecisely predict drug cytotoxicity. Cardiotoxicity and cardiovascular therapeutic drug effects are largely studied on immortalized two-dimensional cultures and animal models. Nonetheless, both systems are inefficient to correctly predict the human response (Thomas et al. 2018). The advantages of tissue engineering have led to the generation of human hearts *in vitro* which, in combination with microfluidic technologies, granted the creation of heart-on-a-chip devices. The use of this technology permits the researcher to probe drug-induced global effects on the cardiac tissue (Zhang et al. 2015). But before understanding hearts-on-a-chip by themselves, it is important to understand the anatomy and physiology of the heart. Cardiac tissue has a mesodermal origin. The myocardium is conformed by cardiomyocytes and this tissue represents the pumping engine of the vascular system. The myocardium cells are disposed of in a parallel fashion, and are joined together by Fascia adherens unions and gap junctions that build up the myocardial fiber that allows muscle contraction. Early in the development, atrial and ventricular cardiomyocytes, pacemaker cells, and Purkinje cells are formed. In the left side of the myocardium, Purkinje cells spawn, conduct the electrical stimuli, and pass it to the right branch of the myocardium. Cardiomyocytes and fibroblasts (60% of cardiac mass) are located under the extracellular matrix of the heart wall. The endocardium cells line the blood vessels and cardiac valves. The epicardium cells form the pericardium parietal wall, which is in the coronary vasculature. Natural crest cells oversee cardiac tissue oxygenation, which also contributes to the outflow of the tract septum. Ultimately, macrophages are in charge of the cardiac immune response (Olson 2006). Besides, in cell-to-cell communication, the heart is subjected to endocrine and paracrine signaling, elicited by cytokines, growth factors, and extracellular matrix molecules that comprise the cardiac microenvironment. Cellular intercommunication with extracellular and signaling molecules set up heart development and maintenance in a coordinated and complex microenvironment. With all these previous considerations, tissues can be engineered. Heart tissue engineering has developed two particular approaches to construct these systems: one that involves the use of biological substitutes, and a scaffold-free system. The latter is developed as cell-sheet engineering, where cells are first obtained by primary isolation, are seeded on temperature-sensitive polymers, and right after cells would be attached, the confluent layer of cells is released by decreasing culture temperature and decreasing cell-substrate affinity. This multiple-layer structure is capable of resembling thick myocardium-like tissues (Hwang et al. 2009). The other approach relies on spheroid formation. In this system, embryonic stem cells are seeded onto anti-adherent wells made of PDMS or PEG. Embryonic stem cells will acquire their phenotype as cardiomyocytes, osteoblasts, or neurons, depending on well cellular density and size. Cardiogenesis can be confirmed by sarcomeric a-actinin and cardiomyocytes genes expression (Hwang et al. 2009). Moreover, natural and human-made polymers are used to build artificial matrices for this purpose. Natural polymers like elastin, hyaluronic acid, alginate, fibrin, gelatin, and collagen are preferred since they express specific ligands that improve cell adhesion. Nevertheless, great variability may exist between batches, and the polymer mechanical properties may not support the tissues. Therefore, biomaterial densification techniques and the use of hybrid systems may be utilized to increase mechanical support. For example, compression of collagen gels has been used successfully as acellular patches for cardiac function improvement. Also, the use of hybrid systems, such as semi-interpenetrating networks of photocrosslinkable hyaluronic acid (MeHa) and collagen, increase significantly the mechanical features when compared with collagen or MeHa alone (Nichol et al. 2010). In contrast, artificial polymers could be simply modified for specific purposes. For example, the polyglycolic acid (PGA) turn-over rate, which is measured by the conversion of hydroxypropanoic acid to alpha-Hydroxyacetic acid, is completed in approximately 1 year, allowing the scaffold to gradually disappear as tissue develops, leaving a polymer-free functional tissue (Prabhakaran et al. 2011). Recently, the development of muscular thin films (MTFs) by the team of Parker et al. was another significant advancement for pharmacological and electrophysical studies. Furthermore, a microfabricated hybrid tissue system was built with cardiomyocytes and elastomeric thin films to measure fiber contraction, cytoskeletal organization, and the influence of diastole and systole MTF maximum stress levels on cellular alignment. Later, this device was modified for drug screening and to address genetic, structural, and functional features of myocardium injury. Finally, the device was improved with a clear cover that allows the identification of MTF deformations (Agarwal et al. 2013). Oppositely, Zhang et al. designed an OOC system for pharmaceutical research. The microfluidic 3D system resembles heart functions such as mechanical contractions, molecular signaling, electrophysical responses, and activation after drug stimulation. Sidorov et al. were able to arrange sarcomeres on an aligned distribution as seen in the heart tissue. Furthermore, cardiac tissue differentiation was completed by using a sustained pulse. This experiment shed light to the contraction as a determinant feature to design a successful model with mature cardiomyocytes (Mathur et al. 2015). Another engineered 3D cardiac tissue construct (ECTC) was developed by Sidorov et al., by using a six-well plate seeded with rat neonatal ventricular cells on a matrix of fibrinogen and thrombin. For cardiac tissue design, different cell types are needed. Thus, they co-cultured cardiomyocytes, vessel-specific cells, and ECM molecules on a device perfused with a culture medium. Synchronic beats were registered as soon as 5 days of culture and fiber formation was detectable within 13–15 days (Schroer et al. 2017). The measure of contractions was achieved with an optical microscope coupled to a cantilever probe fixed on the condenser, allowing an optic measure of probe shifting and possible ECTC twisting. The ECTC is anchored on a motorized stage. The contraction was then stimulated by applying a controlled lateral force. The action potential was measured by using a KCl electrode. Moreover, voltage stimulation was achieved by conduction through titanium wires (Sidorov et al. 2017). Sidorov's research team achieved cardiomyocyte differentiation into a beating cardiac muscle that displayed normal organization cytology. Also, the amplitude of the action potential (AP) was 80–90 mV, and the resting potential was 60–75 mV at a pacing rate of 2000 ms. The results of this experiment were later validated by the Wikswo laboratory by using horizontal force to determine ECTC contractility. Sidorov's team then used the heart-on-chip device to address drug response when β-adrenergic stimulation was applied. The heart rate is highly controlled by adrenergic neurotransmitters. Thus, the addition of isoproterenol was able to increase the force and contraction rate, through the activation of adrenergic receptors. Nevertheless, isoproterenol diminishes the duration of contraction when added to cardiomyocytes *in vivo*. However, it is necessary to translate this research to studies that would focus on human cardiac diseases. Another human 3D microtissue model was built by Marsano et al. by using induced pluripotent stem cells (Vunjak-Novakovic et al. 2010). The device demonstrated a higher cardiomyocyte maturation rate when subjected to cyclic mechanical stimulation. Even more, both the spontaneous and synchronous beating, as well as the cardiomyocyte's stimuli response to the beta-adrenergic-agonist, isoprenaline, were detected. Nevertheless, this device requires further improvement to accurately mimic the complexity of cardiac muscle anatomy (Zhuang et al. 2000). Finally, this device is suitable to develop pharmacologic and research protocols, if contraction and action potentials are registered by using a transversal force throughout the device coupled to a cantilever probe, capable of measuring ECTC responses. Thus, if wires are connected directly to the PDMS mold, the cultured cells can be tracked constantly. The addition of this detection system is cheap when compared with other devices. The development of a device that mimics the anatomy and structure of the human heart represents the first step in the generation of a multi-organ microdevice (Zhang et al. 2015).

Hepatic

The liver is the organ where metabolic and drug biotransformation processes take place. Thus, this organ is more susceptible to display metabolic disorders and drug-induced toxicity. Anatomically, the liver is divided into two lobules that represent the functional surface of the organ. Lobules are organized in sinusoids that are formed of specialized fenestrated endothelial cells (McCuskey et al. 2003) (Figure 3.A). Hepatocytes are the most abundant cells in the liver (80%) and are arranged in plates attached by thigh junctions that form the bile canaliculi, small tubular apical spaces between adjacent hepatocytes. These cells express high levels of detoxification enzymes such as cytochrome CYP450, acute-phase response proteins, and plasmatic proteins. These proteins coordinate nanoparticle clearance and xenobiotic removal through conjugation with molecules such as glycine sulfate, glucuronate, or glutathione, which ultimately diminish their toxicity and make them soluble and easily removable. Hepatocytes are also determinants for glucose and lipid metabolism. The space of Disse forms a reticular basement layer made of collagen I, IV, fibronectin, and laminin located between the hepatocytes and the endothelial cells. Hepatocytes around the sinusoid display abundant microvilli protrusions to the Disse, which allows direct contact with the blood surrounding the sinusoid. The remaining 20% of the liver mass is formed of non-parenchymal cells (NPC) such as liver sinusoidal endothelial cells; Kupffer cells, which lead the response to liver injury; hepatic stellate cells, that oversee the extracellular matrix remodeling after cellular loss; and pit cells. Sinusoid zonation is given by the vascular disposition that spreads throughout the liver, which varies from the periportal to the perivenous area. The most relevant modifications are the extracellular matrix (ECM) framework throughout the sinusoid, cytology, genetic expression, and metabolism. Furthermore, fenestration is different in the periportal region (larger) in contrast to the proximities of the central vein that is marked for oxygen delivery, although recent studies have demonstrated that this zonation could be determined by Wnt signaling driven by endothelial cells. As a consequence of the loss of liver, architecture leads to loss of tissue function (Kaplowitz 2006). The average blood flow that runs throughout the sinusoid has an average rate of 144 µm/s per 1.9s of mean residence time. The hepatocytes that are closer to the sinusoid receive ~ 2 nmol/mL; the oxygen delivery rate given throughout the sinusoid is 72 nmol/(min × 10^6 cells). The oxygen supply of the liver comes from two sources: the hepatic artery, and the hepatic portal vein circulation. Since the oxygen pressures across the sinusoid decrease, the responses to metabolic and/or toxic stimuli are differential depending on the zone. Blood supply alterations could lead to inadequate cell response and the release of hepatocyte factors potency, which is a function of the blood-to-cell volume ratio (20.5 mL blood/ mL tissue) for healthy adult livers. If the hepatocyte volume is 3.4 pL[11], then it is calculated that 14 million hepatocytes are supplied by only 1 mL of blood. This feature of liver blood supply is determinant for the design of *in vivo* and *ex vivo* liver culture systems (Nahmias et al. 2006). In culture, hepatocytes displayed zonation; hence, it is pivotal to maintain this arrangement when a liver is emulated on a chip (Beckwitt et al. 2018, Moradi et al. 2020). Liver Chip design can be performed by standard computer-aided design (CAD) software (AutoCAD, Corel Draw, etc.) where the creation of gradients represents a pivotal feature on microfluidic system development. Thus, for laminar flows, it is necessary to keep Reynolds numbers low to maintain layer integration. Also, shear stress must be fine-tuned to exert cell response upon mechanical cues that will maintain the microenvironment homeostasis. Microfluidics allows fluid control that mimics the physiological environment. In the case of the liver, the manipulation of fluids allows metabolic zonation. Liver research on drug-induced injury and pharmacokinetics is conducted in small animals. Hence, species differences (such as genetics and metabolism) limit the lab bench-to-clinic translation. Even more, a modification on the European cosmetic regulation (EC 1223/2009) prohibited the use of products that contain compounds previously tested on animals, accelerating the necessity of a human *in vitro* model of the liver function system development (Blais et al. 2017). In

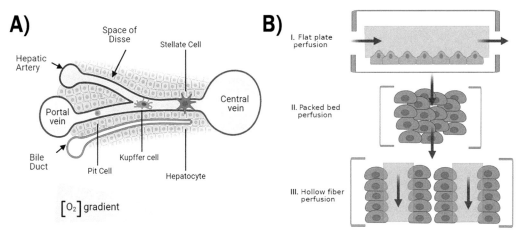

Figure 3. (A) Oxygen gradient throughout the hepatic sinusoid. The figure shows the complexity of oxygen delivery for pericentral cells. Stimuli and drugs that are delivered by the portal vein are distributed in the same manner (B) Different types of bioreactor perfusion strategies. I. Media culture is diffused only on one side of the cell culture. II. Media culture is diffused through the cell culture. III. The media culture is diffused through a channel that mimics tissue vasculature. The figure was designed using BioRender.

tissue engineering, there are different approaches to maintain liver tissue organized and functional. Reperfusion of the rat liver is widely used since the organ maintains the heterotypic cell-to-cell interactions and 3D functional conformation. Nevertheless, isolated livers exhibit a short functional lifespan. Thus, researchers developed precision-cut liver slices (0.2 mm) that displayed increased viability (~ 72 h) when cultured under a highly oxygenated environment. The use of microfluidic devices for the culture of precision-cut slices was successful in detecting unstable drug metabolites via high-performance liquid chromatography, due to increased tissue survival to a few days in culture (Schumacher et al. 2007, Harrill et al. 2009). On the other hand, Kane et al. maintained the viability of hepatocytes in culture by seeding rat liver cells and 3T3-J2 fibroblasts on a system based on fenestrated pores, yielding a 2D Micro-Patterned co-culture. Rat hepatocytes showed optimal albumin synthesis, normal physiology, and active metabolism (Ren et al. 2012).

Finally, the recent development of 3D bioprinters represents an excellent alternative to achieve liver-like architecture from scratch. The Organovo company uses bioprinting to develop a co-culture of human hepatocytes and stellate cells on a 24-well plate format with a lifespan of approximately 35 days *in vitro*, but with a deficient production of albumin that is associated with low functional activity (Nguyen et al. 2016). Also, the bioprinting of human-induced pluripotent stem cells (iPSCs)-derived hepatocytes, adipose-derived cells, and endothelial cells on a hydrogel base of Ma and colleagues resembles the hexagonal architecture of the liver, and it was maintained up to 8 days in culture, showing low albumin production values of 2–8 mg (day $\times$ 10^6 cells) (Ma et al. 2016). Co-culture of NPC in the engineering of 3D liver tissue models increases hepatocyte activity compared to culture alone (Khetani et al. 2015, Beckwitt et al. 2018). Thus, organ-on-a-chip technology made the Liver Chip system a successful model in terms of cellular phenotype preservation and function. There are three types of bioreactors for liver systems: a flat-plate system, a packed-bed design, and a hollow fiber design. The latter is a complex system, in which cells are arranged as lining fibers that deliver oxygen and nutrients, mimicking perfused vasculature (Figure 3B). Cell selection is of primordial importance to develop accurate engineered liver systems. Both NPCs and hepatocytes must be present to set out the pharmacodynamics, drug biotransformation, and liver disease genesis and progression due to thigh intercommunication between hepatic cells. Fresh human isolated hepatocytes are very difficult to obtain, thus there are many commercially available cells such as cells isolated from humanized $Fah^{-/-}$ NOD/SCID mice, or the primary human hepatocytes isolated from FRG® ($Fah^{-/-}/Rag2^{-/-}/Il2rg^{-/-}$) mice (Azuma et al. 2007). Surprisingly, these cells have no advantage in terms of lifespan in culture when compared to cells isolated from standard isolation protocols. Consequently, the best alternative to obtain freshly isolated cells is cryopreserved hepatocytes that are readily available for purchase. The downside of the use of cryopreserved cells is that function can vary drastically based on the freezing process. Furthermore, the HepG2/C3A cancer cell line shows functional resemblance and rapid culture expansion. Nonetheless, these cancer cells rely on glycolytic metabolism and are dedifferentiated. Another HepG2/C3-like cell line is the HepaRG cell line, which was obtained from a hepatic tumor from a patient with hepatitis C. HepaRG grows rapidly in culture and can differentiate into hepatocytes and cholangiocytes after 28 days of 1.7% DMSO treatment. Hepatocytes derived from HepaRG present similar levels of function and less variability. Recently, HepatoCells were developed by Corning and display the same functional activity as HepaRG (Anthérieu et al. 2010). Also, for spheroid formation, a 96-well plate that streamlines hanging drop aggregation was created by InSphero (Schlieren, Switzerland). This system produces hepatocytes spheroids (250-mm diameter) that exhibit very low variability (10%). Liver organoids are very similar to heterotypic 3D spheroids. Ho et al. demonstrated that, by creating a radial electrical field by using electrophoresis, circular cellular arrangements would be designed on polydimethylsiloxane (PDMS) chips. A novel application for this microfluidic chip is the use of immunosensors since these would trace hepatotoxicity biomarkers, and monitor drug effect on skin sensitization by metabolite production quantification and antigen-presenting cells (APCs) activation. This model consisted of three components: cell chambers, flow channels, and endothelial barriers constructed by using lithography technology. Culture media, nutrients, and drugs were delivered by, and endothelial connected to, the cellular compartment. The main function of the endothelial barrier was to avoid direct flow but assuring the transfer of oxygen and nutrients through diffusion. HepG2 cells were also cultured by Baudoin et al., in a microfluidic bioreactor system to investigate HepG2 cell behavior. Microfluidic systems can host different cell types to achieve cellular behavior research, by providing an environment that resembles the tissue architecture more accurately. Interactions between HepG2 and MCF-7 cells, and their respective metabolisms, were studied by Zhang et al. Taken together, the reviewed system represents half of a century of research on bioartificial livers. However, it is still unclear if the collected data from liver-on-a-chip devices is confident enough, or how it should be utilized on physiologically-based pharmacokinetic modeling for safety assessments (Ahmed et al. 2017).

Renal

The kidney contains a myriad of cell types that build up a 3D structure embedded in a complex extracellular matrix and vasculature. Renal xenobiotic excretion is performed by both glomeruli filtration and dynamic secretion/reabsorption by the tubular apparatus. Recently, there is a major effort to understand the physiology and intracellular metabolism of the proximal tube—the most important region within the nephron where filtration takes place. Influx takes place at the basolateral and apical membrane of renal proximal tubule epithelial cells (PTECs). Efflux is carried out at the apical membrane through the ATP-binding cassette (ABC) transporters, and multidrug and toxin extrusion transporters 1 that, besides efflux, also play an important role in kidney detoxification (Wilmer et al. 2016). Drug-induced kidney injury (DIKI) is a major outcome of pharmacotherapy, and a comprehensive study of the molecular mechanisms of nephrotoxicity can improve drug development. Nevertheless, current models are not suitable to study the biological functions of the kidney. The development of cell systems by using improved culturing technologies will eventually result in bioengineered 3D devices, which will allow the assessment of nephrotoxicity events. To achieve this, it is imperative to identify DIKI biomarkers that could be traced after the testing of kidney-on-a-chip devices to understand *in vivo* response (Arany and Safirstein 2003). Many proposed kidney-on-a-chip systems use animal cells such as Madin-Darby canine kidney (MDCK) cells, and the porcine LLC-PK1 cell line, since ciPTEC engineered cells require a better molecular characterization to be used for the creation of these devices. Also, there are some other considerations when using PTEC cells to be considered. These cells are under continuous luminal fluid shear stress (FSS) and a trans-epithelial osmotic gradient *in vivo*. FSS cannot be measured *in vivo* since the urinary flow is decreased along the tubule due to tubular reabsorption. The approximate FSS in humans varies between 0.7 dyne/cm^2 and 1.2 dyne/cm^2 under physiological conditions but can increase to 1.6 dyne/cm^2 during pathology. Modifications on the FSS can disturb signal transduction, iron reabsorption, receptor-mediated endocytosis, cytoskeleton organization, cell-to-cell communication, as well as gene and protein expression (Adler et al. 2016). For example, an antineoplastic drug, Ifosfamide, diminished inflammatory responses, integrin pathway disturbance, and modulation of several cancer therapeutic targets under a microfluidic device but these phenomena were not detected on static conditions. Tubule curvature is another physiological condition to consider when assessing the tubule cell performance *in vitro*. MDKC cells cultured on curved hollow fibers increased transporter activities and highly expressed metabolic enzymes. Furthermore, 3D kidney systems attached to an extracellular matrix displayed more sensitivity to develop DIKI, which allows long-term toxicity monitoring. Taken together, these studies confirm a critical intercommunication between the 3D cellular components and the homeostatic and pathological responses. Nowadays, most kidney-on-a-chip are developed by using renal cells seeded on an extracellular matrix or membranes next to nutrient delivery microchannels, waste clearance, and constant flow. Microfluidic devices are seeded with renal cells and cultured on 2D extracellular matrix-coated membranes, sandwiched between PDMS-microchannels, or etched on a glass substrate. Furthermore, 3D tubular systems were improved with hollow fibers that resemble the tubular structure formed in hydrogels. These systems are now used on FSS studies, and usually consist of a monolayer of human PTECs that show active reabsorption seeded inside a fibrin-coated, polyethersulfone (PES)-polyvinyl pyrrolidone (PVP) hollow structure implanted in a PDMS-glass chamber. Increased detection/expression of the detoxification protein HO-1 following drug administration was compared with data obtained from rat models. Even more, microfluidic systems allow researchers to co-culture multiple cell types, making accessible the investigation of crosslinked signaling, along with cell-to-cell interactions such as immune system cell recruitment and inflammation, which are pivotal features for the study of kidney health and injury (Wilmer et al. 2016). Moreover, outstanding molecular techniques such as transcriptor activator-like effector nuclease (TALEN) and clustered regularity interspaced short palindromic repeats/Cas9 (CRISPR-associated 9) (CRISPR) have been used for the creation of gene mutations observed in renal diseases. The future perspective for kidney-on-a-chip devices relies on the inclusion of more cell types that can mimic several of the proximal tube functions, such as secretion and solute passages. The final goal of this technology is to recreate a model for human kidney disease and thus, identify prevention and effective therapeutics. Lastly, it is important to highlight that the use of microfluidic culture opens a great opportunity to study the interaction of structure and function of the proximal tubule, and apply it to the nephrotoxicity research (Adler et al. 2016).

Neural

The brain is an extremely complex organ, which receives and processes information that is transformed into human body responses. Communication between neurons in the brain is via unidirectional synapses between the terminal axon of one neuron with the dendrites of another one. Neurons also communicate with other cells such as oligodendrocytes, microglial cells, and astrocytes. Brain function is accomplished by electrical and chemical signals, but when this neuronal communication is disrupted, neurodegenerative diseases such as Huntington's disease or Alzheimer's could arise. Neurodegenerative diseases have been a focus of research for decades, although the underlying mechanisms of disease onset and progression are still largely unknown. The main limitation of these studies is the lack of *in vitro* model systems that can resemble complex tissue structures, intercellular interactions, and responses to stimuli (Bang et al. 2019). The advances of brain-on-a-chip technology have not been able to fully engineer the full structure and function of the brain, but specific parts of the brain tissue. Traditional culture of neurons on glass substrates or Petri dishes has evolved to Whiteside's soft lithography, where the culture is micro-organized in channels. Modification to this method was introduced by using the multistep lithography, where soma and axon were compartmentalized, allowing the research to focus on axon's responses to drug treatment, axon regeneration processes, and the reconstruction of the interconnection of several neurons (Mofazzal Jahromi et al. 2019). Myelination is the formation of the myelin sheath around a nerve fiber and is fundamental for the action potential propagation by electrically separating the axon from the environment. Demyelination is the loss of this shield of myelin due to autoimmune responses, brain injury, or neurogenerative diseases. Brain-on-a-chip technology could resemble this process and allow its study more accurately. Indeed, nowadays there are brain-on-a-chip systems that have been shown to be effective for the study of myelinization, although with some limitations such as 2D cell adhesion and stiffness of biomaterials (Cummings et al. 2018). On the other hand, the brain-blood barrier (BBB) is a highly selective semipermeable structure that functions to protect the brain from infections, neurotoxicants, and hydrophilic molecules. Thus, there is an imperative restriction for large molecules that could be used to treat disease, which has led to failure in drug screening stages. To solve this problem, a microfluidic model based on a porous membrane laid between two microchannels was developed (Figure 4). Cultured cells can communicate through the

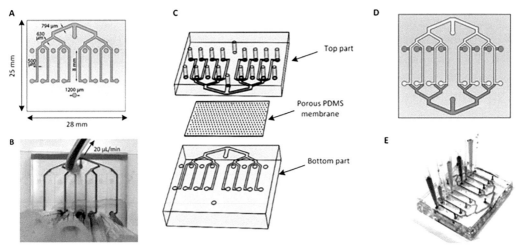

Figure 4. Design of one- and two-layer multiplexed chip. (A) Eight parallel channels branching from a single inlet with independent access points. (B) Eight different conditions in a one-layer device (colored by food dye) (C) Design of a chip with a PDMS membrane in the center of the two layers (top and bottom) (D) Cartoon of a two-layer device (E) Eight different conditions in the two-layer device. (Reprinted from M Zakharova et al. under a Creative Commons Attribution 3.0 Unported Licence.)

membrane pores and secreted cytokines. This model allows to mimic the BBB structure. Furthermore, the addition of a neural cell-laden hydrogel can establish a 3D cell culture. The transendothelial electrical resistance of the BBB can also be measured by the insertion of an electrode in each channel. Nevertheless, cell-to-cell interactions are restricted between different cell types since this communication depends on long-range cytokine secretion. Furthermore, physical interaction can be achieved by increasing pore density, thus increasing intercellular interaction (Bang et al. 2019). Besides the brain-on-a-chip developed system, a cord-on-a-chip has also been created by Sances et al., where induced pluripotent stem cell (iPSC)-derived ventral spinal neurons were co-cultured with brain endothelial cells in the porous-membrane-based device. The culture expressed SST at the same levels as a fetus. These results highlight the importance of endothelial cell–neuron interaction and a 3D culture environment. The technology of brain-on-a-chip can be divided into different platforms, with 3D high-content systems being arguably the most promising *in vitro* models. These systems resemble the whole brain environment, such as the interconnected multichip systems that can respond to external and internal stimuli. These high-throughput systems can allow massive screening (Zakharova et al. 2020). It is important to highlight the need for additional features for brain-on-a-chip development such as cell sources, cell-to-cell interactions, and cell-extracellular matrix interactions. These devices are necessary to implement personalized analysis of individual conditions and for the prediction of the treatment outcome (Mofazzal Jahromi et al. 2019).

12. Challenges and forecast of organs-on-a-chip

So far, organ-on-a-chip technologies have enjoyed great interest, allowing the creation of various biomimetic systems for virtually any relevant organ or tissue. Even so, its insertion in industry and society has not been as disruptive as other techs such as 3D printing itself, machine learning, blockchain, or even e-health technology. Organs-on-a-chip, together with a large part of biomedical and clinical trends, must go through slow but necessary filters that must ensure that their implementation will be the safest, most regulated, and the most beneficial. However, compared to other trends, OOCs have a particular limitation. Due to the enormous dependence that preclinical research has on two-dimensional cell cultures and animal models, the transition to biomimetic models will be (and must be) laborious and time-consuming, since such rigorous changes in the backbone of pharmaceutical research must be accomplished with the utmost caution. Because of this, Zhang et al. concluded in their work that this technology would be "more evolutionary than revolutionary" (Zhang et al. 2018). Even if partial or total adoption of OOC technology is achieved, other valuable challenges are coming from the personalized medicine model, such as the personalization, customization, and democratization of these devices.

Adoption

Even with the number of deficiencies that modern preclinical models present, current chip models are not necessarily finer substitutes. A more complex model does not mean it will be better at meeting the current needs of biomedical research. Therefore, it is necessary to specify important goals and milestones on the path to technological adoption. For this purpose, the work of Allwardt et al. presents a linear roadmap, touching the most important challenges and requirements to achieve the adoption of organs-on-a-chip (Allwardt et al. 2020). However, despite its utility displaying the value proposition of this transition, these emerging technologies almost always are developed in parallel. Accordingly, the homologation and consensus on organ-on-a-chip design perhaps should not lie in the overall process, but concrete challenges. Consequently, this work presents two of the most important breakthroughs to achieve the desirable transition: validation and industrialization.

Validation

Preclinical research models, even with their downsides, are ideal for conducting strict and standardized studies. In consequence, one of the first challenges to consider for validation is the transition from proofs-of-concept to robust and homogeneous models, capable of achieving reproducibility and high fidelity. This, by itself, is composed of various challenges. The first might start by defining the design, materials, and manufacturing standards. Within this, collaborative and serious discussions should begin on the correct anatomical and physiological scaling, the fidelity of the cultured cells, the type of materials to be used, and any other factor that requires homologation to yield analytically coherent and trustworthy testing models. Another valuable detail is the adaption of organs-on-a-chip to the standard analytical assessing systems, such as automated microplate reader systems. For this, a series of researchers have begun to adapt their chip models to this type of equipment (Phan et al. 2017, Gaio et al. 2018). Potentially the biggest and most apparent problem is the difficulty and intricacy of comparing OCCs against other preclinical models, since procedures for evaluating toxicity, metabolic activity, or response to drugs are divergent. With all these considerations, the validation and conversion of OOCs into gold standards may depend on the consensus of different parties, such as biomedical and clinical researchers, scientific societies, the private pharmaceutical industry, and government and regulatory organisms.

Industrialization and commercialization

The adoption success of these models will not only depend on their suitability as testing models, but also their industrial scaling, their market size, and their economic feasibility overall. Due to this, special emphasis has been placed in this work on material and manufacturing process selection. The price and amount of the materials, as well as the ease and simplicity of the production process, should be considered by the researcher if commercialization is desirable. Moreover, being a relatively new and disruptive technology, the global market size of organs-on-a-chip is quite small when compared to other technologies, hovering between $5 and $43 million dollars in recent years. However, the estimated compound annual growth rates are huge, hovering between 30% and 63.2%, thus making the market projection very attractive in the next few years. This can partially be interpreted as the interest that the academy, the pharmaceutical industry, and the new OOC start-ups have in this technology. Hence, despite the considerable challenges to achieve industrialization, economic behavior seems to open valuable opportunities to skyrocket this tech.

Customization

Customization is one of the main pillars of the personalized medicine model, and one that can be extrapolated to tissue engineering and OCCs. The ability to make biomimetic models for "each one" would be distant, but also a very desirable step to follow. Pharmacological profiles can vary from person to person depending on their gender, age, race, ethnicity, genetic polymorphisms, unique mutations, and overall lifestyle, approaching various biomedical trends closer to the needs of the individual. For example, fields like pharmacogenomics and nutrigenomics have already offered several commercial platforms for the development of personalized therapies and diets. However, due to the very flexibility of the technology, engineering an *ad hoc* organ-on-a-chip for a specific individual is not only possible, but it would potentially be the richest and most reliable source of information for pharmacological and toxicological assessment (Van Den Berg et al. 2019).

Democratization

Along with customization, the democratization of OOC technology is a distant, but worthy flagship challenge. Organs-on-a-chip are expensive systems to produce, so increasing their accessibility to both the scientific community and the general population could lead to some of the most significant innovations within biomedical sciences. On one side, the popularization of these models among scientists could greatly enhance the quality of experimental outputs, as well as providing an encouraging opportunity to replace animal models almost entirely. The latter, seen from both a bioethical and scientific perspective, is immensely beneficial. For this, scalability and industrialization must be as efficient as the current cell culture industry is for researchers, institutes, and companies. Another important approach to consider comes from the democratization of information, where the proliferation of databases and research networks could benefit the production of home-run testing models. However, the most prodigious goal lies in the democratization of this technology for the general population. The creation of chips that are not only for "each one", but for "each and everyone" might stand as the most ambitious objective in the future of OOCs. Under ideal circumstances, custom chips could be designed for most of the population. That level of accessibility would not only allow every person to evaluate the efficacy and safety of their medications with reliable and high-throughput screening models, but it could change the paradigm of how pharmacological screening will be assessed in the future.

13. Conclusions

Organ-on-a-chip technology is one of the most disruptive and promising technologies within the biomedical field, by being introduced as the future alternative to 2D cultures and animal testing models. Upon this development, a very appreciable series of biomimetic models have been made with a plethora of integrated materials, techniques, and components. However, this parallel development has not yet been sufficiently uniform, leaving several milestones to reach within the roadmap to clinical adoption, industrialization, and democratization. Therefore, for the optimal and successful designing of future chips, various considerations, ranging from the biological and medical fields to physics and engineering backgrounds, will have to be considered if partial or complete acceptance and insertion of the tech is desirable. Given this interdisciplinarity, much emphasis has been placed in this chapter on approaching organ-on-a-chip design from a comprehensive perspective, since much of the success might not only depend on technical details but compendious considerations. Ultimately, this compendious approach may open a valuable opportunity for consensus and one more step in the transition from proofs-of-concept to more ideal and more robust devices, which simultaneously are expected to be more ethical, more accessible, and more equitable.

References

Adler, M., Ramm, S., Hafner, M., Muhlich, J.L., Gottwald, E.M., Weber, E., Jaklic, A., Ajay, A.K., Svoboda, D., Auerbach S., Kelly, E.J., Himmelfarb, J. and Vaidya, V.S. 2016. A quantitative approach to screen for nephrotoxic compounds *in vitro*. *J. Am. Soc. Nephrol.* [Internet]. [accessed 2021 Feb 8], 27(4): 1015–1028. https://pubmed.ncbi.nlm.nih.gov/26260164/.

Agarwal, A., Goss, J.A., Cho, A., McCain, M.L. and Parker, K.K. 2013. Microfluidic heart on a chip for higher throughput pharmacological studies. *Lab Chip* [Internet]. [accessed 2021 Feb 24], 13(18): 3599–3608. /pmc/articles/PMC3786400/.

Agrawal, G., Aung, A. and Varghese, S. 2017. Skeletal muscle-on-a-chip: An *in vitro* model to evaluate tissue formation and injury. *Lab Chip*, 17(20): 3447–3461.

Ahmed, H.M.M., Salerno, S., Morelli, S., Giorno, L. and De Bartolo, L. 2017. 3D liver membrane system by co-culturing human hepatocytes, sinusoidal endothelial and stellate cells. *Biofabrication* [Internet]. [accessed 2021 Feb 8], 9(2). https://pubmed.ncbi.nlm.nih.gov/28548045/.

Allwardt, V., Ainscough, A.J., Viswanathan, P., Sherrod, S.D., McLean, J.A., Haddrick, M. and Pensabene, V. 2020. Translational roadmap for the organs-on-a-chip industry toward broad adoption. *Bioengineering* [Internet], 7(3): 1–27. www.mdpi.com/journal/bioengineering.

Almany, L. and Seliktar, D. 2005. Biosynthetic hydrogel scaffolds made from fibrinogen and polyethylene glycol for 3D cell cultures. *Biomaterials* [Internet], 26: 2467–2477. www.elsevier.com/locate/biomaterials.

Anthérieu, S., Chesné, C., Li, R., Camus, S., Lahoz, A., Picazo, L., Turpeinen, M., Tolonen, A., Uusitalo, J., Guguen-Guillouzo, C. and Guillouzo, A. 2010. Stable expression, activity, and inducibility of cytochromes P450 in differentiated HepaRG cells. *Drug Metab. Dispos.* [Internet]. [accessed 2021 Feb 24], 38(3): 516–525. https://pubmed.ncbi.nlm.nih.gov/20019244/.

Arany, I. and Safirstein, RL. 2003. Cisplatin nephrotoxicity. *Semin. Nephrol.* [Internet]. [accessed 2021 Feb 24], 23(5): 460–464. https://pubmed.ncbi.nlm.nih.gov/13680535/.

Aziz, A., Geng, C., Fu, M., Yu, X., Qin, K. and Liu, B. 2017. The role of microfluidics for organ on chip simulations. *Bioengineering* [Internet]. [accessed 2021 Feb 5], 4(4): 39. http://www.mdpi.com/2306-5354/4/2/39.

Azizipour, N., Avazpour, R., Rosenzweig, D.H., Sawan, M. and Ajji, A. 2020. Evolution of biochip technology: A review from lab-on-a-chip to organ-on-a-chip. *Micromachines* [Internet]. [accessed 2021 Feb 7], 11(6): 1–15. /pmc/articles/PMC7345732/?report=abstract.

Azuma, H., Paulk, N., Ranade, A., Dorrell, C., Al-Dhalimy, M., Ellis, E., Strom, S., Kay, M.A., Finegold, M. and Grompe, M. 2007. Robust expansion of human hepatocytes in Fah-/-/Rag2 -/-/Il2rg-/- mice. *Nat. Biotechnol.* [Internet]. [accessed 2021 Feb 24], 25(8): 903–910. https://pubmed.ncbi.nlm.nih.gov/17664939/.

Baino, F., Novajra, G. and Vitale-Brovarone, C. 2015. Bioceramics and scaffolds: A winning combination for tissue engineering. *Front Bioeng. Biotechnol.* [Internet], 3. www.frontiersin.org.

Bakhchova, L., Jonušauskas, L., Andrijec, D., Kurachkina, M., Baravykas, T., Eremin, A. and Steinmann, U. 2020. Femtosecond laser-based integration of nano-membranes into organ-on-a-chip systems. *Materials* (Basel) [Internet]. [accessed 2021 Feb 7], 13(14). /pmc/articles/PMC7412128/?report=abstract.

Bang, S., Jeong, S., Choi, N. and Kim, H.N. 2019. Brain-on-a-chip: A history of development and future perspective. *Biomicrofluidics* [Internet]. [accessed 2021 Feb 28], 13(5): 051301. http://aip.scitation.org/doi/10.1063/1.5120555.

Bavli, D., Prill, S., Ezra, E., Levy, G., Cohen, M., Vinken, M., Vanfleteren, J., Jaeger, M. and Nahmias, Y. 2016. Real-time monitoring of metabolic function in liver-onchip microdevices tracks the dynamics of Mitochondrial dysfunction. *Proc. Natl. Acad. Sci. U S A* [Internet]. [accessed 2021 Feb 7], 113(16): E2231–E2240. https://pubmed.ncbi.nlm.nih.gov/27044092/.

Beckwitt, C.H., Clark, A.M., Wheeler, S., Taylor, D.L., Stolz, D.B., Griffith, L. and Wells, A. 2018. Liver 'organ on a chip.' *Exp. Cell. Res.* [Internet]. [accessed 2021 Feb 8], 363(1): 15–25. https://pubmed.ncbi.nlm.nih.gov/29291400/.

Bhatia, S.N. and Ingber, D.E. 2014. Microfluidic organs-on-chips. *Nat. Biotechnol.* [Internet]. [accessed 2021 Feb 5], 32(8): 760–772. https://www.nature.com/articles/nbt.2989.

Bishop, E.S., Mostafa, S., Pakvasa, M., Luu, H.H., Lee, M.J., Wolf, J.M., Ameer, G.A., He, T.C. and Reid, R.R. 2017. 3-D bioprinting technologies in tissue engineering and regenerative medicine: Current and future trends. *Genes Dis.*, 4(4): 185–195.

Blais, E.M., Rawls, K.D., Dougherty, B.V., Li, Z.I., Kolling, G.L., Ye, P., Wallqvist, A. and Papin, J.A. 2017. Reconciled rat and human metabolic networks for comparative toxicogenomics and biomarker predictions. *Nat. Commun.* [Internet]. [accessed 2021 Feb 24], 8(1): 1–15. www.nature.com/naturecommunications.

Bovard, D., Sandoz, A., Luettich, K., Frentzel, S., Iskandar, A., Marescotti, D., Trivedi, K., Guedj, E., Dutertre, Q., Peitsch, M.C. and Hoeng, J. 2018. A lung/liver-on-a-chip platform for acute and chronic toxicity studies. *Lab Chip*, 18(24): 3814–3829.

Carlborg, C.F., Haraldsson, T., Öberg, K., Malkoch, M. and Van Der Wijngaart, W. 2011. Beyond PDMS: Off-stoichiometry thiol-ene (OSTE) based soft lithography for rapid prototyping of microfluidic devices. *Lab Chip*, [Internet]. [accessed 2021 Feb 7], 11(18): 3136–3147. https://pubs.rsc.org/en/content/articlehtml/2011/lc/c1lc20388f.

Chin, C.D., Linder, V. and Sia, S.K. 2007. Lab-on-a-chip devices for global health: Past studies and future opportunities. *Lab Chip* [Internet]. [accessed 2021 Feb 8], 7(1): 41–57. https://pubs.rsc.org/en/content/articlehtml/2007/lc/b611455e.

Cidonio, G., Glinka, M., Dawson, J.I. and Oreffo, R.O.C. 2019. The cell in the ink: Improving biofabrication by printing stem cells for skeletal regenerative medicine. *Biomaterials* [Internet]. [accessed 2021 Feb 8], 209: 10–24. /pmc/articles/PMC6527863/?report=abstract.

Cummings, J., Reiber, C. and Kumar, P. 2018. The price of progress: Funding and financing Alzheimer's disease drug development. *Alzheimer's Dement. Transl. Res. Clin. Interv.* [Internet]. [accessed 2021 Feb 28], 4: 330–343. https://pubmed.ncbi.nlm.nih.gov/30175227/.

D'Souza, S.E., Ginsberg, M.H. and Plow, E.F. 1991. Arginyl-glycyl-aspartic acid (RGD): A cell adhesion motif. *Trends Biochem. Sci.*, 16(C): 246–250.

da Ponte, R.M., Gaio, N., van Zeijl, H., Vollebregt, S., Dijkstra, P., Dekker, R., Serdijn, W.A. and Giagka, V. 2021. Monolithic integration of a smart temperature sensor on a modular silicon-based organ-on-a-chip device. *Sensors Actuators, A Phys.*, 317: 112439.

Dahlin, R.L., Kasper, F.K. and Mikos, A.G. 2011. Polymeric nanofibers in tissue engineering. *Tissue Eng. - Part B Rev.* [Internet]. [accessed 2021 Feb 8], 17(5): 349–364. /pmc/articles/PMC3179616/?report=abstract.

Díaz Lantada, A., de Blas Romero, A., Schwentenwein, M., Jellinek, C., Homa, J. and García-Ruíz, J.P. 2017. Monolithic 3D labs- and organs-on-chips obtained by lithography-based ceramic manufacture. *Int. J. Adv. Manuf. Technol.* [Internet], 93(9–12): 3371–3381. http://www.upm.es.

Ding, H., Illsley, N.P. and Chang, R.C. 2019. 3D Bioprinted GelMA based models for the study of trophoblast cell invasion. *Sci. Rep.* [Internet], 9(1). https://doi.org/10.1038/s41598-019-55052-7.

Fan, X., Chen, K., He, X., Li, N., Huang, J., Tang, K., Li, Y. and Wang, F. 2016. Nano-TiO2/collagen-chitosan porous scaffold for wound repairing. *Int. J. Biol. Macromol.*, 91: 15–22.

Ferrari, E., Palma, C., Vesentini, S., Occhetta, P. and Rasponi, M. 2020. Integrating biosensors in organs-on-chip devices: A perspective on current strategies to monitor microphysiological systems. *Biosensors* [Internet], 10(9). www.mdpi.com/journal/biosensors.

Figeys, D. and Pinto, D. 2000. Lab-on-a-chip: A revolution in biological and medical sciences. *Anal. Chem.* [Internet]. [accessed 2021 Feb 8], 72(9). https://pubs.acs.org/sharingguidelines.

Gaio, N., Waafi, A., Vlaming, M.L.H., Boschman, E., Dijkstra, P., Nacken, P., Braam, S.R., Boucsein, C., Sarro, P.M. and Dekker, R. 2018. A multiwell plate Organ-on-Chip (OOC) device for *in vitro* cell culture stimulation and monitoring. pp. 314–317. *In*: *Proc. IEEE Int. Conf. Micro. Electro. Mech. Syst.* Vol. 2018-January. [place unknown]: Institute of Electrical and Electronics Engineers Inc.

Gencturk, E., Mutlu, S. and Ulgen, K.O. 2017. Advances in microfluidic devices made from thermoplastics used in cell biology and analyses. *Biomicrofluidics* [Internet]. [accessed 2021 Feb 7], 11(5). /pmc/articles/PMC5654984/?report=abstract.

Gorain, B., Choudhury, H., Pandey, M., Kesharwani, P., Abeer, M.M., Tekade, R.K. and Hussain, Z. 2018. Carbon nanotube scaffolds as emerging nanoplatform for myocardial tissue regeneration: A review of recent developments and therapeutic implications. Biomed. Pharmacother., 104: 496–508.

Gudapati, H., Dey, M. and Ozbolat, I. 2016. A comprehensive review on droplet-based bioprinting: Past, present and future. *Biomaterials* [Internet]. [accessed 2021 Feb 8], 102: 20–42. https://pubmed.ncbi.nlm.nih.gov/27318933/.

Guo, B. and Ma, P.X. 2014. Synthetic biodegradable functional polymers for tissue engineering: A brief review. Sci. China Chem., 57(4): 490–500.

Haeberle, S. and Zengerle, R. 2007. Microfluidic platforms for lab-on-a-chip applications. *Lab Chip* [Internet]. [accessed 2021 Feb 8], 7(9): 1094–1110. https://pubs.rsc.org/en/content/articlehtml/2007/lc/b706364b.

Haleem, A. and Javaid, M. 2019. Polyether ether ketone (PEEK) and its 3D printed implants applications in medical field: An overview. *Clin. Epidemiol. Glob. Heal.* [Internet]. [accessed 2021 Feb 7], 7(4): 571–577. https://doi.org/10.1016/j.cegh.2019.01.003.

Haring, A.P., Sontheimer, H. and Johnson, B.N. 2017. Microphysiological human brain and neural systems-on-a-chip: Potential alternatives to small animal models and emerging platforms for drug discovery and personalized medicine. *Stem Cell. Rev. Reports*, 13(3): 381–406.

Harink, B., Le Gac, S., Truckenmüller, R., Van Blitterswijk, C. and Habibovic, P. 2013. Regeneration-on-a-chip? The perspectives on use of microfluidics in regenerative medicine. *Lab Chip* [Internet], 13(18): 3512–3528. www.rsc.org/loc.

Harrill, A.H., Watkins, P.B., Su, S., Ross, P.K., Harbourt, D.E., Stylianou, I.M., Boorman, G.A., Russo, M.W., Sackler, R.S., Harris, S.C., Smith, P.C., Tennant, R., Bogue, M., Paigen, K., Harris, C., Contractor, T., Wiltshire, T., Rusyn, I. and Threadgill, D.W. 2009. Mouse population-guided resequencing reveals that variants in CD44 contribute to acetaminophen-induced liver injury in humans. *Genome. Res.* [Internet]. [accessed 2021 Feb 24]. 19(9): 1507–1515. /pmc/articles/PMC2752130/.

Hingorani, A.D., Kuan, V., Finan, C., Kruger, F.A., Gaulton, A., Chopade, S., Sofat, R., MacAllister, R.J., Overington, J.P., Hemingway, H., Denaxas, S., Prieto, D. and Casas, J.P. 2019. Improving the odds of drug development success through human genomics: Modelling study. *Sci. Rep.* [Internet]. [accessed 2021 Feb 5], 9(1): 18911. https://www.nature.com/articles/s41598-019-54849-w.

Hirama, H., Satoh, T., Sugiura, S., Shin, K., Onuki-Nagasaki, R., Kanamori, T. and Inoue, T. 2019. Glass-based organ-on-a-chip device for restricting small molecular absorption. *J. Biosci. Bioeng.* [Internet]. [accessed 2021 Feb 7], 127(5): 641–646. https://pubmed.ncbi.nlm.nih.gov/30473393/.

Hoffmann, A., Leonards, H., Tobies, N., Pongratz, L., Kreuels, K., Kreimendahl, F., Apel, C., Wehner, M. and Nottrodt, N. 2017. New stereolithographic resin providing functional surfaces for biocompatible three-dimensional printing. *J. Tissue Eng.* [Internet]. [accessed 2021 Feb 8], 8. /pmc/articles/PMC5753888/?report=abstract.

Hölzl, K., Lin, S., Tytgat, L., Van Vlierberghe, S., Gu, L. and Ovsianikov, A. 2016. Bioink properties before, during and after 3D bioprinting. *Biofabrication*, 8(3).

Hua, S., Sukumar, S., Prakash, J., Hu, Y., Zhang, B. and Pang, Z. 2017. Modulating the tumor microenvironment to enhance tumor nanomedicine delivery. *Front Pharmacol.* | www.frontiersin.org [Internet], 8:952. www.frontiersin.org.

Huh, D., Hamilton, G.A. and Ingber, D.E. 2011. From 3D cell culture to organs-on-chips. *Trends Cell Biol.*, 21(12): 745–754.

Huh, D., Kim, H.J., Fraser, J.P., Shea, D.E., Khan, M., Bahinski, A., Hamilton, G.A. and Ingber, D.E. 2013. Microfabrication of human organs-on-chips. *Nat. Protoc.*, [Internet]. [accessed 2021 Feb 7], 8(11): 2135–2157. https://www.nature.com/articles/nprot.2013.137.

Hwang, Y.S., Bong, G.C., Ortmann, D., Hattori, N., Moeller, H.C. and Khademhosseinia, A. 2009. Microwell-mediated control of embryoid body size regulates embryonic stem cell fate via differential expression of WNT5a and WNT11. *Proc. Natl. Acad. Sci. U S A* [Internet]. [accessed 2021 Feb 24], 106(40): 16978–16983. https://pubmed.ncbi.nlm.nih.gov/19805103/.

Ji, Q., Zhang, J.M., Liu, Y., Li, X., Lv, P., Jin, D. and Duan, H. 2018. A modular microfluidic device via multimaterial 3D printing for emulsion generation. *Sci. Rep.* [Internet], 8(1): 1–11. http://dx.doi.org/10.1038/s41598-018-22756-1.

Kaarj, K. and Yoon, J.Y. 2019. Methods of delivering mechanical stimuli to Organ-on-a-Chip. *Micromachines* [Internet], 10(10). www.mdpi.com/journal/micromachines.

Kačarević, Ž.P., Rider, P.M., Alkildani, S., Retnasingh, S., Smeets, R., Jung, O., Ivanišević, Z. and Barbeck, M. 2018. An introduction to 3D bioprinting: Possibilities, challenges and future aspects. *Materials* (Basel) [Internet]. [accessed 2021 Feb 8], 11(11). /pmc/articles/PMC6266989/?report=abstract.

Kaplowitz, N. 2006. Liver biology and pathobiology. *Hepatology* [Internet]. [accessed 2021 Feb 24], 43(2 SUPPL. 1). https://pubmed.ncbi.nlm.nih.gov/16447293/.

Khalid, M.A.U., Kim, Y.S., Ali, M., Lee, B.G., Cho, Y.J. and Choi, K.H. 2020. A lung cancer-on-chip platform with integrated biosensors for physiological monitoring and toxicity assessment. *Biochem. Eng. J.* [Internet], 155. www.elsevier.com/locate/bej.

Khetani, S.R., Berger, D.R., Ballinger, K.R., Davidson, M.D., Lin, C. and Ware, B.R. 2015. Microengineered liver tissues for drug testing. *J. Lab. Autom.* [Internet]. [accessed 2021 Feb 24], 20(3):216–250. https://pubmed.ncbi.nlm.nih.gov/25617027/.

Kim, I.Y., Seo, S.J., Moon, H.S., Yoo, M.K., Park, I.Y., Kim, B.C. and Cho, C.S. 2008. Chitosan and its derivatives for tissue engineering applications. *Biotechnol. Adv.* [Internet]. [accessed 2021 Feb 7], 26(1): 1–21. https://pubmed.ncbi.nlm.nih.gov/17884325/.

Kim, J.H., Park, J.Y., Jin, S., Yoon, S., Kwak, J.Y. and Jeong, Y.H. 2019. A microfluidic chip embracing a nanofiber scaffold for 3D cell culture and real-time monitoring. *Nanomaterials* [Internet]. [accessed 2021 Feb 8], 9(4). /pmc/articles/PMC6523224/?report=abstract.

Kimura, H., Sakai, Y. and Fujii, T. 2018. Organ/body-on-a-chip based on microfluidic technology for drug discovery. *Drug Metab. Pharmacokinet*, 33(1): 43–48.

Kojic, S.P., Stojanovic, G.M. and Radonic, V. 2019. Novel cost-effective microfluidic chip based on hybrid fabrication and its comprehensive characterization. *Sensors.* [Internet]. [accessed 2021 Feb 7], 19(7):1719. https://www.mdpi.com/1424-8220/19/7/1719.

Kondiah, P.P.D., Choonara, Y.E., Marimuthu, T., Kondiah, P.J., du Toit, L.C., Kumar, P. and Pillay, V. 2020. Nanotechnological paradigms for neurodegenerative disease interventions. pp. 277–292. *In*: *Adv 3D-Printed Syst. Nanosyst. Drug Deliv. Tissue. Eng.* [place unknown]: *Elsevier*.

Langhans, S.A. 2018. Three-dimensional *in vitro* cell culture models in drug discovery and drug repositioning. *Front. Pharmacol.* [Internet]. [accessed 2021 Feb 5], 9(JAN): 6. www.frontiersin.org.

Leberfinger, A.N., Ravnic, D.J., Dhawan, A. and Ozbolat, I.T. 2017. Concise review: Bioprinting of stem cells for transplantable tissue fabrication. *Stem. Cells Transl. Med.* [Internet]. [accessed 2021 Feb 8], 6(10): 1940–1948. https://pubmed.ncbi.nlm.nih.gov/28836738/.

Lee, H. and Cho, D.W. 2016. One-step fabrication of an organ-on-a-chip with spatial heterogeneity using a 3D bioprinting technology. *Lab Chip* [Internet]. [accessed 2021 Feb 7], 16(14): 2618–2625. www.rsc.org/loc.

Li, Y.C., Lin, M.W., Yen, M.H., Fan, S.M.Y., Wu, J.T., Young, T.H., Cheng, J.Y. and Lin, S.J. 2015. Programmable laser-assisted surface microfabrication on a Poly(Vinyl Alcohol)-coated glass chip with self-changing cell adhesivity for heterotypic cell patterning. *ACS Appl. Mater. Interfaces* [Internet]. [accessed 2021 Feb 8], 7(40): 22322–22332. https://pubmed.ncbi.nlm.nih.gov/26393271/.

Liu, F., Chen, Q., Liu, C., Ao, Q., Tian, X., Fan, J., Tong, H. and Wang, X. 2018. Natural polymers for organ 3D bioprinting. *Polymers* (Basel) [Internet], 10(11). www.mdpi.com/journal/polymers.

Liu, W., Zhang, Y.S., Heinrich, M.A., De Ferrari, F., Jang, H.L., Bakht, S.M., Alvarez, M.M., Yang, J., Li, Y.C., Trujillo-de Santiago, G., Miri, A.K., Zhu, K., Khoshakhlagh, P., Prakash, G., Cheng, H., Guan, X., Zhong, Z., Ju, J., Zhu, G.H., Jin, X., Shin, S.R., Dokmeci, M.R. and Khademhosseini, A. 2017. Rapid continuous multimaterial extrusion bioprinting. *Adv. Mater.* [Internet]. [accessed 2021 Feb 8], 29(3). https://www.ncbi.nlm.nih.gov/pmc/articles/PMC5235978/.

Liu, Y., Gu, J. and Fan, D. 2020. Fabrication of high-strength and porous hybrid scaffolds based on nano-hydroxyapatite and human-like collagen for bone tissue regeneration. *Polymers* (Basel) [Internet]. [accessed 2021 Feb 7], 12(1). https://pubmed.ncbi.nlm.nih.gov/31906327/.

Ma, X., Qu, X., Zhu, W., Li, Y.S., Yuan, S., Zhang, H., Liu, J., Wang, P., Lai, C.S.E., Zanella, F., Feng, G.S., Sheikh, F., Chien, S. and Chen, S. 2016. Deterministically patterned biomimetic human iPSC-derived hepatic model via rapid 3D bioprinting. *Proc. Natl. Acad. Sci. U S A* [Internet]. [accessed 2021 Feb 24], 113(8): 2206–2211. www.pnas.org/cgi/doi/10.1073/pnas.1524510113.

Maschmeyer, I., Lorenz, A.K., Schimek, K., Hasenberg, T., Ramme, A.P., Hübner, J., Lindner, M., Drewell, C., Bauer, S., Thomas, A., Sambo, N.S., Sonntag, F., Lauster, R. and Marx, U. 2015. A four-organ-chip for interconnected long-term co-culture of human intestine, liver, skin and kidney equivalents. *Lab Chip* [Internet], 15(12): 2688–2699. www.rsc.org/loc.

Mathur, A., Loskill, P., Shao, K., Huebsch, N., Hong, S.G., Marcus, S.G., Marks, N., Mandegar, M., Conklin, B.R., Lee, L.P. and Healy, K.E. 2015. Human iPSC-based cardiac microphysiological system for drug screening applications. *Sci. Rep.* [Internet]. [accessed 2021 Feb 8], 5(1): 1–7. www.nature.com/scientificreports.

McCuskey, R.S., Ekataksin, W., LeBouton, A.V., Nishida, J., McCuskey, M.K., McDonnell, D., Williams, C., Bethea, N.W., Dvorak, B. and Koldovsky, O. 2003. Hepatic microvascular development in relation to the morphogenesis of hepatocellular plates in neonatal rats. *Anat. Rec. - Part A Discov. Mol. Cell Evol. Biol.* [Internet]. [accessed 2021 Feb 24], 275(1): 1019–1030. https://pubmed.ncbi.nlm.nih.gov/14533176/.

Md Ali, M.A., Kayani, A.B.A., Yeo, L.Y., Chrimes, A.F., Ahmad, M.Z., Ostrikov, K. (Ken) and Majlis, B.Y. 2018. Microfluidic dielectrophoretic cell manipulation towards stable cell contact assemblies. *Biomed. Microdevices.* [Internet]. [accessed 2021 Feb 8], 20(4): 1–13. https://doi.org/10.1007/s10544-018-0341-1.

Mehta, P.P. and Pawar, V.S. 2018. Electrospun nanofiber scaffolds: Technology and applications. pp. 509–573. *In*: *Appl. Nanocomposite Mater Drug Deliv.* [place unknown]: Elsevier.

Miao, L., Lin, C.M. and Huang, L. 2015. Stromal barriers and strategies for the delivery of nanomedicine to desmoplastic tumors HHS public access. *J. Control Release*, 219: 192–204.

Misun, P.M., Rothe, J., Schmid, Y.R.F., Hierlemann, A. and Frey, O. 2016. Multi-analyte biosensor interface for real-time monitoring of 3D microtissue spheroids in hanging-drop networks. *Microsystems Nanoeng.* [Internet]. [accessed 2021 Feb 7], 2(1): 1–9. www.nature.com/micronano.

Mofazzal Jahromi, M.A., Abdoli, A., Rahmanian, M., Bardania, H., Bayandori, M., Moosavi Basri, S.M., Kalbasi, A., Aref, A.R., Karimi, M. and Hamblin, M.R. 2019. Microfluidic brain-on-a-chip: perspectives for mimicking neural system disorders. *Mol. Neurobiol.* [Internet]. [accessed 2021 Feb 28], 56(12): 8489–8512. https://pubmed.ncbi.nlm.nih.gov/31264092/.

Moradi, E., Jalili-Firoozinezhad, S. and Solati-Hashjin, M. 2020. Microfluidic organ-on-a-chip models of human liver tissue. *Acta Biomater.*, 116: 67–83.

Moraes, C., Labuz, J.M., Leung, B.M., Inoue, M., Chun, T.-H. and Takayama, S. 2013. On being the right size: Scaling effects in designing a human-on-a-chip. *Integr. Biol.*, 5(9): 1149–1161.

Moya, A., Ortega-Ribera, M., Guimerà, X., Sowade, E., Zea, M., Illa, X., Ramon, E., Villa, R., Gracia-Sancho, J. and Gabriel, G. 2018. Online oxygen monitoring using integrated inkjet-printed sensors in a liver-on-a-chip system. *Lab Chip* [Internet]. [accessed 2021 Feb 7], 18(14): 2023–2035. https://pubs.rsc.org/en/content/articlehtml/2018/lc/c8lc00456k.

Murphy, SV. and Atala, A. 2014. 3D bioprinting of tissues and organs. *Nat. Biotechnol.* [Internet]. [accessed 2021 Feb 8], 32(8): 773–785. https://pubmed.ncbi.nlm.nih.gov/25093879/.

Nahmias, Y., Kramvis, Y., Barbe, L., Casali, M., Berthiaume, F., Yarmush, M.L., Nahmias, Y., Kramvis, Y., Barbe, L., Casali, M., Nahmias, Y., Kramvis, Y., Barbe, L., Casali, M., Berthiaume, F. and Yarmush, L. 2006. A novel formulation of oxygen-carrying matrix enhances liver-specific function of cultured hepatocytes. *FASEB J.* [Internet]. [accessed 2021 Feb 24], 20(14): 2531–2533. https://pubmed.ncbi.nlm.nih.gov/17077286/.

Nguyen, D.G., Funk, J., Robbins, J.B., Crogan-Grundy, C., Presnell, S.C., Singer, T. and Roth, A.B. 2016. Bioprinted 3D primary liver tissues allow assessment of organ-level response to clinical drug induced toxicity *in vitro*. van Grunsven, L.A., editor. *PLoS One* [Internet]. [accessed 2021 Feb 24], 11(7): e0158674. https://dx.plos.org/10.1371/journal.pone.0158674.

Nichol, J.W., Koshy, S.T., Bae, H., Hwang, C.M., Yamanlar, S. and Khademhosseini, A. 2010. Cell-laden microengineered gelatin methacrylate hydrogels. *Biomaterials*, 31(21): 5536–5544.

Nie, J., Gao, Q., Wang, Y., Zeng, J., Zhao, H., Sun, Y., Shen, J., Ramezani, H., Fu, Z., Liu, Z., Xiang, M., Fu, J., Zhao, P., Chen, W. and He, Y. 2018. Vessel-on-a-chip with Hydrogel-based Microfluidics. *Small* [Internet]. [accessed 2021 Feb 7], 14(45): 1802368. http://doi.wiley.com/10.1002/smll.201802368.

Occhetta, P., Sadr, N., Piraino, F., Redaelli, A., Moretti, M. and Rasponi, M. 2013. Fabrication of 3D cell-laden hydrogel microstructures through photo-mold patterning. *Biofabrication* [Internet], 5(3): 035002. https://iopscience.iop.org/article/10.1088/1758-5082/5/3/035002.

Odijk, M., Van Der Meer, A.D., Levner, D., Kim, H.J., Van Der Helm, M.W., Segerink, L.I., Frimat, J.P., Hamilton, G.A., Ingber, D.E. and Van Den Berg, A. 2015. Measuring direct current trans-epithelial electrical resistance in organ-on-a-chip microsystems. *Lab Chip* [Internet], 15(3): 745–752. www.rsc.org/loc.

Olson, E.N. 2006. Gene regulatory networks in the evolution and development of the heart. Science (80-) [Internet]. [accessed 2021 Feb 18], 313(5795): 1922–1927. https://pubmed.ncbi.nlm.nih.gov/17008524/.

Pasqualini, F.S., Emmert, M.Y., Parker, K.K. and Hoerstrup, S.P. 2017. Organ Chips: Quality Assurance Systems in Regenerative Medicine. *Clin. Pharmacol. Ther.*, 101(1): 31–34.

Pensabene, V., Costa, L., Terekhov, A., Gnecco, J.S., Wikswo, J. and Hofmeister, W. 2016. Ultrathin polymer membranes with patterned, micrometric pores for organs-on-chips HHS public access. *ACS. Appl. Mater. Interfaces* [Internet], 8(34): 22629–22636. http://pubs.acs.org.

Pepelanova, I., Kruppa, K., Scheper, T. and Lavrentieva, A. 2018. Gelatin-methacryloyl (GelMA) hydrogels with defined degree of functionalization as a versatile toolkit for 3D cell culture and extrusion bioprinting. *Bioengineering* [Internet], 5(3). www.mdpi.com/journal/bioengineering.

Phan, D.T.T., Wang, X., Craver, B.M., Sobrino, A., Zhao, D., Chen, J.C., Lee, L.Y.N., George, S.C., Lee, A.P. and Hughes, C.C.W. 2017. A vascularized and perfused organ-on-a-chip platform for large-scale drug screening applications. *Lab Chip* [Internet]. [accessed 2021 Feb 8], 17(3): 511–520. /pmc/articles/PMC6695340/?report=abstract.

Prabhakaran, M.P., Kai, D., Ghasemi-Mobarakeh, L. and Ramakrishna, S. 2011. Electrospun biocomposite nanofibrous patch for cardiac tissue engineering. *Biomed. Mater* [Internet]. [accessed 2021 Feb 24], 6(5). https://pubmed.ncbi.nlm.nih.gov/21813957/.

Pridgeon, C.S., Schlott, C., Wong, M.W., Heringa, M.B., Heckel, T., Leedale, J., Launay, L., Gryshkova, V., Przyborski, S., Bearon, R.N., Wilkinson, E.L., Ansari, T., Greenman, J., Hendriks, D.F.G., Gibbs, S., Sidaway, J., Sison-Young, R.L., Walker, P., Cross, M.J., Park, B.K. and Goldring, C.E.P. 2018. Innovative organotypic *in vitro* models for safety assessment: Aligning with regulatory requirements and understanding models of the heart, skin, and liver as paradigms. *Arch. Toxicol.* [Internet], 92(2): 557–569. https://doi.org/10.1007/s00204-018-2152-9.

Quirós-Solano, W.F., Gaio, N., Stassen, O.M.J.A., Arik, Y.B., Silvestri, C., Van Engeland, N.C.A., Van der Meer, A., Passier, R., Sahlgren, C.M., Bouten, C.V.C., van den Berg, A., Dekker, R. and Sarro, P.M. 2018. Microfabricated tuneable and transferable porous PDMS membranes for Organs-on-Chips. *Sci. Rep.* [Internet], 8(1). www.nature.com/scientificreports.

Raj, A., Suthanthiraraj, P.P.A. and Sen, A.K. 2018. Pressure-driven flow through PDMS-based flexible microchannels and their applications in microfluidics. *Microfluid. Nanofluidics.* [Internet], 22(11): 128. https://doi.org/10.1007/s10404-018-2150-5.

Rehm, B., Consultant, D., Haghshenas, A., Paknejad, A.S. and Schubert, J. 2008. Situational Problems in MPD. In: Manag Press Drill. [place unknown]: Elsevier; p. 39–80.

Ren, W., Zhang, A. and Dong, J. 2012. [Liver sinusoidal endothelial cells and liver regeneration]. Zhonghua gan zang bing za zhi = Zhonghua ganzangbing zazhi = *Chinese J. Hepatol.*, 20(9): 715–717.

Rosser, J., Bachmann, B., Jordan, C., Ribitsch, I., Haltmayer, E., Gueltekin, S., Junttila, S., Galik, B., Gyenesei, A., Haddadi, B., Harasek, M., Egerbacher, M., Ertl, P. and Jenner, F. 2019. Microfluidic nutrient gradient–based three-dimensional chondrocyte culture-on-a-chip as an *in vitro* equine arthritis model. *Mater Today Bio.*, 4: 100023.

Sahin, O., Ashokkumar, M. and Ajayan, P.M. 2018. Micro- and nanopatterning of biomaterial surfaces. pp. 67–78. *In: Fundam Biomater Met.* [place unknown]: Elsevier.

Sakai, S., Kamei, H., Mori, T., Hotta, T., Ohi, H., Nakahata, M. and Taya, M. 2018. Visible light-induced hydrogelation of an alginate derivative and application to stereolithographic bioprinting using a visible light projector and acid red. Biomacromolecules [Internet]. [accessed 2021 Feb 8], 19(2): 672–679. https://pubs.acs.org/doi/abs/10.1021/acs.biomac.7b01827.

Sano, E., Mori, C., Matsuoka, N., Ozaki, Y., Yagi, K., Wada, A., Tashima, K., Yamasaki, S., Tanabe, K., Yano, K. and Torisawa, Y.S. 2019. Tetrafluoroethylene-propylene elastomer for fabrication of microfluidic organs-on-chips resistant to drug absorption. *Micromachines* [Internet], 10(11). www.mdpi.com/journal/micromachines.

Schroer, A.K., Shotwell, M.S., Sidorov, V.Y., Wikswo, J.P. and Merryman, W.D. 2017. I-Wire Heart-on-a-Chip II: Biomechanical analysis of contractile, three-dimensional cardiomyocyte tissue constructs. *Acta Biomater.* [Internet]. [accessed 2021 Feb 8], 48: 79–87. https://pubmed.ncbi.nlm.nih.gov/27818306/.

Schumacher, K., Khong, Y.M., Chang, S., Ni, J., Sun, W. and Yu, H. 2007. Perfusion culture improves the maintenance of cultured liver tissue slices. pp. 197–205. *In: Tissue Eng* [Internet]. Vol. 13. [place unknown]: Tissue Eng; [accessed 2021 Feb 24]. https://pubmed.ncbi.nlm.nih.gov/17518593/.

Serra, P., Duocastella, M., Fernández-Pradas, J.M. and Morenza, J.L. 2010. The laser-induced forward transfer technique for microprinting. pp. 367–393. *In: Adv. Laser Mater Process Techno. Res. Appl.* [place unknown]: Elsevier Inc.

Shaegh, S.A.M., De Ferrari, F., Zhang, Y.S., Nabavinia, M., Mohammad, N.B., Ryan, J., Pourmand, A., Laukaitis, E., Sadeghian, R.B., Nadhman, A., hin, S.R., Nezhad, A.S., Khademhosseini, A. and Dokmeci, M.R. 2016. A microfluidic optical platform for real-time monitoring of pH and oxygen in microfluidic bioreactors and organ-on-chip devices. *Biomicrofluidics* [Internet], 10(4). http://dx.doi.org/10.1063/1.4955155].

Shik Mun, K., Arora, K., Huang, Y., Yang, F., Yarlagadda, S., Ramananda, Y., Abu-El-Haija, M., Palermo, J.J., Appakalai, B.N., Nathan, J.D. and Naren, A.P. 2019. Patient-derived pancreas-on-a-chip to model cystic fibrosis-related disorders. *Nat. Commun.* [Internet], 10(1). https://doi.org/10.1038/s41467-019-11178-w.

Sidorov, V.Y., Samson, P.C., Sidorova, T.N., Davidson, J.M., Lim, C.C. and Wikswo, J.P. 2017. I-wire heart-on-a-chip I: Three-dimensional cardiac tissue constructs for physiology and pharmacology. *Acta Biomater.* [Internet]. [accessed 2021 Feb 8], 48: 68–78. https://pubmed.ncbi.nlm.nih.gov/27818308/.

Sieber, S., Wirth, L., Cavak, N., Koenigsmark, M., Marx, U., Lauster, R. and Rosowski, M. 2018. Bone marrow-on-a-chip: Long-term culture of human haematopoietic stem cells in a three-dimensional microfluidic environment. *J. Tissue Eng. Regen. Med.* [Internet]. [accessed 2021 Feb 7], 12(2): 479–489. https://onlinelibrary.wiley.com/doi/abs/10.1002/term.2507.

Singh, Y.P., Dasgupta, S. and Bhaskar, R. 2019. Preparation, characterization and bioactivities of nano anhydrous calcium phosphate added gelatin–chitosan scaffolds for bone tissue engineering. *J. Biomater. Sci. Polym. Ed.* [Internet]. [accessed 2021 Feb 7], 30(18): 1756–1778. https://pubmed.ncbi.nlm.nih.gov/31526176/.

Sontheimer-Phelps, A., Hassell Bryan, A. and Ingber, D.E. 2019. Modelling cancer in microfluidic human organs-on-chips. *Nat. Rev. Cancer.*, 19(2): 65–81.

Sosa-Hernández, J.E., Villalba-Rodríguez, A.M., Romero-Castillo, K.D., Aguilar-Aguila-Isaías, M.A., García-Reyes, I.E., Hernández-Antonio, A., Ahmed, I., Sharma, A., Parra-Saldívar, R. and Iqbal, H.M.N. 2018. Organs-on-a-Chip Module: A review from the development and applications perspective. *Micromachines* [Internet]. [accessed 2021 Feb 5], 9(10): 536. http://www.mdpi.com/2072-666X/9/10/536.

Stratton, S., Shelke, N.B., Hoshino, K., Rudraiah, S. and Kumbar, S.G. 2016. Bioactive polymeric scaffolds for tissue engineering. *Bioact Mater.*, 1(2): 93–108.

Sun, H., Jia, Y., Dong, H., Dong, D. and Zheng, J. 2020. Combining additive manufacturing with microfluidics: An emerging method for developing novel organs-on-chips. *Curr. Opin. Chem. Eng.*, 28: 1–9.

Takahashi, S., Leiss, M., Moser, M., Ohashi, T., Kitao, T., Heckmann, D., Pfeifer, A., Kessler, H., Takagi, J., Erickson, H.P. and Fässler, R. 2007. The RGD motif in fibronectin is essential for development but dispensable for fibril assembly. *J. Cell Biol.* [Internet], 178(1): 167–178. http://www.jcb.org/cgi/doi/10.1083/jcb.200703021.

Takebe, T., Zhang, B. and Radisic, M. 2017. Synergistic engineering: Organoids meet organs-on-a-chip. *Cell Stem Cell.*, 21(3): 297–300.

Tanaka, H.Y. and Kano, M.R. 2018. Stromal barriers to nanomedicine penetration in the pancreatic tumor microenvironment. *Cancer Sci.*, 109(7): 2085–2092.

Taubenberger, A.V., Woodruff, M.A., Bai, H., Muller, D.J. and Hutmacher, D.W. 2010. The effect of unlocking RGD-motifs in collagen I on pre-osteoblast adhesion and differentiation. *Biomaterials.*, 31(10): 2827–2835.

Thomas, D. and Radhakrishnan, P. 2019. Tumor-stromal crosstalk in pancreatic cancer and tissue fibrosis. *Mol. Cancer.* [Internet], 18(1). https://doi.org/10.1186/s12943-018-0927-5.

Thomas, H., Diamond, J., Vieco, A., Chaudhuri, S., Shinnar, E., Cromer, S., Perel, P., Mensah, G.A., Narula, J., Johnson, C.O., Roth, G.A. and Moran, A.E. 2018. Global atlas of cardiovascular disease 2000–2016: The path to prevention and control. *Glob Heart* [Internet]. [accessed 2021 Feb 18], 13(3): 143–163. https://pubmed.ncbi.nlm.nih.gov/30301680/.

Toepke, M.W. and Beebe, D.J. 2006. PDMS absorption of small molecules and consequences in microfluidic applications. *Lab Chip* [Internet]. [accessed 2021 Feb 7], 6(12): 1484–1486. https://pubs.rsc.org/en/content/articlehtml/2006/lc/b612140c.

Torino, S., Corrado, B., Iodice, M. and Coppola, G. 2018. PDMS-Based microfluidic devices for cell culture. *Inventions* [Internet]. [accessed 2021 Feb 7], 3(3): 65. http://www.mdpi.com/2411-5134/3/3/65.

Trapecar, M., Communal, C., Velazquez, J., Maass, C.A., Huang, Y.J., Schneider, K., Wright, C.W., Butty, V., Eng, G., Yilmaz, O., Trumper, D. and Griffith, L.G. 2020. Gut-liver physiomimetics reveal paradoxical modulation of IBD-related inflammation by short-chain fatty acids. *Cell Syst.*, 10(3): 223–239.e9.

Van Berlo, J.H. and Molkentin, J.D. 2014. An emerging consensus on cardiac regeneration. *Nat. Med.* [Internet]. [accessed 2021 Feb 18], 20(12): 1386–1393. https://pubmed.ncbi.nlm.nih.gov/25473919/.

Van Den Berg, A., Mummery, C.L., Passier, R. and Van der Meer, A.D. 2019. Personalised organs-on-chips: Functional testing for precision medicine. *Lab Chip* [Internet]. [accessed 2021 Feb 5], 19(2): 198–205. https://pubs.rsc.org/en/content/articlehtml/2019/lc/c8lc00827b.

Van Norman, G.A. 2019. Limitations of animal studies for predicting toxicity in clinical trials: Is it time to rethink our current approach? *JACC Basic to Transl. Sci.*, 4(7): 845–854.

Ventola, C.L. 2017. Progress in nanomedicine: Approved and investigational nanodrugs. P T., 42(12): 742–755.

Vunjak-Novakovic, G, Tandon, N., Godier, A., Maidhof, R., Marsano, A., Martens, T.P. and Radisic, M. 2010. Challenges in Cardiac Tissue Engineering. Tissue Eng Part B Rev [Internet]. [accessed 2022 Mar 7] 16(2): 169. /pmc/articles/PMC2946883.

Wang, X., Liu, Z. and Pang, Y. 2017. Concentration gradient generation methods based on microfluidic systems. *RSC Adv.*, 7(48): 29966–29984.

Wang, Z., Samanipour, R. and Kim, K. 2015. Organ-on-a-chip platforms for drug screening and tissue engineering. *In: Biomed. Eng. Front. Res. Converging Technol.* [Internet]. Vol. 9. [place unknown]: Springer International Publishing; [accessed 2021 Feb 7]; pp. 209–233. https://link.springer.com/chapter/10.1007/978-3-319-21813-7_10.

Wang, Z., Kumar, H., Tian, Z., Jin, X., Holzman, J.F., Menard, F. and Kim, K. 2018. Visible light photoinitiation of cell-adhesive gelatin methacryloyl hydrogels for stereolithography 3D bioprinting. *ACS Appl. Mater. Interfaces* [Internet]. [accessed 2021 Feb 8], 10(32): 26859–26869. https://pubs.acs.org/doi/abs/10.1021/acsami.8b06607.

Wikswo, J.P., Curtis, E.L., Eagleton, Z.E., Evans, B.C., Kole, A., Hofmeister, L.H. and Matloff, W.J. 2013. Scaling and systems biology for integrating multiple organs-on-a-chip. *Lab Chip* [Internet], 13(18): 3496–3511. www.rsc.org/loc.

Wilmer, M.J., Ng, C.P., Lanz, H.L., Vulto, P., Suter-Dick, L. and Masereeuw, R. 2016. Kidney-on-a-Chip technology for drug-induced nephrotoxicity screening. *Trends Biotechnol.* [Internet]. [accessed 2021 Feb 24] 34(2): 156–170. https://pubmed.ncbi.nlm.nih.gov/26708346/.

Wojciak-Stothard, B. and Ridley, A.J. 2003. Shear stress-induced endothelial cell polarization is mediated by Rho and Rac but not Cdc42 or PI 3-kinases. *J. Cell Biol.* [Internet]. [accessed 2021 Feb 8], 161(2): 429–439. /pmc/articles/PMC2172912/?report=abstract.

Xiao, S., Coppeta, J.R., Rogers, H.B., Isenberg, B.C., Zhu, J., Olalekan, S.A., McKinnon, K.E., Dokic, D., Rashedi, A.S., Haisenleder, D.J., Malpani, S.S., Arnold-Murray, C.A., Chen, K., Jiang, M., Bai, L., Nguyen, C.T., Zhang, J., Laronda, M.M., Hope, T.J., Maniar, K.P., Pavone, M.E., Avram, M.J., Sefton, E.C., Getsios, S., Burdette, J.E., Kim, J.J., Borenstein, J.T. and Woodruff, T.K. 2017. A microfluidic culture model of the human reproductive tract and 28-day menstrual cycle. *Nat. Commun.* [Internet]. 8. www.nature.com/naturecommunications.

Xu, C., Zhang, M., Chen, W., Jiang, L., Chen, C. and Qin, J. 2020. Assessment of air pollutant PM2.5 pulmonary exposure using a 3D lung-on-chip model. *Cite This ACS Biomater. Sci. Eng.* [Internet]. [accessed 2021 Feb 6], 6: 3090. https://dx.doi.org/10.1021/acsbiomaterials.0c00221.

Yang, X., Li, K., Zhang, X., Liu, C., Guo, B., Wen, W. and Gao, X. 2018. Nanofiber membrane supported lung-on-a-chip microdevice for anti-cancer drug testing. *Lab Chip* [Internet]. [accessed 2021 Feb 8], 18(3): 486–495. https://pubs.rsc.org/en/content/articlehtml/2018/lc/c7lc01224a.

Yue, K., Trujillo-de Santiago, G., Alvarez, M.M., Tamayol, A., Annabi, N. and Khademhosseini, A. 2015. Synthesis, properties, and biomedical applications of gelatin methacryloyl (GelMA) hydrogels. *Biomaterials*, 73: 254–271.

Zakharova, M., Palma Do Carmo, M.A., Van Der Helm, M.W., Le-The, H., De Graaf, M.N.S., Orlova, V., Van Den Berg, A., Van Der Meer, A.D., Broersen, K. and Segerink, L.I. 2020. Multiplexed blood-brain barrier organ-on-chip. *Lab Chip.* 20(17): 3132–3143.

Zhang, B., Korolj, A., Lai, B.F.L. and Radisic, M. 2018. Advances in organ-on-a-chip engineering. *Nat. Rev. Mater.* [Internet], 3(8): 257–278. http://dx.doi.org/10.1038/s41578-018-0034-7.

Zhang, Y.S., Aleman, J., Arneri, A., Bersini, S., Piraino, F., Shin, S.R., Dokmeci, M.R. and Khademhosseini, A. 2015. From cardiac tissue engineering to heart-on-a-chip: Beating challenges. *Biomed. Mater.* [Internet]. [accessed 2021 Feb 18], 10(3). https://pubmed.ncbi.nlm.nih.gov/26065674/.

Zhuang, J., Yamada, K.A., Saffitz, J.E. and Kleber, A.G. 2000. Pulsatile stretch remodels cell-to-cell communication in cultured myocytes. *Circ. Res.* [Internet]. [accessed 2021 Feb 8], 87(4): 316–322. https://pubmed.ncbi.nlm.nih.gov/10948066/.

CHAPTER 13

Overview of Nanostructured Carbon-based Catalysts
Recent Advances and Perspectives

Mattia Bartoli,[1,2] *Pravin Jagdale,*[3] *Mauro Giorcelli,*[1,2]
Massimo Rovere[1,2] *and Alberto Tagliaferro*[1,2,*]

1. Introduction

Since the beginning of the industrial age, chemical productions have required the use of catalysts to implement most processes. Among the huge number of possibilities, heterogeneous catalysts supported onto carbonaceous materials have always stood as the most competitive ones. Catalysts such as Lindlar's (Lindlar 1952) marked the history of industrial chemistry reaching new heights of performances.

The use of nanostructured carbonaceous species represented a step forward for carbon supported catalysts and nanostructured carbonaceous materials encompass a great number of heterogeneous compounds. They can be broadly classified in two main families. The first is composed of carbon allotropes such as fullerene (Omacrsawa 2012), graphene (Zhu 2017) and carbon nanotubes (CNTs) (Endo et al. 2013). The morphology of these materials can be tuned leading to the formation of carbon quantum dots (Hola et al. 2014), nanohorns (Pagona et al. 2009), nanofibers (Ko and Wan 2014), nanoribbons (James and Tour 2012), nanocages (Wu et al. 2019) and other morphologies (Coville et al. 2011). All of these materials have sizes in the range of nanometers and a well-defined morphology.

The second group is represented by the fully engineered carbonaceous materials realized by designing hybrid and hierarchical shapes never seen before. These materials are rationally nano-architectured by using pre-assembled or bioinspired scaffolds. This approach introduces a new level of control in the nano- and microscale assembly of the materials.

All these materials combined with nano- and microparticles represent the new frontier for heterogeneous catalysis. In this chapter, we overview graphene, carbon nanotubes (CNTs) and other nanostructured carbons as catalysts for the realization of heterogeneous catalysts. After a brief

[1] Department of Applied Science and Technology, Politecnico di Torino, C.so Duca degli Abruzzi 24, 10129 Turin, Italy.
[2] Consorzio Interuniversitario Nazionale per la Scienza e Tecnologia dei Materiali (INSTM), Via G. Giusti 9, 50121 Florence, Italy.
[3] Center for Sustainable Future, Italian Institute of Technology, Via Livorno 60, 10144 Turin, Italy.
* Corresponding author: alberto.tagliaferro@polito.it

materials' introduction about materials, we discuss the main catalytic procedures where nanocarbon supported catalysts are used, enlightening the advantages and peculiarities of each catalyst herein reported.

2. Nanocarbon materials: Structures and modifications

2.1 Nano and nanostructured carbon

2.1.1 Graphene and related materials

Ideal Graphene is a material composed of an infinite single plane of pure graphitic carbon. Graphene elementary cell structure contains atoms bonded by three in-plane σ bonds with the p orbital system perpendicular to the sp^2 plane. As a consequence, graphene displays a full delocalization of π bonds on its all structure, allowing the electrons to freely move in the graphene plane (Mintmire et al. 1992, Yan et al. 2009, Dresselhaus et al. 2010). The electron mobility in the graphene plane is the reason for its high electrical conductivity (Rhee 2020).

The properties of graphene are leveraged by its difficult handability and scarce availability (Lee et al. 2016). Additionally, a great part of scientific literature uses the term graphene to define few layered materials and even nanographite (Narita et al. 2015, Sun et al. 2020). To overcome the troubles related to the use of real graphene, several materials have been proposed. The first material is graphene oxide (GO). GO is highly oxidized graphene carrying a great amount of oxygen residues (Brisebois and Siaj 2020). The main functionalities of GO basal lattice are epoxy and hydroxyl residues while carboxylic ones are more represented on its edges. Contrary to graphene, GO structure is closely related to the productive way adopted for its study.

Production of GO saw the light over one and half century ago with the pioneering studies of Brodie in 1859. He investigated the oxidation of graphite trying to define the atomic weight of carbon (Brodie 1859) by using a mixture of nitric acid and potassium chlorate and the process time can go upto days. After this long oxidative process, he produced light-yellow powder made by tiny shining flakes. He named the material graphitic acid but nowadays it is known worldwide under the name of GO. Half a century later, Staudenmaier described a new reaction for the production of "graphitic acids" by using an acidic solution of $KClO_3$ (Staudenmaier 1898). Another route to GO was described by Hummers and Offeman in the middle of the 20th century (William et al. 1958). Authors treated graphite with a sulphuric acid solution containing both sodium nitrate and potassium permanganate. Afterwards, Hummers' method became the most used process to produce graphite oxide even if all three above-mentioned protocols are still used to produce GO with different features. On this very same topic, Poh and co-workers (Poh et al. 2012) reported an exhaustive research characterizing reduced GO (rGO) produced from GO. Authors enlighten the relevancy of the concentration and types of oxygen-containing residues of GO used as precursor for the preparation of rGO due to the different production pathways adopted.

Due to the different synthetic routes, GO composition and non-stoichiometric structure are closely dependent on the product steps (Araújo et al. 2017). As a consequence, several studies have described different spatial arrangement for GO, namely the Ruess, Szabo, Hofmann, Nakajima-Matsuo, Lerf-Klinowski and Scholz-Boehm models (Szabó et al. 2006), as reported in Figure 1(b)–(g).

Scientists generally agreed the Lerf-Klinowski model is the most representative one and it matches quite well the real structure of GO as proved by numerous experimental data (Lerf et al. 1997). Based on this model, defective regions (i.e., cracks, holes, wrinkles and impurities) are induced by the oxidative procedure used for the production.

Scientists' interest have been caught not only by GO but also by the previously mentioned rGO. rGO is a rising star in the research world of graphene (Thakur and Karak 2015). Actually, it

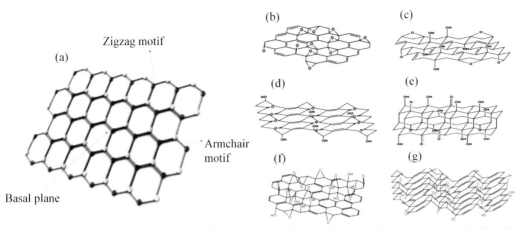

Figure 1. Structures of pristine graphene (a) and its GO derivatives based on Hofmann (b), Ruess (c), Scholz-Boehm (d), Nakajima-Matsuo (e), Lerf-Klinowski (f) and Szabo (g) models. Reproduced, adapted and reprinted with permission from Ref. (Lavagna et al. 2020) under CC license.

is produced through a reduction process of GO by using several reductive agents (Guex et al. 2017, Liu et al. 2011). Through reductive process, the oxygen-based groups of GO are removed forming rGO with a C/O ratio ranging from 12.5 wt.% to 0.4 wt.% (Lee et al. 2019). Both GO and rGO are characterized by high surface area and a good chemical and thermal stability representing a bright new generation of nanosized carbon support for plenty of catalysis applications.

2.1.2 Carbon nanotubes

Carbon Nanotubes (CNTs) are a tube-shaped allotropic form of carbon, with a diameter in the nanometers range and shape similar to those sketched in Figure 2.

CNTs have different structures as they differ in length, thickness, the type of helicity and number of layers. The number of layers defines them as single wall CNT (SWCNT), double wall CNT (DWCNT) and multiwalled CNT (MWCNT) while the terms for their helicity are armchair, zig-zag and chiral. CNTs have a high aspect ratio up to 10^6 (Li et al. 2007) and exhibit unusual mechanical properties such as high toughness and elastic moduli (Treacy et al. 1996). Furthermore,

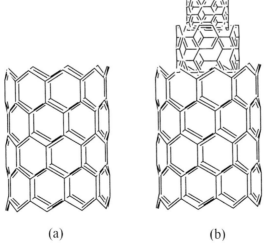

(a)   (b)

Figure 2. Example of "zig-zag" (a) single wall CNT (SWCNT) and (b) multiwalled CNT (MWCNT).

CNTs show very attractive electrical properties as their behaviour ranges from semiconducting to metallic (Lekawa-Raus et al. 2014, Bandaru 2007). Several procedures have been developed to produce CNTs on a large scale such as arc discharge (Arora and Sharma 2014), laser ablation (Scott et al. 2001), chemical vapor deposition (Kumar and Ando 2010) and flame synthesis approach (Hamzah et al. 2017).

2.1.3 Other carbonaceous materials

Other than allotropic carbon, several carbonaceous materials have gained a lot of interest for catalytic application.

Among them, carbon black (CB) has led the field for decades with many applications (Yang et al. 2011). CB particles represent the most common carbon support in industry due to its large availability, low cost and robustness. Nonetheless, other carbon materials have gained a great interest such bioderived activated carbons (ACs) and templated carbon structures. ACs are very interesting due to their sustainable production and large surface area together with residual functionalities such hydroxylic groups. While ACs totally or partially retained the structure of the biomass used for their production, template carbons shapes can be tailored creating completely new architectures. This process based on the creation of a removable support for the carbon growth opens the way for a vast realm of morphologies such as graphene sponges (Zhao et al. 2012) and mesoporous carbons (Kyotani et al. 1997).

2.2 Tailored nanocarbon properties: Functionalization and decoration

Pure carbonaceous structures rarely show a remarkable catalytic activity mainly because of the functionalities on the carbon surface or of deposited structures such as inorganic species. In the next subsections, we report the main features of organic and inorganic tailoring processes and their effects on the catalytic activity of nanocarbonaceous materials.

2.2.1 Organic surface functionalization

Organic functionalization of carbon surface is a well-established practice and could lead to the introduction of plenty of different functionalities, which are summarized in Figure 3.

Introduction of functionalities deeply affects the catalytic performances of graphene as described by Sun and co-workers (Sun et al. 2012) with a decrease of activity due to the defects of carbonaceous structure. However, other studies have proved that the introduction of defects into the sp^2 plane could enhance certain reactivities (Lai et al. 2011). A newcomer to the field is hence faced with mixed opinions on the topic and contradictory results. It is quite hard to find a straight line through the immense literature on graphene and related materials but it is possible to say that promotion and depletion of catalytic activity is related to both functionalities and catalytic reaction.

Accordingly, Lemes and co-workers (Lemes et al. 2019) described the nitrogen doped graphene enhanced activity during the oxygen reduction process and Zhang reported the same results by considering sulphur doped materials (Zhang et al. 2014).

Considering CNTs instead of graphene simplifies the discussion. Functional groups introduced on the CNTs surface magnify their catalytic activity. Cao and co-workers (Cao et al. 2018) reported the role of carboxylic function on the surface of MWCTNs for the active promotion of phenol oxidation. Similarly, Qui et al. (Qui et al. 2011) have shown that all oxygen containing groups participate in the oxidative processes.

On ACs and TCSs, residual groups are abundant and their effect is remarkable. Szymański et al. (Szymański et al. 2004) clearly show this phenomenon during the reduction of NOx to NH_3.

Introduction of functionalities is not the only procedure to enhance the catalytic activity of nanostructured carbons. A simpler approach is based on the increment of surface area by using

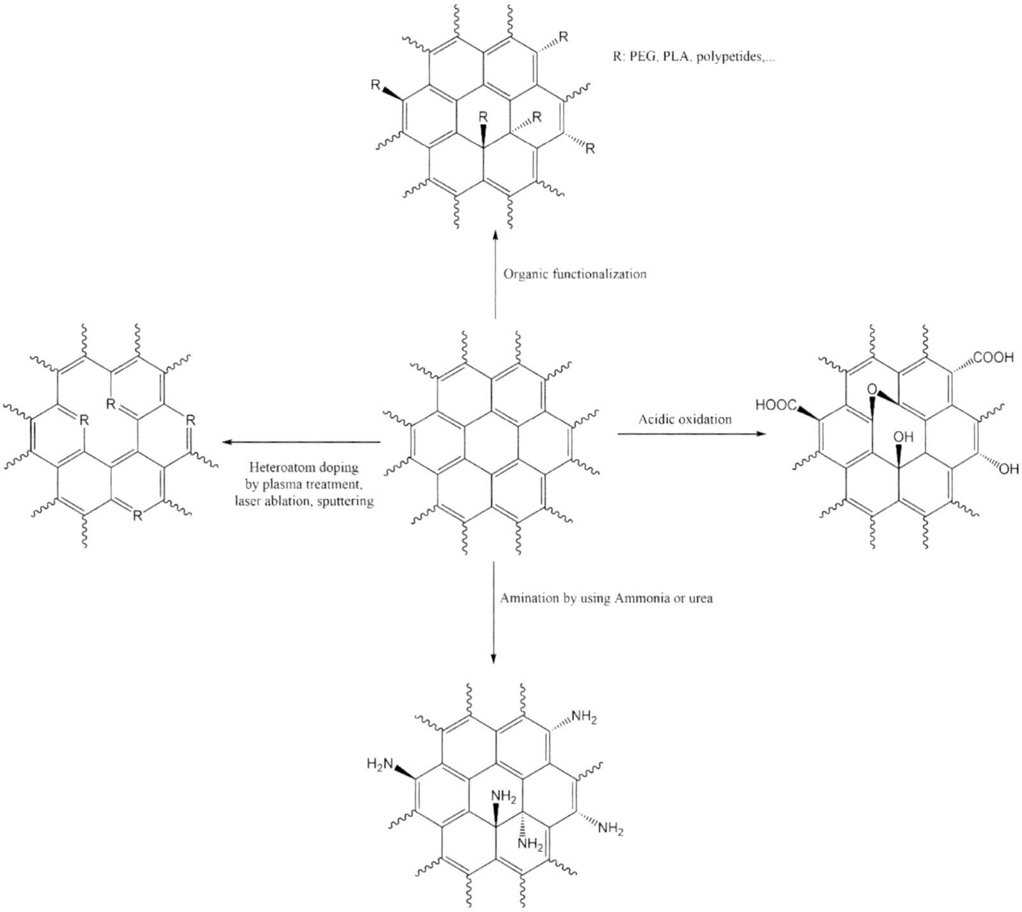

Figure 3. Summary of different functionalities introduced onto carbonaceous structures.

chemical or physical activation. These processes are not used for graphene and CNTs but they are very performing for bioderived carbon and represent the common production route for AC. AC can be obtained through treatment of carbon by using steam or CO_2 as described by many authors (Molina-Sabio et al. 1996, Bouchelta et al. 2008, Zhang et al. 2004). Steam activation is classified as a physical activation process: it removes the tar and increases the pore size. CO_2 activation is based on the reaction:

$$CO_{2(g)} + C_{(S)} \longrightarrow 2\ CO_{(g)} \tag{1}$$

This reaction (Equation 1) removes part of the carbon matrix with an increment of both porosity and surface area and positively affects the catalytic performances as reported for instance by Diaz et al. (Díaz et al. 2007). Similar results can be achieved by using KOH (Lin et al. 2013) or HNO_3 (Liu et al. 2011).

2.2.2 *Inorganic surface tailoring*

Carbon materials are often used as support for inorganic nanostructures in catalytic applications. There are three main routes for tailoring the carbon surface with nanoparticles as summarized in Figure 4: chemical reduction of metal precursors, metal vapour deposition (MVD) and carbothermal processes.

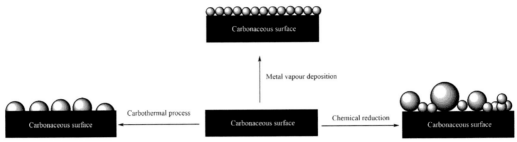

Figure 4. Approaches to nanoparticle tailoring of carbon surfaces.

Chemical reduction of metal precursors is a simple approach based on the use of a reduction agent such as $NaBH_4$ or H_2. The process leads to an inhomogeneous particle size distribution in absence of any additives such as polymers (Chen and Xing 2005).

MVD is based on the simultaneous condensation at very low temperature of a metal vapor with a vapor of a weakly stabilizing organic solvent. After co-deposition, the frozen matrix is allowed to melt, generating solvent-stabilized metal microclusters MVD with a good control on particle size and distribution as described by Oberhauser et al. (Oberhauser et al. 2015). MVD drawback is represented by the unanchored nanostructures that only interact with the carbonaceous matrix though weak forces.

Carbothermal processes represent a good compromise between the high performances of MVD and the facile synthesis through chemical reduction. Carbothermal routes induce the reduction of a metal salt using the carbon support as a reducing agent in an inert atmosphere at high temperature. This method leads to well anchored particles but damages the carbonaceous structure and its use is discouraged in combination with precious carbon nanostructures such as graphene or CNTs related materials. On the contrary, the other two approaches could be used without concern with any kind of nanostructured carbon. The further advantage of carbothermal reduction is the drastic increment of carbon support surface area.

3. Nanostructured and nanosized carbon materials: Catalytic applications

3.1 Selective hydrogenation of unsaturated carbon-carbon bonds

The research for new recyclable hydrogenation catalysts promoted the synthesis of a great deal of heterogeneous catalysts (Nishimura 2001). In this context, metal supported onto carbon structures find plenty of applications due to the unique combination of tunability and robustness.

Selective hydrogenation of carbon–carbon unsaturated bonds is a widely studied topic. The most investigated systems are based on the use of platinum or palladium atoms due to their activity on hydrogenation reactions. A key study on the topic was reported by Efremenko and Sheintuch (Efremenko and Sheintuch 1998). Authors performed a quantum mechanical analysis of the interactions between carbon materials and palladium clusters providing a deep insight into Pd/C bonding, functional-group effects, and cluster growth on carbon surfaces. Efremenko and Sheintuch (Efremenko and Sheintuch 1998) showed that the surface carbon formed much stronger bonds to the basal graphite plane. Furthermore, data collected by studying ACs proved an increment of palladium clusters activity due to a delocalized donor–acceptor interaction with the support π system orbital system.

About the conversion of phenylacetylene to styrene, the effect of modification of graphitic structure to enhance catalytic properties of nanoparticles supported onto carbonaceous materials was investigated by Xia et al. (Xia et al. 2019). Authors produced a nitrogen doped graphene tailored with nanoparticles of platinum and tested it in the process of semi-hydrogenation of phenylacetylene. Through DFT calculations, authors proved that the remarkable activity of the catalyst was due to the confinement effect of doped graphene on styrene. This effect promotes the styrene desorption from nanoparticles surface avoiding its over reduction to benzyl ethanol.

A similar effect was observed by Wang et al. (Wang et al. 2017a) by using a bimetallic Pt-Au nanostructured system supported onto CNT. Authors claimed that the enhanced selectivity obtained by the catalyst was due to a hindered surface dynamic of the nanoscale bimetallic phases and the preference for desorption of partial hydrogenation products.

The high selectivity reached by using metal nanoparticles supported on CNTs was compared with all traditional catalysts by Teschner et al. (Teschner et al. 2006). Authors clearly proved carbon-based nanomaterials outperformed the other carbon supported catalyst in selectivity and turn over frequency.

The other class of hydrogenation reactions is the selective reduction of carbon–carbon unsaturated bonds when carbonyl functions are also present. Among them, the selective reduction of α,β-unsaturated carbonyl compounds is relevant in the manufacture of fine chemical intermediates. The selective reduction of α,β-unsaturated carbonyl species still remains a hot topic even if it has been widely investigated. Chemoselective reduction of cinnamaldehyde represents a perfect case of study for the comparison of different nanocarbon based catalysts.

Truong-Huu et al. (Truong-Huu et al. 2012) supported palladium nanoparticles onto a few layers of graphene. They reported a high selectivity for carbon–carbon double bond hydrogenation due to the very accessible surface together with fast intermediate products' desorption. Peres et al. (Peres et al. 2019) modified a similar graphene structure by tailoring the surface with platinum concave nanocubes stabilized by alkyl amines. The presence of the aminic stabilizer improved the stability of the nanoparticles and at high concentration promoted the chemoselectivity for carbon oxygen unsaturation reduction. The effect of amine was also studied by Dongil and co-workers (Dongil et al. 2011) by grafting ruthenium phosphine complexes on GO and graphite in the presence of amine groups, showing variations in activity and selectivity after the amine-functionalization.

An iridium-based catalyst was described by Li et al. (Li et al. 2019a). Authors supported the iridium cluster onto graphene aerogel made by converting GO into rGO. Also, in this case carbonyl functionality was reduced with a high chemoselectivity over carbon–carbon unsaturated bond. This was due to the weak interaction occurring between cinnamaldehyde and iridium supported materials as reported by Wang and co-workers (Wang et al. 2019).

CNTs have been also studied for similar catalytic reactions. Ma et al. (Ma et al. 2017) studied the confinement effect of CNTs surface and how this phenomenon affects the product distribution by using a ruthenium catalyst. Authors reported a high selectivity for hydrocinnamaldehyde when ruthenium nanostructures were dispersed on the outer walls of CNTs and a high selectivity for hydrocinnamyl alcohol when the particles were confined within CNTs' inner cavity facilitating hydrogenation. Results proved the interactions between the confined nanoparticles are responsible for a crucial role of modulating the product distribution.

This was also reported by several studies showing a high hydrocinnamaldehyde selectivity by using nanoparticles dispersed onto the surface of CNTS (Tessonnier et al. 2005, Vu et al. 2006, Zhu et al. 2019).

Carbon nanofibers (CNFs) were also used for the same reactions as reported by Jung co-workers (Jung et al. 2009). In this study, authors compared several carbon supported catalysts with CNF such as those supported on ACs and CNTs showing a strong influence of carbon nanomaterials' textures on the catalytic performances. Carbon black (CB) was also used as described by Szumełda et al. (Szumełda et al. 2014) by tailoring it with palladium/gold particles by using microemulsion technique showing a tunable selectivity based on gold percentage.

3.2 Fischer-Tropsch Synthesis

Fischer-Tropsch Synthesis (FTS) is the conversion process of a carbon monoxide and hydrogen mixture into (a) saturated or (b) unsaturated liquid hydrocarbons as sketched in the following scheme:

(a) $nCO + (2n+1) H_2 \longrightarrow C_nH_{2n+2} + nH_2O$

(b) $nCO + 2nH_2 \longrightarrow C_nH_{2n} + nH_2O$ (2)

FTS is a pillar of industrial gas-to-liquid technology used for the production of an oil substitute starting from reforming of typically coal, natural gas, or biomass. Cobalt-containing catalysts are very active during FTS by using a H_2:CO ratio of around 2. Nonetheless, H_2/CO produced from cheap feedstocks such as low quality coal or biomass are characterized by lower ratios and in these cases cobalt-iron doped catalysts could be more effective. Supports of these metals are an essential part of the FTS conversion and carbon ones are commonly used due to the tunable pore sizes and high surface area.

In 2015, Karimi co-workers (Karimi et al. 2015) firstly described the use of graphene tailored with 15 wt.% of cobalt as an efficient catalyst for FTS. Their catalyst was able to work for up to 480h with a decrease of only 22% of the initial conversion efficiency. Karimi's research group moved a step forward in improving the stability of cobalt catalysts using graphene nanosheets improving the conversion in the same condition after 480 h of working time to 92% (Karimi et al. 2015). Luo and co-workers (Luo et al. 2020a) studied the effect of oxidation of the graphene support before the cobalt deposition. Pretreatment improved the catalyst activity as a consequence of the increment of defect sites, improvement of support-metal interaction and cobalt dispersion. Taghavi et al. (Taghavi et al. 2017) evaluated the roles of nitrogen-based residual groups on the surface of graphene on the selectivity, activity and stability of cobalt-containing catalyst during FTS with a gain of 4% of conversion compared with non-functionalized one. A more detailed insight on the effect of nitrogen groups on FTS was reported by Chernyak et al. (Chernyak et al. 2019). Authors studied GO and nitrogen doped graphene nanoflakes as supports for cobalt-containing FTS catalysts. Authors observed a decrease of activity with the localization of nitrogen sites on the nitrogen group's particle edges.

Wang and co-workers (Wang et al. 2019) developed a three-dimensional nitrogen doped graphene aerogel with an increased activity compared to classical graphene catalyst.

Zhao et al. (Zhao et al. 2013), using supported nanoparticles of iron oxide on pyrolytic GO, found a close relationship between carbon supports and iron poisoning during FTS.

CNTs have also been studied as in the research by Chernyak et al. (Chernyak et al. 2020). Authors decorated CNTs with up to 30 wt.% of cobalt nanoparticles showing that the catalyst sintering increased its stability, depleting the formation of wax.

A cheaper nanostructured carbon support was used by Asalieva et al. (Asalieva et al. 2020). In their work, exfoliated graphite was combined with cobalt pellets showing a favourable influence of the graphitic additive with a catalyst productivity of 455 Kg/h.

The use of carbon spheres was described by Dlamini et al. (Dlamini et al. 2020) through carbonization of melamine derivatives showing good performances by using just a 10 wt.% of cobalt. Also, hollow carbon spheres could be used with very close results by carbonization of poly(styrene) precursors (Phaahlamohlaka et al. 2020).

Another profitable route for FTS is represented by the combination of AC supported catalysts with inorganic additives such as zeolite (Karre et al. 2013), silica (Pei et al. 2015), zirconia (Wang et al. 2004, 2008) or $KMnO_4$ (Tian et al. 2017).

3.3 Hydrodesulfurization

Hydrodesulfurization (HDS) is a well established procedure for the treatment of high sulphur content oils with the removal of organic sulphur as H_2S under high pressure of hydrogen (Startsev 1995). This process is mediated by inorganic catalyst used as oxide such as CoO, NiO, MoQ and WO_3 or as sulfides (Vasudevan and Fierro 1996). The crucial point for the catalyst's preparation is the dispersion of such species in a very effective way. Nanostructured carbons have good performances during the synthesis of HDS catalysts. Wang and co-workers (Wang et al. 2015a) prepared a Ni-Mo catalyst supported on rGO/mesoporous $TiO2$ and tested it for dibenzothiophene HDS. This study on a model compound showed the incorporation of the acidic catalyst on the rGO, thus promoting H_2S formation. Hajjar et al. (Hajjar et al. 2017) treated naphtha by using a Co-Mo/graphene material produced through spray pyrolysis technique. Again, the catalyst acidity was a key feature with a total acidity of 5 mmol/g. The catalyst showed a desulfurization activity close to 100% that is quite remarkable considering the high sulphur content of the feedstock of up to 2.8 wt.%. Similarly, Saleh and Al-Hammadi (Saleh and Al-Hammadi 2021) use a Ni-Mo catalyst supported onto graphene to enhance the activity of Ni-MO species increasing the desulfurization from 84% to 99%. Also, Co-Mo catalysts were developed and used such as the one described by Xu and co-workers (Xu et al. 2019) and zeolite/graphene hybrids as the one described by Ali et al. (Ali and Saleh 2020).

CNTs have been used by Yang et al. (Yang et al. 2019) to support palladium based materials for the treatment of dibenzothiophenes solutions. Authors claimed an activity three times higher than the commercial NiMoS supported onto alumina Al_2O_3 catalyst and significantly higher than the other traditional metal silicide catalysts. Liu et al. (Liu et al. 2018) used a NiMoS supported onto CNTs, thus proving the better catalytic activity of these well-known species when deposited onto CNTs' surface. Co-Mo based catalysts were studied in combination with CNTs monolithic structures, proving the positive effect of these "nanoscale reactors" (Soghrati et al. 2012, 2014) that can compete with more traditional zeolite based catalysts.

Far from the frontiers on nanostructured carbon materials, CB has played a major role as support for HDS catalyst as overviewed by Gheek in 2007 paper "Carbon black composites—supports of HDS catalysts" (Gheek et al. 2007). Similarly, Wang and co-workers (Wang et al. 2017b) supported a Ni-Mo species on carbon through carbothermal procedures showing good performances on the HDS of model compounds. In this case, the catalyst calcination temperature is a crucial feature for an effective material's production as reported by Abubakar et al. (Abubakar et al. 2019). Authors explored a range of calcination temperatures up to 400°C showing that the catalyst produced at 300°C represents the best combination of physiochemical properties and catalytic activity. Similar considerations were reported by Farag et al. (Farag et al. 1999) that added a further step by comparing carbon support with alumina. Authors found that all the preparation parameters affected the nanostructured carbon supports and they must be tuned according to the different catalysts aimed.

Further studies showed a good enhancement of catalytic performances of carbon-inorganic hybrid support as in the case of carbon mixed with alumina (Nikulshin et al. 2014) or aluminosilicates (Aridi and Al-Daous 2009).

3.4 Hydroformylation

Hydroformylation is a catalytic procedure for the conversion of alkene into carbonyl compounds as sketched as follows:

$$R\diagup\!\!= + CO + H_2 \longrightarrow R\diagup\!\!\diagup\overset{O}{\underset{H}{\diagdown}} \qquad (3)$$

Traditionally, hydroformylation was performed by using phosphine or amine rhodium, ruthenium and cobalt complexes (Van Leeuwen and Claver 2002).

Cobalt was supported onto carbon nanofibers (CNFs) by Qiu et al. (Qiu et al. 2006) by simple reduction of a cobalt precursor. Cobalt nanoparticles supported onto CNFs showed a high homogeneous dispersion with spherical shape and an average diameter around 14 nm. This catalyst displayed a remarkable activity and regioselectivity for 1-octene hydroformylation with conversion of up to 70%.

Zhang and co-workers (Zhang et al. 1999) described the tailoring of CNTs surface with phosphine rhodium species used for the hydroformylation of 1-propene. Authors compared the CNTs supported catalyst with the same rhodium complexes supported onto AC and silica. Authors claimed a superior activity of CNTs based catalysts and a very high regioselectivity to t-butylaldehyde formation. The reason was due to the nanosized CNTs tubular together with their hydrophobic surface made by six membered aromatic rings.

Similar results were described by Tan and co-workers (Tan et al. 2017) using rGO as support during the terminal olefins hydroformylation in the presence of phosphine-based ligand and rhodium. rGO is a very promising support for this process as shown by Tan et al. (Tan et al. 2016) due to its ability to outperform even CNTs supported rhodium catalysts.

3.5 Cross coupling reactions

A cross-coupling reaction is a process where two species are combined together with the formation of a new carbon-heteroatom or carbon–carbon bond promoted by a metal catalyst (Miyaura and Buchwald 2002). In the next subsections, we will describe the carbon–carbon bond formation through mechanisms as the one summarized in Figure 5.

The formation of new carbon bonds through cross-coupling represents a key tool for synthetic chemistry and the development of new efficient catalysts is an open challenge. Heterogeneous catalysts are far less performing than homogeneous one (Pagliaro et al. 2012) but their high stability along multiple catalytic processes counterbalance the inferior activity. Nanocarbons are widely used in cross-coupling processes due to the versatility of surface tailoring that lead to well-exposed and anchored metal particles able to catalyse a wide range of reactions as reported by Cornelio and co-workers (Cornelio et al. 2013). In the following subsections, we are reporting the main achievements related to the use of heterogeneous catalysts applied in the cross-coupling reactions.

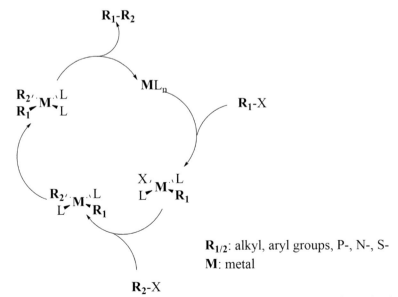

Figure 5. Schematic pathway of cross-coupling reaction for the formation of new carbon-carbon bonds.

3.5.1 Suzuki-miyaura reactions

Suzuki-Miyaura reaction is a cross-coupling reaction where the coupling species are a boronic acid derivative and an organohalide catalysed by a Pd(0) specie as summarized in Figure 6.

A wide diffusion of this synthetic approach has led to the development of various palladium supported onto carbon used for the synthesis of several biaryl and heterocyclic compounds (Maluenda and Navarro 2015, Beletskaya et al. 2019). It could be used for various non-aryl organic moieties coupling such as alkenes, alkynes, or alkanes. The very first example of reaction of Suzuki mediate by a palladium supported onto carbon was described by Marck et al. (Marck et al. 1994) without additional phosphine ligands in aqueous solvent.

In aqueous media, Diler al. (Diler et al. 2020) achieved good results by using mono dispersed palladium nanoparticles supported onto GO. Similarly, Hemmati et al. (Hemmati et al. 2019) supported palladium nanoparticles on metformin modified producing high recyclable catalysts.

The introduction of nitrogen atoms and functionalities on graphene supports has been proved as a very powerful way to improve the catalyst activity as described by many authors (Ibrahim et al. 2020, Liu et al. 2020a, Movahed et al. 2014, Rana et al. 2015, Zarnegaryan et al. 2019) due to the enhancement of basicity of the catalyst itself. Other authors explored more exotic ways to produce catalysts such as the biosynthesis (Salehi et al. 2019) or the production of graphene nanocages (Wang et al. 2014). Tran et al. (Tran et al. 2018) wisely introduced benzene-1,4-diboronic acid as an intercalating agent in a GO supported palladium catalyst for improving activity and durability. Huang et al. (Huang et al. 2018) used a mixture of 3-D structured amine ionic liquid functionalized graphene aerogel to produce tough and easily recoverable, highly recyclable, and stable catalysts.

Another attractive approach is represented by the production of magnetically recoverable catalyst by the addition of Fe_3O_4 nanoparticle to more standard catalysts such as palladium onto graphene materials (Elazab et al. 2015, Feng et al. 2014, Hoseini et al. 2015, Niakan et al. 2020, Rafiee et al. 2019).

Furthermore, other metals' activity was explored as doping agent as in the case of Ni_2O_3 (Nie et al. 2014) or as self-active one as for gold nanoparticles (Candu et al. 2017, Mondal et al. 2015, Thomas et al. 2017) or copper complexes (Ansari and Bhat 2019, Anuma et al. 2019, Kumar et al. 2019).

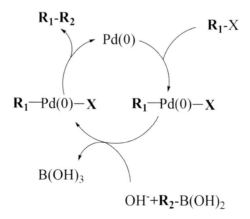

R-X: organic halide

Figure 6. Reaction pathway of Suzuki-Miyaura reaction.

Palladium supported on CNTs was also deeply used in Suzuki couplings (Corma et al. 2005). There are plenty of reports about the use of palladium supported by MWCNTs (Navidi et al. 2013, Sullivan et al. 2009) and SWCNTs (Veisi et al. 2014, Yang et al. 2015).

Also, CNTs surface tailoring with thiols (Lee et al. 2017, Veisi et al. 2019) or amine (Ghorbani-Vaghei et al. 2015) groups has become a well-established practice to tune the solubility and enhance the activity of those catalysts as well as the decoration with magnetic nanoparticles.

3.5.2 Heck reactions

The Heck reaction is one of the most powerful and diffused approaches to couple alkenes with organic moieties containing a halide, a triflate, or a diazonium functionality as shown in the Figure 7.

The first example of palladium supported on to carbon applied in a Heck reaction was the pioneering study by Julia and co-workers (Julia et al. 1973) in 1973. Authors performed a reaction between styrene and aromatic halides with yields that reached 62%. Köhler et al. (Köhler et al. 2002) reported a detailed investigation on all parameters that affect the Heck mechanism, showing the simultaneous effect of the palladium dispersion, oxidation and the conditions of catalyst preparation (impregnation method or pretreatment conditions).

GO was used by Wang and co-workers (Wang et al. 2015b) for palladium nanoparticles' synthesis by reduction of a palladium salt precursor showing good catalytic activity, with neglectable metal leaching and high recyclability. Also, palladium complexes could be anchored to graphene for the Heck reaction between styrene or n-butyl acrylate with plenty of substituted aryl bromides as reported by Fernández-García et al. (Fernández-García et al. 2016). Several authors enhanced the activity of Heck's catalysts by a further modification of GO such as polymer grafting (Kozur et al. 2019). Also, for Heck's catalyst, the combination with ferrites represents a good tool to increment the recyclability as proved by the study of Elazab et al. (Elazab et al. 2015). Authors used ferrite

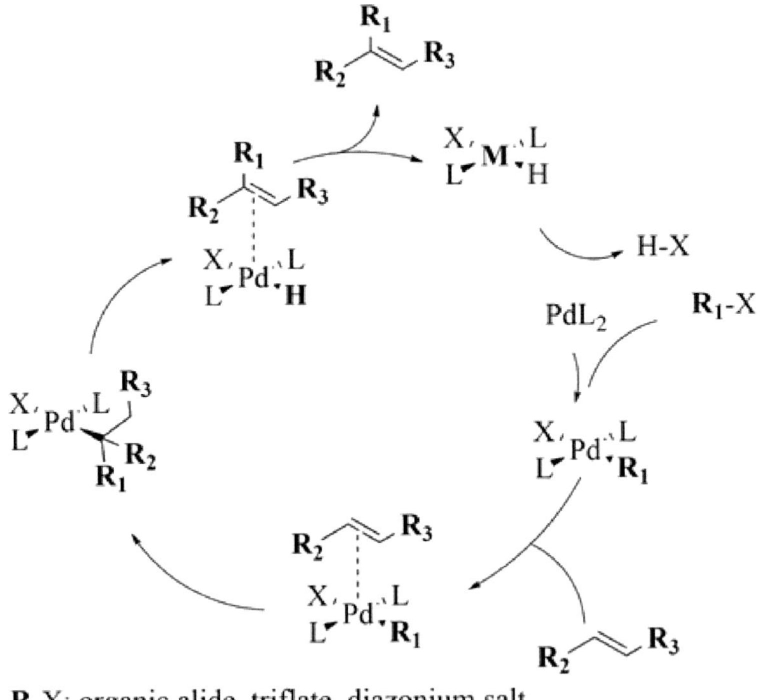

R-X: organic alide, triflate, diazonium salt

Figure 7. Reaction pathway of Heck reaction.

particles of around 16 nm for tailoring a palladium supported onto GO catalysts showing remarkable catalytic activity and recoverability. Similar results were obtained by using a magnetic catalyst in a eutectic solvent by Niakan et al. (Niakan et al. 2021).

A relevant study was reported by Janowska and co-workers (Janowska et al. 2011) that supported palladium nanoparticles on vertically aligned MWCNTs for Heck reactions induced by microwaves heating. The microwave irradiation combined with high loss factor of MWCTNs boosted the reaction kinetics without compromising recyclability and activity. Interestingly, Sun et al. (Sun et al. 2013) proved that defective CNTs allowed CNTs-palladium interaction, avoiding the increment of palladium particle size. Another factor that enhances the CNTs based catalysts activity during the Heck reaction is the presence of nitrogen sites. Cao et al. (Cao et al. 2020) proved nitrogen doped CNTs are a better support for loading active palladium particles. Also, grafting of peptides (Afshari et al. 2020), poly(amidoamines) (Nabid et al. 2011), poly (lactic acid) (Neelgund and Oki 2011) or cyclodextrin nanosponges (Sadjadi et al. 2018) could be better ways to improve the usability of the CNT catalysts without compromising their activity.

Another interesting approach was reported by Zhu and co-workers (Zhu et al. 2009) based on the use of palladium catalyst supported onto carbon nanofibers. Authors claimed a high activity and stability and a low leaching in multi-cycles with a palladium loading of up to 5 wt.%.

3.5.3 Sonogashira reactions

Sonogashira reaction is a cross-coupling reaction based on the use of palladium catalyst and a copper co-catalyst to form a carbon–carbon bond between an aryl or vinyl halide and a terminal alkyne as summarized in Figure 8.

De la Rosa described the palladium supported on charcoal as catalyst for Sonogashira reaction in 1990 using several aryl bromide with excellent results (De La Rosa et al. 1990). Later in 2003, Novák et al. (Novák et al. 2003) reported the very same study but with the addition of phosphine ligands to improve the conversion.

Lee et al. (Lee et al. 2013) reported a comprehensive study on the palladium nanoparticles supported onto GO as effective catalysts for Sonogashira reactions. The catalyst was prepared by reduction of GO impregnated with Pd(OAc)$_2$ at 100ºC under hydrogen atmosphere producing palladium nanoparticles of 2 nm of diameter supported onto rGO. These materials showed a good activity related to the efficient dispersion of nanoparticles on rGO matrix. Feng and co-workers

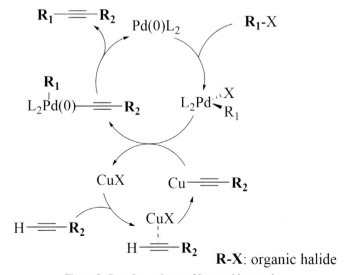

Figure 8. Reaction pathway of Sonogashira reaction.

(Feng et al. 2014) introduced a bimetallic catalyst on graphene by adding cobalt to a more traditional palladium based catalyst. Authors claimed an improvement of catalytic activity for a Pd/Co ratio close to 1. Naimi and Kiani (Naeimi and Kiani 2019) proved the ability of cheap metals such as nickel to catalyse the Sonogashira reaction. Rohani et al. (Rohani et al. 2019) described an efficient cobalt-aluminium catalyst supported onto graphene tailored with amine groups proving the beneficial effect of nitrogen and sulphur doping. In the case of Sonogashira reactions, it is also possible to combine traditional palladium catalyst supported onto nanostructured carbon with ferrites to improve the recyclability (Niakan et al. 2021).

3.6 Immobilized enzyme on to nanostructured carbon supports

Immobilized enzymes are one of the cutting-edge frontiers of catalysis due to the ultra-high efficiency and selectivity of these materials.

Recently, graphene and graphene-like nanomaterials have been largely used as supporting matrices for the enzymatic arrays immobilization targeting bio-catalytic applications (Adeel et al. 2018). The combination of mechanical and electrochemical properties with the functional surface residues' tunability has genereted a great interest (Zhang et al. 2010). Xue and co-workers (Xue et al. 2012) immobilized monomeric hemin on graphene. These bio catalysts showed a catalytic activity of up to 10 times better than that of the hemin immobilized into a hydrogel and of up to 100 times better than enzyme. Huang et al. (Huang et al. 2011) immobilized haemoglobin onto Go for catalysing a peroxidatic reaction in organic solvent benzene-1,2,3-triol. Gong et al. (Gong et al. 2017) further extended the combination of graphene derivatives with enzymes by developing a real nanoscale bioreactor using immobilizing naringinase on graphene. These nano-architectured reactors were able to work in flow condition for the production of isoquercitrin.

Similarly, Xie et al. (Xie et al. 2020) produced a micro reactor system immobilizing a lipase on 4-(1-Pyrenyl)butanoic acid functionalized monolithic nickel-CNTs foam improving enzyme efficiency and long-term catalyst stability. Choi and co-workers (Choi et al. 2015) used the same approach to develop an enzyme modified electrode based on CNTs/Nafion composites for biofuel cell purposes.

3.7 Nanocarbon supported catalysts for electrochemical oxidation

Nanostructured carbon materials are not the best choice as support for wet oxidation due to the reactivity of the carbon matrix but they achieved unreachable results during the electrooxidation process. This was mainly due to the high conductivity of carbon supports that actively play a major role during electrochemical reactions.

In the field of electrochemical oxidation process, there are two main sub-field represented by environmental remediation through organic pollutants degradation and energy storage through fuel cells technologies. In the next subsections, we are summarizing both of them with the most remarkable achievements reported in the scientific literature.

3.7.1 Electrochemical oxidation: Environmental remediation through organic pollutants degradation

Electrochemical oxidation of organic pollutants is a degradative process occurring to organic matter in watery media. This process is due to the *in situ* electro-generated oxygen radicals, as shown in Figure 9, acting as strong oxidant agents for organic substrates.

During the electrooxidation process, metallic active sites are converted into hydroxylic radical species. These last ones are highly unstable and very reactive leading to degradation of organic

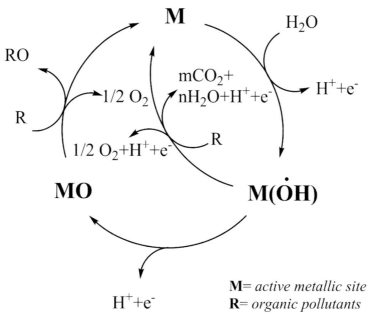

Figure 9. Schematic reaction pathway occurring during electrooxidation of organic (R) pollutants.

matters into carbon dioxide with the formation of metal oxides. Metal oxide can further react with organic matter promoting its partial oxidation.

The more relevant drawback of electrooxidation is the great energy demand (Anglada et al. 2009), but it is compensated by the wide range of applications including sanification of wastewater from chemical (Canizares et al. 2006, Han et al. 2008, Cho et al. 2008, El-Ashtoukhy et al. 2009, Rajkumar and Palanivelu 2004), use in textile (Vlyssides et al. 2000, Chennah et al. 2019) and food (Werkneh et al. 2019, Piya-Areetham et al. 2006) industry, together with the treatment of urban wastewater streams (Rodrigo et al. 2010).

Actually, electrooxidation is performed using expensive metal doped electrodes containing platinum, palladium, iridium or rhodium (Comninellis et al. 2008). This is due to the electrochemical potential required for oxygen evolution (around 1.7 V) together with the great stability (i.e., resistance to corrosion, formation of passivation layers) and the high electrical conductivity required.

Graphene is a perfect case of study for this aim and it has been widely tested. Zhang and co-workers (Zhang et al. 2018) produced a platinum tailored graphene electrode to remove acetaminophene. Similarly, Song et al. (Song et al. 2017) degraded chlorophenol by doped graphene with palladium while Duan et al. (Duan et al. 2017) used lead oxide.

CNTs were used by Jian et al. (Jiang et al. 2014) combined with iron and titanium to degrade nitrophenol with high efficiency. Similar bimetallic doped CNTs electrodes were used by Ferreira et al. (Ferreira et al. 2015) for the degradation of aniline and by Duan and co-workers (Duan et al. 2020) for degradation of emerging pharmaceutical pollutants.

On large scale applications, ACs have plenty of use in purification of wastewater through the electrooxidation process as reported by many authors (Duan et al. 2015, Martínez-Huitle and Panizza 2018, Garcia et al. 2017).

3.7.2 Electrochemical oxidation: Energy storage

A fuel cell is an electrochemical cell able to produce electrical energy by oxidation of a fuel by an oxidizing agent such as oxygen or air into electricity through a pair of redox reactions. Initially fuel cells required hydrogen as fuel and this represented a huge limitation to their spread. During the past decades, plenty of different feedstocks were implemented ranging from alcohols (i.e., methanol, ethanol, glycols) to formic acid trying to circumvent the necessity for hydrogen. The use of these unconventional alimentations has pushed for the production of highly efficient anodic materials.

Methanol direct fuel cells have gained a great attention due to the possibility of methanol production from renewable feedstock. Anyhow, its use required the development of well-architectured anodes. Graphene was efficiently used to create anodic materials by supporting bimetallic catalysts based on silver/palladium (Ghiabi et al. 2018), ruthenium/palladium (Awasthi and Singh 2013), gold/palladium (Luo et al. 2020b) and platinum/palladium nanostructures (Arukula et al. 2019, Li et al. 2014b, 2014).

Liu and co-workers (Liu et al. 2015b) proved that palladium itself could be used to decorate graphene aerogels and applied them in direct methanol fuel cells. A further improvement can be achieved by doping the graphene with nitrogen atoms. As enlightened by Fang et al. (Fang et al. 2020) for rGO and by Ren et al. (Ren et al. 2020) for neat graphene, nitrogen doping prevents metals clusters agglomeration avoiding the deterioration of their electrocatalytic performances. Additionally, Zhu et al. (Zhu et al. 2017) proved that cheaper metals such as nickel could work as replacement of palladium with comparable performances.

Ethanol fuel cells are also based on the palladium catalysts as reported by Kabir et al. (Kabir et al. 2016). Authors assembled a 3-D graphene structure loaded with 30 wt.% of palladium reaching a peak current density of 1568 Ag that outperformed commercial palladium and platinum clusters supported onto AC. Hydrogenated graphene was used by Xiao and co-workers (Xiao et al. 2019) in alkaline media and outperformed the analogue graphene material. Similar to methanol direct fuel cells, nitrogen doping enhances the performances as enlightened by Wang and co-workers (Wang et al. 2020). Multimetallic catalysts containing palladium and silver (Douk et al. 2019), gold (Karuppasamy et al. 2018), platinum/nickel (Ma et al. 2015), cobalt (Rostami et al. 2015), iron (Wei et al. 2018) and lead (Wu et al. 2015) could also work quite well. Surprisingly, nickel nanoparticles supported onto graphene showed a total operation time of up 104 days (Wu et al. 2021) that was quite higher than previously reported research.

Similar to methanol and ethanol direct fuel cells, formic acid direct fuels cells can be implemented by using nanostructured palladium supported graphene (Chen et al. 2015, Liu et al. 2014) and graphene mixed with cheaper carbon support such as carbon black (Lv et al. 2017). Formic acid fuel cells were implemented by using CNTs anodes decorated with cobalt/silver/palladium clusters (Er et al. 2020), antimony (Qu et al. 2014a) or indium tin oxide nanoparticles (Qu et al. 2014b).

Ethylene glycol can also feed fuel cells as reported by Kannan and co-workers (Kannan et al. 2015). Authors decorated graphene with palladium nanoparticles and manganese oxides reaching an anodic peak current density of 3.8 mA/cm^2. Li et al. (Li et al. 2019a) doped graphene with nitrogen and tailored it with palladium/silver nanoparticles reaching a current density of 4208.7 mA/mg.

4. Conclusions and future perspectives

As reported in this chapter, nanostructured carbons have played a crucial role in the production of heterogeneous catalysts for all industrially relevant processes.

The features of nano and nanostructured species are a cutting-edge resource for the realization of new high performance materials that can open the way to innovative and game-changing approaches. Nonetheless, carbon black and AC are a cheaper solution and they are generally preferred for industrial applications. A relevant exception is represented by the electrooxidation process where the performances of graphene and CNTs cannot be reached otherwise.

Balancing performances and affordability of catalysts still remain the starting point of all decisional processes and carbon-based species are struggling against zeolites and silica supported catalysts. We firmly believe the high tunability of nanocarbons represents the real advantage together with their electrical properties. Those two factors are the reasons for nanocarbon competitiveness leading to their bright future.

References

Abubakar, Umar Cheche, Khalid R. Alhooshani, Sagir Adamu, Jameel Al Thagfi and Tawfik A. Saleh. 2019. The effect of calcination temperature on the activity of hydrodesulfurization catalysts supported on mesoporous activated carbon. *Journal of Cleaner Production*, 211: 1567–1575.

Adeel, Muhammad, Muhammad Bilal, Tahir Rasheed, Ashutosh Sharma and Hafiz M.N. Iqbal. 2018. Graphene and graphene oxide: Functionalization and nano-bio-catalytic system for enzyme immobilization and biotechnological perspective. *International Journal of Biological Macromolecules*, 120: 1430–1440.

Afshari, Ronak, Seyyed Emad Hooshmand, Mojtaba Atharnezhad and Ahmad Shaabani. 2020. An insight into the novel covalent functionalization of multi-wall carbon nanotubes with pseudopeptide backbones for palladium nanoparticles immobilization: A versatile catalyst towards diverse cross-coupling reactions in bio-based solvents. *Polyhedron*, 175: 114238.

Ali, Islam and Tawfik A. Saleh. 2020. Zeolite-graphene composite as support for molybdenum-based catalysts and their hydrodesulfurization performance. *Applied Catalysis A: General*, 598: 117542.

Anglada, Angela, Ane Urtiaga and Inmaculada Ortiz. 2009. Contributions of electrochemical oxidation to waste-water treatment: fundamentals and review of applications. *Journal of Chemical Technology & Biotechnology*, 84(12): 1747–1755.

Ansari, Rasheeda M. and Badekai Ramachandra Bhat. 2019. Copper (II) Schiff base-graphene oxide composite as an efficient catalyst for Suzuki-Miyaura reaction. *Chemical Physics*, 517: 155–160.

Anuma, Saroja, Praveen Mishra and Badekai Ramachandra Bhat. 2019. Copper complex with N-,O- architecture grafted graphene oxide nanosheet as a heterogeneous catalyst for Suzuki cross coupling reaction. *Journal of the Taiwan Institute of Chemical Engineers*, 95: 643–651.

Araújo, Mariana P., Soares, O.S.G.P., Fernandes, A.J.S., Pereira, M.F.R. and Freire, C. 2017. Tuning the surface chemistry of graphene flakes: New strategies for selective oxidation. *RSC Advances*, 7(23): 14290–14301.

Aridi, Toufic N. and Mohammed A. Al-Daous. 2009. HDS of 4,6-dimethyldibenzothiophene over MoS2 catalysts supported on macroporous carbon coated with aluminosilicate nanoparticles. *Applied Catalysis A: General*, 359(1): 180–187.

Arora, Neha and Sharma, N.N. 2014. Arc discharge synthesis of carbon nanotubes: Comprehensive review. *Diamond and Related Materials*, 50: 135–150.

Arukula, Ravi, Mohanraj Vinothkannan, Ae Rhan Kim and Dong Jin Yoo. 2019. Cumulative effect of bimetallic alloy, conductive polymer and graphene toward electrooxidation of methanol: An efficient anode catalyst for direct methanol fuel cells. *Journal of Alloys and Compounds*, 771: 477–488.

Asalieva, Ekaterina, Lilia Sineva, Svetlana Sinichkina, Igor Solomonik, Kirill Gryaznov, Ekaterina Pushina, Ekaterina Kulchakovskaya, Andrei Gorshkov, Boris Kulnitskiy, Danila Ovsyannikov, Sergey Zholudev and Vladimir Mordkovich. 2020. Exfoliated graphite as a heat-conductive frame for a new pelletized Fischer–Tropsch synthesis catalyst. *Applied Catalysis A: General*, 601: 117639.

Awasthi, R. and Singh, R.N. 2013. Graphene-supported Pd–Ru nanoparticles with superior methanol electrooxidation activity. *Carbon*, 51: 282–289.

Bandaru, Prabhakar R. 2007. Electrical properties and applications of carbon nanotube structures. *Journal of Nanoscience and Nanotechnology*, 7(4-5): 1239–1267.

Beletskaya, Irina P, Francisco Alonso and Vladimir Tyurin. 2019. The Suzuki-Miyaura reaction after the Nobel prize. *Coordination Chemistry Reviews*, 385: 137–173.

Bouchelta, Chafia, Mohamed Salah Medjram, Odile Bertrand and Jean-Pierre Bellat. 2008. Preparation and characterization of activated carbon from date stones by physical activation with steam. *Journal of Analytical and Applied Pyrolysis*, 82(1): 70–77.

Brisebois, P.P. and Mohamed, Siaj. 2020. Harvesting graphene oxide–years 1859 to 2019: A review of its structure, synthesis, properties and exfoliation. *Journal of Materials Chemistry C*, 8(5): 1517–1547.

Brodie, Benjamin Collins. 1859. XIII. On the atomic weight of graphite. *Philosophical Transactions of the Royal Society of London*, (149): 249–259.

Candu, Natalia, Amarajothi Dhakshinamoorthy, Nicoleta Apostol, Cristian Teodorescu, Avelino Corma, Hermenegildo Garcia, and Vasile I. Parvulescu. 2017. Oriented Au nanoplatelets on graphene promote Suzuki-Miyaura coupling with higher efficiency and different reactivity pattern than supported palladium. *Journal of Catalysis*, 352: 59–66.

Canizares, P., Paz, R., Lobato, J., Sáez, C. and Rodrigo, M.A. 2006. Electrochemical treatment of the effluent of a fine chemical manufacturing plant. *Journal of Hazardous Materials*, 138(1): 173–181.

Cao, Ning, Mao Fei Ran, Yanyan Feng, Wei Chu, Chengfa Jiang, Wenjing Sun and Congmei Chen. 2020. Density functional theory study of N-doping effect on the stability and activity of Pd/NCNT catalysts for heck reaction. *Applied Surface Science*, 506: 144960.

Cao, Yonghai, Bo Li, Guoyu Zhong Yuhang, Li Hongjuan, Wang Hao and Yu Feng Peng. 2018. Catalytic wet air oxidation of phenol over carbon nanotubes: Synergistic effect of carboxyl groups and edge carbons. *Carbon*, 133: 464–473.

Chen, Min and Yangchuan Xing. 2005. Polymer-mediated synthesis of highly dispersed Pt nanoparticles on carbon black. *Langmuir*, 21(20): 9334–9338.

Chen, Qing-Song, Zhong-Ning Xu, Si-Yan Peng, Yu-Min Chen, Dong-Mei Lv, Zhi-Qiao Wang, Jing Sun and Guo-Cong Guo. 2015. One-step electrochemical synthesis of preferentially oriented (111) Pd nanocrystals supported on graphene nanoplatelets for formic acid electrooxidation. *Journal of Power Sources*, 282: 471–478.

Chennah, A., Anfar, Z., Amaterz, E., Taoufyq, A., Bakiz, B., Bazzi, L., Guinneton, F. and Benlhachemi, A. 2019. Ultrasound-assisted electro-oxidation of Methylene blue dye using new Zn3 (PO4) 2 based electrode prepared by electro-deposition. *Materials Today: Proceedings*, 22: 32–34.

Chernyak, S.A., Stolbov, D.N., Ivanov, A.S., Klokov, S.V., Egorova, T.B., Maslakov, K.I., Eliseev, O.L., Maximov, V.V., Savilov, S.V. and Lunin, V.V. 2019. Effect of type and localization of nitrogen in graphene nanoflake support on structure and catalytic performance of Co-based Fischer-Tropsch catalysts. *Catalysis Today*, 357: 193–202.

Chernyak, Sergei, Alexander Burtsev, Sergey Maksimov, Stepan Kupreenko, Konstantin Maslakov and Serguei Savilov. 2020. Structural evolution, stability, deactivation and regeneration of Fischer-Tropsch cobalt-based catalysts supported on carbon nanotubes. *Applied Catalysis A: General*, 603: 117741.

Cho, Seung-Hee, Hong-Joo Lee and Seung-Hyeon Moon. 2008. Integrated electroenzymatic and electrochemical treatment of petrochemical wastewater using a pilot scale membraneless system. *Process Biochemistry*, 43(12): 1371–1376.

Choi, Sung Deuk, Jin Ho Choi, Young Ho Kim, Sung Yeol, Kima Prabhat, Dwivedi Ashutosh Sharma, Sanket Goeld Gyu and Man Kim. 2015. Enzyme immobilization on microelectrode arrays of CNT/Nafion nanocomposites fabricated using hydrogel microstencils. *Microelectronic Engineering*, 141: 193–197.

Comninellis, Christos, Agnieszka Kapalka, Sixto Malato, Simon A. Parsons, Ioannis Poulios and Dionissios Mantzavinos. 2008. Advanced oxidation processes for water treatment: advances and trends for R&D. *Journal of Chemical Technology & Biotechnology: International Research in Process, Environmental & Clean Technology*, 83(6): 769–776.

Corma, Avelino, Hermenegildo Garcia and Antonio Leyva. 2005. Catalytic activity of palladium supported on single wall carbon nanotubes compared to palladium supported on activated carbon: Study of the Heck and Suzuki couplings, aerobic alcohol oxidation and selective hydrogenation. *Journal of Molecular Catalysis A: Chemical*, 230(1): 97–105.

Cornelio, Benedetta, Graham A. Rance, Marie Laronze-Cochard, Antonella Fontana, Janos Sapi and Andrei N. Khlobystov. 2013. Palladium nanoparticles on carbon nanotubes as catalysts of cross-coupling reactions. *Journal of Materials Chemistry A*, 1(31): 8737–8744.

Coville, Neil J., Sabelo D. Mhlanga, Edward N. Nxumalo and Ahmed Shaikjee. 2011. A review of shaped carbon nanomaterials. *South African Journal of Science*, 107(3-4): 01–15.

De La Rosa, Martha, A., Esperanza Velarde and Angel Guzmán. 1990. Cross-coupling reactions of monosubstituted acetylenes and aryl halides catalyzed by palladium on charcoal. *Synthetic Communications*, 20(13): 2059–2064.

Diaz, E., Mohedano, A.F., Calvo, L., Gilarranz, M.A., Casas, J.A. and Rodríguez, J.J. 2007. Hydrogenation of phenol in aqueous phase with palladium on activated carbon catalysts. *Chemical Engineering Journal*, 131(1-3): 65–71.

Diler, Fatma, Hakan Burhan, Hayriye Genc, Esra Kuyuldar, Mustafa Zengin, Kemal Cellat and Fatih Sen. 2020. Efficient preparation and application of monodisperse palladium loaded graphene oxide as a reusable and effective heterogeneous catalyst for suzuki cross-coupling reaction. *Journal of Molecular Liquids*, 298: 111967.

Dlamini, Mbongiseni W., Tumelo N. Phaahlamohlaka, David O. Kumi, Roy Forbes, Linda L. Jewell and Neil J. Coville. 2020. Post doped nitrogen-decorated hollow carbon spheres as a support for Co Fischer-Tropsch catalysts. *Catalysis Today*, 342: 99–110.

Dongil, A.B., Bachiller-Baeza, B., Guerrero-Ruiz, A. and Rodríguez-Ramos, I. 2011. Chemoselective hydrogenation of cinnamaldehyde: A comparison of the immobilization of Ru–phosphine complex on graphite oxide and on graphitic surfaces. *Journal of Catalysis*, 282(2): 299–309.

Douk, Abdollatif Shafaei, Hamideh Saravani, Majid Farsadrooh and Meissam Noroozifar. 2019. An environmentally friendly one-pot synthesis method by the ultrasound assistance for the decoration of ultrasmall Pd-Ag NPs on graphene as high active anode catalyst towards ethanol oxidation. *Ultrasonics Sonochemistry*, 58: 104616.

Dresselhaus, M.S., Jorio, A. and Saito, R. 2010. Characterizing graphene, graphite, and carbon nanotubes by Raman spectroscopy. *Annu. Rev. Condens. Matter Phys.*, 1(1): 89–108.

Duan, Feng, Yuping Li, Hongbin Cao, Yi Wang, John C. Crittenden and Yi Zhang. 2015. Activated carbon electrodes: Electrochemical oxidation coupled with desalination for wastewater treatment. *Chemosphere*, 125: 205–211.

Duan, Pingzhou, Shiheng Gao, Jiawei Lei, Xiang Li and Xiang Hu. 2020. Electrochemical oxidation of ceftazidime with graphite/CNT-Ce/PbO2–Ce anode: Parameter optimization, toxicity analysis and degradation pathway. *Environmental Pollution*: 114436.

Duan, Xiaoyue, Cuimei Zhao, Wei Liu, Xuesong Zhao and Limin Chang. 2017. Fabrication of a novel PbO2 electrode with a graphene nanosheet interlayer for electrochemical oxidation of 2-chlorophenol. *Electrochimica Acta*, 240: 424–436.

Efremenko, Irena and Moshe Sheintuch. 1998. Quantum chemical study of small palladium clusters. *Surface Science*, 414(1): 148–158.

El-Ashtoukhy, E.-S.Z., Amin, N.K. and Abdelwahab, O. 2009. Treatment of paper mill effluents in a batch-stirred electrochemical tank reactor. *Chemical Engineering Journal*, 146(2): 205–210.

Elazab, Hany A., Ali, R. Siamaki, Sherif Moussa, B. Frank Gupton and M. Samy El-Shall. 2015. Highly efficient and magnetically recyclable graphene-supported Pd/Fe3O4 nanoparticle catalysts for Suzuki and Heck cross-coupling reactions. *Applied Catalysis A: General*, 491: 58–69.

Endo, Morinobu, Sumio Iijima and Mildred S. Dresselhaus. 2013. Elsevier Science Ltd, The Boulevard, Langford Lane, Kidlington, Oxford OX5 1GB, U.K. *Carbon Nanotubes*.

Er, Omer Faruk, Aykut Caglar, Berdan Ulas, Hilal Kivrak and Arif Kivrak. 2020. Novel carbon nanotube supported Co@ Ag@Pd formic acid electrooxidation catalysts prepared via sodium borohydride sequential reduction method. *Materials Chemistry and Physics*, 241: 122422.

Fang, Difan, Liming Yang, Guang Yang, Genping Yia, Yufa Feng, Penghui Shao, Hui Shia, Kai Yu, Deng You and Xubiao Luo. 2020. Electrodeposited graphene hybridized graphitic carbon nitride anchoring ultrafine palladium nanoparticles for remarkable methanol electrooxidation. *International Journal of Hydrogen Energy*, 45(41): 21483–21492.

Farag, Hamdy, Whitehurst, D.D., Kinya Sakanishi and Isao Mochida. 1999. Carbon versus alumina as a support for Co–Mo catalysts reactivity towards HDS of dibenzothiophenes and diesel fuel. *Catalysis Today*, 50(1): 9–17.

Feng, Yi-Si, Xin-Yan Lin, Jian Hao and Hua-Jian Xu. 2014. Pd–Co bimetallic nanoparticles supported on graphene as a highly active catalyst for Suzuki–Miyaura and Sonogashira cross-coupling reactions. *Tetrahedron*, 70(34): 5249–5253.

Fernández-García, L., Blanco, M., Blanco, C., Álvarez, P., Granda, M., Santamaría, R. and Menéndez, R. 2016. Graphene anchored palladium complex as efficient and recyclable catalyst in the Heck cross-coupling reaction. *Journal of Molecular Catalysis A: Chemical*, 416: 140–146.

Ferreira, M., Pinto, M.F., Neves, L.C., Fonseca, A.M., Soares, O.S.G.P., Órfão, J.J.M., Pereira, M.F.R., Figueiredo, J.L. and Parpot, P. 2015. Electrochemical oxidation of aniline at mono and bimetallic electrocatalysts supported on carbon nanotubes. *Chemical Engineering Journal*, 260: 309–315.

Garcia, Luane Ferreira, Ana Claudia, Rodrigues Siqueira, Germán Sanz Lobón, Jossano Saldanh Marcuzzo, Benevides Costa Pessel, Eduardo Mendez,Telma Alves Garcia and Ericde Souza Gil. 2017. Bio-electro oxidation of indigo carmine by using microporous activated carbon fiber felt as anode and bioreactor support. *Chemosphere*, 186: 519–526.

Gheek, P., S. Suppan, J. Trawczyński, A. Hynaux, C. Sayag and G. Djega-Mariadssou. 2007. Carbon black composites—supports of HDS catalysts. *Catalysis Today*, 119(1): 19–22.

Ghiabi, Caesar, Ali Ghaffarinejad, Hojjat Kazemi and Razieh Salahandish. 2018. *In situ*, one-step and co-electrodeposition of graphene supported dendritic and spherical nano-palladium-silver bimetallic catalyst on carbon cloth for electrooxidation of methanol in alkaline media. *Renewable Energy*, 126: 1085–1092.

Ghorbani-Vaghei, Ramin, Saba Hemmati, Majid Hashemi and Hojat Veisi. 2015. Diethylenetriamine-functionalized single-walled carbon nanotubes (SWCNTs) to immobilization palladium as a novel recyclable heterogeneous nanocatalyst for the Suzuki–Miyaura coupling reaction in aqueous media. *Comptes Rendus Chimie*, 18(6): 636–643.

Gong, A., Zhu, C.T., Xu, Y., Wang, F.Q., Wu, F.A. and Wang, J. 2017. Moving and unsinkable graphene sheets immobilized enzyme for microfluidic biocatalysis. *Scientific Reports*, 7(1): 4309.

Guex, L.G., Sacchi, B., Peuvot, K.F., Andersson, R.L., Pourrahimi, A.M., Ström, V., Farris, S. and Olsson, R.T. 2017. Experimental review: chemical reduction of graphene oxide (GO) to reduced graphene oxide (rGO) by aqueous chemistry. *Nanoscale*, 9(27): 9562–9571.

Hajjar, Zeinab, Mohammad Kazemeini, Alimorad Rashidi, Saeed Soltanali and Farzad Bahadoran. 2017. Naphtha HDS over Co-Mo/Graphene catalyst synthesized through the spray pyrolysis technique. *Journal of Analytical and Applied Pyrolysis*, 123: 144–151.

Hamzah, N., Mohd M.F. Yasin, Mohd M.Z. Yusop, Saat, A. and Mohd N.A. Subha. 2017. Rapid production of carbon nanotubes: A review on advancement in growth control and morphology manipulations of flame synthesis. *Journal of Materials Chemistry A*, 5(48): 25144–25170.

Han, Wei-Qing, Lian-Jun Wang, Xiu-Yun Sun and Jian-Sheng Li. 2008. Treatment of bactericide wastewater by combined process chemical coagulation, electrochemical oxidation and membrane bioreactor. *Journal of Hazardous Materials*, 151(2-3): 306–315.

Hemmati, Saba, Lida Mehrazin, Mozhgan Pirhayati and Hojat Veisi. 2019. Immobilization of palladium nanoparticles on Metformin-functionalized graphene oxide as a heterogeneous and recyclable nanocatalyst for Suzuki coupling reactions and reduction of 4-nitrophenol. *Polyhedron*, 158: 414–422.

Hola, Katerina, Yu Zhang, Yu Wang, Emmanuel P. Giannelis, Radek Zboril and Andrey L. Rogach. 2014. Carbon dots—Emerging light emitters for bioimaging, cancer therapy and optoelectronics. *Nano Today*, 9(5): 590–603.

Hoseini, S. Jafar, Vahid Heidari and Hasan Nasrabadi. 2015. Magnetic Pd/Fe3O4/reduced-graphene oxide nanohybrid as an efficient and recoverable catalyst for Suzuki–Miyaura coupling reaction in water. *Journal of Molecular Catalysis A: Chemical*, 396: 90–95.

Huang, Cancan, Hua Bai, Chun Li and Gaoquan Shi. 2011. A graphene oxide/hemoglobin composite hydrogel for enzymatic catalysis in organic solvents. *Chemical Communications*, 47(17): 4962–4964.

Huang, Yanli, Qiuli Wei, Yuanyuan Wang and Liyi Dai. 2018. Three-dimensional amine-terminated ionic liquid functionalized graphene/Pd composite aerogel as highly efficient and recyclable catalyst for the Suzuki cross-coupling reactions. *Carbon*, 136: 150–159.

Ibrahim, Amr Awad, Andrew Lin, Mina Shawky Adly and Samy M. El-Shall. 2020. Enhancement of the catalytic activity of Pd nanoparticles in Suzuki coupling by partial functionalization of the reduced graphene oxide support with p-phenylenediamine and benzidine. *Journal of Catalysis*, 385: 194–203.

James, Dustin K. and James M. Tour. 2012. The chemical synthesis of graphene nanoribbons—a tutorial review. *Macromolecular Chemistry and Physics*, 213(10-11): 1033–1050.

Janowska, Izabela, Kambiz Chizari, Jean-Hubert Olivier, Raymond Ziessel, Marc Jacques Ledoux and Cuong Pham-Huu. 2011. A new recyclable Pd catalyst supported on vertically aligned carbon nanotubes for microwaves-assisted Heck reactions. *Comptes Rendus Chimie*, 14(7): 663–670.

Jiang, Yonghai, Zhongxin Hu, Minghua Zhou, Lei Zhou and Beidou Xi. 2014. Efficient degradation of p-nitrophenol by electro-oxidation on Fe doped Ti/TiO2 nanotube/PbO2 anode. *Separation and Purification Technology*, 128: 67–71.

Julia, M., Duteil, M., Grard, C. and Kuntz, E. 1973. Condensation of Aryl halides with Olefins catalyzed by Palladium. *Bulletin De La Societe Chimique De France Partie Ii-Chimie Moleculaire Organique Et Biologique*, (9-10): 2791–2794.

Jung, A., Jess, A., Schubert, T. and Schütz, W. 2009. Performance of carbon nanomaterial (nanotubes and nanofibres) supported platinum and palladium catalysts for the hydrogenation of cinnamaldehyde and of 1-octyne. *Applied Catalysis A: General*, 362(1): 95–105.

Kabir, Sadia, Alexey Serov, Kateryna Artyushkova and Plamen Atanassov. 2016. Design of novel graphene materials as a support for palladium nanoparticles: Highly active catalysts towards ethanol electrooxidation. *Electrochimica Acta*, 203: 144–153.

Kannan, Ramanujam, Ae Rhan Kim, Kee Suk Nahm and Dong Jin Yoo. 2015. One-pot synthesis and electrocatalytic performance of Pd/MnOx/graphene nanocomposite for electrooxidation of ethylene glycol. *International Journal of Hydrogen Energy*, 40(35): 11960–11967.

Karimi, Saba, Ahmad Tavasoli, Yadollah Mortazavi and Ali Karimi. 2015. Cobalt supported on Graphene—A promising novel Fischer–Tropsch synthesis catalyst. *Applied Catalysis A: General*, 499: 188–196.

Karimi, S., Tavasoli, A., Mortazavi, Y. and Karimi, A. 2015. Enhancement of cobalt catalyst stability in Fischer–Tropsch synthesis using graphene nanosheets as catalyst support. *Chemical Engineering Research and Design*, 104: 713–722.

Karre, Avinashkumar V., Alaa Kababji, Edwin L. Kugler and Dady B. Dadyburjor. 2013. Effect of time on stream and temperature on upgraded products from Fischer–Tropsch synthesis when zeolite is added to iron-based activated-carbon-supported catalyst. *Catalysis Today*, 214: 82–89.

Karuppasamy, Lakshmanan, Chin-Yi Chen, Sambandam Anandan and Jerry J. Wu. 2018. Sonochemical fabrication of reduced graphene oxide supported Au nano dendrites for ethanol electrooxidation in alkaline medium. *Catalysis Today*, 307: 308–317.

Ko, Frank K. and Yuqin Wan. 2014. *Introduction to Nanofiber Materials*: Cambridge University Press.

Köhler, Klaus, Roland G. Heidenreich, Jürgen G.E. Krauter and Jörg Pietsch. 2002. Highly active palladium/activated carbon catalysts for Heck reactions: Correlation of activity, catalyst properties, and Pd leaching. *Chemistry–A European Journal*, 8(3): 622–631.

Kozur, Alexander, Laura Burk, Ralf Thomann, Pierre J. Lutz and Rolf Mülhaupt. 2019. Graphene oxide grafted with polyoxazoline as thermoresponsive support for facile catalyst recycling by reversible thermal switching between dispersion and sedimentation. *Polymer*, 178: 121553.

Kumar, Lolakshi Mahesh, Praveen Mishra and Badekai Ramachandra Bhat. 2019. Iron pincer complex and its graphene oxide composite as catalysts for Suzuki coupling reaction. *Journal of Saudi Chemical Society*, 23(3): 307–314.

Kumar, Mukul and Yoshinori Ando. 2010. Chemical vapor deposition of carbon nanotubes: a review on growth mechanism and mass production. *Journal of Nanoscience and Nanotechnology*, 10(6): 3739–3758.

Kyotani, Takashi, Takayuki Nagai, Sanjuro Inoue and Akira Tomita. 1997. Formation of new type of porous carbon by carbonization in zeolite nanochannels. *Chemistry of Materials*, 9(2): 609–615.

Lai, Linfei, Luwei Chen, Da Zhan, Li Sun, Jinping Liu, San Hu, Limb Che, Kok Poh, Zexiang Shen and Jianyi Lin. 2011. One-step synthesis of NH2-graphene from in situ graphene-oxide reduction and its improved electrochemical properties. *Carbon*, 49(10): 3250–3257.

Lavagna, Luca, Giuseppina Meligrana, Claudio Gerbaldi, Alberto Tagliaferro and Mattia Bartoli. 2020. Graphene and lithium-based battery electrodes: A review of recent literature. *Energies*, 13(18): 4867.

Lee, E.K., Park, S.A., Woo, H., Park, K.H., Kang, D.W., Lim, H. and Kim, Y.T. 2017. Platinum single atoms dispersed on carbon nanotubes as reusable catalyst for Suzuki coupling reaction. *Journal of Catalysis*, 352: 388–393.

Lee, H. Cheun, Wei-Wen Liu, Siang-Piao Chai, Abdul Rahman Mohamed, Chin Wei Lai, Cheng-Seong Khee C.H. Voon, Hashim, U. and Hidayah, N.M.S. 2016. Synthesis of single-layer graphene: A review of recent development. *Procedia Chemistry*, 19: 916–921.

Lee, Kyoung Hoon, Sang-Wook Han, Ki-Young Kwon and Joon B. Park. 2013. Systematic analysis of palladium–graphene nanocomposites and their catalytic applications in Sonogashira reaction. *Journal of Colloid and Interface Science*, 403: 127–133.

Lee, X.J., Hiew, B.Y.Z., Lai, K.C., Lee, L.Y., Gan, S., Thangalazhy-Gopakumar, S. and Rigby S. 2019. Review on graphene and its derivatives: Synthesis methods and potential industrial implementation. *Journal of the Taiwan Institute of Chemical Engineers*, 98: 163–180.

Lekawa-Raus, Agnieszka, Jeff Patmore, Lukasz Kurzepa, John Bulmer and Krzysztof Koziol. 2014. Electrical properties of carbon nanotube based fibers and their future use in electrical wiring. *Advanced Functional Materials*, 24(24): 3661–3682.

Lemes, Giovanni, David Sebastián, Elena Pastor and María J Lázaro. 2019. N-doped graphene catalysts with high nitrogen concentration for the oxygen reduction reaction. *Journal of Power Sources*, 438: 227036.

Lerf, Anton, Heyong He, Thomas Riedl, Michael Forster and Jacek Klinowski. 1997. 13C and 1H MAS NMR studies of graphite oxide and its chemically modified derivatives. *Solid State Ionics*, 101: 857–862.

Li, Jing, Peng Cheng Ma, Wing Sze Chow, Chi Kai To, Ben Zhong Tang and J.-K. Kim. 2007. Correlations between percolation threshold, dispersion state, and aspect ratio of carbon nanotubes. *Advanced Functional Materials*, 17(16): 3207–3215.

Li, Ling, Ge Gao, Jia Zheng, Xin Shi and Zhi Liu. 2019a. Three-dimensional graphene aerogel supported Ir nanocomposite as a highly efficient catalyst for chemoselective cinnamaldehyde hydrogenation. *Diamond and Related Materials*, 91: 272–282.

Li, Shan-Shan, Jing-Jing Lv, Yuan-Yuan Hu, Jie-Ning Zheng, Jian-Rong Chen, Ai-Jun Wang and Jiu-Ju Feng. 2014b. Facile synthesis of porous Pt–Pd nanospheres supported on reduced graphene oxide nanosheets for enhanced methanol electrooxidation. *Journal of Power Sources*, 247: 213–218.

Li, Shan-Shan, Jianyan Yu, Yuan-Yuan Hu, Ai-Jun Wang, Jian-Rong Chen, and Jiu-Ju Feng. 2014c. Simple synthesis of hollow Pt–Pd nanospheres supported on reduced graphene oxide for enhanced methanol electrooxidation. *Journal of Power Sources*, 254: 119–125.

Li, Zhao, Xu Zhao, Yixin Zhang and Jianjun Wu. 2019b. One-pot construction of N-doped graphene supported 3D PdAg nanoflower as efficient catalysts for ethylene glycol electrooxidation. *Colloids and Surfaces A: Physicochemical and Engineering Aspects*, 562: 409–415.

Lin, Bingyu, Kemei Wei, Jun Ni and Jianxin Lin. 2013. KOH activation of thermally modified carbon as a support of Ru catalysts for ammonia synthesis. *ChemCatChem*, 5(7): 1941–1947.

Lindlar, Helv. 1952. Ein neuer Katalysator für selektive Hydrierungen. *Helvetica Chimica Acta*, 35(2): 446–450.

Liu, Lei, Qing-Fang Deng, Yu-Ping Liu, Tie-Zhen Ren and Zhong-Yong Yuan. 2011. HNO3-activated mesoporous carbon catalyst for direct dehydrogenation of propane to propylene. *Catalysis Communications*, 16(1): 81–85.

Liu, Mingrui, Cheng Peng, Wenke Yang, Jiaojiao Guo, Yixiong Zheng, Peiqin Chen, Tingting Huang and Jing Xu. 2015. Pd nanoparticles supported on three-dimensional graphene aerogels as highly efficient catalysts for methanol electrooxidation. *Electrochimica Acta*, 178: 838–846.

Liu, Ning, NiNa Qiao, Feng-Shou Liu, ShaoHua Wang and Ying Liang. 2020. Bulky α-diimine palladium complexes supported graphene oxide as heterogeneous catalysts for Suzuki-Miyaura reaction. *Journal of Molecular Structure*, 1218: 128537.

Liu, Sen, Jingqi Tian, Lei Wang and Xuping Sun. 2011. A method for the production of reduced graphene oxide using benzylamine as a reducing and stabilizing agent and its subsequent decoration with Ag nanoparticles for enzymeless hydrogen peroxide detection. *Carbon*, 49(10): 3158–3164.

Liu, Sijia, Qiu Jin, Yuan Xu, Xiangchen Fang, Ning Liua, Jie Zhang, Xin Liang and Biaohua Chen. 2018. The synergistic effect of Ni promoter on Mo-S/CNT catalyst towards hydrodesulfurization and hydrogen evolution reactions. *Fuel*, 232: 36–44.

Liu, Xue-nan, Chao Deng, Ying Gao and Bing Wu. 2014. Preparation of graphene and graphene supported Pd catalysts for formic acid electrooxidation. *Journal of Fuel Chemistry and Technology*, 42(4): 476–480.

Luo, Mingsheng, Shuo Li, Zuoxing Di, Zhi Yang, Weichao Chou and Buchang Shi. 2020a. Fischer-Tropsch synthesis: Effect of nitric acid pretreatment on graphene-supported cobalt catalyst. *Applied Catalysis A: General*, 599: 117608.

Luo, Cong, Jianhua Yang, Jinling Li, Shijie, He, Bowen Meng, Tao Shao, Qiankun Zhang, Dongxia Zhang and Xibin Zhou. 2020b. Green synthesis of Au@N-CQDs@Pd core-shell nanoparticles for enhanced methanol electrooxidation. *Journal of Electroanalytical Chemistry*, 873: 114423.

Lv, M., Li, W., Liu, H., Wen, W., Dong, G., Liu, J. and Peng, K. 2017. Enhancement of the formic acid electrooxidation activity of palladium using graphene/carbon black binary carbon supports. *Chinese Journal of Catalysis*, 38(5): 939–947.

Ma, Hongfei, Tie Yu, Xiulian Pan and Xinhe Bao. 2017. Confinement effect of carbon nanotubes on the product distribution of selective hydrogenation of cinnamaldehyde. *Chinese Journal of Catalysis*, 38(8): 1315–1321.

Ma, J., Wang, J., Zhang, G., Fan, X., Zhang, G., Zhang, F. and Li, Y. 2015. Deoxyribonucleic acid-directed growth of well dispersed nickel–palladium–platinum nanoclusters on graphene as an efficient catalyst for ethanol electrooxidation. *Journal of Power Sources*, 278: 43–49.

Maluenda, Irene and Oscar Navarro. 2015. Recent developments in the Suzuki-Miyaura reaction: 2010–2014. *Molecules*, 20(5): 7528–7557.

Marck, Guy, Alois Villiger and Richard Buchecker. 1994. Aryl couplings with heterogeneous palladium catalysts. *Tetrahedron Letters*, 35(20): 3277–3280.

Martínez-Huitle, Carlos Alberto and Marco Panizza. 2018. Electrochemical oxidation of organic pollutants for wastewater treatment. *Current Opinion in Electrochemistry*, 11: 62–71.

Mintmire, John W., Brett I. Dunlap and Carter T. White. 1992. Are fullerene tubules metallic? *Physical Review Letters*, 68(5): 631.

Miyaura, Norio and Stephen L. Buchwald. 2002. *Cross-coupling Reactions: A Practical Guide*. Vol. 219: Springer-Verlag Berlin Heidelberg 2002. Printed in Germany.

Molina-Sabio, M., Gonzalez, M.T., Rodriguez-Reinoso, F. and Sepúlveda-Escribano, A. 1996. Effect of steam and carbon dioxide activation in the micropore size distribution of activated carbon. *Carbon*, 34(4): 505–509.

Mondal, Paramita, Noor Salam, Avijit Mondal, Kajari Ghosh, K. Tuhina and Sk Manirul Islam. 2015. A highly active recyclable gold–graphene nanocomposite material for oxidative esterification and Suzuki cross-coupling reactions in green pathway. *Journal of Colloid and Interface Science*, 459: 97–106.

Movahed, Siyavash Kazemi, Minoo Dabiri and Ayoob Bazgir. 2014. Palladium nanoparticle decorated high nitrogen-doped graphene with high catalytic activity for Suzuki–Miyaura and Ullmann-type coupling reactions in aqueous media. *Applied Catalysis A: General*, 488: 265–274.

Nabid, Mohammad Reza, Yasamin Bide and Seyed Jamal Tabatabaei Rezaei. 2011. Pd nanoparticles immobilized on PAMAM-grafted MWCNTs hybrid materials as new recyclable catalyst for Mizoraki–Heck cross-coupling reactions. *Applied Catalysis A: General*, 406(1): 124–132.

Naeimi, Hossein and Fatemeh Kiani. 2019. Functionalized graphene oxide anchored to Ni complex as an effective recyclable heterogeneous catalyst for Sonogashira coupling reactions. *Journal of Organometallic Chemistry*, 885: 65–72.

Narita, Akimitsu, Xiao-Ye Wang, Xinliang Feng and Klaus Müllen. 2015. New advances in nanographene chemistry. *Chemical Society Reviews*, 44(18): 6616–6643.

Navidi, Mozhgan, Nasrin Rezaei and Barahman Movassagh. 2013. Palladium(II)–Schiff base complex supported on multi-walled carbon nanotubes: A heterogeneous and reusable catalyst in the Suzuki–Miyaura and copper-free Sonogashira–Hagihara reactions. *Journal of Organometallic Chemistry*, 743: 63–69.

Neelgund, Gururaj M. and Aderemi Oki. 2011. Pd nanoparticles deposited on poly(lactic acid) grafted carbon nanotubes: Synthesis, characterization and application in Heck C–C coupling reaction. *Applied Catalysis A: General*, 399(1): 154–160.

Niakan, Mahsa, Majid Masteri-Farahani, Hemayat Shekaari and Sabah Karimi. 2020. Pd supported on clicked cellulose-modified magnetite-graphene oxide nanocomposite for C-C coupling reactions in deep eutectic solvent. *Carbohydrate Polymers*:117109.

Niakan, M., Masteri-Farahani, M., Shekaari, H. and Karimi, S. 2021. Pd supported on clicked cellulose-modified magnetite-graphene oxide nanocomposite for C-C coupling reactions in deep eutectic solvent. *Carbohydrate Polymers*, 251: 117109.

Nie, Renfeng, Juanjuan Shi, Weichen Du and Zhaoyin Hou. 2014. Ni2O3-around-Pd hybrid on graphene oxide: An efficient catalyst for ligand-free Suzuki–Miyaura coupling reaction. *Applied Catalysis A: General*, 473: 1–6.

Nikulshin, P.A., Salnikov, V.A., Mozhaev, A.V., Minaev, P.P., Kogan, V.M. and Pimerzin, A.A. 2014. Relationship between active phase morphology and catalytic properties of the carbon–alumina-supported Co(Ni)Mo catalysts in HDS and HYD reactions. *Journal of Catalysis*, 309: 386–396.

Nishimura, Shigeo. 2001. *Handbook of Heterogeneous Catalytic Hydrogenation for Organic Synthesis*: Wiley New York.

Novák, Zoltán, András Szabó, József Répási and András Kotschy. 2003. Sonogashira coupling of Aryl halides catalyzed by palladium on charcoal. *The Journal of Organic Chemistry*, 68(8): 3327–3329.

Oberhauser, Werner, Claudio Evangelisti, Ravindra P. Jumde, Rinaldo Psaro, Francesco Vizza, Manuela Bevilacqua, Jonathan Filippi, Bruno F. Machado and Philippe Serp. 2015. Platinum on carbonaceous supports for glycerol hydrogenolysis: Support effect. *Journal of Catalysis*, 325: 111–117.

Omacrsawa, Eiji. 2012. *Perspectives of fullerene nanotechnology*: Springer Science & Business Media.

Pagliaro, Mario, Valerica Pandarus, Rosaria Ciriminna, François Béland and Piera Demma Carà. 2012. Heterogeneous versus homogeneous palladium catalysts for cross-coupling reactions. *ChemCatChem.*, 4(4): 432–445.

Pagona, Georgia, Grigoris Mountrichas, Georgios Rotas, Nikolaos Karousis, Stergios Pispas and Nikos Tagmatarchis. 2009. Properties, applications and functionalisation of carbon nanohorns. *International Journal of Nanotechnology*, 6(1-2): 176–195.

Pei, Yanpeng, Yunjie Ding, Hejun Zhu and Hong Du. 2015. One-step production of C1–C18 alcohols via Fischer-Tropsch reaction over activated carbon-supported cobalt catalysts: Promotional effect of modification by SiO2. *Chinese Journal of Catalysis*, 36(3): 355–361.

Peres, Laurent, M. Rosa Axet, Deliang Yi, Philippe Serp and Katerina Soulantica. 2019. Selective hydrogenation of cinnamaldehyde by unsupported and few layer graphene supported platinum concave nanocubes exposing {110} facets stabilized by a long-chain amine. *Catalysis Today,* 357: 166–175.

Phaahlamohlaka, Tumelo N., Mbongiseni W. Dlamini, David O. Kumi, Roy Forbes, Linda L. Jewell and Neil J. Coville. 2020. Co inside hollow carbon spheres as a Fischer-Tropsch catalyst: Spillover effects from Ru placed inside and outside the HCS. *Applied Catalysis A: General*, 599: 117617.

Piya-Areetham, P., Shenchunthichai, K. and Hunsom, M. 2006. Application of electrooxidation process for treating concentrated wastewater from distillery industry with a voluminous electrode. *Water Research*, 40(15): 2857–2864.

Poh, Hwee Ling, Filip Šaněk, Adriano Ambrosi, Guanjia Zhao, Zdeněk Sofer and Martin Pumera. 2012. Graphenes prepared by Staudenmaier, Hofmann and Hummers methods with consequent thermal exfoliation exhibit very different electrochemical properties. *Nanoscale*, 4(11): 3515–3522.

Qiu, Jieshan, Hongzhe Zhang, Changhai Liang, Jiawei Li and Zongbin Zhao. 2006. Co/CNF catalysts tailored by controlling the deposition of metal colloids onto CNFs: Preparation and catalytic properties. *Chemistry–A European Journal*, 12(8): 2147–2151.

Qu, WeiLi, ZhenBo Wang, XuLei Sui and DaMing Gu. 2014a. An efficient antimony doped tin oxide and carbon nanotubes hybrid support of Pd catalyst for formic acid electrooxidation. *International Journal of Hydrogen Energy*, 39(11): 5678–5688.

Qu, Wei-Li, Da-Ming Gu, Zhen-Bo Wang and Jing-Jia Zhang. 2014b. High stability and high activity Pd/ITO-CNTs electrocatalyst for direct formic acid fuel cell. *Electrochimica Acta*, 137: 676–684.

Qui, N.V., Scholz, P., Krech, T., Keller, T.F., Pollok, K. and Ondruschka, B. 2011. Multiwalled carbon nanotubes oxidized by UV/H2O2 as catalyst for oxidative dehydrogenation of ethylbenzene. *Catalysis Communications*, 12(6): 464–469.

Rafiee, Fatemeh, Parvaneh Khavari, Zahra Payami and Narges Ansari. 2019. Palladium nanoparticles immobilized on the magnetic few layer graphene support as a highly efficient catalyst for ligand free Suzuki cross coupling and homo coupling reactions. *Journal of Organometallic Chemistry*, 883: 78–85.

Rajkumar, Duraiswamy and Palanivelu, K. 2004. Electrochemical treatment of industrial wastewater. *Journal of Hazardous Materials*, 113(1-3): 123–129.

Rana, Surjyakanta, Suresh Maddila, Kotaiah Yalagala and Sreekantha B. Jonnalagadda. 2015. Organo functionalized graphene with Pd nanoparticles and its excellent catalytic activity for Suzuki coupling reaction. *Applied Catalysis A: General*, 505: 539–547.

Ren, J., Zhang, J., Yang, C., Yang, Y., Zhang, Y., Yang, F., Ma, R., Yang, L., He, H. and Huang, H. 2020. Pd nanocrystals anchored on 3D hybrid architectures constructed from nitrogen-doped graphene and low-defect carbon nanotube as high-performance multifunctional electrocatalysts for formic acid and methanol oxidation. *Materials Today Energy*, 16: 100409.

Rhee, Kyong Yop. 2020. Electronic and thermal properties of graphene. *Nanomaterials*, 10.5: 926.

Rodrigo, M.A., Cañizares, P., Buitrón, C. and Sáez, C. 2010. Electrochemical technologies for the regeneration of urban wastewaters. *Electrochimica Acta*, 55(27): 8160–8164.

Rohani, Sahar, Ghodsi Mohammadi Ziarani, Abolfazl Ziarati and Alireza Badiei. 2019. Designer 3D CoAl-layered double hydroxide@N, S doped graphene hollow architecture decorated with Pd nanoparticles for Sonogashira couplings. *Applied Surface Science*, 496: 143599.

Rostami, Hussein, Abbas Ali Rostami and Abdollah Omrani. 2015. Investigation on ethanol electrooxidation via electrodeposited Pd–Co nanostructures supported on graphene oxide. *International Journal of Hydrogen Energy*, 40(33): 10596–10604.

Sadjadi, Samahe, Majid M. Heravi and Maryam Raja. 2018. Combination of carbon nanotube and cyclodextrin nanosponge chemistry to develop a heterogeneous Pd-based catalyst for ligand and copper free C-C coupling reactions. *Carbohydrate Polymers*, 185: 48–55.

Saleh, Tawfik A. and Saddam A. Al-Hammadi. 2021. A novel catalyst of nickel-loaded graphene decorated on molybdenum-alumina for the HDS of liquid fuels. *Chemical Engineering Journal*, 406: 125167.

Salehi, Mirmehdi Hashemi, Mohammad Yousefi, Malak Hekmati and Ebrahim Balali. 2019. *In situ* biosynthesis of palladium nanoparticles on Artemisia abrotanum extract-modified graphene oxide and its catalytic activity for Suzuki coupling reactions. *Polyhedron*, 165: 132–137.

Scott, Carl D., Sivaram Arepalli, Pavel Nikolaev and Richard E. Smalley. 2001. Growth mechanisms for single-wall carbon nanotubes in a laser-ablation process. *Applied Physics A*, 72(5): 573–580.

Soghrati, E., Kazemeini, M., Rashidi, A.M. and Kh Jafari Jozani. 2012. Preparation and characterization of Co-Mo catalyst supported on CNT coated cordierite monoliths utilized for Naphta HDS process. *Procedia Engineering*, 42: 1484–1492.

Soghrati, E., Kazemeini, M., Rashidi, A.M. and Jozani, K.J. 2014. Development of a structured monolithic support with a CNT washcoat for the naphtha HDS process. *Journal of the Taiwan Institute of Chemical Engineers*, 45(3): 887–895.

Song, Xiaozhe, Qin Shi, Hui Wang, Shaolei Liu, Chang Tai and Zhaoyong Bian. 2017. Preparation of Pd-Fe/graphene catalysts by photocatalytic reduction with enhanced electrochemical oxidation-reduction properties for chlorophenols. *Applied Catalysis B: Environmental*, 203: 442–451.

Startsev, Anatolii N. 1995. The mechanism of HDS catalysis. *Catalysis Reviews*, 37(3): 353–423.

Staudenmaier, L. 1898. Verfahren zur darstellung der graphitsäure. *Berichte der deutschen chemischen Gesellschaft*, 31(2): 1481–1487.

Sullivan, James A., Keith A. Flanagan and Holger Hain. 2009. Suzuki coupling activity of an aqueous phase Pd nanoparticle dispersion and a carbon nanotube/Pd nanoparticle composite. *Catalysis Today*, 145(1): 108–113.

Sun, Hongqi, Shizhen Liu, Guanliang Zhou, Ha Ming Ang, Moses O Tadé and Shaobin Wang. 2012. Reduced graphene oxide for catalytic oxidation of aqueous organic pollutants. *ACS Applied Materials & Interfaces*, 4(10): 5466–5471.

Sun, Wenjing, Zhongqing Liu, Chengfa Jiang, Ying Xue, Wei Chu and Xiusong Zhao. 2013. Experimental and theoretical investigation on the interaction between palladium nanoparticles and functionalized carbon nanotubes for Heck synthesis. *Catalysis Today*, 212: 206-–214.

Sun, Zhuxing, Siyuan Fang and Yun Hang Hu. 2020. 3D graphene materials: From understanding to design and synthesis control. *Chemical Reviews,* 120(18): 10336–10453.

Szabó, T., Berkesi, O., Forgó, P., Josepovits, K., Sanakis, Y., Petridis, D. and Dékány, I. 2006. Evolution of surface functional groups in a series of progressively oxidized graphite oxides. *Chemistry of Materials*, 18(11): 2740–2749.

Szumełda, T., Drelinkiewicz, A., Kosydar, R. and Gurgul, J. 2014. Hydrogenation of cinnamaldehyde in the presence of PdAu/C catalysts prepared by the reverse "water-in-oil" microemulsion method. *Applied Catalysis A: General*, 487: 1–15.

Szymański, Grzegorz S., Teresa Grzybek and Helmut Papp. 2004. Influence of nitrogen surface functionalities on the catalytic activity of activated carbon in low temperature SCR of NOx with NH3. *Catalysis Today*, 90(1-2): 51–59.

Taghavi, Somayeh, Alireza Asghari and Ahmad Tavasoli. 2017. Enhancement of performance and stability of Graphene nano sheets supported cobalt catalyst in Fischer–Tropsch synthesis using Graphene functionalization. *Chemical Engineering Research and Design*, 119: 198–208.

Tan, Minghui, Guohui Yang, Tiejun Wang, Tharapong Vitidsant, Jie Li, Qinhong Wei, Peipei Ai, Mingbo Wu, Jingtang Zheng and Noritatsu Tsubaki. 2016. Active and regioselective rhodium catalyst supported on reduced graphene oxide for 1-hexene hydroformylation. *Catalysis Science & Technology*, 6(4): 1162–1172.

Tan, M., Ishikuro, Y., Hosoi, Y., Yamane, N., Ai, P., Zhang, P. and Tsubaki, N. 2017. PPh3 functionalized Rh/rGO catalyst for heterogeneous hydroformylation: Bifunctional reduction of graphene oxide by organic ligand. *Chemical Engineering Journal*, 330: 863–869.

Teschner, Detre, Elaine Vassa, Michael Hävecker, Spiros Zafeiratos, Péter Schnörch, Hermann Sauer, Axel Knop-Gericke, Robert Schlögl, Mounir Chamam, Attila Wootsch, Arran S. Canning, Jonathan J. Gamman, S. David Jackson, James McGregor and Lynn F. Gladden. 2006. Alkyne hydrogenation over Pd catalysts: A new paradigm. *Journal of Catalysis*, 242(1): 26–37.

Tessonnier, Jean-Philippe, Laurie Pesant, Gabrielle Ehret, Marc J. Ledoux and Cuong Pham-Huu. 2005. Pd nanoparticles introduced inside multi-walled carbon nanotubes for selective hydrogenation of cinnamaldehyde into hydrocinnamaldehyde. *Applied Catalysis A: General*, 288(1): 203–210.

Thakur, Suman and Niranjan Karak. 2015. Alternative methods and nature-based reagents for the reduction of graphene oxide: A review. *Carbon*, 94: 224–242.

Thomas, Molly, Mehraj Ud Din Sheikh, Devendra Ahirwar, Mustri Bano and Farid Khan. 2017. Gold nanoparticle and graphene oxide incorporated strontium crosslinked alginate/carboxymethyl cellulose composites for o-nitroaniline reduction and Suzuki-Miyaura cross-coupling reactions. *Journal of Colloid and Interface Science*, 505: 115–129.

Tian, Zhipeng, Chenguang Wang, Zhan Sia, Longlong Ma, Lungang Chen, Qiying Liu, Qi Zhang and Hongyu Huang. 2017. Fischer-Tropsch synthesis to light olefins over iron-based catalysts supported on KMnO4 modified activated carbon by a facile method. *Applied Catalysis A: General*, 541: 50–59.

Tran, Thuy Phuong Nhat, Ashutosh Thakur, Dai Xuan Trinh, Anh Thi Ngoc Dao and Toshiaki Taniike. 2018. Design of Pd@ Graphene oxide framework nanocatalyst with improved activity and recyclability in Suzuki-Miyaura cross-coupling reaction. *Applied Catalysis A: General*, 549: 60–67.

Treacy, M.M. JEBBESSEN, Thomas W. Ebbesen and John M. Gibson. 1996. Exceptionally high Young's modulus observed for individual carbon nanotubes. *Nature*, 381(6584): 678–680.

Truong-Huu, Tri, Kambiz Chizarialzabel Janowska, Maria Simona Moldovan, Ovidiu Ersen, Lam D. Nguyen, Marc J. Ledoux, Cuong Pham-Huu and Dominique Begin. 2012. Few-layer graphene supporting palladium nanoparticles with a fully accessible effective surface for liquid-phase hydrogenation reaction. *Catalysis Today*, 189(1): 77–82.

Van Leeuwen, Piet, W.N.M. and Carmen Claver. 2002. *Rhodium catalyzed hydroformylation*. Vol. 22: Springer Science & Business Media.

Vasudevan, P.T. and García J.L. Fierro. 1996. A review of deep hydrodesulfurization catalysis. *Catalysis Reviews*, 38(2): 161–188.

Veisi, Hojat, Ardeshir Khazaei, Maryam Safaei and Davood Kordestani. 2014. Synthesis of biguanide-functionalized single-walled carbon nanotubes (SWCNTs) hybrid materials to immobilized palladium as new recyclable heterogeneous nanocatalyst for Suzuki–Miyaura coupling reaction. *Journal of Molecular Catalysis A: Chemical*, 382: 106–113.

Veisi, Hojat, Ahmad Nikseresht, Nasim Ahmadi, Kaveh Khosravi and Fatemeh Saeidifar. 2019. Suzuki–Miyaura reaction by heterogeneously supported Pd nanoparticles on thio-modified multi walled carbon nanotubes as efficient nanocatalyst. *Polyhedron*, 162: 240–244.

Vlyssides, A.G., Papaioannou, D., Loizidoy, M., Karlis, P.K. and Zorpas, A.A. 2000. Testing an electrochemical method for treatment of textile dye wastewater. *Waste Management*, 20(7): 569–574.

Vu, Hung, Filomena Gonçalves, Régis Philippe, Emmanuel Lamouroux, Massimiliano Corrias, Yolande Kihn, Dominique Plee, Philippe Kalck and Philippe Serp. 2006. Bimetallic catalysis on carbon nanotubes for the selective hydrogenation of cinnamaldehyde. *Journal of Catalysis*, 240(1): 18–22.

Wang, Bo, Yaoyao Han, Sufang Chen, Yuhua Zhang, Jinlin Li and Jingping Hong. 2019. Construction of three-dimensional nitrogen-doped graphene aerogel (NGA) supported cobalt catalysts for Fischer-Tropsch synthesis. *Catalysis Today*, 355: 10–16.

Wang, Haiyan, Shida Liu, Rubenthran Govindarajan and Kevin J. Smith. 2017. Preparation of Ni-Mo2C/carbon catalysts and their stability in the HDS of dibenzothiophene. *Applied Catalysis A: General*, 539: 114–127.

Wang, H., Xiao, B., Cheng, X., Wang, C., Zhao, L., Zhu, Y and Lu, X. 2015. NiMo catalysts supported on graphene-modified mesoporous TiO2 toward highly efficient hydrodesulfurization of dibenzothiophene. *Applied Catalysis A: General*, 502: 157–165.

Wang, P., Zhang, G., Jiao, H., Deng, X., Chen, Y. and Zheng, X. 2015. Pd/graphene nanocomposite as highly active catalyst for the Heck reactions. *Applied Catalysis A: General*, 489: 188–192.

Wang, Shenghua, Zhiling Xin, Xing Huang, Weizhen Yu, Shuo Niu and Lidong Shao. 2017. Nanosized Pd–Au bimetallic phases on carbon nanotubes for selective phenylacetylene hydrogenation. *Physical Chemistry Chemical Physics*, 19(8): 6164–6168.

Wang, T., Ding, Y.J., Xiong, J.M., Chen, W.M., Pan, Z.D., Lu, Y. and Lin, L.W. 2004. Fischer-Tropsch reaction over cobalt catalysts supported on zirconia-modified activated carbon. pp. 349–354. *In*: Bao, X. and Xu, Y. (eds.). *Studies in Surface Science and Catalysis*: Elsevier, Vol. 147.

Wang, Tao, Yunjie Ding, Yuan Lü, Hejun Zhu and Liwu Lin. 2008. Influence of lanthanum on the performance of Zr-Co/activated carbon catalysts in Fischer-Tropsch synthesis. *Journal of Natural Gas Chemistry*, 17(2): 153–158.

Wang, Xizheng, Wufeng Chen and Lifeng Yan. 2014. Three-dimensional reduced graphene oxide architecture embedded palladium nanoparticles as highly active catalyst for the Suzuki–Miyaura coupling reaction. *Materials Chemistry and Physics*, 148(1): 103–109.

Wang, Ying, Liujun Jin, Caiqin Wang and Yukou Du. 2020. Nitrogen-doped graphene nanosheets supported assembled Pd nanoflowers for efficient ethanol electrooxidation. *Colloids and Surfaces A: Physicochemical and Engineering Aspects*, 587: 124257.

Wang, Y., Rong, X., Wang, T., Wu, S., Rong, Z., Wang, Y. and Qu, J. 2019. Influence of graphene surface chemistry on Ir-catalyzed hydrogenation of p-chloronitrobenzene and cinnamaldehyde: Weak molecule-support interactions. *Journal of Catalysis*, 377: 524–533.

Wei, Meng, Lili Zhang, Di Luo, Liang-Xin Ding, Suqing Wang and Haihui Wang. 2018. Graphene-assisted synthesis of PdFe-embedded porous carbon nanofibers for efficient ethanol electrooxidation. *Electrochimica Acta*, 289: 311–318.

Werkneh, Adhena Ayaliew, Hayelom Dargo Beyene and Abduljeleel A. Osunkunle. 2019. Recent advances in brewery wastewater treatment; approaches for water reuse and energy recovery: A review. *Environmental Sustainability*, pp. 1–11.

William, S., Hummers, J.R. and Richard E. Offeman. 1958. Preparation of graphitic oxide. *J. Am. Chem. Soc.*, 80(6): 1339–1339.

Wu, Peng, Yiyin Huang, Liqun Zhou, Yaobing Wang, Yakun Bu and Jiannian Yao. 2015. Nitrogen-doped graphene supported highly dispersed palladium-lead nanoparticles for synergetic enhancement of ethanol electrooxidation in alkaline medium. *Electrochimica Acta*, 152: 68–74.

Wu, Qiang, Lijun Yang, Xizhang Wang and Zheng Hu. 2019. Carbon-based nanocages: A new platform for advanced energy storage and conversion. *Advanced Materials*: 1904177.

Wu, T., Wang, X., Emre, A. E., Fan, J., Min, Y., Xu, Q. and Sun, S. 2021. Graphene-nickel nitride hybrids supporting palladium nanoparticles for enhanced ethanol electrooxidation. *Journal of Energy Chemistry*, 55: 48–54.

Xi, Lixin, Dan Li, Jun Long, Fei Huang, Lini Yang, Yushu Guo, Zhimin Jia, Jianping Xiao and Hongyang Liu. 2019. N-doped graphene confined Pt nanoparticles for efficient semi-hydrogenation of phenylacetylene. *Carbon*, 145: 47–52.

Xiao, He, Jingjuan Zhang, Man Zhao, Tianjun Hu, Jianfeng Jia and Haishun Wu. 2019. Hydrogenated graphene as support of Pd nanoparticles with improved electrocatalytic activity for ethanol oxidation reaction in alkaline media. *Electrochimica Acta*, 297: 856–863.

Xie, Wenqin, Jun Xiong and Guangyan Xiang. 2020. Enzyme immobilization on functionalized monolithic CNTs-Ni foam composite for highly active and stable biocatalysis in organic solvent. *Molecular Catalysis*, 483: 110714.

Xu, Jundong, Yunfeng Guo, Tingting Huang and Yu Fan. 2019. Hexamethonium bromide-assisted synthesis of CoMo/graphene catalysts for selective hydrodesulfurization. *Applied Catalysis B: Environmental*, 244: 385–395.

Xue, Teng, Shan Jiang, Yongquan Qu, Qiao Su, Rui Cheng, Sergey Dubin, Chin-Yi Chiu,Richard Kaner, Yu Huang and Xiangfeng Duan. 2012. Graphene-supported hemin as a highly active biomimetic oxidation catalyst. *Angewandte Chemie International Edition*, 51(16): 3822–3825.

Yan, Jia-An, W.Y. Ruan and Chou, M.Y. 2009. Electron-phonon interactions for optical-phonon modes in few-layer graphene: First-principles calculations. *Physical Review B*, 79(11): 115443.

Yang, Fan, Cheng Chi, Sen Dong Chunxia Wang, Xilai Jia, Liang Ren, Yunhan Zhang, Liqiang Zhang and Yongfeng Li. 2015. Pd/PdO nanoparticles supported on carbon nanotubes: A highly effective catalyst for promoting Suzuki reaction in water. *Catalysis Today*, 256: 186–192.

Yang, Kaixuan, Xiao Chen, Ji Qi, Zongxuan Bai, Liangliang Zhang and Changhai Liang. 2019. A highly efficient and sulfur-tolerant Pd2Si/CNTs catalyst for hydrodesulfurization of dibenzothiophenes. *Journal of Catalysis*, 369: 363–371.

Yang, Yunxia, Ken Chiang and Nick Burke. 2011. Porous carbon-supported catalysts for energy and environmental applications: A short review. *Catalysis Today*, 178(1): 197–205.

Zarnegaryan, Ali, Zahra Dehbanipour and Dawood Elhamifar. 2019. Graphene oxide supported Schiff-base/palladium complex: An efficient and recoverable catalyst for Suzuki–Miyaura coupling reaction. *Polyhedron*, 170: 530–536.

Zhang, Jiali, J., Zhang, F., Yang, H., Huang, X., Liu, H., Zhang, J. and Guo, S. 2010. Graphene oxide as a matrix for enzyme immobilization. *Langmuir*, 26(9): 6083–6085.

Zhang, Lipeng, Jianbing Niu, Mingtao Li and Zhenhai Xia. 2014. Catalytic mechanisms of sulfur-doped graphene as efficient oxygen reduction reaction catalysts for fuel cells. *The Journal of Physical Chemistry C*, 118(7): 3545–3553.

Zhang, Qian, Wan Huang, Jun-ming Hong and Bor-Yann Chen. 2018. Deciphering acetaminophen electrical catalytic degradation using single-form S doped graphene/Pt/TiO2. *Chemical Engineering Journal*, 343: 662–675.

Zhang, Tengyan, Walter P. Walawender, Fan, L.T., Maohong Fan, Daren Daugaard and Brown, R.C. 2004. Preparation of activated carbon from forest and agricultural residues through CO2 activation. *Chemical Engineering Journal*, 105(1-2): 53–59.

Zhang, Yu, Hong-Bin Zhang, Guo-Dong Lin, Ping Chen, You-Zhu Yuan and Tsai, K.R. 1999. Preparation, characterization and catalytic hydroformylation properties of carbon nanotubes-supported Rh–phosphine catalyst. *Applied Catalysis A: General*, 187(2): 213–224.

Zhao, Huabo, Qingjun Zhu, Yongjun Gao, Peng Zhai and Ding Ma. 2013. Iron oxide nanoparticles supported on pyrolytic graphene oxide as model catalysts for Fischer Tropsch synthesis. *Applied Catalysis A: General*, 456: 233–239.

Zhao, Jinping, Wencai Ren and Hui-Ming Cheng. 2012. Graphene sponge for efficient and repeatable adsorption and desorption of water contaminations. *Journal of Materials Chemistry*, 22(38): 20197–20202.

Zhu, Hai, Juntao Wang, Xiaoling Liu and Xiaoming Zhu. 2017. Three-dimensional porous graphene supported Ni nanoparticles with enhanced catalytic performance for Methanol electrooxidation. *International Journal of Hydrogen Energy*, 42(16): 11206–11214.

Zhu, Hongwei. 2017. *Graphene: Fabrication, Characterizations, Properties and Applications*: Academic Press. The Boulevard, Langford Lane, Kidlington, Oxford OX5 1GB, U.K.

Zhu, Jie, Mengdi Dou, Mohong Lu, Xu Xiang, Xuejie Ding, Wenxin Liu and Mingshi Li. 2019. Thermo-responsive polymer grafted carbon nanotubes as the catalyst support for selective hydrogenation of cinnamaldehyde: Effects of surface chemistry on catalytic performance. *Applied Catalysis A: General*, 575: 11–19.

Zhu, Jun, Jinghong Zhou, Tiejun Zhao, Xinggui Zhou, De Chen and Weikang Yuan. 2009. Carbon nanofiber-supported palladium nanoparticles as potential recyclable catalysts for the Heck reaction. *Applied Catalysis A: General*, 352(1): 243–250.

CHAPTER 14

Synthesis of Graphene onto Semi-insulating Substrates
Epitaxial Graphene on SiC and CVD Graphene on Sapphire

Neeraj Mishra[1,2,*] and *Domenica Convertino*[1,2,*]

1. Introduction

Graphene is a 2D material extracted foremost in 2004 from a mass of crystalline graphite by mechanical exfoliation by two scientists working at the University of Manchester, Geim and Novoselov (Novoselov et al. 2004). Graphene is the 1st man-made 2D material and is well known for its limitless property. The story of the discovery of graphene is very interesting. Every Friday after dinner, the group of Geim and Novoselov used to work on different projects. One Friday night they were exfoliating graphite with the help of scotch tape, sticking the flakes on a cleaned silicon dioxide (SiO_2), and looking at the microscope. Sometimes, to achieve a result, your luck should favor you and you need to be lucky. Fortunately, they used 300 nm SiO_2/Si to place the exfoliated flakes. Due to this thickness, the contrast of graphene is very visible at the microscope and it is easy to distinguish the thickness of graphene (Novoselov et al. 2004). The increased attention on Graphene by the community is ascribed to its peculiar features: loftier physical-mechanical properties, both in strength and flexibility, elevated electron mobility, thermal conductivity and surface area, impermeability to gases (even the He atom), and transparency (Banszerus et al. 2015, Bao and Loh 2012, Castro Neto et al. 2009, Lee et al. 2008a). This interest has given rise to new production methods, including chemical-based approaches and large-scale techniques, whose choice depends on the required properties and applications (Bonaccorso et al. 2012).

2. Graphene growth

Since its first isolation in 2004, a great effort has been made by the scientific community to produce graphene with different methods, whose choice depends on the applications, ranging from micromechanical exfoliation, to chemical-based approaches, to annealing of silicon carbide (SiC)

[1] Center for Nanotechnology Innovation @NEST, Istituto Italiano di Tecnologia, Piazza San Silvestro 12, 56127 Pisa, Italy.
[2] Graphene Labs, Istituto Italiano di Tecnologia, Via Morego 30, 16163 Genova, Italy.
* Corresponding authors: neeraj.mishra@iit.it; Domenica.Convertino@iit.it

at high temperature and chemical vapor deposition (CVD) on metals (Cu, Ni, Pt, and Au), silicon or sapphire (Figure 1).

Micromechanical exfoliation involves the use of repeated peeling of graphite flakes using sticky tape (Figure 1(a)) (Novoselov et al. 2004). The weak interplanar interactions allow for the isolation of monolayer graphene, transferred to another substrate by simply pressing the tape onto it. This exfoliated graphene presents perfect crystallinity and low density of defects and is usually preferred for fundamental studies. However, the limited size of the exfoliated flakes makes them unsuited for large-scale production and applications.

In liquid-phase exfoliated (LPE) graphene, the exfoliation process is scaled up. The graphite is dispersed in a solvent, exfoliated using ultrasound and ultracentrifugation to remove the un-exfoliated flakes (Figure 1(b)) (Bonaccorso et al. 2012). The principle of LPE can also be used not only to exfoliate pristine graphite but also for the exfoliation of graphite interpolated compounds and graphite oxide (GO). The GO can be reduced using thermal or chemical treatments. However, these graphene-like structures have both sp^2 and sp^3 hybridized carbon atoms, resulting in degraded quality and purity, modified electronic structure and physical properties, and are usually preferred in the production of conducting films and composites (Park and Ruoff 2009).

The two main fabrication methods for large-scale (up to 8-inch) production of high-grade monolayer graphene are the CVD technique on transition metals like copper (Cu) (Figure 1(d)) (Li et al. 2009, Miseikis et al. 2015, Yu et al. 2008) and thermal decomposition of SiC (Figure 1(c)) (First et al. 2010, Starke et al. 2012). Cu is usually preferred to grow single-layer graphene. Thanks to the reduced solubleness of carbon in copper, graphene growth is self-limited and ceases as soon as a homogeneous graphene layer covers the copper surface (Li et al. 2009). One of the shortcomings of this technique is that graphene needs to be transferred, since most applications require graphene

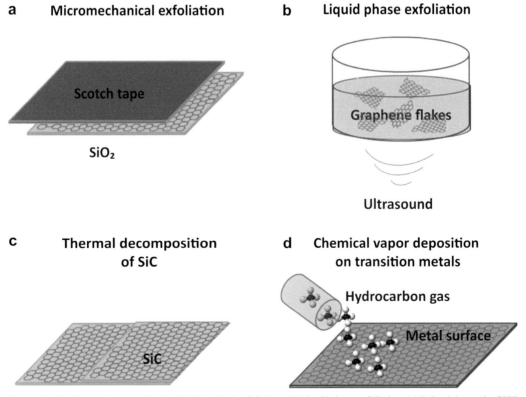

Figure 1. Graphene synthesis methods: (a) Mechanical exfoliation. (b) Liquid phase exfoliation. (c) Epitaxial growth of SiC. (d) Chemical vapor deposition on transition metals. (Original figure)

on an insulating substrate. This limitation can be overcome by using graphene grown via thermal decomposition of SiC (Emtsev et al. 2009) or CVD growth on silicon or sapphire (Mishra et al. 2019, Song et al. 2012, Su et al. 2011). The CVD approach is a surface-catalyzed method that exploits transition metals to synthesize both single-crystal and polycrystalline graphene, which are subsequently transferred to arbitrary substrates. Nickel is usually preferred to synthesize multilayer graphene. Due to the significant carbon solubility, carbon diffuses into the nickel and segregates forming a non-homogeneous multilayer graphene film (Yu et al. 2008), differently from copper where the low carbon solubility makes the growth process self-limiting, resulting in a homogeneous single-layer film (Deng et al. 2019, Li et al. 2009, Miseikis et al. 2015). The CVD technique is one of the most used scalable approaches to grow graphene. Indeed, controlling the H_2 and CH_4 gas mixture and the growth parameters, one can control both the nucleation density and graphene crystal dimension. Moreover, just a short time ago it has been shown that graphene crystal location can be controlled by using chromium-based nucleation seeds to create a pattern on the copper substrates (Miseikis et al. 2017a) or polymers (Song et al. 2016), facilitating the use of CVD graphene in wafer-chip manufacturing. However, a downside of the CVD approach is the transfer of graphene from metals to insulating substrates, which may induce contaminations, unintentional doping, and mechanical stress to the graphene film, that negatively affects its physical integrity and electrical performance (Bonaccorso et al. 2012, Lupina et al. 2015). The classic transfer method from copper involves the spinning of a polymer on graphene/copper foil, to help the manipulation of the thin graphene membrane after the chemical etching of the metal substrate (Miseikis et al. 2017b). Generally, copper is etched using a copper etchant solution like iron chloride ($FeCl_3$), ammonium persulphate (APS), or other commercial copper etchants. APS is usually preferred to have clean and polymer-free graphene surfaces (De Fazio et al. 2019). Figure 2 exhibits a schematic of the transfer process of graphene from copper (G/Cu) on the desired substrate. A layer of polymethyl methacrylate (PMMA) is spin-coated on the G/Cu foil. The Cu foil is then etched using a copper etchant solution, leaving the floating PMMA/G. The membrane is rinsed many times in deionized (DI) water to reduce the amount of contaminants of the etchant. The PMMA/G layer is then placed onto the target substrate and let dry. Subsequently, the substrate is kept in acetone for more than 2 hrs to remove the PMMA, rinsed for few minutes in isopropanol (IPA), and dried with nitrogen.

A cleaner transfer can be obtained using bubbling transfer (Gao et al. 2012) or semi-dry transfer (Miseikis et al. 2017b), but these methods have limitations when a continuous graphene

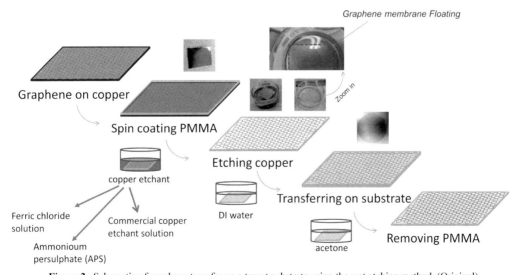

Figure 2. Schematic of graphene transfer on a target substrate using the wet etching method. (Original)

film is needed since they can prompt the formation of holes in graphene. In addition, the level of the metallic contaminants in transferred graphene is typically not satisfactory for industrial-scale back-end-of-line (BEOL) integration (Lupina et al. 2015).

To overcome the limitation of CVD graphene on metals, hard work has been done to develop a transfer-free synthesis of graphene directly on insulating supports. The use of epitaxial graphene on SiC is one the most known method (Berger et al. 2004, Emtsev et al. 2008). However, the enormous cost of both the SiC wafer and the growth process have encouraged the research community to find alternative ways to obtain metal-free graphene on insulating substrates that can be integrated with mature device-processing technology, such as graphene growth on sapphire (Fanton et al. 2011, Mishra et al. 2019).

3. Graphene characterization

Graphene is typically characterized with standard surface techniques like atomic force microscopy (AFM), scanning electron microscopy (SEM), as well as Raman spectroscopy, to understand its fundamental properties. AFM and SEM are usually used to investigate the large-scale topography, while Raman spectroscopy enables researchers to assess graphene crystallinity and its electronic behavior in a non-invasive way.

Atomic force microscopy

AFM is a scanning probe technique used for morphological characterization, force measurement, and sample manipulation with atomic resolution. The standard operation is to scan a flexible cantilever integrated with a needle-pointed tip over the sample. A laser beam is centered on the cantilever and is reflected onto a photodetector. During the scanning, the synergy between the cantilever and the sample forces the cantilever to bend, and a photodiode detects the laser beam deflection. Typically, the AFM is operated in three modes that diverge from each other by the prick and sample synergy forces: non-contact mode (attractive force), contact mode (repulsive force), and tapping mode (the cantilever oscillates around its resonant frequency lightly tapping at the surface). AFM is usually employed to analyze topographic features of the surface with a nanometric resolution. In addition to topography, the sample's material properties and frictional forces can be detected by phase signal and lateral force signal analysis, respectively. Lateral force microscopy has been used for spatially mapping the frictional properties of partially graphitized SiC(0001), providing high frictional contrast between the buffer layer and monolayer graphene (Kellar et al. 2010). Moreover, AFM phase sensitivity to material changes allows one to distinguish the numbers of graphene layers from the contrast change (Kruskopf et al. 2015, Tamayo and García 2002).

Raman spectroscopy

Raman spectroscopy is a spectroscopic method typically utilized to measure vibrational states related to interaction and symmetry in the molecules, by measuring the inelastic scattering of light. Each band in a Raman spectrum derives from a molecular or lattice vibration, giving information about the material's molecular structure, crystallinity, and residual stress. Raman spectroscopy is prevalently used to assess graphene thickness, strain, and doping (Ferrari et al. 2006). The characteristic graphene Raman spectrum has two main peaks or bands associated with phonon vibrational modes: the graphitic G band (~ 1580 cm^{-1}), originated from a first-order Raman scattering, and the 2D band (~ 2700 cm^{-1}), associated with a double resonance electron-phonon inelastic scattering process. When the process involves a phonon and a defect, there is also the D band (~ 1350 cm^{-1}) (Ferrari 2007, Malard et al. 2009).

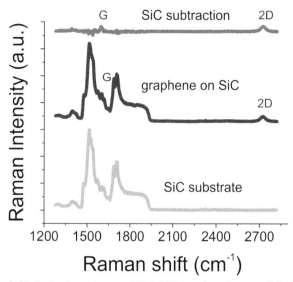

Figure 3. Raman spectra of SiC (bottom), graphene on SiC (middle), and graphene on SiC after background subtraction (top). (Original).

The position and profile of the 2D band and the ratio between I(2D) over I(G) give an indication of the graphene thickness. In particular, a sharp and symmetric 2D band, fitted with a single Lorentzian curve with a full width at half maximum (FWHM) lower than 30 cm^{-1} is characteristic of monolayer graphene (ML). In bilayer and trilayer graphene, the 2D band is wider and the fitting requires multiple Lorentzian curves. Moreover, the 2D band energy increases significantly with increasing layer number (Lee et al. 2008b). In monolayer graphene, the 2D peak is two times more intense than the G band. Furthermore, the G to 2D band intensity ratio increases with the number of graphene layers (Das et al. 2008a, Hao et al. 2010), even if considering the ratio alone is not sufficient because it depends also on graphene doping (Das et al. 2008b).

A shift in the G and 2D bands can also be associated with doping, strain, or their combination (Das et al. 2008b, Lee et al. 2012, Neumann et al. 2015). The energy of the 2D band increases with p-type doping and decreases with n-type doping, with respect to the case of pure undoped or neutral graphene. The G-band of graphene shifts towards higher wavenumbers for both n and p-type doping (Casiraghi et al. 2007, Das et al. 2008a).

In graphene on SiC, the G and D bands are difficult to identify due to the surrounding Raman bands from the SiC substrate between 1000 and 2000 cm^{-1}. On the contrary, the 2D band is well visible around ~ 2700–2750 cm^{-1} since it is isolated from the SiC peaks. The substrate contribution is usually removed by subtracting a SiC reference spectrum, making both the G and D bands visible (Röhrl et al. 2008), as depicted in Figure 3.

When compared to micro mechanically cleaved graphene, the 2D band FWHM of graphene on SiC is typically broader and both G and 2D Raman bands are strongly blue-shifted (G band of ~ 10 cm^{-1}, 2D band ~ 39 cm^{-1}), due to the compressive strain triggered by the different thermal expansion coefficient of SiC substrate and graphene (Lee et al. 2008b, Wang et al. 2008). Differently from the 2D band, the G peak position and width are not influenced by the graphene thickness (Lee et al. 2008b).

Scanning electron microscopy

A scanning electron microscope (SEM) is an instrument that exercises a focused high-energy electron beam to probe a target surface. The interaction between the beam and the sample generates electrons (e.g., either excited from the atoms of the sample or backscattered) that are used to analyze the chemical composition and morphology of wide areas with a resolution of tens of nanometers. Indeed, secondary electrons and backscattered electrons, generated respectively by inelastic and elastic scattering interactions with the incident electron beam, are typically detected. In our case, to see the graphene on SiC and sapphire, we used the Inlens signal, 5eV, and 97 pA of current.

4. Epitaxial growth on Silicon Carbide (SiC)

Epitaxial growth of graphene on SiC is achieved by thermal decomposition of SiC wafers at elevated temperature (1300–1800ºC) in ultra-high vacuum or under high pressure of Argon (800 sccm). This treatment leads to silicon (Si atoms) sublimation from the surface and the rearrangement of carbon atoms behind forming the graphene layers (First et al. 2010). Graphene can thus be grown directly on a commercially available semiconducting (SC) or semi-insulating (SI) substrate, obviating the need to transfer it to another substrate. Among its crystalline forms, the most used ones are the cubic 3C, the hexagonal 4H, and 6H polytypes. In the polytype name, the letter C or H indicates the cubic or hexagonal symmetry, while the numbers 3, 4, or 6 represent how many SiC bilayers are present within one stacking unit. The most used polytypes are the 4H- and 6H-SiC because they have an ideal crystallographic symmetry to grow graphene (Figure 4). By cutting a hexagonal SiC crystal along a plane perpendicular to the c-axis, two surfaces are formed, one terminated by Si atoms, known as (0001) or the Si-face, and the other terminated with C atoms, known as SiC(0001⁻) or the C-face (Figure 4(c)). Graphene can be synthesized both on the Si-face and the C-face surfaces of the SiC crystal showing different graphitization behavior (First et al. 2010).

Indeed, thermal decomposition of SiC crystal is not a self-restrictive process that generates a reduced graphene thickness uniformity. Especially on the C-terminated surface where there is a poor thickness control in the growth process, rotationally disordered multilayer graphene films are produced (Emtsev et al. 2009). Prior to growth, both polarity surfaces are hydrogen etched (H-etched) to form wide, highly uniform, atomically flat terraces (Ramachandran et al. 1998).

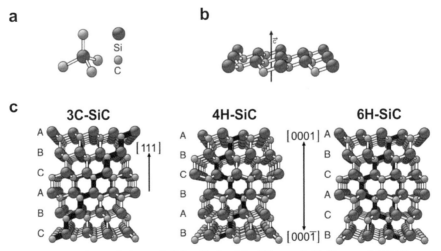

Figure 4. SiC crystal. (a) SiC tetrahedron, a building block of the SiC crystal. (b) Hexagonal SiC bilayer arranges in parallel planes of regular hexagonal networks whose stacking defines the different polytypes of SiC. (c) Crystal structure of three different SiC polytypes: cubic 3C-SiC, with an ABC stacking sequence, hexagonal 4H-SiC, with ABCB stacking, and hexagonal 6H-SiC, with ABC ACB stacking. One bilayer of Si and C has a 'c' dimension of 0.252 nm, which means that 4H-SiC has a unit cell of 1 nm and 6H-SiC of 1.5 nm. Adapted from (Forti 2014).

5. Hydrogen etching

The crystalline quality of graphene grown on SiC is strongly dependent on the quality of the SiC surface. As-received optically polished wafers have many residual polishing scratches (Figure 5(a)). The crystalline defects can be eliminated by etching the SiC in hydrogen. The process consists of annealing the substrate under a flow of molecular hydrogen and can be performed at both atmospheric and low pressure (Bianco et al. 2015a, Convertino et al. 2016, Frewin et al. 2009). The high temperature causes silicon atom sublimation, exposing the underlying carbon atoms that react with the hydrogen atoms forming hydrocarbons that leave the surface. The process continues until several hundred nanometers of SiC are removed, obtaining large atomically flat surfaces, thus improving the surface morphology (Ramachandran et al. 1998). Due to the unintentional miscut of the surface even for on-axis substrates, the etching process produces regular step structures (i.e., reveals the atomic steps of the SiC surface with unit-cell height), as shown in Figure 5(b). These atomically planar surfaces are used to synthesize high crystalline graphene (Starke et al. 2012).

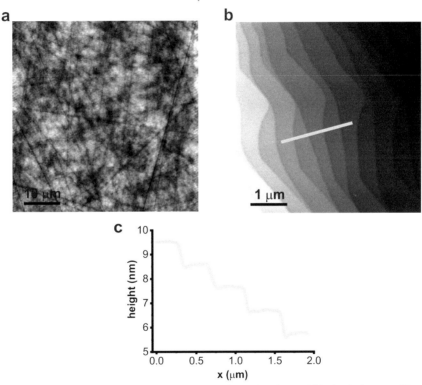

Figure 5. AFM images of on-axis 4H-SiC wafer surface morphology. (a) Before and (b) after hydrogen etching, carried out by heating the samples under a flow of molecular hydrogen at 450 mbar and 1300°C. (a) Optically polished C-face presents damage and deep scratches. (b) Hydrogen etched Si-face surface with atomically flat micrometric terraces separated by an atomic step. (c) Step profile taken along the step flow direction reveals steps of unit cell height (1 nm) with an average distance of 0.5 μm. (Original)

6. Graphene obtained via thermal decomposition of SiC

As mentioned above, graphene grows differently on the two basal planes of the SiC polytypes with the hexagonal crystal structure (First et al. 2010). The graphitization process on the Si-face produces a boundary between the initial graphene layer (monolayer graphene on SiC) and the underlying SiC. This interface layer is named the zero layer or buffer layer, and 30% of its C atoms are covalently bound to the Si atoms of the top SiC layer, with a periodicity of $(6\sqrt{3} \times 6\sqrt{3})R30°$. Graphene on the Si-face grows via step-edge nucleation, due to increased Si desorption at the terrace step edges (Hupalo et al. 2009). Following Si sublimation, the buffer layer is released from SiC as new graphene and another buffer layer is formed. Thus, each graphene layer maintains the 30° rotation relative to the substrate imprinted by the buffer layer (Emtsev et al. 2008). Moreover, its influence on the above graphene layers strongly affects their electronic properties, reducing graphene mobility and inducing an intrinsic electron doping ($n \approx 10^{13}$ cm^{-2}) (Emtsev et al. 2009).

While on SiC(0001) it is possible to grow homogeneous mono and few-layer graphene, on the SiC(0001$^-$) monolayer growth of epitaxial graphene is difficult to achieve, since the thermal decomposition takes place quite fast and in an uncontrolled way (First et al. 2010). Moreover, since there is no interface layer, the obtained graphene layers are electronically decoupled and do not have an azimuthal orientation (i.e., turbostratic graphene) so that each layer is isolated, much less sensitive to SiC surface defects. For these reasons, on C-face graphene carrier mobilities are typically higher than on Si-face (Strupinski et al. 2011).

Graphene on the C-face of SiC

Knowing the difficulties to control the graphene layers that are formed on the C-face via conventional thermal decomposition growth, numerous strategies have been used to improve the control over the growth rate to reduce the silicon sublimation rate (Camara et al. 2009, Strupinski et al. 2011). Recently, a tailored CVD approach has been developed to improve the graphene surface morphology and the layer thickness uniformity (Convertino et al. 2016). The structural, chemical and electronic properties of C-face graphene obtained by using this CVD approach have been also compared with the conventional thermal decomposition growth.

Multilayer graphene was synthesized on an H-etched C-face SiC wafer via both thermal decomposition and CVD in a resistively-heated cold-wall reactor (HT-BM Aixtron reactor) (Convertino et al. 2016). Thermal decomposition was obtained by heating the samples in argon at 1350°C and 780 mbar for 15 minutes (Candini et al. 2015, 2017, Emtsev et al. 2008, Haghighian et al. 2018, Meng et al. 2016). CVD growth was obtained using methane (CH$_4$) as a carbon precursor. The wafer was heated at about 1350°C within a mixture of 50% of argon and 50% hydrogen held at 780 mbar, to improve the consistency of the layer thickness and reduce the defects, while flowing the carbon precursor (3 sccm of CH$_4$) for 5 minutes.

AFM analysis of the obtained graphene revealed a relative inhomogeneous surface morphology similar to previous investigations (de Heer et al. 2011, Prakash et al. 2010). As shown in Figure 6(a), ridges of few nanometers were observed. Such features are characteristic of C-face graphene and are ascribed to the distinct expansion coefficients of graphene and SiC and to weaker graphene-substrate coupling (de Heer et al. 2011). Moreover, the surface-displayed step bunching of tens of nanometers as shown in a typical AFM image in panel (a). Adjacent micrometric domains of different heights were also made evident by the distinct contrast in the SEM images (panel (b)) and were found to be similar to those in previous works (Johansson et al. 2011, Razado-Colambo et al. 2015). These results confirm that the growth process on C-face is quick and difficult to control. Indeed, the low surface energy of the C-face, and the absence of an enclosure (Razado-Colambo et al. 2015), probably contributed to produce this canyon-like morphology.

Different from the thermal decomposition graphene, in the CVD-grown samples, the nanometric height of the steps was well-preserved, as reported in Figure 6(c). Locally, one could observe domain inclusions with an augmented roughness (even though still within the nanometric range) (darker areas in SEM micrograph in Figure 6(d)). These inclusions probably arise by an incomplete

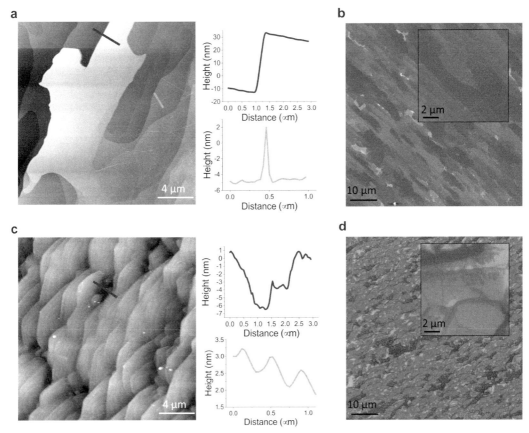

Figure 6. Thermal decomposition graphene (a,b) and CVD graphene (c,d) on 4H-SiC(000-1) (a) AFM image. Inset: typical high step (top) and a ridge (bottom) profile. (b) SEM image. Inset: high magnification of the same sample. (c) AFM micrograph. Right insets: typical profile of an inhomogeneous area (top) and steps (bottom). (d) SEM image. Inset: high magnification of the same sample. Adapted from (Convertino et al. 2016).

growth causing the development of submicrometric single-crystal domains, and might potentially be improved by increasing the partial pressure of hydrogen during growth.

The measure of the attenuation of a characteristic SiC Raman peak (~ 1516 cm^{-1}) was utilized to evaluate graphene thickness. This peak is an overtone of the L point optical phonon and presents a reduced intensity after graphene growth (Shivaraman et al. 2009). This method was employed to map graphene thickness over the thermal decomposition sample and CVD sample. As expected from the morphological analysis, despite the good crystallinity and the lack of the disorder-induced D band, in these samples the thickness varies between a few and about twenty layers. For the CVD-grown samples, the attenuation of the SiC Raman signal reveals a more homogeneous thickness distribution, between 3 and 7 layers (Convertino et al. 2016).

The CVD method can be adapted to grow multilayer graphene (up to 90) with good crystallinity, which was exploited to investigate the absorption in the THz range (Bianco et al. 2015b, Zanotto et al. 2017) and to realize innovative graphene mode-locked THz lasers (Bianco et al. 2015b).

Graphene on the Si-face of SiC

The thermal decomposition growth on the Si-face of SiC could be obtained by heating Si-face SiC in ultra-high vacuum (UHV) or under Ar atmospheric pressure (First et al. 2010, Hibino et al. 2008, Starke et al. 2012). Different from UHV-grown graphene that was notoriously obtained in small

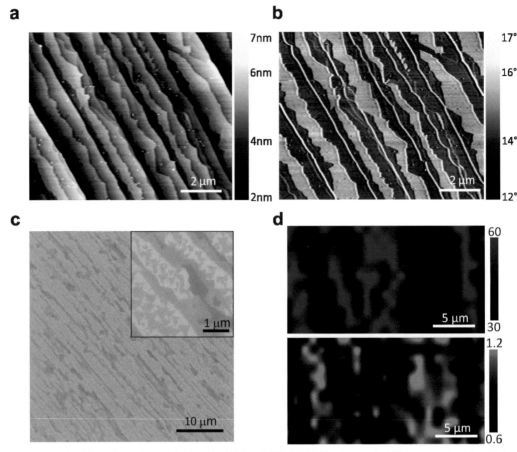

Figure 7. Epitaxial graphene characterization via AFM and SEM. (a,b) Tapping mode AFM topography (a) and phase signal (b) of graphene on 4H-SiC(0001). The contrast in the phase signal is due to monolayer graphene (brighter) or bilayer graphene (darker). (c) SEM image of mono-bilayer graphene with a higher magnification (inset). (d) 2D bandwidth (top) reveals single-layer graphene inside the terraces (darker areas) and bilayer inclusions at the step edges (lighter areas). The bilayer inclusions are confirmed by an increase in the ratio of G and 2D peaks intensity (bottom).

domains of few hundreds of nanometers, Ar-grown graphene presented much larger domain sizes of several tens of microns, thanks to more controlled silicon sublimation (Emtsev et al. 2009).

By changing the temperature and time, it is possible to obtain mono, bi, and few-layer graphene in a controlled way. The buffer layer is usually grown at lower temperatures, ranging from 1200°C to 1250°C and a pressure of 780 mbar for 5–10 min. Monolayer graphene is obtained by raising the temperature to 1300–1330°C (Bianco et al. 2015a, Chen et al. 2016, Convertino et al. 2018, Mashoff et al. 2015).

Figure 7 shows a representative epitaxial graphene sample. The graphene substrate was characterized by AFM (Figure 7(a,b)) and SEM (Figure 7(c)) that showed the characteristic atomic terraces and steps of the graphitized SiC surface with graphene with different thicknesses inside the terraces. Raman spectroscopy was then used to assess the distribution of the number of layers. The AFM micrograph correlates with the spatial distribution of the FWHM of the 2D Raman band (Figure 7(d, top)), and the ratio of 2D and G band intensities (Figure 7(d, bottom)), revealing the monolayer character of the graphene on the terraces. Bilayer inclusions were typically located at the terrace edges (Bianco et al. 2015a, Coletti et al. 2010).

Hydrogen intercalation

Epitaxial graphene on Si-face suffers from the strong interaction with the buffer layer that induces n-doping of graphene (Starke et al. 2012). The intrinsic negative charge can be compensated via transfer doping, by molecular functionalization of graphene surface with electronegative molecules (Coletti et al. 2010). As an alternative, the undesired effect of the interface layer can be repressed by removing the covalent bonds using hydrogen intercalation, to obtain floating graphene layers on SiC (Riedl et al. 2009). The technique of hydrogen intercalation produces quasi-free standing epitaxial graphene by annealing a buffer layer under molecular hydrogen to break the covalent bonds between carbon and silicon atoms and saturate the Si atoms. Thus, the buffer layer (Figure 8(a)) is released from SiC and turns into quasi-free standing monolayer graphene (QFMLG) (Figure 8(c)), while monolayer graphene (Figure 8(b)) turns into a decoupled bilayer (Figure 8(d)) (Riedl et al. 2009).

The process is typically carried out by exposing the grown buffer layer samples to molecular hydrogen at 780 mbar and temperature ranging from 600°C to 1000°C, typically for 10 minutes (Convertino et al. 2018, Kang et al. 2016).

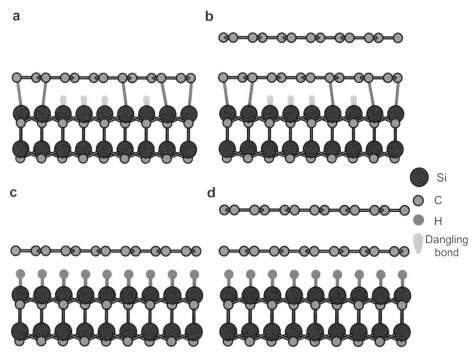

Figure 8. Side-view model of the buffer layer, monolayer, and quasi-free standing graphene. Schematic side view of (a) buffer layer and (b) monolayer graphene. C atoms are partially covalently bonded to Si atoms of SiC. After hydrogen intercalation, hydrogen atoms saturate the Si atoms of the Si-face SiC, releasing the buffer layer and the monolayer into (c) quasi-free standing monolayer graphene and (d) quasi-free standing bilayer graphene. Adapted from (Riedl et al. 2009).

7. Wafer-scale synthesis of graphene on sapphire

Direct growth of graphene on insulator

Lots of hard work has been made to synthesize graphene directly on SiO_2/Si (Su et al. 2011) and glass substrate (Chen et al. 2019), but usually, graphene was nano-crystalline. Ismach et al. has grown graphene on a thin copper film (100–450 nm) pre-deposited on different insulators, like quartz, SiO_2, and fused silica (Ismach et al. 2010). However, at high temperatures, the copper melts, and the surface is not archetyped for continuous graphene, desirable in CMOS application. Shin et al. synthesized graphene unswervingly on an insulator by vertical dispersal of carbon precursors through the interface between the two grains of copper, to the substrate in contact with the catalyst copper film (Shin et al. 2018). However, this method requires wet etching for copper removal.

For the enlargement of graphene device technology, the direct growth of graphene on an insulator with a metal-catalyst-free method is still desirable. Wafer-scale metal-catalyst-free growth of graphene has been obtained at low temperatures (< 1000°C) (Song et al. 2012). However, graphene quality was reduced, compared to the one synthesized using a metal catalyst. Fanton et al. showed good quality graphene on sapphire, growth with a catalyst-free approach, with mobility at room temperature around 3000 cm^2/Vs (Fanton et al. 2011). However, a significant disadvantage of this method is the high synthesis temperature above 1500°C that is not compatible with the conventional device processes. Hwang et al. showed van der Waals epitaxy synthesis of graphene directly on the sapphire substrate, with good quality and mobility around 2000 cm^2/V.s. However, the growth temperature was still high (above 1600°C) (Hwang et al. 2013). Recently, Kim et al. were able to grow millimeter-sized graphene by a novel capping technique using an overlaid Cu thin film, in direct contact with the substrate (Kim et al. 2018).

Zhang et al. proposed a three-step technique to grow graphene of the copper-assisted sapphire substrate (Zhang et al. 2018). Copper and a carbon precursor were co-deposited on sapphire by metalorganic chemical vapor deposition (MOCVD); subsequently, the substrates were annealed to grow graphene from the carbon precursor.

However, there is still a lot of space for improvement in the direct growth of graphene on sapphire using a metal-free, low temperature, and single-step approach. Therefore, it is important to investigate in more detail the growth mechanism and surface preparation and reconstruction of the sapphire substrate during the process.

Recently, the growth of high-quality monolayer graphene via CVD on the c-plane of an H-etched Al_2O_3 (0001) substrate with a catalyst-free method in a commercial CVD reactor (HT-BM, Aixtron) has been demonstrated (Mishra et al. 2019). The growth temperature of 1200°C was much lower compared to what was used in previous works (Hwang et al. 2013).

In this part, we will discuss the reconstruction of the sapphire substrate to grow homogenous and high-quality graphene. The structural and chemical properties of graphene were assessed by Raman spectroscopy, AFM, low-energy electron diffraction (LEED), and scanning-tunneling microscopy (STM). The graphene on the sapphire growth process was successfully scaled up from batch scale to wafer-scale (up to 8-inch wafer), keeping the same quality and uniformity. The carrier mobility measured at room temperature was above 2200 cm^2/V.s. In addition, if needed, an entire graphene wafer can be easily transferred from sapphire to the desired substrate by a polymer-assisted technique. The direct and metal-free CVD approach is certainly interesting thanks to its implementation in a commercial reactor, that can be easily scaled-up, leading to applications in photodetectors, terahertz spectroscopy, optoelectronics, and many more.

Experimental set up for synthesis of graphene on the sapphire

A single c-axis, single-crystal HEMCOR, and double-side polished sapphire (0001) substrate was used for graphene synthesis (Alfa Aesar, batch: 45020, LOT: U20E038). Before growth, sapphire was cleaned by ultrasonication in acetone, isopropanol, and de-ionized (DI) water. Subsequently, the substrate was treated with piranha solution (1:3, H_2O_2:H_2SO_4) for 15 min, cleaned in DI water, and finally dried with nitrogen. After the cleaning, sapphire was hydrogen etched in Aixtron high-temperature black magic (HT-BM) cold wall CVD system for 5 min at 1180°C in H_2 (Figure 9)

Figure 9. HT-BM Aixtron reactor used to grow graphene on sapphire, with its internal view. The sample holder or susceptor is made of graphite and can hold a substrate of a maximum of 1 cm². The reactor can reach up to 1400°C.

(Mishra et al. 2019). Then, the substrate was annealed for 10 min at 1200°C in Ar (1000 sccm). After the annealing procedure, graphene was grown in a flow of Ar, H_2, and CH_4. CH_4 was introduced as a carbon precursor in a mixture of Ar and H_2 (1000 sccm of Ar and 100 sccm of H_2) for 30 min, with a ratio of CH_4:H_2 of 1:20 (Mishra et al. 2019). After the growth, the system was cooled in Ar. Graphene was characterized by Raman spectroscopy, AFM, LEED, and STM.

H-etching of the sapphire substrate

As-received sapphire substrate has polishing scratches and contaminations, similar to SiC wafers (see Figure 5) (Mishra et al. 2016). Therefore, a hydrogen etching step was performed to remove the polishing scratches and produce nanometric terraces to grow graphene (Mishra et al. 2016). The hydrogen etching step required a temperature of 1180–1200°C for 5 min in H_2 at 450 mbar. The etching step forms nice terraces and reconstruct the sapphire substrate. Terraces are typically 2–4 nm high with an RMS value of 0.5–0.7 nm (Figure 10) (Mishra et al. 2019).

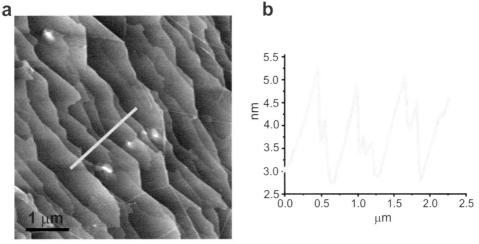

Figure 10. The H-etched surface of sapphire substrate (a) and the step profile taken along the step flow direction (b).

Graphene characterization

Raman spectroscopy

Graphene on sapphire was characterized with a Renishaw inVia system, using a 532 nm laser and a 100 X objective, with a spot size of ~ 1 μm (5% power) and an exposure time of 1s. The G and 2D bands were at 1586 cm^{-1} and 2689 cm^{-1}, respectively (see Figure 11(a)) (Das et al. 2008b). The sharp 2D band had an FWHM of around 36 cm^{-1} and the I(2D)/I(G) was higher than 2 (Figure 11(b) and 11(c)), which indicate the presence of monolayer graphene. Moreover, the I(D)/I(G) around 0.25 indicates low defects (Neumann et al. 2015) (Figure 11(d)).

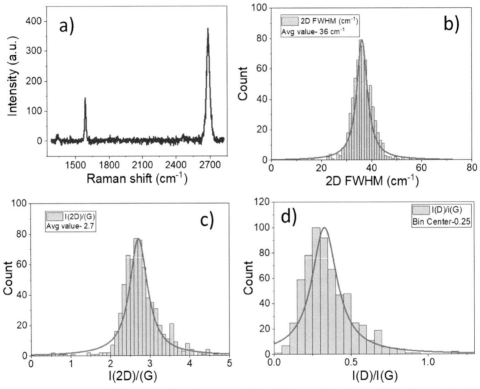

Figure 11. Raman spectra and histograms of graphene on sapphire. (a) Spectra of graphene with the characteristic 2D, G, and D bands. Histogram of (b) 2D FWHM, (c) I(2D)/I(G) and (d) I(D)/I(G) obtained from 30 × 30 μm^2 Raman maps.

Atomic force microscopy

The topography of graphene on sapphire was analyzed by Bruker AFM. Raw data were analyzed by Gwyddion software. Figure 12 shows the topography of a sapphire substrate after graphene growth. The white lines are ridges caused by the distinct thermal expansion coefficients of graphene and sapphire and are 1–2 nm high (Song et al. 2012). Intense bright spots are defects formed during the growth at high temperatures.

Scanning electron Microscopy

Zeiss merlin SEM was performed on graphene on sapphire to investigate its morphology. Low current (76 pA) and voltage (2 kV) were used to perform the imaging (Figure 13). The dark areas confirmed the presence of a continuous graphene film. The white thin connections are the wrinkle (same observed with AFM, see Figure 12) caused by the different thermal expansion coefficients

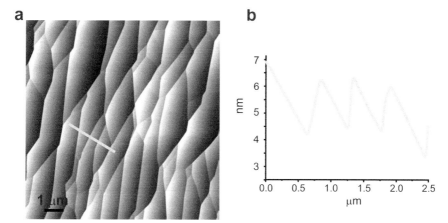

Figure 12. (a) AFM micrograph of a typical graphene on the sapphire sample, with ridges of 1–2 nm (Song et al. 2012). (b) Step profile taken along the step flow direction reveals steps of 2–3 nm.

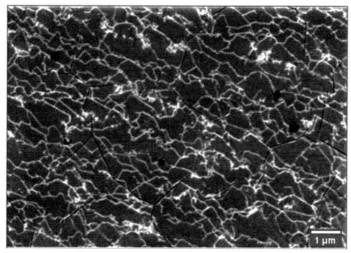

Figure 13. SEM image of sapphire after graphene growth. The white lines indicate graphene wrinkles formed due to thermal expansion coefficient (Song et al. 2012). The dark lines are the grain boundaries of the sapphire substrate.

of graphene and sapphire (Garcia et al. 2012). Agglomerated white lines are defects present during the growth. In addition, the dark connected lines are the grain boundaries of the sapphire substrate.

Low-energy electron diffraction (LEED) and scanning-tunneling microscopy (STM)

To understand better the growth mechanism and the structural properties at the graphene/sapphire interface, LEED and STM analyses were performed. LEED was conducted on as received and hydrogen etched sapphire substrates, respectively. The as-received surface was very difficult to measure, due to its insulting nature, and no LEED pattern was collected. In contrast, we were able to see a $(\sqrt{31} \times \sqrt{31})\,R9°$ reconstruction on hydrogen etched sapphire, even at 60 eV which is a quite low voltage for insulating substrate (Figure 14(a,b)). This can be explained by the Al-rich surface induced by the H-etching treatment. Despite a reasonably long history of disagreement and discussions regarding its definite atomic structure (Barth and Reichling 2001, Gautier et al. 1991, Jarvis and Carter 2001, Lauritsen et al. 2009, Renaud et al. 1994), in 2009 Lauritsen et al. were able to convince the scientific community that Al-rich (111) layer is formed on the Al-terminated c-plane of the substrate (Lauritsen et al. 2009). At present, it is known that heating the sapphire substrate of Al_2O_3 (0001) at high

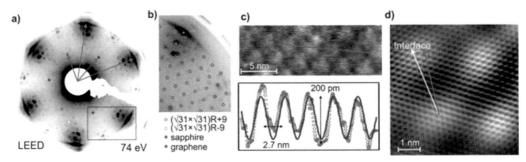

Figure 14. (a) LEED configuration of graphene on hydrogen etched sapphire, measured at 74 eV. (b) Zoomed-in of the insert in panel (a), revealing the Al_2O_3 reconstruction place over the theoretical diffraction spots. (c) STM appearance and the line profile of a part of graphene over the ($\sqrt{31} \times \sqrt{31}$)R ± 9°. (d) 2D-FFT-filtered STM image from a portion of (c) (Mishra et al. 2019).

temperature well above 1200°C in ultra-high vacuum (UHV) (Barth and Reichling 2001, Gautier et al. 1991, Lauritsen et al. 2009, Renaud et al. 1994) causes a loss of surface oxygen which leads to surface reconstruction. Our process can encourage the ($\sqrt{31} \times \sqrt{31}$)R9° reconstruction at lower temperatures and higher pressures than those previously published (Barth and Reichling 2001, Gautier et al. 1991, Lauritsen et al. 2009, Renaud et al. 1994), due to the H_2-etching, i.e., oxygen is replaced by hydrogen. Figure 14(a) shows the LEED pattern of graphene on H_2-etched sapphire and, as reported in Figure 14(b), diffraction spots measured at different locations show two rotating areas of the ($\sqrt{31} \times \sqrt{31}$)R9° reconstruction (i.e., ($\sqrt{31} \times \sqrt{31}$)R ± 9°). Figure 14(c) reports an atomic-scale resolved STM image of graphene over the H-etched ($\sqrt{31} \times \sqrt{31}$)R ± 9° reconstruction. The multifaceted arrangement of the underlying reassembled sapphire layer is visible as an intervallic arrangement of unevenly rhombus-shaped domains. In the bottommost insert of the panel (c), the dark line designates the spacing between the utmost to be (27 ± 1) Å, consistent with the insignificant length of the re-establishment of 26.5 Å. Pane (d) displays a 2-dimensional Fast Fourier Transform (2D FFT) filtered part of the image in panel (c) along with the basic vector of graphene (dark) and the reconstruction (white). The reciprocated orientation between graphene and sapphire reconstruction is 21° confirming that graphene aligns along the R30 direction with respect to the Al_2O_3 (0001) (1 × 1), as also evident from the LEED pattern in panel (a). The fundamental archetypal of the Al-rich reconstruction on sapphire is complicated by the changing of the lattice arrangement between the Al atoms in the interior of the unit cell and relies on the joint organization of the Al-rich atoms with respect to the substrate archive (Lauritsen et al. 2009). Despite the actual resolution of the STM scans preventing us from reaching a decision on the atomic prearrangement between graphene and the Al-rich atoms at the boundary, we were able to see that the graphene framework imitates well the intervallic corrugation forced by the ($\sqrt{31} \times \sqrt{31}$)R ± 9°. Al-rich reconstructed surface act as a metal catalyst and at elevated temperature, carbon decomposes easily and forms graphene. During growth, the Al-rich surface acts like strong Lewis acids that disassociate the methane molecules, and thus activate the graphene growth.

Mobility measurements at room temperature

The room temperature transport parameters (density, mobility) of the graphene film were extracted through magneto-transport measurements in the magnetic field range –0.37 T to 0.37 T (Mishra et al. 2019). To this end, several Hall bar and Van der Pauw geometry were fabricated out of each chip by first depositing Cr/Au electrodes at randomly chosen areas of the chip and by defining their shape through an aligned dry etching process, as shown in Figure 15(a) for a representative Hall bar device (Mishra et al. 2019). The light area in the optical image is the etch mask region used

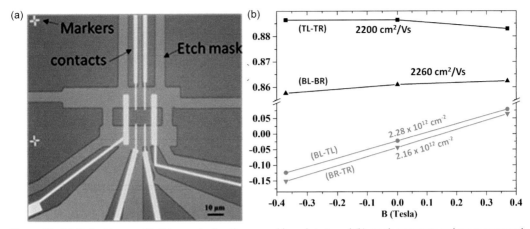

Figure 15. (a) Optical image of hall bar made directly on sapphire substrate and (b) graph represents resistance measured in magnetic field ± 0.37 T. Resistance measured between S-D, TL-TR, BL-BR, and TL-BL, whereas mobility is measured between TL-TR and BL-BR and carrier densities measured between BL-TL and BR-TR, respectively. Where abbreviations are S-source, D-Drain, TL: Top left, TR-Top right, BL-Bottom left, and BR-Bottom Right.

to remove the graphene with oxygen plasma to avoid short circuits. Moreover, the bright area is contacts evaporated with Au/Cr layer.

Electron beam lithography (EBL) was used to make hall bars. Since sapphire is an insulating substrate, a conductive film, easily soluble in water, was spun over the PMMA layer. A constant current passed between source and drain and the resistance was measured. Figure 15(b) shows the graph of the resistance vs applied magnetic field and carrier density measured between different combinations inside the channel (Figure 15(b)). The connections were checked by using the 2-probe resistance method. The resistance was measured by passing a perpetual current, and the voltage was measured between all the possible combinations (Figure 16). Figure 16 shows different hall bars prepared in one sample and named A0, A1, and B0. Graphene is p-doped, with hole densities and mobilities in the range $1.5 - 6.5 \times 10^{12}$ cm^{-2} and 1000–2500 cm^2/vs, respectively (Figure 15(b)). As a general trend, we find that lower carrier densities correspond to higher hole mobility. Finally, we speculate that the electrical performance of our chips is most probably limited by the grain-boundary-induced disorder. In this scenario, the higher carrier mobility of graphene of H-etched substrates is conceivably linked to the larger, observed graphene grain size (1 μm) (Mishra et al. 2019).

Maps of 2-probe resistances

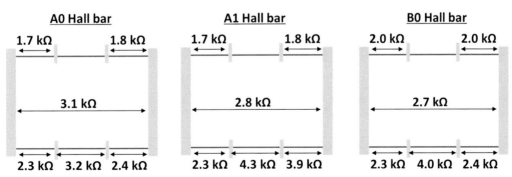

Figure 16. Two probe resistances were measured between the source, drain, and all the channels. (original)

Transfer of graphene grown on sapphire wafers

To broaden the applicative range, graphene can be transferred from sapphire to a target substrate using the poly-vinyl alcohol (PVA) lamination approach (Shivayogimath et al. 2019). For this purpose, Gr/sapphire was kept overnight in water, then the substrate was spun with polyvinyl alcohol (PVA) polymer and baked at 110°C for 1 min (Mishra et al. 2019, Shivayogimath et al. 2019). PVA film was used due to its high solubility in water. The weak bonding between graphene and sapphire (van der Waals) makes it easy to peel the graphene from the substrate with PVA film on top. The peeled film was placed onto the desired substrate (290 nm SiO$_2$/Si substrate) and heated for few minutes at 110°C for better adhesion. To remove the PVA film, the sample was kept for a couple of hours in DI H$_2$O. Figure 17 shows an optical image of graphene transferred from sapphire and the Raman spectra before and after the graphene transfer, revealing no changes in the peaks position and graphene quality.

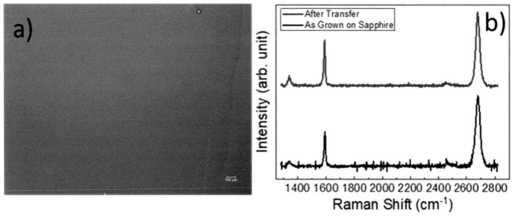

Figure 17. (a) Graphene transferred to SiO$_2$/Si (290 nm) using the PVA-assisted transfer method. (b) Raman spectra before and after graphene transfer.

Scalable approach

It is always a dream of every project to transfer the technology from the lab to the industry for large-scale applications. Graphene grown in our lab got the attention of the CEO of AIXTRON Company, Prof. Ken Teo. AIXTRON Company is the worldwide manufacturer of CVD and MOCVD reactors. Tying up with AIXTRON led to the scaling up of this project, and following the optimization of the parameters we were able to grow from 2-inch up to 8-inch wafers in an MOCVD reactor (Mishra et al. 2019). In addition, AIXTRON manufactured a production scale reactor (CCS 2D MOCVD) for the wafer-scale growth of graphene and other 2D materials. The system is easy to use, well-controlled in temperature, and flexible for 2D materials growth. In the CCS 2D MOCVD reactor (Figure 18), the susceptor is made of graphite and can be easily swapped as needed. In this reactor, we can easily grow 19 wafers of 2-inch, 5 wafers of 4-inch, 1 wafer of 6-inch, or 1 wafer of 8-inch, as can be seen from Fig. 18.

Figure 18. AIXTRON CCS 2D MOCVD reactor for graphene and h-BN growth. The susceptor is easy to swap from 2-inch to 8-inch wafers (Mishra et al. 2019).

References

Banszerus, L., Schmitz, M., Engels, S., Dauber, J., Oellers, M., Haupt, F., Watanabe, K., Taniguchi, T., Beschoten, B. and Stampfer, C. 2015. Ultrahigh-mobility graphene devices from chemical vapor deposition on reusable copper. *Science Advances*, 1(6): e1500222.

Bao, Qiaoliang and Kian Ping Loh. 2012. Graphene photonics, plasmonics, and broadband optoelectronic devices. *ACS Nano*, 6(5): 3677–94.

Barth, C. and M. Reichling. 2001. Imaging the atomic arrangements on the high-temperature reconstructed α-Al2O3(0001) Surface. *Nature*, 414(6859): 54–57.

Berger, C., Song, Z., Li, T., Li, X., Ogbazghi, A.Y., Feng, R., Dai, Z., Marchenkov, A.N., Conrad, E.H., First, P.N. and de Heer, W.A. 2004. Ultrathin epitaxial graphite: 2D electron gas properties and a route toward graphene-based nanoelectronics. *The Journal of Physical Chemistry B*, 108(52): 19912–16. https://pubs.acs.org/doi/10.1021/jp040650f.

Bianco, F., Perenzoni, D., Convertino, D., De Bonis, S. L., Spirito, D., Perenzoni, M., Coletti, C., Vitiello, M.S. and Tredicucci, A. 2015a. Terahertz detection by epitaxial-graphene field-effect-transistors on silicon carbide. *Applied Physics Letters*, 107(13): 131104.

Bianco, F., Miseikis, V., Convertino, D., Xu, J.-H., Castellano, F., Beere, H. E., Ritchie, D. a., Vitiello, M. S., Tredicucci, A. and Coletti, C. 2015b. THz saturable absorption in turbostratic multilayer graphene on silicon carbide. *Optics Express*, 23(9): 11632. http://www.opticsexpress.org/abstract.cfm?URI=oe-23-9-11632.

Bonaccorso, F., Lombardo, A., Hasan, T., Sun, Z., Colombo, L. and Ferrari, A.C. 2012. Production and processing of graphene and 2d crystals. *Materials Today*, 15(12): 564–89. http://dx.doi.org/10.1016/S1369-7021(13)70014-2.

Camara, N., Huntzinger, J.R., Rius, G., Tiberj, A., Mestres, N., Pérez-Murano, F., Godignon, P. and Camassel, J. 2009. Anisotropic growth of long isolated graphene ribbons on the C face of graphite-capped 6H-SiC. *Physical Review B - Condensed Matter and Materials Physics*, 80(12): 1–8.

Candini, A., Martini, L., Chen, Z., Mishra, N., Convertino, D., Coletti, C., Narita, A., Feng, X., Müllen, K. and Affronte, M. 2015. Electroburning of few-layer graphene flakes, epitaxial graphene, and turbostratic graphene discs in air and under vacuum. *Beilstein Journal of Nanotechnology*, 6(1): 711–19.

Candini, A., Martini, L., Chen, Z., Mishra, N., Convertino, D., Coletti, C., Narita, A., Feng, X., Müllen, K. and Affronte, M. 2017. High photoresponsivity in graphene nanoribbon field-effect transistor devices contacted with graphene electrodes. *Journal of Physical Chemistry C*, 121(19): 10620–25.

Casiraghi, C., Pisana, S., Novoselov, K.S., Geim, A.K. and Ferrari, A.C. 2007. Raman fingerprint of charged impurities in graphene. *Applied Physics Letters*, 91(23): 1–4.

Castro Neto, A.H., Guinea, F., Peres, N.M.R., Novoselov, K.S. and Geim, A.K 2009. The electronic properties of graphene. *Reviews of Modern Physics*, 81(1): 109–62. https://link.aps.org/doi/10.1103/RevModPhys.81.109.

Chen, J.W., Huang, H.C., Convertino, D., Coletti, C., Chang, L.Y., Shiu, H.W., Cheng, C.M., Lin, M.F., Heun, S., Chien, F.S. Sen, Chen, Y.C., Chen, C.H. and Wu, C.L. 2016. Efficient N-type doping in epitaxial graphene through strong lateral orbital hybridization of Ti adsorbate. *Carbon*, 109: 300–305.

Chen, Z., Qi, Y., Chen, X., Zhang, Y. and Liu, Z. 2019. Direct CVD growth of graphene on traditional glass: Methods and mechanisms. *Advanced Materials*, 31(9): 1803639.

Coletti, C., Riedl, C., Lee, D.S., Krauss, B., Patthey, L., von Klitzing, K., Smet, J.H. and Starke, U. 2010. Charge neutrality and band-gap tuning of epitaxial graphene on SiC by molecular doping. *Physical Review B - Condensed Matter and Materials Physics*, 81(23): 1–8.

Convertino, D., Rossi, A., Miseikis, V., Piazza, V. and Coletti, C . 2016. Thermal decomposition and chemical vapor deposition: A comparative study of multi-layer growth of graphene on SiC(000-1). *MRS Advances*, 1(55): 3667–72. http://www.journals.cambridge.org/abstract_S2059852116000037.

Convertino, Domenica, Stefano Luin, Laura Marchetti and Camilla Coletti. 2018. Peripheral neuron survival and outgrowth on graphene. *Frontiers in Neuroscience*, 12(JAN): 1–8.

Das, Anindya, Biswanath Chakraborty and Sood, A.K. 2008a. Raman spectroscopy of graphene on different substrates and influence of defects. *Bulletin of Materials Science*, 31(3): 579–84.

Das, A., Pisana, S., Chakraborty, B., Piscanec, S., Saha, S.K., Waghmare, U.V., Novoselov, K.S., Krishnamurthy, H.R., Geim, A.K., Ferrari, A.C. and Sood, A.K. 2008b. Monitoring dopants by raman scattering in an electrochemically top-gated graphene transistor. *Nature Nanotechnology*, 3(4): 210–15.

Deng, Bing, Zhongfan Liu and Hailin Peng. 2019. Toward mass production of CVD graphene films. *Advanced Materials*, 31(9): 1800996.

De Fazio, D., Purdie, D.G., Ott, A.K., Braeuninger-Weimer, P., Khodkov, T., Goossens, S., Taniguchi, T., Watanabe, K., Livreri, P., Koppens, F.H.L., Hofmann, S., Goykhman, I., Ferrari, A.C. and Lombardo, A. 2019. High-mobility, wet-transferred graphene grown by chemical vapor deposition. *ACS Nano*, 13(8): 8926–35.

de Heer, W.A., Berger, C., Ruan, M., Sprinkle, M., Li, X., Hu, Y., Zhang, B., Hankinson, J. and Conrad, E. 2011. Large area and structured epitaxial graphene produced by confinement controlled sublimation of silicon carbide. *Proceedings of the National Academy of Sciences of the United States of America*, 108(41): 16900–905. http://www.ncbi.nlm.nih.gov/pubmed/21960446%0Ahttp://www.pubmedcentral.nih.gov/articlerender. fcgi?artid=PMC3193246.

Emtsev, K. V., Speck, F., Seyller, T., Ley, L. and Riley, J.D. 2008. Interaction, growth, and ordering of epitaxial graphene on SiC{0001} surfaces: A comparative photoelectron spectroscopy study. *Physical Review B - Condensed Matter and Materials Physics*, 77(15): 1–10.

Emtsev, Konstantin, V., Bostwick, A., Horn, K., Jobst, J., Kellogg, G.L., Ley, L., McChesney, J.L., Ohta, T., Reshanov, S.A., Röhrl, J., Rotenberg, E., Schmid, A.K., Waldmann, D., Weber, H.B. and Seyller, T. 2009. Towards wafer-size graphene layers by atmospheric pressure graphitization of silicon carbide. *Nature Materials*, 8(3): 203–7. http://dx.doi. org/10.1038/nmat2382.

Fanton, M.A., Robinson, J.A., Puls, C., Liu, Y., Hollander, M.J., Weiland, B.E., Labella, M., Trumbull, K., Kasarda, R., Howsare, C., Stitt, J. and Snyder, D.W. 2011. Characterization of graphene films and transistors grown on sapphire by metal-free chemical vapor deposition. *ACS Nano*, 5(10): 8062–69.

Ferrari, A.C., Meyer, J.C., Scardaci, V., Casiraghi, C., Lazzeri, M., Mauri, F., Piscanec, S., Jiang, D., Novoselov, K.S., Roth, S. and Geim, A.K. 2006. Raman spectrum of graphene and graphene layers. *Physical Review Letters*, 97(18): 1–4.

Ferrari, Andrea C. 2007. Raman spectroscopy of graphene and graphite: Disorder, electron-phonon coupling, doping and nonadiabatic effects. *Solid State Communications*, 143(1-2): 47–57.

First, P.N., de Heer, W.A., Seyller, T., Berger, C., Stroscio, J.A. and Moon, J.-S. 2010. Epitaxial graphenes on silicon carbide. *MRS Bulletin*, 35(4): 296–305.

Forti, Stiven. 2014. Large-area epitaxial graphene on SiC(0001): From decoupling to interface engineering. http://opus4. kobv.de/opus4-fau/frontdoor/index/index/docId/4933.

Frewin, C.L., Coletti, C., Riedl, C., Starke, U. and Saddow, S.E. 2009. A comprehensive study of hydrogen etching on the major SiC polytypes and crystal orientations. *Materials Science Forum*, 615–617(July 2015): 589–92.

Gao, L., Ren, W., Xu, H., Jin, L., Wang, Z., Ma, T., Ma, L.P., Zhang, Z., Fu, Q., Peng, L.M., Bao, X. and Cheng, H.M. 2012. Repeated growth and bubbling transfer of graphene with millimetre-size single-crystal grains using platinum. *Nature Communications*, 3.

Garcia, A.G.F., Neumann, M., Williams, J.R., Watanabe, K., Taniguchi, T. and Goldhaber-gordon, D. 2012. Effective Cleaning of Hexagonal Boron Nitride for Graphene Devices. *Nano Letters*, 12(9): 4449–4454.

Gautier, M., Duraud, J.P., Pham L. Van and Guittet, M.J. 1991. Modifications of α-Al2O3(0001) surfaces induced by thermal treatments or ion bombardment. *Surface Science*, 250(1–3): 71–80.

Haghighian, N., Convertino, D., Miseikis, V., Bisio, F., Morgante, A., Coletti, C., Canepa, M. and Cavalleri, O. 2018. Rippling of graphitic surfaces: A comparison between few-layer graphene and HOPG. *Physical Chemistry Chemical Physics*, 20(19): 13322–30.

Hao, Y., Wang, Y., Wang, L., Ni, Z., Wang, Z., Wang, R., Koo, C.K., Shen, Z. and Thong, J.T.L. 2010. Probing layer number and stacking order of few-layer graphene by raman spectroscopy. *Small*, 6(2): 195–200.

Hibino, H., Kageshima, H., Maeda, F., Nagase, M., Kobayashi, Y. and Yamaguchi, H. 2008. Microscopic thickness determination of thin graphite films formed on SiC from quantized oscillation in reflectivity of low-energy electrons. *Physical Review B - Condensed Matter and Materials Physics*, 77(7): 1–7.

Hupalo, M., Conrad, E.H. and Tringides, M.C. 2009. Growth mechanism for epitaxial graphene on vicinal 6H-SiC (0001) surfaces: A scanning tunneling microscopy study. *Physical Review B - Condensed Matter and Materials Physics*, 80(4): 1–4.

Hwang, J., Kim, M., Campbell, D., Alsalman, H.A., Kwak, J.Y., Shivaraman, S., Woll, A.R., Singh, A.K., Hennig, R.G., Gorantla, S., Rümmeli, M.H. and Spencer, M.G. 2013. Van Der Waals epitaxial growth of graphene on sapphire by chemical vapor deposition without a metal catalyst. *ACS Nano*, 7(1): 385–95.

Ismach, A., Druzgalski, C., Penwell, S., Schwartzberg, A., Zheng, M., Javey, A., Bokor, J. and Zhang, Y. 2010. Direct chemical vapor deposition of graphene on dielectric surfaces. *Nano Letters*, 10(5): 1542–48.

Jarvis, Emily A.A. and Emily A. Carter. 2001. Metallic character of the Al 2 O 3 (0001)–($\sqrt{31} \times \sqrt{31}$) R ± 9° surface reconstruction. *The Journal of Physical Chemistry B*, 105(18): 4045–52. https://pubs.acs.org/doi/10.1021/jp003587c.

Johansson, L.I., Watcharinyanon, S., Zakharov, A.A., Iakimov, T., Yakimova, R. and Virojanadara, C. 2011. Stacking of adjacent graphene layers grown on C-Face SiC. *Physical Review B - Condensed Matter and Materials Physics*, 84(12): 1–8.

Kang, Jung-hyun, Yuval Ronen, Yonatan Cohen and Domenica Convertino. 2016. MBE growth of self—assisted in as nanowires on graphene. *Semiconductor Science and Technology*, 31(11): 115005. http://dx.doi.org/10.1088/0268-1242/31/11/115005.

Kellar, Joshua A., Justice M.P. Alaboson, Qing Hua Wang and Mark C. Hersam. 2010. Identifying and characterizing epitaxial graphene domains on partially graphitized SiC(0001) surfaces using scanning probe microscopy. *Applied Physics Letters*, 96(14): 1–4.

Kim, Y., Moyen, E., Yi, H., Avila, J., Chen, C., Asensio, M.C., Lee, Y.H. and Pribat, D. 2018. Synthesis of high quality graphene on capped (1 1 1) Cu thin films obtained by high temperature secondary grain growth on c -plane sapphire substrates. *2D Materials*, 5(3): 035008.

Kruskopf, M., Pierz, K., Wundrack, S., Stosch, R., Dziomba, T., Kalmbach, C.C., Müller, A., Baringhaus, J., Tegenkamp, C., Ahlers, F.J. and Schumacher, H.W. 2015. Epitaxial graphene on SiC: Modification of structural and electron transport properties by substrate pretreatment. *Journal of Physics Condensed Matter*, 27(18).

Lauritsen, J.V., Jensen, M.C.R., Venkataramani, K., Hinnemann, B., Helveg, S., Clausen, B.S. and Besenbacher, F. 2009. Atomic-scale structure and stability of the 31×31R9° surface of Al2O3(0001). *Physical Review Letters*, 103(7): 8–11.

Lee, Changgu, Xiaoding Wei, Jeffrey W Kysar and James Hone. 2008a. Measurement of the elastic properties and intrinsic strength of monolayer graphene. *Science*, 321(5887): 385–88. http://www.sciencemag.org/cgi/doi/10.1126/science.1157996.

Lee, D.S., Riedl, C., Krauss, B., von Klitzing, K., Starke, U. and Smet, J.H. 2008b. Raman spectra of epitaxial graphene on SiC and of epitaxial graphene transferred to SiO 2. *Nano Letters*, 8(12): 4320–25. https://pubs.acs.org/doi/10.1021/nl802156w.

Lee, J.E., Ahn, G., Shim, J., Lee, Y.S. and Ryu, S. 2012. Optical separation of mechanical strain from charge doping in graphene. *Nature Communications*, 3(May): 1024–28. http://dx.doi.org/10.1038/ncomms2022.

Li, X., Cai, W., An, J., Kim, S., Nah, J., Yang, D., Piner, R., Velamakanni, A., Jung, I., Tutuc, E., Banerjee, S.K., Colombo, L. and Ruoff, R.S. 2009. Large area synthesis of high quality and uniform graphene films on copper foils. *Science*, 324(5932): 1312–14.

Lupina, G., Kitzmann, J., Costina, I., Lukosius, M., Wenger, C., Wolff, A., Vaziri, S., Östling, M., Pasternak, I., Krajewska, A., Strupinski, W., Kataria, S., Gahoi, A., Lemme, M.C., Ruhl, G., Zoth, G., Luxenhofer, O. and Mehr, W. 2015. Residual metallic contamination of transferred chemical vapor deposited graphene. *ACS Nano*, 9(5): 4776–85.

Malard, L.M., Pimenta, M.A., Dresselhaus, G. and Dresselhaus, M.S. 2009. Raman spectroscopy in graphene. *Physics Reports*, 473(5-6): 51–87. http://dx.doi.org/10.1016/j.physrep.2009.02.003.

Mashoff, T., Convertino, D., Miseikis, V., Coletti, C., Piazza, V., Tozzini, V., Beltram, F. and Heun, S. 2015. Increasing the active surface of titanium islands on graphene by nitrogen sputtering. *Applied Physics Letters*, 106(8).

Meng, F., Thomson, M.D., Bianco, F., Rossi, A., Convertino, D., Tredicucci, A., Coletti, C. and Roskos, H.G. 2016. Saturable absorption of femtosecond optical pulses in multilayer turbostratic graphene. *Optics Express*, 24(14): 15261. http://www.opticsexpress.org/abstract.cfm?URI=oe-24-14-15261.

Miseikis, V, Convertino, D., Mishra, N., Gemmi, M., Mashoff, T., Heun, S., Haghighian, N., Bisio, F., Canepa, M., Piazza, V. and Coletti, C. 2015. Rapid CVD growth of millimetre-sized single crystal graphene using a cold-wall reactor. *2D Materials*, 2(1): 014006. http://stacks.iop.org/2053-1583/2/i=1/a=014006?key=crossref.ed1d1af3c6014f3c03dfbbeccafe3504.

Miseikis, V., Xiang, S., Roddaro, S., Heun, S. and Coletti, C. 2017a. Perfecting the growth and transfer of large single-crystal CVD graphene: A platform material for optoelectronic applications. *Carbon Nanostructures*, 0(9783319581323): 113–24.

Miseikis, Vaidotas, Bianco, F., David, J., Gemmi, M., Pellegrini, V., Romagnoli, M. and Coletti, C. 2017b. Deterministic patterned growth of high-mobility large-crystal graphene: A path towards wafer scale integration. *2D Materials*, 4(2).

Mishra, Neeraj, Miseikis, V., Convertino, D., Gemmi, M., Piazza, V. and Coletti, C. 2016. Rapid and catalyst-free van der waals epitaxy of graphene on hexagonal boron nitride. *Carbon*, 96: 497–502. http://dx.doi.org/10.1016/j.carbon.2015.09.100.

Mishra, N., Forti, S., Fabbri, F., Martini, L., McAleese, C., Conran, B., Whelan, P.R., Shivayogimath, A., Buß, L., Falta, J., Aliaj, I., Roddaro, S., Flege, J.I., Bøggild, P., Teo, K.B.K. and Coletti, C. 2019. Going beyond copper: Wafer-scale synthesis of graphene on sapphire. 1904906: 1–8. http://arxiv.org/abs/1907.01610.

Mishra, Neeraj, Forti, S., Fabbri, F., Martini, L., McAleese, C., Conran, B.R., Whelan, P.R., Shivayogimath, A., Jessen, B.S., Buß, L., Falta, J., Aliaj, I., Roddaro, S., Flege, J.I., Bøggild, P., Teo, K.B.K. and Coletti, C. 2019. Wafer-scale synthesis of graphene on sapphire: Toward fab-compatible graphene. *Small*, 15(50).

Neumann, C., Reichardt, S., Venezuela, P., Drögeler, M., Banszerus, L., Schmitz, M., Watanabe, K., Taniguchi, T., Mauri, F., Beschoten, B., Rotkin, S.V. and Stampfer, C. 2015. Raman spectroscopy as probe of nanometre-scale strain variations in graphene. *Nature Communications*, 6(May): 1–7.

Novoselov, K.S., Geim, A. K., Morozov, S. V, Jiang, D., Zhang, Y., Dubonos, S. V, Grigorieva, I.V. and Firsov, A.A. 2004. Electric field effect in atomically thin carbon films. *Science*, 306(5696): 666–69.

Park, Sungjin and Rodney S. Ruoff. 2009. Chemical methods for the production of graphenes. *Nature Nanotechnology*, (4): 217–224.

Prakash, G., Capano, M. A., Bolen, M.L., Zemlyanov, D. and Reifenberger, R.G. 2010. AFM study of ridges in few-layer epitaxial graphene grown on the carbon-face of 4H-SiC (000Γ). *Carbon*, 48(9): 2383–93. http://dx.doi.org/10.1016/j.carbon.2010.02.026.

Qingkai Yu, Jie Lian, Sujitra Siriponglert, Hao Li, Yong P. Chenq and Pei, S.-S. 2008. Graphene segregated on Ni surfaces and transferred to insulators. *Applied Physics Letters*, 93(June 2008): 113103.

Ramachandran, V., Brady, M.F., Smith, a.R., Feenstra, R.M. and Greve, D.W. 1998. Preparation of atomically flat surfaces on silicon carbide using hydrogen etching. *Journal of Electronic Materials*, 27(4): 308–12.

Razado-Colombo, I., Avila, J., Chen, C., Nys, J.P., Wallart, X., Asensio, M.C. and Vignaud, D. 2015. Probing the electronic properties of graphene on C-Face SiC down to single domains by nanoresolved photoelectron spectroscopies. *Physical Review B - Condensed Matter and Materials Physics*, 92(3): 1–10.

Renaud, G., Villette, B., Vilfan, I. and Bourret, A. 1994. Atomic structure of the α-Al2O3(0001)(31 × 31) R ± 9° reconstruction. *Physical Review Letters*, 73(13): 1825–28.

Riedl, C., Coletti, C., Iwasaki, T., Zakharov, A.A. and Starke, U. 2009. Quasi-free-standing epitaxial graphene on SiC obtained by hydrogen intercalation. *Physical Review Letters*, 103(24): 1–4.

Röhrl, J., Hundhausen, M., Emtsev, K.V., Seyller, T., Graupner, R. and Ley, L. 2008. Raman spectra of epitaxial graphene on SiC(0001). *Applied Physics Letters*, 92(20): 20–23.

Shin, Jae Hyeok, Su Han Kim, Sun Sang Kwon and Won Il Park. 2018. Direct CVD growth of graphene on three-dimensionally-shaped dielectric substrates. *Carbon*, 129: 785–89.

Shivaraman, Shriram, M.V.S. Chandrashekhar, John J. Boeckl and Michael G. Spencer. 2009. Thickness estimation of epitaxial graphene on sic using attenuation of substrate raman intensity. *Journal of Electronic Materials*, 38(6): 725–30.

Shivayogimath, A., Whelan, P.R., MacKenzie, D.M.A., Luo, B., Huang, D., Luo, D., Wang, M., Gammelgaard, L., Shi, H., Ruoff, R.S., Bøggild, P. and Booth, T.J. 2019. Do-it-yourself transfer of large-area graphene using an office laminator and water. *Chemistry of Materials*, 31(7): 2328–36.

Song, H.J., Son, M., Park, C., Lim, H., Levendorf, M.P., Tsen, A.W., Park, J. and Choi, H.C. 2012. Large scale metal-free synthesis of graphene on sapphire and transfer-free device fabrication. *Nanoscale*, 4(10): 3050.

Song, X., Gao, T., Nie, Y., Zhuang, J., Sun, J., Ma, D., Shi, J., Lin, Y., Ding, F., Zhang, Y. and Liu, Z. 2016. Seed-assisted growth of single-crystalline patterned graphene domains on hexagonal boron nitride by chemical vapor deposition. *Nano Letters*, 16(10): 6109–16.

Starke, U., Forti, S., Emtsev, K.V. and Coletti, C. 2012. Engineering the electronic structure of epitaxial graphene by transfer doping and atomic intercalation. *MRS Bulletin*, 37(12): 1177–86.

Strupinski, W., Grodecki, K., Wysmolek, A., Stepniewski, R., Szkopek, T., Gaskell, P. E., Grüneis, A., Haberer, D., Bozek, R., Krupka, J. and Baranowski, J.M. 2011. Graphene epitaxy by chemical vapor deposition on SiC. *Nano Letters*, 11(4): 1786–91. https://pubs.acs.org/doi/10.1021/nl200390e.

Su, C.-Y., Lu, A.-Y., Wu, C.-Y., Li, Y.-T., Liu, K.-K., Zhang, W., Lin, S.-Y., Juang, Z.-Y., Zhong, Y.-L., Chen, F.-R. and Li, L.-J. 2011. Direct formation of wafer scale graphene thin layers on insulating substrates by chemical vapor deposition. *Nano Letters*, 11(9): 3612–16.

Tamayo, J. and García, R. 2002. Deformation, contact time, and phase contrast in tapping mode scanning force microscopy. *Langmuir*, 12(18): 4430–35.

Wang, Y., hua Ni, Z., Yu, T., Xiang Shen, Z., min Wang, H., hong Wu, Y., Chen, W. and Thye Shen Wee, A. 2008. Raman studies of monolayer graphene: The substrate effect. *Journal of Physical Chemistry C*, 112(29): 10637–40. https://pubs.acs.org/doi/pdf/10.1021/jp8008404.

Zanotto, S., Bianco, F., Miseikis, V., Convertino, D., Coletti, C. and Tredicucci, A. 2017. Coherent absorption of light by graphene and other optically conducting surfaces in realistic on-substrate configurations. *APL Photonics*, 2(1): 016101. http://dx.doi.org/10.1063/1.4967802%5Cnhttp://scitation.aip.org/content/aip/journal/app/2/1/10.1063/1.4967802.

Zhang, C., Huang, J., Tu, R., Zhang, S., Yang, M., Li, Q., Shi, J., Li, H., Zhang, L., Goto, T. and Ohmori, H. 2018. Transfer-free growth of graphene on Al2O3 (0001) using a three-step method. *Carbon*, 131: 10–17.

CHAPTER 15

Nanostructures for Hydrogen Storage

Bholanath Mukherjee,[1,*] *Vikaskumar Gupta*[2]
and *Suyash Agnihotri*[3]

1. Introduction

Hydrogen was discovered by scientist Henry Cavendish FRS (10th Oct 1731–24th Oct 1810) in 1766 which he termed "flammable air". Hydrogen is a Greek word meaning "maker of water". The name Hydrogen was put forward by Antoine Lavoisier in 1783 after reproducing Cavendish's experiment and realizing that it produces water on combustion. Hydrogen is the smallest known element having atomic number 1 and is found as three different isotopes viz. hydrogen, deuterium and tritium having mass number 1, 2 and 3, respectively.

A hydrogen molecule is diatomic and the atoms are bonded together by a covalent bond. The most abundant isotope of hydrogen atom nucleus contains one proton and one electron. The protons have a magnetic moment and the proton's spin associated with it is 1/2. Therefore, the molecular hydrogen exists as two isomers. The molecule which has its two proton spins aligned parallel form a triplet state, designated as ortho-hydrogen, while the other which has its two proton spins aligned antiparallel form a singlet state, designated as para-hydrogen. The ratio of ortho-hydrogen and para-hydrogen at thermal equilibrium is found to be 75 to 25. These states are identical (physically and chemically), but differ nominally in their specific heats and spectrograph. Interestingly, when hydrogen is cooled, it eventually becomes pure para-hydrogen (McCarty et al. 1981).

Hydrogen is an odourless, colourless, tasteless gas, which does not react at ambient temperature. At normal pressure, volume occupied by 1 g of hydrogen gas is about 11 liters. The melting point and boiling point of hydrogen are 14 K and 20.4 K, respectively. It is highly inflammable (at 6.2% concentration in air) and is sparingly soluble in water (1.93 ml of hydrogen dissolves in 100 ml of water at 0°C).

Extensive research attention is being rendered towards the utilization and exploitation of sustainable energy sources including hydrogen energy due to excessive demand of various fossil fuels and destruction of environment. The researchers have been putting their efforts for designing and developing advanced devices for energy conversion and storage. The ever-increasing energy requirements has created the situation to explore various resources of sustainable energy. For instance, in 2018, the total electricity consumption worldwide was found to be 22315 Terawatt-hour

[1] Department of Chemistry, K. V. Pendharkar College, Dombivli, India.
[2] Department of Physical Science and Engineering, Nagoya Institute of Technology, Nagoya, Aichi, Japan.
[3] Department of Physics, K. V. Pendharkar College, Dombivli, India.
* Corresponding author: btmukherjee@gmail.com

(TWh), which is 4.0% higher than it was in 2017. Thus, due to increase in demand, more efficient and reliable components are being developed to meet the growing demands.

Due to the inherent properties of hydrogen, this gas is considered as the future fuel. This hydrogen energy is easily convertible into other forms like mechanical energy, and electrical energy with a fair amount of efficiency. In addition, greenhouse gases like carbon dioxide and such other pollutants are not emitted in this process. Francis Bacon (1959) had developed a 5-kW fuel cell system and demonstrated a mixture of hydrogen and air to be a viable energy source. Water, one of the most important source of hydrogen, faces no the danger of getting depleted as in case of fossil fuels. These alternative sources should not use different fossil fuels and should be carbon neutral, that is, non-polluting. Hydrogen has become an interesting alternative fuel in recent times due to continuous increase in global energy requirement and concern for the environmental safety. The main roadblock towards hydrogen being used as a commercial fuel is its on-board storage in adequate quantity. High storage capacity of hydrogen is required when it acts as an energy carrier in case of high-energy density rechargeable batteries or in fuel cells. Thus, in the present scenario of energy requirements, hydrogen seems to be the key for clean, green, reliable and sustainable energy source.

2. Advantages and disadvantages of hydrogen gas as a future fuel

Abundance of hydrogen is what makes it a future fuel as it can be easily generated by different techniques from water (Winter and Nitsch 1988).

The energy considerations show that hydrogen is most efficient when compared with other available fuels. One kilogram of hydrogen contains energy equivalent to 2.1 kilogram of natural gas or equivalent to 2.8 kilogram of gasoline. Hydrogen energy can be converted into different forms of energy like mechanical energy, electrical energy among others with 50–60% high efficiencies. Hydrogen can be used in a fuel cell (Schlapbach and Zuettel 2001, Ogden et al. 2002).

Hydrogen gas behaves as a truly sustainable and clean fuel in an internal combustion engine very much like other alternative fuels.

It is imperative to consider hydrogen to be a green fuel. It is perfectly non-polluting, as the exhaust formed on combustion is water, whereas gasoline and oils are hazardous and highly toxic to plants and animals. Also, hydrogen diffuses easily in air and in case of accidental spillage, it evaporates rapidly. Oil and gasoline spillage calls for elaborate clean-up efforts. The measures have been found to be inadequate as this spilled oil and gasoline penetrates into the immediate ecosystem causing long term and more often irreparable damage (Phillips et al. 1972).

Notwithstanding the high inflammability, the ability of hydrogen gas to disperse easily into the surrounding, in case of an accidental leakage, makes it a safer fuel. On the other hand, propane and gasoline have relatively high densities due to which the dispersal rate become slow, thus increasing the chances of explosion. Only 1% concentration of gasoline is sufficient in atmosphere to cause explosion while concentration of hydrogen should be about 4% to cause the same effect (Kalyanam et al. 1987). Thus, hydrogen is an excellent candidate for clean energy source, unlike fossil fuels.

There are definitely a few disadvantages of hydrogen to be a clean energy and significant work needs to be carried out to make this a viable source. The hydrogen cylinders are too heavy to be fitted in a vehicle. The most worrisome is that the flame of hydrogen runs back into the cylinder to cause an explosion.

The volume occupied by one kilogram of hydrogen gas at NTP conditions is about 10 m^3. Hence, to drive a hydrogen-powered car up to 600 km, it needs at least 4 kg of hydrogen. This volume needs to be stored in a small space for any practical utility. Moreover, the hydrogen cylinders should be so designed to be safe, compact, lasting, light in weight, and economical and in addition, should not contain any toxic pollutant (Das et al. 2002).

The relative energy content of hydrogen (liquid), CNG and gasoline in kWh/liter unit are 2.36, 5.8 and 8.76, respectively. Therefore, in comparison to gasoline or natural gas, storing hydrogen in liquid form is not economical. These facts are relevant to determine the size of a hydrogen cylinder and that of a gasoline tank in addition to the explosion hazard associated with the hydrogen. Other properties of hydrogen gas are that it is very light in comparison to air (about 14 times), and has high flammability and low flash point. Hydrogen has 20.4 K as boiling point and 14 K as melting point (Greenwood et al. 1997). Another hurdle is that the kinetic molecular diameter of hydrogen is 0.289 nm, which is much lesser than 0.335–0.350 nm, the interlayer distance found in graphite like materials (Bacon 1951).

Hence, the necessity to find a suitable system for storing significant amount of the hydrogen gas in a container of small size becomes more pertinent.

3. Various methods of hydrogen storage

3.1 Gaseous hydrogen

Different techniques of hydrogen storage in gaseous forms are discussed here:

3.1.1 Metal hydrides

Metal hydrides (MHs) composed from different alloys functions by adsorption of hydrogen gas and then desorption later on under different possibilities of ambient temperature, elevated temperature or when the external pressure is reduced. Magnesium hydride shows up to 7.6 wt% reversible hydrogen storage capacity for on board applications. MHs absorb a relatively small amount of gas. The advantage of MHs is that hydrogen can be safely delivered at a constant pressure. The purity of hydrogen gas is responsible for the life of a MH container in which it is stored. Accordingly, as the MH container is used repeatedly, impurities get adsorbed by the MHs, possibly acting as poisons. This reduces the hydrogen storing capacity as well as the lifetime of the MH container gets compromised as the adsorbed impurities do not get desorbed. The kinetics as well as the reversibility being very slow limits the use of MHs as hydrogen storage facility. Also, the systems are heavy and expensive for regular usage. Balde et al. (2008), in their work on hydrogen storage by sodium aluminum hydride ($NaAlH_4$), proved that if MHs used are of nano size particles, then it reduces desorption temperature of hydrogen and also its activation energy. In addition, the reloading pressure is also lowered by use of smaller particle size.

Sodium borohydride ($NaBH_4$) is another favourable option as hydrogen storage material, as its adsorption capacity is 10.8 wt%. In spite of various efforts made by researchers, the Department of Energy, USA, denied permission to $NaBH_4$ to be used in automotives (Demirci et al. 2009). Jain et al. (2010) made efforts towards decreasing the desorption temperature and simultaneously increase the kinetics, and thereby cycle life of Mg based hydrides. The kinetics has been improved by the addition of suitable catalysts into the system and by introducing defects and surface modifications by ball milling. Severa et al. (2010) found specific conditions to store more than 11 wt% hydrogen, reversibly. This was achieved by converting magnesium diboride to magnesium borohydride by direct hydrogenation process. Zhang et al. (2012) carried out direct synthesis of Mg-B-H compounds and studied its storage characteristics. A remarkable feat of 6.7 wt% reversible hydrogen adsorption capacity was achieved at 303 K by employing thermodynamic destabilization, thereby decreasing the kinetic barriers, which had never been achieved earlier (Zhang et al. 2021).

3.2 Compressed hydrogen gas in high pressure tank

Hydrogen is usually stored in tanks by compression. The storage capacity of a compressed gas tank may be increased gravimetrically and volumetrically by two different methods. The first method involves cryogenic compressed tanks, which is based on Charles's Law. At constant pressure, the volumetric capacity of a gas tank increases with the decrease in the tank temperature. The energy requirement to achieve this is very large, the volume occupied by the gas is also very large which reduces the energy density. The hydrogen storage tanks can be manufactured to withstand very high pressure to the tune of $6,000 \, Kgm^{-2}$ but the molecular diameter of hydrogen molecules being very small, these get diffused through the matrix of the metallic container at $200 \, Kgm^{-2}$ and above. This process makes the metal brittle. Accordingly, the cylinder to be mounted in the vehicle has to be a thick-walled cylinder. Hence, the disadvantage will be that the actual weight of the hydrogen gas will be much less compared to the weight of the hydrogen cylinder. Hence, the vehicle will carry more weight, which makes the system uneconomical. Compressed gas cylinder may be suitable where the container is kept stationary like a household cooking gas cylinder. Considering the shortage of conventional electrical power and the problems which are being faced with the load shedding of electrical power and having to use diesel generators, compressed hydrogen gas cylinders instead of diesel could be the answer.

3.3 Glass micro-sphere

Hollow glass micro-microspheres (HGMs) have diameter range of 25–500 μm while the thickness of the walls ranges between 1–8 μm. In HGMs, the hydrogen gas is stored by the diffusion of the molecules in the porous wall of the HGMs. Hydrogen gas is filled in these HGMs at definite pressure and at a temperature range of 473–673 K. These glass spheres are cooled to almost room temperature so that the hydrogen gas remains trapped inside these spheres (Morgan and Sissine 1995). Dalai et al. (2015) has reported hydrogen storage of 3.26 wt% in HGMs at 473 K and at a pressure of 1MPa. Some researchers have reported 2.2 wt% in HGMs at 673K and 34.5 MPa (Shelby et al. 2009). These spheres show hydrogen storage capacity of about 5–6 wt% at pressure of 20–49 MPa (Eklund et al. 1983). The hydrogen gas from these spheres is released by heating or crushing them. Reuse of these spheres is uneconomical due to crushing so it is not a favourable option.

4. In chemicals

Different chemicals viz. methanol (CH_3OH), ammonia (NH_3), and methyl cyclohexane ($CH_3C_6H_{11}$) (Strickland et al. 1984) containing hydrogen can also be considered for this purpose. These chemicals exist as liquid at NTP condition, the infrastructure used for transportation and storage of gasoline can also be used for these chemicals (Padro et al. 1999). Another possible candidate for hydrogen storage is aluminum hydride (AlH_3), having hydrogen density 10 wt% (Sandrock et al. 2005, Graetz 2009, Graetz et al. 2011). The advantage of this method over compressed hydrogen gas cylinders is that in this system leak-proof systems are not required. The hydrogen storage capacity of such chemicals is relatively high viz. methanol 8.9 wt%, ammonia 15.1 wt%, etc. In addition, the containers of such chemicals should be made of composites or plastics at least, in some cases to decrease the dead-weight of the container.

The major drawback of this method is that it is non-reversible, that is, regeneration of the compounds by hydrogen is not possible. It requires the production of the compounds in a centralized unit and recycling of the reaction products. Another disadvantage of this method is with compounds like ammonia that produces environmentally toxic waste products such as nitrogen oxides. Besides, it additionally produces carbon monoxide gas that is further unfavourable (Strickland et al. 1984).

Various other compounds like lithium borohydride, sodium borohydride, and hydrazine have been seriously investigated for their hydrogen storage capacities.

4.1 Metal organic framework (MOF)

MOFs display unique features, viz. relatively large pore volumes associated with surface area, variable pore size, tailored metal selection, ligand functionalization and framework–adsorbate interaction. Such materials can easily replace adsorbents like activated carbon as these materials too exhibit comparable hydrogen adsorption at significantly low temperature; unfortunately, the affinity for hydrogen decreases as the temperature rises to room temperature. While metal hydrides have higher hydrogen adsorption property, the desorption kinetics and thermodynamics are not favourable (Collins et al. 2007).

The molecules of hydrogen and the walls of the MOFs suffer weak interaction between them; hence, the void space present in the cages of MOFs is not completely utilized. This unutilized void space can be optimally used by introduction of mesoporous or microporous particles like carbon nanomaterials into these void spaces and suitably alter the pore size, and thereby the pore volume for greater hydrogen adsorption (Prasanth et al. 2011). Hydrogen adsorption by such type of materials increased at cryogenic temperatures and also at room temperature.

In recent years, researcher groups have figured out various strategies for optimizing interaction of hydrogen with these materials utilizing the dangling bonds or unsaturated centers for binding of hydrogen.

Monte Carlo simulations show that large hydrogen storage by MOFs is due to the physical characteristics, which are responsible for the strong interactions between hydrogen molecules and the MOF. These interactions between molecules of hydrogen and MOF are mainly due to Van der Waal's forces, charge-quadrupole and induction.

Physisorption of hydrogen mainly takes place due to the polarization effects and charged framework having narrow pores further increases the effects. The presence of these features in MOF makes it an excellent candidate for hydrogen storage (Belof et al. 2007).

Li et al. (2010) developed a system in which ammonia borane (AB) was captivated in a MOF and was labeled as JUC-32-Y. This captivation assisted by metallic catalysis enhanced the kinetics of release of hydrogen gas, thus preventing ammonia formation. The AB captured in the MOF JUC-32-Y released the captivated hydrogen gas even at 50ºC. It can release 8.2 wt% of hydrogen in 3 min at 95ºC and 10.2 wt% of hydrogen within 10 min at 85ºC. Wang et al. (2020) synthesized a microporous aluminum-based MOF labeled as BUT-22 for hydrogen storage at 296 K and 8 MPa. The capacity of hydrogen storage of BUT-22 was reported to be 12 wt% at 10 MPa and 77 K.

It can be thus concluded that MOFs are good materials to be used for hydrogen storage.

5. In Carbon

Activated carbon (AC) and various other allotropes of carbon like fullerenes, graphene, CNTs, nano fibers having large SSA are considered to be good sorbates and adsorption takes place by physisorption. Naturally, this characteristic of carbon materials has been correlated with hydrogen storage since long. AC having SSA 3000 m^2g^{-1} has been synthesized (Chahine et al. 1994). Dillon et al. (1999a) studied various adsorption capacities of such materials while Baughman et al. (2002) suggested that in addition to hydrogen storage, these materials can be used in fuel cell and automobiles. There is also the need to understand the geometric and electronic features required for activation of hydrogen molecules on these surfaces.

Theoretical calculations supported by experimental works have indicated the adsorption of hydrogen by carbon nanomaterials to be reversible and significant. The possibilities of achieving target set by the Department of Energy, USA seems unlikely as these materials have low bulk density.

In spite of the existence of certain doubts, researchers have claimed varying degrees of possibilities of carbon allotropes being used for hydrogen storage at ambient temperature and possible use in automobiles and other modes of transportation.

Contributions made by some of the researchers on hydrogen storage using various CNMs are as follows:

ACs being highly porous materials and large SSA (more than 3000 m^2g^{-1}) are favourable features for being good materials for hydrogen storage and for 'Hydrogen Economy'. The concept of AC being an energy material has been acknowledged widely (Sevilla and Mokaya 2014). Earlier researchers, Chahine et al. (1994) demystified the relationship between the capacity of hydrogen storage of AC and micro pore volume. They reported 2 wt% of hydrogen adsorptions at –196°C and Rzepka (1998) obtained 1.3 wt% of hydrogen adsorption using similar materials, while Strobel et al. (1999) obtained a hydrogen adsorption of 1.5 wt% by using a microbalance in isothermal gravimetric method, pressures of 125 bars and at 238°C. Darkrim et al. (2002) proved that adsorption hydrogen depends on micro pore volume and micro pore size distribution. Zhan et al. (2002) calculated the adsorption of hydrogen on ACs to be 23.76 wt%. According to them, the hydrogen adsorption potential becomes 0kJ.mol^{-1} when the hydrogen adsorption reaches about 800 bars; hence, no more hydrogen molecules could be adsorbed into the micro pores. At such high pressure, the porous nature of the adsorbent is likely to be destroyed and part of the hydrogen adsorbed may get desorbed. Shindo et al. (2003) prepared mechanically milled activated carbon; they suggested that the mechanism of adsorption by ACs was because of adsorption of molecules of hydrogen into the pores in the carbon as hydrogen molecules and assisted by the dangling bonds formed due to ball-milling. In 2004, Challet et al. found an increase in chemisorption of hydrogen gas with Li and K doped micro porous activated carbons, though the isotherms of these doped AC did not exhibit any alteration in sorption ratio compared to the as obtained AC at 77 K. Kojima and Suzuki (2004) obtained similar results, but the hydrogen adsorbed by the K-doped AC was desorbed at temperature range of about 440–1370 K, implying chemisorption. Various researchers have reported that hydrogen uptake by such carbons was proportional to the SSA and pore volume. Metal decorated Activated carbon fibres (ACF) have better hydrogen adsorption properties, as observed by Takagi et al. (2004). Lee et al. (2007) further substantiated this finding and reported that such fibres are more efficient than as obtained fibres, possibly due to chemisorption. ACF were oxidized by ammonium persulphate and another aliquot was reduced by hydrogen, and their hydrogen adsorption capacities were investigated over a range of temperature and pressure. Hydrogen adsorption capacity was increased by reduction and decreased by oxidation. Shang Li in 2005 reported the electrochemical hydrogen adsorption on AC–Copper electrode; the discharge capacity was 510 mAh/g after 384 cycles, such high capacities have not been observed in CNT–Cu electrodes. According to Strobel et al. (2006) 6 wt% of hydrogen adsorption would be possible if the SSA of the samples exceeds 4000 m^2g^{-1}. Potassium doped AC synthesized from coconut shells exhibited hydrogen adsorption of 0.85 wt% under pressure of 10 MPa and 298 K (Hangkyo et al. 2007). AC obtained from anthracite, a type of coal, by activating with potassium hydroxide had SSA of 3183 m^2g^{-1} and hydrogen adsorption reported was 3.2 wt% at 200 bars and ambient temperature (Jorda-Beneyto et al. 2007). The theoretical calculations made by Zhan for hydrogen adsorption was reported to be up to 23wt% but this high capacity is never reached by anyone. Jorda-Beneyto reached up to 3.2 wt% hydrogen adsorption with a SSA of 3183 m^2g^{-1}. Zielinski et al. in 2007 studied hydrogen storage in AC decorated with nickel particles. They studied the impact of some of the factors which influence the hydrogen storage viz. factors like metal salt used as precursor, extent of metal present, and also the different methods of synthesis were analyzed. The investigation summarized that the adsorption of hydrogen on carbon was due to chemisorption by *spill-over* of hydrogen molecule. The molecules dissociate on the decorated metal particles and then migrate on to the carbon surface. A number of researchers used this technique to increase the extent of hydrogen adsorption with varying degree of success. Wang et al. (2009) demonstrated that potassium hydroxide activation could create remarkable porosity to the carbon material. Accordingly, the SSA of the material was increased to remarkable extent (3190 m^2g^{-1}) and hydrogen adsorption reported was 7.08 wt% at 77 K and pressure of 2 MPa.

Nitrogen containing ACs, activated by potassium hydroxide with high SSA of 500–2400 m^2g^{-1} and pore volume range 0.26–1.16 cm^3g^{-1}, have shown 2.94 wt% of hydrogen storage at 77 K and at 0.1 MPa. The hydrogen uptake was found to be linearly dependent on ultra-pore volume 0.5–0.7 nm, but non linearly with SSA and pore volume (Sethia and Sayari 2016). Carbon synthesized from rice straw on activation with potassium hydroxide and potassium carbonate has shown large surface area of 2000–2100 m^2/g, and 0.65 wt% and 0.55 wt% of hydrogen adsorption capacity, respectively, at ambient temperature and at a pressure of 100 bars (Schaefer et al. 2017).

Plerdsranoy et al. (2019) has shown that transition metal halide provides catalytic effect for dehydrogenation process, improved thermal conductivity of ACs, increased hydrogen permeability and in addition, prevention of particle agglomeration during cycling.

Thus, it can be concluded that porous ACs containing substantial number of active sites, very high SSA, and large pore volume is necessary for high gravimetric hydrogen storage. Though theoretical analysis indicates SSA as a relevant factor for hydrogen storage, many researchers have found it to be less important than micropore volume. (Jimenez et al. 2010)

6. Carbon nano materials

6.1 Graphitic nano-fibres—Graphitic nanofibres exhibit high hydrogen adsorption capacity and it is more than 20 L at STP conditions for each gram of carbon fibres at 120 KPa and 25°C (Chambers et al. 1996, 1998). A near similar observation was made in case of Vapour Grown Carbon nanofibres (VGCNF). These fibres adsorb more than 10 wt% of hydrogen and the process is fairly spontaneous (Fan et al. 1999). Browning et al. (2002) prepared graphite carbon fibres from the decomposition of ethylene using various alloys of Fe-Ni-Cu and investigated their use in hydrogen storage. Mao et al. (2000) has reported maximum hydrogen adsorption of 9.99 wt%. They too reported that carbon nanofibres (CNF) can rapidly release hydrogen at room temp. Poirier et al. (2001) found that CNFs adsorb hydrogen to about 0.7 wt% at 10.5 MPa at room temperature. Nickel-copper alloy when used as catalyst in different ratios altered the morphology as well as SSA of the fibres. Hwang et al. (2002) found hydrogen adsorption of 1.4 wt% at 120 KPa and 25°C with CNFs prepared by pyrolysis of methane gas in presence of Ni-MgO as catalyst at 1200°C in nitrogen atmosphere. Awasthi et al. (2002) tried to ball-mill graphite to obtain graphitic carbon but this did not help positively in hydrogen adsorption. Hanada et al. (2003) detected up to 5.1 wt% hydrogen adsorption by CNFs at a temperature of 303–873 K by a TGA on CNFs doped with potassium nitrate. Yoon et al. (2004) warned that activation of CNFs under severe conditions can destroy its morphology. According to Blackman et al. (2006), activation of CNFs using potassium hydroxide was a better choice than carbon dioxide as the SSA was about 1000 m^2g^{-1}. This was much higher than when activated by carbon dioxide. An interesting finding was that the increase in SSA offered no proportional rise in hydrogen adsorption. Young et al. (2007) modified ACFs by doping with Ni and fluorination to increase the SSA and pore volume. It was found that due to fluorination, the hydrogen adsorption increased though the pore volume was found to decrease, while due to doping with nickel, pore volume remained unchanged but hydrogen adsorption was significantly enhanced as the result of catalytic activation of nickel. Im et al. (2008) concluded that in activated CNFs prepared from polyacrylonitrile and activated by potassium hydroxide and zinc chloride, if the diameter of the pores were in the range of 0.6–0.7 nm, then hydrogen adsorption was found to be more effective. Pillared Graphene—a three-dimensional carbon nanostructure having tunable pore sizes by design and high SSA—was used by Dimitrakaki et al. (2008) for hydrogen adsorption via grand canonical Monte Carlo and *ab initio* calculations. When this carbon material was decorated with lithium nanoparticles, its hydrogen storage was raised to 41 g hydrogen per liter at ambient conditions. Diaz et al. (2010) modified different CNF based materials and studied their hydrogen storage capacity at ambient pressure. They found that for impregnation of metals like palladium, its aqueous precursors were more effective than organic precursors and functionalized CNF were found have about 100% more hydrogen adsorption than as obtained CNF. According to Wu et al. (2010), the important factors in increasing the hydrogen adsorption in turbostatic CNFs was due to the defects formed on the carbon materials and other graphitic substances.

Lee et al. (2010) synthesized calcium particles decorated Zigzag Graphene Nano-Ribbon (ZGNR) which could adsorb hydrogen up to 5 wt%. Du et al. (2010) observed that lithium particles decorated porous graphene to be better materials for hydrogen adsorption than non-porous graphene, with hydrogen adsorption of 12 wt%. When aluminum particles decorated on both the sides of graphene layers were used for hydrogen storage, it adsorbed 13.79 wt% and adsorption energy was found to be –0.193 eV per hydrogen molecule, according to density-functional calculations (Ao and Peeters 2010). A simulated process was used to study chemisorption of hydrogen using corrugated graphene sheets and discharge by mechanical deformations. The corrugation was found to be out-of-plane up to ± 0.2 Å yielding fast adsorption and desorption of hydrogen. These systems offer hydrogen storage up to 8 wt% (Tozzini and Pellegrini 2011).

Bimetallic decorated CNFs containing Nickel and Ceria were tested for hydrogen adsorption capacity by George et al. (2021) and 1.858 wt% hydrogen adsorption was reported at the charge-discharge density of 500 mA/g.

The graphitic carbon fibres were expected to show high hydrogen storage capacity but very high degree of variation is observed. It was proved that the process of synthesis of CNFs was responsible for its hydrogen adsorption capacity.

6.2 Single Walled Carbon Nano Tubes (SWCNTs)—Dillon et al. (1997) investigated the hydrogen storage capacity using Temperature—Programmed Desorption (TPD) measurements. Their finding was that desorption of hydrogen was at ~ 133 K for both the samples. However, a second peak was observed at 290 K when heated in vacuum, which indicates second type of hydrogen adsorption, possibly in the interior part of SWCNTs. The desorption process at high temperature was absent in ACs. Dillon et al. (1999b) achieved hydrogen storage adsorption ~ 3.5–4.5 wt% at room temperature and at pressure of about 0.6 bars by introducing cutting method to synthesize short SWCNTs in high concentration with open ends. Later on, Liu et al. (1999) observed 8 wt% of hydrogen adsorptions on crystalline ropes of SWCNTs at 80 K and 40 bars. Ye et al. (1999) broadened the diameter of SWCNTs by rupturing the crystalline rope by sonication and achieved the hydrogen adsorption to 8.25 wt% at 80 K and 7 MPa. Ding et al. (2001) increased the specific surface area by splitting the ropes into individual tubes. Wang et al. (2002) found hydrogen adsorption of 8 wt% for CNTs.

Chen et al. (2001) proposed that there must be sites at which adsorption of hydrogen is favoured in SWCNTs. Accordingly, SWCNTs with open ends occupy hydrogen in two manners: endohedral adsorption and exohedral adsorption. They synthesized CNTs of 50–100 nm diameters capable of storing hydrogen upto 5–7 wt% later improvised to 13 wt% by pretreatments. Another suggestion made by Ding et al. (2001) was that by thermal treatment catalyst tips were removed, which helped in increasing the hydrogen adsorption.

Surprisingly, when Dai et al. (2002) synthesized macroscopic ropes of SWCNTs with a diameter of 1.72 nm, they used a semi-continuous arc discharge method and reported 503 mAhg[-1] of discharge capacity, equivalent to 1.84 wt% hydrogen adsorption. Lawrence and Xu (2004) studied the monolayer saturation plateau on purified SWCNTs at ambient temperature and reported 0.9 wt% of hydrogen adsorption. This indicates that the results may not be reproducible. Schimmel et al. (2004) suggested the reason that due to bundling of SWCNT and opened-SWCNT, they possess a low SSA. Efremenko and Sheintuch (2005) predicted low adsorption (~ 1 wt%) at 10 MPa in Langmuir isotherms. Few others like Nobuyuki et al. (2002), Shen et al. (2004) and Shaijumon et al. (2005) have synthesized SWCNTs by various methods; unfortunately, these samples adsorbed not more than 3 wt% of hydrogen. Takagi et al. (2006, 2007) found that activation by nitric acid enhances the adsorption of hydrogen by changing bundle structure by binding out functional group introduced during acid treatment. Lee et al. (2006) has reported that Ni decorated SWCNTs are capable of releasing nearly 10 wt% hydrogen at room temperature. Nikitin et al. (2008) studied chemisorption of hydrogen in SWCNTs using atomic hydrogen for hydrogenation. Their study showed specific carbon nanotubes form reversible C−H bonds, due to which the adsorption increases to beyond 7 wt%.

Rashidi et al. (2010) prepared SWCNT by chemical vapour deposition (CVD) at 900ºC over cobalt-molybdenum nanoparticles and reported maximum adsorption of 0.8 wt%. They concluded that there was no desired adsorption due to insufficient binding between carbon materials and hydrogen. Zhou and Williams (2011) made similar observations for platinum decorated CNTs.

SWCNT bundles and aligned CNTs are expected to show very high hydrogen adsorption capacity because of the availability of two side surfaces of the CNTs. At the same time, large variations in the results are obtained, i.e., 8% in case of SWCNT bundles and 13% in case of aligned CNTs.

Vellingiri et al. (2019) synthesized composite of SWCNT-LiBH$_4$ by facile ultra-sonication assisted impregnation and then oxidation at 573 K. During adsorption/desorption study, highest hydrogen adsorption was 4 wt% under pressure 5 bar at 373 K.

The above-mentioned results indicate that various methods of synthesis and treatments are responsible in controlling the hydrogen storage. Therefore, standardization of the method of synthesis of SWCNTs is a necessity.

6.3 Multi-Walled Carbon Nano Tubes (MWCNTs)—Hou et al.'s (2002) findings suggested that MWCNTs may have more hydrogen storage capacity than SWCNTs due to its opened tips, simple chemical state and large pore volume. Hydrogen storage capacity was 6.3 wt%. Accordingly, Liu et al. (2003) synthesized MWCNTs with open ends and reported 0.69 wt% adsorption. The adsorption of hydrogen, 3.7 wt%, and desorption of 3.6 wt% at 69 bar was reported by Lueking et al. (2003) on MWCNT/NiMgO system. According to Huang et al. (2003), annealing of CNT in nitrogen at 500ºC and soaking in a 1.0 M potassium nitrate solution increases hydrogen adsorption, but hydrogen adsorption capacity achieved was only 3.2 wt% at a moderate pressure of 120 bars. Peng-Xiang et al. (2003) observed that the adsorption capacity was proportional to the diameter of the nano tube and there was no complete desorption at standard conditions. Additionally, structural morphology reveals that the surface modification and porous structures enhance the hydrogen uptake capacity. Zheng et al. (2005) investigated hydrogen storage on MWCNTs volumetrically and calculated the intermolecular interaction energy, and found no evidence of adsorption. Shaijumon et al. (2005) synthesized MWCNTs from acetylene and achieved 3.2 wt% hydrogen storage at 60 bars and 125 K temperature. Naab et al. (2006) was highly disappointed to find that SWCNT, MWCNTs and carbon nanofibres show < 1 wt% adsorption. Chen et al. (2007) used potassium hydroxide to modify MWCNTs and annealed those at high temperatures. They observed that capacity of KOH treated and untreated CNTs to store hydrogen was 4.47 and 0.71 wt%, respectively. Interestingly, Rakhi et al. (2008) also synthesized MWCNT acetylene as precursor, Dy-Ni alloy as the catalyst and found the adsorption capacity to be 3.5 wt% at 143 K and 7.5 MPa. Park et al. (2010) enhanced the adsorption property of MWCNT by microwave treatment. The treatment makes the surface microporous. Park and Lee (2010) reported that platinum particles decorated on MWCNT showed better hydrogen storage capacity compared to the as obtained MWCNTs. In a similar work, Reyhani et al. (2011) observed that decoration of MWCNT by transition metal particles enhances its hydrogen storage capacity. Palladium (Pd) seems to be a better metal as it dissociates hydrogen faster. Aghababaei et al. (2020) carried out surface activation of MWCNTs by potassium hydroxide and studied the effect. Li and Co metal nanoparticles were decorated on the nanotubes. At 278 K, the Li-decorated MWCNT adsorbed 1.33 wt% of hydrogen while Co-decorated MWCNTs adsorbed 1.06 wt% at the same temperature.

Thus, the hydrogen storage by SWCNTs was detected as less than MWCNTs as hydrogen atom can occupy the empty space between the layers of carbon and be enhanced by transition metal decoration and other chemical treatments.

6.4 Metal Doped Carbon Nano Tubes—Chen et al. (1999) reported 20 wt% gravimetric storage of hydrogen with Li-doped CNTs while undoped CNTs showed virtually negligible amount (0.4 wt%). For the process, doping was done by reaction with Li containing compound such as nitrates and carbonates. Froudakis (2001) used QM/MM model to study adsorption of molecular hydrogen on CNTs doped with alkali-metal and suggested that such nanotubes exhibit greater adsorption of hydrogen in comparison to as obtained SWCNTs due to charge-induced dipole interaction. Hydrogen adsorption capacities were investigated in different materials like doped porous carbons, AC (1600 m^2g^{-1}) and SWCNTs by Challet et al. (2004) and observed increase in adsorption at ambient condition but not at 77 K. Kojima and Suzuki (2004) found that due to high SSA, K-doped AC adsorbs more hydrogen than K-doped graphite.

Hu et al. (2005) reported 4.21 wt% and 3.95 wt % adsorption of hydrogen for Li and K doped SWCNTs. Giraudet and Zhu (2011) found that doping of carbon by nickel nanoparticles increases hydrogen storage at room temperature as well as at elevated temperatures.

Thus, it can be commented that CNTs doped with metal particles show greater adsorption of hydrogen than primal CNTs and doping by lithium metal is a better option than other alkali metals. In these findings, surface area also plays an important role. Other than alkali metals, doping of carbon by nickel improves adsorption of hydrogen even at ambient temperature. Dou et al. (2019) has reported that 3.35 wt% reversible hydrogen storage capacity of Mg-Ni-Cu/CNTs can be achieved at 583 K and 3.0 atm pressure. In this experiment, isothermal hydrogen adsorption and desorption of doped-CNTs were studied and the findings were analyzed by shrinking core model method. Magnesium metal doped reduced graphene oxide efficiently promoted hydrogen adsorption and desorption processes up to 6.5 wt% (Cho et al. 2017). Huang et al. (2017) prepared carbon wrapped transition metal (Ni/C, Co/C) nano-particles and used as catalyst in magnesium hydride system. Magnesium hydride doped with Ni/C shows 6.1 wt% of hydrogen adsorption and release at 523 K. Also, magnesium hydride Ni/C is able to store hydrogen gas at 373 K within 30s.

7. In pyrolyzed plant/animal materials

In 1989, Laine et al. prepared ACs from coconut shells catalyzed by potassium. They studied the different factors affecting these syntheses. Sharon et al. (2008) synthesized the carbon materials with different conditions of temperature, atmosphere, activation, etc. These materials were prepared from plants-based precursors, and for them hydrogen adsorption was substantially below 6.4 wt%. Suzuki et al. (2007) used rice bran as precursor for hydrogen storage and synthesized activated mesoporous carbon having SSA of 652-m^2 g^{-1} and 0.137 cm^3 g^{-1} as the pore volume. Balathanigaimani et al. (2009) performed study on corn grain-carbon monoliths (CG-CMs) at about 1 bar pressure and 77 K and also at 50 bars and 298 K. The findings showed that the CG-CMs and carbon monoliths have similar hydrogen adsorption capacities. Takahata et al. (2009) reported 0.6 wt% hydrogen storage by carbon activated by potassium hydroxide synthesized from waste coffee beans having very high SSA of 2070 m^2g^{-1} with pore size 0.6 to 1.1 nm. Akasaka et al. (2011) performed an identical study on AC from waste coffee beans. Takahata et al. (2009) found that, the morphology of the pores and the adsorption capacity of the carbon material to be proportional. It is also proportional to the SSA. Senoz and wool (2011) carried out an innovative work on pyrolyzed chicken feather fibre (PCFF) which showed very poor hydrogen adsorption capacity. The biomass of Posidonia Oceanica and Wood Chips were pyrolyzed at 600ºC in an inert atmosphere, which showed hydrogen uptake up to 5 wt% (Pedicini et al. 2020). Im et al. (2010) studied the pyrolysis of mesoporous carbon nanofibres

(MCNFs). The MCNF pyrolyzed at 1200ºC, at 77 K and 1 bar pressure shows hydrogen uptake of 0.73 wt%. This work was based on dependence of SSA of MCNFs, carbonization temperature and hydrogen storage capacity.

Though different carbon materials have been synthesized from plant and animal-based precursors since a long time, the idea to use such materials to study hydrogen adsorption cropped up only in 2007, courtesy Sharon et al. These materials have high SSA, but still the extent of hydrogen adsorption is not up to the mark. Thus, the study of hydrogen adsorption by these remains a challenge and open to further investigation and research.

8. Theoretical models

Based on the observations of various workers, two theoretical models of types of CNM for adsorption of hydrogen can be proposed:

1. Surface adsorption of hydrogen by the dangling bonds of CNTs.
2. Storage of hydrogen in the lumen of CNT as well as between the walls of MWCNTs.

8.1 Surface adsorption of hydrogen by CNT

The earliest attempt of storing hydrogen in ACs and CNTs was performed by molecular simulation of hydrogen storage in both ACs and CNTs (Rzepka et al. 1998, Wang and Johnson 1999). Their findings, however, did not support either CNT or activated carbon suitable for hydrogen storage, though they presented MO calculations, which showed superiority of CNT over activated carbon as far as hydrogen storage is concerned. Later, various workers performed experimental observations using CNT. Storage of hydrogen atom takes place in the nanotube by rupturing the carbon-carbon bond but maintaining the stability of the walls of nanotubes (Lee et al. 2001). Simultaneously, Froudakis et al. (2001) suggested that due to insertion of the atoms of hydrogen to the walls of tube, the binding of hydrogen atoms will occur in a zigzag manner, which leads to alteration of the shape of the tube resulting in 15% increase in the tube volume.

Dodziuk et al. (2002) presented calculations on hydrogen adsorption by CNTs and suggested that large quantities of hydrogen adsorption by CNTs cannot be achieved through physisorption. Li et al. (2003) proposed that due to slow kinetics, theoretically there is an upper limit for chemisorption of 7.7 wt%, but in practice this could be difficult to achieve. Moreover, during desorption, due to attachment of dangling bonds on the surface of CNT, the CNT loses its metallic nature and strong C–H covalent bonding slows down the kinetics of hydrogen recombination. There was a trivial Raman shift seen in the Q branch; thus, this could help in clear corroboration that the mechanism of interaction with the adsorbing material is simply physisorption while there is no charge transfer involved in the reaction. Energy barrier was found of 2.7 eV for hydrogen dissociation on the surface of tube. However, once there is occurrence of chemisorption, the hydrogenated tube is metastable only against ripping (Chen et al. 2005).

8.2 Uptake of hydrogen in lumen or internal spaces

Cracknell (2002) suggested that due to space limitation, hydrogen molecules, which were absorbed in the interstitial spaces, were not able to form stable clusters. The enthalpy of adsorption, which supports stable cluster formation in the internal spaces, is < 303 K (Murata et al. 2002).

It must be mentioned here that uptake of hydrogen is greater in the optimal graphitic nanofibres with slit-like pores than predicted in case of the internal space of a CNT. Moreover, the hydrogen that enters into the tube is difficult to release because the lumen diameter is very small and it would require very high energy to come out. Therefore, when release of hydrogen was recorded, it was suggested that hydrogen atoms must have been bound to the tube walls instead of being part of interior.

From the discussion, it can be concluded that hydrogen, which is adsorbed on tube surface, can escape easily but those, which enter the tube (endohedral), cannot do so easily. The covalent bond formed between carbon and hydrogen as well as the orientation of the tubes restricts the removal by increasing the desorption energy. The studies performed so far reveal three features that need elaborated study and understanding prior to contemplating CNMs as hydrogen storage devices. These features are:

Rather slow uptake of hydrogen by CNM

Less desorption of adsorbed hydrogen

Possibility of promoting adsorption of hydrogen by using transition metals

9. Hydrogen storage measurement

Hydrogen storage capacity of different materials is determined by three methods, i.e., by

Gravimetric,

Volumetric

Temperature Programmed Desorption (TPD).

For the present work, volumetric method has been used, as this facility was available. Hence, here the volumetric method is being discussed briefly using Sievert's apparatus.

In case of volumetric method, the sample is kept in a sample holder having definite volume. It is degassed and then subjected to hydrogen gas at known pressure, so adsorption of hydrogen gas causes reduction in pressure; this difference between pressure values is used for calculation of the volume of adsorbed hydrogen.

Different scientists (Sandrock et al. 1981, Chambers et al. 1998, Sivakumar et al. 1999, Zacharia et al. 2005, Reddy and Ramaprabhu 2008, Sharon et al. 2007, 2008, 2011) made use of Sievert's apparatus for investigating hydrogen uptake on CNMs at different pressures. Calculation of adsorption capacity is done by using ideal gas law, i.e., $PV = nRT$. Van der Waal's correction factors 'a' and 'b' have been applied considering non-ideal behavior of hydrogen molecules (a—for the inter molecular force, b—for molecular size) for improving precision of the result. Accordingly, the modified form of equation used for the calculation of the extent of hydrogen adsorbed due to the reduction in the pressure and the Sievert's apparatus is

$$\left[P_r + a \left(\frac{n}{V_r} \right)^2 \right] (V_r - nb) = nRT$$

10. Summary

This chapter is a depiction and overview of variety of promising possibilities for storage of hydrogen in diverse conditions (physical) and in combination with various forms of carbon along with different kinds of materials. It is corroborated that specific properties possessed by a creditable hydrogen storage material be listed as high capacity to store hydrogen per unit mass and volume to ascertain the quantity of energy available, lower enthalpy of formation to reduce the energy required for release of hydrogen as well as lower heat of dissipation for hydride formation which is exothermic process, reversibility, moderate dissociation pressure, low dissociation temperature, negligible energy loss for the charging and discharging of hydrogen, optimum kinetics, significantly long cycle life, maintaining high degree of safety and low recycling cost.

Metal hydrides are also good alternative for hydrogen storage but are heavy and their shorter lifetime is a major constraint. High-pressure tank storage of compressed hydrogen is uneconomical, mainly due to energy considerations. Hydrogen storage in glass microspheres or in an array of glass microtubules is also promising but highly uneconomical. Different chemicals have been explored to store hydrogen but the processes are found to be irreversible. Substances like metal organic framework systems have shown promising results as hydrogen storage materials but their practical viability is a distant possibility.

After so much of efforts, even now there is no best hydrogen storage material and still there is a long way to go. Authors have considered different parameters like the morphology of CNMs, SSA, pore size, pore volume, doping and the activation of CNMs by nano metals. The reported hydrogen storage capacity values have a very wide range. Hence, hydrogen adsorption by CNMs is a wide open issue, which needs to be solved by keeping in mind various factors controlling uptake of hydrogen and obviously the solution for which world is eagerly waiting should be able to give required and reproducible results.

References

Aghababaei, M., Ghoreyshi, A.A. and Esfandiari, K. 2020. Optimizing the conditions of multi-walled carbon nanotubes surface activation and loading metal nanoparticles for enhanced hydrogen storage. *International Journal of Hydrogen Energy*, 45(43): 23112–23121.

Akasaka, H., Takahata, T., Toda, I., Ono, H., Ohshio, S., Himeno, S. and Saitoh, H. 2011. Hydrogen storage ability of porous carbon material fabricated from coffee bean wastes. *International Journal of Hydrogen Energy*, 36(1): 580–585.

Ao, Z.M. and Peeters, F.M. 2010. High-capacity hydrogen storage in Al-adsorbed graphene. *Physical Review B*, 81(20).

Awasthi, K. 2002. Ball-milled carbon and hydrogen storage. *International Journal of Hydrogen Energy*, 27(4): 425–432.

Bacon, G.E. 1951. The interlayer spacing of graphite *Acta Cryst.*, 4: 558–561.

Balathanigaimani, M.S., Shim, W.-G., Kim, T.-H., Cho, S.-J., Lee, J.-W. and Moon, H. 2009. Hydrogen storage on highly porous novel corn grain-based carbon monoliths. *Catalysis Today.*, 146(1-2): 234–240.

Balde Cornelis, P., Hereijgers P.C. Bart, Bitter H. Johannes and de Jong Krijn, P. 2008. Sodium alanate nanoparticles—linking size to hydrogen storage properties. *J. Am. Chem. Soc.*, 130(21): 6761–6765.

Baughman, R.H., Zakhidov, A.A. and Heer, W.A. 2002. Carbon Nanotubes—the Route toward Applications. *Science*, 2 August, 297(5582): 787–792.

Belof, J.L., Stern, A.C., Eddaoudi, M. and Space, B. 2007. On the mechanism of hydrogen storage in a metal–organic framework material. *Journal of the American Chemical Society*, 129(49): 15202–15210.

Blackman, J.M., Patrick, J.W., Arenillas, A., Shi, W. and Snape, C.E. 2006. Activation of carbon nanofibres for hydrogen storage. *Carbon*, 44(8): 1376–1385.

Browning, D.J., Gerrard, M.L., Lakeman, J.B., Mellor, I.M., Mortimer, R.J. and Turpin, M.C. 2002 Studies into the storage of hydrogen in carbon nanofibers: Proposal of a possible reaction mechanism. *Nano Letters*, 2: 201–205.

Chahine, R. and Bose, T.K. 1994. Low-pressure adsorption storage of hydrogen. *International Journal of Hydrogen Energy*, 19: 161–164.

Challet, S., Azais, P., Pellenq, R.J.-M., Isnard, O., Soubeyroux, J.-L. and Duclaux, L. 2004. Hydrogen adsorption in microporous alkali-doped carbons (activated carbon and single wall nanotubes). *Journal of Physics and Chemistry of Solids*, 65(2-3): 541–544.

Chambers, A., Rodriguez, N.M. and Baker, R.T.K. 1996. Influence of copper on the structural characteristics of carbon nanofibers produced from the cobalt catalyzed decomposition of ethylene. *Journal of Materials Research*, 11: 430–438.

Chambers, A., Park, C., Baker, R.T.K. and Rodriguez, N.M. 1998. Hydrogen storage in graphite nanofibers. *The Journal of Physical Chemistry B*, 102(22): 4253–4256.

Chen, C. and Huang, C. 2007. Hydrogen storage by KOH-modified multi-walled carbon nanotubes. *International Journal of Hydrogen Energy*, 32(2): 237–246.

Chen, P. 1999. High H_2 uptake by alkali-doped carbon nanotubes under ambient pressure and moderate temperatures. *Science.* 285(5424): 91–93.

Chen, Y., Shaw, D.T., Bai, X.D., Wang, E.G., Lund, C., Lu, W.M. and Chung, D.D.L. 2001. Hydrogen storage in aligned carbon nanotubes. *Applied Physics Letters*, 78(15): 2128–2130.

Chen, Z., Appenzeller, J., Knoch, J., Lin, Y. and Avouris, P. 2005. The role of metal–nanotube contact in the performance of carbon nanotube field-effect transistors. *Nano Letters*, 5(7): 1497–1502.

Cho, E., Ruminski, A,. Liu, Y., Shea, P., Kang, S., Zaia, E., Park, J.,

Chuang, Y., Yuk, J., Zhou, X., Heo, T., Guo, J., Wood, B. and Urban, J. 2017. Hierarchically controlled inside-out doping of Mg nanocomposites for moderate temperature hydrogen storage. *Advanced Functional Materials*, 27(47): 1704316.

Collins David, J. and Zhou Hong-Cai. 2007. Hydrogen storage in metal–organic frameworks. *J. Mater. Chem.*, 17: 3154–3160.

Cracknell, R.F. 2002. Simulation of hydrogen adsorption in carbon nanotubes. *Molecular Physics*, 100(13): 2079–2086.

Dai, G.-P., Liu, C., Liu, M., Wang, M.-Z. and Cheng, H.-M. 2002. Electrochemical hydrogen storage behavior of ropes of aligned single-walled carbon nanotubes. *Nano Letters*, 2: 503–506.

Dalai, S., Savithri, V., Shrivastava, P., Param Sivam, S. and Sharma, P. 2015. Fabrication of zinc-loaded hollow glass microspheres (HGMs) for hydrogen storage. *International Journal of Energy Research*, 39(5): 717–726.

Darkrim, F.L., Malbrunot, P. and Tartaglia, G.P. 2002. Review of hydrogen storage by adsorption in carbon nanotubes. *International Journal of Hydrogen Energy*, 27: 193–202.

Das, L. 2002. Near-term introduction of hydrogen engines for automotive and agricultural application. *International Journal of Hydrogen Energy*, 27(5): 479–487.

Demirci, U.B., Akdim, O. and Miele, P. 2009. Ten-year efforts and a no-go recommendation for sodium borohydride for on-board automotive hydrogen storage. *International Journal of Hydrogen Energy*, 34(6): 2638–2645.

Díaz Eva, León Marta and Ordóñez Salvador. 2010. Hydrogen adsorption on Pd-modified carbon nanofibres: Influence of CNF surface chemistry and impregnation procedure. *International Journal of Hydrogen Energy*, 35(10): 4576–4581.

Dillon, A.C., Jones, K.M., Bekkedahl, T.A., Kiang, C.H., Bethune, D.S. and Heben, M.J. 1997. Storage of hydrogen in single-walled carbon nanotubes. *Nature*, 386(6623): 377–379.

Dillon, A.C., Gennet, T., Alleman, J.L., Jones, K.M., Parilla, P.A. and Heben, M.J. 1999a. Proceedings of the U. S. DOE Hydrogen Program Review. pp. 421–440.

Dillon, A.C., Gennet, T., Alleman, J.L., Jones, K.M., Parilla, P.A. and Heben M.J. 1999b. A simple and complete purification of single-walled carbon nanotube materials. *Advanced Materials*, 11(16): 1354–1358.

Dimitrakakis, G.K., Tylianakis, E. and Froudakis, G.E. 2008. Pillared graphene: A Nnew 3-D network nanostructure for enhanced hydrogen storage. *Nano Letters*, 8(10).

Ding, M.S., Xu, K., Zhang, S.S., Amine, K., Henriksen, G.L. and Jow, T.R. 2001. Change of conductivity with salt content, solvent composition, and temperature for electrolytes of LiPF[sub 6] in ethylene carbonate-ethyl methyl carbonate. *Journal of The Electrochemical Society*, 148(10): A1196.

Dodziuk, H. and Dolgonos, G. 2002. Molecular modeling study of hydrogen storage in carbon nanotubes. *Chemical Physics Letters*, 356(1-2): 79–83.

Dou, B., Zhang, H., Cui, G., He, M., Ruan, C., Wang, Z., Chen, H., Xu, Y., Jiang, B. and Wu, C. 2019. Hydrogen sorption and desorption behaviors of Mg-Ni-Cu doped carbon nanotubes at high temperature. *Energy*, 167: 1097–1106.

Du Aijun, Zhu Zhonghua and Smith Sean, C. 2010. Multifunctional porous graphene for nanoelectronics and hydrogen storage: New properties revealed by first principle calculations. *J. Am. Chem. Soc.*, 132(9): 2876–2877.

Efremenko, I. and Sheintuch, M. 2005. Enthalpy and entropy effects in hydrogen adsorption on carbon nanotubes. *Langmuir*, 21(14): 6282–8.

Eklund, G. and Vonkruenstierna, O. 1983. Storage and transportation of merchant hydrogen. *International Journal of Hydrogen Energy*, 8(6): 463–470.

Fan, Y.Y., Liao, B., Liu, M., Wei, Y.L., Lu, M.Q., Cheng, H.M. 1999. Hydrogen uptake in vapor-grown carbon nanofibers. *Carbon*, 37: 1649–1652.

Froudakis, G.E. 2001. Why alkali-metal-doped carbon nanotubes possess high hydrogen uptake. *Nano Letters*, 1(10): 531–533.

Geng, H.-Z., Kim, T.H., Lim, S.C., Jeong, H.-K., Jin, M.H., Jo, Y.W. and Lee, Y.H. 2010. Hydrogen storage in microwave-treated multi-walled carbon nanotubes. *International Journal of Hydrogen Energy*, 35(5): 2073–2082.

George, J.K., Yadav, A. and Verma, N. 2021. Electrochemical hydrogen storage behavior of Ni-Ceria impregnated carbon micro-nanofibers. *International Journal of Hydrogen Energy*, 46(2): 2491–2502.

Giraudet, S. and Zhu, Z. 2011. Hydrogen adsorption in nitrogen enriched ordered mesoporous carbons doped with nickel nanoparticles. *Carbon*, 49(2): 398–405.

Graetz, J. 2009. New approaches to hydrogen storage. *Chem. Soc. Rev.*, 38(1): 73–82.

Graetz, J., Reilly, J.J., Yartys, V.A., Maehlen, J.P., Bulychev, B.M., Antonov, V.E., Tarasov, B.P. and Gabis, I. 2011. Aluminum hydride as a hydrogen and energy storage material: Past, present and future. *Journal of Alloys and Compounds*, 509: S517–S528.

Greenwood, N.N. and Earnshaw, A. 1997. Chemistry of the Elements, Oxford; Boston, Butterworth-Heinemann.

Hanada, K., Shiono, H. and Matsuzaki, K. 2003. Hydrogen uptake of carbon nanofiber under moderate temperature and low pressure. *Diamond and Related Materials*, 12(3–7): 874–877.

Hou, P.X., Bai, S. Yang, Q.H., Liu, C. and Cheng, H.M. 2002. Multi-step purification of carbon nanotubes Carbon 40(1): 81–85.

Hu, N., Sun, X. and Hsu, A. 2005. Monte Carlo simulations of hydrogen adsorption in alkali-doped single-walled carbon nanotubes. *The Journal of Chemical Physics*, 123(4): 044708.

Hwang, J.Y., Lee, S.H., Sim, K.S. and Kim, J.W. 2002. Synthesis and hydrogen storage of carbon nanofibers. *Synthetic Metals*, 126(1): 81–85.

Im J.S., Park S.J., Kim T.J., Kim, Y.H. and Lee, Y.S. 2008. The study of controlling pore size on electrospun carbon nanofibers for hydrogen adsorption. *Journal of Colloid and Interface Science*, 318(1): 42–49.

Im, J.-E., Oh, S.-L., Choi, K.-H., Wang, K.-K., Jung, S., Cho, W., Oh, M. and Kim, Y.-R. 2010. Hydrogen uptake efficiency of mesoporous carbon nanofiber and its structural factors to determine the uptake efficiency. *Surface and Coatings Technology*, 205: S99–S103.

Jain, I.P., Lal, C. and Jain, A. 2010. Hydrogen storage in Mg: A most promising material. *International Journal of Hydrogen Energy*, 35(10): 5133–5144.

James, E. Shelby, Matthew M. Hall, Michael J. Snyder and Peter B. Wachtel. 2009. A radically new method for hydrogen storage in hollow glass microspheres. *DOE Report*, 958673.

Jimenez, V., Sanchez, P., Díaz, J.A., Valverde, J.L. and Romero, A. 2010. Hydrogen storage capacity on different carbon materials. *Chemical Physics Letters*, 485(1-3): 152–155.

Jin, H., Lee, Y.S. and Hong, I. 2007. Hydrogen adsorption characteristics of activated carbon. *Catalysis Today*, 120(3-4): 399–406.

Jordá-Beneyto, M., Suárez-García, F., Lozano-Castelló, D., Cazorla-Amorós, D. and Linares-Solano, A. 2007. Hydrogen storage on chemically activated carbons and carbon nanomaterials at high pressures. *Carbon*, 45(2): 293–303.

Kalyanam, K.M. and Robert, H.D. 1987. Safety guide for hydrogen Ottawa. *National Research Council Canada*, 124 p.

Kojima, Y. and Suzuki, N. 2004. Hydrogen adsorption and desorption by potassium-doped superactivated carbon. *Applied Physics Letters*, 84(20): 4113–4115.

Laine, J., Calafat, A. and labady, M. 1989. Preparation and characterization of activated carbons from coconut shell impregnated with phosphoric acid. *Carbon*, 27(2): 191–195.

Lawrence, J. and Xu, G. 2004. High pressure saturation of hydrogen stored by single-wall carbon nanotubes. *Applied Physics Letters*, 84(6): 918–920.

Lee, H., Ihm, J., Cohen, M.L. and Louie, S.G. 2010. Calcium-decorated graphene-based nanostructures for hydrogen storage. *Nano Letters*, 10(3): 793–798.

Lee, S.M., An, K.H., Lee, Y.H., Seifert, G. and Frauenheim, T. 2001. A hydrogen storage mechanism in single-walled carbon nanotubes. *J. Am. Chem. Soc.*, 123: 5059–5063.

Lee, Y.S., Kim, Y.H., Hong, J.S., Suh, J.K. and Cho, G.J. 2007. The adsorption properties of surface modified activated carbon fibers for hydrogen storages. *Catalysis Today*, 120(3-4): 420–425.

Leela Mohana Reddy, A. 2008. Hydrogen adsorption properties of single-walled carbon nanotube—Nanocrystalline platinum composites. *International Journal of Hydrogen Energy*, 33(3): 1028–1034.

Li, S., Pan, W. and Mao, Z. 2005. A comparative study of the electrochemical hydrogen storage properties of activated carbon and well-aligned carbon nanotubes mixed with copper. *International Journal of Hydrogen Energy*, 30(6): 643–648.

Li, Z., Zhu, G., Lu, G., Qiu, S. and Yao, X. 2010. Ammonia borane confined by a metal−organic framework for chemical hydrogen storage: Enhancing kinetics and eliminating ammonia. *Journal of the American Chemical Society*, 132(5): 1490–1491.

Liu, C. 1999. Hydrogen storage in single-walled carbon nanotubes at room temperature. *Science*, 286(5442): 1127–1129.

Liu, Fu, Zhang Xiaobin, Cheng Jipeng, Tu Jiangpin, Kong Fanzhi, Huang Wanzhen and Chen Changpin. 2003. Preparation of short carbon nanotubes by mechanical ball milling and their hydrogen adsorption behavior. *Carbon*, 41(13): 2527–2532.

Lueking, A. and Yang, R.T. 2003. Hydrogen storage in carbon nanotubes: Residual metal content and pretreatment temperature. *AIChE Journal*, 49(6): 1556–1568.

Maheshwar, S., Madhuri, S., Golap, K. and Mukherjee, B. 2011. Hydrogen storage by carbon fibers synthesized by pyrolysis of cotton fibers. *Carbon Letters*, 12(1): 39–43.

Mao, Z-Q., Xu, C., Yan, J., Liang, J., Sun, G., Wei, B. and Wu, D. 2000. Preliminary study on hydrogen storage in carbon nanofibers. *New Carbon Materials*, 15: 64–67.

McCarty, R.D., Hord, J. and Roder, H.M. 1981. Selected properties of hydrogen (engineering design data) Washington, D.C.: U.S. Dept. of Commerce, National Bureau of Standards 1981, E-Resource U.S. Federal Government Document Book.

Morgan, D. and Sissine, F. 1995. Hydrogen: Technology and Policy, Congressional Research Service, Report for Congress, USA.

Murata, K., Kaneko, K., Kanoh, H., Kasuya, D., Takahashi, K., Kokai, F., Yudasaka, M. and Iijima, S. 2002. Adsorption mechanism of supercritical hydrogen in internal and interstitial nanospaces of single-wall carbon nanohorn assembly. *The Journal of Physical Chemistry B*, 106(43): 11132–11138.

Naab, F.U., Dhoubhadel, M., Gilbert, J.R., Gilbert, M.C., Savage, L.K., Holland, O.W., Duggan, J.L. and McDaniel, F.D. 2006. Direct measurement of hydrogen adsorption in carbon nanotubes/nanofibers by elastic recoil detection. *Physics Letters A*, 356(2): 152–155.

Nikitin Anton, Li Xiaolin, Zhang Zhiyong, Ogasawara Hirohito, Dai Hongjie and Nilsson Anders. 2008. Hydrogen storage in carbon nanotubes through the formation of stable C−H bonds. *Nano Lett*, 8(1): 162–167.

Nishimiya, N., Ishigaki, K., Takikawa, H., Ikeda, M., Hibi, Y., Sakakibara, T., Matsumoto, A. and Tsutsumi, K. 2002. Hydrogen sorption by single-walled carbon nanotubes prepared by a torch arc method. *Journal of Alloys and Compounds*, 339(1-2): 275–282.

Ogden, J.M. 2002. Hydrogen: The fuel of the future? *Physics Today*, 69–74.

Padro, C. and Putsche, V. 1999. Survey of the Economics of Hydrogen Technologies, Technical Report, National Renewable Energy Laboratory, Colorado, USA.

Park, S.-J. and Lee, S.-Y. 2010. Hydrogen storage behaviors of platinum-supported multi-walled carbon nanotubes. *International Journal of Hydrogen Energy*, 35(23): 13048–13054.

Pedicini, R., Maisano, S., Chiodo, V., Conte, G., Policicchio, A. and Agostino, R.G. 2020. Posidonia Oceanica and Wood chips activated carbon as interesting materials for hydrogen storage. *International Journal of Hydrogen Energy*, 45(27): 14038–14047.

Peng-Xiang, H., Shi-Tao, Xu, Z.Y., Quan-Hong, Y., Liu, C. and Hui-Ming, C. 2003. Hydrogen adsorption/desorption behavior of multi-walled carbon nanotubes with different diameters. *Carbon*, 41(13): 2471–2476.

Phillips. 1972. Unique Fuel Therefor Patent number: 3818875 Filing date: 30 Nov 1972 Issue date: Jun 1974.

Plerdsranoy, P., Thiangviriya, S., Dansirima, P., Thongtan, P., Kaewsuwan, D., Chanlek, N. and Utke, R. 2019. Synergistic effects of transition metal halides and activated carbon nanofibers on kinetics and reversibility of MgH$_2$. *Journal of Physics and Chemistry of Solids*, 124: 81–88.

Poirier, E., Chahine, R. and Bose, T.K. 2001. Hydrogen adsorption in carbon nanostructures. *International Journal of Hydrogen Energy*, 26(8): 831–835. August 2001.

Prasanth, K.P., Rallapalli, P., Raj, M.C., Bajaj, H.C. and Jasra, R.V. 2011. Enhanced hydrogen sorption in single walled carbon nanotube incorporated MIL-101 composite metal–organic framework. *International Journal of Hydrogen Energy*, 36(13): 7594–7601.

Qikun, W. 2002. Hydrogen storage by carbon nanotube and their films under ambient pressure. *International Journal of Hydrogen Energy*, 27(5): 497–500.

Rakhi, R., Setupathi, K. and Ramaprabhu, S. 2008. Synthesis and hydrogen storage properties of carbon nanotubes. *International Journal of Hydrogen Energy*, 33(1): 381–386.

Rashidi, A.M., Nouralishahi, A., Khodadadi, A.A., Mortazavi, Y., Karimi, A. and Kashefi, K. 2010. Modification of single wall carbon nanotubes (SWNT) for hydrogen storage. *International Journal of Hydrogen Energy*, 35(17): 9489–9495, September 2010.

Reyhani, A., Mortazavi, S.Z., Mirershadi, S., Moshfegh, A.Z., Parvin, P. and Nozad, G.A. 2011. Hydrogen Storage in decorated multiwalled carbon nanotubes by Ca, Co, Fe, Ni, and Pd nanoparticles under ambient conditions. *The Journal of Physical Chemistry C*, 115(14): 6994–7001.

Rzepka, M., Lamp, P. and de la Casa-Lillo, M.A. 1998. Physisorption of hydrogen on microporous carbon and carbon nanotubes. *The Journal of Physical Chemistry B*, 102(52): 10894–10898.

Sandrock, G.D. and Huston, E.L. 1981. How metals store hydrogen. *Chemtech*, 11: 754–762.

Sandrock, G., Reilly, J., Graetz, J., Zhou, W.-M., Johnson, J. and Wegrzyn, J. 2005. Accelerated thermal decomposition of AlH3 for hydrogen-fueled vehicles. *Applied Physics A*, 80(4): 687–690.

Schaefer, S., Muniz, G., Izquierdo, M.T., Mathieu, S., Ballinas-Casarrubias, M.L., Gonzalez-Sanchez, G., Celzard, A. and Fierro, V. 2017. Rice straw-based activated carbons doped with SiC for enhanced hydrogen adsorption. *International Journal of Hydrogen Energy*, 42(16): 11534–11540.

Schimmel, H.G., Nijkamp, G., Kearley, G.J., Rivera, de Jong K.P. and Mulder, F.M. 2004. Hydrogen adsorption in carbon nanostructures compared. *Materials Science and Engineering B*, 108(1-2): 124–129.

Schlapbach, L. and Zuettel, A. 2001. Hydrogen-storage materials for mobile applications, *Nature*, 414: 353–358.

Senoz, E. and Wool, R.P. 2011. Hydrogen storage on pyrolyzed chicken feather fibers. *International Journal of Hydrogen Energy*, 36(12): 7122–7127.

Sethia, G. and Sayari, A. 2016. Activated carbon with optimum pore size distribution for hydrogen storage. *Carbon*, 99: 289–294.

Severa, G., Rönnebro, E. and Jensen, C.M. 2010. Direct hydrogenation of magnesium boride to magnesium borohydride: Demonstration of > 11 weight percent reversible hydrogen storage. *Chem. Commun*, 46(3): 421–423.

Sevilla, M. and Mokaya, R. 2014. Energy storage applications of activated carbons: supercapacitors and hydrogen storage. *Energy Environ. Sci*, 7(4): 1250–1280.

Shaijumon, M.M., Rajalakshmi, N., Ryu, H. and Ramaprabhu, S. 2005. Synthesis of multi-walled carbon nanotubes in high yield using Mm based AB2 alloy hydride catalysts and the effect of purification on their hydrogen adsorption properties. *Nanotechnology*, 16(4): 518–524.

Sharon, M., Soga, T., Afre, R., Sathiyamoorthy, D., Dasgupta, K., Bhardwaj, S., Sharon, M. and Jaybhaye S., 2007. Hydrogen storage by carbon materials synthesized from oil seeds and fibrous plant materials. *International Journal of Hydrogen Energy*, 32: 4238–4249.

Sharon, M., Bhardwaj, S., Jaybhaye, S., Sathiyamoorthy, D., Dasgupta, K. and Sharon, M. 2008. Hydrogen adsorption by carbon nanomaterials from natural source. *Asian J. Exp. Sci.*, 22(2): 75–88.

Shen, K., Xu, H., Jiang, Y. and Pietra, T. 2004. The role of carbon nanotube structure in purification and hydrogen adsorption. *Carbon*, 42(11): 2315–2322.

Shindo, K., Kondo, T., Arakawa, M. and Sakurai, Y. 2003. Hydrogen adsorption/desorption properties of mechanically milled activated carbon. *Journal of Alloys and Compounds*, 359(1-2): 267–271.

Sivakumar, R., Ramaprabhu, S., Rama Rao, K.V.S., Anton, H. and Schmidt, P. C. 1999. Hydrogen absorption characteristics in the Tb1−xZrxFe3 (x=0.1, 0.2, 0.3) system. *Journal of Alloys and Compounds*, 285(1-2): 143–149.

Strickland, G. 1984. Hydrogen derived from ammonia: Small-scale costs. *International Journal of Hydrogen Energy*, 9(9): 759–766.

Strobel, R., Jorissen, L., Schliermann, T., Trapp, V., Schutz, W., Bohmhammel, K., Wolf, G. and Garche, J. 1999. Hydrogen adsorption on carbon materials. *Journal of Power Sources*, 84(2): 221–224.

Strobel, R., Garche, J., Moseley, P.T., Jorissen, L. and Wolf, G. 2006. Hydrogen storage by carbon materials. *Journal of Power Sources*, 159(2): 781–801.

Suzuki, R.M., Andrade, A.D., Sousa, J.C. and Rollemberg, M.C. 2007. Preparation and characterization of activated carbon from rice bran. *Bioresource Technology*, 98(10): 1985–1991.

Takagi, H., Hatori, H., Yamada, Y., Matsuo, S. and Shiraishi, M. 2004. Hydrogen adsorption properties of activated carbons with modified surfaces. *Journal of Alloys and Compounds*, 385(1-2): 257–263.

Takagi, H., Soneda, Y., Hatori, H., Zhu, Z.H. and Lu, G.Q. 2006. Hydrogen adsorption properties of single-walled carbon nanotubes treated with nitric acid. *Nanoscience and Nanotechnology*, Volume, Issue, 3–7 July 2006.

Takagi, H., Soneda, Y., Hatori, H., Zhu, Z.H. and Lu, G.Q. 2007. Effects of nitric acid and heat treatment on hydrogen adsorption of single-walled carbon nanotubes. *Australian Journal of Chemistry*, 60(7): 519–523.

Takahata, T., Toda, I., Ono, H., Ohshio, S., Akasaka, H. and Himeno, S. 2009. Detailed structural analyses of KOH activated carbon from waste coffee beans. *Japanese Journal of Applied Physics*, 48(11): 117001.

Tozzini, V. and Pellegrini, V. 2011. Reversible hydrogen storage by controlled buckling of graphene layers. *The Journal of Physical Chemistry C*, 115(51): 25523–25528.

Vellingiri, L., Annamalai, K., Kandasamy, R. and Kombiah, I. 2019. Single-walled carbon nanotubes/lithium borohydride composites for hydrogen storage: role of in situ formed LiB (OH) 4, Li2CO3 and LiBO2 by oxidation and nitrogen annealing. *RSC Advances*, 9(54): 31483–31496.

Wang, B., Zhang, X., Huang, H, Zhang, Z., Yildirim, T., Zhou, W., Xiang, S. and Chen, B. 2020. A microporous aluminum-based metal-organic framework for high methane, hydrogen, and carbon dioxide storage. *Nano Research*, 14(2): 507–511.

Wang, H., Gao, Q. and Hu, J. 2009. High hydrogen storage capacity of porous carbons prepared by using activated carbon. *Journal of the American Chemical Society*, 131(20): 7016–7022.

Wang, Q. and Johnson, J.K. 1999. Molecular simulation of hydrogen adsorption in single-walled carbon nanotubes and idealized carbon slit pores. *The Journal of Chemical Physics*, 110(1): 577–586.

Winter, C.J. and Nitsch, J. 1988. Hydrogen as an energy carrier. *OSTI ID*: 5621380.

Wu, H.-C., Li, Y.-Y. and Sakoda, A. 2010. Synthesis and hydrogen storage capacity of exfoliated turbostratic carbon nanofibers. *International Journal of Hydrogen Energy*, 35(9): 4123–4130.

Ye, Y., Ahn, C.C., Witham, C., Fultz, B., Liu, J., Rinzler, A.G., Colbert, D., Smith, K.A. and Smalley, R.E. 1999. Hydrogen adsorption and cohesive energy of single-walled carbon nanotubes. *Applied Physics Letters*, 74(16): 2307–2309.

Yoon, S., Lim, S., Song, Y., Ota, Y., Oiao, W. and Tanaka, A. 2004. KOH activation of carbon nanofibers. *Carbon*, 42(8-9): 1723–1729.

Young Seak Lee, Young Ho Kim, Ji Sook Hong, Jeong Kwon Suh and Gyou Jin Cho. 2007. The adsorption properties of surface modified activated carbon fibers for hydrogen storages. *Catalysis Today*, 120(3-4): 420–425.

Zacharia, R., Kim, K.Y., Fazle Kibria, A.K.M. and Nahm, K.S. 2005. Enhancement of hydrogen storage capacity of carbon nanotubes via spill-over from vanadium and palladium nanoparticles. *Chemical Physics Letters*, 412(4-6): 369–375.

Zhan, L., Li, K., Zhu, X., Lv, C. and Ling, L. 2002. Adsorption limit of supercritical hydrogen on super-activated carbon. *Carbon*, 40(3): 455–457.

Zhang, X., Cao, D. and Chen, J. 2003. Hydrogen adsorption storage on single-walled carbon nanotube arrays by a combination of classical potential and density functional theory. *The Journal of Physical Chemistry B*, 107(21): 4942–4950.

Zhang, X.,Liu, Y., Ren, Z., Zhang, X., Hu, J., Huang, Z., Lu, Y., Gao, M. and Pan, H. 2021. Realizing 6.7 wt% reversible storage of hydrogen at ambient temperature with non-confined ultrafine magnesium hydrides. *Energy & Environmental Science*, 14(4): 2302–2313.

Zhang, Z.G., Luo, F.P., Wang, H., Liu, J.W. and Zhu, M. 2012. Direct synthesis and hydrogen storage characteristics of Mg–B–H compounds. *International Journal of Hydrogen Energy*, 37(1): 926–931.

Zheng, Q.R., Gu, A.Z., Lu, X.S. and Lin, W.S. 2005. Adsorption equilibrium of supercritical hydrogen on multi-walled carbon nanotubes. *The Journal of Supercritical Fluids*, 34(1): 71–79, May 2005.

Zhou, Jian-Ge and Williams Quinton L. 2011. Hydrogen storage on platinum-decorated carbon nanotubes with boron, nitrogen dopants or sidewall vacancies. *Carbon Nanotubes*, Pages: 12.

Zielinski, M., Wojcieszak, R., Monteverdi, S., Mercy, M. and Bettahar, M.M. 2007. Hydrogen storage in nickel catalysts supported on activated carbon. *International Journal of Hydrogen Energy*, 32(8): 1024–1032.

CHAPTER 16

Electrochemical Water Splitting

Bhushan Patil,[1,] A. Martinez-Lázaro,[2] R. Escalona-Villalpando,[2]*
Mayra Polett Gurrola,[3] J. Ledesma-García[2] and L.G. Arriaga[4]

1. Introduction

1.1 Introduction to water splitting

Hydrogen is one of the clean source of energy with the gravimetric energy density ca. 120 MJ Kg, which is more than 2.5 times than gasoline (44 MJ kg). Its clean production does not lead to the CO_2 emission or greenhouse gases, which makes it one of the ideal eco-friendly energy source that can replace the non-renewable energy sources like fossil fuels (Wang et al. 2020). Oxygen is another important gas which is needed for various energy devices (Song et al. 2017, Nørskov et al. 2004); it is also needed in the medical facilities like in the Covid-19 situation in the year 2020 and 2021 where many deaths occurred due to a shortage of oxygen supply (Bikkina et al. 2021).

Water is abundant and easily available on most parts of the Earth. Chemically, it is made up of 2 molecules of hydrogen and one molecule of oxygen forming H_2O molecule. Thus, it can act as the source to supply hydrogen and oxygen if the elements are separated or bond is broken in the H_2O molecule ($H_2O \longrightarrow H_2 + O_2$). The process of breaking the water molecule is water splitting. In other words, water splitting is the chemical reaction where water molecule is broken down into hydrogen and oxygen with some additional energy like chemical, with solar light, electric, thermal, nuclear, and so on. The photochemical process is the slowest method among all of the methods. Steam forming mostly generates hazardous gases like CO, CO_2, and oxides of sulfur. Conditions to operate most of these methods require elevated temperature and pressure. The instability of catalysts in the photoelectrochemical cells limits prolonging the application of such catalysts in water splitting (Anantharaj and Noda 2020). Water splitting by utilization of combined energies of the electric and chemical energy is referred to as Electrochemical Water Splitting (EWS).

[1] CNR-SPIN, Corso F.M. Perrone, 24, 16152 Genoa, Italy.
[2] División de Investigación y Posgrado, Facultad de Ingeniería, Universidad Autónoma de Querétaro, 76010, Santiago de Querétaro, México. Emails: aleem.lazaro@live.com; escalona.qfb@gmail.com; janet.ledesma@uaq.mx
[3] Tecnológico Nacional de México/Instituto Tecnológico de Chetumal. Av. Insurgentes 330, David Gustavo Gutiérrez, 77013, Chetumal, Quintana Roo, México and Cátedra Consejo Nacional de Ciencia y Tecnología-Tecnológico Nacional de México/ Instituto Tecnológico de Chetumal. Av. Insurgentes 330, David Gustavo Gutiérrez, 77013, Chetumal, Quintana Roo, México. mayra.pg@chetumal.tecnm.mx
[4] Laboratorio Nacional de Micro y Nanofluidica (LABMyN), Centro de Investigación y Desarrollo Tecnológico en Electroquímica, Pedro Escobedo, Qro., C.P. 76703, México. larriaga@cideteq.mx
* Corresponding author: bhushanpatil25@gmail.com

Benefits of EWS: EWS is a clean source of hydrogen and oxygen production. Thus, it is one of the highest energy generation sources to produce pure hydrogen without any greenhouse gases, which makes it a sustainable, pollution-free, eco-friendly method. It is easy for setup and can be used from small scale to large level generation of hydrogen and oxygen (Yan et al. 2016). EWS can be used as and when required or as on-demand energy source. It is possible to make commercial units with a low cost of investment. It is an energy source which can be used in stationary as well as mobile devices.

1.2 Electrochemical fundamentals of water splitting

Electrolyte: Use of pure water for electrolysis is constrained due to its high resistance of 18 MΩcm whereas tap water and seawater have a low resistance of ca. 5 and 20 Ωcm, respectively. However, contamination of redox active elements restricts their applicability in the EWS. Therefore, highly acidic and highly alkaline solutions are used for the EWS. Water splitting obeys Nernstian behavior, i.e., reaction strongly depends on the pH of electrolyte whereas unit increase in pH has the cathodic redox potential shift of 0.059V.

Electrode/Electrocatalysts: The ideal catalyst for the EWS can be selected based on the thermodynamic and kinetic catalytic activity, stability under operation and during stationary state, and high selectivity towards hydrogen and oxygen adsorption. It is accepted as a standard to relate the overpotential requisite to achieve 10 mA cm^{-2} current density with respect to geometric area at standard temperature and pressure (O_2: 1 atm), which is ca. 10 % effectual water splitting by the solar radiations of 1 sun. Thus, the figure describing overpotential to achieve 10 mA cm^{-2} is the figure of merit in the EWS (McCrory et al. 2013). However, there are many aspects which can alter the real electrocatalytic results like catalyst design, porosity, spillover, hydrophobicity, and hydrophilicity; thus, some researchers demand the overpotential normalized with electrochemical active surface area (i.e., ECSA) (Anantharaj and Noda 2020). To determine the rate of reaction on the specific electrocatalyst, kinetics is determined using Tafel slop, exchange current density, and turnover frequency (Anantharaj and Noda 2020).

Amorphous and crystalline nature of the catalysts has a high impact on the EWS. Anantharaj and Noda 2020 reviewed an influence of the crystallinity of the electrocatalysts towards the EWS and concluded that the amorphous nature of the electrocatalysts increases the electrocatalytic active surface area (ESCA) due to the irregular crystal structure which further helps in the diffusion of reactant, intermediate products and mass transport of gases evolved by electrolysis enhances the surface-confined reaction. Furthermore, the flexibility of electrodes with the amorphous catalysts is more than the crystalline electrocatalysts which also provides room for the catalyst to alter the crystal structure during electrolysis (Anantharaj and Noda 2020). Low cost electrocatalysts replacing the high cost state-of-the-art electrocatalysts have been experimentally proved as an efficient electrocatalysts such as transition metal (oxy)hydroxides, layer double hydroxides (LDH), chalcogenides, phosphides, nitrides, borides, carbides, and multimetallic alloys.

Set up of EWS: In general, for the half-cell reaction, 3-electrode cell is used as represented in Figure 1 and for the overall water splitting, 2-electrode cell is fabricated (Figure 1A). As usual, there are electrodes, working electrode on which electrocatalyst is deposited for the testing, counter or auxiliary electrode which provides the surface for the counter reaction and it is selected in such a way that the reaction should not be limited by the slow reaction that occurs at the counter electrode. The third electrode is the reference electrode (RE); it is preferable that the RE is in the same solvent as used in the electrolyte (Figure 1B). For all the experiments and comparisons, same electrode surface area, the spacing in-between the electrodes, quantity, and concentration of electrolyte should be kept constant for accuracy and reproducibility. Generally, acidic electrolyte, for example,

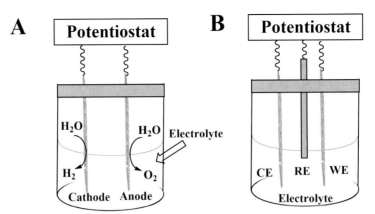

Figure 1. A: two-electrode system and B: three-electrode system.

0.5 M H_2SO_4, is used for hydrogen evolution reaction (HER) because of the high concentration of hydrogen ions while an alkaline solution is used for oxygen evolution reaction (OER) due to the high concentration of hydroxyl ions promoting easy adsorption on the electrocatalysts (Ramaswamy and Mukerjee 2012). However, for half-cell reactions, it can be selected while for commercial use one electrolyte should be used for both reactions. Thus, it is challenging to perform HER in alkaline and OER in acidic conditions (Gu et al. 2020a, Mahmood et al. 2018).

1.3 State-of-the-art catalysts Pt (HER) and Ru, Ir (OER)

The catalyst for water splitting is selected based on the ΔG (free energy to adsorb hydrogen, hydroxyl, or oxygen on the catalyst surface). The ΔG close to zero with high current density is the criteria to select the superior electrocatalyst. There are volcano plots plotted according to the potential or current densities versus ΔG (Cook et al. 2010, Man et al. 2011, Trasatti 1972). Based on these, Pt is referred to as a state-of-the-art catalyst for the HER, while Ir and Ru are for the OER. The important criteria for the selection of catalyst is to adsorb the hydrogen, hydroxyl, or oxygen and to detach the H_2 and O_2 evolved or any byproduct attached to the surface of the catalyst. Thus, in addition to element, its morphology, size, crystallinity, hydrophilicity, porosity, and surface area are important parameters influencing the EWS. High cost of the state-of-the-art catalysts demands exploring alternative cheap and stable catalysts.

1.4 Applications of EWS

Water splitting is a simple electrochemical process that has had many applications in recent years since through electrolysis, it is possible to decompose different compounds; the only condition is that these melted or dissolved compounds release ions. Electrolysis is often used in electroplating metals such as copper, gold, tin, silver and chromium to avoid or reduce corrosion or to enhance beauty of the materials. EWS opts as a tool for the immobilization of biomolecules on the electrodes by changing the surface pH in the vicinity of electrodes, for example, ascorbate oxidase immobilized on the gold electrode by hydrogen evolution (Patil et al. 2012). Many devices have been improved based on this simple electrochemical technique like metal-air batteries and hydrogen fuel cells (Schmidt et al. 2003).

Metal-air batteries: The storage of hydrogen obtained by EWS has enormous potential for portable devices for electric vehicles and extensive energy-storage networks. Among the batteries, the metal-air battery has promising future due to its high theoretical energy density 40.1 MJ/kg, equivalent to gasoline (40.6 MJ/kg) (Tahir et al. 2017). In the non-aqueous metal-air batteries, oxygen evolution mechanism and process is different than the aqueous one. EWS is applicable in the aqueous metal-air barriers in the cathode where OER is carried out by electrocatalysts, for example, zinc-air, lithium-air, aluminium-air batteries, and so on (Liu et al. 2020).

Fuel cells: The fuel cell is an energy device based on the electrochemical reactions whose concept is similar to that of a battery and it can be considered as the reverse process of EWS where hydrogen and oxygen are used and water is produced through overall reaction $2H_2 + O_2 \rightarrow 2H_2O$. It consists of the production of electricity through the use of chemicals, which are usually hydrogen (oxidized at anode through $2H_2 \rightarrow 4H^+ + 4e^-$) and oxygen (reduced at cathode as $4H^+ + 4e^- + O_2 \rightarrow 2H_2O$), where hydrogen acts as a fuel element and oxygen is obtained directly from the air. Thus, EWS can be used as the clean energy source for this purpose (Ströbel et al. 2002).

CO_2 reduction: During CO_2 reduction in the aqueous electrolyte, one of the major challenges is HER interference, particularly altering the local pH near the vicinity of the electrocatalyst surface, thus indirectly proving its importance to avoid or minimize HER during the CO_2 reduction. Therefore, it is important to reveal the fundamentals of the HER on the selective electrocatalysts (Ooka et al. 2017).

2. Electrochemical water splitting mechanism

The theoretical standard potential of EWS is 1.23 V vs. reversible hydrogen electrode (RHE) at 25ºC and 1 atm irrespective of the reaction media. Overall, EWS is the resultant of two half-cell processes, which are HER and OER. Thus, OER should occur at the anode (1.23 V/RHE), while HER at the cathode (0.0 V/RHE). As a matter of fact, for breaking the intrinsic activation barriers like the contact and the solution resistances, a higher potential is required for the practical or experimental water splitting than the thermodynamic potentials. Such an additional potential required for water splitting after deducting thermodynamic standard potential is described as an overpotential (η). Thus, minimizing overpotential required for water splitting is the key aspect, which can be achieved by selecting proper catalyst or catalysts and optimize electrolyzer design. The sluggish nature of OER due to multiple steps and intermediates formation increases the overpotential from 1.23 to usually 1.5 and more anodic.

Water splitting mechanism is different in different electrolytes; however, the overall total reaction remains the same $2 H_2O \longrightarrow 2H_2 + O_2$.

Reactions involved in the EWS:

HER and OER in acidic medium (Equations 1–2) and neutral/alkaline medium (Equations 3–4) (Anantharaj et al. 2016a)

$$4H^+ + 4 \text{ electrons} \longrightarrow 2H_{2(g)} \quad E^0 = 0.0 \text{ V} \tag{1}$$

$$2H_2O \longrightarrow O_{2(g)} + 4H^+ + 4 \text{ electrons} \quad E^0 = +1.23 \text{ V} \tag{2}$$

$$4H_2O + 4 \text{ electrons} \longrightarrow 2H_{2(g)} + 4OH^- \quad E^0 = -0.828 \text{ V} \tag{3}$$

$$4OH^- \longrightarrow O_{2(g)} + H_2O + 4 \text{ electrons} \quad E^0 = +0.401 \text{ V} \tag{4}$$

2.1 HER mechanism

HER mechanism in acidic medium (Yang et al. 2016, Tavares et al. 2001)

The first process in the HER is Volmer process where a proton from hydronium cation (H_3O^+) is attracted towards the electron rich electrocatalysts, resulting into an adsorbed hydrogen intermediate (H*) on the electrocatalysts.

These adsorbed hydrogen intermediates (H*) can couple with the hydronium ions to form hydrogen by Heyrovsky reaction or by coupling of two adjacent hydrogen intermediates (H*) through Tafel reaction.

HER mechanism in neutral/alkaline medium (Darband et al. 2019, Anantharaj et al. 2016a)

In the neutral or alkaline medium, a proton from the water molecule (H_2O) is attracted towards negatively charged electrocatalysts and forms the adsorbed layer of hydrogen intermediate (H*). Thus, the only difference in the Volmer reactions between the acidic and alkaline or neutral medium is the source of protons, which is adsorbed on the electrocatalysts and initiates the reaction.

$H_2O + e^- \rightleftarrows H_{adsorbed} + OH^-$ (Volmer adsorption process)

$H_{adsorbed} + H_2O + electron \rightleftarrows H_{2(g)} + OH^-$ (Heyrovsky process)

$2\ H_{adsorbed} \rightleftarrows H_{2(g)}$ (Tafel process)

The hydrogen production follows two pathways like in acidic medium by Heyrovsky reaction by an attraction between hydrogen intermediate (H*) and water molecule or by Tafel reactions where hydrogen is produced by the combination of two adjacent hydrogen intermediates. The possible pathways of the reaction can be interpreted experimentally using Tafel slopes. For the accurate interpretation of Tafel slopes, it is important to plot IR corrected plots as well as precise temperature and applied potential during the experiments. The Tafel slopes 118 mV dec^{-1}, 39 mV dec^{-1}, and 29 mV dec^{-1} corresponds to the Volmer, Heyrovsky, and Tafel reactions (Li et al. 2016). Therefore, the rate-determining step can be found out using the Tafel slopes. For example, the rate determining step in the HER at the Pt (100) plane electrode is Heyrovsky while at the Pt (110) plane electrode, it is Tafel reaction (Schmidt et al. 2003).

In the HER, the first process is adsorbed intermediate modified electrocatalysts formed by the proton adsorption (from the acidic electrolyte) and H_2O (in neutral or alkaline electrolyte) from the electrolyte solution ($H_{adsorbed}$). The adsorbed proton or water molecule can produce hydrogen by two pathways: Volmer-Heyrovsky reaction or Volmer-Tafel reaction mechanism (Patil et al. 2019).

2.2 OER mechanism

Numerous studies reveal that OER mechanism is not clear and occur on the anode with sluggish and complex proposed mechanisms than HER with multistep 4 electron transfer. OER occurs at a slower rate in comparison to HER. Like HER, instead of H^+ ions, OH^- first forms the adsorbed layer on the electrocatalyst followed by the series of reactions as shown in the Equations (5–13).

OER in neutral/alkaline media (Hu et al. 2019a)

$$E + OH^- \longrightarrow EOH + e^- \tag{5}$$

$$EOH + OH^- \longrightarrow EO + H_2O + e^- \tag{6}$$

$$EO + OH^- \longrightarrow EOOH + e^- \tag{7}$$

$$EOOH + OH^- \longrightarrow E + O_{2(g)} + H_2O + e^- \tag{8}$$

OER in acidic media (Shi et al. 2020, Reier et al. 2017)

$$H_2O + cat^\square \longleftrightarrow H^+ + OH^{cat\square} + electron \tag{9}$$

$$HO^{cat\square} \longleftrightarrow H^+ + O^{cat\square} + electron \tag{10}$$

$$O^{cat\square} + H_2O \longleftrightarrow H^+ + HOO^{cat\square} + electron \tag{11}$$

$$HOO^{cat\square} \longleftrightarrow H^+ + O_{2(g)}{}^{cat\square} + electron + {}^{cat\square} \tag{12}$$

$$2\,H_2O \longleftrightarrow 4H^+ + O_{2(g)} + 4\,electrons \tag{13}$$

where E is the electrode or electrocatalyst surface and $cat^\square$ is the catalytically active site. The above-mentioned mechanisms are general mechanisms while they may vary based on the electrocatalysts used. Also, during the measurements or applied potentials, resistances may increase due to cathodic and anodic reactions, gas bubble formation, or intermediate products. Electrocatalyst properties may also transform during catalysis. All these variations may change the required potentials for the reaction and the mechanism may also alter periodically.

3. Calculations and data interpretation

3.1 Onset potential

It is the redox potential required to initiate the HER or OER. From the CV or LSV, initial potential deviating from the non-Faradaic current (decrease in the case of HER while increase in the case of OER) excluding the redox peaks of catalysts (if electrocatalyst is oxidize and reduce in the selected potential window). Measurement of onset potential is a tricky method and there are various opinions about the correct method for estimating the onset potential. One method is to draw two tangents—one parallel to capacitance and another with the inclined or declined current of OER and HER; the junction of these two lines is considered as the onset potential. Another method is considering the potential necessary to obtain the specific current density, for example, 0.5 mV above the non-Faradaic current. An alternate method is considering the potential at which the current inclines or declines due to OER and HER. For the HER, since it is the cathodic reaction, the higher the anodic shift of the onset potential as a result of electrocatalytic activity, the better the electrocatalytic efficiency towards HER. On the contrary, OER is an anodic reaction higher the cathodic shift of onset potential better the electrocatalytic activity towards OER.

3.2 Tafel slope

The relation between the Tafel slope and charge transfer coefficient (α) is Tafel slope = $1/\alpha$. The Tafel slope mainly helps to interpret the rate of limiting reactions. Tafel slope proposes the additional required overpotential to upsurge the current density by 10-folds.

The Tafel slope can be estimated from the LSV, chronoamperometry, and impedance spectroscopy. Commonly, Tafel slope is obtained using the LSV data (iR-compensated overpotential and non-Faradaic capacitance correction) based on the slope obtained by plotting the log (j) versus η.

$$\eta = a + (2.3RT/\alpha nF) \log (j)$$

where, number of electrons require to complete the reactions is n, i.e., 4 for the OER and 2 for the HER while others are the constants with their usual standard meaning. Due to the inverse correlation between Tafel slope and α, the smaller the Tafel slope, higher the charge transferability or faster the kinetics. To obtain a precise slope with the least experimental inaccuracy, slow scan rate like 1 mV s^{-1} is preferred.

Due to the limitation of obtaining the steady-state current in LSV, the chronoamperometric analysis is preferred for Tafel slope estimation. In the chronoamperometry, after few seconds of initiation of reaction, steady-state current is obtained. Thus, current versus time plots at small potential intervals like 5 mV s^{-1} measured at the IrTiO$_2$ shows linear region even at the lower overpotential region as compared to the Tafel slope obtained from CV (Fabbri et al. 2014). Therefore, for accuracy in Tafel slope, correction with the capacitance is an important step (although it is tricky and needs to be done very carefully to avoid misinterpretation of data) while estimating from the LSV. Another well accepted method of Tafel slope calculation is from the electrochemical impedance spectroscopy (EIS). For using EIS, the Nyquist plots at a small interval of overpotentials are measured and Tafel slope is estimated from the log (R$_{ct}$) vs. η. R$_{ct}$ is the charge transfer resistance. As the solution resistance Rs (the real axis resistance value at the high-frequency intercept) is exempted and only depends on the electrocatalytic resistance, it is considered as one of the accurate methods of Tafel slope calculations (Anantharaj et al. 2016a, b).

The ideal catalyst must show a low Tafel slope for the efficient reaction kinetics.

3.3 Exchange current density

Exchange current density generally denoted by i_0 or j_0 is a key parameter which emphasizes the rate of reaction at the zero/equilibrium potential (Patil et al. 2019).

The Buttler-Volmer equation at higher overpotential ($\eta > 0.05$ V) can be rewritten as

$$\eta = a + (2.3RT/\alpha nF) \log (j) = (-2.3RT/\alpha nF) \log (j_0) + (2.3RT/\alpha nF) \log (j)$$

Due to sluggish OER reaction, i_0 or j_0 is frequently utilized to determine the HER catalysis (Anantharaj et al. 2016a). Higher the i_0 or j_0, better the electrocatalytic activity.

3.4 Stability of electrocatalysts

Stability of the electrocatalysts towards HER and OER are mainly investigated by the continuous CV at higher scan rate, i.e., 50 or 100 mV s^{-1} (generally, till present date, 1000 cycles are considered as a stable catalyst). Another method of evaluation of electrocatalytic stability is measuring the change in the current when applied potential is fixed (chronoamperometry) or by measuring the change in the potential when applied current is constant (chronopotentiomentry), preferably at 10 mA cm^{-1} from few minutes till some hours (generally, till present date, 12 hours are considered as a stable catalyst) (Anantharaj et al. 2016a). It is always better to analyze the catalyst by LSV to understand the shift in the onset potential, η_{10}, and Tafel slope after the stability testing, whether by continuous CV, chronoamperometry or chronopotentiometry.

3.5 Turn over frequency (TOF)

The hydrogen or oxygen molecules evolved per surface electrocatalyst atom is measured by turn over frequency. In other words, TOF can be referred to the rate of electron delivered per surface electrocatalyst atom per second.

TOF = $IN_A/(AFn\Gamma)$ (Anantharaj and Kundu 2019)

TOF = $I/(4 \times F \times m)$ (Lu and Zhao 2015)

where I denotes the measured current in Amperes, N_A and A are the Avogadro constant and geometrical surface area, F is Faraday constant, Γ is the number of atoms determined from the concentration of catalysts either on the surface or in total atoms, and m is the number of active moles of electrocatalysts. The concentration of atoms can be estimated from the average particle diameter of electrocatalyst (Anantharaj et al. 2016c), using another elemental characterization like ICP-MS (Khalily et al. 2019), based on the charge determined from the redox activity of electrocatalyst (Patil et al. 2019). During the estimation of m, mostly it is assumed that all the electrocatalyst molecules are active towards the HER and OER. Some studies assume that the electrocatalyst is forming a monolayer or forms a flat smooth surface (Anantharaj et al. 2016a,b). Due to these assumptions, chances of error cannot be ruled out because it is well known that not all the electrocatalyst area is active and to obtain and retain flat surface during electrocatalysis is challenging.

3.6 Faradaic efficiency

One method for Faradaic efficiency is through rotating ring disk electrode (RRDE) relevant mostly in the OER (Guo et al. 2014). In general, RRDE is performed with the glassy carbon as a disk and Pt as the ring. After cleaning and polishing, prior to catalytic measurements, the empirical collection efficiency of RRDE ($N_{empirical}$) is determined using the conventional ferrous/ferric redox response with respect to selected rotational speed and slow scan rate (~ 50 mV s^{-1}) by the ratio of $I_{limiting,ring}/I_{limiting,disk}$. The catalyst material is deposited only on the disk. The measurements should be performed by sweeping the potential of the disc while keeping the ring potential at the constant potential at which ORR catalyzes on the ring (Anantharaj et al. 2016a). Thus, the Faradaic efficiency is estimated by

FE = $(I_r \times n_d)/(I_d \times n_r \times N_{empirical})$

where I_d, I_r, n_d, and n_r are disc current, ring current, number of electrons transferred at disc and ring, respectively. While measuring FE using RRDE, side reactions or another catalytic reaction may interfere, so one has to be attentive to avoid misinterpretation of the data and FE. An alternative method is by measuring the hydrogen and oxygen evolved by conventional water gas displacement method, gas chromatography, spectroscopic techniques for OER, and by the data interpretation based on the ideal gas equation.

3.7 Mass and specific activities

The mass activity is referred to an experimental current divided by the weight of the electrocatalysts measured in amperes per gram. It is an important property especially for making lightweight batteries or portable device energy sources. The specific activity of the catalyst is estimated by the current normalized with the real surface area of the electrocatalyst measured either electrochemically or through BET. However, the surface area measured by BET may not be completely active in electrocatalysis; thus, it is always preferable that specific activity is measured based on ECSA.

ECSA is estimated by the non-Faradaic capacitance, impedance spectroscopy, and from the charge measured from the redox peaks. In the case of Pt, and Pd, hydrogen adsorption-desorption, CO adsorption, etc., can also be useful for ECSA calculations. For estimating the figure of merit, η_{10}, geometric area is used and conventionally accepted for the catalytic efficiency measurements; however, for the real fundamental catalytic efficiency should be revealed based on the specific capacity.

4. Electrochemical water splitting summary until 2020

Several proposals have emerged in recent years through which various combinations of electrocatalysts have been evaluated in a half cell under equilibrium conditions (1 atm and 25°C) to demonstrate that electron transfer in the OER and HER can occur at the surfaces with different chemical and morphological compositions. The noble metals like Pt, Pd, Au have been extensively studied and proved to be efficient electrocatalysts, the attempts to replace or reduce these high cost catalysts with the various combinations of transition metals and organic compounds is the topic of interest in the recent era. Even noble metal-based catalysts decorated on organic membranes have been tested as electrolyte within electrolyzers (Cruz et al. 2011a, Ornelas et al. 2009).

Although bifunctional electrocatalysts have been reported, the ideal way is to evaluate materials for both processes separately due to differences in the reaction potentials and the conditions of the reactions themselves. It is important to recognize that although water oxidation can occur in both basic and acidic media, basic electrolytes are mostly used for their different advantages such as easy handling, anti-corrosion of materials, and high efficiency in oxidation processes.

4.1 Transition metals

Abundant availability, strong activity, economically cheap, and substantial electrocatalytic activities of transition metals make them an alternative to replace the noble EWS electrocatalysts. They are explored preliminary towards HER in acidic and alkaline media. Some transition metals are superior to the state-of-the-art electrocatalysts for EWS. Transition metals are in the forms of chalcogenides, phosphides, nitrides, carbides, and oxides. Transition metal dichalcogenides (TMDs) are the utmost researched 2-dimensional material towards HER. Active edges, easy separation of 2D layers which increase surface area, the probability of altering the band gap of material, and easy to incorporate heteroatoms make it suitable for the EWS electrocatalysts. Co_3S_4 due to octahedron planes and adoptive spin states improve its OER catalysis, which increases as the number of nanosheets decreases. Atomic nanosheet of the Co_3S_4 is more active than the bulk (Liu et al. 2015). Transition metal selenides like $NiSe_2$, $CoSe_2$, or a combination of different transition metals and selenides show excellent HER catalysis due to very small intrinsic resistivity, the possibility of engineering by adjusting the stoichiometry of transition metals in the lattice (Huang et al. 2018a). Transition metal carbides (TMCs) reform the structure during the EWS (Peng et al. 2020). TMCs can be easily oxidized to form heterostructures such as $CoSe_xS_{2-x}@Co(OH)_2$ (Han et al. 2016). Transition metal phosphides (TMPs) are widely studied material in the full pH range including the neutral pH. Weak delocalization between transition metal and P facilitates H_2O adsorption and their decomposition. In addition, results by addition of two or more transition metals in the single-metal phosphides boost the electrocatalytic activities of TMPs (Shi and Zhang 2016, Li et al. 2018, Huang et al. 2018a). Transition metal nitrides, which are metallic in nature, encourage electrocatalysis towards EWS (Paquin et al. 2015). Transition metal oxides are poor in conductivity; however, the enormous oxygen vacancies generated by electrocatalytic reduction promotes the adsorption/desorption process in the HER (Luo et al. 2019). Therefore, transition metal oxides have a high rate of hydrogen production in acidic and neutral solutions (Peng et al. 2020). Electrocatalysts from the first line transition metal oxides and hydroxides show outstanding OER activity.

4.2 Transition metal complexes/Alloys/Heteroatoms for Pt-free HER

Fe exhibits relatively better HER catalysis with the overpotential around 260 mV at 10 mA cm^{-2}. Also, its different alloys proved to be better catalyst than Fe, like using Co in the alloy has increased real surface area and decreased the overpotential (Cabello et al. 2017). Alloys with three and four metals have been reported with Fe varying its stoichiometric relationship $Fe_{82}B_{18}$, $Fe_{80}Si_{10}B_{10}$, $Fe_{60}Co_{20}Si_{10}B_{10}$ alloys and investigated their catalytic activity in order to increase its stability and electrocatalysis towards HER in alkaline solution (Müller et al. 2014, Ďurovič et al. 2019). Alternative to iron, Co and its alloys are other materials used to replace Pt because of excellent electrical conductivity and durable catalysis (Subramania et al. 2007). Co-Mo and Co-Ni-Mo showed comparable results to the Fe (Laszczyńska and Szczygieł 2020). Hollow nanoparticles (Co-HNP) with catalytic activity at neutral pH (pH 7) keeping an overpotential of 85 mV at 10 mA cm^{-2} and 330 mV for 20 h at 150 mA cm^{-2} have been reported (Liu et al. 2016a). Another material and its respective alloys most used is Ni in medium alkaline, which depends on its morphology and surface area, supporting 217 mV, 180 mV, 275 mV at 10 mA cm^{-2} (Abbas et al. 2017, Hu et al. 2019b).

In addition to Ni in different structural configurations, it has been reported supported in reduced graphene oxide (rGO) obtaining an overpotential of 97 mV at 10 mA cm^{-2} with a durability time of 30 days in alkaline conditions del Ni-foam (Gutic et al. 2019). While Ni alloys have been diverse to modify their catalytic properties, stability, and surface, NiAlMo powder on nickel sheet presents a low overpotential of 82 mV at 200 mA cm^{-2} and a Tafel slope of 36 mV dec^{-1} (Razmjooei et al. 2020). An interesting compound has been reported with different steps in its synthesis of $MoNi_4$/MoO_2@Ni where it reports a low overpotential and a high stability time of 10 h (Zhang et al. 2017). The report on the synthesis of Ni-P-La was interesting due to its long stability up to 2000 h (Madram et al. 2020).

Other types of materials have been used combining a transition metal and a non-metal in order to replace noble metals to reduce costs, maintain catalytic activity and apply at a higher pH range. An example is the sulfides such as MoS_2 that presents catalytic activity at the edges of 2-D sheets in the acid medium for HER overpotential of 10 mA cm^{-2} around 180 mV (Lukowski et al. 2013). The compounds MoS_2 doped with heteroatom Co,O to form Co,O–MoS_2 resulted in a lower overpotential of 113 mV but under alkaline conditions (Cao et al. 2019). Meanwhile, the derivate of NiS, NiS_2, N_2S_3 have certain activity in the HER in alkaline and acidic conditions, and the performances are lower compared to molecules that do not contain precious metals or derivatives of Pt (Jiang et al. 2016). On the other hand, with the addition of CoS_2 to the Ni crystal lattice, it gives it catalytic properties at a neutral pH with comparable results with respect to an alkaline and acid medium (Zheng et al. 2015).

4.3 Metal-Organic Frameworks (MOFs)

MOFs are made by the metal nodes and organic linkers with the fabulous features of highly porous nature, which ultimately increases the surface area, makes it flexible, easy for structural modifications and so on. The micro and nano morphologies of MOFs have served as templates and also as precursors to manufacture numerous materials for OER. These types of materials are widely used in water oxidation, mainly due to the abundant triphasic surface regions that they have for the reaction to occur since their composition and structure ensure rapid transport of electrons and masses. MOFs are one of the alternatives against noble metals since they have four main benefits: (i) the alternate replication of organic molecules and metallic cores as active electrocatalytic sites improves the stability of metal active site; (ii) the crystalline structure of MOFs; (iii) an intrinsic porosity enhances exposure towards the metallic active site, which makes it more accessible and

Table 1. Literature summary of the electrocatalysts towards OER.

Catalysts	Electrolyte	Overpotential (η_{10}) At 10 mA cm^{-2}	Tafel slope mV dec^{-1}	TOF (s^{-1})	Mass loading (mg cm^{-2})	References
IrO$_2$	1 M KOH	338	47	-	0.21	(Tahir et al. 2017)
IrO$_2$	Solid Polymer Electrolyte	220	-	-	3	(Cruz et al. 2011b)
IrO$_2$/C	0.1 M KOH	370	-	-	0.2	(Tahir et al. 2017)
IrO$_2$-Pt	Nafion 117	270	-	-	6	(Ornelas et al. 2009)
IrO$_2$-Ta$_2$O$_5$	Nafion 117	270			6	(Ornelas et al. 2009)
RuO$_2$	Solid Polymer Electrolyte	170	65		3	(Cruz et al. 2011a)
FeMOFs-SO$_3$	1M KOH	218	36.2	-	-	(Feng et al. 2020)
FeCo$_2$-MOF	1 M KOH	335	36.2	0.0609 (200 mV)	-	(Gu et al. 2020b)
Fe/NiBTC@NF thin layer	1 M KOH	270	47	468 h^{-1} (300 mV)	-	(Wang et al. 2016)
NiCo-UMOFNs ultra thin nanosheets	1 M KOH	250	42	0.86 (300 mV)	-	(Zhao et al. 2016)
NiFe-MOF-74 Rhombic crystals	1 M KOH	223	71.6	-	-	(Xing et al. 2018)
FeNi-MOF nanoflowers	1 M KOH	150	38.7	3.1 (250 mV)		(Luo et al. 2020)
FeCoNi(OH)$_3$(BDC1.5/ NF) nano slab morphology	1 M KOH	196	29.5	0.865(250 mV)	-	(Senthil Raja et al. 2020)
Co$_3$O$_4$	1 M KOH	339	82	-	1.5	(Liu et al. 2008)
NiO QDs	0.5 M KOH	320	40	-		(Fominykh et al. 2014)
MnO	0.1 M KOH	540	-	-	0.028	(Gorlin and Jaramillo 2010)
NiCo nanosheets	1 M KOH	332	41	-	0.07	(Song and Hu 2014)
Ni-Co oxide	1 M NaOH	340	51	-	-	(Song and Hu 2014)
NiCo$_2$O$_4$ microcuboids	1 M NaOH	290	53	-	-	(Gao et al. 2016)
NiFe$_2$O$_4$ hollow spheres	0.1 M KOH	370	85	-	-	(Martínez-Lázaro et al. 2019)
CoP	1 M KOH	300	65	-	0.1	(Yang et al. 2015)
Co$_4$N	1M KOH	257	44	-	0.82	(Yu et al. 2016)

Table 1 contd. ...

...Table 1 contd.

Catalysts	Electrolyte	Overpotential (η_{10}) At 10 mA cm^{-2}	Tafel slope mV dec^{-1}	TOF (s^{-1})	Mass loading (mg cm^{-2})	References
Ni-P	1 M KOH	300	64	-	0.2	(Chen et al. 2015)
Co$_2$B	0.1 M KOH	360	45	-	0.21	(Masa et al. 2016)
CoSn(OH)$_6$ nanocubes	1 M KOH	274	-	-	0.283	(Song et al. 2016)
Co$_3$O$_4$@C	0.1 M KOH	310	69	-	-	(Liu 2016b)
Co$_3$O$_4$-NW/CC	1 M KOH	320	72	-	0.82	(Chen et al. 2015)
ZnCo$_2$O$_4$/N-CNT	0.1 M	350	70.6	-	0.2	(Liu et al. 2016)
CoOx@NC	1 M	260		-	1	(Haiyan et al. 2015)
N-CG-CoO	1 M KOH	340	71	-	-	(Mao et al. 2014)

feasible for the electrocatalysis; (iv) the adjustability of MOFs due to their easy structural tuning and hybridization with supplementary metals can boost its electrocatalysis towards EWS (Maleki 2016).

Variants of MOFs have been made, the most important being those that encompass the Fe group in their structure such as FeMOFs-SO$_3$ spindle-like (Feng et al. 2020) and FeCo$_2$-MOF (Gu et al. 2020b) that showed Tafel slopes ca. 36.2 mV dec^{-1}. The most widely used variants of these materials have been thin films with defined nanostructures such as Fe/NiBTC@NF thin layer (Wang et al. 2016), reported NiCo-UMOFNs ultra-thin nanosheets (Zhao et al. 2016) and NiFe-MOF-74 Rhombic crystals (Xing et al. 2018) with low operating overpotentials. Other alternatives like FeNi-MOF nanoflowers (Luoa19).

Several materials for electrocatalysis activity towards the HER have been studied, finding their strong influence on the Me-H junction that is expressed by the "volcano" plot, which can be searched as volcano-type reliance among the i_0 or j_0 and metal-hydrogen bond energy. Speaking of catalysts with electrocatalytic properties for HER, platinum has conventionally been used, but due to its high price it is not very profitable in industrial applications. One of the alternatives has been the use of Pt in addition to the other metals (Pt-M) such as Fe, Co, Ru, Cu, Au to reduce the mass of Pt with important results. MoPt$_2$ has presented an overpotential at 100 mA cm^{-2} of 210 mV (Stojić et al. 2006) and with the addition of (Na-molybdate and tris(ethylenediamine)Co(III) chloride) which is reduced to 196 mV (Marčeta Kaninski et al. 2007). Pt-Ni combination also proved to be enhancing electrocatalytic activity (Zhang et al. 2019).

In the same way, supports with metals C, Mo or Ni have been used to increase their electroactive area. For example, the Pt with C (around 20 wt % more used) the overpotential is ca. 46 mV at 10 mA cm^{-2} under alkaline medium (Zhang et al. 2019), with the Tafel slope change with electrolyte concentration of 180, 94 and 30 mV dec^{-1} for 0.01, 0.1 and 1 M KOH, respectively (Wang et al. 2019). Also, Pt deposited on the Ni–Fe layered double hydroxide has been reported overpotential of 101 mV at the 10 mA cm^{-2} (Anantharaj et al. 2017). N-Mo$_2$C has also been used as a support (Qiu et al. 2019) and the NiRu which has considerably improved accelerate water disassociation and the immobilization of Pt (Li et al. 2020).

Table 2. Literature summary of the electrocatalysts towards HER.

Catalysts	Electrolyte	Overpotential (η_{40}) At 10 mA cm^{-2}	Tafel slope mV dec^{-1}	Mass activity	Stability (h^{-1})	References
Pt/C	KOH (1 M)	46	30	20 %		(Wang et al. 2019)
Pt/Ni-Fe	KOH	101				(Anantharaj et al. 2017)
Pt/NMo$_2$C		100	108	1.08		(Qiu et al. 2019)
Pt/NiRu	KOH	38				(Li et al. 2020)
MoPt$_2$	alkaline	210	39			(Stojić et al. 2006)
MoPt$_2$-NaMoEt-Co	alkaline	196	34			(Marčeta Kaninski et al. 2007)
Pt-Ni	alkaline		61			(Zhang et al. 2019)
Fe	alkaline	260				(Pentland et al. 1957)
Fe-Co	alkaline	145	68			(Cabello et al. 2017)
Co-Mo	alkaline	145	103	45%		(Laszczyńska and Szczygieł 2020)
Co-Ni-Mo	alkaline	110				(Subramania et al. 2007)
Co-HNP	PBS pH 7	85			20	(Liu et al. 2016)
Ni-foam	alkaline	217				(Hu et al. 2019)
Ni-mesh	alkaline	275				(Hu et al. 2019)
Ni-Urchin	alkaline	180				(Abbas et al. 2017)
Ni-rGO	alkaline	97				(Gutic et al. 2019)
NiAlMo		82	36			(Razmjooei et al. 2020)
MoNi$_4$/MoO$_2$@Ni	alkaline	15	30		10	(Zhang et al. 2017)
Ni-P-La		139	93		2000	(Madram et al. 2020)
MoS$_2$	Acid	180	43			(Lukowski et al. 2013)
Co,O–MoS$_2$	alkaline	113	50		50	(Cao et al. 2019)
NiCoS$_2$	Neutral Alkaline acid	72	68 118 44			(Zheng et al. 2015)
MoSe$_2$	alkaline	364	68			(Zhao et al. 2020)
NiSe	alkaline	163				(Song et al. 2019)
CoSe/MoSe$_2$	Alkaline acid	115 190	54 62			(Song et al. 2019)
Ni$_2$P–NiSe$_2$	alkaline	66			12	(Liu et al. 2020)
Ni$_3$N/Ni	alkaline	121				(Xing et al. 2016)
Ni$_3$N	neutral	112			12	(Paquin et al. 2015)
Mo-Ni$_3$FeN	alkaline	69				(Liu et al. 2020c)
Co–Mo$_2$N@NC		47	43			(Lang et al. 2019)
Co$_4$N	alkaline	62	37		72	(Yuan et al. 2020)

Table 2 contd. ...

...Table 2 contd.

Catalysts	Electrolyte	Overpotential (η_{40}) At 10 mA cm^{-2}	Tafel slope mV dec^{-1}	Mass activity	Stability (h^{-1})	References
WC/WC$_2$	alkaline	75			480	(Chen et al. 2020)
Mo$_2$C@2D-NPCs	alkaline	45			20	(Lu et al. 2017)
Co-Mo		46	46		500	(Liu et al. 2019)
NiC	Alkaline	200				(You et al. 2017)
Mo$_2$C	alkaline	204				(Yu et al. 2017)
Cu@WC nanowires	acid Alkaline neutral	92 119 173				(Yao et al. 2021)
NiFeCoP/NM	alkaline	33	71.1			(Pan et al. 2016)
N–NiCoP NWs/CFP	alkaline	162.5				(Zhang et al. 2020)
Mo-Co-P	alkaline	30–35			24	(Thenuwara et al. 2018)
FeP-TiO$_2$	neutral	102				(Callejas et al. 2014)

References

Abbas, S.A., Iqbal, M.I., Kim, S.H. and Jung, K.D. 2017. Catalytic activity of urchin-like Ni nanoparticles prepared by solvothermal method for hydrogen evolution reaction in alkaline solution. *Electrochim. Acta*, 227: 382–390.

Anantharaj, S., Ede, S.R., Sakthikumar, K., Karthick, K., Mishra, S. and Kundu, S. 2016a. Recent trends and perspectives in electrochemical water splitting with an emphasis on sulfide, selenide, and phosphide catalysts of Fe, Co, and Ni: A Review. *ACS Catal.*, 6: 8069–8097.

Anantharaj, S., Jayachandran, M. and Kundu, S. 2016b. Unprotected and interconnected Ru0 nano-chain networks: Advantages of unprotected surfaces in catalysis and electrocatalysis. *Chem. Sci.*, 7: 3188–3205.

Anantharaj, S., Karthik, P.E., Subramanian, B. and Kundu, S. 2016c. Pt nanoparticle anchored molecular self-assemblies of DNA: An extremely stable and efficient HER electrocatalyst with Ultralow Pt Content. *ACS Catal.*, 6: 4660–4672.

Anantharaj, S., Karthick, K., Venkatesh, M., Simha T.V.S.V., Salunke, A.S., Ma, L., Liang, H. and Kundu, S. 2017. Enhancing electrocatalytic total water splitting at few layer Pt-NiFe layered double hydroxide interfaces. *Nano Energy*, 39: 30–43.

Anantharaj, S. and Kundu, S. 2019. Do the evaluation parameters reflect intrinsic activity of electrocatalysts in electrochemical water splitting? *ACS Energy Lett.*, 4(6): 1260–64.

Anantharaj, S. and Noda, S. 2020. Amorphous catalysts and electrochemical water splitting: An untold story of harmony. *Small*, 16: 1–24.

Bikkina, S., Manda, V.K. and Rao, U.V.A. 2021. Medical oxygen supply during COVID-19: A study with specific reference to state of Andhra Pradesh, India. *Mater. Today Proc.*, 1–5.

Cabello, G., Gromboni, M.F., Pereira, E.C., Mascaro, L.H. and Marken, F. 2017. Microwave-electrochemical deposition of a Fe-Co alloy with catalytic ability in hydrogen evolution. *Electrochim. Acta*, 235: 480–487.

Callejas, J.F., McEnaney, J.M., Read, C.G., Crompton, J.C., Biacchi, A.J., Popczun, E.J.Gordon, T.R., Lewis, N.S. and Schaak, R.E. 2014. Electrocatalytic and photocatalytic hydrogen production from acidic and neutral-PH aqueous solutions using iron phosphide nanoparticles. *ACS Nano*, 8: 11101–11107.

Cao, D., Ye, K., Moses, O.A., Xu, W., Liu, D., Song, P.Wu, C., Wang, C., Ding, S., Chen, S., Ge, B., Jiang, J. and Song, L. 2019. Engineering the in-plane structure of metallic phase molybdenum disulfide via Co and O dopants toward efficient alkaline hydrogen evolution. *ACS Nano*, 13: 11733–11740.

Chen, P., Xu, K., Fang, Z., Tong. Y., Wu, J., Lu, X., Peng, X., Ding, H., Wu, C. and Xie, Y. 2015. Metallic Co$_4$N porous nanowire arrays activated by surface oxidation as electrocatalysts for the oxygen evolution reaction. *Angew. Chemie.*, 127: 14923–14927.

Chen, Z., Gong, W., Cong, S., Wang, Z., Song, G., Pan, T., Tang, X., Chen, J., Lu, W. and Zhao, Z. 2020. Eutectoid-structured WC/W2C heterostructures: A new platform for long-term alkaline hydrogen evolution reaction at low overpotentials. *Nano Energy*, 68: 104335.

Cook, T.R., Dogutan, D.K., Reece, S.Y., Surendranath, Y., Teets, T.S. and Nocera, D.G. 2010. Solar energy supply and storage for the legacy and nonlegacy Worlds. *Chemical Reviews*, 110: 6474–6502.

Cruz, J.C., Baglio, V., Siracusano, S., Antonucci, V., Aricò, A.S. and Ornelas, R. 2011a. Preparation and characterization of RuO2 catalysts for oxygen evolution in a solid polymer electrolyte. *Int. J. Electrochem. Sci.*, 6: 6607–19.

Cruz, J.C., Baglio, V., Siracusano, S., Ornelas, R., Arriaga, L.G., Antonucci, V. and Arico, A.S. 2011b. Nanosized IrO2 Electrocatalysts for Oxygen Evolution Reaction in an SPE Electrolyzer, 1639–46.

Darband, G.B., Aliofkhazraei, M. and Shanmugam, S. 2019. Recent advances in methods and technologies for enhancing bubble detachment during electrochemical water splitting. *Renew. Sustain. Energy Rev.*, 114: 109300.

Ďurovič, M., Hnát, J., Bernäcker, C.I., Rauscher, T., Röntzsch, L., Paidar, M. and Bouzek, K.. 2019. Nanocrystalline $Fe_{60}Co_{20}Si_{10}B_{10}$ as a cathode catalyst for alkaline water electrolysis: Impact of surface activation. *Electrochim. Acta*, 306: 688–97.

Fabbri, E., Habereder, A., Waltar, K., Kötz, R. and Schmidt, T.J. 2014. Developments and perspectives of oxide-based catalysts for the oxygen evolution reaction. *Catal. Sci. Technol.*, 4: 3800–3821.

Feng, K., Zhang, D., Liu, F., Li, H., Xu, J., Xia, Y., Li, Y., Lin, H., Wang, S., Shao, M., Kang, Z. and Zhong, J. 2020. Highly efficient oxygen evolution by a thermocatalytic process cascaded electrocatalysis over sulfur-treated fe-based metal–organic-frameworks. *Adv. Energy Mater.*, 10: 1–8.

Fominykh, K., Feckl, J.M., Sicklinger, J., Döblinger, M., Böcklein, S., Ziegler, J., Peter, L., Rathousky, J., Scheidt, E.W., Bein, T. and Fattakhova-Rohlfing, D. 2014. Ultrasmall dispersible crystalline nickel oxide nanoparticles as high-performance catalysts for electrochemical water splitting. *Adv. Funct. Mater.*, 24: 3123–3129.

Gao, X., Zhang, H., Li, Q., Yu, X., Hong, Z., Zhang, X., Liang, C. and Lin, Z. 2016. Hierarchical NiCo2O4 hollow microcuboids as bifunctional electrocatalysts for overall water-splitting. *Angew. Chemie - Int. Ed.*, 55: 6290–6294.

Gorlin, Y. and Jaramillo, T.F. 2010. A bifunctional nonprecious metal catalyst for oxygen reduction and water oxidation. *J. Am. Chem. Soc.*, 132: 13612–13614.

Gu, X.K., Carl, J., Camayang, A., Samira, S. and Nikolla, E. 2020a. Oxygen evolution electrocatalysis using mixed metal oxides under acidic conditions: Challenges and opportunities. *J. Catal.*, 388: 130–140.

Gu, M., Wang, S.C., Chen, C., Xiong, D. and Yi, F.Y. 2020b. Iron-based metal-organic framework system as an efficient bifunctional electrocatalyst for oxygen evolution and hydrogen evolution reactions. *Inorg. Chem.*, 59: 6078–6086.

Guo, S.X., Liu, Y., Bond, A.M., Zhang, J., Karthik, P.E., Maheshwaran, I., Senthil S. Kumar and Phani, K.L.N. 2014. Facile electrochemical Co-deposition of a graphene-cobalt nanocomposite for highly efficient water oxidation in alkaline media: Direct detection of underlying electron transfer reactions under catalytic turnover conditions. *Phys. Chem. Chem. Phys.*, 16: 19035–19045.

Gutic, S.J., Šabanovic, M., Metarapi, D., Pašti, I.A., Korac, F. and Mentus, S.V. 2019. Electrochemically synthesized Ni@ reduced graphene oxide composite catalysts for hydrogen evolution in alkaline media—the effects of graphene oxide support. *Int. J. Electrochem. Sci.*, 14: 8532–8543.

Haiyan, J., Jing, W., Diefeng, S., Zhongzhe, W., Zhenfeng, P. and Yong, W. 2015. *In situ* cobalt-cobalt oxide/N-doped carbon hybrids as superior bifunctional electrocatalysts for hydrogen and oxygen evolution. *J. Am. Chem. Soc.*, 137: 2688–2694.

Han, L., Meng, Q., Wang, D., Zhu, Y., Wang, J., Du, X., Stach, E.A. and Xin, H.L. 2016. Interrogation of bimetallic particle oxidation in three dimensions at the nanoscale. *Nat. Commun.*, 7: 1–9.

Hu, C., Zhang, L. and Gong, J. 2019a. Recent progress made in the mechanism comprehension and design of electrocatalysts for alkaline water splitting. *Energy Environ. Sci.*, 12: 2620–2645.

Hu, X., Tian, X., Lin, Y.W. and Wang, Z. 2019b. Nickel foam and stainless steel mesh as electrocatalysts for hydrogen evolution reaction, oxygen evolution reaction and overall water splitting in alkaline media. *RSC Adv.*, 9: 31563–31571.

Huang, C., Pi, C., Zhang, X., Ding, K., Qin, P., Fu, J., Peng, X., Gao, B., Chu, P.K. and Huo, K. 2018a. *In situ* synthesis of MoP nanoflakes intercalated N-doped graphene nanobelts from MoO3–amine hybrid for high-efficient hydrogen evolution reaction. *Small.*, 14: 1800667.

Huang, Z., Liu, J., Xiao, Z., Fu, H., Fan, W., Xu, B., Dong, B., Liu, D., Dai, F. and Sun, D. 2018b. A MOF-Derived Coral-like NiSe@NC Nanohybrid: An efficient electrocatalyst for the hydrogen evolution reaction at All PH values. *Nanoscale*, 10: 22758–22765.

Jiang, N., Tang, Q., Sheng, M., You, B., Jiang, D.E. and Sun, Y. 2016. Nickel sulfides for electrocatalytic hydrogen evolution under alkaline conditions: A case study of crystalline NiS, NiS2, and Ni3S2 nanoparticles. *Catal. Sci. Technol.*, 6: 1077–1084.

Khalily, M.A., Patil, B., Yilmaz, E. and Uyar, T. 2019. Atomic layer deposition of Co_3O_4 nanocrystals on N-doped electrospun carbon nanofibers for oxygen reduction and oxygen evolution reactions. *Nanoscale Adv.*, 1: 1224–1231.

Lang, X., Qadeer, M.A., Shen, G., Zhang, R., Yang, S., An, J., Pan, L. and Zou, J.J. 2019. A Co-Mo_2N composite on a nitrogen-doped carbon matrix with hydrogen evolution activity comparable to that of Pt/C in alkaline media. *J. Mater. Chem. A*, 7: 20579–20583.

Laszczyńska, A. and Szczygieł, I. 2020. Electrocatalytic activity for the hydrogen evolution of the electrodeposited Co–Ni–Mo, Co–Ni and Co–Mo alloy coatings. *Int. J. Hydrogen Energy.*, 45: 508–520.

Li, D., Chen, X., Lv, Y., Zhang, G., Huang, Y., Liu, W., Li, Y., Chen, R., Nuckolls, C. and Ni, H. 2020. An effective hybrid electrocatalyst for the alkaline HER: Highly dispersed Pt sites immobilized by a functionalized NiRu-Hydroxide. *Appl. Catal. B Environ.*, 269: 118824.

Li, G., Sun, Y., Rao, J., Wu, J., Kumar, A., Xu, Q.N., Fu, C., Liu, E., Blake, G.R., Werner, P., Shao, B., Liu, K., Parkin, S., Liu, X., Fahlman, M., Liou, S.C., Auffermann, G., Zhang, J., Felser, C. and Feng, X. 2018. Carbon-tailored semimetal MoP as an efficient hydrogen evolution electrocatalyst in both alkaline and acid media. *Adv. Energy Mater.*, 8: 1801258.

Li, X., Hao, X., Abudula, A. and Guan, G. 2016. Nanostructured catalysts for electrochemical water splitting: Current state and prospects. *J. Mater. Chem. A*, 4: 11973–12000.

Liu, B., Zhang, L., Xiong, W. and Ma, M. 2016a. Cobalt-nanocrystal-assembled hollow nanoparticles for electrocatalytic hydrogen generation from neutral-PH water. *Angew. Chemie - Int. Ed.*, 55: 6725–29.

Liu, C., Gong, T., Zhang, J., Zheng, X., Mao, J., Liu, H., Li, Y. and Hao, Q. 2020b. Engineering Ni_2P-$NiSe_2$ heterostructure interface for highly efficient alkaline hydrogen evolution. *Appl. Catal. B Environ.*, 262: 118245.

Liu, F., Zheng, H.L., Chen, S.C., Jing, J.J., Chen, D.H., Chen, G., Wang, W., Zhu, M.H. and Bin Shi, J. 2008. Surgical treatment for well-differentiated thyroid carcinoma invading the larynx, trachea, esophagus and hypopharynx. *Acad. J. Second Mil. Med. Univ.*, 29: 213–1216.

Liu, G., Bai, H., Ji, Y., Wang, L., Wen, Y., Lin, H., Zheng, L., Li, Y., Zhang, B. and Peng, H. 2019. A highly efficient alkaline HER Co-Mo bimetallic carbide catalyst with an optimized Mo d-orbital electronic state. *J. Mater. Chem. A*, 7: 12434–12439.

Liu, Q., Pan, Z., Wang, E., An, L. and Sun, G. 2020a. Aqueous metal-air batteries: Fundamentals and applications. *Energy Storage Mater.*, 27: 478–505.

Liu, X., Lv, X., Wang, P., Zhang, Q., Huang, B., Wang, Z., Liu, Y., Zheng, Z. and Dai, Y. 2020c. Improving the HER activity of Ni_3FeN to convert the superior OER electrocatalyst to an efficient bifunctional electrocatalyst for overall water splitting by doping with molybdenum. *Electrochim. Acta.*, 333: 135488.

Liu, Y., Xiao, C., Lyu, M., Lin, Y., Cai, W., Huang, P., Tong, W., Zou, Y. and Xie, Y. 2015. Ultrathin Co_3S_4 nanosheets that synergistically engineer spin states and exposed polyhedra that promote water oxidation under neutral conditions. *Angew. Chemie - Int. Ed.*, 54: 11231–11235.

Liu, Z.Q., Cheng, H., Li, N., Ma, T.Y. and Su, Y.-Z. 2016b. $ZnCo_2O_4$ quantum dots anchored on nitrogen-doped carbon nanotubes as reversible oxygen reduction evolution electrocatalysts. *Adv. Mat.*, 28: 3777–3784.

Lu, C., Tranca, D., Zhang, J., Hernández, F.R., Su, Y., Zhuang, X., Zhang, F., Seifert, G. and Feng, X. 2017. Molybdenum carbide-embedded nitrogen-doped porous carbon nanosheets as electrocatalysts for water splitting in alkaline media. *ACS Nano*, 11: 3933–3942.

Lu, X. and Zhao, C. 2015. Electrodeposition of hierarchically structured three-dimensional nickel-iron electrodes for efficient oxygen evolution at high current densities. *Nature Communications*, 6: 1–7.

Lukowski, M.A., Daniel, A.S., Meng, F., Forticaux, A., Li, L. and Jin, S. 2013. Enhanced hydrogen evolution catalysis from chemically exfoliated metallic MoS2 nanosheets. *J. Am. Chem. Soc.*, 135: 10274–10277.

Luo, F., Xu, R., Ma, S., Zhang, Q., Hu, H., Qu, K., Xiao, S., Yang, Z. and Cai, W. 2019. Engineering oxygen vacancies of cobalt tungstate nanoparticles enable efficient water splitting in alkaline medium. *Appl. Catal. B Environ.*, 259: 118090.

Luo, S.W., Gu, R., Shi, P., Fan, J., Xu, Q.J. and Min, Y.L. 2020. Π-Π interaction boosts catalytic oxygen evolution by self-supporting metal-organic frameworks. *J. Power Sources*, 448: 227406.

Madram, A., Mohebbi, M., Nasiri, M. and Sovizi, M.R. 2020. Preparation of Ni–P–La alloy as a novel electrocatalysts for hydrogen evolution reaction. *Int. J. Hydrogen Energy*, 45: 3940–47.

Mahmood, N., Yao, Y., Zhang, J.W., Pan, L., Zhang, X. and Zou, J.J. 2018. Electrocatalysts for hydrogen evolution in alkaline electrolytes: Mechanisms, challenges, and prospective solutions. *Adv. Sci.*, 5: 1700464

Maleki, H. 2016. Recent advances in aerogels for environmental remediation applications: A review. *Chem. Eng. J.*, 300: 98–118.

Man, I.C., Su, H.Y., Calle-Vallejo, F., Hansen, H.A., Martínez, J.I., Inoglu, N.G., Kitchin, J., Jaramillo, T.F., Nørskov, J.K. and Rossmeisl, J. 2011. Universality in oxygen evolution electrocatalysis on oxide surfaces. *ChemCatChem*, 3: 1159–1165.

Mao, S., Wen, Z., Huang, T., Hou, Y. and Chen, J. 2014. High-performance Bi-functional electrocatalysts of 3D crumpled graphene-cobalt oxide nanohybrids for oxygen reduction and evolution reactions. *Energy Environ. Sci.*, 7: 609–616.

Marčeta Kaninski, M.P., Nikolić, V.M., Potkonjak, T.N., Simonović, B.R. and Potkonjak, N.I. 2007. Catalytic activity of Pt-based intermetallics for the hydrogen production-influence of ionic activator. *Appl. Catal. A Gen.*, 321: 93–99.

Martínez-Lázaro, A., Rico-Zavala, A., Espinosa-Lagunes, F.I., Torres-González, J., Álvarez-Contreras, L., Gurrola, M.P., Arriaga, L.G., Ledesma-García, J. and Ortiz-Ortega, E. 2019. Microfluidic water splitting cell using 3D $NiFe_2O_4$ hollow spheres. *J. Power Sources*, 412: 505–513.

Masa, J., Weide, P., Peeters, D., Sinev, I., Xia, W., Sun, Z., Somsen, C., Muhler, M. and Schuhmann, W. 2016. Amorphous cobalt boride (Co2B) as a highly efficient nonprecious catalyst for electrochemical water splitting: Oxygen and hydrogen evolution. *Adv. Energy Mater.*, 6: 1–10.

McCrory, C.C.L., Jung, S., Peters, J.C. and Jaramillo, T.F. 2013. Benchmarking heterogeneous electrocatalysts for the oxygen evolution reaction. *J. Am. Chem. Soc.*, 135: 16977–16987.

Müller, C.I., Rauscher, T., Schmidt, A., Schubert, T., Weißgärber, T., Kieback, B. and Röntzsch, L. 2014. Electrochemical investigations on amorphous Fe-base alloys for alkaline water electrolysis. *Int. J. Hydrogen Energy*, 39: 8926–8937.

Nørskov, J.K., Rossmeisl, J., Logadottir, A., Lindqvist, L., Kitchin, J.R., Bligaard, T. and Jónsson, H. 2004. Origin of the overpotential for oxygen reduction at a fuel-cell cathode. *J. Phys. Chem. B.*, 108: 17886–17892.

Ooka, H., Figueiredo, M.C. and Koper, M.T.M. 2017. Competition between hydrogen evolution and carbon dioxide reduction on copper electrodes in mildly acidic media. *Langmuir*, 33: 9307–9313.

Ornelas, R., Arico, A.S., Matteucci, R.F., Orozco, G., Beltran, D., Meas, Y. and Arriaga, L.G. 2009. Preparation and evaluation of $RuO_2 – IrO_2$, $IrO_2 – Pt$ and $IrO_2–Ta_2O_5$ catalysts for the oxygen evolution reaction in an SPE electrolyzer. *J. Appl. Electrochem.*, 39: 191–96.

Pan, Y., Lin, Y., Chen, Y., Liu, Y. and Liu, C. 2016. Cobalt phosphide-based electrocatalysts: Synthesis and phase catalytic activity comparison for hydrogen evolution. *J. Mater. Chem. A*, 4: 4745–4754.

Paquin, F., Rivnay, J., Salleo, A., Stingelin, N. and Silva, C. 2015. Multi-phase semicrystalline microstructures drive exciton dissociation in neat plastic semiconductors. *J. Mater. Chem. C*, 3: 10715–10722.

Patil, B., Fujikawa, S., Okajima, T. and Ohsaka, T. 2012. Enzymatic direct electron transfer at ascorbate oxidase-modified gold electrode prepared by one-step galvanostatic method. *Int. J. Electrochem. Sci.*, 7: 5012–5019.

Patil, B., Satilmis, B., Khalily, M.A. and Uyar, T. 2019. Atomic layer deposition of $NiOOH/Ni(OH)_2$ on PIM-1-based N-doped carbon nanofibers for electrochemical water splitting in alkaline medium. *ChemSusChem*, 1469–1477.

Peng, J., Dong, W., Wang, Z., Meng, Y., Liu, W., Song, P. and Liu, Z. 2020. Recent advances in 2D transition metal compounds for electrocatalytic full water splitting in neutral media. *Mater. Today Adv.*, 8: 100081.

Pentland, N., J. O'M. Bockris and Sheldon, E. 1957. Hydrogen evolution reaction on copper, gold, molybdenum, palladium, rhodium, and iron. *J. Electrochem. Soc.*, 104: 182.

Qiu, Y., Wen, Z., Jiang, C., Wu, X., Si, R., Bao, J., Zhang, Q., Gu, L., Tang, J. and Guo, X. 2019. Rational design of atomic layers of Pt anchored on Mo_2C nanorods for efficient hydrogen evolution over a wide PH range. *Small*, 15: 1–9.

Ramaswamy, N. and Mukerjee, S. 2012. Fundamental mechanistic understanding of electrocatalysis of oxygen reduction on Pt and Non-Pt surfaces: Acid versus alkaline media. *Adv. Phys. Chem.*, 2012: 1–18.

Razmjooei, F., Liu, T., Azevedo, D.A., Hadjixenophontos, E., Reissner, R., Schiller, G., Ansar, S.A. and Friedrich, K.A. 2020. Improving plasma sprayed raney-type nickel–molybdenum electrodes towards high-performance hydrogen evolution in alkaline medium. *Sci. Rep.*, 10: 1–13.

Reier, T., Nong, H.N., Teschner, D., Schlögl, R. and Strasser, P. 2017. Electrocatalytic oxygen evolution reaction in acidic environments – reaction mechanisms and catalysts. *Adv. Energy Mater.*, 7.

Schmidt, T.J., Stamenkovic, V., Ross, P.N. and Markovic, N.M. 2003. Temperature dependent surface electrochemistry on Pt single crystals in alkaline electrolyte: Part 3. The oxygen reduction reaction. *Physical Chemistry Chemical Physics*, 5: 400–406.

Senthil Raja, D., Huang, C.L., Chen, Y.A., Choi, Y.M. and Lu, S.Y. 2020. Composition-balanced trimetallic MOFs as ultra-efficient electrocatalysts for oxygen evolution reaction at high current densities. *Appl. Catal. B Environ.*, 279: 119375.

Shi, Y. and Zhang, B. 2016. Recent advances in transition metal phosphide nanomaterials: Synthesis and applications in hydrogen evolution reaction. *Chem. Soc. Rev.*, 45: 1529–1541.

Shi, Z., Wang, X., Ge, J., Liu, C. and Xing, W. 2020. Fundamental understanding of the acidic oxygen evolution reaction: Mechanism study and state-of-the-art catalysts. *Nanoscale*, 12: 13249–13275.

Song, F. and Hu, X. 2014. Exfoliation of layered double hydroxides for enhanced oxygen evolution catalysis. *Nat. Commun.*, 5: 1–9.

Song, F., Schenk, K. and Hu, X. 2016. A nanoporous oxygen evolution catalyst synthesized by selective electrochemical etching of perovskite hydroxide $CoSn(OH)_6$ nanocubes. *Energy Environ. Sci.*, 9: 473–477.

Song, K., Agyeman, D.A., Park, M., Yang, J. and Kang, Y.M. 2017. High-energy-density metal–oxygen batteries: lithium–oxygen batteries vs sodium–oxygen batteries. *Adv. Mater.*, 29: 1–31.

Song, W., Wang, K., Jin, G., Wang, Z., Li, C., Yang, X. and Chen, C. 2019. Two-step hydrothermal synthesis of CoSe/MoSe2 as hydrogen evolution electrocatalysts in acid and alkaline electrolytes. *ChemElectroChem*, 6: 4842–4847.

Stojić, D.L., Grozdić, T.D., Kaninski, M.P.M., Maksić, A.D. and Simić, N.D. 2006. Intermetallics as advanced cathode materials in hydrogen production via electrolysis. *Int. J. Hydrogen Energy*, 31: 841–846.

Ströbel, R., Oszcipok, M., Fasil, M., Rohland, B., Jörissen, L. and Garche, J. 2002. The compression of hydrogen in an electrochemical cell based on a PE fuel cell design. *J. Power Sources*, 105: 208–215.

Subramania, A., Sathiya A.R. Priya and Muralidharan, V.S. 2007. Electrocatalytic cobalt-molybdenum alloy deposits. *Int. J. Hydrogen Energy*, 32: 2843–2847.

Tahir, M., Pan, L., Idrees, F., Zhang, X., Wang, L., Zou, J.J. and Wang, Z.L. 2017. Electrocatalytic oxygen evolution reaction for energy conversion and storage: A comprehensive review. *Nano Energy*, 37: 136–157.

Tavares, M.C., Machado, S.A.S. and Mazo, L.H. 2001. Study of hydrogen evolution reaction in acid medium on Pt microelectrodes. *Electrochim. Acta*, 46: 4359–4369.

Thenuwara, A.C., Dheer, L., Attanayake, N.H., Yan, Q., Waghmare, U.V. and Strongin, D.R. 2018. Co-Mo-P based electrocatalyst for superior reactivity in the alkaline hydrogen evolution reaction. *ChemCatChem*, 10: 4846–4851.

Trasatti, S. 1972. Work function, electronegativity, and electrochemical behaviour of metals. III. Electrolytic hydrogen evolution in acid solutions. *J. Electroanal. Chem.*, 39: 163–184.

Wang, J., Yue, X., Yang, Y., Sirisomboonchai, S., Wang, P., Ma, X., Abudula, A. and Guan, G. 2020. Earth-abundant transition-metal-based bifunctional catalysts for overall electrochemical water splitting: A review. *J. Alloys Compd.*, 819: 153346.

Wang, L., Wu, Y., Cao, R., Ren, L., Chen, M., Feng, X., Zhou, J. and Wang, B. 2016. Fe/Ni metal-organic frameworks and their binder-free thin films for efficient oxygen evolution with low overpotential. *ACS Appl. Mater. Interfaces*, 8: 16736–16743.

Wang, X., Xu, C., Jaroniec, M., Zheng, Y. and Qiao, S.Z. 2019. Anomalous hydrogen evolution behavior in high-PH environment induced by locally generated hydronium ions. *Nat. Commun.*, 10: 1–8.

Xing, J., Guo, K., Zou, Z., Cai, M., Du, J. and Xu, C. 2018. *In situ* growth of well-ordered NiFe-MOF-74 on Ni foam by Fe^{2+} induction as an efficient and stable electrocatalyst for water oxidation. *Chem. Comm.*, 54: 7046–7049.

Xing, Z., Li, Q., Wang, D., Yang, X. and Sun, X. 2016. Self-supported nickel nitride as an efficient high-performance three-dimensional cathode for the alkaline hydrogen evolution reaction. *Electrochim. Acta*, 191: 841–845.

Yan, Y., Xia, B.Y., Zhao, B. and Wang, X. 2016. A review on noble-metal-free bifunctional heterogeneous catalysts for overall electrochemical water splitting. *J. Mater. Chem. A*, 4: 17587–17603.

Yang, X., Li, H., Lu, A.Y., Min, S., Idriss, Z., Hedhili, M.N., Huang, K.W., Idriss, H. and Li, L.J. 2016. Highly acid-durable carbon coated Co_3O_4 nanoarrays as efficient oxygen evolution electrocatalysts. *Nano Energy*, 25: 42–50.

Yang, Y., Fei, H., Ruan, G. and Tour, J.M. 2015. Porous cobalt-based thin film as a bifunctional catalyst for hydrogen generation and oxygen generation. *Adv. Mater.*, 27: 3175–3180.

Yao, M., Wang, B., Sun, B., Luo, L., Chen, Y., Wang, J., Wang, N., Komarneni, S., Niu, X. and Hu, W. 2021. Rational design of self-supported Cu@WC core-shell mesoporous nanowires for PH-universal hydrogen evolution reaction. *Appl. Catal. B Environ.*, 280: 119451.

You, B., Liu, X., Hu, G., Gul, S., Yano, J., Jiang, D.E. and Sun, Y. 2017. Universal surface engineering of transition metals for superior electrocatalytic hydrogen evolution in neutral water. *J. Am. Chem. Soc.*, 139: 12283–12290.

Yu, H., Fan, H., Wang, J., Zheng, Y., Dai, Z., Lu, Y., Kong, J., Wang, X., Kim, Y.J., Yan, Q. and Lee, J.M. 2017. 3D ordered porous MoxC (x = 1 or 2) for advanced hydrogen evolution and li storage. *Nanoscale*, 9: 7260–7267.

Yu, X.Y., Feng, Y., Guan, B., Lou, X.W.D. and Paik, U. 2016. Carbon coated porous nickel phosphides nanoplates for highly efficient oxygen evolution reaction. *Energy Environ. Sci.*, 9: 1246–1250.

Yuan, W., Wang, S., Ma, Y., Qiu, Y., An, Y. and Cheng, L. 2020. Interfacial engineering of cobalt nitrides and mesoporous nitrogen-doped carbon: Toward efficient overall water-splitting activity with enhanced charge-transfer efficiency. *ACS Energy Lett.*, 5: 692–700.

Zhang, C., Chen, B., Mei, D. and Liang, X. 2019. The OH^--driven synthesis of Pt-Ni nanocatalysts with atomic segregation for alkaline hydrogen evolution reaction. *J. Mater. Chem. A*, 7: 5475–5481.

Zhang, J., Wang, T., Liu, P., Liao, Z., Liu, S., Zhuang, X., Chen, M., Zschech, E. and Feng, X. 2017. Efficient hydrogen production on $MoNi_4$ electrocatalysts with fast water dissociation kinetics. *Nat. Commun.*, 8: 1–8.

Zhang, L., Qi, Y., Sun, L., Chen, G., Wang, L., Zhang, M., Zeng, D., Chen, Y., Wang, X., Xu, K. and Ma, F. 2020. Facile route of nitrogen doping in nickel cobalt phosphide for highly efficient hydrogen evolution in both acid and alkaline electrolytes. *Appl. Surf. Sci.*, 512: 145715.

Zhao, S., Wang, Y., Dong, J., He, C.T., Yin, H., An, P., Zhao, K., Zhang, X., Gao, C., Zhang, L., Lv, J., Wang, J., Zhang, J., Khattak, A.M., Khan, N.A., Wei, Z., Zhang, J., Liu, S., Zhao, H. and Tang, Z. 2016. Ultrathin metal-organic framework nanosheets for electrocatalytic oxygen evolution. *Nat. Energy*, 1: 1–10.

Zhao, X., Zhao, Y., Huang, B., Cai, W., Sui, J., Yang, Z. and Wang, H.E. 2020. $MoSe_2$ nanoplatelets with enriched active edge sites for superior sodium-ion storage and enhanced alkaline hydrogen evolution activity. *Chem. Eng. J.*, 382: 123047.

Zheng, G., Peng, Z., Jia, D., Al-Enizi, A.M. and Elzatahry, A.A. 2015. From water oxidation to reduction: Homologous Ni-Co based nanowires as complementary water splitting electrocatalysts. *Adv. Energy Mater.*, 5: 1–7.

CHAPTER 17

Nanomaterials Applied in the Construction Industry

D.L. Trejo-Arroyo,[1] D. Pech-Núñez,[2] J.C. Cruz,[2] L.G. Arriaga,[3] R.E. Vega-Azamar[2] and M.P. Gurrola[1],*

1. Introduction

Historically, nanotechnology has been adopted in various industrial fields such as microbiology, medicine, engineering, computer science, chemistry, and materials (Ho and Kjeang 2011, Gurrola et al. 2016, Francesko et al. 2018), which contributes to the improvement of the quality of life and greater welfare. Some of these industries are associated to the construction sector, for which it is important to focus on the demands of nanotechnology and nanomaterials that can stimulate the sector's overall productivity. The developments and implementation of nanotechnology are promising in terms of improving traditional materials, making the construction industry benefit by opening a new vision of the materials needed to generate more sophisticated, innovative, modern and environmental-friendly constructions (Papadaki et al. 2018, Vishwakarma 2020). The growth of nanotechnology can be related to the development of nanomaterials with a wide variety of particles that can be used to generate both functional and structural properties in cement-based composites, such as cementitious structures with the ability to self-heal, improved mechanical properties, that are lighter, reduce heat transfer, and have high solar reflectance, with antibacterial and self-cleaning surfaces, with greater energy and environmental efficiency. Nanocoatings can also provide fire protection due to their insulating capabilities and protection against the phenomenon of corrosion of steel structures, either in structures exposed to the environment or as reinforcement of concrete (Askarian et al. 2019). In general, nanotechnology provides unique features for construction and building through a variety of applications (Spitzmiller et al. 2013, Huseien et al. 2019, Ardalan et al. 2020).

[1] CONACYT-Tecnológico Nacional de México/Instituto Tecnológico de Chetumal. Av. Insurgentes 330, David Gustavo Gutiérrez, Chetumal 77013, Mexico. Email: danna.ta@chetumal.tecnm.mx
[2] Tecnológico Nacional de México/Instituto Tecnológico de Chetumal. Av. Insurgentes 330, David Gustavo Gutiérrez, 77013, Chetumal, Quintana Roo, México. Emails: M12390315@chetumal.tecnm.mx; julio.ca@chetumal.tecnm.mx; ricardo.va@chetumal.tecnm.mx
[3] Laboratorio Nacional de Micro y Nanofluidica (LABMyN), Centro de Investigación y Desarrollo Tecnológico en Electroquímica, Pedro Escobedo, Qro., C.P. 76703, México. Email: larriaga@cideteq.mx
* Corresponding author: mayra.pg@chetumal.tecnm.mx

Besides, it is well known that the current development in the field of nanotechnology, and therefore in the increase in the use of nanomaterials, would not be viable without characterization through different analysis and identification techniques of materials, given that it is imperative to understand the relationship between the initial composition, which can be controlled, with the resulting structure and properties of a material designed at the nanoscale that can present a higher and unique performance. In terms of synthesis, the properties of these materials depend not only on the properties of the starting materials, but also on their morphology and interfacial characteristics, so their determination is very important.

2. Nanomaterials and applications in cementitious materials

One of the typical current barriers in cement and concrete research is the improvement of strength and durability properties; therefore, a key point required is the extension of the lifespan by modifying the traditional materials used in construction and building, and generating new mixtures that integrate other cementitious binder components with a low CO_2 footprint, such as geopolymers (Duxson et al. 2007, Juenger et al. 2011). The cement-based binder is the main active component of concrete used in most constructions and buildings, in addition to water and fine and coarse aggregates (Ardalan et al. 2020, Hatem Nawar 2021). Portland cement is a hydraulic binder made from ground 'clinker', which is obtained by firing to sintering, along with small amounts of calcium sulfate, and can also contain silicon-based or/and aluminous pozzolanic materials that in contact with water chemically react with calcium hydroxide (Ardalan et al. 2017, Askarian et al. 2019, Motahari et al. 2019). In this regard, and in order to save energy, as well as to reduce the carbon emissions generated in the cement manufacturing process, industrial by-products called supplementary cementitious materials (SCM) have been used, incorporating them either as additions or as partial substitution of cement, improving their rheological, mechanical and durability properties through their hydraulic or pozzolanic activity in the same way (Siddique and Khan 2011).

Between the cement and the aggregates used to produce cement-based composites, such as mortars and concretes, there is an Interfacial Transition Zone (ITZ), which is the area most prone to the generation of cracks and/or voids filled with air on a nanometric scale that can significantly affect the mechanical properties of the composite (Qudoos et al. 2018). There is a fairly wide field for the improvement of cementitious composites, such as mortars and concretes, by incorporating nanomaterials within the cementitious matrix (Reches 2018). In this sense, nanoparticles improve the microstructural properties of concrete, since they intervene in the formation of hydration products and/or fill the existing voids, achieving a more closed packing to the aggregates' surface in the ITZ (Scrivener et al. 2004, Nazari and Riahi 2011). The hydration process between the aggregates and the cementitious paste is improved, helping to dissipate the growth of cracks and/or to enhance the sealing effect, improving toughness and resistance to bending (Wang et al. 2020). Some nanoparticles also help accelerate cement hydration and can inhibit the growth of $Ca(OH)_2$ crystals, generating more quantity (Arefi and Rezaei-Zarchi 2012).

The aggregates' characteristics, in terms of shape, texture and gradation, notably influence the properties of fresh concrete such as workability, cohesiveness, segregation, among others; they can directly affect the strength of concrete as well as its durability in the hardened state (Alexander and Mindess 2005). It has been reported that the shape of the aggregates does not affect the strength of concrete in the hardened state, but significantly influences workability (León and Ramírez 2010). For its part, the texture has a significant influence on the ITZ—a rough surface produces higher microhardness values in the area closest to the aggregate surface and decreases as it distances (Qudoos et al. 2018). In turn, porosity tends to decrease near the aggregates' surface as the surface roughness increases and, in addition, concretes made with coarse aggregates with rough surfaces present an improvement in mechanical properties as well as compressive and tensile strength, elastic modulus and Poisson's ratio (Hong et al. 2014). It has also been reported that aggregates produced with stratified sediments may vary in their characteristics, due to the conditions under which they consolidate (Chan Yam et al. 2003).

Several studies have focused on the mechanical response of concrete and durability, to the addition of different nanostructured materials (nanotubes, nanofibers, nanoparticles) and SCM, for which, various reinforcement mechanisms have been proposed focusing on the ITZ and its characterization by different methods. With the addition of nanoparticles such as silicon dioxide (SiO_2) and titanium dioxide (TiO_2), it has been confirmed that the hydration products of the cementitious paste and the durability of concrete are improved, which is attributed to different mechanisms such as nucleation sites and/or fill effect (Meng et al. 2012, Wang et al. 2018). Ren et al. evaluated the influence of the addition of SiO_2 and TiO_2 nanoparticles, in different percentages, on the compressive strength of concretes in the curing stages of 3, 7, 21 and 28 days of curing. They determined that the optimum percentage of addition was 3% by weight of cement and that both type of nanoparticles improved the compressive strength at 28 days of curing. The TiO_2 nanoparticles do not contribute to the formation of the C-S-H gel; however, they can fill the pores in the C-S-H gel and also help to control the growth of C-H crystals at the interface between the aggregate and the cementing paste. For their part, the SiO_2 nanoparticles react with the C-H to form C-S-H; both contribute to the obtention of denser concrete and better mechanical properties (Ren et al. 2018).

On the contrary, Meng et al. assessed the influence of 5 and 10% of TiO_2 nanoparticles on the flexural and compression strength of cement-based mortars at 1, 3, 7 and 28 days. They reported that the mechanical properties of mortars added with TiO_2 nanoparticles increased at early stages and decreased at 28 days; they indicated that this was due to a variation in the orientation of the crystal C-H by nucleation effect, and not by an increase in its amount. However, by adding slag powders to the same mixtures, these consume the C-H during the hydration process, helping to improve compressive strength at late stages (Meng et al. 2012).

Wang et al. evaluated the influence of the addition of different nanostructured materials or nano-fillers, such as nano-SiO_2, nano-TiO_2, nano-ZrO_2, both non-functionalized and functionalized multi-walled carbon nanotubes (MWCNTs), carbon nanotubes covered with nickel, multilayer graphenes (MLGs) and boron nano-nitride, on the bond strength and microstructural development between the cementitious paste and the aggregates. They determined that all the added nanomaterials helped in the improvement of the ITZ hydration and compaction products. Nanostructured materials, for example 0D nanoparticles, can form structures called core-shell and inhibit the growth of cracks (pinning effect), and 1D and 2D can bridge the hydration products generated in the ITZ and in the bulk of the cementitious matrix, thus avoiding the appearance and development of cracks (Wang et al. 2020). Only if the bonding in the matrix is controlled, substantial mechanical reinforcement can occur at the various scales of the material (Stynoski et al. 2015). The nano-core effect was proposed by Han et al., which integrates the core effect with the nano effect and it is closely linked to the intrinsic properties of nano-fillers, and to the composition and processing of cementitious composites (Han et al. 2017b). On the other hand, the influence of nanoparticles on the mechanical response of Reactive Powder Concrete (RPC) has also been assessed, which is an ultra-high-performance concrete. It has been reported that bending, compression and splitting strength were increased after 28 days of curing in RPC with the addition of ZrO_2 nanoparticles. The addition of this kind of nanoparticles did not interfere in the hydration process; however, it inhibited the growth space of C-H crystals by restricting their size and also reduced the crystal orientation, obtaining a denser microstructure than that of conventional RPC (Han et al. 2017a). Likewise, Wang et al. reported that the addition of TiO_2, SiO_2 and ZrO_2 nanoparticles improved the properties of wear resistance and resistance to chloride ion penetration, thus improving the microstructure of RPC (Wang et al. 2018).

As for mortars and concretes made with limestone aggregates, their initial and final properties both in fresh and hardened state may differ when compared to composites made with natural sand of siliceous origin. The demand for water is increased due to the higher degree of absorption, so in practice an overdose of cement could be needed to achieve the desired strengths. The effect of the addition of ZrO_2 nanoparticles has also been evaluated in cementitious composites made with limestone aggregates and it has been reported that the reinforcement in mortars added with this kind of nanoparticles was attributed to the nucleation sites generated by the nanoparticles and the inhibition of the growth of C-H crystals in addition to the filler effect, with obtaining of a better microstructural development and a denser mortar, which was reflected in the increase in compressive strength, when compared to conventional mortar, proposing that the onset of failure of the composite occurred inside the limestone aggregate and not at the interface ITZ (Trejo-Arroyo et al. 2019).

Materials with photocatalytic properties have also been evaluated for cementitious composites that seek to develop a self-cleaning and/or inhibitory function of microorganisms such as fungi and bacteria generated mainly by the high humidity in the air, which are more prone in tropical climates and mostly grow in large exposed areas of cementing nature, specifically for the case of TiO_2 and ZrO_2.

3. Nanomaterials used as coatings

In the last decade, the use of photocatalysts in concrete technology has been researched and developed with the integration of semiconductors in construction materials, with great promising benefits for good environmental quality, and applications have been expanding for self-cleaning facades, roads and pavements, conservation of stone-based architectural heritage, etc. However, there are still barriers, or a certain degree of skepticism, probably regarding its cost-benefit related to the efficiency of the photocatalyst. Photocatalysis is associated to NO_x abatement, to mitigate the impact of airborne particles and nitrogen oxides that deteriorate air quality in urban environments and, as already mentioned, it provides self-cleaning, antimicrobial and water purifying capabilities (Chen and Poon 2009, Zhong and Haghighat 2015, Ganguly et al. 2018). The most researched and commercialized photocatalysts is TiO_2.

Research focused on the use of antimicrobial and antifungal materials in the form of surface coatings with stable and inhibitory capabilities is of great importance. In the literature, there is a large amount of information about the mechanisms of photocalytic activity and the self-cleaning effect, as well as research works focused on improving the photoactive effect in the visible light range by means of theoretical approaches, experimental methods, modeling of electronic structures, starting from the use of non-metals, mono-doped transition metals and co-doping, among others (Yan et al. 2013, Etacheri et al. 2015). In the same way, just as materials with improved photocatalytic properties have been developed, the implications on their durability properties must also be widely evaluated, as a function of the exposure time to the environment and under real working conditions.

Implementation of TiO_2 is a relatively new technology; its effect has been evaluated in cementitious composites when applied by different techniques, either in the form of coatings or directly integrated into the matrix in bulk. Mainly, the self-cleaning property is evaluated through the degradation of organic colorants such as Rhodamine B (Rh-B), methyl orange, methylene blue, among others, and the decontaminating effect in the air through the degradation of NOX, as mentioned above (Ramirez et al. 2010, Zhao et al. 2015, Cerro-Prada et al. 2016). The use of TiO_2 as part of coatings for mortars or cement has been tested in different studies; these coatings are usually prepared in solutions based on water, alcohol or resins, starting with simple applications such as spray or brush, up to a little more specialized applications such as immersion, vacuum saturation or others (Folli et al. 2012, Guo et al. 2015, Atta-ur-Rehman et al. 2019). In the tests, the coatings have demonstrated photocalytic effectiveness in terms of self-cleaning and for air decontamination, they have a high degree of efficiency. Nevertheless, this is overshadowed by the low resistance offered by the coatings when exposed to wear and tear due to environmental exposure, for which the photocalytic effectiveness is affected over time (Guo et al. 2015, 2016, Gherardi et al. 2016, Cerro-Prada et al. 2018, Luo et al. 2020).

In 2009, Ruot et al. reported in their research the influence of the addition of TiO_2 in cementitious pastes (white cement) and mortars. The samples were cured and stored for 4 years at 23°C and relative humidity of 55 ± 5%; they were tested for density, porosity, X-ray diffraction and degradation of a Rh-B solution (0.05 gL^{-1}). In the results, it was stated that both the pastes and mortars with the addition of TiO_2 showed photocalytic abilities, which increased as the percentage of addition rose and, also, the porosity increased in the mortars when compared to that of the pastes (Ruot et al. 2009). In 2010, Ramirez et al., developed a study with TiO_2 classified into two groups: group 1—decorative cement, concrete slab, green plaster with limestone sand and white cement mortar with limestone sand; group 2—cellular cement and a cement slab with three finishes. Both groups were covered by immersion in different solutions. Both groups were exposed to a 3-phase attrition test; the first phase without abrasive process on the surface, the second phase under medium wear conditions, and the third phase under high wear conditions. The samples from the first group retained the highest amount of TiO_2 and presented good results in decontamination; however, the material that showed the greatest photocalytic effectiveness was cellular cement (Ramirez et al. 2010). Also, effectiveness of TiO_2 has been reported as a function of size, evaluating the effect of TiO_2 on a micrometric and nanometric scale on the anatase phase in cement mortar samples (Folli et al. 2012). The aesthetic durability was determined by the degradation of Rh-B dye in a solution and NO_X decontamination tests were carried out. The results showed a greater photocalytic effect with TiO_2 at the nanometric scale that rised with the increase of the addition, which resulted in an effective strategy to achieve self-cleaning and air-decontaminating facades.

Yang et al., in their research, evaluated the photocatalytic activity of TiO_2 supported in siliceous aggregates as a substrate and mounted on a concrete surface, against the conventional method used, which is dispersion in a layer of surface mortar, depending on photon efficiency. They reported a considerable rise in the photocatalytic activity in these composites supported with TiO_2 when compared to the conventional method; however, they only presented a model to illustrate the impact in the air near the surface of the composites, but they did not predict the impact on air quality (Yang et al. 2018).

Treatment of TiO_2 based on acids such as HNO_3 and H_2SO_3 is a method proposed by Paolini et al. to develop pigments for building materials with improved optical properties (Paolini et al. 2016). It has been pointed out in the literature that a pretreatment of TiO_2 nanoparticles, with different acids before being added to the selected construction material, can considerably improve their reflectance as well as maintain the photocatalytic activity for a longer time, that is, the energy and environmental efficiency of these self-cleaning materials can be improved (Paolini et al. 2018).

The use of steel elements in the construction industry has been of great relevance, especially due to the great advantages that this material offers, such as high resistance, ductility and uniformity, which makes it suitable for building countless structures of different sizes and shapes (Brockenbrough and Merritt 2006, Aghayere and Vigil 2020). Despite these advantages, steel elements present problems in aggressive environments since they are prone to being affected by corrosion as well as being thermodynamically unstable.

On the other hand, due to the great impact that corrosion has on the economy of industrialized countries, protection against corrosion has become a topic of primary relevance in the modern metal industry (Chang et al. 2012). Currently, it is possible to use several methods to control corrosion; however, in the construction industry the use of coatings made by electrodeposition and polymeric coatings stand out, due to easiness of application, especially when it comes to protecting large structures. Nonetheless, most of these, specifically the polymeric ones, tend to have low resistance to the penetration of corrosive solutions, which considerably reduces the protective action of the coating.

3.1 Electrodeposition

One of the techniques for the production of coatings is electrodeposition, since it presents a useful and versatile option. The properties of the coatings obtained by this technique, such as hardness, resistance to wear and electrical resistivity, largely depend on particle size of the nanomaterial used for their manufacture. The materials most used for this technique are Zn, ZrO_2 and Ni. From this, it has been verified that when comparing coatings with Zn and ZrO_2, the latter acquires greater hardness and improvements in their anticorrosive properties (Vathsala and Venkatesha 2011). In this type of coatings, combinations of the most used materials with others, such as Zn with carbon nanotubes (CNTs) and Ni with CNTs, have also been made, for which electrochemical studies have shown that the union of these materials has improved resistance to corrosion of the compounds because, with the use of these nanomaterials, grain size is much smaller, which in turn covers the micro-holes of the surface in the metals, generating homogenization and making coatings strong and resistant to corrosion cracking (Praveen et al. 2007, Saji and Thomas 2007, Praveen and Venkatesha 2009).

3.2 Organic Polymeric Coatings

Organic coatings can be defined as a relatively stable dispersion of one or more finely divided pigments in a well-tested solution that, when applied and dried, results in a film that forms a flexible and adherent barrier that protects against corrosion (Pons Frias 1984). The main components of this type of coatings are shown in Figure 1.

Despite the fact that these coatings form a protective barrier, most of them are usually porous to oxygen, water and chlorides, which favor the initiation of the corrosion process. In recent years, several options have been investigated that can help solve this problem and nanomaterials have been used to try to specifically improve chemical and mechanical properties. The use of nanoparticles

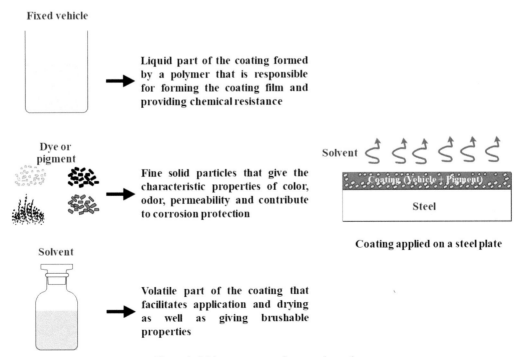

Fixed vehicle

Liquid part of the coating formed by a polymer that is responsible for forming the coating film and providing chemical resistance

Dye or pigment

Fine solid particles that give the characteristic properties of color, odor, permeability and contribute to corrosion protection

Solvent

Coating (Vehicle + Pigment)

Steel

Coating applied on a steel plate

Solvent

Volatile part of the coating that facilitates application and drying as well as giving brushable properties

Figure 1. Main components of an organic coating.

in coatings has been able to effectively reduce the impact of corrosive environments due to the different properties they can improve, such as surface hardness, adhesion, drying time, barrier and tribological properties, to mention a few (Farag 2020). In addition to the above mentioned, coatings with nanomaterials allow adaptability on various surfaces, as well as reduction in maintenance costs. Specifically, for this type of coatings, the difference between the particle sizes is an important element to take into account, since it has been proven that smaller particle sizes translate into greater resistance to corrosion, due to reduction in the coating's porosity, thus decreasing the passage of external agents (Aglan et al. 2007).

Among the nanomaterials that have been studied, there are epoxy coatings added with particles of SiO_2, Zn, Fe_2O_3, clays, TiO_2, silica, zirconium, graphene, CNTs, to name a few, whose use has considerably improved different properties of the coatings and, specifically, resistance to corrosion (Li et al. 2016). Of all the above, carbon nanomaterials stand out significantly. Specifically, CNTs are of particular interest due to the structural, mechanical, electrical and thermal properties to be incorporated in polymeric matrices (Osorio et al. 2008), because they can withstand high temperatures, abrasive conditions and corrosive environments (Harb et al. 2016). In coatings, it has been shown that the use of CNTs provides mechanical properties such as resistance, rigidity, strengthening of the adhesion and cohesion properties of the film, as well as the improvement of conductivity (Esposito et al. 2014); likewise, their ability to spread over large surfaces has made them an excellent material for use in functional coatings (De Volder et al. 2013). Coatings with presence of CNTs that have undergone impedance measurements have been shown to have a significant increase in their resistance to charge transfer (Aglan et al. 2007), which translates into an increase in corrosion resistance. Other options that have also been studied are the mixture of different nanomaterials; an example of this is the mixture of MWCNTs and Zn, which has been shown to delay reactions, reducing the passage of H_2O, O_2 and other pollutants towards the metal. This phenomenon can be attributed to the anticorrosive properties of Zn, which shows two protection mechanisms, cathodic protection and barrier protection that is reinforced by the inclusion of MWCNTs, Figure 2, since the use of these provides a physical barrier to the corrosive medium preventing pores from being active sites for metal dissolution (Praveen et al. 2007).

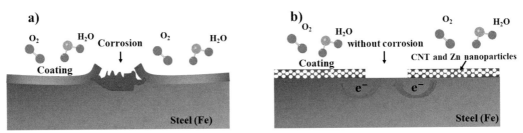

Figure 2. Schematic representation of coatings (a) coating barrier protection, film damage, (b) coating with nanomaterials, cathodic protection mechanism.

4. Nanomaterials characterization

Concrete is a reliable, durable, and strong building material, so it has not undergone many changes over the years. Nevertheless, with the properties exhibited by concrete components at the nanoscale and the interest in improving the properties of this material, nowadays, the incorporation of nanomaterials into the construction industry is a reality. In the case of CNTs, these are incorporated into concrete mixes to inhibit crack propagation and improve their mechanical and thermal response. On the other hand, the development of small sensors at the micro or nanometric scale, based on a combination of CNT and polycarbonate that can be integrated into concrete structural elements, has also been reported, since when subjected to some gradient of stress, temperature, humidity, etc., they may register a transitory change in electrical conductivity as an indicator of some initial damage to the structure. Additionally, Fe_2O_3 nanoparticles can also be used as a means of reinforcement in concrete, generating a greater degree of packing in the construction elements (Lee et al. 2010). Thus, a number of nanomaterials have been used in the construction industry, so that today one of the most important challenges in the area of nanotechnology applied to construction is to understand the relationship between the initial composition, with the resulting structure and properties of a nanoscale engineered material that has unique performance. In this sense, the characterization of nanoparticles is important, since it is possible to obtain parameters that include surface area and porosity, shape and size, particle distribution, solubility, wettability, among others (Schmid 2011). The following can be used as characterization techniques.

4.1 SEM-EDS

In general, electron microscopy is one of the most effective tools for the characterization and investigation of nanostructures. Specifically, Scanning Electron Microscopy (SEM) has versatility in its various imaging modes, excellent spatial resolution, relative ease of imaging, and accessibility of associated spectroscopic and diffraction techniques. This technique is used to analyze the morphological, chemical and structural characteristics of materials at the micro and nanoscale due to its high resolution and integrated techniques. It is also equipped with an Energy Dispersive X-ray Spectroscopy (EDS) that is capable of identifying the elements of the periodic table. Chemical composition EDS can analyze concrete in its mineral phase, while SEM provides element analysis for morphology. The microstructure of cement is complex and, in many respects, it has not been quantitatively characterized. The main hydration product, hydrated calcium silicates (C-H-S), has a complex porous structure (Gallucci et al. 2013). SEM technique can be used to analyze cementitious composites such as mortar and concretes; it is generally necessary that the surfaces of the samples be totally parallel, on some occasions can be mirror polished, which depends on the analysis that is required. Different authors have used this technique to determine the penetration of chloride ions within concrete (Neville et al. 1995). By means of this technique, the characteristics of concrete can be evaluated. For example, microstructural development can be analyzed as a function of curing time; identify the formation of hydration products; analyze the ITZ between the surfaces of the aggregates and the cementitious matrix; pore formation, their distribution, porosity and reaction kinetics of different pozzolanic materials and the deterioration mechanism of the concrete (Villain and Thiery 2005, Tang et al. 2011, Islam et al. 2016). By analyzing backscattered electron images (BSE) of polished sections of concrete, it is possible to measure the 'average' microstructure in the ITZ (Scrivener et al. 2004).

4.2 TEM

Transmission Electron Microscopy (TEM) is a tool used in various fields, including materials science and engineering. All types of materials can be analyzed at the nanoscale, in some cases at the atomic scale (High Resolution Transmission Electron Microscopy, HRTEM), which allows to directly determine the chemical, crystalline and electronic structure of a single nanocrystal. This technique, as in SEM, makes use of the interactions of the electron beam with the specimen. For the incident electrons to penetrate the sample, this must be very thin, around 5 to 10 nm. The thickness required for the analysis of the sample is a function of its density and elemental composition, as well as the desired resolution. In addition to being used for imaging, in a magnification range of 10^3 to 10^6, TEM can also be used for the acquisition of electron diffraction patterns (Donald et al. 2000). With an electron beam finely focused on a selected area, information on the phases and/or crystals in a concrete sample can be generated, as well as the finely fibrous morphology, typical of C-S-H, which is the main hydration product of cement Portland (Gallucci et al. 2013).

4.3 AFM

An alternative and very useful technique for the analysis of cementitious composites is Atomic Force Microscopy (AFM). This technique allows the exploration of cementitious samples under ambient conditions, without being subjected to controlled atmospheres such as high vacuum as in SEM and TEM. Modifying the surface microstructure generated by high vacuum dehydration is avoided and the natural state of the hydrated cement can be analyzed (Papadakis et al. 1999). AFM is an exceptional technique not only because it generates high-resolution microstructure topography, but also because it combines the advantages of three-dimensional digital morphological information at room atmosphere, no vacuum requirements, real-time imaging during reaction, being able to observe, for example, the early stage of hydration of the hydrated calcium silicate (C-S-H) (Yang et al. 2003). To generate the magnified image, this technique uses a mechanical probe instead of using electron beams as in SEM or TEM (Mitchell et al. 1996). Using this technique, it is possible to study the hydration of cement in its natural environment in real time.

4.4 UV-Visible and microindentation

UV-VS is used for the detection of functional groups and qualitative and quantitative analysis. It uses electromagnetic radiation, the source of light is generally a Xenon lamp, since it provides a strong illumination, from the ultraviolet UV region, to near infrared NIR. When the wavelength emitted in that region corresponds to an energy level in which an electron is promoted to a higher orbital level, the energy is absorbed. The equipment determines the amount of ultraviolet light absorbed at a certain wavelength, and this wavelength is recorded as the one with the highest absorbance of ultraviolet light. This technique has been used to evaluate the reflectance of TiO_2 powders pre-treated with different acids, as well as to evaluate the photocatalytic activity of paints containing pre-treated TiO_2 powders and to assess the degradation of Rh-B dye, which is representative of organic contaminants (Paolini et al. 2018).

For their part, microindentation and nanoindentation tests have been used to study the nano and micromechanical properties of cementitious composites, as well as the gradients in the main concrete phases, which are the bulk matrix and the ITZ. Wang et al. used the microhardness test as well as the backscattered electronic microscopy (BSEM) technique to study the mechanisms that influence the curing temperature of steam-cured concretes, which are used during the manufacture of precast components for high-speed railways; the hydrated cement paste as well as the ITZ between the cement paste and the aggregates were studied (Wang et al. 2019).

4.5 XRD

One of the most accurate and highly reproducible methods to identify the phases of nanomaterials within cement, concrete and cement pastes is the X-Ray Diffraction (XRD) technique (Santillán et al. 2016). Amorphous phases in a material do not present narrow and intense reflections (peaks) as in the crystalline phases; on the other hand, they appear as bulge-like in a very wide range of the distance in 2theta on the diffractogram (L'Hôpital et al. 2016). When an X-ray beam falls on a crystal, it causes the atoms that make it up to scatter the incident wave in such a way that each one of them produces an interference phenomenon, where, for certain directions, the incidence will be destructive and, for others, it will be constructive, thus suggesting the phenomenon of diffraction. An X-ray diffractogram presents the intensity data as a function of the diffraction angle (2θ), obtaining a series of peaks (Kitano et al. 1987). The most important data obtained from a diffractogram are:

(a) Peak position expressed in θ, 2θ and distance values; (b) Peak intensity, the intensities can be taken as peak heights or, for work with more precise areas, as the calculation of average crystal sizes; (c) Peak profile. Although it is used less than the previous ones, the shape of the peaks also provides useful information about the analyzed sample.

XRD is important in the study of cementitious materials and/or cement-based composites such as concrete. Since this technique is capable of determining the composition of the phases in the development of the hydration process, as well as to define the crystallite size, the degree of crystallinity, and the orientation of the crystals, the phase changes as a function of time or temperature of cured. It also makes it possible to identify unknown materials or minerals, and the intensity of the peaks is compared to the standard JCPDS database.

4.6 TGA-DSC

Differential Scanning Calorimetry (DSC)-Thermogravimetric Analysis (TGA): this technique records all the thermal transitions of a material (enthalpy and melting temperature, etc.), simultaneously with the record of the weight change and the speed at which it occurs in the sample as a function of temperature and/or time. TGA of cement pastes, mortars and concretes allows to quantify the different compounds due to the decomposition that occurs in the temperature curves. The decomposition of pastes, mortars or concrete is divided into four processes: the first one associated with evaporable water in a temperature range of 20°C to 110°C; the second process, where water is lost from the hydrated calcium silicate gel (C-S-H) and from the minor phases of aluminates and sulfoaluminates from 120°C to 400°C; the third transformation corresponds to calcium hydroxide (Ca $(OH)_2$) between 410°C–530°C; finally, the decomposition process, between 550°C and 950°C, corresponding to the calcium carbonate ($CaCO_3$) present in the cementitious material and to that formed during carbonation, releasing CO_2 in the form of gas (Traversa et al. 2013). Differential Scanning Calorimetry (DSC) techniques are based on the energy differences required to keep the sample and the reference at an identical temperature. The BET (Brunauer-Emmett-Teller) method measures the specific surface area of the materials as well as the shape, volume and pore size distribution. This is important in engineering elements formed by cementitious composites as most present porous microstructures.

5. Electrochemical techniques as a tool for materials characterization in the construction industry

By means of electrochemical techniques, for example Electrochemical Impedance Spectroscopy (EIS), it is possible to characterize coatings or monitor corrosion phenomena, for example steel in reinforced concrete, since the microstructure of the material directly influences the transport properties. The pore network within the concrete structure serves as the ionic flow path, which is directly related to its conductivity. On the other hand, for the evaluation of the corrosion phenomenon and the determination of reaction rates, it is possible to use electrochemical techniques which, besides, can be used both in the field and in the laboratory. In comparison to the methods of weight loss and visual observation that are the most common to evaluate this process, electrochemistry offers advantages since the kinetics of corrosion can be quantitatively studied.

To perform electrochemical studies, it is necessary to use an electrochemical cell, which is a system composed of three electrodes and an electrolyte solution, which is described below:

- Reference electrode: It is characterized by having a constant and known potential value; therefore, it allows to know at what potential the reduction or oxidation process studied occurs.

- Working electrode: The reaction of interest occurs in this electrode, that is, the reaction that is going to be the object of study.

- Counter electrode: Electrode that is used only to make an electrical connection to the electrolyte, in order to be able to apply a current to the working electrode. Since the process that occurs on this electrode is not important, an inert material (noble metal or graphite) is usually used to avoid its dissolution. It is also known as an auxiliary electrode.
- Electrolytic solution: Solution that contains free ions, conductive of electric current.

Figure 3 represents the basic design of an electrochemical cell, as well as the equipment needed to obtain and process the information obtained in each electrochemical characterization technique.

In electrochemical techniques, an electrical signal is introduced into the system under study and information is obtained from it by analyzing the electrical response generated by this signal; with the same equipment, it is possible to perform different techniques and measurements, among which Linear Polarization Resistance (LPR), Tafel Analysis and EIS stand out.

LPR is a quick way to measure uniform corrosion rate. The technique uses small ranges of potential (relative to corrosion potential) applied to metal samples. Linear polarization measurements start at approximately –20 mV relative to corrosion potential (E_{corr}) and end at + 20 mV relative to E_{corr}, so the test is non-destructive. Graphically, the current values are represented as a function of the applied potential, but they can also be presented as polarization curves (E_{corr} as a function of log i), Figure 4a, since in this way the values of the Tafel slopes can be obtained, which are used to calculate the corrosion current density (i_{corr}), Figure 4b. The i_{corr} value is related to the resistance to polarization and high results of this indicate that the corrosion rate of a material is high, and the values of i_{corr} are calculated by means of the Equation 1.1 developed in 1957 by Stern and Geary.

$$i_{corr} = \frac{\beta_a \beta_c}{2.303 \, R_p \, (\beta_a + \beta_c)} \tag{1}$$

where, i_{corr} is the corrosion current density in A/cm², R_p is the corrosion resistance in ohm*cm², $\beta a \, y \, \beta c$ are the Tafel slopes in V/decade or in mV/decade of current density.

The materials that are susceptible to be evaluated by this method must be placed in conductive electrolytes and be free of resistive films, since it is an inclusive technique that globalizes all resistive contributions and does not manage to discriminate them. When dealing with highly resistive systems, the use of EIS is recommended.

As already mentioned, EIS is an electrochemical method used in corrosion studies. Specifically for coatings, it is possible to study both the deterioration and the stability of these due to the action of an electrolyte and the increase in the corrosion rate of the substrate, either due to the damage of the coating or due to the attack of the electrolyte (Olaya-Flores and Torres-Luque 2012). The most widely used technique is the application of a small signal of potential E to an electrode and its response in current (i) at different frequencies is measured. However, under certain circumstances, it is possible to apply a small current signal and measure the response of the system's potential. Thus, the electronic equipment processes the measurements of potential—time and current—displaying the outcomes in a series of impedance values corresponding to each studied frequency. This relationship of impedance and frequency values is called the 'impedance spectrum'. Impedance spectra are usually analyzed using electrical circuits, made up of resistors (R), capacitances (C), inductances (L), etc., components representing some physical phenomenon of the corrosive process of the studied material, providing valuable information about these and their variation in relation to the immersion time.

There are two ways to graphically represent the impedance results obtained from an EIS test:

(1) Nyquist diagram: where the imaginary part multiplied by –1 (–Z") is represented, compared to the real part (Z'). It is the most widely used representation system and the information obtained from it is based on the form that spectra take.

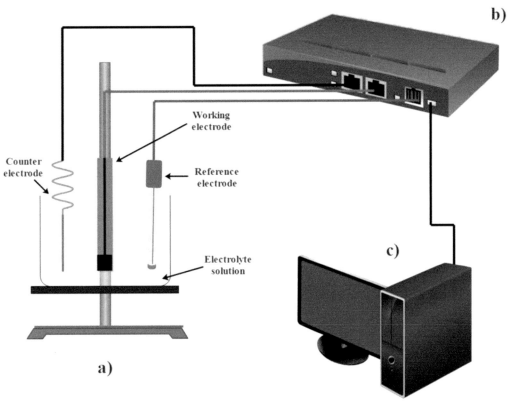

Figure 3. Schematic representation of an electrochemical test, (a) electrochemical cell, (b) potentiostat, (c) computer equipment.

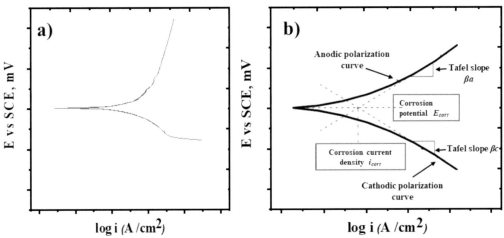

Figure 4. (a) E_{corr} polarization curve as a function of log i, (b) Representation of the calculation of the Tafel slopes, obtention of i_{corr} and E_{corr}

(2) Bode diagrams: where the logarithm of the impedance modulus ($\log |Z|$ /ohm) and the phase angle are represented as a function of the logarithm of the frequency. The information obtained from this type of representation is aimed mainly at performance as a function of frequency.

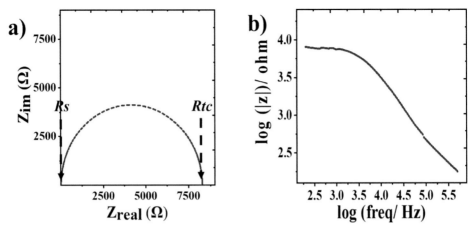

Figure 5. EIS for an organic coating, (a) Nyquist diagram, (b) Bode diagram.

As shown in Figure 5, Nyquist diagrams are the most used, since by means of them it is possible to quickly identify different parameters. For example, from the cuts with the real axis, it is possible to obtain the electrical resistance of the working electrolyte (Rs), and the resistance to charge transfer (Rtc), which is related to the resistance of the film that the coating offers; finally, from the value of the frequency at the maximum point, it is possible to calculate the capacitance of the electrochemical double layer (Cp) that is associated to the insulation property that a coating may have (Zoski 2007, Frankel 2016).

6. Conclusions

Despite the undoubted contribution of the construction industry, both throughout history and to this day, to the configuration of the World, this industry remains among the main concerns regarding its impact on climate change and sustainability. Therefore, one of the most important challenges in this industry is the search for environmental-friendly materials, especially when it is taken into account that two of the most widely used materials, not only in construction activities but when compared with any other type of activity, are cement and steel, which are also major contributors to global warming.

In response to the above mentioned, the addition of nanomaterials to essential construction materials has shown to improve their fundamental properties, such as strength and durability, which allows the use of less materials, a longer useful life and fewer maintenance needs. The application of nanomaterials in construction materials has proved to improve mechanical properties, energy efficiency, heat transfer, self-cleaning features and corrosion, among others.

The research avenues for the use of nanomaterials in this field continue to offer great opportunities for advancement, not only in terms of fundamental technical aspects, but also in terms of applicability under real working conditions and of economic-environmental cost benefit assessment when compared against other constructive alternatives.

Declaration of competing interest

The authors declare no conflict of interest.

Acknowledgments

Authors would like to thank the Mexican Council of Science and Technology (CONACYT) for its financial support through the Chairs-CONACYT-Project Number 746 and the National Laboratories 2021–315871 for their technical support.

References

Aghayere, A.O. and Vigil, J. 2020. Structural Steel Design. Third Edit. Mercury Learning and Information, Virgina, USA.

Aglan, A., Allie, A., Ludwick, A. and Koons, L. 2007. Formulation and evaluation of nano-structured polymeric coatings for corrosion protection. *Surface and Coatings Technology*, 202: 370–378.

Alexander, M. and Mindess, S. 2005. Aggregates in Concrete. CRC Press, Ney York.

Ardalan, R.B., Jamshidi, N., Arabameri, H., Joshaghani, A., Mehrinejad, M. and Sharafi, P. 2017. Enhancing the permeability and abrasion resistance of concrete using colloidal nano-SiO2 oxide and spraying nanosilicon practices. *Construction and Building Materials*, 146: 128–135.

Ardalan, R.B., Emamzadeh, Z.N., Rasekh, H., Joshaghani, A. and Samali, B. 2020. Physical and mechanical properties of polymer modified self-compacting concrete (SCC) using natural and recycled aggregates. *Journal of Sustainable Cement-Based Materials*, 9: 1–16.

Arefi, M.R. and Rezaei-Zarchi, S. 2012. Synthesis of zinc oxide nanoparticles and their effect on the compressive strength and setting time of self-compacted concrete paste as cementitious composites. *International Journal of Molecular Sciences*, 13: 4340–4350.

Askarian, M., Aval, S.F. and Joshaghani, A. 2019. A comprehensive experimental study on the performance of pumice powder in self-compacting concrete (SCC). *Journal of Sustainable Cement-Based Materials*, 7: 340–356.

Atta-ur-Rehman, Kim, J.H., Kim, H.G., Qudoos, A. and Ryou, J.S. 2019. Effect of leaching on the hardened, microstructural and self-cleaning characteristics of titanium dioxide containing cement mortars. *Construction and Building Materials*, 207: 640–650.

Brockenbrough, R.L. and Merritt, F.S. 2006. Structural Steel Designer's Handbook. McGraw-Hill Education, New York.

Cerro-Prada, E., Manso, M., Torres, V. and Soriano, J. 2016. Microstructural and photocatalytic characterization of cement-paste sol-gel synthesized titanium dioxide. *Frontiers of Structural and Civil Engineering*, 10: 189–197.

Cerro-Prada, E., García-Salgado, S., Quijano, M.A. and Varela, F. 2018. Controlled synthesis and microstructural properties of sol-gel TiO2 nanoparticles for photocatalytic cement composites. *Nanomaterials*, 9: 1–26.

Chan Yam, J.L., Solís Carcaño, R. and Moreno, É.I. 2003. Influencia de los agregados pétreos en las características del concreto. *Ingeniería*, 7: 39–46.

Chang, C.-H., Huang, T.-C., Peng, C.-W., Yeh, T.-C., Lu, H.-I., Hung, W.-I., Weng, C.-J., Yang, T.-I. and Yeh, J.-M. 2012. Novel anticorrosion coatings prepared from polyaniline/graphene composites. *Carbon*, 50: 5044–5051.

Chen, J. and Poon, C. 2009. Photocatalytic construction and building materials: From fundamentals to applications. *Building and Environment*, 44: 1899–1906.

Cynthia, G. Zoski, 2007. Handbook of Electrochemistry. Elsevier, Kidlington, Oxford.

De Volder, M.F.L., Tawfick, S.H., Baughman, R.H. and Hart, A.J. 2013. Carbon nanotubes: Present and future commercial applications. *Science*, 339: 535–539.

Donald, A.M., He, C., Royall, C.P., Sferrazza, M., Stelmashenko, N.A. and Thiel, B.L. 2000. Applications of environmental scanning electron microscopy to colloidal aggregation and film formation. *Colloids and Surfaces A: Physicochemical and Engineering Aspects*, 174: 37–53.

Duxson, P., Provis, J.L., Lukey, G.C. and Deventer, J.S.J. Van. 2007. The role of inorganic polymer technology in the development of 'green concrete'. *Cement and Concrete Research*, 37: 1590–1597.

Esposito, L.H., Ramos, J.A. and Kortaberria, G. 2014. Dispersion of carbon nanotubes in nanostructured epoxy systems for coating application. *Progress in Organic Coatings*, 77: 1452–1458.

Etacheri, V., Di Valentin, C., Schneider, J., Bahnemann, D. and Pillai, S.C. 2015. Visible-light activation of TiO2 photocatalysts: Advances in theory and experiments. *Journal of Photochemistry and Photobiology C: Photochemistry Reviews*, 25: 1–29.

Farag, A.A. 2020. Applications of nanomaterials in corrosion protection coatings and inhibitors. *Corrosion Reviews*, 38: 67–86.

Folli, A., Pade, C., Hansen, T.B., De Marco, T. and Macphee, D.E. 2012. TiO$_2$ photocatalysis in cementitious systems: Insights into self-cleaning and depollution chemistry. *Cement and Concrete Research*, 42: 539–548.

Francesko, A., Cardoso, V.F. and Lanceros-Méndez, S. 2018. Lab-on-a-chip technology and microfluidics. pp. 3–36. *In*: Élder, A. Santos, Dongfei Liu and Hongbo Zhang (eds.). *Microfluidics for Pharmaceutical Applications: From Nano/ Micro Systems Fabrication to Controlled Drug Delivery*. William Andrew.

Frankel, G.S. 2016. Fundamentals of corrosion kinetics. pp. 17–32. *In*: Anthony, E. Hughes, Johannes M.C. Mol, Mikhail L. Zheludkevich and Rudolph G. Buchheit (eds.). *Active Protective Catings*. Springer, Dordrecht.

Gallucci, E., Zhang, X. and Scrivener, K.L. 2013. Effect of temperature on the microstructure of calcium silicate hydrate (C-S-H). *Cement and Concrete Research*, 53: 185–195.

Ganguly, P., Byrne, C., Breen, A. and Pillai, S.C. 2018. Antimicrobial activity of photocatalysts: Fundamentals, mechanisms, kinetics and recent advances. *Applied Catalysis B: Environmental*, 225: 51–75.

Gherardi, F., Colombo, A., D'Arienzo, M., Di Credico, B., Goidanich, S., Morazzoni, F., Simonutti, R. and Toniolo, L. 2016. Efficient self-cleaning treatments for built heritage based on highly photo-active and well-dispersible $TiO2$ nanocrystals. *Microchemical Journal*, 126: 54–62.

Gómez-Zamorano, L.Y., Iñiguez-Sánchez, C.A. and Lothenbach, B. 2015. Microestructura y propiedades mecánicas de cementos compuestos: Efecto de la reactividad de adiciones puzolánicas e hidráulicas. *Revista ALCONPAT*, 5: 18–30.

Guo, M.-Z., Maury-Ramirez, A. and Poon, C.S. 2015. Versatile photocatalytic functions of self-compacting architectural glass mortars and their inter-relationship. *Materials & Design*, 88: 1260–1268.

Guo, M.-Z., Maury-Ramirez, A. and Poon, C.S. 2016. Self-cleaning ability of titanium dioxide clear paint coated architectural mortar and its potential in field application. *Journal of Cleaner Production*, 112: 3583–3588.

Gurrola, M.P., Ortiz-Ortega, E., Farias-Zuñiga, C., Chávez-Ramírez, A.U., Ledesma-García, J. and Arriaga, L.G. 2016. Evaluation and coupling of a membraneless nanofluidic device for low-power applications. *Journal of Power Sources*, 307: 244–250.

Han, B., Wang, Z., Zeng, S., Zhou, D., Yu, X., Cui, X., Ou, J. 2017a. Properties and modification mechanisms of nano-zirconia filled reactive powder concrete. *Construction and Building Materials*, 141: 426–434.

Han, B., Zhang, L., Zeng, S., Dong, S., Yu, X., Yang, R., Ou, J. 2017b. Nano-core effect in nano-engineered cementitious composites. *Composites Part A: Applied Science and Manufacturing*, 95: 100–109.

Harb, S.V., Pulcinelli, S.H., Santilli, C.V., Knowles, K.M. and Hammer, P. 2016. A comparative study on graphene oxide and carbon nanotube reinforcement of PMMA-Siloxane-Silica anticorrosive coatings. *ACS Applied Materials & Interfaces*, 8: 16339–16350.

Hatem Nawar A. Nano-technologies and Nano-materials for civil engineering construction works applications, Materials Today: Proceedings, https://doi.org/10.1016/j.matpr.2021.01.497.

Ho, B. and Kjeang, E. 2011. Microfluidic fuel cell systems. *Open Engineering*, 1: 123–131.

Hong, L., Gu, X. and Lin, F. 2014. Influence of aggregate surface roughness on mechanical properties of interface and concrete. *Construction and Building Materials*, 65: 338–349.

Huseien, G.F., Shah, K.W. and Sam, A.R.M. 2019. Sustainability of nanomaterials based self-healing concrete: An all-inclusive insight. *Journal of Building Engineering*, 23: 155–171.

Islam, J., Collins, F., Aldridge, L.P. and Gates, W. 2016. Effect of pore characteristics and chloride binding on time-dependent chloride diffusion into cementitious material. In Proceedings of the Annual Conference of the Australasian Corrosion Association 2016: Corrosion and Prevention 2016: Corrosion and Prevention 2016. 513–522.

Juenger, M.C.G., Winnefeld, F., Provis, J.L. and Ideker, J.H. 2011. Cement and concrete research advances in alternative cementitious binders. *Cement and Concrete Research*, 41: 1232–1243.

Kitano, H., Hirai, Y., Okada, Y., Nakamura, K. and Ise, N. 1987. Pulse injection analysis of the binding of serum proteins to porous polymer gels modified with formyl groups. *Journal of Chromatography B: Biomedical Sciences and Applications*, 420: 13–24.

L'Hôpital, E., Lothenbach, B., Kulik, D.A. and Scrivener, K. 2016. Influence of calcium to silica ratio on aluminium uptake in calcium silicate hydrate. *Cement and Concrete Research*, 85: 111–121.

Lee, J., Mahendra, S. and Alvarez, P.J.J. 2010. Nanomaterials in the construction industry: A review of their applications and environmental health and safety considerations. *ACS Nano*, 4: 3580–3590.

León, M.P. and Ramírez, F. 2010. Caracterización morfológica de agregados para concreto mediante el análisis de imágenes. *Revista ingeniería de construcción*, 25: 215–240.

Li, J., Cui, J., Yang, J., Li, Y., Qiu, H. and Yang, J. 2016. Reinforcement of graphene and its derivatives on the anticorrosive properties of waterborne polyurethane coatings. *Composites Science and Technology*, 129: 30–37.

Luo, H., Dimitrov, S., Daboczi, M., Kim, J.-S., Guo, Q., Fang, Y., Stoeckel, M.-A., Samorì, P., Fenwick, O., Sobrido, A.-B.-J., Wang, X. and Titirici, M.-M. 2020. Nitrogen-doped carbon Dots/TiO_2 nanoparticle composites for photoelectrochemical water oxidation. *ACS Applied Nano Materials*, 3: 3371–3381.

Meng, T., Yu, Y., Qian, X., Zhan, S. and Qian, K. 2012. Effect of nano-TiO_2 on the mechanical properties of cement mortar. *Construction and Building Materials*, 29: 241–245.

Mitchell, L.D., Prica, M. and Birchall, J.D. 1996. Aspects of portland cement hydration studied using atomic force microscopy. *Journal of Materials Science*, 31(16): 4207–4212.

Motahari, S.M., Balapour, M. and Karakouzian, M. 2019. Improving the hardened and transport properties of perlite incorporated mixture through different solutions: Surface area increase, nanosilica incorporation or both. *Construction and Building Materials*, 209: 187–194.

Nazari, A. and Riahi, S. 2011. The effects of ZnO_2 nanoparticles on strength assessments and water permeability of concrete in different curing media. *Materials Research*, 14: 178–188.

Neville, A. 1995. Chloride attack of reinforced concrete : An overview. *Materials and Structures*, 28: 63–70.

Olaya-Flores, J. and Torres-Luque, M.M. 2012. Resistencia a la corrosión de recubrimientos orgánicos por medio de espectroscopía de impedancia. *Ingenieria y Universidad*, 16: 43–58.

Osorio, A.G., Silveira, I.C.L., Bueno, V.L. and Bergmann, C.P. 2008. H_2SO_4/HNO_3/HCl—Functionalization and its effect on dispersion of carbon nanotubes in aqueous media. *Applied Surface Science*, 255: 2485–2489.

Paolini, R., Sleiman, M., Pedeferri, M. and Diamanti, M.V. 2016. TiO_2 alterations with natural aging: Unveiling the role of nitric acid on NIR reflectance. *Solar Energy Materials and Solar Cells*, 157: 791–797.

Paolini, R., Borroni, D., Pedeferri, M. and Diamanti, M.V. 2018. Self-cleaning building materials: The multifaceted effects of titanium dioxide. *Construction and Building Materials*, 182: 126–133.

Papadaki, D., Kiriakidis, G. and Tsoutsos, T. 2018. Applications of nanotechnology in construction industry. pp. 347–370. *In*: Ahmed Barhoum and Abdel Salam Hamdy Makhlouf (eds.). *Fundamentals of Nanoparticles*. William Andrew.

Papadakis, V.G., Pedersen, E.J. and Lindgreen, H. 1999. AFM-SEM investigation of the effect of silica fume and fly ash on cement paste microstructure. *Journal of Materials Science*, 34: 683–690.

Pons Frias, H. 1984. Recubrimientos anticorrosivos con zinc y su aplicación. Instituto Politecnico Nacional, México D.F.

Praveen, B.M., Venkatesha, T.V., Arthoba Naik, Y. and Prashantha, K. 2007. Corrosion studies of carbon nanotubes-Zn composite coating. *Surface and Coatings Technology*, 201: 5836–5842.

Praveen, B.M. and Venkatesha, T.V. 2009. Electrodeposition and properties of Zn–Ni–CNT composite coatings. *Journal of Alloys and Compounds*, 482: 53–57.

Qudoos, A., Atta-ur-Rehman, Kim, H.G. and Ryou, J.-S. 2018. Influence of the surface roughness of crushed natural aggregates on the microhardness of the interfacial transition zone of concrete with mineral admixtures and polymer latex. *Construction and Building Materials*, 168: 946–957.

Ramirez, A.M., Demeestere, K., De Belie, N., Mäntylä, T. and Levänen, E. 2010. Titanium dioxide coated cementitious materials for air purifying purposes: Preparation, characterization and toluene removal potential. *Building and Environment*, 45: 832–838.

Reches, Y. 2018. Nanoparticles as concrete additives: Review and perspectives. *Construction and Building Materials*, 175: 483–495.

Ren, J., Lai, Y. and Gao, J. 2018. Exploring the influence of SiO_2 and TiO_2 nanoparticles on the mechanical properties of concrete. *Construction and Building Materials*, 175: 277–285.

Ruot, B., Plassais, A., Olive, F., Guillot, L. and Bonafous, L. 2009. TiO_2-containing cement pastes and mortars: Measurements of the photocatalytic efficiency using a rhodamine B-based colourimetric test. *Solar Energy*, 83: 1794–1801.

Saji, V.S. and Thomas, J. 2007. Nanomaterials for corrosion control. *Current Science*, 92: 51–55.

Santillán, L., Villagrán Zaccardi, Y.A., Alderete, N., Zega, C.J. and De Belie, N. 2016. Cuantificación mineralógica de cementos mediante análisis Rietveld de DRX. Análisis cruzado de difractogramas experimentales y programas de refinamiento. VII Congreso Internacional-21ª Reunión Técnica de la AATH. Salta, Argentina, 529–536.

Schmid, G. 2011. Nanoparticles: from theory to application. John Wiley & Sons.

Scrivener, K.L., Crumbie, A.K. and Laugesen, P. 2004. The Interfacial Transition Zone (ITZ) between cement paste and aggregate in concrete. *Interface Science*, 12: 411–421.

Siddique, R. and Khan, M.I. 2011. Supplementary cementing materials. Springer Science & Business Media, Berlin, Heidelberg.

Spitzmiller, M., Mahendra, S. and Damoiseaux, R. 2013. Safety issues relating to nanomaterials for construction applications. pp. 127–158. *In*: Pacheco-Torgal, F., Diamanti, M.V., Nazari, A. and Granqvist, C.-G. (eds.). *Nanotechnology in Eco-Efficient Construction*. Woodhead Publishing.

Stynoski, P., Mondal, P. and Marsh, C. 2015. Effects of silica additives on fracture properties of carbon nanotube and carbon fiber reinforced Portland cement mortar. *Cement and Concrete Composites*, 55: 232–240.

Tang, L., Nilsson, L.-O. and Basheer, P.A.M. 2011. Resistance of concrete to chloride ingress: Testing and modelling. CRC Press.

Traversa, L.P., Iloro, F.H. and Benito, D.E. 2013. Determinación mediante ensayos térmicos del CO_2 absorbido por morteros de cemento. *Ciencia y Tecnología de los Materiales*, 33–41.

Trejo-Arroyo, D., Acosta, K., Cruz, J., Valenzuela-Muñiz, A., Vega-Azamar, R. and Jiménez, L. 2019. Influence of ZrO2 nanoparticles on the microstructural development of cement mortars with limestone aggregates. *Applied Sciences*, 9: 598.

Vathsala, K. and Venkatesha, T.V. 2011. Zn-ZrO2 nanocomposite coatings: Elecrodeposition and evaluation of corrosion resistance. *Applied Surface Science*, 257: 8929–8936.

Villain, G. and Thiery, M. 2005. Impact of carbonation on microstructure and transport properties of concrete. Conference Proceedings 10DBMC International Conférence On Durability of Building Materials and Components. 1–8.

Vishwakarma, V. 2020. Synthesis, design, and characterization of controlled nanoparticles for the construction industry. pp. 323–341. *In*: Chaudhery Mustansar Hussain (ed.). *Handbook of Nanomaterials for Manufacturing Applications.* Elsevier.

Wang, D., Zhang, W., Ruan, Y., Yu, X. and Han, B. 2018. Enhancements and mechanisms of nanoparticles on wear resistance and chloride penetration resistance of reactive powder concrete. *Construction and Building Materials*, 189: 487–497.

Wang, M., Xie, Y., Long, G., Ma, C. and Zeng, X. 2019. Microhardness characteristics of high-strength cement paste and interfacial transition zone at different curing regimes. *Construction and Building Materials*, 221: 151–162.

Wang, X., Dong, S., Ashour, A., Zhang, W. and Han, B. 2020. Effect and mechanisms of nanomaterials on interface between aggregates and cement mortars. *Construction and Building Materials*, 240: 117942.

Yan, H., Wang, X., Yao, M. and Yao, X., 2013. Band structure design of semiconductors for enhanced photocatalytic activity: The case of TiO_2. *Progress in Natural Science: Materials International*, 23: 402–407.

Yang, L., Hakki, A., Wang, F. and Macphee, D.E. 2018. Photocatalyst efficiencies in concrete technology: The effect of photocatalyst placement. *Applied Catalysis B: Environmental*, 222: 200–208.

Yang, T., Keller, B., Magyari, E., Hametner, K. and Günther, D. 2003. Direct observation of the carbonation process on the surface of calcium hydroxide crystals in hardened cement paste using an atomic force microscope. *Journal of Materials Science*, 38: 1909–1916.

Zhao, A., Yang, J. and Yang, E.-H. 2015. Self-cleaning engineered cementitious composites. *Cement and Concrete Composites*, 64: 74–83.

Zhong, L. and Haghighat, F. 2015. Photocatalytic air cleaners and materials technologies—Abilities and limitations. *Building and Environment*, 91: 191–203.

CHAPTER 18

The Era of Human-Machine Hybrid
Medical Advances in Biomimetic Devices

Mauricio A. Medina,[1,*] *Amanda V. Haglund,*[2]
José Manuel Hernández Hernández[3] and *José Tapia Ramírez*[4,*]

1. Introduction: Life is Tissue

The world around us is a space fraught with dangers: no one is exempt from suffering unexpected accidents and therefore physical trauma. Roughly, 1.3 million people die every year on the highways worldwide and 50 million people suffer nonfatal injuries because of motor vehicle accidents (Salam 2017). In addition, everyday life is plagued by other dangers such as acute and chronic diseases like metabolic syndrome, kidney failure, or acute myocardial infarction (Perrone-Filardi et al. 2015, Raikou and Gavriil 2018). The consequences of these coronary and metabolic diseases could lead to the loss of a limb or the latest epidemic, heart failure, currently affecting 5 million Americans (Wang et al. 2014). The need for organ transplants in patients is increasing alarmingly across the world, while the previous demand for transplants has never been satisfied (Nguan 2012). Also, the loss of limbs in accidents due to high-energy trauma or vascular metabolic disorders leads to a decrease in the quality of life of patients, increasing co-morbidity. In the United States, around 135,000 amputations are performed each year (not counting war casualties) (James and Laurencin 2015). It is expected that 3.6 million people in the same country will be living with limb loss by 2050 (Ziegler-Graham et al. 2008).

Because of the growing demand and suffering in this type of patient, cybernetics, nanomedicine, and robotics have come to the rescue in an interesting fusion with synthetic biology, artificial intelligence (Delgado and Porcar 2013) and bionics-microsurgery to improve the classic prostheses that current medicine offers (Aman et al. 2019). Under this new paradigm, the patient who has lost a limb can have an intelligent prosthesis with nano-sensors, able to detect even touch (Touch Bionics™) (Schweitzer et al. 2018), which allows the patient in her or his new *"every-day-cyborg-status"* (Quigley and Ayihongbe 2018) to live without any physical disability, enabling integration back into regular life (Aman et al. 2019).

[1] Nanoscience and Nanotechnology Department, Centro de Investigación y de Estudios Avanzados-IPN, México City.
[2] Department of Materials Science and Engineering, University of Tennessee, Knoxville, Tennessee, USA, 37996.
[3] Department of Cell Biology, Centro de Investigación y de Estudios Avanzados-IPN, México City.
[4] Department of Genetics and Molecular Biology, Centro de Investigación y de Estudios Avanzados-IPN, México City, Mexico C.P.07360.
* Corresponding authors: mauricio.medina@cinvestav.mx; jtapia@cinvestav.mx

2. Nano-bionics and nano-cybernetics

Nano-electronic devices are currently used in the aeronautical and aerospace industry (Zhang et al. 2017b) as well as in biomedical applications (nano-bioelectronics) (Dai et al. 2018). These nanomaterials have improved and increased optoelectronic properties compared to micrometric scale materials (Jeevanandam et al. 2018). Thus, electromagnetic devices such as microchips have potential for optimization as well. In addition, the classic results of electronic transport and heat transfer such as Fourier's law or Stefan-Boltzmann's law change at the nanoscale, representing unique properties, which could enhance medical devices and are suitable for several applications (Salata 2004). Therefore, nanoelectronics has found great applications in bionics (Wallace et al. 2012), which is the science that studies the creation and development of technological procedures that help the natural functions of living beings (Vincent et al. 2006). As an example, bionic engineering helps improve the impaired functions of biological systems through the creation of prostheses (Sensky 1980, Williams and Walter 2015), artificial organs such as synthetic exo-bionic cardiac valve replacements (Ashrafian et al. 2010), brain-computer interfaces (BCI) (Shih et al. 2012), and artificial retinas (Bloch et al. 2019), etc.

It is logical that nano-electronic devices are part of the scrutiny of cybernetics, which is the science that studies system theories of communication and autonomous control in living beings and machines (Vasil'ev 2013). Nanocircuits are currently being integrated into bionics and into the gap between artificial intelligence (AI) and synthetic biology, creating outstanding promise not only in the field of microsurgery and smart biomimetic prostheses for those patients who have lost a limb (Williams and Walter 2015) but also in biomedical devices to study metabolic diseases (Yun et al. 2009). Nano-cybernetics will be necessary to understand the dynamic processes that occur at nanoscale.

Electronic devices have been continuously reducing in size as Moore's law predicted. The latest developments in technology such as valleytronic devices (Peng et al. 2017) and quantum computers (Solenov et al. 2018) are playing an important role in miniaturization. Consequently, the latest advances could increase the feasibility of electrode-cellular interface technologies, which are currently able to transfer information across nanostructured materials to create interactions with living systems such as cells or tissue (Wallace et al. 2012). These human-machine interface technologies can initially be built in 2D and then integrated into a 3D printed scaffold (Kong et al. 2019). Applications include not only integrated 3D biomimetic prostheses compatible with human tissue without undesired effects such as scar tissue or inflammation, but also smart organ-on-a-chip systems for real time theragnostic information (Esch et al. 2015) and superior neural interfaces (Kumar et al. 2017).

Biomechanics incorporates nano-bionics and nano-cybernetics modeling to predict patients' experiences in real scenarios; this so-called 4D modeling can predict how nanomaterials and nanodevices for *in vitro* tissue engineering (Devillard et al. 2018) will behave under hard conditions such as heat, rain, pressure, and daily life impact through months and years. The nano-electronic device applications that can benefit from this modeling are biosensors, bio-actuators, tissue engineering and bio-robotics (Ashammakhi et al. 2018a). It is expected that a smart prosthesis using these technologies can readapt and readjust itself, giving the patient intuitive proportional control and sensing by using artificial intelligence-tacit learning systems and computer machine learning (Oyama et al. 2016) to deal with unexpected weather changes or other unpredictable scenarios.

3. Brain-computer interface (BCI) implants

BCI research has demonstrated to be particularly important for the next generation of advanced prostheses where subjects will have neuro-prosthetic myoelectrical control through the thought processes (Lee et al. 2009). A BCI is a computer system able to analyze, record and interpret brain signals, such as the voltage produced by the ion-gated channels in the neuronal postsynaptic membrane, and decodify those signals by detecting movement-related cortical potentials (MRCP) and relaying them to an output device which acts as a bridge between cell tissue (neurons) and electronic devices (robotic arms, hardware, etc.) (Jeong et al. 2015). Early studies in monkeys revealed in the 1960s the promises of BCIs to move objects through decodified brain cortical signaling (Fetz 1969). By 2006, a disabled patient with C3-C4 cervical injury had received a BCI micro-implant inserted into the primary motor cortex. As a result, the BCI system allowed the patient to have prosthetic hand neural control (Hochberg et al. 2006). Moreover, BCI has been used to treat amyotrophic lateral sclerosis and spinal cord injury (Shih et al. 2012).

Current BCIs have been created as an effort to solve the challenges regarding the necessary precision to fuse living cell tissue (microglia, astrocytes or cardio-myocytes) (Dai et al. 2018) with a nano-electronic chip-interface (Dai et al. 2016). Unfortunately, stimulation and sensing can cause inflammation by immune system activation; therefore, innocuous nano-materials are being explored with enhanced biomimetic electronic properties better able to avoid immunological response (Tian et al. 2012). Thus, the nano brain-implant histocompatibility is now more suitable for the brain's architecture (Kim et al. 2018).

4. Advanced smart prosthesis (ASP)

Some of the applications of BCI technology are: 1. Neurofeedback rehabilitation, 2. The treatment of neuromuscular disorders, and 3. The rehabilitation of quadriplegia and tetraplegia (Huggins et al. 2017). BCIs are essential components of advanced smart prostheses (ASP) (Muller-Putz and Pfurtscheller 2008). Unlike conventional prostheses, an ASP must be integrated with a sophisticated electronic-BCI supported by Artificial Intelligence to decode the brain activity (Bouton et al. 2016). Thus, an ASP is an adaptable dynamic device, which allows patient's environmental temporal-space control (Gao et al. 2003). In this context, the patient's intention signals (the thoughts) are the equivalent of a software. These thoughts (neuron activity) must have a strong correlation with the intention of the user; that is why electroencephalogram (EEG) devices are used to discard noise such as electromyographic signals or residual signaling. Subsequently, a BCI transforms the useful intentioned electrical activity such as the thoughts of a tetraplegic patient to restore hand-grasp function to an output command able to execute the order (Pfurtscheller et al. 2003). Therefore, translated algorithms allow the user to manipulate hardware such as a wheelchair (Galan et al. 2008), or smart prosthesis (Muller-Putz et al. 2005) such as a bionic arm or leg; See Figure 1.

However, BCIs must be implanted in the patient through surgery to create a successful bond with the ASP, and there is always an electronic materials immune system rejection risk (Murphy et al. 2015). Since nano-electronic devices have greater surface capacity for ion exchange or electrical signal detection (Timko et al. 2010) and are also more histo-compatible, they could be the key to solve current microelectronic device problems such as inflammation and infection (Kwong et al. 2011, Rezaei et al. 2019). There are numerous reviews where nano-electronic devices have been successfully inserted into cortical neurons without dramatic activation of the immune system response (Timko et al. 2010, Wang et al. 2007). Nevertheless, current work is aimed towards ASPs operating by using non-invasive methods (Gao et al. 2003) such as external electrodes and nano-sensors, conferring to the self-transcending human-machine hybrids dependence and autonomy.

5. Organ-on-a-chip transplant

The greatest ever scientific breakthrough in the field of therapeutics of different pathologies is the development of *Organ-on-a-chip* that biomimetically simulates human organs outside the body. *Organ-on-a-chip* is a novel technology that simulates the physiological conditions (pH, temperature, cationic proteins, function, homeostasis, metabolism), mechanics and biochemistry

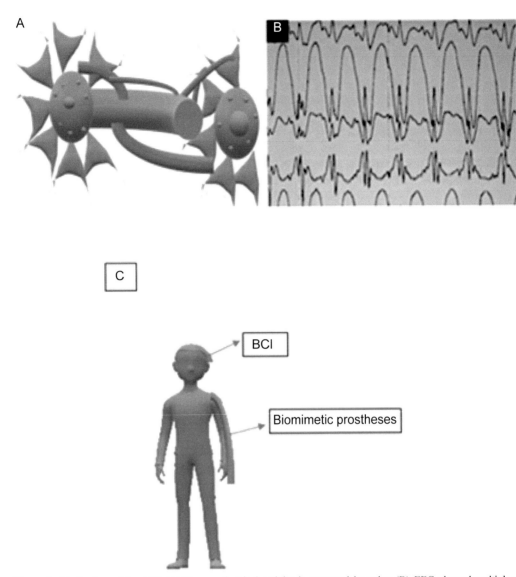

Figure 1. Mechanism of the ASP. (A) Neuron electrical activity is measured by using (B) EEG channel multiplexed recordings. The signals (brain waves) are codified by a BCI to algorithms, so then (C) the patient's thoughts can be translated to commands of intention to operate hardware such as bionic devices and advanced smart biomimetic prostheses.

of a specific organ (heart, kidney, pancreas) on a chip (Rothbauer et al. 2019). This kind of chip has many applications in the area of medical investigation; some examples of these applications are the comprehension of tissues metabolism, fabrication of different kinds of tissue models to completely replace animal models, the study of drug effects in terms of toxicity as well as cure, and organ transplantation research (Sosa-Hernandez et al. 2018). The donation of organs has increased over time, but there are still not enough for all the people who need it.

Ashammakhi et al. (Ashammakhi et al. 2018b) introduced artificially developed organ-supporting devices and machineries that are integrated with an array of organ-on-a-chip platforms. Such machineries will be useful in supporting organs externally for increased efficiency of the organs so that they become functional and are ready for transplantation for better outcomes.

Scientists from Massachusetts Institute of Technology (MIT), as well as NSF-funded Synthetic Biology Engineering Research Center (Synberc), innovated a technique to program human stem cells towards the development of a desired tissue using an organ-on-a-chip platform. This will be a new technique introduced for transplant patients. The tissues grown from the patient's own stem cells will reduce the risk of their immune systems, while there will also be no need to wait for donors for a long time (Beckwitt et al. 2018).

Mayo Clinic researchers are engineering the liver-on-a-chip; toxic substances trigger liver fibrosis and hepatic insufficiency. What is intended is to use systems where various types of cells can be cultured within microfluidic compartments. These microfluidic devices allow to recreate real and dynamic mycourse that simulate the signaling interactions between the different groups of cells that are studied (Zhou et al. 2015). Currently, micro-electromechanical systems (MEMS) are allowing scientist to create highly specialized organs (organ-on-a-chip) by understanding the dynamic cell environment (Sun et al. 2019). The applications for this technology are ideal for the study and evaluation of drug medication (Osaki et al. 2018), pharmacokinetics (PK), median lethal dose, etc. (Pasqualini et al. 2017).

The chips can simulate the cellular microenvironment (Esch et al. 2015) (including chemical mediators, metabolites, biological molecules, etc.) (Sackmann et al. 2014) in order to replicate organ function. Interestingly, these chips can connect to each other building the so-called *multi-organ-on-a-chip* system (for example lung organ-on-a-chip connected to heart organ-on-a-chip) so then, *"in vitro"* models (unreal compared to an *"in vivo"* system) would be obsolete, allowing scientists to dismiss animal models for metabolic studies.

Current work has been done integrating carbon electrodes to offer an accurate controllable physiological mechanical condition to create cardiac patches for both replacement of cardiac tissues and the creation of *heart-on-a-chip* (Xiao et al. 2014) for research purposes. Other organ-on-a-chip models available are: vessel-on-a-chip, liver-on-a-chip, kidney-on-a-chip and a novel microfluidic device fabricated by Shamloo and coworkers called *neuron-on-a-chip* (An et al. 2015). Chief organ-on-a-chip advantages are: (1) They emulate the real conditions of the organism (presence of macrophages, nutrients, hormones, myeloperoxidases, hydrolases and drug permeability). (2) They dispense of the necessity to sacrifice animals, which do not correctly simulate human organisms. The main disadvantage is that, like biomimetic limbs, they are still quite expensive.

Recently, Dr. Anthony Atala and his team from Harvard University carried out a successful transplant of a bladder cultivated from the patient's cells. All of these advances in regenerative medicine and cell tissue 3D bioprinting can save the lives of patients who cannot wait for an organ donation, a serious problem in contemporary health care systems (Atala 2016, Pincock 2011).

Despite MEMS (Whitesides 2006), biomimetic tissue structures made by 3-D printing (Yi et al. 2017) and organ-on-a-chip being still in their infancy, it is expected that in the future, medical scientists and researchers will base their experimental models on these devices or, perhaps physicians may begin transplanting bionic human organs at hospitals soon.

6. Transhumanism

Transhumanism is a contemporary quasi-medical ideology (McNamee and Edwards 2006), a cultural, sociopolitical, and intellectual trend that supports not only the use of current technology to improve the beauty, lifespan, and resistance to disease in human beings, but also cutting-edge technologies, such as nanotechnology, genetics (Bostrom 2003), artificial organs, robotics (Zeilig et al. 2012, Zhang et al. 2017a), cybernetics, quantum computing, synthetic biology and neurological chips that allow linking human brains to computers (BCI) (Daly and Huggins 2015), to radically modify both the human body and mind (McNamee and Edwards 2006).

Numerous papers describe this progressive technological trend, and *The Journal of Medicine and Philosophy* has nurtured the debate between bio-conservatives and transhumanists concerning ethics in genetic engineering, memory and thought transfection. The manuscript entitled "*Human Mind Control of Rat Cyborg's Continuous Locomotion with Wireless Brain-to-Brain Interface*" discusses thought to thought transfection ideas (Zhang et al. 2019).

A social problem often associated with the emergence of transhumanism is its inevitable effect on social equality—the rich will have access to these technologies, but the poor will most likely be excluded. Contemporary plastic surgery is an example of this same situation: people who can afford a surgical procedure to look more beautiful or healthy can increase their chances to obtain a job. Another modern example is the use of single nucleotide polymorphisms, which can act as biological markers to predict an individual's risk of developing human eye disorders (Shastry 2009), susceptibility to environmental factors such as toxins, and response to medications and drugs, all of which provide the user who can afford the treatment with advantages against illness compared to those who cannot pay. In a transhumanist future, rich people could secure an artificial heart (Cook et al. 2015), an artificial lung (Wrenn et al. 2018), and exoskeletons (Cerasa et al. 2018) for physical therapy or to replace lost limbs, whereas the poor would die long before being able to afford the technology.

We are already living the benefits and consequences of transhumanism. Failing to acknowledge this fact would be equivalent to denying the opportunity of using a robotic prosthesis or a pacemaker to a patient who needs it.

7. Cyborgs among us

Unlike a robot, which is entirely machine, a cyborg is a human hybrid who has machine devices integrated into her or his biological body (Lapum et al. 2012). There is a collection of ten essays written between 1978 and 1989 exploring the cyborg ideology based on concept, definition, classification, ethics, morality and philosophy. One of these essays is a controversial manuscript published by Haraway in 1985 (Haddow et al. 2015) which considers these ideas in a construct entitled *The Cyborg Manifesto*. The manifesto describes a world with a realistic cyber-society, and found a deep resonance with those who, by that time, were already living with devices surgically implanted for medical therapy. Currently, devices such as exoskeletons are being used to rehabilitate people (Fukaya et al. 2017), even in extreme cases such as spinal cord injury (Kandilakis and Sasso-Lance 2019), while other electronic devices such as brain implants, electronic pacemakers, bionic limbs, organ-on-a-chip, drug implant systems and artificial retina are not only a reality but also, usually the only resource to keep alive or restore those patients with a status known as *"Everyday Cyborg"* (Haddow et al. 2015).

Nevertheless, even when there are serious technical and ethical aspects to be solved, BCIs (Ngan et al. 2019), actuators and bionic devices are still promising extraordinary applications in healthcare (Kogel et al. 2019). A few years ago, the Hugh M. Herr team tested and compared a passive prosthesis versus a bionic ankle prosthesis to provide enhanced joint position and torque control in a patient who suffers from a traumatic transtibial amputation (Rouse et al. 2015). The results were outstanding: kinetics and kinematics were enhanced in comparison with passive prosthesis and the patient was able to dance rumba by using an advanced bionic ankle prosthesis.

However, biomimetic prostheses must still be customized for a given task that requires fine movements, torque or a certain impedance. Depending on the part of the body, the physical forces need to be calculated and then engineered into the prosthesis motor-actuators (Pasquina et al. 2015). In 2018, researchers successfully developed a novel biomimetic robotic elbow joint using calculations performed through Monte Carlo software; new applications due to this achievement range from a humanoid arm to rehabilitation (Cui et al. 2018). In a different study, researchers developed a myo-electrically robotic prosthesis for a patient who had suffered a transradial amputation (Stillfried et al. 2018). On this occasion, the team successfully designed a customized robotic limb able to play the drums since the patient is a musician. Finally, Elon Musk *Neuralink*™ brain interface technologies are under exhaustive-ongoing clinical trials, not only for the restoration of sensory and motor function in patients but also for human brain integration with artificial intelligence software (Musk and Neuralink 2019).

8. Artificial intelligence at the service of an emerging transhuman society

AI applications range from marketing to healthcare, but our interest in this document focuses on AI for medical applications such as drug development (Fujiwara et al. 2018) and biomimetic devices (Winkler 2017). The same as with autonomous cars, a biomimetic device which works with AI should be able to make the user feel as comfortable as possible. A very common syndrome after losing a limb is the so-called *phantom limb syndrome* (Chahine and Kanazi 2007), where patients experience sensations in a limb that does not exist anymore. An ideal machine learning AI-ASP, integrated into the patient's neurological network, would modulate the BCI signals so the patient would feel a prosthesis as familiar as if the bionic limb were a natural limb (Valeriani et al. 2019).

A trans-human society also demands novel methods to fight pandemics such as SARS-CoV-2 COVID-19, and other viral diseases such as hepatitis C and Dengue virus. Since viruses have the capacity to be so mutable and modify their DNA or RNA, it is difficult to develop antivirals to keep pace with their constant structure modifications. An AI machine learning software can select novel molecules that could be found in antineoplastic drugs by searching at digital drug library. Consequently, AI identifies chemical structure activity through design and process algorithms to develop novel antivirals (Stebbing et al. 2020). In addition, scientists are using AI to target protein-receptor bonds by developing powerful and novel synthetic antibiotics (Zoffmann et al. 2019).

9. New technocracy paradigms: Risks and perspectives

Adaptation processes between human-machine EEG-ASP-BCI have successfully advanced in the last years (Tariq et al. 2018). However, there are some concerns regarding security. For example, the integration of other devices such as phones, cameras, weapons, Wi-Fi, Bluetooth, etc., could make a prosthesis vulnerable to hacking (Landhuis 2016); criminals could take control of the bionic arm and execute a crime. The idea exists in nature, the *Dicrocoelium dendriticum* fungus species hacks ants' (*Formica fusca*) navigational skills (Libersat et al. 2018). Protecting robotic limbs or bionic devices from hackers will be an additional challenge for cyberpolice, since brain-hacking leads to the manipulation of neural information (Lenca 2016). A new technocracy will have to accept or not the risk, benefits and consequences of a new *every-day-cyborg-status* (Quigley and Ayihongbe 2018). Self-transcending human-machine hybrids could be endowed with enhanced military-grade enhancements such as strength or superior hand-eye coordination; thus, they could displace non-cyborg humans from employ possibilities, just like what is happening nowadays with robots in factories.

Current times have indicated that future human evolution will be by "human direction" instead of natural direction. Quantum computing-artificial intelligence working together with the BCI neuralink™ can modify the human brain, enabling it to connect to the internet and other machines ("Human-robotic interfaces to shape the future of prosthetics," 2019). Therefore, cyborg-transhuman rights will also be required for a user's privacy policy as well as ontological (Lapum et al. 2012) and moral support (Quigley and Ayihongbe 2018). The creation of new laws will be necessary, and a global service support for technological maintenance as well.

The future could sound ominous, but the possibilities are infinite: from a future without disabilities to the next generation of an amazing degree-provided human-machines much better adapted to explore and colonize extraterrestrial space. Maybe we could say a modern babel tower is in progress.

References

Aman, M., Festin, C., Sporer, M.E., Gstoettner, C., Prahm, C., Bergmeister, K.D. and Aszmann, O.C. 2019. Bionic reconstruction: Restoration of extremity function with osseointegrated and mind-controlled prostheses. *Wien. Klin. Wochenschr*, 131(23-24): 599–607. doi: 10.1007/s00508-019-1518-1.

An, F., Qu, Y., Liu, X., Zhong, R. and Luo, Y. 2015. Organ-on-a-Chip: New platform for biological analysis. *Anal. Chem. Insights*, 10: 39–45. doi: 10.4137/ACI.S28905.

Anthony Atala, M.D. Building Organs for the Future. 2016. *Transplantation*, 100(8): 1595–1596. doi: 10.1097/TP.0000000000001301.

Ashammakhi, N., Ahadian, S., Zengjie, F., Suthiwanich, K., Lorestani, F., Orive, G., Ostrovidov, S. and Khademhosseini, A. 2018. Advances and future perspectives in 4D bioprinting. *Biotechnol. J.*, 13(12): e1800148. doi: 10.1002/biot.201800148.

Ashammakhi, N., Elkhammas, E. and Answarul, H. 2018. Translating advances in organ-on-a-chip technology for supporting organs. *Journal of Biomedical Materials Research Part B: Applied Biomaterials*, 107(6).

Ashrafian, H., Darzi, A. and Athanasiou, T. 2010. Autobionics: A new paradigm in regenerative medicine and surgery. *Regen. Med.*, 5(2): 279–288. doi: 10.2217/rme.10.2.

Beckwitt, C.H., Clark, A.M., Wheeler, S., Taylor, D.L., Stolz, D.B., Griffith, L. and Wells, A. 2018. Liver 'organ on a chip'. *Exp. Cell. Res.*, 363(1): 15–25. doi: 10.1016/j.yexcr.2017.12.023.

Bloch, E., Luo, Y. and da Cruz, L. 2019. Advances in retinal prosthesis systems. *Ther. Adv. Ophthalmol.*, 11: 2515841418817501. doi: 10.1177/2515841418817501.

Bostrom, N. 2003. Human genetic enhancements: A transhumanist perspective. *J. Value. Inq.*, 37(4): 493–506.

Bouton, C.E., Shaikhouni, A., Annetta, N.V., Bockbrader, M.A., Friedenberg, D.A., Nielson, D.M., Sharma, G., Sederberg, P.B., Glenn, B.C., Mysiw, W.J, Morgan, A.G., Deogaonkar, M., and Rezai, A.R. 2016. Restoring cortical control of functional movement in a human with quadriplegia. *Nature*, 533(7602): 247–250. doi: 10.1038/nature17435.

Cerasa, A., Pignolo, L., Gramigna, V., Serra, S., Olivadese, G., Rocca, F., Perrotta, P., Dolce, G. Quattrone, A. and Tonin, P. 2018. Exoskeleton-robot assisted therapy in stroke patients: A lesion mapping study. *Front. Neuroinform.*, 12: 44. doi: 10.3389/fninf.2018.00044.

Cook, J.A., Shah, K.B., Quader, M.A., Cooke, R.H., Kasirajan, V., Rao, K.K., Smallfield, M.C., Tchoukina, I. and Tang, D.G. 2015. The total artificial heart. *J. Thorac. Dis.*, 7(12): 2172–2180. doi: 10.3978/j.issn.2072-1439.2015.10.70.

Cui, B., Chen, L., Xie, Y. and Wang, Z. 2018. Kinematic decoupling analysis and design of a biomimetic robotic elbow joint. *Appl. Bionics. Biomech.*, 2018: 4613230. doi: 10.1155/2018/4613230.

Chahine, L. and Kanazi, G. 2007. Phantom limb syndrome: A review. *Middle East J. Anaesthesiol.*, 19(2): 345–355.

Dai, X., Zhou, W., Gao, T., Liu, J. and Lieber, C.M. 2016. Three-dimensional mapping and regulation of action potential propagation in nanoelectronics-innervated tissues. *Nat. Nanotechnol.*, 11(9): 776–782. doi: 10.1038/nnano.2016.96.

Dai, X., Hong, G., Gao, T. and Lieber, C.M. 2018. Mesh nanoelectronics: Seamless integration of electronics with tissues. *Acc. Chem. Res.*, 51(2): 309–318. doi: 10.1021/acs.accounts.7b00547.

Daly, J.J. and Huggins, J.E. 2015. Brain-computer interface: Current and emerging rehabilitation applications. *Arch. Phys. Med. Rehabil.*, 96(3 Suppl.): S1–7. doi: 10.1016/j.apmr.2015.01.007.

Delgado, A. and Porcar, M. 2013. Designing *de novo*: Interdisciplinary debates in synthetic biology. *Syst. Synth. Biol.*, 7(1-2): 41–50. doi: 10.1007/s11693-013-9106-6.

Devillard, C.D., Mandon, C.A., Lambert, S.A., Blum, L.J. and Marquette, C.A. 2018. Bioinspired multi-activities 4D printing objects: A new approach toward complex tissue engineering. *Biotechnol. J.*, 13(12): e1800098. doi: 10.1002/biot.201800098.

Esch, E.W., Bahinski, A. and Huh, D. 2015. Organs-on-chips at the frontiers of drug discovery. *Nat. Rev. Drug. Discov.*, 14(4): 248–260. doi: 10.1038/nrd4539.

Fetz, E.E. 1969. Operant conditioning of cortical unit activity. *Science*, 163(3870): 955–958. doi: 10.1126/science.163.3870.955.

Fujiwara, T., Kamada, M. and Okuno, Y. 2018. [Artificial Intelligence in Drug Discovery]. *Gan To Kagaku Ryoho*, 45(4): 593–596.

Fukaya, T., Mutsuzaki, H., Yoshikawa, K., Sano, A., Mizukami, M. and Yamazaki, M. 2017. The training effect of early intervention with a hybrid assistive limb after total knee arthroplasty. *Case. Rep. Orthop.*, 2017: 6912706. doi: 10.1155/2017/6912706.

Galan, F., Nuttin, M., Lew, E., Ferrez, P.W., Vanacker, G., Philips, J. and Millan Jdel, R. 2008. A brain-actuated wheelchair: asynchronous and non-invasive Brain-computer interfaces for continuous control of robots. *Clin. Neurophysiol.*, 119(9): 2159–2169. doi: 10.1016/j.clinph.2008.06.001.

Gao, X., Xu, D., Cheng, M. and Gao, S. 2003. A BCI-based environmental controller for the motion-disabled. *IEEE Trans. Neural. Syst. Rehabil. Eng.*, 11(2): 137–140. doi: 10.1109/TNSRE.2003.814449.

Haddow, G., King, E., Kunkler, I. and McLaren, D. 2015. Cyborgs in the everyday: Masculinity and biosensing prostate cancer. *Sci. Cult. (Lond)*, 24(4): 484–506. doi: 10.1080/09505431.2015.1063597.

Hochberg, L.R., Serruya, M.D., Friehs, G.M., Mukand, J.A., Saleh, M., Caplan, A.H., Branner, A., Chen, D., Penn, R.D. and Donoghue, J.P. 2006. Neuronal ensemble control of prosthetic devices by a human with tetraplegia. *Nature*, 442(7099): 164–171. doi: 10.1038/nature04970.

Huggins, J.E., Guger, C., Ziat, M., Zander, T.O., Taylor, D., Tangermann, M., Soria-Frisch, A., Simeral, J., Scherer, R., Rupp, R., Ruffini, G., Robinson, D.K.R., Ramsey, N.F., Nijholt, A., Müller-Putz, G., McFarland, D.J., Mattia, D., Lance, B.J., Kindermans, P.-J., Iturrate, I., Herff, C., Gupta, D., Do, A.H., Collinger, J.L., Chavarriaga, R., Chase, S.M., Bleichner, M.G., Batista, A., Anderson, C.W. and Aarnoutse, E.J. 2017. Workshops of the sixth international brain-computer interface meeting: Brain-computer interfaces past, present, and future. *Brain Comput. Interfaces (Abingdon)*, 4(1-2): 3–36. doi: 10.1080/2326263X.2016.1275488.

Human-robotic interfaces to shape the future of prosthetics. 2019. *EBioMedicine*, 46: 1. doi: 10.1016/j.ebiom.2019.08.018.

James, R. and Laurencin, C.T. 2015. Regenerative engineering and bionic limbs. *Rare Metals*, 34(3): 143–155. doi: 10.1007/s12598-015-0446-0.

Jeevanandam, J., Barhoum, A., Chan, Y.S., Dufresne, A. and Danquah, M.K. 2018. Review on nanoparticles and nanostructured materials: History, sources, toxicity and regulations. *Beilstein. J. Nanotechnol.*, 9: 1050–1074. doi: 10.3762/bjnano.9.98.

Jeong, J.W., Shin, G., Park, S.I., Yu, K.J., Xu, L. and Rogers, J.A. 2015. Soft materials in neuroengineering for hard problems in neuroscience. *Neuron*, 86(1): 175–186. doi: 10.1016/j.neuron.2014.12.035.

Kandilakis, C. and Sasso-Lance, E. 2019. Exoskeletons for personal use after spinal cord injury. *Arch. Phys. Med. Rehabil.*, doi: 10.1016/j.apmr.2019.05.028.

Kim, Y., Meade, S.M., Chen, K., Feng, H., Rayyan, J., Hess-Dunning, A. and Ereifej, E.S. 2018. Nano-architectural approaches for improved intracortical interface technologies. *Front. Neurosci.*, 12: 456. doi: 10.3389/fnins.2018.00456.

Kogel, J., Schmid, J.R., Jox, R.J. and Friedrich, O. 2019. Using brain-computer interfaces: a scoping review of studies employing social research methods. *BMC Med. Ethics*, 20(1): 18. doi: 10.1186/s12910-019-0354-1.

Kong, Y.L., Gupta, M.K., Johnson, B.N. and McAlpine, M.C. 2019. Corrigendum to "3D printed bionic nanodevices" [Nano Today 11(2016): 330–350]. *Nano. Today*, 25: 156. doi: 10.1016/j.nantod.2019.03.009.

Kumar, A., Tan, A., Wong, J., Spagnoli, J.C., Lam, J., Blevins, B.D., Natasha, G., Thorne, L., Ashkan, K., Xie, J. and Liu, H. 2017. Nanotechnology for neuroscience: Promising approaches for diagnostics, therapeutics and brain activity mapping. *Adv. Funct. Mater.*, 27(39). doi: 10.1002/adfm.201700489.

Kwong, B., Liu, H. and Irvine, D.J. 2011. Induction of potent anti-tumor responses while eliminating systemic side effects via liposome-anchored combinatorial immunotherapy. *Biomaterials*, 32(22): 5134–5147. doi: 10.1016/j.biomaterials.2011.03.067.

Landhuis, E. 2016. Science and culture: Crafting prostheses with form, function, and flair. *Proc. Natl. Acad. Sci. U S A*, 113(47): 13258–13259. doi: 10.1073/pnas.1616194113.

Lapum, J., Fredericks, S., Beanlands, H., McCay, E., Schwind, J. and Romaniuk, D. 2012. A cyborg ontology in health care: Traversing into the liminal space between technology and person-centred practice. *Nurs. Philos.*, 13(4): 276–288. doi: 10.1111/j.1466-769X.2012.00543.x.

Lee, J.H., Ryu, J., Jolesz, F.A., Cho, Z.H. and Yoo, S.S. 2009. Brain-machine interface via real-time fMRI: Preliminary study on thought-controlled robotic arm. *Neurosci. Lett.*, 450(1): 1–6. doi: 10.1016/j.neulet.2008.11.024.

Lenca, M. 2016. Hacking the brain: Brain–computer interfacing technology and the ethics of neurosecurity. *Ethics and Information Technology*, 18: 117–129.

Libersat, F., Kaiser, M. and Emanuel, S. 2018. Mind control: How parasites manipulate cognitive functions in their insect hosts. *Front. Psychol.*, 9: 572. doi: 10.3389/fpsyg.2018.00572.

McNamee, M.J. and Edwards, S.D. 2006. Transhumanism, medical technology and slippery slopes. *J. Med. Ethics.*, 32(9): 513–518. doi: 10.1136/jme.2005.013789.

Muller-Putz, G.R., Scherer, R., Pfurtscheller, G. and Rupp, R. 2005. EEG-based neuroprosthesis control: A step towards clinical practice. *Neurosci. Lett.*, 382(1-2): 169–174. doi: 10.1016/j.neulet.2005.03.021.

Muller-Putz, G.R. and Pfurtscheller, G. 2008. Control of an electrical prosthesis with an SSVEP-based BCI. *IEEE Trans. Biomed. Eng.*, 55(1): 361–364. doi: 10.1109/TBME.2007.897815.

Murphy, M.D., Guggenmos, D.J., Bundy, D.T. and Nudo, R.J. 2015. Current challenges facing the translation of brain computer interfaces from preclinical trials to use in human patients. *Front. Cell Neurosci.*, 9: 497. doi: 10.3389/fncel.2015.00497.

Musk, E. and Neuralink. 2019. An integrated brain-machine interface platform with thousands of channels. *J. Med. Internet. Res.*, 21(10): e16194. doi: 10.2196/16194.

Ngan, C.G.Y., Kapsa, R.M.I. and Choong, P.F.M. 2019. Strategies for neural control of prosthetic limbs: From electrode interfacing to 3D printing. *Materials (Basel)*, 12(12). doi: 10.3390/ma12121927.

Nguan, C. 2012. Trying to meet the demands in organ transplantation. *Can. Urol. Assoc. J.*, 6(6): 453–454. doi: 10.5489/cuaj.12339.

Osaki, T., Sivathanu, V. and Kamm, R.D. 2018. Vascularized microfluidic organ-chips for drug screening, disease models and tissue engineering. *Curr. Opin. Biotechnol.*, 52: 116–123. doi: 10.1016/j.copbio.2018.03.011.

Oyama, S., Shimoda, S., Alnajjar, F.S., Iwatsuki, K., Hoshiyama, M., Tanaka, H. and Hirata, H. 2016. Biomechanical reconstruction using the tacit learning system: intuitive control of prosthetic hand rotation. *Front. Neurorobot.*, 10: 19. doi: 10.3389/fnbot.2016.00019.

Pasqualini, F.S., Emmert, M.Y., Parker, K.K. and Hoerstrup, S.P. 2017. Organ chips: Quality assurance systems in regenerative medicine. *Clin. Pharmacol. Ther.*, 101(1): 31–34. doi: 10.1002/cpt.527.

Pasquina, P.F., Perry, B.N., Miller, M.E., Ling, G.S.F. and Tsao, J.W. 2015. Recent advances in bioelectric prostheses. *Neurol. Clin. Pract.*, 5(2): 164–170. doi: 10.1212/CPJ.0000000000000132.

Peng, B., Li, Q., Liang, X., Song, P., Li, J., He, K., Fu, D., Li, Y., Shen, C., Wang, H., Wang, C., Liu, T., Zhang, L., Lu, H., Wang, X., Zhao, J., Xie, J., Wu, M., Bi, L., Deng, L. and Loh, K.P. 2017. Valley polarization of trions and magnetoresistance in heterostructures of MoS2 and yttrium iron garnet. *ACS Nano.*, 11(12): 12257–12265. doi: 10.1021/acsnano.7b05743.

Perrone-Filardi, P., Paolillo, S., Costanzo, P., Savarese, G., Trimarco, B. and Bonow, R.O. 2015. The role of metabolic syndrome in heart failure. *Eur. Heart. J.*, 36(39): 2630–2634. doi: 10.1093/eurheartj/ehv350.

Pfurtscheller, G., Muller, G.R., Pfurtscheller, J., Gerner, H.J. and Rupp, R. 2003. 'Thought'—control of functional electrical stimulation to restore hand grasp in a patient with tetraplegia. *Neurosci. Lett.*, 351(1): 33–36. doi: 10.1016/s0304-3940(03)00947-9.

Pincock, S. 2011. Anthony atala: At the cutting edge of regenerative surgery. *Lancet*, 378(9800): 1371. doi: 10.1016/S0140-6736(11)61600-0.

Quigley, M. and Ayihongbe, S. 2018. Everyday cyborgs: On integrated persons and integrated goods. *Med. Law Rev.*, 26(2): 276–308. doi: 10.1093/medlaw/fwy003.

Raikou, V.D. and Gavriil, S. 2018. Metabolic syndrome and chronic renal disease. *Diseases*, 6(1). doi: 10.3390/diseases6010012.

Rezaei, R., Safaei, M., Mozaffari, H.R., Moradpoor, H., Karami, S., Golshah, A., Salimi, H. and Karami, H. 2019. The role of nanomaterials in the treatment of diseases and their effects on the immune system. *Open Access. Maced. J. Med. Sci.*, 7(11): 1884–1890. doi: 10.3889/oamjms.2019.486.

Rothbauer, M., Rosser, J.M., Zirath, H. and Ertl, P. 2019. Tomorrow today: Organ-on-a-chip advances towards clinically relevant pharmaceutical and medical *in vitro* models. *Curr. Opin. Biotechnol.*, 55: 81–86. doi: 10.1016/j.copbio.2018.08.009.

Rouse, E.J., Villagaray-Carski, N.C., Emerson, R.W. and Herr, H.M. 2015. Design and testing of a bionic dancing prosthesis. *PLoS One*, 10(8): e0135148. doi: 10.1371/journal.pone.0135148.

Sackmann, E.K., Fulton, A.L. and Beebe, D.J. 2014. The present and future role of microfluidics in biomedical research. *Nature*, 507(7491): 181–189. doi: 10.1038/nature13118.

Salam, M.M. 2017. Motor vehicle accidents: The physical versus the psychological trauma. *J. Emerg. Trauma. Shock*, 10(2): 82–83. doi: 10.4103/0974-2700.201584.

Salata, O. 2004. Applications of nanoparticles in biology and medicine. *J. Nanobiotechnology*, 2(1): 3. doi: 10.1186/1477-3155-2-3.

Schweitzer, W., Thali, M.J. and Egger, D. 2018. Case-study of a user-driven prosthetic arm design: Bionic hand versus customized body-powered technology in a highly demanding work environment. *J. Neuroeng. Rehabil.*, 15(1): 1. doi: 10.1186/s12984-017-0340-0.

Sensky, T. 1980. A consumer's guide to "bionic arms". *Br. Med. J.*, 281(6233): 126–127. doi: 10.1136/bmj.281.6233.126.

Shastry, B.S. 2009. SNPs: Impact on gene function and phenotype. *Methods Mol. Biol.*, 578: 3–22. doi: 10.1007/978-1-60327-411-1_1.

Shih, J.J., Krusienski, D.J. and Wolpaw, J.R. 2012. Brain-computer interfaces in medicine. *Mayo Clin. Proc.*, 87(3): 268–279. doi: 10.1016/j.mayocp.2011.12.008.

Solenov, D., Brieler, J. and Scherrer, J.F. 2018. The potential of quantum computing and machine learning to advance clinical research and change the practice of medicine. *Mo. Med.*, 115(5): 463–467.

Sosa-Hernandez, J.E., Villalba-Rodriguez, A.M., Romero-Castillo, K.D., Aguilar-Aguila-Isaias, M.A., Garcia-Reyes, I.E., Hernandez-Antonio, A., Ahmed, I., Sharma, A., Parra-Saldívar, R. and Iqbal, H.M.N. 2018. Organs-on-a-chip module: A review from the development and applications perspective. *Micromachines (Basel)*, 9(10). doi: 10.3390/mi9100536.

Stebbing, J., Phelan, A., Griffin, I., Tucker, C., Oechsle, O., Smith, D. and Richardson, P. 2020. COVID-19: Combining antiviral and anti-inflammatory treatments. *The Lancet*, 20(4): 400–402.

Stillfried, G., Stepper, J., Neppl, H., Vogel, J. and Hoppner, H. 2018. Elastic elements in a wrist prosthesis for drumming reduce muscular effort, but increase imprecision and perceived stress. *Front. Neurorobot.*, 12: 9. doi: 10.3389/fnbot.2018.00009.

Sun, W., Luo, Z., Lee, J., Kim, H. J., Lee, K., Tebon, P., Feng, Y., Dokmeci, M.R., Sengupta, S. and Khademhosseini, A. 2019. Organ-on-a-chip for cancer and immune organs modeling. *Adv. Healthc. Mater.*, 8(4): e1801363. doi: 10.1002/adhm.201801363.

Tariq, M., Trivailo, P.M. and Simic, M. 2018. EEG-Based BCI control schemes for lower-limb assistive-robots. *Front. Hum. Neurosci.*, 12: 312. doi: 10.3389/fnhum.2018.00312.

Tian, B., Liu, J., Dvir, T., Jin, L., Tsui, J. H., Qing, Q., Suo, Z., Langer, R., Kohane, D.S. and Lieber, C.M. 2012. Macroporous nanowire nanoelectronic scaffolds for synthetic tissues. *Nat. Mater.*, 11(11): 986–994. doi: 10.1038/nmat3404.

Timko, B.P., Cohen-Karni, T., Qing, Q., Tian, B. and Lieber, C.M. 2010. Design and implementation of functional nanoelectronic interfaces with biomolecules, cells, and tissue using nanowire device arrays. *IEEE Trans. Nanotechnol.*, 9(3): 269–280. doi: 10.1109/TNANO.2009.2031807.

Valeriani, D., Cinel, C. and Poli, R. 2019. Brain(-)computer interfaces for human augmentation. *Brain Sci.*, 9(2). doi: 10.3390/brainsci9020022.

Vasil'ev, G.F. 2013. [Cybernetics and biology]. *Biofizika*, 58(4): 732–736.

Vincent, J.F., Bogatyreva, O.A., Bogatyrev, N.R., Bowyer, A. and Pahl, A.K. 2006. Biomimetics: Its practice and theory. *J. R. Soc. Interface.*, 3(9): 471–482. doi: 10.1098/rsif.2006.0127.

Wallace, G.G., Higgins, M.J., Moulton, S.E. and Wang, C. 2012. Nanobionics: The impact of nanotechnology on implantable medical bionic devices. *Nanoscale*, 4(15): 4327–4347. doi: 10.1039/c2nr30758h.

Wang, C.W., Pan, C.Y., Wu, H.C., Shih, P.Y., Tsai, C.C., Liao, K.T., Lu, L.L., Hsieh, W.H., Chen, C.D. and Chen, Y.T. 2007. *In situ* detection of chromogranin a released from living neurons with a single-walled carbon-nanotube field-effect transistor. *Small*, 3(8): 1350–1355. doi: 10.1002/smll.200600723.

Wang, Z.V., Li, D.L. and Hill, J.A. 2014. Heart failure and loss of metabolic control. *J. Cardiovasc. Pharmacol.*, 63(4): 302–313. doi: 10.1097/FJC.0000000000000054.

Whitesides, G.M. 2006. The origins and the future of microfluidics. *Nature*, 442(7101): 368–373. doi: 10.1038/nature05058.

Williams, M.R. and Walter, W. 2015. Development of a prototype over-actuated biomimetic prosthetic hand. *PLoS One*, 10(3), e0118817. doi: 10.1371/journal.pone.0118817.

Winkler, D.A. 2017. Biomimetic molecular design tools that learn, evolve, and adapt. *Beilstein. J. Org. Chem.*, 13: 1288–1302. doi: 10.3762/bjoc.13.125.

Wrenn, S.M., Griswold, E.D., Uhl, F.E., Uriarte, J.J., Park, H.E., Coffey, A.L., Dearborn, J.S., Ahlers, B.A., Deng, B., Lam, Y.W., Huston, D.R., Lee, P.C., Wagner, D.E. and Weiss, D.J. 2018. Avian lungs: A novel scaffold for lung bioengineering. *PLoS One*, 13(6): e0198956. doi: 10.1371/journal.pone.0198956.

Xiao, Y., Zhang, B., Liu, H., Miklas, J.W., Gagliardi, M., Pahnke, A., Thavandiran, N., Sun, Y., Simmons, C., Keller, G. and Radisic, M. 2014. Microfabricated perfusable cardiac biowire: A platform that mimics native cardiac bundle. *Lab Chip*, 14(5): 869–882. doi: 10.1039/c3lc51123e.

Yi, H.G., Lee, H. and Cho, D.W. 2017. 3D printing of organs-on-chips. *Bioengineering (Basel)*, 4(1). doi: 10.3390/bioengineering4010010.

Yun, Y.H., Eteshola, E., Bhattacharya, A., Dong, Z., Shim, J.S., Conforti, L., Kim, D., Schulz, M.J., Ahn, C.H. and Watts, N. 2009. Tiny medicine: Nanomaterial-based biosensors. *Sensors (Basel)*, 9(11): 9275–9299. doi: 10.3390/s91109275.

Zeilig, G., Weingarden, H., Zwecker, M., Dudkiewicz, I., Bloch, A. and Esquenazi, A. 2012. Safety and tolerance of the ReWalk exoskeleton suit for ambulation by people with complete spinal cord injury: A pilot study. *J. Spinal. Cord. Med.*, 35(2): 96–101. doi: 10.1179/2045772312Y.0000000003.

Zhang, H., Wu, R., Li, C., Zang, X., Zhang, X., Jin, H. and Zhao, J. 2017. A force-sensing system on legs for biomimetic hexapod robots interacting with unstructured terrain. *Sensors (Basel)*, 17(7). doi: 10.3390/s17071514.

Zhang, R., Zhang, Y. and Wei, F. 2017. Controlled synthesis of ultralong carbon nanotubes with perfect structures and extraordinary properties. *Acc. Chem. Res.*, 50(2): 179–189. doi: 10.1021/acs.accounts.6b00430.

Zhang, S., Yuan, S., Huang, L., Zheng, X., Wu, Z., Xu, K. and Pan, G. 2019. Human mind control of rat cyborg's continuous locomotion with wireless brain-to-brain interface. *Sci. Rep.*, 9(1): 1321. doi: 10.1038/s41598-018-36885-0.

Zhou, Q., Patel, D., Kwa, T., Haque, A., Matharu, Z., Stybayeva, G., Gao, Y., Diehl, A.M. and Revzin, A. 2015. Liver injury-on-a-chip: Microfluidic co-cultures with integrated biosensors for monitoring liver cell signaling during injury. *Lab. Chip*, 15(23): 4467–4478. doi: 10.1039/c5lc00874c.

Ziegler-Graham, K., MacKenzie, E.J., Ephraim, P.L., Travison, T.G. and Brookmeyer, R. 2008. Estimating the prevalence of limb loss in the United States: 2005 to 2050. *Arch. Phys. Med. Rehabil.*, 89(3): 422–429. doi: 10.1016/j.apmr.2007.11.005.

Zoffmann, S., Vercruysse, M., Benmansour, F., Maunz, A., Wolf, L., Blum Marti, R., Heckel, T., Ding, H., Truong, H.H., Prummer, M., Schmucki, R., Mason, C.S., Bradley, K., Jacob, A.I., Lerner, C., Araujo del Rosario, A., Burcin, M., Amrein, K.E. and Prunotto, M. 2019. Machine learning-powered antibiotics phenotypic drug discovery. *Sci. Rep.*, 9(1): 5013. doi: 10.1038/s41598-019-39387-9.

Index